# What Is LIFE?

SECOND EDITION

## A GUIDE TO BIOLOGY

# What Is LIFE?

SECOND EDITION

## A GUIDE TO BIOLOGY

## Jay Phelan

University of California, Los Angeles

W. H. FREEMAN AND COMPANY
NEW YORK

*Publisher:* Peter Marshall

*Senior Development Editor:* Elizabeth Howe

*Development Editors:* Jane Tufts, Alicia Moretti

*Assistant Editor:* Marni Rolfes

*Associate Director of Marketing:* Debbie Clare

*Art Development Editor:* Tommy Moorman

*Illustrations:* Tommy Moorman

*Managing Editor:* Philip McCaffrey

*Senior Project Editor:* Mary Louise Byrd

*Art Director:* Diana Blume

*Text Design:* Tom Carling, Carling Design Inc

*Photo Editor:* Christine Buese

*Photo Researchers:* Deborah Anderson, Julia Phelan

*Production Manager:* Susan Wein

*Composition and Layout:* Sheridan Sellers

*Printing and Binding:* Quad/Graphics Dubuque

Library of Congress Control Number: 2011938007

Student Edition:    ISBN-13: 978-1-4641-0720-7

ISBN-10: 1-4641-0720-3

Loose-leaf Edition:   ISBN-13: 978-1-4641-0587-6

ISBN-10: 1-4641-0587-1

Printed in the United States of America

Second Printing

W. H. Freeman and Company

41 Madison Avenue, New York, NY 10010

Houndsmills, Basingstoke RG21 6XS, England

www.whfreeman.com

# BRIEF CONTENTS

# CONTENTS

# 3 · Cells
(81)

## The smallest part of you

# 4 · Energy

(129)

## From the sun to you in just two steps

# PART 2 Genetics, Evolution, and Behavior

# 5 · DNA, Gene Expression, and Biotechnology  169

What is the genetic code, and how is it harnessed?

# 6 · Chromosomes and Cell Division  221

Continuity and variety

# 7 · Genes and Inheritance

Family resemblance: how traits are inherited

# 8 · Evolution and Natural Selection

## Darwin's dangerous idea

# 9 • Evolution and Behavior

Communication, cooperation, and conflict in the animal world

## PART 3 Evolution and the Diversity of Life

# 10 · The Origin and Diversification of Life on Earth

**387**

Understanding biodiversity

# 11 · Animal Diversification

**427**

Visibility in motion

# 12 · Plant and Fungi Diversification

(473)

Where did all the plants and fungi come from?

# 13 · Evolution and Diversity Among the Microbes

Bacteria, archaea, protists, and viruses: the unseen world

**511**

# PART 4 Ecology and the Environment

# 14 · Population Ecology  547
## Planet at capacity: patterns of population growth

**StreetBIO: Knowledge you can use**
Life history trade-offs and a mini-fountain of youth: what is the relationship between reproduction and longevity?  576

# 15 · Ecosystems and Communities  583
## Organisms and their environments

# 16 · Conservation and Biodiversity

**623**

## Human influences on the environment.

# Dear Reader,

If you can learn anything from reading this book, I hope it is this: *Biology is about you, and it touches your life every day, in dozens of ways. It's creative. It's fun.*

In these pages, you'll find an overview of the key themes in biology as well as detailed information and stories about meaningful topics. I hope you will find answers to questions you're curious about, and will be spurred to ask many more, whether you are a future lawyer, teacher, entrepreneur, parent, consumer, citizen, or all of the above.

Consider these:

- Does it rain more on weekends than during the week?
- Do megadoses of vitamin C reduce cancer risk?
- Why doesn't natural selection lead to the production of perfect organisms?
- Why are big, fierce animal species so rare in the world?

As you read, you'll see big red **Q**s, identifying questions like these, often with real relevance to your own life. They point toward passages where you can uncover the answers. If you don't think the answer is found there, look again and think some more. Sometimes you know things that you don't realize you know. Sometimes, as you read, you are learning about things you're not actually reading about. Recognizing and developing these abilities will help you reason your way through novel problems and will serve you well long after you have forgotten this or that specific fact.

At the end of each chapter, you'll find a section called **StreetBIO: Knowledge You Can Use**. These sections unpack some questions and issues that are particularly practical, such as *How clean is that food you just dropped?* (see Chapter 13).

There's much more to biology than just words. Flip through *What Is Life?* and look at the **photographs**. Images can inspire and provide an alternative hook for remembering and understanding ideas. I have hand-picked every photo, with a goal of provoking, engaging, and even entertaining you, always while helping you make connections between complex ideas.

You'll also notice brief quotes from a variety of literary sources. There is a rich tradition of scientific imagery, references, and metaphors throughout literature. With these, I hope to illustrate how, as your scientific literacy is increased, so too will be the richness of your experience and appreciation of literature.

In a world of information overload, it is more important than ever to learn how to distill ideas, examples, and implications, forming hierarchies of importance. I don't want you to lose sight of the big picture. In organizing each chapter, I have broken down the topics into **discrete sections** (I think of them as nuggets). And at the end of each, I highlight the **Take-Home Message** that concisely and precisely highlights and reinforces the section's most important ideas. There is much material here, but in these short chunks you can master it. Packaged with every new copy of the book are laminated, illustrated **To-Go** summaries of each chapter.

Increasingly, the information you consume includes graphs. It's essential to understand how to read and interpret such figures. To help you, I've included an exercise at the end of each chapter called **Graphic Content**. This critical thinking challenge will help you become adept at reading and analyzing visual displays of information, while identifying subtle assumptions, biases, and even manipulations.

At the end of each chapter, you'll also find review questions with a wide range of difficulties indicated by a **difficulty thermometer**. It's okay if you get some of these questions wrong. By noting the approximate difficulty level of those giving you trouble, you'll be able to more accurately assess where you stand as you gain proficiency with the material. And there are thousands more questions available for easy and efficient practice online.

This is just a sampling of some of the features provided in this textbook. I really hope that you gain as much satisfaction from reading this book as I have received in putting it together for you.

Sincerely,

Jay Phelan

P.S. About the cover: I wanted to convey that biology isn't something that exists far away, separate from our personal lives. Rather, it intersects with our lives and is a central part of our world.

# About the Author

Jay Phelan teaches biology at UCLA, where he has taught introductory biology to more than 9,000 majors and non-majors students over the past fourteen years. He is the recipient of more than a dozen teaching awards, including UCLA's highest teaching honor, the Distinguished Teaching Award, in 2011. He received his Ph.D. in evolutionary biology from Harvard in 1995, and his master's and bachelor's degrees from Yale and UCLA. His primary area of research is evolutionary genetics, and his original research has been published in *Evolution, Experimental Gerontology*, and the *Journal of Integrative and Comparative Biology*, among other journals. His research has been featured on *Nightline*, CNN, the BBC, and National Public Radio; in *Science Times* and *Elle*; and in more than a hundred newspapers.

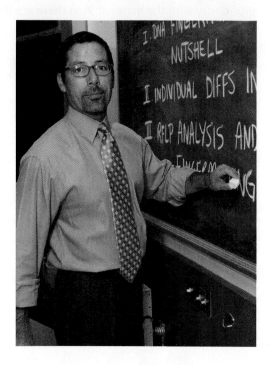

Jay lectures frequently on a variety of topics in education, including the nurturing of critical thinking skills in undergraduate students, and the use and efficacy of online adaptive assessment software. His research in these areas has been published in the *International Encyclopedia of Education* and other journals.

With economist Terry Burnham, Jay is co-author of the international best-seller *Mean Genes: From Sex to Money to Food—Taming Our Primal Instincts*. Written for the general reader, *Mean Genes* explains in simple terms how knowledge of the genetic basis of human nature can empower individuals to lead more satisfying lives.

To Julia

**Q** **HOW DOES *WHAT IS LIFE?* SO THOROUGHLY CAPTIVATE NON-MAJORS?**

**IT WAS CREATED WITH THEM IN MIND.**

## Engaging Examples

*What Is Life? A Guide to Biology* threads fascinating, relevant, contemporary examples throughout each chapter.

## Brief Sections

Each chapter is broken down into a series of short, accessible sections.

## Clear, Consistent Illustrations

Fresh and easy-to-understand figures bring the concepts to life. Collaboratively developed by the author and scientific illustrator, the text and illustrations are seamlessly integrated, effective learning tools.

## Vivid Photos

Striking images appear as unit openers and are combined with illustrations of biological processes, concepts, and experimental techniques to engage the imagination of the student.

 *Once breeders recognized the existence of heredity, selective breeding—such as for reduced size in horses—became possible.*

## Intriguing, Often Surprising Q Questions

Q questions spark students' interest and encourage critical thinking.

**Q** Drinking diet soda can be deadly if you carry a single bad gene. What gene is it and why is it so deadly?

## Take-Home Messages

Each section of the chapter includes a concise, memorable summary of key ideas.

**TAKE-HOME MESSAGE 7·11**

Many traits, including continuously varying traits such as height, eye color, and skin color, are influenced by multiple genes.

## End-of-Chapter Study Tools

Each chapter includes Big Ideas recaps of the key ideas from each section, Key Terms that are page-referenced for quick review, and Check Your Knowledge and short-answer questions.

## StreetBIOs

**STREETBIO: KNOWLEDGE YOU CAN USE** features are found at the end of every chapter, and demonstrate the practicality and fun of biology. 25 percent are new!

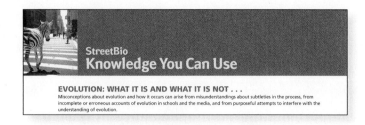

**StreetBio**
**Knowledge You Can Use**

**EVOLUTION: WHAT IT IS AND WHAT IT IS NOT . . .**
Misconceptions about evolution and how it occurs can arise from misunderstandings about subtleties in the process, from incomplete or erroneous accounts of evolution in schools and the media, and from purposeful attempts to interfere with the understanding of evolution.

## Graphic Content Helps Students Analyze Data and Think Critically

Each chapter features a graph or chart accompanied by a series of critical-thinking questions that requires students to evaluate the data presented.

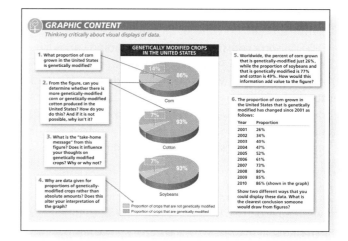

## Prep-U Thermometers Rate the Question Difficulty

Thermometers appear next to each end-of-chapter Check Your Knowledge question to help students understand their current level of proficiency.

## To-Go Guides Provide a Visual Summary of Each Chapter

These laminated guides offer a visual summary for each chapter and serve as a great study tool. They can be packaged with the text at no additional cost.

# Acknowledgments

As a new graduate student at Harvard, I heard from experienced teaching fellows that if you were interested in learning how to be an effective teacher, it was essential to seek out extraordinary mentors. Based on word of mouth, I became involved with E. O. Wilson's course in Evolutionary Biology and Irven DeVore's course in Human Behavioral Biology. Both were known to be unusually provocative, challenging, and entertaining classes for non-science majors. I aggressively pursued teaching positions in both classes—which I held onto tightly for twelve semesters. Working under these legendary instructors, I was set on a course that inspired and prepared me to write this book.

The two courses were quite different from each other, but at their core both were built on two beliefs that are central to this book and to my thinking about education: (1) Biology is creative, interesting, and fun. (2) Biology is relevant to the daily life of every person. There was a palpable sense that, in teaching non-science majors especially, we had a responsibility to provide our students with the tools to thrive in a society increasingly permeated by scientific ideas and issues, and that one of our most effective strategies would be to convey the excitement we felt for biology and the enormous practical value it has to help us understand the world. I thank Professors Wilson and DeVore for all that they have shared with me.

My development as a scientist and, particularly, my appreciation for rigorous and methodical critical thinking have been shaped by the kind support and wise guidance of Richard Lewontin. I have also been fortunate to have as a long-time mentor and collaborator Michael Rose, who has instilled in me a healthy skepticism about any observation in life that is not fivefold replicated. And for almost daily insightful input on matters relating to scientific content, teaching, writing, and more, I thank Terry Burnham.

There are many other friends and colleagues I wish to thank for helping me with *What Is Life?*

In researching and writing the book and in developing the numerous courses I teach, I have benefited from more than a decade of perceptive and valuable contributions, too numerous to list, from Glenn Adelson, Alon Ziv, Michael Cooperson, and Alicia Moretti. I am tremendously appreciative of all they have done for me.

For a project covering so many topics and years, it is essential to have a close group of trusted, tolerant, and knowledgeable colleagues, I am grateful to Harold Owens, Greg Graffin, Brian Swartz, Greg Laden, Jeff Egger, Andy Tobias, Elisabeth Tobias, Joshua Malina, Melissa Merwin-Malina, Bill U'ren, Chris Bruno, Michelle Richmond, and Meredith Dutton, who have offered advice, guidance, and support, far beyond the call of duty.

Numerous colleagues at UCLA provided assistance and support, including Steve Strand, Cliff Brunk, Fred Eiserling, Emil Reisler, Victoria Sork, Deb Pires, Lianna Johnson, Gaston Pfluegl, Bob Simons, Frank Laski, Jeff Thomas, and Tracy Newman. As a result of their commitment to excellence in the UCLA Life Science Core Curriculum, I have been able to acquire a wealth of experiences that have helped me continue improving as a teacher.

I owe a tremendous debt to Sara Tenney, without whose encouragement and support this project could never have been begun or completed.

W. H. Freeman is an extremely author-centric publisher. Throughout the process of creating this book, from the first inception of the idea through the production of all the supplementary materials, Liz Widdicombe, Brian Napack, and John Sargent have been tremendously supportive. I am grateful for their welcoming me into their publishing family. Publisher Peter Marshall has been a tenacious, versatile, and skillful manager of the entire team. I am very fortunate to have such a wise leader overseeing all aspects of this project.

The team of editors that worked with me on this book—and two people in particular—improved it immeasurably. Development editor Beth Howe, oversaw every aspect of the writing and production of the book, attending to issues of content, production, and design while making insightful contributions throughout and expertly managing the thousand details necessary to put everything together. I could not have completed this book without Beth's commitment and guidance. And I cannot adequately convey my gratitude to development editor Jane Tufts, whose meticulous attention to detail, commitment to accuracy, and almost obsessive drive to create a thorough and readable book are apparent on every page.

It is impossible to teach biology without illustrations. My deepest gratitude goes to Tommy Moorman for creating such innovative and effective figures for the book. Tommy's vision for an elegant and beautiful art program completely integrated with the text is apparent on every page. Working with him to develop each illustration in this book has been (and continues to be) one of my most enjoyable and satisfying professional collaborations. Thanks also go to Alison Kendall and to Erin Daniel for assisting with the creation of the illustrations and for work on the To Go guides. For the design of the book, I thank Tom Carling. And for excellent assistance with photo research thanks to Julia Phelan, Deborah Anderson, and Christine Buese.

I wish to thank Harvey Pough for his assistance with the first edition of What Is Life? for which he provided excellent drafts of Chapters 11–13 and contributions to Chapter 16.

For creating the innovative media and print materials that accompany the book, I am thankful for the contributions of Patrick Shriner, Mike Jones, and Beth McHenry, and for the extensive input of supplements editors Amanda Dunning and Marni Rolfes. I thank all of the contributors and advisors who helped create the student and instructor resources; your efforts have been invaluable. I thank Jennifer Warner and Meredith Norris for their work on the Student Success Guide, and for Jennifer's willingness to share her classroom experiences with the sales force. I also appreciate the contributions of Troy Williams and the Prep-U team. Sheri Snavely provided significant input in developing pedagogical strategies throughout the book; I also appreciate her thoughtful and smart advice at nearly every step in the publishing process.

Copyeditor Linda Strange helped to ensure consistency and readability throughout the text. I thank Chris Hunt for compiling the thorough index. The rest of the life sciences editorial team at W. H. Freeman, too, have been knowledgeable and supportive, particularly Kate Parker, Marc Mazzoni, Jerry Correa, Susan Winslow, Elaine Palucki, and Lisa Samols.

For their efficiency and commitment to producing a beautiful book, I am most grateful to the W. H. Freeman production team: Sheridan Sellers, Mary Louise Byrd, Diana Blume, Susan Wein, Philip McCaffrey, and Ellen Cash.

The people on the marketing team at W. H. Freeman have contributed enormously in helping with the challenging task of introducing a new book to students and instructors across the country. Debbie Clare, Steve Rigolosi, Lindsey Veautour, and John Britch have been enthusiastic and dedicated in creating materials and strategies to assist instructors in evaluating the ways in which What Is Life? can aid them as they develop their own courses and strategies for success.

Finally, I thank my family—Kevin Phelan, Patrick Phelan, Erin Enderlin, and my parents—for their unwavering support and interest as I wrote this book. Reading draft after draft and following each revision, they made valuable contributions at every stage. I thank Jack, Charlie, and Sam, too. Most of all, for her generous and passionate support of this project from day one, her substantive contributions to both the content and presentation of ideas, and so much more, I thank Julia.

## Contact the author with your feedback.

The content of this book has been greatly improved through the comments of reviewers and students. Your comments, suggestions, and criticism are also welcome; they are essential in guiding its ongoing evolution. Please contact the author at **jay@jayphelan.com.** I'm serious.

# We thank the many reviewers that aided in the development of this text.

## Second Edition Text Development

Jessica Adams, Southwestern Illinois College
Sylvester Allred, Northern Arizona University
Nina Baghai-Riding, Delta State University
E. A. Bailey, Alamance Community College
Donna Becker, Northern Michigan University
Donna Bivans, Pitt Community College
TJ Bliss, Brigham Young University
Barbara Blonder, Flagler College
Judy Bluemer, Morton College
Lisa Boggs, Southwestern Oklahoma State University
Bruno Borsari, Winona State University
Laura Bradshaw, Illinois Central College
Judy Brooks, Blinn College
Wendy Brown, Danville Area Community College
Bruce Carroll, Lone Star College–North Harris
Michelle Cawthorn, Georgia Southern University
Jianguo Chen, Claflin University
Richard Cheney, Christopher Newport University
Genevieve Chung, Broward College
Elizabeth Co, Diablo Valley College
Robert Conover, Glendale College
Erica Corbett, Southeastern Oklahoma State University
Cathy Cornett, University of Wisconsin–Platteville
Sarah Cotton, Chaffey College
James Courtright, Marquette University
Richard Cowart, University of Dubuque
Hattie Dambroski, Normandale Community College
Jim Daugherty, Glendale Community College
Buffany DeBoer, Wayne State College
Jennifer Dechaine, Central Washington University
Tracy Deem, Bridgewater College
Craig Denesha, Spartanburg Community College
Hartmut Doebel, George Washington University
Brett Dooley, Patrick Henry Community College
Hawkins Dowis, Fresno City College
Dani DuCharme, Waubonsee Community College
Jacquelyn Duke, Baylor University
Susan Epperson, University of Colorado, Colorado Springs
Marirose Ethington, Genesee Community College
Arnaldo Ferreira, Chaffey College
Ryan Fisher, Salem State University
Reza Forough, Bellevue College
Brandon Foster, Wake Technical Community College
Cynthia Galloway, Texas A&M University–Kingsville
Joseph Gar, West Kentucky Community and Technical College
Alexandros Georgakilas, East Carolina University

Patricia Geppert, University of Texas at San Antonio
Julie Gibbs, College of DuPage
Caitlin Gille, Pasco Hernando Community College
Becky Graham, Ohio Dominican University
Sherri Graves, Sacramento City College
Jeff Green, Nashville State Community College
Cindy Gustafson-Brown, University of California, San Diego
Sue Habeck, Tacoma Community College
Mark Haefele, Community College of Denver
Thomas Horvath, State University of New York–Oneonta
Anne-Marie Hoskinson, Gustavus Adolphus College
Laurie Host, Harford Community College
Catherine Hurlbut, Florida State College at Jacksonville
Godwin Ifere, Clark Atlanta University
Virginia Irintcheva, Black Hawk College
Robert Iwan, Inver Hills Community College
Mark Jackson, Central Connecticut State University
Elisabeth Javazon, Morehouse College
John Jenkin, Blinn College
Mitrick Johns, Northern Illinois University
Scott Johnson, Wake Technical Community College
Richard Karp, University of Cincinnati
Elena Keeling, California Polytechnic State University
Joe Keen, Patrick Henry Community College
Amanda Kin, Jefferson State Community College
Melissa Kosinski-Collins, Brandeis University
Joe Kremer, Alvernia University
Dana Kurpius, Elgin Community College
Karen Kurvink, Moravian College
Kim Lackey, University of Alabama
Jordana LaFantasie, Fort Hays State University
Ellen Lamb, University of North Carolina at Greensboro
Lorraine Leiser, Southeast Community College
Laurie Len, El Camino College
Suzanne Long, Monroe Community College
Mark Manteuffel, St. Louis Community College
Brian Maricle, Fort Hays State University
John Markham, University of Manitoba
Jennifer Maupin, Santa Barbara City College
Maura McGrail, Iowa State University
Michael McLeod, Belmont Abbey College
Patricia Melloy, Fairleigh Dickinson University
Diane Melroy, University of North Carolina at Wilmington
Rachel Mintell, Manchester Community College
June Minter, Patrick Henry Community College
Kiran Misra, Edinboro University of Pennsylvania
Alice Monroe, St. Petersburg College
Frances Moore, Patrick Henry Community College

Ulrike Muller, California State University–Fresno
Necia Nicholas, Calhoun Community College
Zia Nisani, Antelope Valley College
Fran Norflus, Clayton State University
Chuma Okere, Clark Atlanta University
Marie Panec, Moorpark College
Wiline Pangle, The Ohio State University
Randi Papke, Southwestern Illinois College
Micah Perkins, Owensboro Community and Technical College
Lori Pitkofsky, Ramapo College of New Jersey
Anita Plagge, San Diego Mesa College
Michael Plotkin, Mt. San Jacinto College
Heather Prestridge, Blinn College
Stuart Reichler, University of Texas at Austin
Sheila Reilly, Belmont Abbey College
Carol Rhodes, Canada College
Bill Rogers, Ball State University
John Rousseau, Whatcom Community College
Matthew Rowe, Sam Houston State University
Edward Saiff, Ramapo College of New Jersey
Eduardo Salazar, Temple College
John Schampel, Phoenix College
Roger Seeber, West Liberty University
Brian Seymour, Edward Waters College
Dave Sheldon, St. Clair County Community College
Cara Shillington, Eastern Michigan University
Jack Shurley, Idaho State University
Dale Smoak, Piedmont Technical College
Suzanne Sousa, Middlesex Community College
Agnes Southgate, College of Charleston
Richard Stiehl, Pima Community College
Lisa Strain, Northeast Lakeview College
Jackie Swanik, Wake Technical Community College
Robert Swatski, Harrisburg Area Community College–York Campus
Ignatius Tan, New York University
John Taylor, Southern Utah University
Pamela Thinesen, Century College
Douglas Thrower, University of California, Santa Barbara
Nina Thumser, Edinboro University of Pennsylvania
Elizabeth Tomlin, University of North Carolina at Greensboro
Edward Tucker, Baruch College, City University of New York
Virginia Turner, Harper College
Kimberly Vietti, Illinois Central College
Thomas Vogel, Western Illinois University
Joannna Vondrasek, Piedmont Virginia Community College
Dennis Walsh, MassBay Community College
Yunqiu Wang, University of Miami
Katherine Warpeha, University of Illinois at Chicago
David Webb, St. Clair County Community College
Michael Wenzel, California State University–Sacramento
Tina Whitney, Governors State University
Robert Whyte, California University of Pennsylvania

Rachel Wiechman, West Liberty University
Donald Wood, Odessa College
Dulce Zayas-Alvarez, Southwestern Community College
David Zeigler, University of North Carolina at Pembroke
Jacquelyn Zevenbergen, Kent State University, Stark Campus
Brenda Zink, Northeastern Junior College
Gerald Zuercher, University of Dubuque

## Second Edition Focus Group Participants

### (2010, Albuquerque)
Tonya Bates, University of North Carolina at Charlotte
Craig Clifford, Northeastern State University, Oklahoma
Danielle DuCharme, Waubonsee Community College
Ana Escandon, Los Angeles Harbor College
Mark Hens, University of North Carolina at Greensboro
Jessica Hopkins, University of Akron
Lorraine Leiser, Southeast Community College
Ashley Ramer, University of Akron
Sydha Salihu, West Virginia University
Jennifer Stueckle, West Virginia University
Jennifer Warner, University of North Carolina at Charlotte

### (2010, New York City)
Joseph Ahlander, Northeastern State University
Tami Asplin, North Dakota State University
Caroline "Lena" Ballard, Rock Valley College
Charlotte Borgeson, University of Nevada–Reno
Susan Bower, Pasadena College
Greg Dalhem, North Kentucky University
Moon Draper, University of Texas at Austin
Gloria Hoffman, Morgan State University
Craig Jordan, University of Texas at San Antonio
Michael Koban, Morgan State University
Josephine Kurdziel, University of Michigan
Ellen Lamb, University of North Carolina at Greensboro
Jennifer Scoby, Illinois Central College

### (2011, New York City)
Jennie Bever, Arizona State University
Michelle Cawthorn, Georgia Southern University
John Constable, California State University–Fresno
Elisabeth Javazon, Morehouse College
Brian Kram, Prince George's Community College
Lorraine Leiser, Southeast Community College
Cassandra Moore-Crawford, Prince George's Community College
Ulrike Müller, California State University–Fresno
Wiline Pangle, The Ohio State University
Jennifer Warner, University of North Carolina at Charlotte
Michael Wenzel, California State University–Sacramento
  and Folsom Lake Community College

## First Edition
## Text Development

Stephanie J. Aamodt, University of Louisiana–Shreveport
M. Stephen Ailstock, Anne Arundel Community College
Sylvester Allred, Northern Arizona University
Jessica K. Baack, Montgomery Community College–Rockville
Sarah Barlow, Middle Tennessee State University
Rebecca A. Bartow, Western Kentucky University
Brian J. Baumgartner, Trinity Valley Community College
Jose Bava, El Camino Community College
Lisa L. Boggs, Southwestern Oklahoma State University
Charlotte Borgeson, University of Nevada–Reno
Peggy Brickman, University of Georgia
Jason Brown, Young Harris College
Carole L. Browne, Wake Forest University
Neil J. Buckley, State University of New York–Plattsburgh
Anne E. Bunnell, East Carolina University
Joseph Burdo, Bridgewater State University
David Byres, Florida Community College at Jacksonville
William Caire, University of Central Oklahoma
Michael S. Carr, Oakton Community College
Kelly Carrillo Burke, College of the Canyons
Thomas Chen, Santa Monica College
Thomas F. Chubb, Villanova University
Nira Clark, Southwestern College
Erica Cline, University of Washington
George Clokey, University of Wisconsin–Whitewater
Anthony D. Cornett, Valencia Community College, Osceola
Karen Curto, University of Pittsburgh
Don C. Dailey, Austin Peay State University
Douglas W. Darnowski, Indiana University–Southeast
Garry Davies, University of Alaska
Renee Dawson, University of Utah
Lewis E. Deaton, University of Louisiana–Lafayette
Tom Deaton, Auburn University
Elizabeth Anne Desy, Southwestern Minnesota State University
Doreen R. Dewell, Whatcom Community College
Mary E. Dominiecki, Slippery Rock University
Charles Dunn, Austin Community College
Frank DuRoy, Essex County College
Kari Eamma, Tarrant County Community College–Northeast
Heidi Engelhardt, Brandon University
Marirose T. Ethington, Genesee Community College
Rebecca A. Fetherson, University of Dayton
Robert C. Frankis, University of New Mexico
Monica C. Frazier, Columbus State University
Anne Galbraith, University of Wisconsin–La Crosse
Sandra Gibbons, Moraine Valley College
Philip G. Gibson, Gwinnett Technical College
Andrew Goliszek, North Carolina A &T State University
Stephen M. Gómez, Central New Mexico Community College
Becky C. Graham, University of West Alabama
Cara Gubbins, Butte College
Charles (Billy) Gunnels IV, Florida Gulf Coast University

Janelle M. Hare, Morehead State University
Carla Hass, Pennsylvania State University
Colleen Hatfield, California State University–Chico
J. L. Henriksen, Bellevue Community College
John A. Hnida, Peru State College
Thomas Horvath, State University of New York–Oneonta
Adam W. Hrincevich, Louisiana State University
Dianne Jennings, Virginia Commonwealth University
Mitrick A. Johns, Northern Illinois University
Robert D. Johnson Jr., Pierce College
Martin A. Kapper, Central Connecticut State University
Arnold J. Karpoff, University of Louisville
Todd Kelson, Brigham Young University–Idaho
Stephen T. Kenny, Yakima Valley Community College
Jennifer M. Kilbourne, Community College of Baltimore County
Joanne Kilpatrick, Auburn University–Montgomery
Brenda Knotts, Eastern Illinois University
Michael Koban, Morgan State University
Catherine Koo, Caldwell College
Jerome A. Krueger, South Dakota State University
Jim Krupa, University of Kentucky
Ellen Shepard Lamb, University of North Carolina at Greensboro
Thomas Landefeld, California State University–Dominquez Hills
Leah LaPerle Larkin, University of New Mexico
Lynn Larsen, Portland Community College
Kathleen H. Lavoie, State University of New York–Plattsburgh
Gregory P. Lewis, Furman University
Harvey Liftin, Broward Community College
Tammy J. Liles, Lexington Community College
Cynthia W. Littlejohn, University of Southern Mississippi
Jonathan Lochamy, Georgia Perimeter College
Melanie Loo, California State University–Sacramento
Juan M. Lopez-Bautista, University of Alabama
Blasé Maffia, University of Miami
Jeffrey M. Marcus, Western Kentucky University
Betsy Maxim, Austin Community College
Karen McFarland, Austin Peay State University
Mark A. McGinley, Texas Technical University
John Mersfelder, Sinclair Community College
Shahroukh Mistry, Westminster College
Jeanne Mitchell, Truman State University
Beth A. Montelone, Kansas State University
Cynthia L. Morin, Lincoln University of Missouri
Royden Nakamura, California Polytechnic–San Luis Obispo
Rocky Nation, Southern Wesleyan University
Allan D. Nelson, Tarleton State University
Tania Nezrick, Kishwaukee College
David Niebuhr, College of William and Mary
Meredith Sommerville Norris, University of North Carolina at Charlotte
Alexander E. Olvido, Virginia State University
Randi Papke, Southwestern Illinois College
Mary Paulson, Central Washington University
John M. Pleasants, Iowa State University
Gregory J. Podgorski, Utah State University
Melissa Presch, California State University–Fullerton

Eric Rabitoy, Citrus College
Karen Raines, Colorado State University
Dawn Ranish, Nova Southeastern University
Erin Rempala, San Diego Mesa College
Melody Ricci, Victor Valley
Kathleen Richardson, Portland Community College
Dave Rintoul, Kansas State University
Steven Rissing, The Ohio State University
Robert R. Robbins, Utah Valley State College
Bill Rogers, Ball State University
Troy T. Rohn, Boise State University
Susan T. Rouse, Southern Wesleyan University
Lynette Rushton, South Puget Sound Community College
Michael L. Rutledge, Middle Tennessee State University
Shamili Ajgaonkar Sandiford, College of DuPage
Fayla Schwartz, Everett Community College
Erik P. Scully, Towson University
Juanita C. Sharpe, Chicago State University
Marilyn Shopper, Johnson County Community College
Michele Shuster, New Mexico State University
Neil E. Simister, Brandeis University
Anu Singh-Cundy, Western Washington University
Jennifer Skillen, College of Southern Nevada
Kerri M. Skinner, University of Nebraska–Kearny
Marc Allen Smith, Sinclair Community College
Nancy G. Solomon, Miami University of Ohio
Roberta Lynn Soltz, Central Washington University
Bryan G. Spohn, Florida Community College at Jacksonville–Kent Campus
Amanda Starnes, Emory University
Leo Sternberg, University of Miami
Joan C. Stover, South Seattle Community College
Sukanya V. Subramanian, Collin County Community College
Franklyn Tan Te, Miami Dade Community College
William J. Thieman, Ventura College
Pamela Thinesen, Century College
Michael W. Thompson, Middle Tennessee State University
Martin A. Vaughan, Indiana University–Purdue University Indianapolis
Jack Waber, West Chester University of Pennsylvania
Jennifer M. Warner, University of North Carolina at Charlotte
John E. Whitlock, Hillsborough Community College
Elizabeth Willott, University of Arizona
Clifford Wilson, Chicago State University
Mark Woelfle, Vanderbilt University
Calvin Young, Fullerton College
Brenda Zink, Northeastern Junior College

## First Edition
## Art Development

Ann Aguanno, Marymount Manhattan College
Kelly Carrillo Burke, College of the Canyons
William Caire, University of Central Oklahoma
Debra Chapman, Wilkes University
Genevieve C. Chung, Broward Community College
Jeffrey Scott Coker, Elon University
Michael S. Dann, Pennsylvania State University
Renee Dawson, University of Utah

Francisco Delgado, Pima Community College
Doreen R. Dewell, Whatcom Community College
Hartmut G. Doebel, George Washington University
Diane Lynn Doidge, Grand View College
James Doyle, College of the Holy Cross
Dave Eakin, Eastern Kentucky University
Paul Farnsworth, University of New Mexico
Brandon L. Foster, Wake Technical and Community College
Patrick Galliart, North Iowa Community College
David Gordon, Pittsburg State University
Charles (Billy) Gunnels IV, Florida Gulf Coast University
Bernard Hauser, University of Florida
Jane J. Henry, Baton Rouge Community College
Mark A. Holland, Salisbury University
Virginia E. Irintcheva, Black Hawk College–Quad City Campus
Philip Jardim, City College of San Francisco
Denim M. Jochimsen, University of Idaho
Judy Kaufman, Monroe Community College
Ariel R. Krakowski, Laney College
Dan E. Krane, Wright State University
Maria Kretzman, Glendale College
Kathleen H. Lavoie, State University of New York–Plattsburgh
Mary LeFever, Columbus State Community College
Stephen Matheson, Calvin College
James McCaughern-Carucci, Quinnipiac University
Lori A. Pitkofsky, Rampano College
Michael P. Robinson, Miami University
Wendy F. Rothwell, University of California–Santa Cruz
Jason F. Schreer, State University of New York–Potsdam
Jennifer J. Scoby, Illinois Central College
Dawn Sherry, Macon State University
Cara Shillington, Eastern Michigan University
Patricia L. Smith, Valencia Community College
Jeffrey Thomas, California State University–Northridge
Bruce Tomlinson, State University of New York–Fredonia
Joy B. Trauth, Arkansas State University
Alan R. Wasmoen, Metropolitan Community College
Michael Windelspecht, Appalachian State University
David E. Wolfe, American River College
Kerry Yurewicz, Plymouth State University

# First Edition
# Accuracy Reviewers

Christine Bezotte, Elmira College
Michael Carr, Oakton Community College
Erica Champion
William Crampton, University of Central Florida
Michael S. Dann, Pennsylvania State University
Bernard Hauser, University of Florida
John A. Hnida, Peru State College
Kelly Hogan, University of North Carolina at Chapel Hill
Anne Houtman, California State University–Fullerton
Ariel R. Krakowski, Laney College
Michael Silva, El Paso Community College
Pamela Thinesen, Century College
Jeffrey Thomas, California State University–Northridge

Michael Wenzel, California State University–Sacramento
John E. Whitlock, Hillsborough Community College

# First Edition
# Focus Group Attendees
## Text and Content Development

**(2004, Los Angeles)**
Garen Baghdasarian, Santa Monica College
Thomas Chen, Santa Monica College
Elizabeth Ciletti, Pasadena City College
Mary Colavito, Santa Monica College
Ana Ester Escandon, Los Angeles Harbor College
Phyllis Hirsh, East Los Angeles College
Melody Ricci, Victor Valley College

**(2004, Memphis)**
T. Wayne Barger, Tennessee Technical University
Don Baud, University of Memphis
Jay Blundon, Rhodes College
Martha P. Brown, University of Memphis
Dave Eakin, Eastern Kentucky University
Stan Eisen, Christian Brothers University
Jack Grubaugh, University of Memphis
Min-Ken Lao, Furman University

**(2004, Chicago)**
Sandra Bobick, Community College of Alleghany
Christopher Dobson, Grand Valley State University
Sondra Dubowsky, Allen County Community College
Eileen Gregory, Rollins College
Tracy Harris, Florida Keys Community College
Robert Krasner, Providence College
Bernard A. Marcus, Genesee Community College
Laurel B. Roberts, University of Pittsburgh
Lyndell P. Robinson, Lincoln Land Community College
Brian Schmaefsky, Kingwood College
Janet Vigna, Grand Valley State University
Daniel W. Ward, Waubonsee Community College

**(2006, Florida)**
Peggy Brickman, University of Georgia
Anne E. Bunnell, East Carolina University
Cara Gubbins, Butte College
Janelle M. Hare, Morehead State University
John A. Hnida, Peru State College
Mitrick A. Johns, Northern Illinois University
Kathleen H. Lavoie, State University of New York–Plattsburgh
Blasé Maffia, University of Miami
Gregory J. Podgorski, Utah State University
Susan T. Rouse, Southern Wesleyan University
Jennifer M. Warner, University of North Carolina at Charlotte

## Art Development

**(2007, Atlanta)**
Renee Dawson, University of Utah
Charles (Billy) Gunnels IV, Florida Gulf Coast University
Bernard Hauser, University of Florida

Arnold J. Karpoff, University of Louisville
Stephen R. Kelso, University of Illinois at Chicago
Meredith Sommerville Norris, University of North Carolina at Charlotte
Therese M. Poole, Georgia State University
Mary Celeste Reese, Mississippi State University
Bill Rogers, Ball State University
Kim Cleary Sadler, Middle Tennessee State University
Amy S. Wernette, Hazard Community College

## Media and Supplements

**(2007, Santa Monica)**
Michael Bucher, College of San Mateo
Kelly Carrillo Burke, College of the Canyons
John A. Hnida, Peru State College
Anne Houtman, California State University–Fullerton
Hinrich Kaiser, Victor Valley College
Melanie Loo, California State University–Sacramento
Gregory J. Podgorski, Utah State University
Melissa Presch, California State University–Fullerton
Calvin Young, Fullerton College

**(2008, Miami)**
Barbara Blonder, Flagler College
Charlotte Borgeson, University of Nevada–Reno
Kris Curran, University of Wisconsin–Whitewater
Kari Eamma, Tarrant Community College
Evelyn Frazier, Florida Atlantic University
John Janovy, University of Nebraska–Lincoln
Jennifer Schramm, Chemeketa Community College
Greg Sievert, Emporia University
Heather Vance Chalcraft, East Carolina University
Martin Vaughn, Indiana University–Purdue University Indianapolis

# First Edition
# Comparative Reviewers

Zulfiqar Ahmad, East Tennessee State University
Julie H. Aires, Florida Community College at Jacksonville
Mahmood Anwar, Palm Beach Community College
William Anyonge, Xavier University
Tami Asplin, North Dakota State University
Christine Barrow, Prince George's Community College
Joressia Beyer, John Taylor Community College
Randy Brewton, University of Tennessee–Knoxville
Dustin Brisson, University of Pennsylvania
Jeannie Chapman, University of South Carolina–Upstate
Thomas Chen, Santa Monica College
Cary Chevalier, Missouri Western State University
Craig Clifford, Northeastern State University
Vince Connors, University of South Carolina–Upstate
William Gareth Richard Crampton, University of Central Florida
Chris Davison, Long Beach City College
Kevin Dees, Wharton City Junior College
Doreen Dewell, Whatcom Community College
Tiffany Doan, Central Connecticut State University
Ernest F. DuBrul, University of Toledo

Eugene Fenster, Metropolitan Community College–Longview

Rebecca Fetherson, University of Dayton

Teresa G. Fischer, Indian River State College

Anita Pardue Flick, North Carolina State University

Paul Florence, Jefferson Community and Technical College

Evelyn Frazier, Florida Atlantic University

Diane Wilkening Fritz, Northern Kentucky University

Dennis W. Fulbright, Michigan State University

Douglas C. Gayou, University of Missouri–Columbia

Bagie George, Georgia Gwinnett College

Betsy Gerbec, University of Wisconsin–River Falls

Oliver Ghobrial, Santa Fe College

Julie Gibbs, College of DuPage

Florence K. Gleason, University of Minnesota

Bruce Griffis, Kentucky State University

Tim Grogan, Valencia Community College

Charles J. Grossman, Xavier University

Luis S. Guerra, South Texas College

Pieter de Haan, Berkeley City College

Pamela Hanratty, Indiana University at Bloomington

Mary F. Haskins, Rockhurst University

Keith R. Hench, Kirkwood Community College

Sherry Hickman, Hillsborough Community College

Mark Hollier, Georgia Perimeter College

Robert A. Holmes, Hutchinson Community College

Laurie Host, Harford Community College

Tamalene Imbierowicz, Harford Community College

Robert Iwan, Inver Hills Community College

Philip Jardim, City College of San Francisco

Dianne Jennings, Virginia Commonwealth University

Greg Jones, Santa Fe College

David Knowles, East Carolina University

Janice Krumm, Widener University

Kim Lackey, University of Alabama

Ellen Lamb, University of North Carolina at Greensboro

David Loring, Johnson County Community College

Bill Mackay, Edinboro University of Pennsylvania

Christi Magrath, Troy University

Barbara Mania-Farnell, Purdue University–Calumet

Tom J. McConnell, Ball State University

Wallace M. Meyer, College of the Redwoods

Thelma Miller-Anderson, Hillsborough Community College

Michael R. Millward, University of Cincinnati

Shahroukh Mistry, Westminster College

Brenda Moore, Truman State University

Michael Muller, University of Illinois at Chicago

Zia Nisani, Antelope Valley College

Tanya Noel, York University

James Nolan, Georgia Gwinnett College

Laura Palmer, Pennsylvania State University–Altoona

Louis Pech, University of Wisconsin Centers–Marathon County

Joel B. Piperberg, Millersville University

Karen Plucinski, Missouri Southern State University

Aggie Posthumus, Olivet Nazarene University

Bob Pozos, San Diego State University

Raul Ramirez, Oklahoma City Community College

Bonnie Ripley, Grossmont College

Troy Rohn, Boise State University

Matthew Rowe, Sam Houston State University

Yelena Rudayeva, Palm Beach Community College

Arthur Sandquist, Washburn University

Craig Scott, Clarion University of Pennsylvania

Wallace Sharif, Morehouse College

Donald Slish, Plattsburgh State University

Anthony J. Stancampiano, Oklahoma City Community College

Christine Stracey, University of Florida

Tim Strakosh, State University of New York–Fredonia

Richard P. Stringer, Harrisburg Area Community College

Mark Sturtevant, Oakland University

Sukanya Subramanian, Collin County Community College

Rob Swatski, Harrisburg Area Community College

Kimberly Taugher, Diablo Valley College

Jeffrey Thornsberry, Northwest Missouri State

Nina Thumser, Indiana University of Pennsylvania

Jonathan Titus, State University of New York–Fredonia

Mike Tveten, Pima Community College Northwest

Linda Tyson, Santa Fe College

Katherine M. Van de Wal, Community College of Baltimore County–Essex Campus

Leslie VanderMolen, Humboldt State University

Fred Vogt, Elgin Community College

William A. Wehbi, University of Pennsylvania

John E. Whitlock, Hillsborough Community College

Jennifer Wiatrowski, Pasco-Hernando Community College

Christina Willis, Rockhurst University

Donald S. Wood, Odessa College

Clarence Wright, Tulsa Community College–Metro Campus

Aimee Wurst, Lincoln University of Missouri

Lan Xu, South Dakota State University

# First Edition Class Testers

## Q Questions

Borough of Manhattan Community College
Broward Community College
Central New Mexico Community College
Colorado Mountain College
Florida Community College at Jacksonville
Heartland Community College
Johnson County Community College
Kutztown University of Pennsylvania
Lincoln Land Community College
Metropolitan Community College
Mt. San Jacinto College
Portland State University

Rowan University
Seattle Central Community College
Suffolk County Community College–Ammerman
Tarrant County Community College–Northeast
University of Northern Colorado
Wake Technical Community College

## Chapters

### (2007)

College of San Mateo
Georgia Perimeter College
Georgia State University
Hazard Community College
Southwestern College
University of Arizona
University of Louisville
University of North Carolina at Charlotte
University of Utah
Utah State University
Virginia Commonwealth University

### (2008)

Bluegrass Community College
California State University–Northridge
Chaffey College
Clark Atlanta University
Columbus State University
Daytona Community College
Edinboro University of Pennsylvania
Genesee Community College
George Washington University
Glendale College
Lake Land College
Long Beach Community College
Los Angeles Harbor College
Macon State University
Marshall University
McPherson College
Millersville University of Pennsylvania
Milwaukee Area Technical College
Owensboro Community College
Oxnard College
Philadelphia University
Prince George's Community College
Quinnipiac University
Rock Valley
Sacramento City College
Salem State College
San Jose State University
Sinclair Community College
Southwestern College
St. Clair Community College
University of New Mexico
University of Rhode Island
Valencia Community College

# Contributors and Advisors

## Instructor's Manual—Classroom Catalysts

Verona A. Barr, Heartland Community College
Stephen M. Gómez, Central New Mexico Community College
Paul H. Marshall, Northern Essex Community College

## eBook and Non-Major's Biology Study Tools

Amy Combs
Robert Iwan, Inver Hills Community College
Eric Stavney, Pierce College

## Exploring Biology: Case Studies

Michelle Cawthorn, Georgia Southern University
Jennifer L. Holzman, Emory University

## Figure Conversion Engine

Ann Aguanno, Marymount Manhattan College

## Hands-On Biology: Laboratories for Distance Learning

Mimi Bres, Cassandra Moore-Crawford, and Arnold Weisshaar, Prince George's Community College (all)

## Image Bank Advisor

Jane J. Henry, Baton Rouge Community College

## Keynote Lecture Presentation

Michael C. Bucher, College of San Mateo

## PowerPoint Lecture Outlines

Kristen L. Curran, University of Wisconsin–Whitewater
Danielle DuCharme, Waubonsee Community College
Jennifer Lange, Chabot College
Mark S. Manteuffel, St. Louis Community College

## Prep-U Instructor's Test Bank and Prep-U for Students

Glenn Adelson, Lake Forrest College
Jay Phelan, University of California, Los Angeles
Alon Ziv, Prep-U

## Q Animations

Anne Bunnell, East Carolina University
G. Chung, Broward College
Johnny El-Rady, University of South Florida
Julie V. Gibbs, College of DuPage
Eric Stavney, North Seattle Community College

## Student Worksheets

Michael C. Bucher, College of San Mateo

# Reviewers

Ann Aguanno, Marymount Manhattan College
Charlotte E. Borgeson, University of Nevada–Reno
Kelly Burke, College of the Canyons
Michelle Cawthorn, Georgia Southern University
Michael S. Dann, Pennsylvania State University
Debra B. Folsom, Pasadena City College
Julie V. Gibbs, College of DuPage
Stephen M. Gómez, Central New Mexico Community College
Jane J. Henry, Baton Rouge Community College
Jane Horlings, Saddleback College
Carina Endres Howell, Lock Haven University
John Janovy, Jr., University of Nebraska–Lincoln
Hinrich Kaiser, Victor Valley College
Judy Kaufman, Monroe Community College
Charlease Kelly-Jackson, Claflin University
Patrick J. Lewis, Sam Houston State University
Cindy S. Malone, California State University–Northridge
Meredith S. Norris, University of North Carolina at Charlotte
Greg Podgorski, Utah State University
Lawrence F. Roberge, Florida Community College at Jacksonville
William D. Rogers, Ball State University
Georgianna Saunders, Missouri State University
Jason Schreer, State University of New York–Potsdam
Marilyn Shopper, Johnson County Community College
Carol St. Angelo, Hofstra University
Eric Stavney, Pierce College
Alicia Steinhardt, Hartnell College
Jamey Thompson, Hudson Valley Community College
Helen Walter, Diablo Valley College
Jennifer M. Warner, University of North Carolina at Charlotte
Michael Windelspecht, Appalachian State University

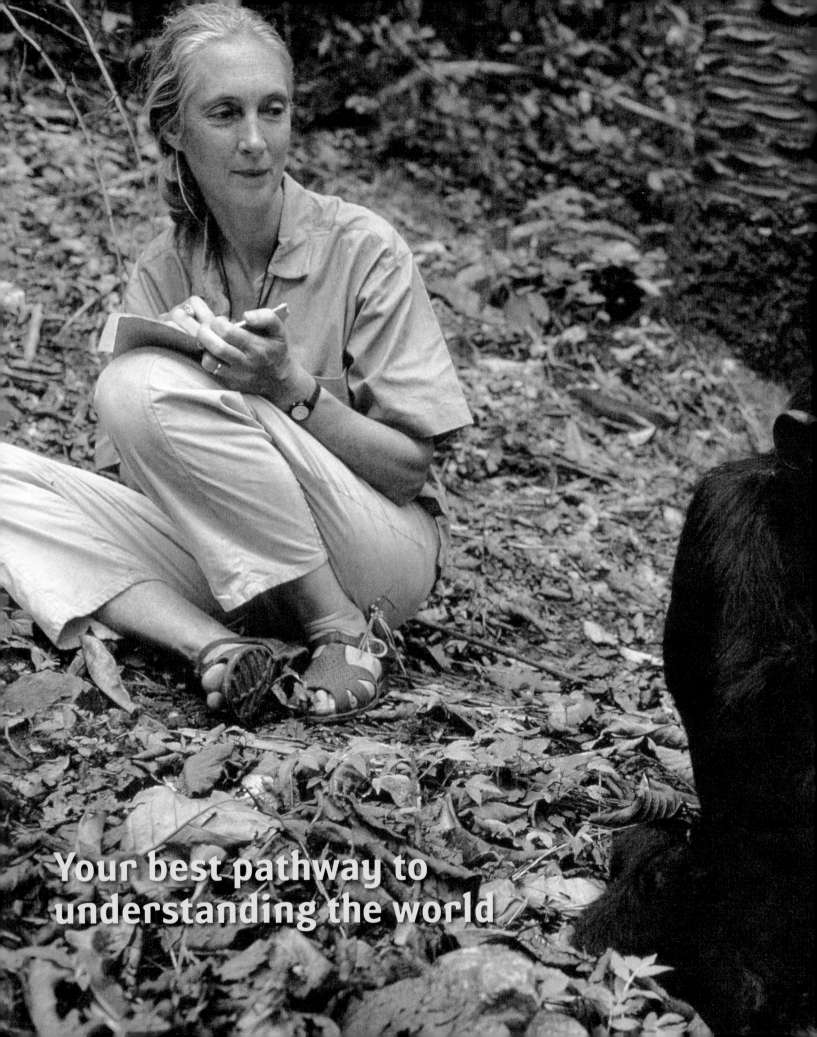

Your best pathway to
understanding the world

# 1 Scientific Thinking

# 1·1–1·3 Science is a collection of facts and a process for understanding the world.

*Already a scientist? It starts with curiosity.*

## 1·1  What is science? What is biology?

You are already a scientist. You may not have realized this yet, but it's true. Because humans are curious, you have no doubt asked yourself or others questions about how the world works and wondered how you might find the answers.

- Does the radiation released by cell phones cause brain tumors?

- Do large doses of vitamin C reduce the likelihood of getting a cold?

These are important and serious questions. But you've probably also pondered some less weighty issues, too.

- Why is morning breath so stinky? And can you do anything to prevent it?

- Why is it easier to remember gossip than physics equations?

And if you really put your mind to the task, you will start to find questions all around you whose answers you might like to know (and some whose answers you'll learn as you read this book).

- Which parent determines a baby's sex? Why?

- Why do so few women get into barroom brawls?

- What is "blood doping," and does it really improve athletic performance?

- Why is it so much easier for an infant to learn a complex language than it is for a college student to learn biology?

Still not convinced you're a scientist? Here's something important to know: science doesn't require advanced degrees or secret knowledge dispensed over years of technical training. It does, however, require an important feature of our species: a big brain, as well as curiosity and a desire to learn. But curiosity, casual observations, and desire can take you only so far. Many pressing issues require some understanding of science, what it can and can't do, and what it can and can't explain. Consider, for example, nutritional claims on foods and dietary supplements, human behavior, health and disease, the interactions of plants, animals, and their environments, global climate change, the continuity and diversity of life in all its forms. By learning more about science, you can understand your life and the world around you.

Explaining how something works or why something happens requires methodical, objective, and rational observations and analysis that are not clouded with emotions or preconceptions. **Science** is not simply a body of knowledge or a list of facts to be remembered. It is an intellectual activity, encompassing observation, description, experimentation, and explanation of natural phenomena. Put another way, science is a pathway by which we can come to discover and better understand our world.

Later in this chapter, we explore specific ways in which we can most effectively use scientific thinking in our lives. But first let's look at a single powerful question that underlies scientific thinking:

**How do you know that is true?**

Once you begin asking this question—of others and of yourself—you are on the road to a better understanding of the world.

The following two stories about popular and successful products show the importance of questioning the truth of many "scientific" claims you see on merchandise packages or read in a newspaper or on the internet.

**Dannon yogurt.** According to the Federal Trade Commission (FTC), a U.S. government agency with the mission of consumer protection, the Dannon Company claimed in nationwide advertisements that its Activia yogurt relieves irregularity and helps with "slow intestinal transit time." Dannon also claimed that its DanActive dairy drink helps prevent colds and flu (**FIGURE 1-1**). The FTC charged that the ads were deceptive because there was no substantiation for the claims and, further, that the claims had been clinically proven to be false. In an agreement finalized in 2011, Dannon agreed to pay $21 million in fines and to stop making those claims unless the company gets reliable scientific evidence demonstrating that the claims are true.

**Airborne.** For almost 15 years, the product Airborne has been marketed and sold to millions of customers. On the packaging and in advertisements, the makers originally asserted that Airborne tablets could ward off colds and boost your immune system (see Figure 1-1). Not surprisingly, Airborne quickly became a great success; it has generated more than $200 million in revenue. Then some consumers posed a reasonable question to the makers of Airborne: *How do you know that it wards off colds?*

**Q** *Can we trust the packaging claims that companies make?*

To prove their claims, the makers of Airborne pointed to the results of a "double-blind, placebo-controlled study" conducted by a company specializing in clinical drug trials. We'll discuss exactly what those terms mean later in the chapter; for now we just need to note that as a result of a class-action lawsuit, it became clear that no such study had been conducted and that there was *no* evidence to back up Airborne's claims. The Airborne company removed the claims from the packaging and agreed to refund the purchase price to anyone who had bought Airborne. It also removed any reference to its "clinical trials," with the company's CEO saying that people "are really not scientifically minded enough to be able to understand a clinical study."

Are you insulted by the CEO's assumption about your intelligence? You should be. Did you or your parents fall for Airborne's false claims? Possibly. But here's some more good news: you can learn to be skeptical and suspicious (in a good way) of product claims. You can learn exactly what it means to have scientific proof or evidence that something is absolutely true. And you can learn this by learning what it means to think scientifically.

Scientific thinking is important in the study of a wide variety of topics: it can help you understand economics, psychology, history, and many other subjects. Our focus in

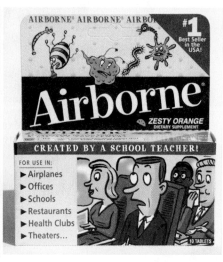

**FIGURE 1-1  Some products claim to improve our health, but how do we know whether they work?**

this book is **biology,** the study of living things. Taking a scientific approach, we investigate the facts and ideas in biology that are already known and study the process by which we come to learn new things. As we move through the four parts of the book, we explore the most important questions in biology.

- What is the chemical and physical basis for life and its maintenance?

- How do organisms use genetic information to build themselves and to reproduce?

- What are the diverse forms that life on earth takes, and how has that diversity arisen?

- How do organisms interact with each other and with their environment?

In this chapter, we explore how to think scientifically and how to use the knowledge we gain to make wise decisions. Although we generally restrict our focus to biology, scientific

thinking can be applied to nearly every endeavor, so in this chapter we use a wide range of examples—including some from beyond biology—as we learn how to think scientifically. Although the examples vary greatly, they all convey a message that is key to scientific thinking: it's okay to be skeptical.

Fortunately, learning to think scientifically is not difficult— and it can be fun, particularly because it is so empowering. **Scientific literacy,** a general, fact-based understanding of the basics of biology and other sciences, is increasingly important in our lives, and literacy in matters of biology is especially essential.

---

### TAKE-HOME MESSAGE 1·1

Through its emphasis on objective observation, description, and experimentation, science is a pathway by which we can discover and better understand the world around us.

---

## 1·2 Biological literacy is essential in the modern world.

A brief glance at any magazine or newspaper will reveal just how much scientific literacy has become a necessity (**FIGURE 1-2**). Many important health, social, medical, political, economic, and legal issues pivot on complex scientific data and theories. For example, why are unsaturated fats healthier for you than saturated fats? And why do allergies strike

children from clean homes more than children from dirty homes? And why do new agricultural pests appear faster than new pesticides?

As you read and study this book, you will be developing **biological literacy,** the ability to (1) use the process of

**July 3, 2008**

**Texas Man Freed By DNA After 15 Years In Prison**

By JEFF CARLTON, Associated Press

A Texas man who spent more than 15 years in prison after being wrongly convicted of kidnapping and robbery raised both arms skyward and collapsed in his mothers embrace Thursday after being told he was a free man.

**FIGURE 1-2  In the news.** Every day, news sources report on social, political, medical, and legal issues related to science.

scientific inquiry to think creatively about real-world issues that have a biological component, (2) communicate these thoughts to others, and (3) integrate these ideas into your decision making. Biological literacy doesn't involve just the big issues facing society or just abstract ideas. It also matters to you personally. Should you take aspirin when you have a fever? Are you using the wrong approach if you try to lose weight and, after some initial success, you find your rate of weight loss diminishing? Is it a good idea to consume moderate amounts of alcohol? Lack of biological literacy will put you at the mercy of "experts" who may try to confuse you or convince you of things in the interest of (their) personal gain. Scientific thinking will help you make wise decisions for yourself and for society.

> **TAKE-HOME MESSAGE 1·2**
>
> Biological issues permeate all aspects of our lives. To make wise decisions, it is essential for individuals and societies to attain biological literacy.

# 1·3 The scientific method is a powerful approach to understanding the world.

It's a brand new age, and science, particularly biology, is everywhere. To illustrate the value of scientific thinking in understanding the world, let's look at what happens in its absence, by considering some unusual behaviors in the common laboratory rat.

Rats can be trained, without much difficulty, to push a lever to receive a food pellet from a feeding mechanism (**FIGURE 1-3**). When the mechanism is altered so that there is a 10-second delay between the lever being pushed and the food pellet being dispensed, however, strange things start to happen. In one cage, the rat will push the lever and then, very methodically, run and push its nose into one corner of the cage. Then it moves to another corner and again pushes its nose against the cage. It repeats this behavior at the third and fourth corners of the cage, after which the rat stands in front of the feeder and the pellet is dispensed. Each time the rat pushes the lever it repeats the nose-in-the-corner sequence before moving to the food tray.

In another cage, with the same 10-second delay before the food pellet is dispensed, a rat pushes the lever and then proceeds to do three quick back-flips in succession. It then moves to the food tray for the food pellet when the 10 seconds have elapsed. Like the nose-in-the-corner rat, the back-flip rat will repeat this exact behavior each time it pushes the lever.

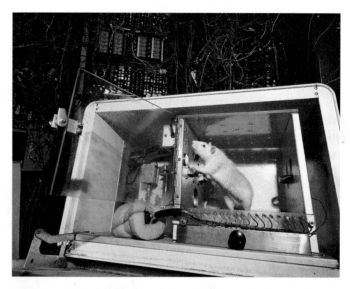

FIGURE 1-3 **"In the absence of the scientific method . . ."** Rats develop strange, superstition-like behaviors.if there is a 10-second delay between when they push a lever and when food is delivered.

associate two events—pushing the lever and engaging in some sequence of behaviors—with another: receiving food. In a sense, they have taken a step toward understanding their world, even though the events are not actually related to each other.

Humans can also mistakenly associate actions with outcomes in an attempt to understand and control their world. The irrational belief that actions that are not logically related to a course of events can influence its outcome is called **superstition.** In the absence of scientific thinking, individuals can develop incorrect ideas, such as superstitions about how the world works (**FIGURE 1-4**). For example, Nomar Garciaparra, a former major league baseball player, always engaged in a precise series of toe taps and adjustments to his batting gloves before he would bat.

Thousands of different narratives, legends, fairy tales, and epics from all around the globe exist to help people understand the

**Q** *Why do people develop superstitions? Can animals be superstitious?*

In cage after cage of rats with these 10-second-delay food levers, each rat eventually develops its own peculiar series of behaviors before moving to the food dish to receive the pellet. Why do they do this? Because it seems to work! They have discovered a method by which they can get a food pellet. To some extent, the rats' behaviors are reasonable. They world around them. These stories explain everything from birth and death to disease and healing.

As helpful and comforting as stories and superstitions may be (or seem to be), they are no substitute for really understanding how the world works. This kind of understanding begins when someone wonders about why something is the way it is and then decides to try to find out the answer. This process of examination and discovery is called the **scientific method.**

The scientific method usually begins with someone observing a phenomenon and proposing an explanation for it. Next, the proposed explanation is tested through a series of experiments. If the experiments reveal that the explanation is accurate, and if the experiments can be done by others with the same result, then the explanation is considered to be valid. If the experiments do not support the proposed explanation, then the explanation must be revised or alternative explanations that more closely reflect the experimental results must be proposed and tested. This process continues as better, more accurate explanations are found.

While the scientific method reveals much about the world around us, it doesn't explain everything. There are many other methods through which we can gain an understanding of the world. For example, much of our knowledge about plants and animals does not come from the use of the scientific method, but rather comes from systematic, orderly observation, without the testing of any explicit hypotheses. Other disciplines also involve understandings of the world based on non-scientific processes. Knowledge about history, for example, comes from the systematic examination of past events as they relate to humans, while the "truths" in other fields, such as religion, ethics, and even politics, often are based on personal faith, traditions, and mythology.

Scientific thinking can be distinguished from these alternative ways of acquiring knowledge about the world in that it is **empirical.** Empirical knowledge is based on experience and observations that are rational, testable, and repeatable. The empirical nature of the scientific approach makes it self-correcting: in the process of analyzing a topic, event, or phenomenon with the scientific method, incorrect ideas are discarded in favor of more accurate explanations. In the next sections, we look at how to put the scientific method into practice.

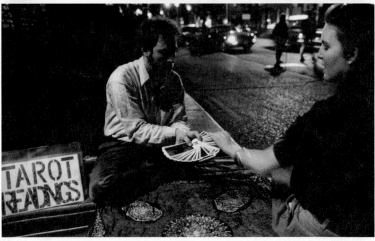

**FIGURE 1-4  Superstitions abound.** As comforting as myths and superstitions may be, they are no substitute for really understanding how the world works.

### TAKE-HOME MESSAGE 1·3

There are numerous ways of gaining an understanding of the world. Because it is empirical, rational, testable, repeatable, and self-correcting, the scientific method is a particularly effective approach.

# 1·4–1·10 A beginner's guide: what are the steps of the scientific method?

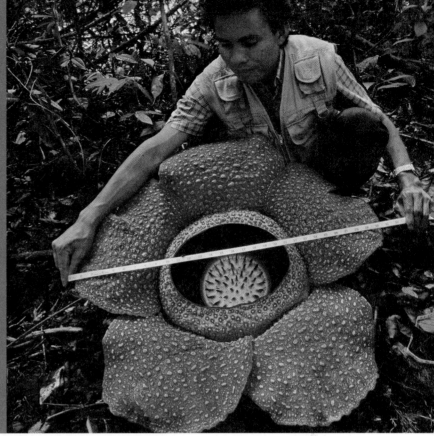

*Scientific thinking relies on rational, testable, and repeatable observations. (Shown here: A botanist in Borneo measures* Rafflesia arnoldii, *which produces the largest flowers of any species.)*

## 1·4  Thinking like a scientist: how do you use the scientific method?

"Scientific method"—this term sounds like a rigid process to follow, much like following a recipe. In practice, however, the scientific method is an adaptable process that can be done effectively in numerous ways. This flexibility makes the scientific method a powerful process that can be used to explore a wide variety of thoughts, events, or phenomena, not only in science but in other areas as well.

The basic steps in the scientific method are:

**Step 1.** Make observations.

**Step 2.** Formulate a hypothesis.

**Step 3.** Devise a testable prediction.

**Step 4.** Conduct a critical experiment.

**Step 5.** Draw conclusions and make revisions.

Once begun, though, the process doesn't necessarily continue linearly through the five steps until it is concluded (**FIGURE 1-5**). Sometimes, observations made in the first step can lead to more than one hypothesis and several testable predictions and experiments. And the conclusions drawn from experiments often suggest new observations, refinements to hypotheses, and, ultimately, increasingly precise conclusions.

### THE SCIENTIFIC METHOD

| STEP 1 | STEP 2 | STEP 3 | STEP 4 | STEP 5 |
| Make observations. | Formulate a hypothesis. | Devise a testable prediction. | Conduct a critical experiment. | Draw conclusions and make revisions. |

 *The scientific method rarely proceeds in a straight line. Conclusions, for example, often lead to new observations and refined hypotheses.*

**FIGURE 1-5  The scientific method: five basic steps and one flexible process.**

What should you
do when
something you
believe in turns
out to be wrong?

**Q** An especially important feature of the scientific method is that its steps are self-correcting. As we continue to make new observations, a hypothesis about how the world works might change (**FIGURE 1-6**). If our observations do not support our current hypothesis, that hypothesis must be given up in favor of one that is not contradicted by any observations. This may be the most important feature of the scientific method: *it tells us when we should change our minds.*

> "If science proves some belief of Buddhism wrong, then Buddhism will have to change. "
> —THE 14TH DALAI LAMA,
> *New York Times,* December 2005

Because the scientific method is a general strategy for learning, it needn't be used solely to learn about nature or scientific things. In fact, we can analyze an important criminal justice question using the scientific method:

- How reliable is eyewitness testimony in criminal courts?

For more than 200 years, courts in the United States have viewed eyewitness testimony as unassailable. Few things are seen as more convincing to a jury than an individual testifying that she can identify the person she saw commit a crime (**FIGURE 1-7**). But is eyewitness identification always right?

**FIGURE 1-6 Hold the fries.** We apply an understanding of science when we choose foods from the menu that have fewer calories and less saturated fat.

**FIGURE 1-7 "With your own two eyes . . ."?** How reliable is eyewitness testimony in criminal courts?

Can the scientific method tell us if whether this perception—or some other commonly held idea—is true? As we describe how to use the scientific method to answer questions about the world, it will become clear that the answer is a resounding *yes.* In the coming sections of this chapter, we also look at how the scientific method can be used to address a variety of issues. In addition to our criminal justice question, we'll answer three additional questions:

- Does echinacea reduce the intensity or duration of the common cold?

- Does chemical runoff give rise to hermaphrodite fish?

- Does shaving hair from your face, legs, or anywhere else cause it to grow back coarser or darker?

## TAKE-HOME MESSAGE 1·4

**The scientific method (observation, hypothesis, prediction, test, and conclusion) is a flexible, adaptable, and efficient pathway to understanding the world, because it tells us when we must change our beliefs.**

# 1·5  Step 1: Make observations.

Scientific study always begins with observations: we simply look for interesting patterns or cause-and-effect relationships. This is where a great deal of the creativity of science comes from. In the case of eyewitness testimony, DNA technologies have made it possible to assess whether tissue such as hair or blood from a crime scene came from a particular suspect. Armed with these tools, the U.S. Justice Department recently reviewed 28 criminal convictions that had been overturned by DNA evidence. It found that in most of the cases, the strongest evidence against the defendant had been eyewitness identification. The observation here is that many defendants who are later found to be innocent were initially convicted based on eyewitness testimony.

Opportunities for other interesting observations are unlimited. Using the scientific method, we can (and will) also answer our three other questions.

Many people have claimed that consuming extracts of the herb echinacea can reduce the intensity or duration of symptoms of the common cold (**FIGURE 1-8**). We can ask: how do you know this is true?

- Does taking echinacea reduce the intensity or duration of the common cold?

Some people have noted that chemicals in sewage runoff—particularly those related to the hormone estrogen—seem to cause male fish to turn into hermaphrodites, organisms that have the reproductive organs of both sexes. Is this true?

- Does chemical runoff give rise to hermaphrodite fish?

And finally, some people have suggested that shaving hair from your face, legs, or anywhere else causes the hair to grow back coarser and darker. Is this true?

- Does hair that is shaved grow back coarser or darker?

STEP 1   STEP 2   STEP 3   STEP 4   STEP 5

**STEP 1: MAKE OBSERVATIONS**

OBSERVATION
To many people, consuming echinacea extract seems to reduce the intensity or duration of symptoms of the common cold.

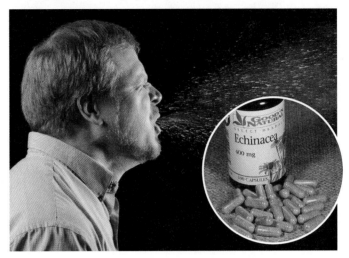

**FIGURE 1-8** The first step of science: making observations about the world.

Using the scientific method, we can answer all of these questions.

## TAKE-HOME MESSAGE 1·5

The scientific method begins by making observations about the world, noting apparent patterns or cause-and-effect relationships.

# 1·6  Step 2: Formulate a hypothesis.

Based on observations, we can develop a **hypothesis** (*pl.* **hypotheses**), a proposed explanation for observed phenomena. What hypotheses could we make about the eyewitness-testimony observations described in the previous section? We could start with the hypothesis

"Eyewitness testimony is always accurate." We may need to modify our hypothesis later, but this is a good start. At this point, we can't draw any conclusions. All we have done is summarize some interesting patterns we've seen in a possible explanation.

STEP 1    STEP 2    STEP 3    STEP 4    STEP 5

## STEP 2: FORMULATE A HYPOTHESIS

**HYPOTHESIS**
Echinacea reduces the duration and severity of the common cold.

**FIGURE 1-9  Hypothesis: the proposed explanation for a phenomenon.**

To be most useful, a hypothesis must accomplish two things.

**1.** It must clearly establish mutually exclusive alternative explanations for a phenomenon. That is, it must be clear that if the proposed explanation is not supported by evidence or further observations, a different hypothesis is a more likely explanation.

**2.** It must generate testable predictions (**FIGURE 1-9**). This characteristic is important because we can evaluate the validity of a hypothesis only by putting it to the test. For example, we could disprove the "Eyewitness testimony is always accurate" hypothesis by demonstrating that, in certain circumstances, individuals who have witnessed a crime might misidentify the criminal when asked to select the suspect from a lineup.

Researchers often pose a hypothesis as a negative statement, proposing that there is no relationship between two factors, such as "Echinacea has no effect on the duration and severity of cold symptoms." Or "There is no difference in the coarseness or darkness of hair that grows after shaving." A hypothesis that states a *lack* of relationship between two

factors is called a **null hypothesis.** Both types of hypothesis are equally valid, but a null hypothesis is easier to disprove. This is because a single piece of evidence or a single new observation that contradicts a null hypothesis is sufficient for us to reject it and conclude that an alternative hypothesis must be considered. So, once you have one piece of solid evidence that your null hypothesis is not true, you gain little by collecting further data. Conversely, it is impossible to prove that a hypothesis is absolutely and permanently true: all evidence or further observations that support a hypothesis are valuable, but they do not rule out the possibility that some future evidence or observation might show that the hypothesis is not true.

> ❝Scientific issues permeate the law. I believe [that] in this age of science we must build legal foundations that are sound in science as well as in law. The result, in my view, will further not only the interests of truth but also those of justice. ❞
> — U.S. SUPREME COURT JUSTICE STEPHEN BREYER, at the annual meeting of the American Association for the Advancement of Science, February 1998

For our three additional observations, we could state each of our hypotheses in two different ways:

*Hypothesis:* Echinacea reduces the duration and severity of the symptoms of the common cold.

*Null hypothesis:* Echinacea has no effect on the duration or severity of the symptoms of the common cold.

*Hypothesis:* Estrogens in sewage runoff turn male fish into hermaphrodites.

*Null hypothesis:* Estrogens in sewage runoff have no effect in turning male fish into hermaphrodites.

*Hypothesis:* Hair that is shaved grows back coarser and darker.

*Null hypothesis:* There is no difference in the coarseness or color of hair that is shaved relative to hair that is not shaved.

### TAKE-HOME MESSAGE 1·6

**A hypothesis is a proposed explanation for an observed phenomenon.**

# 1·7 Step 3: Devise a testable prediction.

Not all hypotheses are created equal. For a hypothesis to be useful, it must generate a prediction. That is, it must suggest that under certain conditions we will be able to observe certain outcomes. Put another way, a good hypothesis helps us make predictions about novel situations. This is a powerful feature of a good hypothesis: it guides us to knowledge about new situations.

All of this is rather abstract. Let's get more concrete with the four hypotheses we are considering. Keep in mind that when you do not understand some aspect of the world, any one of several possible explanations could be true. In devising a testable prediction from a hypothesis, the goal is to propose a situation that will give a particular outcome if your hypothesis is true, but will give a different outcome if your hypothesis is not true.

*Hypothesis:* Eyewitness testimony is always accurate.

*Prediction:* Individuals who have witnessed a crime will correctly identify the criminal regardless of whether multiple suspects are presented one at a time or all at the same time in a lineup.

This is a good, testable prediction because, if our hypothesis is true, then our prediction will always be true. On the other hand, if one method of presenting suspects consistently causes incorrect identification of the criminal, our hypothesis cannot be true and must be revised or discarded.

*Hypothesis:* Echinacea reduces the duration and severity of the symptoms of the common cold.

*Prediction:* If echinacea reduces the duration and severity of the symptoms of the common cold, then individuals taking echinacea should get sick less frequently than those not taking it, and when they do get sick, their illness should not last as long (**FIGURE 1-10**).

*Hypothesis:* Estrogens in sewage runoff turn male fish into hermaphrodites.

*Prediction:* If estrogens in sewage runoff turn male fish into hermaphrodites, then exposing fish to estrogens should cause more fish to turn into hermaphrodites than when fish are not exposed to estrogens.

*Hypothesis:* Hair that is shaved grows back coarser and darker.

*Prediction:* If shaving leads to coarser, darker hair growing back, then if individuals shaved one leg (or eyebrow) only, the

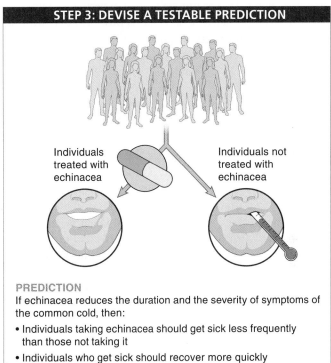

**STEP 3: DEVISE A TESTABLE PREDICTION**

Individuals treated with echinacea

Individuals not treated with echinacea

**PREDICTION**
If echinacea reduces the duration and the severity of symptoms of the common cold, then:
• Individuals taking echinacea should get sick less frequently than those not taking it
• Individuals who get sick should recover more quickly

**FIGURE 1-10 Coming up with a testable prediction.** For a hypothesis to be useful, it must generate a testable prediction.

hair that re-grows on that leg should become darker and coarser than the hair growing on the other leg.

As you begin to think scientifically, you will find yourself making a lot of "if . . . then" types of statements: "*If* that is true," referring to some hypothesis or assertion someone makes, or perhaps a claim made by the manufacturer of a new health product, "*then* I would expect . . . ," proposing your own prediction about a related situation. Once you've made a testable prediction, the next step is to go ahead and test it.

## TAKE-HOME MESSAGE 1·7

**For a hypothesis to be useful, it must generate a testable prediction.**

# 1·8 Step 4: Conduct a critical experiment.

Once we have formulated a hypothesis that generates a testable prediction, we conduct a **critical experiment,** an experiment that makes it possible to decisively determine whether a particular hypothesis is correct. There are many crucial elements in designing a critical experiment, and Section 1-11 covers the details of this process. For now, it is important just to understand that, with a critical experiment, if the hypothesis being tested is not true, we will make observations that compel us to reject that hypothesis.

In this step of the scientific method, a bit of cleverness can come in handy. Suppose we were to devise a critical experiment to test our hypothesis "Eyewitness testimony is always accurate." First, we stage a mock crime such as a purse snatching in front of a group of observers who do not know the crime is staged. Next, we ask observers to identify the criminal. To one group of observers we might present six "suspects" all at once in a lineup. To another group of observers we might present the six suspects one at a time. The beauty of this experiment is that we actually know who the "criminal" is. With this knowledge, we can evaluate with certainty whether an eyewitness's identification is correct or not.

This exact experiment was done, but with an additional, devious little twist: the researcher did not include the actual "criminal" in any of the lineups or one-by-one presentations of the six "suspects." If eyewitness testimony is always accurate, however, this slight variation should not matter. We would predict that the observers in both groups (the lineup group and the one-at-a-time group) would indicate that the criminal was not present. In the next section, we'll see what happened. Right now, let's devise critical experiments for our other hypotheses.

*Hypothesis:* Echinacea reduces the duration and severity of the symptoms of the common cold. The critical experiment for the echinacea hypothesis has been performed, and it is about as close to a perfect experiment as possible. Researchers began with 437 people who volunteered to be exposed to viruses that cause the common cold. Exposure to the cold-causing viruses was a bit unpleasant: all of the volunteers had cold viruses (in a watery solution) dripped into their noses. The research subjects were then secluded in hotel rooms for five days, and doctors examined them for the presence of the cold virus in their nasal cavity and for any cold symptoms (**FIGURE 1-11**).

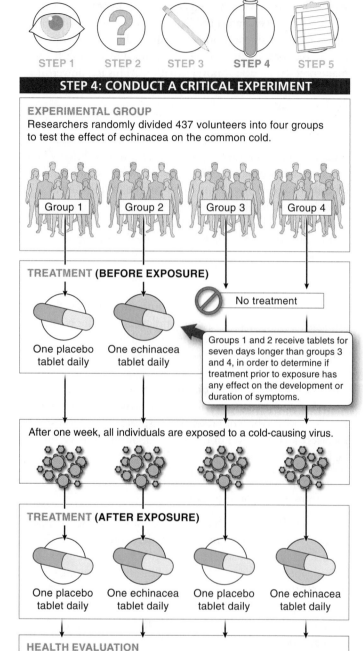

**FIGURE 1-11 "When you need to know . . ."** A critical experiment makes it possible to conclusively determine whether a hypothesis is correct.

The subjects were randomly divided into four groups. In two of the groups, each individual began taking a pill each day for a week prior to exposure to the cold virus. Those in one group received echinacea tablets, while those in the other took a **placebo,** a pill that looked identical to the echinacea tablet but contained no echinacea or other active ingredient. Neither the subject nor the doctor administering the tablets (and later checking for cold symptoms) knew what they contained. In the other two groups, the individuals did not begin taking the tablets until the day on which they were exposed to the cold virus. Again, one group got the echinacea and the other group got the placebo.

*Hypothesis:* Estrogens in sewage runoff turn male fish into hermaphrodites. To try to determine the validity of this hypothesis, researchers placed cages containing male trout downstream from sewage treatment plants on several rivers in England. They then examined the fish for the presence of eggs in their testes and tested their blood for the presence of an egg-production hormone that is not normally found in males.

**Q** Does shaving or cutting hair make it grow back more thickly?

*Hypothesis:* Hair that is shaved grows back coarser and darker. A critical experiment does not have to be complex or high-tech. All that matters is that it can decisively determine whether or not a hypothesis is correct (**FIGURE 1-12**). This point is illustrated by an experiment published in the scholarly journal *Archives of Facial Plastic Surgery.* Researchers decided to test the hypothesis that shaving hair causes it to grow back coarser or darker. A group of volunteers had one of their eyebrows, selected at random, completely shaved off. The subjects were then evaluated for eyebrow re-growth over the course of the next six months, with observers also analyzing photographs of the individuals.

In the next section we'll see how our hypotheses survive being confronted with the results of these critical experiments and how we can move toward drawing conclusions.

**FIGURE 1-12** Does shaving hair cause it to grow back coarser or darker?

> ## TAKE-HOME MESSAGE 1·8
> A critical experiment is one that makes it possible to decisively determine whether a particular hypothesis is correct.

# 1·9 Step 5: Draw conclusions, make revisions.

Once the results of the critical experiment are in, they are pulled apart, examined, and analyzed. Researchers look for patterns and relationships in the evidence they've gathered from their experiments; they draw conclusions and see whether their findings and conclusions support their hypotheses. If an experimental result is not what you expected, that does not make it a "wrong answer." Science includes a great deal of trial

**Q** Is eyewitness testimony in courts always right?

and error, and if the conclusions do not support the hypothesis, then you must revise your hypothesis, which often spurs you to conduct more experiments. This step is a cornerstone of the scientific method because it demands that you must be open-minded and ready to change what you think.

The results of the purse-snatching experiment were surprising. When the suspects (which, as you'll recall, did not include the actual "criminal") were viewed together in a lineup, the observers/witnesses erroneously identified someone as the purse snatcher about a third of the time. When the suspects were viewed one at a time, the observers made a mistaken identification less than 10% of the time.

In this or any other experiment, it does not matter whether we can imagine a reason for the discrepancy between our hypothesis and our results. What is important is that we have demonstrated that our initial hypothesis— "Eyewitness testimony is always accurate"—is not supported by the data. Our observations suggest that, at the very least, the accuracy of an eyewitness's testimony depends on the method used to present the suspects. Based on this result, we might adjust our hypothesis to: "Eyewitness testimony is more accurate when suspects are presented to witnesses one at a time."

We can then devise new and more specific testable predictions in an attempt to further refine our hypothesis. In the case of eyewitness testimony, further investigation suggests that when suspects or pictures of suspects are placed side by side, witnesses compare them and tend to choose the suspect that *most resembles* the person they remember committing the crime. When viewed one at a

time, suspects can't be compared in this way, and witnesses are less likely to make misidentifications.

**Does echinacea help prevent the common cold?**

*Hypothesis:* Echinacea reduces the duration and severity of the symptoms of the common cold. In the echinacea study, the results were definitive (**FIGURE 1-13**). Those who took the echinacea were just as likely to catch a cold as those who took placebo, and, once they caught the cold, the symptoms lasted for the same amount of time. In short, echinacea had no effect at all. Several similar studies have been conducted, all of which show that echinacea does not have any beneficial effect. As one of the researchers commented afterward, "We've got to stop attributing any efficacy to echinacea."

Although it seems clear that our initial hypothesis that echinacea stops people from catching colds and reduces the severity and duration of cold symptoms is not correct, further experimentation might involve altering the amount of echinacea given to the research subjects or the length of time they take echinacea before exposure to the cold-causing viruses.

*Hypothesis:* Estrogens in sewage runoff turn male fish into hermaphrodites. The results of the sewage experiment showed that, in every single river tested, male fish downstream from

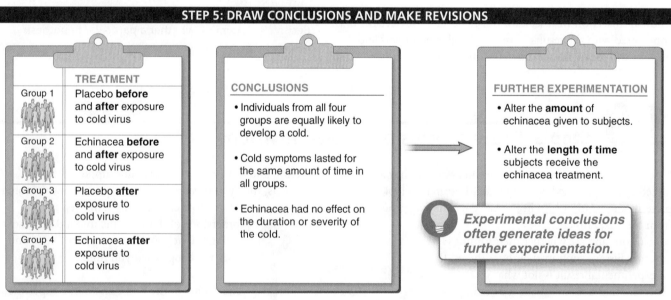

**FIGURE 1-13 Using experimental conclusions to make revisions.**

the runoff developed eggs in their testes and had significant levels of the egg-production hormone in their blood. Further research, stimulated by these results, has investigated a variety of common industrial chemicals to see whether they can stimulate production of the egg-development hormone in male fish under laboratory conditions, in an attempt to identify which particular chemical component in the sewage causes the male fish to change.

*Hypothesis:* Hair that is shaved grows back coarser and darker. In the hair-shaving experiment, the observers discovered that, after six months, it was impossible to distinguish which eyebrow had been shaved. This was not a surprise to dermatologists evaluating the study, because all of the living parts involved in hair growth are below the surface of the skin. (Plucking hairs can damage the root of the hair, potentially affecting future hair growth, but that is another story that awaits further study.)

The outcomes in all of these studies show that after the results of a critical experiment have been gathered and interpreted, it is important not just to evaluate the initial hypothesis but also to consider any necessary revisions or refinements to it. This revision is an important step; by revising a hypothesis, based on the results of experimental tests, we can explain the observable world with greater and greater accuracy.

### TAKE-HOME MESSAGE 1·9

**Based on the results of experimental tests, we can revise a hypothesis and explain the observable world with increasing accuracy. A great strength of scientific thinking, therefore, is that it helps us understand when we should change our minds.**

# 1·10 When do hypotheses become theories, and what are theories?

It's an unfortunate source of confusion that, among the general public, the word "theory" is often used to refer to a hunch or a guess or speculation—that is, something we are not certain about. In fact, to scientists, the word means nearly the opposite: a hypothesis of which they are most certain. To reduce these common misunderstandings, we examine here the ways in which scientists describe our knowledge about natural phenomena.

**Hypothesis** As we have seen, hypotheses are at the very heart of scientific thinking. A hypothesis is a proposed explanation for a phenomenon. A good hypothesis leads to testable predictions. Commonly, when non-scientists use the word "theory"—as in, "I've got a theory about why there's less traffic on Friday mornings than on Thursday mornings"— they actually mean that they have a hypothesis.

**Theory** A **theory** is an explanatory hypothesis for natural phenomena that is exceptionally well supported by the empirical data. A theory can be thought of as a hypothesis that has withstood the test of time and is unlikely to be altered by any new evidence. Like a hypothesis, a theory is testable; but because it has already been repeatedly tested and no observations or experimental results have contradicted it, a

theory is viewed by the scientific community with nearly the same confidence as a fact. For this reason, it is inappropriate to describe something as "just a theory" as a way of asserting that it is not likely to be true.

Theories in science also tend to be broader in scope than hypotheses. In biology, two of the most important theories (which we explore in more detail in Chapters 3 and 8) are the *cell theory,* that all organisms are composed of cells and all cells come from preexisting cells, and the *theory of evolution by natural selection,* that species can change over time and all species are related to one another through common ancestry.

### TAKE-HOME MESSAGE 1·10

**Scientific theories do not represent speculation or guesses about the natural world. Rather, they are hypotheses— proposed explanations for natural phenomena—that have been so strongly and persuasively supported by empirical observation that the scientific community views them as very unlikely to be altered by new evidence.**

# 1·11–1·13 Well-designed experiments are essential to testing hypotheses.

*Controlled experiments increase the power of our observations. Here, laboratory-grown wheat plants are compared.*

## 1·11 Controlling variables makes experiments more powerful.

From our earlier discussion of critical experiments, you have a sense of how important it is to have a well-planned, well-designed experiment. In performing experiments, our goal is to figure out whether one thing influences another thing: if an experiment enables us to draw a correct conclusion about that cause-and-effect relationship, it is a good experiment.

In our initial discussion of experiments, we just described what the experiment was without examining why the researchers chose to perform the experiment the way they did. In this section, we explore some of the ways to maximize an experiment's power, and we'll find that, with careful planning, it is possible to increase an experiment's ability to discern causes and effects.

First, let's consider some elements common to most experiments.

**1. Treatment:** any experimental condition applied to the research subjects. It might be the shaving of one of an individual's eyebrows, or the pattern used to show "suspects" (all at once or one at a time) to the witness of a staged crime, or a dosage of echinacea given to an individual.

**2. Experimental group:** a group of subjects who are exposed to a particular treatment—for example, the

individuals given echinacea rather than placebo in the experiment described above. It is sometimes referred to as the "treatment group."

**3. Control group:** a group of subjects who are treated identically to the experimental group, with one exception—they are not exposed to the treatment. An example would be the individuals given placebo rather than echinacea.

**4. Variables:** the characteristics of an experimental system that are subject to change. They might be, for example, the amount of echinacea a person is given, or a measure of the coarseness of an individual's hair. When we speak of "controlling" variables—the most important feature of a good experiment—we are describing the attempt to minimize any differences (which are also called "variables") between a control group and an experimental group other than the treatment itself. That way, any differences between the groups in the outcomes we observe are most likely due to the treatment.

Let's look at an example that illustrates the importance of considering all these elements when designing an experiment.

Stomach ulcers are erosions of the stomach lining that can be very painful. In the late 1950s, a doctor reported in the *Journal of the American Medical Association* that stomach ulcers could be effectively treated by having a patient swallow a balloon connected to some tubes that circulated a refrigerated fluid. He argued that by super-cooling the stomach, acid production was reduced and the ulcer symptoms relieved. He had convincing data to back up his claim: in all 24 of his patients who received this "gastric freezing" treatment, their condition improved (**FIGURE 1-14**). As a result, the treatment became widespread for many years.

Although there was a clear hypothesis ("Gastric cooling reduces the severity of ulcers") and some compelling

observations (all 24 patients experienced relief), this experiment was not designed well. In particular, there was no clear group with whom to compare the patients who received the treatment. In other words, who is to say that just going to the doctor or having a balloon put into your stomach doesn't improve ulcers? The results of this doctor's experiment do not rule out these interpretations.

A few years later, another researcher decided to do a more carefully controlled study. He recruited 160 ulcer patients and gave 82 of them the gastric freezing treatment. The

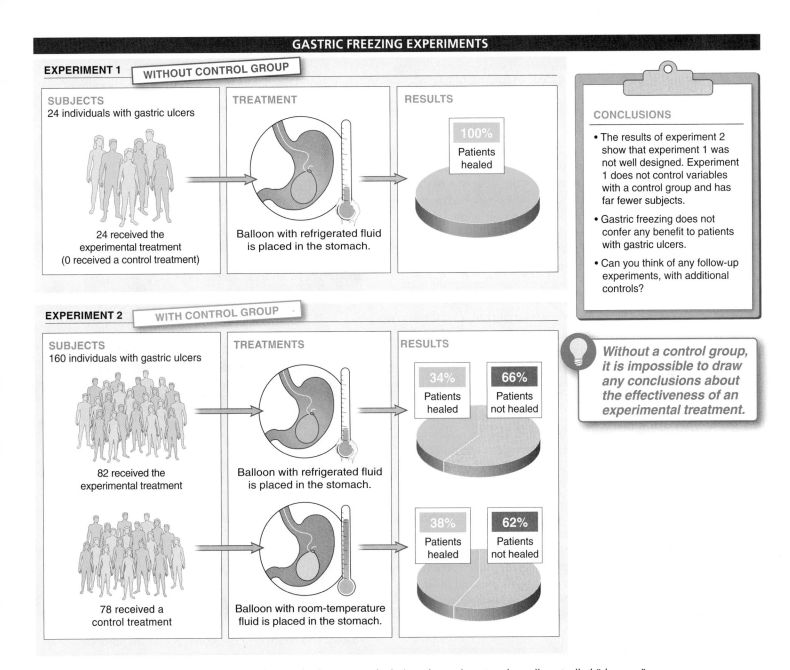

## GASTRIC FREEZING EXPERIMENTS

**EXPERIMENT 1** WITHOUT CONTROL GROUP

SUBJECTS
24 individuals with gastric ulcers

24 received the experimental treatment
(0 received a control treatment)

TREATMENT

Balloon with refrigerated fluid is placed in the stomach.

RESULTS

100%
Patients healed

CONCLUSIONS

• The results of experiment 2 show that experiment 1 was not well designed. Experiment 1 does not control variables with a control group and has far fewer subjects.

• Gastric freezing does not confer any benefit to patients with gastric ulcers.

• Can you think of any follow-up experiments, with additional controls?

**EXPERIMENT 2** WITH CONTROL GROUP

SUBJECTS
160 individuals with gastric ulcers

82 received the experimental treatment

78 received a control treatment

TREATMENTS

Balloon with refrigerated fluid is placed in the stomach.

Balloon with room-temperature fluid is placed in the stomach.

RESULTS

34%
Patients healed

66%
Patients not healed

38%
Patients healed

62%
Patients not healed

*Without a control group, it is impossible to draw any conclusions about the effectiveness of an experimental treatment.*

**FIGURE 1-14  No controls.** Gastric freezing and stomach ulcers: a poorly designed experiment and a well-controlled "do-over."

other 78 received a similar treatment in which they swallowed the balloon but had room-temperature water pumped in. The latter was an appropriate control group because the subjects were treated exactly like the experimental group, with the exception of only a single difference between the groups—whether they experienced gastric freezing or not. The new experiment could test for an effect of the gastric freezing, while controlling for the effects of other, lurking variables that might affect the outcome.

Surprisingly, although the researcher found that for 34% of those in the gastric freezing group their condition improved, he also found that 38% of those in the control group improved. These results indicated that gastric freezing didn't actually confer any benefit when compared with a treatment that did not involve gastric freezing. Not surprisingly, the practice was abandoned. (We'll discuss below why the control group might have had such a high rate of improvement.)

**Q**

**Is arthroscopic surgery for arthritis beneficial for the 300,000 people in the United States who have it each year? How do we know?**

More recently, a well-controlled experiment demonstrated that arthroscopic knee surgery, a treatment for osteoarthritis performed on more than 300,000 people in the United States each year, produced moderate benefits for the patient. But the study demonstrated even greater benefits for patients in a control group who were subjected not to the arthroscopic procedure but rather to "sham" surgery, in which the surgeon made three superficial incisions in the knee, manipulated the knee a bit, and clanged the instruments around in the operating room. As a consequence of this study, many doctors have begun abandoning the use of arthroscopic surgery for the treatment of osteoarthritis—which once accounted for more than a billion dollars in surgical procedures each year. (This type of surgery is still useful for the treatment of some other knee problems, including the repair of ligament and cartilage damage.)

A surprising result from both the gastric freezing studies and the arthroscopic surgery study was that they demonstrate the **placebo effect,** the frequently observed, poorly understood phenomenon in which people respond favorably to *any* treatment. The placebo effect highlights the need for an appropriate control group. We want to know whether the treatment is actually responsible for any effect seen; if the control group receiving the placebo or sham treatment has an outcome much like that of the experimental group, we can conclude that the treatment itself does not have an effect.

Another pitfall to be aware of in designing an experiment is to ensure that the persons conducting the experiment don't influence the experiment's outcome. An experimenter can often unwittingly influence the results of an experiment. This phenomenon is seen in the story of a horse named Clever Hans. Hans was considered clever because his owner claimed that Hans could perform remarkable intellectual feats, including multiplication and division. When given a numerical problem, the horse would tap out the answer number with his foot. Controlled experiments, however, demonstrated that Hans was only able to solve problems when he could see the person asking the question and when that person knew the answer (**FIGURE 1-15**). It turned out that the questioners, unintentionally and through very subtle body language, revealed the answers.

The Clever Hans phenomenon highlights the benefits of instituting even greater controls when designing an experiment. In particular, it highlights the value of **blind experimental design,** in which the experimental subjects do not know which treatment (if any) they are receiving, and

**FIGURE 1-15  Math whiz or ordinary horse?** The horse Clever Hans was said to be capable of mathematical calculations, until a controlled experiment demonstrated otherwise.

**double-blind experimental design,** in which neither the experimental subjects nor the experimenter know which treatment a subject is receiving.

Another hallmark of an extremely well-designed experiment is that it combines the blind/double-blind strategies we've just described in a **randomized,** controlled, double-blind study. In this context, "randomized" refers to the fact that, as in the echinacea study described above, the subjects are randomly assigned into experimental and control groups. In this way, researchers and subjects have no influence on the composition of the two groups.

The use of randomized, controlled, double-blind experimental design can be thought of as an attempt to imagine all the possible ways that someone might criticize an experiment and to design the experiment so that the results cannot be explained by anything other than the effect of the treatment. In this way, the experimenter's results either support the hypothesis or invalidate it—in which case, the hypothesis must be rejected. If multiple explanations can be offered for the observations and evidence from an experiment, then it has not succeeded as a critical experiment.

Suppose you want to know whether a new drug is effective in fighting the human immunodeficiency virus (HIV), the virus that leads to AIDS. Which experiment would be better, one in which the drug is added to HIV-infected cells in a test tube under carefully controlled laboratory conditions, or one in which the drug is given to a large number of HIV-infected individuals? There is no definitive answer. In laboratory studies, it is possible to control nearly every environmental variable. In their simplicity, however, lab studies may introduce difficulties in coming to broader conclusions about a drug's effectiveness under real-world conditions. For example, complex factors in human subjects, such as nutrition and stress, may interact with the experimental drug and influence its effectiveness. These interactions will not be present and taken into account in the controlled lab study.

Good experimental design is more complex than simply following a single recipe. The only way to determine the quality of an experiment is to assess how well the variables that were not of interest were controlled, and how well the experimental treatment tested the relationship of interest.

> ## TAKE-HOME MESSAGE 1·11
>
> To draw clear conclusions from experiments, it is essential to hold constant all those variables we are not interested in. Control and experimental groups should differ only with respect to the treatment of interest. Differences in outcomes between the groups can then be attributed to the treatment.

# 1·12 Repeatable experiments increase our confidence.

**Q**

*Can science be misleading? How can we know?*

In 2005, a study showed that some patients being treated for HIV infection who were also taking an epilepsy drug called valproic acid had significantly reduced numbers of HIV particles in their blood. Newspaper headlines announced this finding—"AIDS Cure Possible, Study Suggests"—and raised many people's hopes. Two years later, however, a study of people who were already taking valproic acid and anti-HIV drugs concluded that those taking valproic acid had not benefited at all from the drug (**FIGURE 1-16**). It's not certain why the later study didn't produce the same results as the first study, but the second study did address two serious shortcomings of the first study—namely, that the first study was much smaller, involving only four patients, and that no control group was used. This pair of studies reveals the importance of repeatability in science.

A powerful way to demonstrate that observed differences between a treatment group and a control group truly reflect the effect of the treatment is for the researchers to conduct the experiment over and over again. Even better is to have other research groups repeat the experiment and get the same results. Researchers describe this desired characteristic of experiments by saying that an experiment must be "reproducible" and "repeatable."

An experiment that can be done over and over again by a variety of researchers to give the same results is an effective defense against biases (which we discuss in the next

**Q**

*Do megadoses of vitamin C reduce cancer risk?*

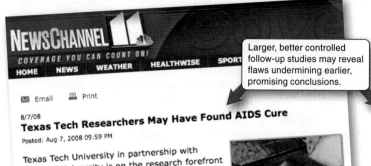

✉ Email    🖨 Print

8/7/08
**Texas Tech Researchers May Have Found AIDS Cure**

Posted: Aug 7, 2008 09:59 PM

Texas Tech University in partnership with Harvard University is on the research forefront of a possible vaccine for HIV, the virus that causes AIDS.

To give you some idea of how big this story is, on Thursday it was the number one story on Newsweek's website and it will be the lead in their next magazine.

Larger, better controlled follow-up studies may reveal flaws undermining earlier, promising conclusions.

*Repeatability is essential! Scientific conclusions are more reliable when experiments have been repeated (and modified, if necessary).*

## Valproic Acid Does Not Reduce Latent HIV Reservoir in Resting CD4 Cells

By Liz Highleyman

Some prior studies have suggested that valproic acid (Depakote), a drug used to treat epilepsy and bipolar disorder, might help eradicate HIV from the body. Two studies published in the June 19, 2008 issue of *AIDS*, however, indicate that it has little long-term effect.

Valproic Acid (Depakote)

### Study 1

David Margolis and colleagues from the University of North Carolina previously reported that valproic acid plus intensified antiretroviral therapy could "flush" HIV out of long-lived latently infected CD4 T-cells and potentially reduce the resting CD4 cell reservoir. They later conducted a study to test the ability of valproic acid to deplete resting CD4 cell infection in patients receiving standard HAART.

Infection of resting CD4 cells was measured in 11 patients with stable undetectable HIV viral load using a standard HIV RNA assay twice rior to, and twice after, adding 1000 mg valproic acid to their standard antiretroviral regimen. Resting CD4 T-cell infection frequency was measured by outgrowth assay, and low-level viremia was quantified in 9 patients using a highly sensitive single-copy plasma HIV RNA assay.

**FIGURE 1-16  Once is not enough.** Experiments and their outcomes must be repeatable for their conclusions to be valid and widely accepted.

section) and reflects a well-designed experiment. Experiments whose results cannot be confirmed by repeated experiments or by experiments performed by other researchers are the downfall of many dramatic claims. Even though chemist Linus Pauling had not one, but two Nobel prizes, he was never able to convince the scientific community that megadoses of vitamin C are an effective treatment for cancer. Every time other individuals tried to repeat Pauling's studies, properly matching the control and treatment groups, they found no difference. Vitamin C just isn't effective against cancer—a conclusion that the vast majority of the medical community now believes. The scientific method is profoundly egalitarian: more important than a scientist's credentials are sound and reproducible results.

When a study is repeated (also referred to as "replicating" a study), sometimes a tiny variation in the experimental design can lead to a different outcome; this can help us isolate the variable that is primarily responsible for the outcome of the experiment. Alternatively, when experiments are repeated and the same results are obtained, our confidence in them is increased.

> ### TAKE-HOME MESSAGE 1·12
>
> **Experiments and their outcomes must be repeatable for their conclusions to be considered valid and widely accepted.**

# 1·13    We've got to watch out for our biases.

**Q** Can scientists be sexist? How would we know?

In 2001, the journal *Behavioral Ecology* changed its policy for reviewing manuscripts that were submitted for publication. Its new policy instituted a double-blind process, whereby neither the reviewers' nor the authors' identities were revealed. Previously, the policy had been a single-blind process in which reviewers' identities were kept secret but the authors' identities were known to the reviewers. In an analysis of papers published between 1997 and 2005, it turned out that after 2001, when the double-blind policy took effect, there was a significant increase in the number of published papers in which the first author was female (**FIGURE 1-17**). Analysis of papers published in a similar journal that maintained the single-blind process over that period revealed no such increase.

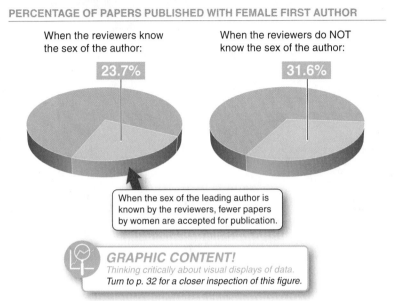

PERCENTAGE OF PAPERS PUBLISHED WITH FEMALE FIRST AUTHOR

When the reviewers know the sex of the author: **23.7%**

When the reviewers do NOT know the sex of the author: **31.6%**

When the sex of the leading author is known by the reviewers, fewer papers by women are accepted for publication.

**GRAPHIC CONTENT!**
*Thinking critically about visual displays of data.*
*Turn to p. 32 for a closer inspection of this figure.*

**FIGURE 1-17  Bias against female scientists?** The journal *Behavioral Ecology* accepted more papers from female authors when the reviewers were not aware of the author's sex.

This study reveals that people, including scientists, may have biases—sometimes subconscious—that influence their behavior. It also serves as a reminder of the importance of proper controls in experiments. If knowing the sex of the author influences a reviewer's decision on whether a paper should be published, it is also possible that researchers' biases can creep in and influence their collection of data and analysis of results. (Even the decision about what—and what not—to study can be influenced by our biases.)

It can be hard to avoid biases. Consider a study that required precise measuring of the fingers of the left and right hands—comparing the extent to which people in different groups were physically symmetrical. The researcher noted that when she measured individuals from the group she predicted would be more symmetrical, she felt a need to re-measure if the reading on her digital ruler indicated a big asymmetry. Because she was thinking that the person's left- and right-hand fingers *should* be symmetrical, she assumed she had made an error. She felt no such need to re-measure for subjects in the other group, for whom asymmetries confirmed her hypothesis. To control for this bias, the researcher connected her digital ruler to a computer and, without ever having the number displayed, transmitted the measurement directly to the computer when she pushed a button. In this way, she was able to make each measurement without introducing any regular bias.

---

### TAKE-HOME MESSAGE 1·13

Biases can influence our behavior, including our collection and interpretation of data. With careful controls, it is possible to minimize such biases.

# 1·14–1·17 Scientific thinking can help us make wise decisions.

*What can you believe? Reading labels is essential to evaluating products and the claims about them.*

# 1·14
## Visual displays of data can help us understand and explain phenomena.

"Let's look at the data." Using points and lines and symbols and a variety of other graphic elements to display measured quantities is a powerful tool commonly used throughout the sciences. Such visual displays of data can serve a range of purposes, which relate to both the presentation of and the exploration of the data.

Whether making a point, illustrating an idea, or facilitating the testing of a hypothesis, visual displays of data typically have one feature in common: they condense large amounts of information into a more easily digested form. In doing so, they can help readers think about and compare data, ultimately helping them to synthesize the information and see useful patterns.

There is an almost infinite variety of ways to display data, including maps, tables, charts, and graphs. Graphs are particularly prevalent in biology, and a few forms are used most frequently. These include bar graphs, line graphs, and pie charts (**FIGURE 1-18**).

Visual displays of data generally have a few common elements. Most have a title, for example, which usually appears at the top and describes the content of the display. Bar graphs and line graphs include axes, usually a horizontal axis, also called

**COMMON VISUAL DISPLAYS OF DATA USED IN BIOLOGY**

**BAR GRAPH**
Rectangular bars are used to represent data, each with a height that is proportional to the value being represented.

**LINE GRAPH**
A line or curve may be used to connect data points or to illustrate trends across many data points.

**PIE CHART**
"Slices" are used to represent data, in which each slice is a proportion of the whole.

**FIGURE 1-18 Presenting what we have observed, precisely and concisely.**

the x-axis, and a vertical axis, called the y- axis. Each axis has a scale, generally labeled with some gradations, indicating one dimension by which the data can be described.

The x-axis of a line graph, for example, might describe the number of hours a student spends studying for a class, in which case it would be labeled "Time spent studying each day (hrs)," while the y-axis might be labeled "Performance on midterm exams (%)." Depending on the size of the data set (the total number of observations collected in the experiment that relate an individual's performance on a midterm and the number of hours that student spent studying each day), individual data points are included on the graph, with additional information conveyed by the shape, color, or pattern of the data points (**FIGURE 1-19**). One data point might reflect that one student spent 2 hours studying per day and scored 85% on a midterm exam, while another spent 1 hour per day and scored 94% on the midterm exam. A line or curve may be used to connect data points or to illustrate a relationship between the two variables.

Rather than displaying individual data points, a bar graph has rectangular bars, each with a height proportional to the value being represented—maybe hours spent studying in a particular class. In a pie chart, each "slice" is a proportion of the whole—for instance, with each slice representing the portion of a student's total number of study hours spent on each different topic every day. A legend may be included for

the graph, identifying which information is represented by which bar or data point or pie slice.

One of the most common functions of visual displays of information is to present the relationship between two variables, such as in a graph. An **independent variable** is some measurable entity that is available (i.e., can be measured) at the start of a process and whose value can be changed as required. A **dependent variable** is one that is created by the process being observed and whose value cannot be controlled. The dependent variable is generally represented by the y-axis and is expected to change in response to a change in the independent variable, represented on the x-axis. The number of hours of sleep a student gets each night, for example, could be thought of as an independent variable, while some measure of academic performance—maybe grade point average—would be a dependent variable.

Visual displays of data can be simple and straightforward, but they can also have features that reduce their effectiveness, or even cause them to be downright misleading. These difficulties can arise from ambiguity in the axis labels or scales, or incomplete information on how each data point was collected (and how the points might have varied), or biases or hidden assumptions in the presentation or grouping of data, or unknown or unreliable sources of data, or an insufficient or inappropriate context given for the data presentation.

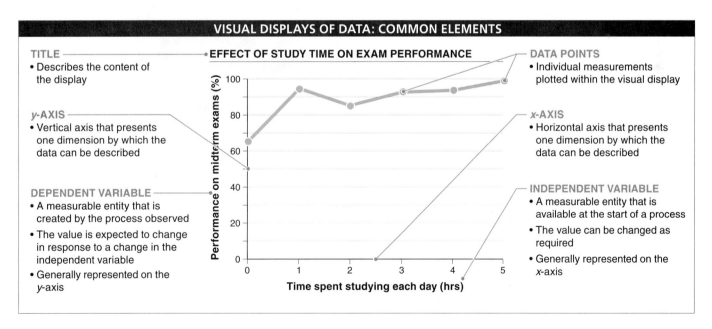

**VISUAL DISPLAYS OF DATA: COMMON ELEMENTS**

**TITLE**
• Describes the content of the display

**y-AXIS**
• Vertical axis that presents one dimension by which the data can be described

**DEPENDENT VARIABLE**
• A measurable entity that is created by the process observed
• The value is expected to change in response to a change in the independent variable
• Generally represented on the y-axis

**EFFECT OF STUDY TIME ON EXAM PERFORMANCE**

**DATA POINTS**
• Individual measurements plotted within the visual display

**x-AXIS**
• Horizontal axis that presents one dimension by which the data can be described

**INDEPENDENT VARIABLE**
• A measurable entity that is available at the start of a process
• The value can be changed as required
• Generally represented on the x-axis

**FIGURE 1-19  Common elements of effective graphical presentations of data.** Shown here: an example of a line graph.

In each chapter of this book, you will see that one of the visual displays of data is labeled with the "Graphic Content" icon (you'll have noticed this on Figure 1-17). For each of these figures, in the Check Your Knowledge section at the end of the chapter you'll find several questions about the figure's content. These questions will guide you in evaluating aspects of the figure (such as: "What is its main point?") and in extracting information from it. They will also help you understand when it might be warranted to approach a data display with more than the usual amount of skepticism, and when greater trust is reasonable.

# 1·15 Statistics can help us in making decisions.

In Section 1-12 we saw that researchers repeatedly found, through experimentation, that megadoses of vitamin C do not reduce cancer risk. If you put yourself in the researchers' shoes, you might wonder how you'd figure out that the vitamin C did not reduce cancer risk. Perhaps there were 100 individuals in the group receiving megadoses of vitamin C: some of them developed cancer and some of them did not. And among the 100 subjects in the group not receiving the megadoses, some of the individuals developed cancer and some did not. How do you decide whether the vitamin C actually had an effect? This knowledge comes from a branch of mathematics called **statistics,** a set of analytical and mathematical tools designed to help researchers gain understanding from the data they gather. To understand statistics, let's start with a simple situation.

Suppose you measured the height of two people. One is a woman who is 5 feet 10 inches tall. The other is a man who is 5 feet 6 inches tall. If these were your only two observations of human height, you might conclude that female humans are taller than males. But suppose you measured the height of 100 women and 100 men chosen randomly from a population. Then you can say, "of the 100 men, the average man is 5 feet 9.5 inches, and of the 100 women, the average woman is 5 feet 4 inches." Better still, the data can illuminate for you not only the average but also some measure of how much variation there is from one individual to another. Statistical analysis can tell you not only that the average man in this study is 5 feet 9.5 inches tall but that two-thirds of the men are between 5 feet 6.5 inches tall (3 inches less than the average) and 6 feet 0.5 inches tall (3 inches more than the average). You will often see this type of range stated as "5 feet 9.5 inches ± 3 inches" ("plus or minus 3 inches"). Similarly, the data might show that the female subjects in the study are "5 feet 6 inches ± 3 inches," indicating that two-thirds of the women are between 5 feet 3 inches and 5 feet 9 inches tall. As we discussed in Section 1-11, on experimental design, and as this example shows, larger numbers of participants are better than smaller numbers if you want to draw general conclusions

about natural phenomena such as the height of men and women (**FIGURE 1-20**).

**HUMAN HEIGHT AND STATISTICS**

*Don't get fooled! One or two observations isn't enough to draw general conclusions about natural phenomena, such as height.*

| AVERAGE FEMALE | AVERAGE MALE |
| 5′4″ ± 3″ | 5′9.5″ ± 3″ |

**FIGURE 1-20 Drawing conclusions based on limited observations is risky.** Measuring a greater number of people will generally help us draw more accurate conclusions about human height.

Using data to describe the characteristics of individuals participating in a study is useful, but often we want to know whether data support (or do not support) a hypothesis. If the scientific method is to be effective in helping us understand the world, it must help us make wise decisions about concrete things. For example, suppose we want to know whether having access to a textbook helps a student perform better in a biology class. Statistics can help us answer this question. After conducting a study, let's say we find that students who had access to a textbook scored an average of 81% ± 8% on their exams, while those who did not scored an average of 76% ± 7%.

In this example, it is difficult to distinguish between the two possible conclusions.

*Possibility 1:* Students having access to a textbook *do perform better* in biology classes. In other words, our sampling of this class reveals a true relationship between the two variables, textbook access and class performance.

*Possibility 2:* Students having access to a textbook *do not perform better* in biology classes. The variation in the scores for the two groups may be too large to allow us to conclude that there's any effect of having access to a textbook. Instead, the difference in average scores for the two groups may mean that more of the high-performing students just happened, through random chance, to be in the group given access to a textbook.

But what if the students with access to a textbook scored, on average, 95% ± 5%, while those without access scored only 60% ± 5%? In this case, we would be much more confident that there is a significant effect of having access to a textbook (**FIGURE 1-21**), because even with the large variation in the scores seen in each group, the two averages are still very different from each other.

Statistical methods help us to decide between these two possibilities and, equally important, to state how confident we are that one or the other is true. The greater the difference between two groups (95% vs. 60% is a greater difference than 81% vs. 76%), and the smaller the variation in each group (± 5% in the 95% and 60% groups, vs. ± 8% and ± 7% in the 81% and 76% groups), the more confident we are of the conclusion that there is a significant effect of the treatment (having access to a textbook, in this case). In other words, in the case where the groups of students scored 95% or 60%, depending on whether they did or did not have access to a textbook, it is possible that having a textbook did not actually improve performance and that this observed difference was just the result of chance. But this conclusion is very, very unlikely.

Statistics can also help us identify relationships (or lack of relationships) between variables. For example, we might note that when there are more firefighters at a fire, the fire is larger and causes more damage. This is a **positive correlation,**

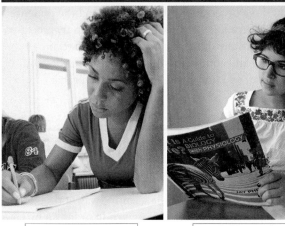

**INTERPRETING STATISTICS: A HYPOTHETICAL STUDY**

Students without textbooks score an average of 60% ± 5%

Students with textbooks score an average of 95% ± 5%

Proportion of students in population

Average exam scores (%)

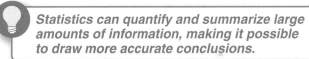

*Statistics can quantify and summarize large amounts of information, making it possible to draw more accurate conclusions.*

**FIGURE 1-21 Drawing conclusions based on statistics.** Statistical analyses can help us organize and summarize the observations that we make and the evidence we gather in an experiment.

meaning that when one variable (the number of firefighters) increases, so does the other (the severity of the fire). Should we conclude that firefighters make fires worse? No. While correlations can reveal relationships between variables, they don't tell us *how* the variables are related or whether change in one variable causes change in another. (You may have heard or read the phrase "correlation is not causation," which refers to this sort of situation.) Before drawing any conclusions about more firefighters causing larger fires, we need to know about the type of fire and its size when the firefighters arrived, because those factors will significantly influence the ultimate amount of damage. To estimate the effect of the number of firefighters on the amount of damage, we would need to compare the amount of damage from fires of similar size that are fought by different numbers of firefighters.

Ultimately, statistical analyses can help us organize and summarize the observations that we make and the evidence we gather in an experiment. These analyses can then help us decide whether any differences we measure between experimental and control groups are likely to be the result of the treatment, and how confident we can be in that conclusion.

# 1·16 Pseudoscience and misleading anecdotal evidence can obscure the truth.

One of the major benefits of the scientific way of thinking is that it can prevent you from being taken in or fooled by false claims. There are two types of "scientific evidence" that frequently are cited in the popular media and are responsible for people erroneously believing that links between two things exist, when in fact they do not.

1. **Pseudoscience,** in which individuals make scientific-sounding claims that are not supported by trustworthy, methodical scientific studies.

2. **Anecdotal observations,** in which, based on just one or a few observations, people conclude that there is or is not a link between two things.

Pseudoscience is all around us, particularly in the claims made on the packaging of consumer products and food (**FIGURE 1-22**). Beginning in the 1960s, for example, consumers encountered the assertion by the makers of a sugarless gum that "4 out of 5 dentists surveyed recommend sugarless gum for their patients who chew gum." Maybe the statement is factually true, but the general relationship it implies may not be. How many dentists were surveyed? If the gum makers surveyed only five dentists, then the statement may not represent the proportion of *all* dentists who would make such a recommendation. And how were the dentists sampled? Were they at a shareholders meeting for the sugarless gum company? What alternatives were given—perhaps gargling with a tooth-destroying acid? You just don't know. That's what makes it pseudoscience.

Pseudoscience capitalizes on a belief shared by most people: that scientific thinking is a powerful method for learning about the world. The problem with pseudoscience is that the scientific bases for a scientific-sounding claim are not clear. The claims generally sound reasonable, and they are persuasive in convincing people to purchase one product over another. But one of the beauties of real science is that you never have to just take someone's word about something. Rather, you are free to evaluate people's research methods and results, and then decide for yourself whether their conclusions and claims are appropriate.

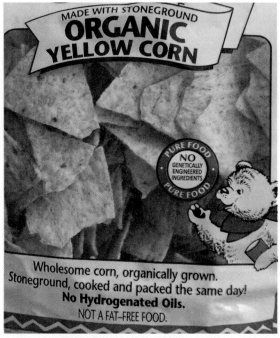

**FIGURE 1-22 Pseudoscientific claims are often found on food products.**

We are all familiar with anecdotal evidence. Striking stories that we read or hear, or our own experiences, can shape our views of cause and effect. We may find compelling parallels between suggestions made in horoscopes and events in our lives, or we may think that we have a lucky shirt, or we may be moved by a child whose cancer went into remission following treatment involving eating apricot seeds. Yet, despite lacking a human face, data are more reliable than anecdotes, primarily because they can illustrate a broader range of observations, capturing the big picture.

Anecdotal observations can seem harmless and can be emotionally powerful. But because they do not include a sufficiently large and representative set of observations of the world, they can lead people to draw erroneous conclusions, often with disastrous consequences. One important case of anecdotal evidence being used to draw general conclusions involves autism, a developmental disorder that impairs social interaction and communication, and the vaccination for measles, mumps, and rubella (commonly called the MMR vaccine) that is given to most children.

**Does the measles, mumps, and rubella vaccine cause autism?**

In 1998, the prestigious medical journal *The Lancet* published a report by a group of researchers that described a set of symptoms (diarrhea, abdominal pain, bloating) related to bowel inflammation in 12 children who exhibited the symptoms of autism. The parents or physicians of 8 of the children in the study said that the behavioral symptoms of autism appeared shortly after the children received MMR vaccination. For this reason, the authors of the report recommended further study of a possible link between the MMR vaccine, the bowel problems, and autism.

In a press conference, one of the paper's authors suggested a link between autism and the MMR vaccine, recommending single vaccines rather than the MMR triple vaccine until it could be proved that the MMR vaccine did not trigger autism. This statement caused a major health scare that only recently has begun to subside. Noting that there had been a significant increase in the incidence of autism in the 1990s and early 2000s, the press and many people took the claims made by the researcher at the press conference as evidence that the MMR vaccine causes autism. Over the course of the next few years, the number of children getting the MMR vaccine dropped significantly, as parents sought to reduce the risk of autism in their children (**FIGURE 1-23**).

Unfortunately, this is a notable case of poor implementation of the scientific method, with numerous flaws leading to an incorrect conclusion. Most important among these is that the study was small (only 12 children), the sample was carefully selected rather than randomized (that is, the researchers selected the study participants based on the symptoms they showed), and no control group was included for comparison (for example, children who had been vaccinated but did not exhibit autism symptoms). On top of this, the study, even flawed as it was, did not actually find or even report a link between autism and the MMR vaccine. That purported link came only from the unsupported, suggestive statements made by one of the study's authors at the press conference.

On their own, the design flaws in the study would be sufficient to invalidate the one author's claims about the link between autism and the MMR vaccine, but there are even more weaknesses in the research. It later was discovered that before the paper's publication, the study's lead author (the one who spoke at the press conference) had received large sums of money from lawyers seeking evidence to use in lawsuits against the MMR vaccine manufacturers. It also came to light that he had applied

 *A misguided fear of "catching" autism has caused some parents to decline immunizations for their children.*

**FIGURE 1-23  Bad science became headline news.**

for a patent for a vaccine that was a rival to the most commonly used MMR vaccine. In the light of these initially undisclosed biases, 10 of the paper's 12 authors published a retraction of their original interpretation of their results. And in 2011, the *BMJ* (*British Medical Journal*) published an article declaring that the "article linking MMR vaccine and autism was fraudulent."

In the years since the original paper was published, many well-controlled, large-scale studies have conducted critical experiments on this subject, and all have been definitive in their conclusion that there is no link between the MMR vaccine and autism. Here are a few of these studies.

**1.** A study of all children born in Denmark between 1991 and 1998 found no difference in the incidence of autism among the 440,655 children who were vaccinated with the MMR vaccine and the 96,648 children who were not vaccinated.

**2.** A 2005 study in Japan showed that after use of the MMR vaccine was stopped in 1993, the incidence of autism continued to increase.

**3.** A study of 1.8 million children in Finland who were followed up for 14 years after getting the MMR vaccine found no link at all between the occurrence of autism and the vaccine.

At this point, the consensus of the international scientific community is that there is no scientific evidence for a link between the MMR vaccine and autism.

> " Science is a way to call the bluff of those who only pretend to knowledge. It is a bulwark against mysticism mysticism, against superstition, against religion misapplied to where it has no business being. If we're true to its values, it can tell us when we're being lied to. "
>
> —CARL SAGAN,
> in *The Demon-Haunted World:*
> *Science as a Candle in the Dark,* 1997

So what explains the observation that there are more autism cases now than in the past? It seems that this is a function of the better identification of autism by doctors and changes in the process by which autism is diagnosed. Another reason for the perceived link between autism and the MMR vaccine was simply the coincidence that most children receive the vaccine at around 18–19 months of age, which happens to be the age at which the first symptoms of autism are usually noticed. In the end, what we learn from this is that we must be wary that we do not generalize from anecdotal observations or let poorly designed studies obscure the truth.

**TAKE-HOME MESSAGE 1·16**

Pseudoscience and anecdotal observations often lead people to believe that links between two phenomena exist, when in fact there are no such links.

# 1·17 There are limits to what science can do.

The scientific method is a framework that helps us make sense of what we see, hear, and read in our lives. There are limits, however, to what science can do.

The scientific method will never prove or disprove the existence of God. Nor is it likely to help us understand the mathematical elegance of Fermat's last theorem or the beauty of Shakespeare's sonnets. As one of several approaches to the acquisition of knowledge, the scientific method is, above all, empirical. It differs from non-scientific approaches such as mathematics and logic, history, music, and the study of artistic expression in that it relies on *measuring* phenomena in some way. The generation of value judgments and other types of non-quantifiable, subjective information—such as religious assertions of faith—falls outside the realm of science. Despite all of the intellectual analyses that the scientific method gives rise to and the objective conclusions it makes possible, it does not, for example, generate moral statements and it cannot give us insight into ethical problems. What "is" (i.e., what we observe in the natural world) is not necessarily what "ought" to be (i.e., what is morally right). It may or may not be.

Further, much of what is commonly considered to be science, such as the construction of new engineering marvels or the heroic surgical separation of conjoined twins, is not scientific at all. Rather, these are technical innovations and developments. While they frequently rely on sophisticated scientific research, they represent the application of research findings to varied fields such as manufacturing and medicine to solve problems (**FIGURE 1-24**).

**FIGURE 1-24 The application of science.** Applying scientific findings to solve a problem can result in sophisticated technical innovations, like this thought-controlled prosthetic arm.

As we begin approaching the world from a more scientific perspective, we can gain important insights into the facts of life, yet must remain mindful of the limits to science.

**TAKE-HOME MESSAGE 1·17**

Although the scientific method may be the most effective path toward understanding the observable world, it cannot give us insights into the generation of value judgments and other types of non-quantifiable, subjective information.

# 1·18 On the road to biological literacy: what are the major themes in biology?

*Contemplating unity and diversity. An entomologist organizes and categorizes specimens.*

## 1·18 A few important themes tie together the diverse topics in biology.

Driven forward by the power of the scientific method, modern biology is in the midst of an explosion of new and exciting developments. Now more than ever, biology is influencing and even transforming our lives. Its increasing relevance outside the world of scientists is felt in a multitude of areas, including breakthroughs in agriculture, innovative developments in criminology, and increased understanding and awareness of environmental issues and the need for widespread changes in behavior.

In this guide to biology, as we explore the many facets of biology and its relevance to life in the modern world, you will find two unifying themes recurring throughout.

**Hierarchical organization.** Life is organized on many levels within individual organisms, including atoms, cells, tissues, and organs. And in the larger world, organisms themselves are organized into many levels: populations, communities, and ecosystems within the biosphere.

**The power of evolution.** Evolution, the change in genetic characteristics of individuals within populations over time, accounts for the diversity of organisms, but also explains the unity among them.

These two major themes connect the diverse topics, as we make our way through four chief areas of focus:

1. The chemical, cellular, and energetic foundations of life

2. The genetics, evolution, and behavior of individuals

3. The staggering diversity of life and the unity underlying it

4. Ecology, the environment, and the subtle and important links between organisms and the world they inhabit

Armed with an appreciation for the scientific method and the broader themes you'll encounter in your study of biology, let's continue our exploration of life!

> **TAKE-HOME MESSAGE 1·18**
>
> **Although the diversity of life on earth is tremendous, the study of life is unified by the themes of hierarchical organization and the power of evolution.**

## RAINY DAYS AND MONDAYS.

**Q:** **WHAT'S THE MOST IMPORTANT OUTCOME OF SCIENTIFIC THINKING?** "Scientific thinking tells us when to change our mind about the natural world." Let's consider a question that might seem silly–even trivial–and watch our minds open.

**Q:** **IT OFTEN SEEMS TO RAIN MORE ON WEEKENDS. COULD IT BE TRUE?** Before you read on, write down your one word answer (and a justification for it, if you'd like).

**Q:** **WELL, DOES IT RAIN MORE ON WEEKENDS?** At first, it seems that the answer must be "NO." How could the weather even "know" what day it is? Weather, after all, existed long before humans, calendars, and Monday mornings.

**Q:** **RATHER THAN RELYING ON COMMON SENSE, LET'S THINK SCIENTIFICALLY.**

*My hypothesis:* It rains more on weekends.

*Restated as a null hypothesis:* The amount of rainfall does not differ across the days of the week.

*A testable prediction:* The amount of rain rainfall should not differ depending on what day of the week it is.

And now the data . . .

In a 1998 study published in the journal *Nature,* researchers using 17 years of data analyzed rainfall along the eastern coast of North America. Here's what they found:

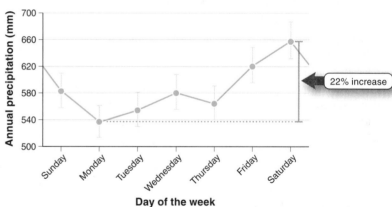

There was 22% more rain on Saturdays than on Mondays! (And this difference was statistically significant.)

**Q:** **HOW CAN THIS BE?** The researchers also tracked the pollutants carbon monoxide and ozone and found them both to gradually increase from Monday through Friday. This pattern, they suggested, is probably related to accumulations from driving, which is greater during the week than on weekends. They suggested that the weekly pollution cycle causes increased cloud formation and rain on weekends (while the reduced pollution on Mondays leads to reduced rainfall). In other words human activities are influencing the weather!

**Q:** **ONE GOOD HYPOTHESIS DESERVES ANOTHER.** If their suggestion is true, we should expect the "Rainy Saturday/Dry Monday" phenomenon to occur only in the vicinity of human activities. As a test of this hypothesis, the researchers measured daily rainfall patterns over the oceans in the northern hemisphere, away from large human populations. Here they found that one-seventh of the rain fell on each day, with no days rainier than others.

**Q:** **THE TAKE HOME MESSAGE HERE?** Scientific thinking rewards open minds with satisfying answers.

# Big Ideas in ... Scientific Thinking

## 1·1–1·3 Science is a collection of facts and a process for understanding the world.

- Through objective observation, description, and experimentation, science is a pathway by which we can understand the world around us. Biological issues permeate all aspects of our lives.
- To make wise decisions, individuals and societies must attain biological literacy.
- There are numerous ways of gaining an understanding of the world. Because it is empirical, rational, testable, repeatable, and self-correcting, the scientific method is a particularly effective approach.

## 1·4–1·10 A beginner's guide: what are the steps of the scientific method?

- The scientific method (observation, hypothesis, prediction, test, and conclusion) is a flexible, adaptable, and efficient pathway to understanding the world, because it tells us when we must change our beliefs.
- The scientific method begins by making observations about the world.
- A hypothesis is a proposed explanation for a phenomenon. For a hypothesis to be useful, it must generate a testable prediction.
- A critical experiment is one that makes it possible to decisively determine whether a particular hypothesis is correct. Based on the results of experimental tests, we can revise our hypothesis and explain the world with increasing accuracy.
- A great strength of scientific thinking is that it helps us understand when we should change our minds.
- Scientific theories are hypotheses—proposed explanations for natural phenomena—that have been so strongly supported by empirical observations that the scientific community views them as very unlikely to be altered by new evidence.

## 1·11–1·13 Well-designed experiments are essential to testing hypotheses.

- To draw clear conclusions from experiments, it is essential to hold constant all those variables we are not interested in.
- Control and experimental groups should vary only with respect to the treatment of interest. Differences in outcomes between the groups can then be attributed to the treatment.
- Experiments and their outcomes must be repeatable for their conclusions to be valid and widely accepted.
- Biases can influence our behavior, including our collection and interpretation of data. With careful controls, we can minimize such biases.

## 1·14–1·17 Scientific thinking can help us make wise decisions.

- Visual displays of data condense large amounts of information and can aid in the presentation and exploration of the data, but care must be taken to evaluate the presentation of the data and their underlying truth.
- Because much variation exists in the world, statistics can help us evaluate whether differences between a treatment group and control group can be attributed to the treatment rather than random chance.
- Pseudoscience and anecdotal observations often lead people to believe that links between two phenomena exist, when there are no such links.
- Although the scientific method may be the most effective path toward understanding the observable world, it cannot give us insights into non-quantifiable, subjective information.

## 1·18 On the road to biological literacy: what are the major themes in biology?

- Although the diversity of life on earth is tremendous, the study of life is unified by the themes of hierarchical organization and the power of evolution.

See the To-Go Guide (packaged with every new copy of this textbook) for an illustrated summary of the chapter.

# Check Your Knowledge

## GRAPHIC CONTENT

*Thinking critically about visual displays of data.*

**1.** What does the green portion of the pie chart represent? What does the blue portion represent? What does the "whole pie" represent?

**2.** Why are there two pie charts?

**3.** What can you conclude from this figure?

**4.** What might be an alternative explanation (other than bias against female scientists) for the phenomenon shown here?

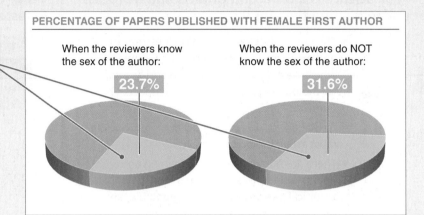

PERCENTAGE OF PAPERS PUBLISHED WITH FEMALE FIRST AUTHOR

When the reviewers know the sex of the author: 23.7%

When the reviewers do NOT know the sex of the author: 31.6%

**5.** How big was the increase in number of papers published with female first author following the institution of a double-blind review policy?

**6.** Why isn't the percentage of papers published with female first author 50%, even after the new review policy? Is that cause for concern? What data would help you to answer that question?

**7.** Do these graphs prove a general bias against female scientists?

See Answers at the back of the book.

## Key Terms in Scientific Thinking

anecdotal observation, p. 26
biological literacy, p. 4
biology, p. 4
blind experimental design, p. 18

control group, p. 16
critical experiment, p. 12
dependent variable, p. 23
double-blind experimental design, p. 19
empirical, p. 6

experimental group, p. 16
hypothesis (*pl.* hypotheses), p. 9
independent variable, p. 23
null hypothesis, p. 10
placebo, p. 13

placebo effect, p. 18
positive correlation, p. 25
pseudoscience, p. 26
randomized, p. 19
science, p. 2
scientific literacy, p. 4

scientific method, p. 6
statistics, p. 24
superstition, p. 5
theory, p. 15
treatment, p. 16
variable, p. 16

# Multiple-Choice Questions

1. **Science is:**

   a) a field of study that requires certain "laws of nature" to be taken on faith.

   b) both a body of knowledge and an intellectual activity encompassing observation, description, experimentation, and explanation of natural phenomena.

   c) a process that can be applied only within the scientific disciplines, such as biology, chemistry, and physics.

   d) the only way to understand the natural world.

   e) None of the above are correct.

2. **All of the following are elements of biological literacy except:**

   a) the ability to use the process of scientific inquiry to think creatively about real-world issues having a biological component.

   b) reading the most important books in biology.

   c) the ability to integrate into your decision making a consideration of issues having a biological component.

   d) the ability to communicate with others about issues having a biological component.

   e) All of the above are elements of biological literacy.

3. **Superstitions are:**

   a) held by many humans, but not by any non-human species.

   b) just one of many possible forms of scientific thinking.

   c) true beliefs that have yet to be fully understood.

   d) irrational beliefs that actions not logically related to a course of events influence its outcome.

   e) proof that the scientific method is not perfect.

4. **In a recent study, patients treated with a genetically engineered heart drug were able to walk on a treadmill 26 seconds longer than those not receiving the drug, and they showed no side effects from taking the drug. Can we conclude that this drug is an effective treatment for heart disease?**

   a) No. Genetically engineered drugs cannot be tested by the scientific method; they require comparative observations.

   b) No. It is not clear how many subjects were in the study.

   c) No. It is not clear that the drug is not a placebo.

   d) Yes.

   e) No. It is not clear that the proper controls were included.

5. **Empirical results:**

   a) rely on intuition.

   b) are generated by theories.

   c) are based on observation.

   d) cannot be replicated.

   e) must support a tested hypothesis.

6. **In a well-designed experiment:**

   a) the prediction will be highly probable if the experiment shows the explanation is correct.

   b) the prediction will be highly improbable if the experiment shows the explanation is incorrect.

   c) the null hypothesis will not be tested.

   d) the prediction will most likely be correct.

   e) Both a) and b) are correct.

7. **Which of the following statements is correct?**

   a) A hypothesis that does not generate a testable prediction is not useful.

   b) Common sense is usually a good substitute for the scientific method when trying to understand the world.

   c) The scientific method can be used only to understand scientific phenomena.

   d) It is not necessary to make observations as part of the scientific method.

   e) All of the above are correct.

**8. The placebo effect:**

a) is the frequently observed, poorly understood phenomenon that people tend to respond favorably to any treatment.

b) reveals that sugar pills are generally as effective as actual medications in fighting illness.

c) reveals that experimental treatments cannot be proven effective.

d) demonstrates that most scientific studies cannot be replicated.

e) is an urban legend.

**9. Before experimental drugs can be brought to market, they must undergo many rigorous trials to ensure that they deliver their medical benefits effectively and safely. One method commonly used in this process is to compare the effect of a drug with that of a placebo in double-blind tests. Which of the following correctly describes a double-blind test?**

a) The researchers apply two-layered blindfolds to the study participants so that they cannot see whether they are receiving the drug or a placebo.

b) Neither the researchers nor the study participants know who is receiving the drug and who is receiving the placebo.

c) The researchers know who is receiving the drug and who is receiving the placebo, but they do not know what the supposed effects of the drug should be.

d) The researchers do not know who receives the drug or the placebo, but the participants know and tell them later.

e) None of the above is correct.

**10. In controlled experiments:**

a) one variable is manipulated while others are held constant.

b) all variables are dependent on each other.

c) all variables are held constant.

d) all variables are independent of each other.

e) all critical variables are manipulated.

**11. If a researcher collects data by using the same experimental setup as in another study, but using different research subjects, the process is considered:**

a) an uncontrolled experiment.

b) intuitive reasoning.

c) extrapolation.

d) replication.

e) exploration.

**12. Upon hearing the males of a particular bird species calling in their natural environment, you wonder why they call. In order to answer your question, you cage one male bird and record his calling rate under several different conditions:**

- in the presence of no conspecific (i.e., of the same species) birds.
- in the presence of 10 conspecific males.
- in the presence of 10 conspecific females.
- in the presence of 5 conspecific males and 5 conspecific females.

**Which of the following null hypotheses could be falsified using this experimental design?**

a) Males of the species do not call.

b) Males of the species call more often than do females.

c) Males of the species call at a rate that is independent of the sex composition of the audience.

d) Males of the species call less often than do females.

e) Males of the species call when called to.

**13. An independent variable:**

a) can cause a change in a dependent variable.

b) is generally less variable than a dependent variable.

c) is plotted on the y-axis in a line graph.

d) can be controlled less well than a dependent variable.

e) is typically of greater interest to the experimenter than a dependent variable.

**14. Statistical methods make it possible to:**

a) prove that any hypothesis is true.

b) determine how likely it is that certain results have occurred by chance.

c) unambiguously learn the truth.

d) reject any hypothesis.

e) test non-falsifiable hypotheses.

**15. Anecdotal evidence:**

a) is a more efficient method for understanding the world than the scientific method.

b) tends to be more reliable than data based on observations of large numbers of diverse individuals.

c) is a necessary part of the scientific method.

d) is often the only way to prove important causal links between two phenomena.

e) can seem to reveal links between two phenomena, but the links do not actually exist.

**16.** A relationship between phenomena that has been established based on large amounts of observational and experimental data is referred to as:

a) a theory.
b) a fact.
c) an assumption.
d) a conjecture.
e) a hypothesis.

**17.** Which of the following issues would be least helped by application of the scientific method?

a) developing more effective high school curricula
b) evaluating the relationship between violence in videogames and criminal behavior in teens
c) determining the most effective safety products for automobiles
d) formulating public policy on euthanasia
e) comparing the effectiveness of two potential antibiotics

**18.** What is the meaning of the statement "Correlation does not imply causation"?

a) Just because two variables vary in a similar pattern does not mean that changing one variable causes a change in the other.
b) It is not possible to demonstrate a correlation between two variables.
c) When a change in one variable causes a change in another variable, the two variables are not necessarily related to each other in any way.
d) It is not possible to prove the cause of any naturally occurring phenomenon.
e) Just because two variables vary in a similar pattern does not mean that they have any relationship to each other.

## Short-Answer Questions

**1.** A pharmaceutical company plans to test a potential anti-cancer drug on human subjects. The drug will be administered in pill form. How should this study be designed so that appropriate controls are in place?

**2.** Lately there have been many claims concerning the health benefits of green tea. Suppose you read a claim that alleges drinking green tea causes weight loss. You are provided with the following information about the studies that led to this claim.

> People were weighed at the beginning of the study.
>
> People were asked to drink two cups of green tea every day for 6 weeks.
>
> People were weighed at the end of the study.
>
> People who drank green tea for 6 weeks lost some weight by the end of the study.
>
> It was concluded that green tea is helpful for weight loss.

This study obviously had some holes in its design. Assuming no information other than that provided above, indicate at least four things that could be done to improve the experimental design.

See Answers at the back of the book.
For additional study questions, go to www.prep-u.com.

### ABOUT THE CHAPTER OPENING PHOTO

Beginning in 1960 (and continuing for more than 45 years), Jane Goodall studied the social and family interactions of chimpanzees in the wild, in Gombe Stream National Park, Tanzania.

Raw materials and fuel
for our bodies

# Chemistry

37

# 2·1–2·3 Atoms form molecules through bonding.

Classical atomic models: a nucleus as a packed cluster of protons and neutrons, orbited by electrons.

## 2·1 Everything is made of atoms.

A little bit of chemistry goes a long way in the study of biology and in understanding a great deal about your everyday life. Will eating those beans in your soup keep you up all night in gastrointestinal distress? Just knowing whether the beans are lentils or lima beans—each contains slightly different types of sugar molecules—will give you an answer. Will the butter you spread on your toast sabotage your efforts to lose weight? Understanding something about the carbon-hydrogen connections in the fat molecules can help you decide.

The chemistry that is most important in biology revolves around a few important elements, which are introduced at the end of this section. An **element** is a substance that cannot be broken down chemically into any other substances. Gold, carbon, and copper are elements you might be familiar with. Whatever the element, if you keep cutting it into ever smaller pieces, each of the pieces behaves exactly the same as any other piece. The smallest piece of pure gold will still have the softness, reflectivity, and malleability characteristic of that element (**FIGURE 2-1**).

Gold

Carbon

Copper

**FIGURE 2-1 Familiar elements.** Elements are substances that cannot be broken down chemically into any other substances.

If you could continue cutting, you would eventually separate the gold into tiny pieces that could no longer be divided without losing their gold-like properties. These individual component pieces of an element are called **atoms.** An atom is a bit of matter that cannot be subdivided any further without losing its essential properties. The word "atom" is from the Greek for "indivisible."

Everything around us, living or not, can be reduced to atoms. All atoms—whether of the element gold or some other element such as oxygen or aluminum or calcium—have the same basic structure. At the center of an atom is a **nucleus,** which is usually made up of two types of particles, called protons and neutrons. **Protons** are particles that have a positive electrical charge and **neutrons** are particles that have no electrical charge. The amount of matter in a particle is its **mass;** protons and neutrons have approximately the same mass (**FIGURE 2-2**).

Whirling in a cloud around the nucleus of every atom are negatively charged particles called **electrons.** An electron weighs almost nothing—less than one-twentieth of one percent of the weight of a proton. (The mass of an atom—its **atomic mass**—is made up of the combined mass of all of its protons and neutrons; for our purposes here, electrons are so light that their mass can be ignored.)

Particles that have the same charge repel each other; those with opposite charges are attracted to each other. Because all electrons have the same charge, the electrons in an atom repel each other. But because they are negatively charged, they are attracted to the positively charged protons in the nucleus. This attraction holds electrons close enough to the nucleus to keep them from flying away, while the energy of their fast movement keeps them from collapsing into the nucleus.

Atoms are tiny. Enlarge an atom by a billion times and it would only be the size of a grapefruit. Paradoxically, most of the space taken up by an atom is empty. That is, because the nucleus is very small and compact, the electrons zip about relatively far from the nucleus. If the nucleus were the size of a golf ball, the electrons would be anywhere from half a mile to six miles away.

What distinguishes one element, such as chlorine, from another, such as neon or oxygen? The number of protons in an atom's nucleus determines what element it is. As a rule, atoms of different elements have a different number of protons in the nucleus. A chlorine atom has 17 protons, a neon atom has 10 protons, and an oxygen atom has 8 protons. Each element is given a name (and an abbreviation, such as O for oxygen and C for carbon) and an **atomic number** that corresponds to how many protons it has (**FIGURE 2-3**).

**THE ATOM: BASIC STRUCTURE**

Nucleus

**HYDROGEN ATOM**
1 Proton
0 Neutrons
1 Electron

**CARBON ATOM**
6 Protons
6 Neutrons
6 Electrons

Nucleus:

Proton    Neutron    Electron

Forces of attraction between positive and negative charges hold the fast-moving electrons (negative) close to the nucleus (positive).

**FIGURE 2-2 The atom.** At the center of an atom is a nucleus containing protons and (in all elements except hydrogen) neutrons. The nucleus is surrounded by electrons whirling about in a cloud.

The mass of an atom is usually about double the element's atomic number. This is because the number of neutrons in the nucleus is usually equal to the number of protons, and

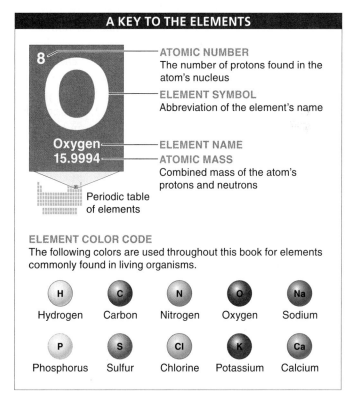

**A KEY TO THE ELEMENTS**

8
O
Oxygen
15.9994

Periodic table of elements

**ATOMIC NUMBER**
The number of protons found in the atom's nucleus

**ELEMENT SYMBOL**
Abbreviation of the element's name

**ELEMENT NAME**
**ATOMIC MASS**
Combined mass of the atom's protons and neutrons

**ELEMENT COLOR CODE**
The following colors are used throughout this book for elements commonly found in living organisms.

H — Hydrogen
C — Carbon
N — Nitrogen
O — Oxygen
Na — Sodium

P — Phosphorus
S — Sulfur
Cl — Chlorine
K — Potassium
Ca — Calcium

**FIGURE 2-3 The vital statistics of atoms.** This key explains how to read a periodic table of elements. A full periodic table of the elements is at the back of the book.

**ISOTOPES**

Atoms with the same atomic number (i.e., the same number of protons) but different atomic mass are isotopes.

|  | CARBON-12 | CARBON-13 | CARBON-14 |
|---|---|---|---|
| ⊕ Protons | 6 | 6 | 6 |
| ○ Neutrons | 6 | 7 | 8 |
| • Electrons | 6 | 6 | 6 |
| Atomic mass | **12** | **13** | **14** |

**FIGURE 2-4 Isotopes are atoms that have extra neutrons or missing neutrons.**

protons and neutrons have approximately the same mass. The element oxygen, for example, has the atomic number 8 because it has 8 protons, so we expect it to have an atomic mass of 16, simply the mass of the 8 protons and 8 neutrons added together—which it generally does. (As noted above, we can ignore the mass of the electrons.)

Atoms don't always have the exact same number of protons and neutrons in their nucleus and electrons circling around it. They sometimes acquire or lose components. An atom that has extra neutrons or fewer neutrons than the number of protons is called an **isotope.** An atom's charge doesn't change in an isotope, because neutrons have no electrical charge, but the atom's mass changes with the loss or addition of another particle in the nucleus (**FIGURE 2-4**). Carbon, for example, has six protons and so usually has a mass of 12. Occasionally, though, carbon has an extra neutron or two and an atomic mass of 13 or 14. This is rare in nature, but it does occur. These isotopes are called carbon-13 ($^{13}$C) and carbon-14 ($^{14}$C) and are referred to as "heavy" carbon. In nature, we frequently see mixtures of several isotopes for a given element. So although a sample of pure carbon is predominantly $^{12}$C atoms (with 6 protons and 6 neutrons), there are some $^{13}$C and $^{14}$C atoms in the sample, too.

Most elements and their isotopes have perfectly stable nuclei that remain unchanged virtually forever, never losing or gaining neutrons, protons, or electrons. A few atomic nuclei are not so stable, however, and break down spontaneously sometime after they are created. These atoms are **radioactive,** and in the process of decomposition they release, at a constant rate, a tiny, high-speed particle carrying a lot of energy. (The particle may be a proton, neutron, electron, or just energy.) For example, uranium-238 (which has 92 protons and 146 neutrons in its nucleus) is a radioactive

element. It spontaneously loses a particle containing 2 protons and 2 neutrons, turning it into an isotope of a different element altogether—thorium-236, in this case. The thorium is radioactive as well, one in a long chain of isotopes, each decaying into another radioactive element, until finally producing the stable element lead (with an atomic mass of 206). Radioactive atoms turn out to be useful in determining the age of fossils (see Section 8-17), in medical imaging and cancer treatment, and in the generation of vast amounts of energy.

All the known elements can be arranged in a scheme, in the order of their atomic number, called the **periodic table** (see Figure 2-3). A copy of the periodic table is provided at the back of the book. So far, about 90 elements have been discovered that are present in nature, and about 25 others can be made in the laboratory. Everything you see around you is made up of some combination of the 90 or so naturally occurring elements.

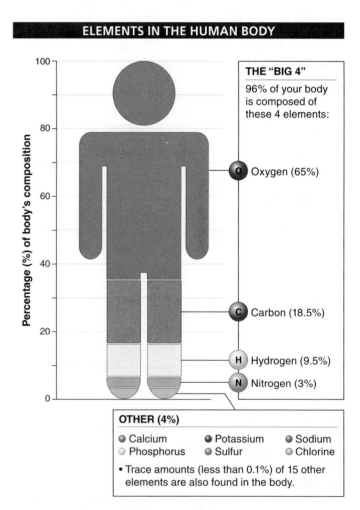

**ELEMENTS IN THE HUMAN BODY**

**THE "BIG 4"**

96% of your body is composed of these 4 elements:

O Oxygen (65%)

C Carbon (18.5%)

H Hydrogen (9.5%)

N Nitrogen (3%)

**OTHER (4%)**
- ○ Calcium  ● Potassium  ● Sodium
- ○ Phosphorus  ● Sulfur  ○ Chlorine
- • Trace amounts (less than 0.1%) of 15 other elements are also found in the body.

**FIGURE 2-5 The body's chemistry.** Just 10 elements make up 99.9% of your body. Four of these—oxygen, carbon, hydrogen, and nitrogen—are most abundant and make up about 96% of your body.

Of all the elements found on earth, only 25 are found in your body. The "Top 10" most common of these make up 99.9% of your body mass, but they are not present in equal measures (**FIGURE 2-5**). The "Big 4" elements, oxygen, carbon, hydrogen, and nitrogen, predominate and make up more than 96% of your body mass. With knowledge about the Big 4, you can understand a huge amount about nutrition and physiology (how your body works), so we'll focus on the properties of these four elements later in this chapter.

# 2·2 An atom's electrons determine how (and whether) the atom will bond with other atoms.

Although the number of protons identifies an element, it is an atom's electrons that determine how (and whether) it bonds with other atoms. Electrons move so quickly that it is impossible to determine, at any given moment, exactly where an electron is. Electrons are not just moving about haphazardly, though. Speeding around the nucleus, they tend to stay within a prescribed area. This area is called an "electron shell," and an atom may have several shells, each shell occupied by its own set of electrons. Within a shell, the electrons stay far apart because their negative charges repel one another. (Electron shells aren't actual physical tracks but, rather, are simplified placeholders for the different possible energy levels of the electrons.)

The first electron shell is closest to the nucleus and can hold two electrons (**FIGURE 2-6**). If an atom has more than two electrons, as most atoms do, the other electrons are arranged in other shells. The second shell is a bit farther away from the nucleus and can hold as many as eight electrons. There can be up to seven shells in total, holding varying numbers of electrons.

Atoms become stable when their outermost shell is filled to capacity. Atoms with this configuration behave like loners, neither reacting nor combining with other atoms. On the other hand, when atoms have outer shells with vacancies, they are likely to interact with other atoms, giving, taking, or sharing electrons to achieve that desirable state: a full outer shell of electrons. In fact, based on the number of electron vacancies in the outermost shell of an atom, it's possible to predict how amenable to bonding that atom will be, and even which other atoms its likely bonding partners will be.

**FIGURE 2-6 Electron interactions.** The chemical characteristics of an atom depend upon the number of electrons in its outermost shell.

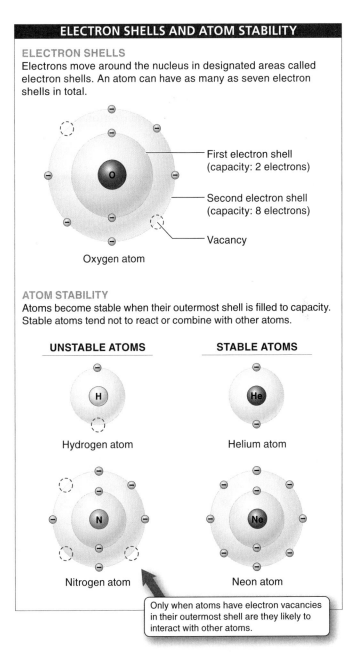

**ELECTRON SHELLS AND ATOM STABILITY**

**ELECTRON SHELLS**
Electrons move around the nucleus in designated areas called electron shells. An atom can have as many as seven electron shells in total.

First electron shell (capacity: 2 electrons)

Second electron shell (capacity: 8 electrons)

Vacancy

Oxygen atom

**ATOM STABILITY**
Atoms become stable when their outermost shell is filled to capacity. Stable atoms tend not to react or combine with other atoms.

UNSTABLE ATOMS | STABLE ATOMS

H — Hydrogen atom

He — Helium atom

N — Nitrogen atom

Ne — Neon atom

Only when atoms have electron vacancies in their outermost shell are they likely to interact with other atoms.

Let's take a brief look at one element—carbon—whose electron configuration gives it considerable versatility when it comes to bonding with other atoms and making important compounds. Carbon has 6 electrons overall: 2 in the first electron shell and 4 in the second electron shell. Because the second electron shell has a capacity of 8 electrons, carbon can share its 4 electrons in its outermost shell. This gives it an ability to bond with other atoms in a large number of different ways—including in four different directions—and makes a huge variety of complex molecules possible (**FIGURE 2-7**).

Carbon most commonly bonds with oxygen, hydrogen, and nitrogen. And because carbon atoms often bind to other carbon atoms, carbon chains are very common—and each carbon bound to two other carbon atoms can still bind with one or two additional atoms. These carbon chains are present in most organic molecules.

Normally, an atom has the same number of electrons as protons. Sometimes an atom may have one or more extra electrons or may lack one or more electrons relative to the number of protons. An atom with extra electrons becomes

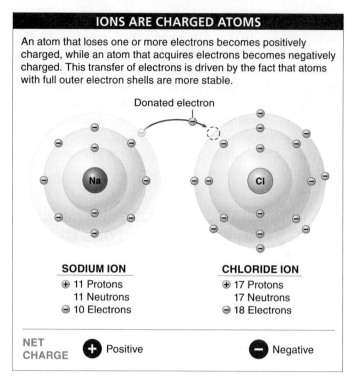

## IONS ARE CHARGED ATOMS

An atom that loses one or more electrons becomes positively charged, while an atom that acquires electrons becomes negatively charged. This transfer of electrons is driven by the fact that atoms with full outer electron shells are more stable.

Donated electron

**SODIUM ION**
⊕ 11 Protons
  11 Neutrons
⊖ 10 Electrons

**CHLORIDE ION**
⊕ 17 Protons
  17 Neutrons
⊖ 18 Electrons

NET CHARGE   ➕ Positive   ➖ Negative

**FIGURE 2-8  Ions are electrically charged atoms.**

negatively charged, and an atom lacking one or more electrons is positively charged. Such a charged atom is called an **ion** (**FIGURE 2-8**). Due to their electrical charge, ions behave very differently from the atoms that give rise to them. As we see later in the chapter, ions are more likely to interact with other, oppositely charged ions.

## TAKE-HOME MESSAGE 2·2

The chemical characteristics of an atom depend on the number of electrons in its outermost shell. Atoms are most stable and least likely to bond with other atoms when their outermost electron shell is filled to capacity.

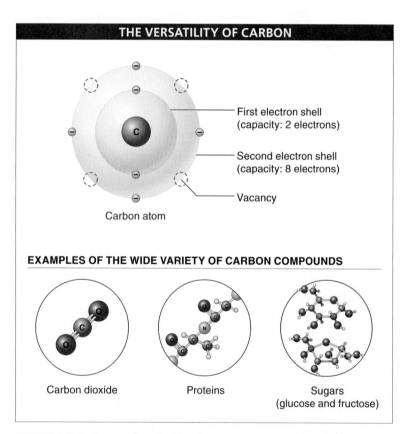

## THE VERSATILITY OF CARBON

First electron shell (capacity: 2 electrons)

Second electron shell (capacity: 8 electrons)

Vacancy

Carbon atom

### EXAMPLES OF THE WIDE VARIETY OF CARBON COMPOUNDS

Carbon dioxide

Proteins

Sugars (glucose and fructose)

**FIGURE 2-7  Carbon can bond with other atoms in many different ways.**

# 2·3 Atoms can bond together to form molecules or compounds.

When you eat a meal, it is like filling your car's tank with gasoline; you have a source of energy that can be used to fuel activities like running, thinking, building muscle, and maintaining the machinery of life. That energy initially comes from the sun and is captured and stored by plants. When we eat plants or eat other animals that eat plants, we ingest the energy stored in the plant material. But how exactly is energy stored in plants?

Groups of atoms held together by bonds are called **molecules.** It takes a certain amount of energy to break a bond between two atoms. The amount of this energy, called the **bond energy,** depends on the atoms involved. When chemical reactions occur—such as when animals (including humans) consume molecules in their diet—some existing bonds are usually broken and some new bonds are usually formed. If the bond energy of the new bonds formed is less than the bond energy of the starting materials, the excess energy is released—and can be used to fuel the body's activities. In a sense, molecules are created as a short-term storage of energy that can be harnessed later.

Before looking at specific types of bonds that hold molecules together, let's look at how molecules are illustrated. In this book you will most often see molecular structures represented by the ball-and-stick and space-filling models.

### DIFFERENT WAYS OF REPRESENTING MOLECULAR STRUCTURE

"Lewis" model    "Ball-and-stick" model    "Space-filling" model

There are three principal types of bonds that hold atoms together. The type of bonding that any atom is likely to take part in depends almost entirely on the number of electrons in its outermost shell.

**Covalent bonds** When two atoms share electrons, a strong **covalent bond** is formed. The simplest example of a covalent bond is the bonding of two hydrogen atoms to form a hydrogen molecule, $H_2$ (**FIGURE 2-9**), the simplest of all molecules. A hydrogen atom has an atomic number of 1: it has a single proton in its nucleus and a single electron circling around the nucleus in the first shell. Because the atom is most stable when the first shell has two electrons, two hydrogen atoms can each achieve a complete outermost shell by sharing electrons. The nuclei come close together (but not

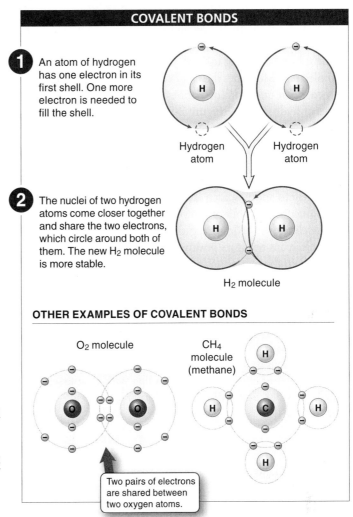

**COVALENT BONDS**

1. An atom of hydrogen has one electron in its first shell. One more electron is needed to fill the shell.

Hydrogen atom    Hydrogen atom

2. The nuclei of two hydrogen atoms come closer together and share the two electrons, which circle around both of them. The new $H_2$ molecule is more stable.

$H_2$ molecule

**OTHER EXAMPLES OF COVALENT BONDS**

$O_2$ molecule    $CH_4$ molecule (methane)

Two pairs of electrons are shared between two oxygen atoms.

**FIGURE 2-9  Covalent bonds: strong chemical bonds formed through electron sharing.**

too close, because they are positively charged and their mutual repulsion would destabilize the molecule), and the two electrons circle around both of the nuclei, almost in a figure 8. The new $H_2$ molecule is very stable because, now that both atoms have two electrons in their outermost shell, they are no longer likely to bond with other atoms.

Oxygen, with an atomic number of 8, has two electrons in its innermost shell and six in its outermost shell. Consequently it needs to form two covalent bonds to fill its outermost shell. Sometimes two oxygen atoms join together to form $O_2$. Each oxygen atom shares a pair of electrons with the other, filling the outermost shell of both. The sharing of two electrons between two atoms is called a

**double bond** (see Figure 2-9). $O_2$ is the most common form in which we find oxygen in the world.

Carbon is a particularly "extroverted" atom. It has an atomic number of 6, meaning that two electrons fill its first shell and four remain to occupy the second shell. Because four electron vacancies are left, carbon can, and frequently does, form four covalent bonds, joining up with other atoms in a wide variety of molecules. Methane, the chief component of natural gas, is formed when one atom of carbon covalently bonds with four atoms of hydrogen (see Figure 2-9). It has the chemical formula $CH_4$.

**Ionic bonds** Atoms can also bond together without sharing electrons. When one atom transfers one or more of its electrons completely to another, each atom becomes an ion, since each has an unequal number of protons and electrons. The atom gaining electrons becomes negatively charged, while the atom losing electrons becomes positively charged. An **ionic bond** occurs when the two oppositely charged ions attract each other and form a **compound,** a molecule made up of two or more elements. Ionic bonds are generally similar in strength to covalent bonds. Unlike covalent bonds, in ionic bonds each electron circles around a single nucleus. Ions of equal and opposite charges are attracted to each other and the compound is neutral—that is, it has no charge (**FIGURE 2-10**). A common ionic compound that you may be familiar with is table salt (NaCl), composed of sodium and chloride ions.

**IONIC BONDS**

**1** One atom transfers one or more electrons completely to another.

Donated electron

**2** The result is two oppositely charged ions.

Sodium ion ⊕    ⊖ Chloride ion

**OPPOSITE CHARGES ATTRACT**

**3** The two oppositely charged ions attract each other, forming a compound.

NaCl compound

NaCl crystals (table salt)    NaCl crystal structure

Cl⁻  Na⁺  Cl⁻
Na⁺  Cl⁻  Na⁺
Cl⁻  Na⁺  Cl⁻

When multiple Na⁺ ions and Cl⁻ ions are attracted to each other, a NaCl crystal structure forms.

**FIGURE 2-10 Ionic bonds: transfer of electrons from one atom to another.** The resulting charged atoms (ions) attract each other.

> **"**In the last third of his life, there came over Laszlo Jamf . . . a hostility, a strangely personal hatred, for the covalent bond . . . That something so mutable, so soft, as a sharing of electrons by atoms of carbon should be at the core of life, *his* life, struck Jamf as a cosmic humiliation. *Sharing?* How much stronger, how everlasting was the ionic bond—where electrons are not shared, but *captured. Seized!* and held! polarized plus and minus, these atoms, no ambiguities . . . how he came to love that clarity: how stable it was, in such mineral stubbornness!**"**

— **THOMAS PYNCHON,** in *Gravity's Rainbow*
[Was Laszlo Jamf mistaken in his understanding of the relative strengths of ionic and covalent bonds?]

**Hydrogen bonds** Ionic and covalent bonds link two or more atoms together. **Hydrogen bonds,** on the other hand, are important in bonding molecules together. A hydrogen bond is formed between a hydrogen atom in one molecule and another atom, often an oxygen or nitrogen atom, in another molecule (or even in another part of the same molecule). This bond is based on the attraction between positive and negative charges.

The atoms taking part in hydrogen bonds are not necessarily ions, so where do the electrical charges come from? The hydrogen atom is already covalently bonded to another atom in the same molecule and shares its electron. That electron circles both the hydrogen nucleus and the nucleus of the other atom, but the electron is not shared equally. Because the other atom always has more than the one proton found in the hydrogen nucleus, it is more positively charged. As a

**1** In a water molecule, oxygen's eight positively charged protons attract electrons more readily than does the single proton in the hydrogen atoms. As a result, the electrons are pulled toward the oxygen side of the molecule, making it slightly negative in charge, while the hydrogen side is slightly positive.

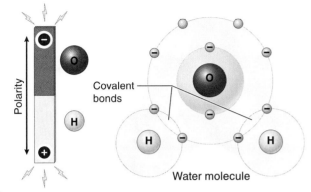

Covalent bonds

Water molecule

**2** Hydrogen bonds are formed between the slightly positively charged hydrogen atoms of one molecule and the slightly negatively charged oxygen atom of another.

Hydrogen bond

**FIGURE 2-11 Hydrogen bonds: attraction between a polar atom or molecule and a hydrogen atom.** As we'll see later in the chapter, the hydrogen bonding between H$_2$O molecules gives water some of its characteristic properties.

result, the electron contributed by hydrogen spends more of its time near the other, more positively charged nucleus than near the hydrogen nucleus. Having an extra electron nearby causes the larger atom to be slightly negatively charged, while the hydrogen atom becomes slightly positively charged (**FIGURE 2-11**).

In a sense, the covalently bonded molecules become like a magnet, with distinct positive and negative sides (see Figure 2-11), and are said to be "polar." Polar molecules are attracted to other polar molecules, lining up in particular orientations such that the positive regions of one molecule are near the negative regions of another. These attractions between polar molecules and hydrogen atoms are called hydrogen bonds. Although they are only about one-thirtieth as strong as covalent or ionic bonds, hydrogen bonds are responsible for many of the unique characteristics of water that make it one

### COVALENT BOND
A strong bond formed when atoms share electrons in order to become more stable, forming a molecule.
**BOND STRENGTH: STRONG**

H$_2$ molecule

### IONIC BOND
An attraction between two oppositely charged ions, forming a compound.
**BOND STRENGTH: STRONG**

NaCl compound

### HYDROGEN BOND
An attraction between the slightly positively charged hydrogen atom of one molecule and the slightly negatively charged atom of another.
**BOND STRENGTH: WEAK**

H$_2$O molecule          H$_2$O molecule

**FIGURE 2-12 The properties of covalent, ionic, and hydrogen bonds compared.**

of the most important molecules for life on earth. See **FIGURE 2-12** for a review of the three types of bonds discussed in this section.

### TAKE-HOME MESSAGE 2·3

Atoms can be bound together in three different ways. Covalent bonds occur when atoms share electrons. In ionic bonds, one atom transfers its electrons to another and the two oppositely charged ions are attracted to each other, forming a compound. Hydrogen bonds, which are weaker than covalent and ionic bonds, involve the attraction between a hydrogen atom and a polar molecule.

# 2·4–2·6 Water has features that enable it to support all life.

*Water flowing down a Japanese waterfall.*

# 2·4

## Hydrogen bonds make water cohesive.

Every so often, a tadpole or small fish gets an unexpected—and life-ending—surprise from above as a giant spider moves quickly across the surface of the water, reaches down with its two front legs, and plucks the animal from the water. The spider, known as the fishing spider (*Dolomedes triton*), injects its prey with venom, carries it back to shore—moving across the water's surface as if across solid ground—and eats it. How can these spiders walk on water?

The fishing spider, like numerous other insects such as the water strider, makes use of the fact that water molecules have tremendous cohesion. That is, they stick together with

**WATER: HIGH SURFACE TENSION**

Pressure applied to water surface

Hydrogen bond

V-shaped water molecules are held together by hydrogen bonds. The bonds are just strong enough to give water a surface tension with net-like properties.

**FIGURE 2-13 Walking on water!** Hydrogen bonds make this possible for some animals.

unusual strength. This molecular cohesiveness is due to hydrogen bonds between the water molecules.

Each water molecule is V-shaped (**FIGURE 2-13**; see also Figure 2-11). The hydrogen atoms are at the ends of the two arms and the oxygen is at the bottom end of the V, between the two hydrogen atoms. Oxygen's strongly positively charged nucleus pulls the circling electrons toward itself and holds on to them for more than its fair share of the time. Consequently, the oxygen at the bottom of the V has a slight negative charge and the end of the water molecule containing the hydrogen atoms has a slight positive charge.

Because of their unequally shared electrons, water molecules are polar (see Figure 2-11). Consequently, large numbers of water molecules orient themselves so that the negative side of one molecule is near the positive side of another. Hydrogen bonds form between the relatively positively charged hydrogen atoms and the relatively negatively charged oxygen atoms of *adjacent* water molecules.

Hydrogen bonds are much weaker than covalent and ionic bonds and they don't last very long. Nonetheless, to the fishing spider, the cumulative effect of all the hydrogen bonds in water is to link together all the water molecules in the stream just enough to give the water a surface tension with some net-like properties.

> ### TAKE-HOME MESSAGE 2·4
>
> **Water molecules easily form hydrogen bonds, giving water great cohesiveness.**

# 2·5 Water has unusual properties that make it critical to life.

All life on earth depends on water; organisms are made up mostly of water and require it more than any other molecule. Hydrogen bonding among water molecules gives water several important properties that contribute to its crucial role in the biology of all organisms.

**1. Cohesion.** We saw in the previous section how the connection of water molecules through hydrogen bonds makes water cohesive, resulting in, for example, high surface tension. The cohesiveness of water molecules also makes it possible for tall trees to exist (**FIGURE 2-14**). Leaves need water. Molecules of water in the leaves are continually lost to the atmosphere through evaporation or are used up in the process of photosynthesis. In order to get more water, plants must pull it up from the soil. The problem for many plants, such as the giant sequoia trees, is that the soil may be 300 feet below the leaf. Hydrogen bonds, however, allow water molecules to pull up adjacent water molecules to which they have hydrogen-bonded. The chain of linked molecules extends all the way down to the soil, where another water molecule is pulled in via the roots each time a water molecule evaporates from a leaf far above.

**FIGURE 2-14 Like a giant straw.** Hydrogen bonds cause water molecules to "stick" together, so they can be pulled up through the giant sequoia.

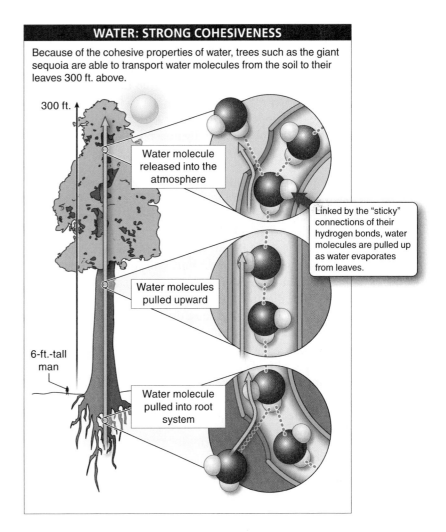

**WATER: STRONG COHESIVENESS**

Because of the cohesive properties of water, trees such as the giant sequoia are able to transport water molecules from the soil to their leaves 300 ft. above.

300 ft.

Water molecule released into the atmosphere

Linked by the "sticky" connections of their hydrogen bonds, water molecules are pulled up as water evaporates from leaves.

Water molecules pulled upward

6-ft.-tall man

Water molecule pulled into root system

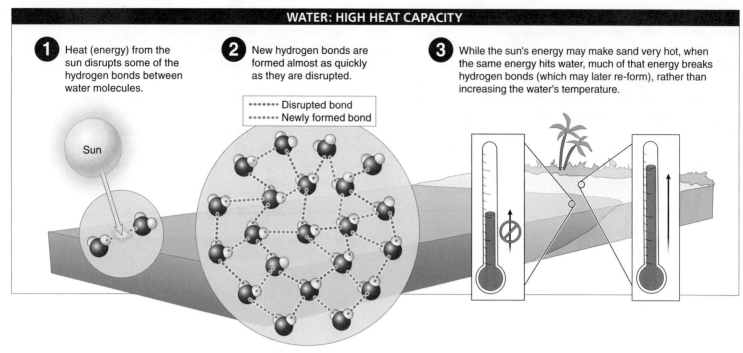

**1** Heat (energy) from the sun disrupts some of the hydrogen bonds between water molecules.

**2** New hydrogen bonds are formed almost as quickly as they are disrupted.

**3** While the sun's energy may make sand very hot, when the same energy hits water, much of that energy breaks hydrogen bonds (which may later re-form), rather than increasing the water's temperature.

······· Disrupted bond
······· Newly formed bond

Sun

**FIGURE 2-15 Water as a moderator of temperature change.** Hydrogen bonds help water resist heating.

**2. Large heat capacity.** Walking across a sandy beach on a hot day, you can feel how easily sand heats up. By comparison, stepping from the beach into the cooler ocean reveals that water resists warming. It takes a lot of energy to change the temperature of water even a small amount. Why? Again, we must look to hydrogen bonding for our answer.

The temperature of a substance is a measure of how quickly all of the molecules are moving. The molecules move more quickly when energy is added in the form of heat. When we heat water, the added energy doesn't immediately increase the movement of the individual water molecules. Rather, it disrupts some of the hydrogen bonds between the molecules (**FIGURE 2-15**). As quickly as they can be disrupted, though, hydrogen bonds form again somewhere else. And since the water molecules themselves don't increase their movement, the temperature doesn't increase. The net effect is that even if you release a lot of energy into water, the temperature doesn't change much. For this reason, because so much of your body is water, you are able to maintain a relatively constant body temperature.

Large bodies of water, especially oceans, can absorb huge amounts of heat from the sun

**Q** Why do coastal areas have milder, less variable climates than inland areas?

during warm times of the year, reducing temperature increases on the land. Similarly, during cold times of the year, the ocean slowly cools, giving off heat that reduces the temperature drop on shore.

**3. Low density as a solid.** Ice floats. This is unusual, since most substances *increase* in density when frozen; as the molecules slow down, they pack together more and more efficiently—and densely. Consequently, the solid sinks. Water, however, becomes less dense and, as you might expect by now, this is due to hydrogen bonding. As the temperature drops and water molecules slow down, rather than becoming more and more tightly packed, they become less so. Each V-shaped water molecule bonds with four partners, via hydrogen bonds, forming a crystalline lattice in which the molecules are held slightly farther apart than in the liquid, causing ice to be less dense than water (**FIGURE 2-16**).

**4. Good solvent.** If you put a pinch of table salt into a glass of water, it will quickly dissolve. This means that all the charged sodium ($Na^+$) and chloride ($Cl^-$) ions that were ionically bonded together become separated from one another. The sodium and chloride ions were initially attracted to each other because they are polar molecules

## WATER: LOWER DENSITY WHEN FROZEN

**FROZEN WATER**
Hydrogen bonding arranges water molecules into a crystalline lattice, keeping them slightly farther apart and, therefore, less dense.

**LIQUID WATER**
Water molecules move about freely, allowing them to be closer to one another.

**RELATIVE AREA OCCUPIED BY THE SAME NUMBER OF H₂O MOLECULES:**

Frozen water      Liquid water

**FIGURE 2-16** **The lattice structure of ice allows it to float.**

## WATER: A GOOD SOLVENT

NaCl crystal (table salt)

When placed in water, ionic compounds such as NaCl are broken down into separate Na⁺ and Cl⁻ ions.

Dissolved Na⁺ and Cl⁻ ions

Positively charged Na⁺ ions are attracted to the negatively charged side of the H₂O molecule, while negatively charged Cl⁻ ions are attracted to the positively charged side.

**FIGURE 2-17** **Solutions.** Water pries apart ionic bonds, dissolving ionic compounds.

(ions), each carrying a slight charge. Water is able to pry them apart because, as a polar molecule, it, too, carries charges. The positively charged sodium ions are attracted to the negatively charged side of the water molecule, and the negatively charged chloride ions are attracted to the positively charged side (**FIGURE 2-17**). Many substances are, like water, polar and consequently dissolve easily in water.

Nonpolar molecules (such as oils) have neither positively charged regions nor negatively charged regions. Consequently, the polar water molecules are not attracted to them. Instead, when an oil is poured into a container of water, the water molecules distance themselves from the oil, leaving the oil molecules in clumps that never dissolve.

*Why don't oceans freeze as easily as freshwater lakes?*

 Because so much salt is dissolved in the oceans, many of the water molecules have their positively charged sides facing Cl⁻ ions. Simultaneously, many molecules of water are turned the other way, with their negatively charged sides facing Na⁺ ions. Consequently, the orderly lattices of hydrogen bonds found in ice cannot form in salt water, and it does not freeze well.

### TAKE-HOME MESSAGE 2·5

The hydrogen bonds between water molecules give water several of its most important characteristics, including cohesiveness, reduced density as a solid, the ability to resist temperature changes, and broad effectiveness as a solvent for ionic and polar substances.

ATOMS   **WATER**   CARBOHYDRATES   LIPIDS   PROTEINS   NUCLEIC ACIDS

# 2·6 Living systems are highly sensitive to acidic and basic conditions.

There's a lot more going on in water than meets the eye. Most of the molecules are present as $H_2O$, but at any instant some of them break into two parts: $H^+$ and $OH^-$. In pure water, the amount of $H^+$ and $OH^-$ must be exactly the same, since every time a molecule splits, one of each type of ion is produced. But in some fluids containing other dissolved materials, this balance is lost: the fluid can have more $H^+$ or more $OH^-$.

### HYDROGEN IONS and HYDROXIDE IONS

| Non-ionized Water Molecule ($H_2O$) | Ionized Hydrogen Molecule ($H^+$) | Ionized Hydroxide Molecule ($OH^-$) |

Donated electron

The amount of $H^+$ or $OH^-$ in a fluid gives it some important properties. In particular, the amount of $H^+$ in a solution is a measure of its acidity and is called **pH.** The greater the number of free hydrogen ions floating around, the more acidic the solution is.

Pure water is in the middle of the pH scale, with a pH of 7.0. Any fluid with a pH below 7.0 has more $H^+$ ions (and fewer $OH^-$ ions) and is considered an **acid.** Any fluid with a pH above 7.0 has fewer $H^+$ ions (and more $OH^-$ ions) and is considered a **base.** The pH scale, like the Richter scale for earthquakes, is logarithmic, although in the case of pH, the *lower* the number the *greater* the acidity: a decrease of 1 on the pH scale represents a 10-fold increase in the hydrogen ion concentration (**FIGURE 2-18**). A decrease of 2 represents a 100-fold increase. This means that a cola, with a pH of about 3.0, is 10,000 times (!) more acidic than a glass of water, with a pH of 7.0.

$H^+$ ions are essentially free-floating protons. Acids can donate their $H^+$ ions to other chemicals. In fact, $H^+$ is a very reactive little ion. Its presence gives acids some unique properties. For instance, the hydrogen ions in acids can bind with atoms in metals, causing them to corrode. That's why you can dissolve nails by dumping them in a bucket of acid (or cola).

Your stomach produces large amounts of hydrochloric acid (HCl dissolved in water) and has a pH between 1 and 3. (HCl is acidic because most of the hydrogen ions split off

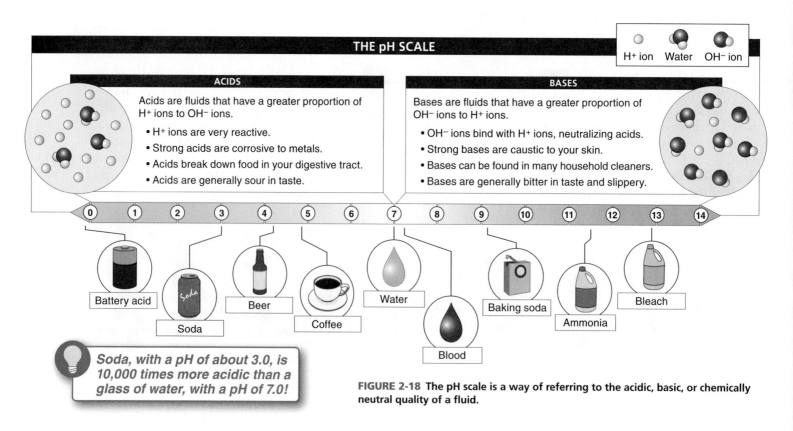

**THE pH SCALE**

H⁺ ion    Water    OH⁻ ion

**ACIDS**

Acids are fluids that have a greater proportion of $H^+$ ions to $OH^-$ ions.

- $H^+$ ions are very reactive.
- Strong acids are corrosive to metals.
- Acids break down food in your digestive tract.
- Acids are generally sour in taste.

**BASES**

Bases are fluids that have a greater proportion of $OH^-$ ions to $H^+$ ions.

- $OH^-$ ions bind with $H^+$ ions, neutralizing acids.
- Strong bases are caustic to your skin.
- Bases can be found in many household cleaners.
- Bases are generally bitter in taste and slippery.

0  1  2  3  4  5  6  7  8  9  10  11  12  13  14

Battery acid  Soda  Beer  Coffee  Water  Blood  Baking soda  Ammonia  Bleach

*Soda, with a pH of about 3.0, is 10,000 times more acidic than a glass of water, with a pH of 7.0!*

**FIGURE 2-18** The pH scale is a way of referring to the acidic, basic, or chemically neutral quality of a fluid.

**Q**

Is acid rain likely to have any impact on populations of microbes living in lakes, streams, and soil? Why?

from the chlorine, raising the H⁺ concentration of the fluid.) The acid in your stomach helps to kill most bacteria that you ingest. It also greatly enhances the breakdown of the chemicals in the food you eat and the efficiency of digestion and absorption. You may have learned firsthand of the high acidity of your stomach fluids if you have experienced heartburn or the sour taste of vomit.

Acids have a higher concentration of H⁺ than of OH⁻. Conversely, bases have a higher concentration of OH⁻ than of H⁺. Baking soda is a common basic substance. Some bases are called "antacids" because the OH⁻ ions in bases can bind with excess H⁺ ions in acidic solutions, neutralizing the acid. Base-containing products such as Alka-Seltzer and Milk of Magnesia reduce the unpleasant feeling of heartburn and acid indigestion that sometimes arise from the overproduction of acids by the stomach. Bases are commonly used in household cleaning products and generally have a bitter taste and a soapy, slippery feel.

The pH of blood is usually 7.4. Given that most cellular reactions produce or consume H⁺ ions, there ought to be great swings in the pH of our blood. Unfortunately, our bodies can't tolerate such swings. Most of the chemicals that aid in the chemical reactions in our blood or cells stop functioning properly if the pH swings up or down by less than half a point. Fortunately, our bodies contain some chemicals that act like bank accounts for H⁺ ions (**FIGURE 2-19**). Called **buffers,** these chemicals can quickly absorb excess H⁺ ions to keep a solution from becoming too acidic, and they can quickly release H⁺ ions to counteract any increases in OH⁻ concentration. Therefore, buffers are chemicals that act to resist changes in pH.

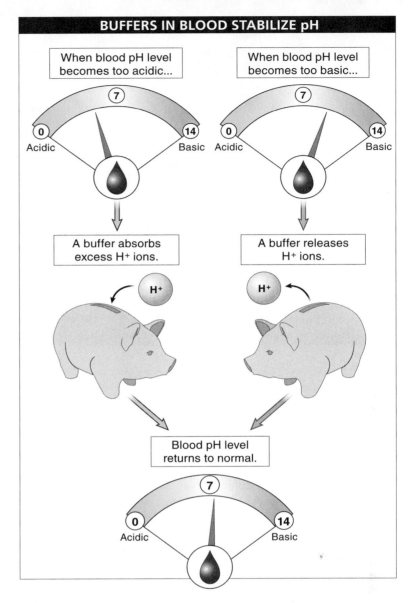

**BUFFERS IN BLOOD STABILIZE pH**

When blood pH level becomes too acidic...

A buffer absorbs excess H⁺ ions.

When blood pH level becomes too basic...

A buffer releases H⁺ ions.

Blood pH level returns to normal.

**FIGURE 2-19 Maintaining a constant internal environment.** Buffers in blood prevent potentially damaging pH swings.

**TAKE-HOME MESSAGE 2·6**

The pH of a fluid is a measure of how acidic or basic the solution is and depends on the concentration of dissolved H⁺ ions present; the lower the pH, the more acidic the solution. Acids, such as vinegar, can donate protons to other chemicals; bases, including baking soda, bind with free protons.

# 2·7–2·10 Carbohydrates are fuel for living machines.

*Worldwide, more corn is produced (by weight) than any other grain.*

## 2·7 Carbohydrates include macromolecules that function as fuel.

Hardly a day goes by without an item in a magazine or newspaper or on TV talking about whether carbohydrates are good or bad for us. Sports drinks such as Gatorade are filled with carbs, and in the days before a big game or race, athletes often "carbo-load" by eating bowls of pasta. Yet other dietary supplements and "power" bars tout that they are low-carb and effective in weight loss by causing the body to resort to using fat for energy. Meanwhile, nutritionists, doctors, and many diet programs exhort people to increase the amount of fiber—another type of carbohydrate—in their diet. What exactly are they all talking about?

Four types of **macromolecules**—large molecules made up from smaller building blocks or subunits—are essential to the building and functioning of living organisms: carbohydrates, lipids, proteins, and nucleic acids. **Carbohydrates** are molecules that contain mostly carbon, hydrogen, and oxygen: they are the primary fuel for running all of the cellular machinery and also form much of the structure of cells in all life forms. Sometimes they contain atoms of other elements, but they must have carbon, hydrogen, and oxygen to be considered a carbohydrate (**FIGURE 2-20**). Additionally, a carbohydrate generally has approximately the same number of carbon atoms as it does $H_2O$ units. For instance, the best-known carbohydrate, glucose, has the composition $C_6H_{12}O_6$ (6 carbons and, as a little math will show us, 6 $H_2O$ units; notice that $6 \times H_2 = H_{12}$ and $6 \times O = O_6$). A carbohydrate called maltose has the composition $C_{12}H_{22}O_{11}$.

Carbohydrates are classified into several categories, based on their size and their composition. The simplest carbohydrates are the **monosaccharides,** or **simple sugars.** The simple sugars contain anywhere from three to six carbon atoms and, when they are broken down, the products usually are not carbohydrates. Two common monosaccharides are glucose, found in the sap and fruit of many plants, and fructose, found primarily in fruits and vegetables, as well as in honey. Fructose is the sweetest of all naturally occurring sugars. The suffix -*ose* tells us that a substance is a carbohydrate.

**SOME COMMON MONOSACCHARIDES**

Glucose
$C_6H_{12}O_6$

Fructose
$C_6H_{12}O_6$

Galactose
$C_6H_{12}O_6$

Carbohydrates function well as fuels because of their many carbon-hydrogen bonds; as those bonds are broken down and other, more stable bonds (primarily between carbon and oxygen) are formed, a great deal of energy is released that organisms can utilize.

In the next section we investigate glucose, the chief carbohydrate used by organisms to fuel their activities.

## CARBOHYDRATES

The carbon-hydrogen bonds of carbohydrates store a great deal of potential energy and are easily broken down by organisms.

H — C
ENERGY

Carbohydrates, such as glucose (shown above), are composed primarily of three elements:

C   H   O

Glucose
$C_6H_{12}O_6$

**FIGURE 2-20** All carbohydrates have a similar structure and function.

Carbohydrates are the primary fuel source for cellular mechanisms.

### TAKE-HOME MESSAGE 2·7

Carbohydrates are the primary fuel for running all cellular machinery and also form much of the structure of cells in all life forms. Carbohydrates contain carbon, hydrogen, and oxygen, and generally have the same number of carbon atoms as they do $H_2O$ units. The simplest carbohydrates, including glucose, are monosaccharides or simple sugars. They contain from three to six carbon atoms. As the chemical bonds of carbohydrates are broken down and other more stable bonds are formed, a great deal of energy is released that can be used by organisms.

# 2·8  Glucose provides energy for the body's cells.

The carbohydrate of most importance to living organisms is glucose. This simple sugar is found naturally in most fruits, but most of the carbohydrates that you eat, including table sugar (called sucrose) and the starchy carbohydrates found in bread and potatoes, are converted into glucose in your digestive system. The glucose then circulates in your blood at a concentration of about 0.1%. Circulating glucose, also called "blood sugar," has one of three fates (**FIGURE 2-21**).

**1. Fuel for cellular activity.** Once it arrives at and enters a cell, glucose can be used as an energy source. Through a series of chemical reactions, the glucose molecule is converted into other molecules with stronger bonds (a process explained in detail in Chapter 4). The change from weaker bonds to stronger bonds releases energy, and organisms use the released energy to fuel cellular activity, including the muscle contractions that enable you to move and the nerve activities that enable you to think.

**2. Stored temporarily as glycogen.** If there is more glucose circulating in your bloodstream than is necessary to meet your body's current energy needs, the excess glucose can be temporarily stored in various tissues, primarily your muscles and liver. The stored glucose molecules are linked together to form a large web of molecules called **glycogen.** When you need energy later, the glycogen can easily be broken down to release glucose molecules back into your

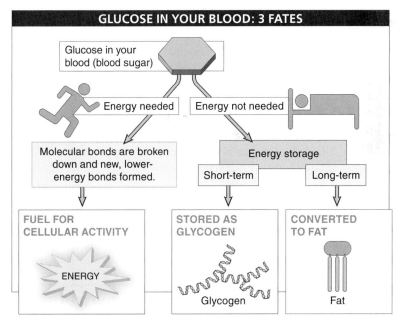

**FIGURE 2-21 What happens to sugar in your blood?** Depending upon whether energy is needed, glucose can be released for immediate use or stored on a short-term or long-term basis.

bloodstream. Glycogen is the primary form of short-term energy storage in animals.

**3. Converted to fat.** Finally, glucose circulating in your bloodstream can be converted into fat, a form of long-term energy storage.

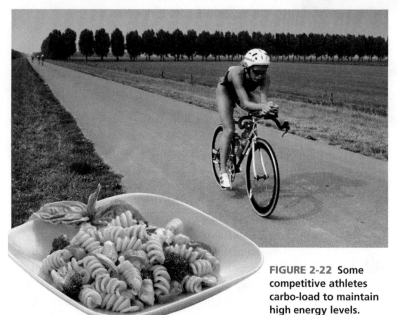

FIGURE 2-22 **Some competitive athletes carbo-load to maintain high energy levels.**

**Because a pound of glycogen has about four pounds of water molecules bound to it, dieters lose a lot of "water weight" during the first days of a diet.**

Glycogen

Water molecules

FIGURE 2-23 **Water weight.** Water molecules bound to glycogen account for much of the weight lost early in a diet.

"Carbo-loading" is a method by which athletes can, for a short time, double or triple the usual amount of glycogen stored in their muscles and liver, increasing the store of fuel available for extended exertion and delaying the onset of fatigue during an endurance event (**FIGURE 2-22**).

**Q** What is "carbo-loading"?

Carbo-loading is usually done in a depletion phase and a loading phase. Six or seven days before competition, the athlete depletes glycogen in the muscles through super-low carbohydrate intake and exhaustive exercise. Then two days before the competition, the athlete eats a super-high carbohydrate diet and reduces exercise to achieve a higher blood glucose than is necessary, so that much of the excess glucose is stored as glycogen.

Glycogen also plays a role in the initial rapid weight loss people experience when dieting. If you reduce your caloric intake such that your body is burning more calories than you are consuming, your body must use stored energy. The first, most accessible molecules that can be broken down for energy in the absence of sufficient sugar in your bloodstream are glycogen molecules in your muscles and liver. Large amounts of water are bound to glycogen. As that glycogen is removed from your tissue, so, too, is the water. This leads to the initial dramatic weight loss that occurs before your body resorts to using stored fat, and the rate of weight loss slows considerably (**FIGURE 2-23**).

**TAKE-HOME MESSAGE 2·8**

Glucose is the most important carbohydrate to living organisms. Glucose in the bloodstream can be used as an energy source, can be stored as glycogen in the muscles and liver for later use, or can be converted to fat.

# 2·9 Many complex carbohydrates are time-release packets of energy.

In contrast to the simple sugars, some carbohydrates contain more than one sugar unit or building block. When two simple sugars are bonded together they form a **disaccharide;** the joining of glucose and galactose, for example, makes the disaccharide lactose, which is the sugar found in milk. When large numbers of simple sugars are joined together—sometimes as many as 10,000 may be linked—the resulting carbohydrate is called a **polysaccharide,** or a **complex carbohydrate** (**FIGURE 2-24**). Depending on how the simple

sugars are bonded together, polysaccharides may function as "time-release" stores of energy or as structural materials that may be completely indigestible by most animals. An example of such a structural material is the polysaccharide cellulose—the primary component of plant cell walls.

Like simple sugars, many disaccharides and polysaccharides are important sources of fuel for cells. Unlike simple sugars, however, disaccharides and polysaccharides must undergo some preliminary processing before the energy can be released from their bonds. Let's look at what happens when we eat sucrose, common table sugar. Sucrose is the primary carbohydrate in plant sap and is composed of two simple sugars, glucose and fructose, linked together. To use sucrose, the body must first break the bond linking the glucose and fructose. Only then can the individual monosaccharides be broken down into their component atoms and the energy that was stored in their chemical bonds be harvested and used.

Energy can also be stored in a complex carbohydrate called **starch,** which consists of a hundred or more glucose molecules joined together in a line. In plants, starch is the primary form of energy storage, found in roots and other tissues (see Figure 2-24). Grains such as barley, wheat, and rye are high in starch content, and corn and rice are more than 70% starch. Although it is composed of glucose molecules linked together, starch does not taste sweet. Because of its molecular shape, it does not stimulate the sweetness receptors on the tongue. The glycogen that stores energy in your muscles and liver is also a complex carbohydrate, so it is sometimes referred to as "animal starch" (although it has a more branched structure than starch and carries more glucose units linked together).

Depending on their structure, dietary carbohydrates can lead to quick-but-brief or slow-but-persistent increases in blood sugar.

**FIGURE 2-25 Short-term versus long-term energy.** Simple sugars and complex carbohydrates differ in the way they make energy available to you.

Many complex carbohydrates are like "time-release" fuel pellets. The glucose molecules become available slowly, as the bonds between glucose units are broken. As a demonstration, put a piece of potato or bread into your mouth and let it rest on your tongue. Initially, it does not taste sweet. After a few minutes, you will begin to taste some sweetness as the chemicals in your saliva break the starch down into glucose.

The relative amounts of complex carbohydrates and simple sugars in foods cause them to have different effects when you eat them. Oatmeal (along with rice and pasta), for example, is rich in complex carbohydrates. Fresh fruits, on the other hand, are rich in simple sugars, such as fructose. Consequently, although fruit will give a quick burst of energy as the sugars are almost immediately available, the fuel will soon be gone from the bloodstream. Simple sugars in the oatmeal will become available only gradually, as the complex carbohydrates of the oats are broken down into their simple sugar components (**FIGURE 2-25**).

**Q**

Before heading to the library for a long study session, students would be wise to consume oatmeal rather than fresh fruits. Why?

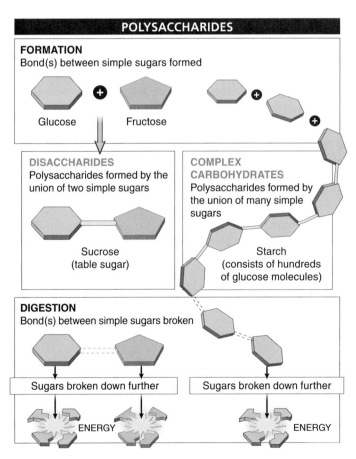

**POLYSACCHARIDES**

**FORMATION**
Bond(s) between simple sugars formed

Glucose ⊕ Fructose ⊕ ⊕

**DISACCHARIDES**
Polysaccharides formed by the union of two simple sugars

Sucrose
(table sugar)

**COMPLEX CARBOHYDRATES**
Polysaccharides formed by the union of many simple sugars

Starch
(consists of hundreds of glucose molecules)

**DIGESTION**
Bond(s) between simple sugars broken

Sugars broken down further

Sugars broken down further

ENERGY

ENERGY

**FIGURE 2-24 Chains of sugars.** Polysaccharides are made from simple sugars bonded together.

### TAKE-HOME MESSAGE 2·9

**Multiple simple carbohydrates are sometimes linked together into more complex carbohydrates. Types of complex carbohydrates include starch, which is the primary form of energy storage in plants, and glycogen, which is a primary form of energy storage in animals.**

# 2·10  Not all carbohydrates are digestible.

Despite their general importance as a fuel source for humans, not all carbohydrates can be broken down by our digestive system. Two different complex carbohydrates—both indigestible by humans—serve as structural materials for invertebrate animals and plants: **chitin** (pronounced KITE-in) and **cellulose** (**FIGURE 2-26**). Chitin forms the rigid outer skeleton of most insects and crustaceans (such as lobsters and crabs). Cellulose forms a huge variety of plant structures that are visible all around us. We find cellulose in trees and the wooden structures we build from them, in cotton and the clothes we make from it, in leaves and in grasses. In fact, it is the single most prevalent compound on earth.

Surprisingly, cellulose is almost identical in composition to starch. Nonetheless, because of one small difference in the chemical bond between the simple sugar units, cellulose has a slightly different three-dimensional structure. Even tiny differences in the shape of a molecule can have a huge effect on its behavior. In this case, the difference in shape makes it impossible for humans to digest cellulose as they can starch. Consequently, the cellulose we eat passes right through our digestive system unused.

Cellulose gives trees the rigidity to grow tall (and makes wood a desirable building material).

Chitin offers protection for the lobster.

**FIGURE 2-26 Carbohydrates can serve as structural materials.**

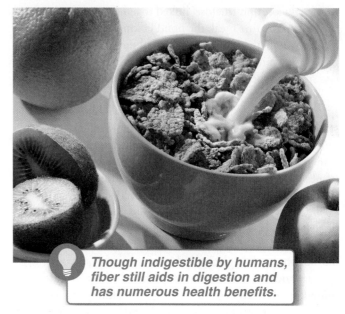

Though indigestible by humans, fiber still aids in digestion and has numerous health benefits.

**FIGURE 2-27 Fiber.** It's not digestible, but it's still important for our diet.

Although it is not digestible, cellulose is still important to human diets. The cellulose in our diet is known as "fiber" (**FIGURE 2-27**). It is also appropriately called "roughage" because, as the cellulose of celery stalks and lettuce leaves passes through our digestive system, it scrapes the wall of the digestive tract. Its bulk and the scraping stimulate the more rapid passage of food and the unwanted, possibly harmful products of digestion through our intestines. That is why fiber reduces the risk of colon cancer and other diseases (but it is also why too much fiber can lead to diarrhea).

Unlike humans, termites have some microorganisms living in their gut that are able to break down cellulose. That's why they can chew on wood and, with the help of the cellulose-digesting boarders in their gut, can break down the cellulose and extract usable energy from the freed glucose molecules.

### TAKE-HOME MESSAGE 2·10

Some complex carbohydrates, including chitin and cellulose, cannot be digested by most animals. Such indigestible carbohydrates in the diet, called fiber, aid in digestion and have many health benefits.

# 2·11–2·13 Lipids store energy for a rainy day.

*A well-insulated harp seal, in Canada.*

## 2·11 Lipids are macromolecules with several functions, including energy storage.

**Lipids** are a second group of macromolecules important to all living organisms. Lipids, just like carbohydrates, are made primarily from atoms of carbon, hydrogen, and oxygen, but the atoms are in different proportions. The lipids in your diet, for example, tend to have significantly more energy-rich carbon-hydrogen bonds than carbohydrates, and so contain significantly more stored energy.

What exactly is a lipid? That's not as easy to answer as you might expect. Lipids come in a wide variety of structures. They don't have any unique subunits (such as the simple sugars that make up disaccharides and polysaccharides) or particular ratios of atoms that serve as defining features. Consequently, lipids

*Why does a salad dressing made with vinegar and oil separate into two layers shortly after you shake it?*

are defined based on their physical characteristics. Most notably, lipids do not dissolve in water and are greasy to the touch—think of salad dressings.

**Q** Lipids are insoluble in water because they tend to have long chains consisting only of carbon and hydrogen atoms, which, in contrast to water, are nonpolar. Nonpolar molecules (or parts of molecules) tend to minimize their contact with water and are considered **hydrophobic** ("water-fearing"). Lipids cluster together when mixed with water, never fully dissolving. Molecules that readily form hydrogen bonds with water, on the other hand, are considered **hydrophilic** ("water-loving").

One familiar type of lipid is *fat*, the type most important in long-term energy storage and insulation

**TYPICAL FEATURES OF LIPIDS**
• Nonpolar molecules that do not dissolve in water
• Greasy to the touch
• Can be significant source of energy storage

**THREE TYPES OF LIPIDS**

| FATS | STEROLS | PHOSPHOLIPIDS |
|---|---|---|
| **FUNCTION** Long-term energy storage and insulation | **FUNCTION** Regulate growth and development | **FUNCTION** Form cellular membranes |

The Emperor penguin (*Aptenodytes forsteri*) benefits from the insulating effect of fat and is able to thrive in the extreme cold of Antarctica.

**FIGURE 2-28 Lipids serve many roles in the body.**

(**FIGURE 2-28**). (Penguins and walruses can maintain relatively high body temperatures, despite living in very cold habitats, due to their thick layer of insulating fat.) Lipids also include *sterols,* which include *cholesterol* and many of the sex hormones that play regulatory roles in animals, and *phospholipids,* which form the membranes that enclose cells.

> **TAKE-HOME MESSAGE 2·11**
>
> Lipids are insoluble in water and greasy to the touch. They are valuable to organisms for long-term energy storage and insulation, in membrane formation, and as hormones.

# 2·12 — Fats are tasty molecules too plentiful in our diets.

All fats have two distinct components: they have a "head" region and two or three long "tails" (**FIGURE 2-29**). The head region is a small molecule called **glycerol.** It is linked to "tail" molecules known as **fatty acids.** A fatty acid is simply a long hydrocarbon—that is, a chain of carbon molecules, often a dozen or more, linked together with one or two hydrogen atoms attached to each carbon.

The fats in most foods we eat are **triglycerides,** which are fats having three fatty acids linked to the glycerol molecule. For this reason, the terms "fats" and "triglycerides" are often used interchangeably. Triglycerides that are solid at room temperature are generally called "fats," while those that are liquid at room temperature are called "oils."

Fat molecules contain much more stored energy than do carbohydrate molecules. That is, the chemical breakdown of fat molecules releases significantly more energy. A single gram of carbohydrate stores about 4 calories of energy, while the same amount of fat stores about 9 calories—not unlike the difference between a $5 bill and a $10 bill. Because fats store

**STRUCTURE OF FATS (TRIGLYCERIDES)**

**HEAD:** Glycerol

**TAILS:** Fatty acids

Fats are composed primarily of three elements: C  H  O

Energy is stored in the many carbon-hydrogen bonds.

C — H

ENERGY

Symbol for fat (triglyceride) used in this book

**FIGURE 2-29 Triglycerides have glycerol heads and fatty acid tails.**

FIGURE 2-30 Animals (including humans!) prefer the taste of fats.

*Because fats store such large amounts of energy, a strong taste preference for fats over other energy sources has evolved in animals.*

such a large amount of energy, a strong taste preference for fats over other energy sources has evolved in animals (**FIGURE 2-30**). Organisms evolving in an environment of uncertain food supply will build the largest surplus by consuming molecules that hold the most amount of energy in the smallest mass. This feature helped the earliest humans to survive, millions of years ago, but today puts us in danger from the health risks of obesity now that fats are all too readily available.

An important distinction is made between "saturated" and "unsaturated" fats (**FIGURE 2-31**). These terms refer to the hydrocarbon chain in the fatty acids. If each carbon atom

| SATURATED FATS | UNSATURATED FATS |
| --- | --- |

In saturated fats, each carbon in the hydrocarbon chain is bound to two hydrogen atoms.

In unsaturated fats, at least one carbon in the hydrocarbon chain is bound to just one hydrogen, causing the fatty acid to have a crooked shape.

Double bond

Straight fatty acids can be packed together tightly. As a result, saturated fats are solid at room temperature.

Crooked fatty acids cannot be packed together tightly. As a result, unsaturated fats on their own (such as olive oil) are liquid at room temperature.

**FIGURE 2-31 Degrees of saturation.** Fatty acids (and thus the fats that contain them) can be unsaturated or saturated.

in the hydrocarbon chain of a fatty acid is bonded to two hydrogen atoms, the fat molecule carries the maximum number of hydrogen atoms and is said to be a **saturated fat.** Most animal fats, including those found in meat and eggs, are saturated. They are not essential to your health and, because they accumulate in your bloodstream and can narrow the vessel walls, they can cause heart disease and strokes.

An **unsaturated fat** is one in which some of the carbon atoms are bound to only a single hydrogen. Most plant fats are unsaturated. Unsaturated fats may be *mono-unsaturated* (if a fatty acid hydrocarbon chain has only one pair of neighboring carbon atoms in an unsaturated state—that is, has only one double bond) or *polyunsaturated* (if more than one pair of carbons is unsaturated—there's more than one double bond). Unsaturated fats are still high in calories, but because they can lower cholesterol, they are generally preferable to saturated fats. Foods high in unsaturated fats include avocados, peanuts, and olive oil. Relative to other animals, fish tend to have less saturated fat.

**Q** How will the "chewy-ness" of a cookie differ depending on whether you make it with butter or vegetable oil as the lipid? Which is better for your health?

The shapes of unsaturated fat molecules and saturated fat molecules are different. When saturated, the hydrocarbon tails of the fatty acids all line up very straight and the fat molecules can be packed together tightly. The tight packing causes the fats, such as butter, to be solid at room temperature. When unsaturated, the fatty acids have kinks in the hydrocarbon tails and the fat molecules cannot be packed together as tightly (see Figure 2-31). Consequently, unsaturated fats, such as canola oil and vegetable oil, do not solidify as easily and are liquid at room temperature.

The ingredient list for many snack foods includes "partially hydrogenated" vegetable oils. The hydrogenation of an oil means that a liquid, unsaturated fat has had hydrogen atoms added to it so that it becomes more saturated. This can be useful in creating a food with a more desirable texture, since increasing a fat's degree of saturation changes its consistency and makes it more solid at room temperatures. By attaining just the right degree of saturation, it is possible to create foods, such as chocolate, that are near the border of solid and liquid and

**Q** Many snack foods contain "partially hydrogenated" vegetable oils. Why might food producers add hydrogen atoms to a vegetable oil?

### HYDROGENATION

Hydrogenation is the artificial addition of hydrogen atoms to an unsaturated fat. This can improve a food's taste, texture, and shelf-life.

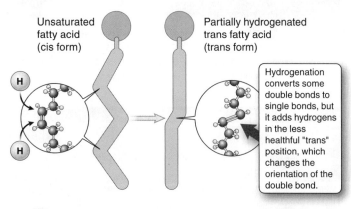

Unsaturated fatty acid (cis form)

Partially hydrogenated trans fatty acid (trans form)

Hydrogenation converts some double bonds to single bonds, but it adds hydrogens in the less healthful "trans" position, which changes the orientation of the double bond.

 *Hydrogenation can increase the saturation of fats—for better and worse.*

**FIGURE 2-32 Hydrogenation improves a food's taste, texture, and shelf-life (but at a cost).**

"melt in your mouth" (**FIGURE 2-32**). Unfortunately, hydrogenation also makes the food less healthful because saturated fats are less reactive—your body is less likely to break them down—and so are more likely to accumulate in your blood vessels, increasing the risk of heart disease.

Hydrogenation of unsaturated fats is doubly problematic from a health perspective because it also creates **trans fats,** the "trans" referring to the unusual orientation of some or all of the double bonds that remain following the addition of hydrogen atoms. This orientation differs from that in other unsaturated dietary fats—which have their double bonds in an orientation called "cis." Trans fats in your diet cause your body to produce more cholesterol, further raising the risk of heart disease, and they also reduce your body's production of a type of cholesterol that protects against heart disease.

Because of the well-documented links between dietary fats and heart disease, many people are trying to reduce their fat intake. "Fake fats" make this possible. They are designed to be similar to fats in taste and texture, but have one big difference: they cannot be digested by humans. One such "fake fat" is olestra. Olestra, instead of being a triglyceride fat with three fatty acid tails linked to a glycerol molecule, has eight separate hydrocarbon fatty acids attached to a molecule of sucrose. This octopus-like molecule allows the fatty acids to stimulate their usual taste buds on your tongue, telling your brain that you are eating a fat. The complex shape of the molecule, however,

prevents your body's digestive chemicals from grabbing onto it and breaking it down. As a consequence, it passes through your digestive system without being digested. It's not a perfect solution, however. Olestra reduces absorption of some vitamins, and in some people causes abdominal cramping.

**Q** Olestra is a recently developed "fake fat" chemical that gives foods the taste of fat, without adding the calories of fats. What chemical structure might make this possible?

## TAKE-HOME MESSAGE 2·12

Fats, including the triglycerides common in the food we eat, are one type of lipid. Characterized by long hydrocarbon tails, fats effectively store energy in the many carbon-hydrogen and carbon-carbon bonds. Their caloric density is responsible for humans' preferring fats to other macromolecules in the diet, and is also responsible for their association with obesity and illness in the modern world.

# 2·13 Cholesterol and phospholipids are used to build sex hormones and membranes.

Not all lipids are fats, nor do lipids necessarily function in energy storage. A second group of lipids, called the **sterols,** plays an important role in regulating growth and development (**FIGURE 2-33**). This group includes some very familiar lipids: cholesterol and the steroid hormones such as testosterone and estrogen. These molecules are all variations on one basic structure formed from four interlinked rings of carbon atoms.

**Cholesterol** is an important component of most cell membranes. For this reason, it is an essential molecule for living organisms. Cells in our liver produce almost 90% of the circulating cholesterol by transforming the saturated fats in our diet. Cholesterol is also present in foods and has a bad reputation in most Western cultures that is mostly well deserved. When we ingest too much cholesterol (present in animal-based foods such as egg yolks, red meat, and cream) and high levels of cholesterol circulate in our bloodstream, the cholesterol can attach to blood vessel walls and cause them to thicken. In turn, this thickening can lead to high blood pressure, a major contributor to strokes and heart attacks. For these reasons, nutritionists advise limiting the consumption of foods high in cholesterol and saturated fats.

STEROLS

CHOLESTEROL
• Important component of cell membranes in animals.
• Dietary cholesterol can attach to and thicken vessel walls and may cause serious health problems.

Sterols are all based on a structure featuring four fused carbon rings.

STEROID HORMONES
• Regulate sexual development, maturation, and sex cell production.
• Estrogen influences memory and mood.
• Testosterone stimulates muscle growth.

Estrogen

Testosterone

**FIGURE 2-33 Not all lipids are for energy storage.** Cholesterol, estrogen, and testosterone are all lipids.

The steroid hormones estrogen and testosterone are built through slight chemical modifications to cholesterol. These hormones are among the primary molecules that direct and regulate sexual development, maturation, and sperm and egg production. In both males and females, estrogen influences memory and mood, among other traits. Testosterone has numerous effects, one of which is to stimulate muscle growth. As a consequence, athletes (particularly bodybuilders) have often been found to take synthetic variants of testosterone to increase their muscularity. But the use of these supplements is often accompanied by dangerous side effects, including extreme aggressiveness ("'roid rage"), high cholesterol, and, following long-term use, cancer. As a consequence, nearly all athletic organizations have banned their use (**FIGURE 2-34**).

Phospholipids and waxes are also lipids. **Phospholipids** are the major component of the membrane that surrounds the contents of a cell and controls the flow of chemicals into and out of the cell (**FIGURE 2-35**). They have a structure similar to

**PHOSPHOLIPIDS**

**HYDROPHILIC HEAD** (attracted to water)

Phosphate group

**HYDROPHOBIC TAILS** (not attracted to water)

Fatty acids

**PHOSPHOLIPIDS IN WATER**

Hydrophilic heads

Water

Hydrophobic tails

Phospholipids align so that their hydrophilic heads extend toward the water, while their hydrophobic tails are directed away from the water.

Symbol for phospholipid used in this book

**FIGURE 2-35 Dual nature.** Phospholipids have a head region that is attracted to water and a tail that is not.

fats, but with two differences: they contain a phosphorus atom (hence *phospho*lipids) and they have two fatty acid chains rather than three. We explore the significant role of phospholipids in cell membranes in the next chapter.

**Waxes** resemble fats but have only one long-chain fatty acid linked to the glycerol head of the molecule. Because the fatty acid chain is highly nonpolar, waxes are strongly hydrophobic; that is, these molecules do not mix with water but repel it. Their water resistance accounts for their presence as a natural coating on the surface of many plants and in the outer coverings of many insects. In both cases, the waxes prevent the plants and animals from losing the water essential to their life processes. Many birds, too, have a waxy coating on their wings, keeping them from becoming water-logged when they get wet.

**FIGURE 2-34 Dangerous bulk.** Steroids can increase muscularity, but with serious health consequences.

### TAKE-HOME MESSAGE 2·13

Cholesterol and phospholipids are lipids that are not fats. Both are important components in cell membranes. Cholesterol also serves as a precursor to steroid hormones, important regulators of growth and development.

# 2·14–2·18 Proteins are versatile macromolecules that serve as building blocks.

*Hair and feathers are built from proteins.*

# 2·14 Proteins are bodybuilding macromolecules.

You can't look at a living organism and not see proteins (**FIGURE 2-36**). Inside and out, **proteins** are the chief building blocks of all life. They make up skin and feathers and horns. They make up bones and muscles. In your bloodstream, proteins fight invading microorganisms and stop you from bleeding to death from a shaving cut. Proteins control the levels of sugar and other chemicals in your bloodstream and carry oxygen from one place in your body to another. And in just about every cell in every living organism, proteins called **enzymes** initiate and assist every chemical reaction that occurs.

Although proteins perform several very different types of functions, all are built in the same way and from the same raw materials in all organisms. In the English language, every sentence is made up of words and every word is formed from one or more of the 26 letters of the alphabet. With 26 letters we can write anything, from sonnets to cookbooks to biology textbooks. Proteins, too, are constructed from a sort of alphabet. Instead of 26 letters there are 20 molecules, known as **amino acids.** Unique combinations of these 20 amino acids are strung together, like beads on a string, and the resulting protein has a unique structure and chemical behavior.

Let's look more closely at the structure of the amino acids in the protein alphabet. They all have the same basic two-part structure: one part is the same in all 20 amino acids, and the other part is unique, differing in each of the 20 amino acids.

Proteins contain the same familiar atoms as carbohydrates and lipids—carbon, hydrogen, and oxygen—but differ in an

Proteins perform a variety of different functions. They all, however, are built the same way and from the same raw materials in organisms.

**STRUCTURAL**
Hair, fingernails, **feathers**, horns, cartilage, tendons

**PROTECTIVE**
Help fight invading microorganisms, coagulate **blood**

**REGULATORY**
Control cell activity, constitute some **hormones**

**CONTRACTILE**
Allow **muscles** to contract, heart to pump, sperm to swim

**TRANSPORT**
Carry molecules such as **oxygen** around your body

**FIGURE 2-36 Proteins everywhere!** Proteins are the chief building blocks of all organisms.

AMINO GROUP

SIDE CHAIN

The side chain is the unique part of each of the 20 amino acids, varying in size, shape, and charge.

CARBOXYL GROUP

SIDE CHAIN EXAMPLES

Alanine    Glycine

Tryptophan

**FIGURE 2-37 Amino acid structure.** Amino acids are made up of an amino group, a carboxyl group, and a side chain.

nitrogen atom bonded to hydrogen atoms (usually two or three). These components—the central carbon with its attached hydrogen atom, carboxyl group, and amino group—are the foundation that identifies a molecule as an amino acid and, as multiple amino acids are joined together, forms the "backbone" of the protein.

The fourth bond of the central carbon attaches to a functional group or side chain. This side chain is the unique part of each of the 20 amino acids. In the simplest amino acid, glycine, for example, the side chain is simply a hydrogen atom. In other amino acids, the side chain is a single $CH_3$ group or three or four such groups. Most of the side chains include both hydrogen and carbon, and a few include nitrogen or sulfur atoms. The side chain determines an amino acid's chemical properties, such as whether the amino acid molecule is polar or nonpolar.

important way: they also contain nitrogen. At the center of every amino acid is a carbon atom, with its four covalent bonds (**FIGURE 2-37**). One bond attaches the carbon to something called a **carboxyl group,** which is a carbon bonded to two oxygen atoms (one by a single bond and the other by a double bond). The second bond attaches the central carbon to a single hydrogen atom. The third bond attaches the central carbon to an **amino group,** which is a

## TAKE-HOME MESSAGE 2·14

Unique combinations of 20 amino acids give rise to proteins, the chief building blocks of the physical structures that make up all organisms. Proteins perform myriad functions, from assisting chemical reactions to causing blood clotting to building bones to fighting microorganisms.

# 2·15 Proteins are an essential dietary component.

The atoms present in the proteins we eat—especially the nitrogen atoms—are essential to the growth, repair, and replacement that take place in our bodies. As we digest protein, breaking it down into its amino acids, our bodies use them for various building projects. Proteins also store energy in their bonds and, like carbohydrates and lipids, they can be used to fuel living processes.

The amount of protein we need depends on the extent of the building projects under way. Most individuals need 40–80 grams of protein per day. Bodybuilders, however, may need 150 grams or more a day to achieve the extensive muscle growth stimulated by their training; similarly, the protein needs of pregnant or nursing women are very high.

Contrary to the impression you might get from food labels, all proteins are not created equal. Every different protein has a different composition of amino acids. And while our bodies can manufacture certain amino acids as they are needed, many other amino acids must come from our diet. Those that we must get from our diet—about half of the 20 amino acids—are called "essential amino acids." For this reason, we shouldn't just speak of needing "x grams of

**Q** Food labels indicate how many grams of protein are contained in a food item. Why is this information only partially helpful for effectively guiding your protein intake?

protein per day." We need to consume all of the essential amino acids every day.

Many foods, containing "complete proteins," have all of the essential amino acids. Animal products such as milk, eggs, fish, chicken, and beef tend to provide complete proteins. Vegetables, fruits, and grains often contain "incomplete proteins," which do not have all the essential amino acids. If you consume only one type of incomplete protein in your diet, you may be deficient in one or more of the essential amino acids. But two incomplete proteins that are "complementary proteins," when eaten together, can provide all the essential amino acids.

Traditional dishes in many cultures often include such pairings. Examples are corn and beans in Mexico and rice and lentils in India (**FIGURE 2-38**).

---

## TAKE-HOME MESSAGE 2·15

Twenty amino acids make up all the proteins necessary for growth, repair, and replacement of tissue in living organisms. Of these amino acids, about half are essential for humans: they cannot be synthesized by the body so must be consumed in the diet. Complete proteins contain all essential amino acids, while incomplete proteins do not.

---

## COMPLETE vs. INCOMPLETE PROTEINS

**ESSENTIAL AMINO ACID CONTENT OF COMMON FOODS**

| | Apples | Almonds | White rice | Lentils | Milk - 2% fat | Tofu - firm (coagulated soy milk) | Chicken | Beef - T-bone steak |
|---|---|---|---|---|---|---|---|---|
| Histidine | ● | ● | ● | ● | ● | ● | ● | ● |
| Isoleucine | ○ | ● | ● | ● | ● | ● | ● | ● |
| Leucine | ● | ● | ● | ● | ● | ● | ● | ● |
| Lysine | ● | ○ | ○ | ● | ● | ● | ● | ● |
| Methionine + Cysteine | ○ | ○ | ● | ○ | ● | ● | ● | ● |
| Phenylalanine + Tyrosine | ○ | ● | ● | ● | ● | ● | ● | ● |
| Threonine | ○ | ● | ● | ● | ● | ● | ● | ● |
| Tryptophan | ○ | ● | ● | ● | ● | ● | ● | ● |
| Valine | ● | ● | ● | ● | ● | ● | ● | ● |

● Present at optimal levels
○ Not present at optimal levels

Optimal levels determined by the Institute of Medicine's Food and Nutrition Board

*GRAPHIC CONTENT!*
*Thinking critically about visual displays of data.*
*Turn to p. 76 for a closer inspection of this figure.*

*Traditional dishes in many cultures combine proteins, bringing together all essential amino acids.*

**FIGURE 2-38 All proteins are not created equal.** Some foods have "complete proteins" with all the essential amino acids. Other foods have "incomplete proteins," and we must consume proteins from multiple sources to get all the essential amino acids.

# 2·16 A protein's function is influenced by its three-dimensional shape.

Proteins are formed by linking individual amino acids together with a **peptide bond,** in which the amino group of one amino acid is bonded to the carboxyl group of another. Two amino acids joined together form a *dipeptide,* and several amino acids joined together form a *polypeptide.* The sequence of amino acids in the polypeptide chain is called the **primary structure** of the protein and can be compared to the sequence of letters that spells a specific word (**FIGURE 2-39**).

Amino acids in a polypeptide chain don't remain in a simple straight line like beads on a string. The chain begins to fold as side chains come together and hydrogen bonds form between various atoms in the chain. The two most common patterns of hydrogen bonding between amino acids cause the chain to either twist in a corkscrew-like shape or form a zigzag folding pattern. This hydrogen bonding between amino acids gives a protein its **secondary structure.**

The protein eventually folds and bends upon itself, and additional bonds continue to form between atoms in the side chains of amino acids that are near each other. Eventually, the protein folds into a unique and complex three-dimensional shape called its **tertiary structure.** The exact form comes about as the secondary structure folds and bends, bringing together amino acids that then form bonds such as hydrogen bonds or covalent sulfur-sulfur bonds (see Figure 2-37).

Some protein molecules have a **quaternary structure** in which two or more polypeptide chains are held together by bonds between amino acids in the different chains. Hemoglobin, the protein molecule that carries oxygen from the lungs to the cells where it is needed, is made from four polypeptide chains, two "alpha" chains and two "beta" chains.

Some proteins are attached to other types of macromolecules. *Lipoproteins,* for example, circulate in the bloodstream carrying fats. They are formed when cholesterol and a triglyceride (both lipids) combine with a protein. *Glycoproteins* are combinations of carbohydrates and proteins. These are found on the surfaces of nearly all animal cells and play a role in helping the immune system distinguish between your own cells and foreign cells. (We learn more about glycoproteins in the next chapter, which discusses cells.)

The overall shape of a protein molecule determines its function—how it behaves and the other molecules it interacts with. For proteins to function properly, they must retain their three-dimensional shape. When their shapes are deformed, they usually lose their ability to function. We can see

**Q** Egg whites contain a lot of protein. Why does beating them change their texture, making them stiff?

---

## STRUCTURE OF PROTEINS

**PRIMARY STRUCTURE**
The sequence of amino acids in a polypeptide chain, similar to the sequence of letters that spell out a specific word

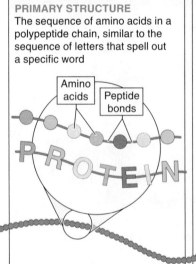

**SECONDARY STRUCTURE**
The corkscrew-like twists or pleated folds formed by hydrogen bonds between amino acids in the polypeptide chain

**TERTIARY STRUCTURE**
The complex three-dimensional shape formed by multiple twists and bends in the polypeptide chain, based on the side chains' interactions with each other and with the aqueous solvent.

**QUATERNARY STRUCTURE**
Two or more polypeptide chains bonded together

**FIGURE 2-39 The several levels of protein structure.** The functions of proteins are influenced by their three-dimensional shape.

proteins deforming when we fry an egg. The heat breaks the hydrogen bonds that give the proteins their shape. The proteins in the clear egg white unfold, losing their secondary and tertiary structure. This disruption of protein folding is called **denaturation** (FIGURE 2-40).

Almost any extreme environment will denature a protein. Take a raw egg, for instance, and crack it into a dish containing baking soda or rubbing alcohol. Both chemicals are sufficiently extreme to turn the clear protein opaque white, as in fried egg whites.

Hair is a protein whose shape most of us have modified at one time or another. Styling hair—whether curling or straightening it—involves altering some of the hydrogen bonds between the amino acids that make up the hair protein, changing its tertiary structure. When your hair gets wet, the water is able to disrupt some of the hydrogen bonds, causing some amino acids in the protein to form hydrogen bonds with the water molecules instead. This enables you to change your hair's shape—making it straighter or, if you manipulate it around curlers, making it curlier—if you style it while it's wet. The hair can then hold this shape when it dries, as the hydrogen bonds to water are replaced by other hydrogen bonds between amino acids of the hair protein as the water evaporates. Once your hair gets wet again, however, unless it is combed, brushed, or wrapped in a different style, it will return to its natural shape.

Whether your hair is straight or curly or somewhere in between depends on your hair protein's amino acid sequence and the three-dimensional shape it confers (FIGURE 2-41).

FIGURE 2-41 **Curly or straight?** Proteins determine it!

**Q** Why *do some people have curly hair and others have straight hair?*

This amino acid sequence is something you're born with (that is, it's genetically determined). The chains are more or less coiled, depending on the extent of covalent and hydrogen bonding between different parts of the coil. Many hair salons make use of the ability to alter covalent bonds to change hair texture semi-permanently. They are able to do this in three simple steps. First, the bonds are broken chemically. Second, the hair is wrapped around curlers to hold the polypeptide chains in a different position. And third, chemicals are put on the hair to create new covalent bonds between parts of the polypeptide chains. The hair thus becomes locked in a new position. (New hair will continue to grow with its genetically determined texture, of course, requiring the procedure to be repeated regularly.)

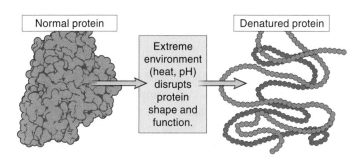

FIGURE 2-40 **Denaturation.** When proteins are unfolded, they lose their function.

### TAKE-HOME MESSAGE 2·16

The particular amino acid sequence of a protein determines how it folds into a particular three-dimensional shape. This shape determines many of the protein's features, such as which molecules it will interact with. When a protein's shape is deformed, the protein usually loses its ability to function.

# 2·17 Enzymes are proteins that speed up chemical reactions.

Protein shape is particularly critical in **enzymes,** molecules that help initiate and accelerate the chemical reactions in our bodies. Enzymes emerge unchanged—in their original form—when the reaction is complete and thus can be used again and again. Here's how they work.

Think of an enzyme as a big piece of popcorn. Its tertiary or quaternary structure gives it a complex shape with lots of nooks and crannies. Within one of those nooks is a small area called the **active site** (**FIGURE 2-42**). Based on the chemical properties of the atoms lining this pocket, the active site provides a place for the participants in a chemical reaction, the reactants or **substrate** molecules, to nestle briefly.

Enzymes are very choosy: they bind only with their appropriate substrate molecules, much like a lock that can be opened with only one key (see Figure 2-42). The exposed

atoms in the active site have electrical charges that attract rather than repel the substrate molecules, and only the substrate molecules can fit into the active-site groove. Once the substrate is bound to the active site, a reaction can take place—and usually does so very quickly.

The chemical reactions that occur in organisms can either release energy or consume energy. But in either case, there is a certain minimum energy—a little "push"—needed to initiate the reaction, called **activation energy.** And although enzymes don't alter the amount of energy released by a reaction, they act as catalysts by lowering the activation energy, which causes the reaction to occur more quickly.

An enzyme can reduce the activation energy and initiate a chemical reaction in a variety of ways. These include:

1.  Stressing critical chemical bonds in the substrate molecule, thus increasing the likelihood of their breaking.

2.  Directly participating in the reaction, increasing the substrate's ability to make or break other bonds.

3.  Creating a "micro-habitat" that favors the reaction. For instance, the active site might be a nonpolar environment, or it might have a different pH than the surrounding fluid. These slight alterations might increase the likelihood that a particular reaction occurs.

4.  Simply orienting or holding substrate molecules in place so that they can be modified.

By virtue of their catalytic capacities, enzymes are at the heart of the chemistry of living organisms. Taken together, all of the chemical reactions in a living organism are its *metabolism.*

These reactions are generally arranged in sequential "metabolic pathways" by which increasingly complex molecules are synthesized or degraded. Each step of these metabolic pathways is catalyzed by an enzyme produced in the body. Proteins are the building blocks with which living organisms are built, but since nearly all enzymes are proteins, proteins can also be thought of as the *builders* of bodies, too.

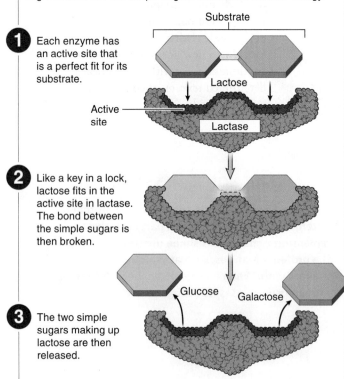

**ENZYMES FACILITATE CHEMICAL REACTIONS**

Enzymes can help to bring about chemical reactions in a variety of ways. The enzyme lactase, for example, breaks down the milk sugar lactose into two simple sugars that can be used for energy.

Substrate

1 Each enzyme has an active site that is a perfect fit for its substrate.

Lactose

Active site

Lactase

2 Like a key in a lock, lactose fits in the active site in lactase. The bond between the simple sugars is then broken.

Glucose    Galactose

3 The two simple sugars making up lactose are then released.

**FIGURE 2-42 Lock and key.** Enzymes are *very* specific about which molecules and reactions they will catalyze.

> ### TAKE-HOME MESSAGE 2·17
>
> **Enzymes are proteins that help initiate and speed up chemical reactions. They aren't permanently altered in the process, but rather can be used again and again.**

# 2·18 Enzymes regulate reactions in several ways (but malformed enzymes can cause problems).

If not for enzymes, it is possible that nothing would ever get done. At least not in living organisms. Enzymes don't alter the outcome of reactions, but without the chemical "nudge" they supply—often increasing reaction rates to millions of times their uncatalyzed rate—the processes necessary to sustain life could not occur.

The rate at which an enzyme catalyzes a reaction is influenced by several chemical and physical factors (**FIGURE 2-43**). These include:

**1. Enzyme and substrate concentration.** For a given amount of substrate, an increase in the amount of enzyme increases the rate at which the reaction occurs. Similarly, for a given amount of enzyme, an increase in the substrate concentration increases the reaction rate. In both cases, the increases occur only up to the point at which all of the enzyme molecules are bound to substrate, or vice versa, at which point additional enzyme or substrate no longer increases the reaction rate.

**2. Temperature.** Because increasing the temperature increases the movement of molecules, reaction rates generally increase at higher temperatures. This occurs only up to the optimum temperature for an enzyme. At temperatures above the optimum, reaction rates decrease as enzymes lose their shape or even denature. Enzymes from different species can have widely differing optimum temperatures.

**3. pH.** As with temperature, enzymes have an optimum pH. Above or below this pH, interactions between excess hydrogen or hydroxide ions and amino acid side chains in the active site disrupt enzyme function (and sometimes structure) and decrease reaction rates.

**4. Presence of inhibitors or activators.** One of the most common ways that cells regulate their metabolic pathways is through the binding of other chemicals to enzymes. This can alter enzyme shape in a way that increases or decreases the enzyme's activity. **Inhibitors** reduce enzyme activity and come in two types: **competitive inhibitors** bind to the active site, blocking substrate molecules from the site and thus from taking part in the reaction. **Noncompetitive inhibitors** do not compete for the active site but rather bind to another

FIGURE 2-43 **Enzyme activity is influenced by physical factors such as temperature and pH, and by chemical factors such as enzyme and substrate concentrations.**

## ENZYME ACTIVITY

The rate at which an enzyme catalyzes a reaction is influenced by several chemical and physical factors.

### ENZYME AND SUBSTRATE CONCENTRATION
Reaction rates increase with increased amounts of enzyme (or substrate), but only up to the point at which all of the enzyme molecules are bound to substrate. At that point, additional enzyme (or substrate) no longer increases the reaction rate.

### TEMPERATURE
Reaction rates generally increase at higher temperatures, but only up to the optimum temperature for an enzyme. At temperatures above the optimum, reaction rates decrease as enzymes lose their shape or even denature.

### pH
Reaction rates generally increase as pH nears the optimum level for an enzyme. Above or below this pH, enzyme function can be disrupted and reaction rates decrease.

### PRESENCE OF INHIBITORS OR ACTIVATORS
Reaction rates increase in the presence of activators and decrease in the presence of inhibitors.

part of the enzyme, altering its shape in a way that changes the structure of the active site, reducing or blocking its ability to bind with substrate. Often, it is the very product of a metabolic pathway that acts as an inhibitor of enzymes early in the pathway, effectively shutting off the pathway when enough of its end product has been produced.

Just as a molecule can bind to an enzyme and inhibit the enzyme's activity, some cellular chemicals act as **activators.** Instead of their binding to an enzyme "turning it off," their binding to the enzyme "turns it on," altering the enzyme's shape or structure so that it can now catalyze a reaction.

Sometimes a cell produces a protein "word" that is misspelled—that is, the sequence of amino acids is incorrect. If an enzyme is altered even slightly, the active site may change and the enzyme no longer functions (**FIGURE 2-44**). Slightly

modified, non-functioning enzymes are responsible for a large number of diseases and physiological problems (see Section 5-9). An example is the body's inability to break down the amino acid phenylalanine (in a condition known as phenylketonuria).

**Q** Why *do some adults get sick when they drink milk?*

One health issue influenced by enzyme function is the condition called lactose intolerance. Normally, during digestion, the lactose in milk is broken down into its component parts, glucose and galactose (see Figure 2-42). These simple sugars are then used for energy. But some people, when they become adults, are unable to break the bond linking the two simple sugars because they no longer produce the enzyme lactase that assists in this process. Consequently, any lactose in their diet passes through their stomach and small intestine undigested. Then, when it reaches the large intestine, bacteria living there consume the lactose. The problem is that, as the bacteria break down the lactose, they produce some carbon dioxide and other gases. These gases are trapped in the intestine and lead to severe discomfort. Interestingly, in regions of the world with long traditions of pastoralism (raising and consuming livestock), lactose intolerance is much rarer than in other parts of the world. Only about 10% of people from Denmark or Sweden have lactose intolerance, but among people from China, which has historically been largely non-pastoral, about 99% are lactose intolerant.

The unpleasant symptoms of lactose intolerance can be avoided by not consuming milk, cheese, yogurt, ice cream, or any other dairy products, but they can also be avoided by taking a pill containing the enzyme lactase. It doesn't matter how the enzyme gets into your digestive system; as long as it's there, the lactose in the milk can be broken down.

---

## NON-FUNCTIONING ENZYMES

Sometimes a protein "word" is misspelled—that is, the sequence of amino acids is incorrect. If an enzyme is altered even slightly, the active site may change, which can cause the enzyme to no longer function.

**FUNCTIONING ENZYME**　　　**NON-FUNCTIONING ENZYME**

When functioning normally, the enzyme phenylalanine hydroxylase converts the amino acid phenylalanine into another amino acid, tyrosine.

If the shape of phenylalanine hydroxylase is altered, it may not be able to convert phenylalanine into tyrosine.

Phenylalanine

Phenylalanine hydroxylase

Tyrosine

💡 *Non-functioning enzymes are responsible for a large number of diseases and physiological problems.*

**FIGURE 2-44 When a protein is "misspelled."**

---

**TAKE-HOME MESSAGE 2·18**

Enzyme activity is influenced by physical factors such as temperature and pH, as well as chemical factors, including enzyme and substrate concentrations. Inhibitors and activators are chemicals that bind to enzymes and by blocking the active site or altering the shape or structure of the enzyme can change the rate at which the enzyme catalyzes reactions.

# 2·19–2·21 Nucleic acids store information on how to build and run a body.

*Three turns of the DNA double helix (from a scanning tunneling micrograph).*

# 2·19

## Nucleic acids are macromolecules that store information.

We have examined three of life's macromolecules: carbohydrates, lipids, and proteins. We turn our attention now to the fourth: **nucleic acids,** macromolecules that store information and are made up of individual units called **nucleotides.** All nucleotides have three components: a molecule of sugar, a phosphate group (containing a phosphorus atom bound to four oxygen atoms), and a nitrogen-containing molecule (**FIGURE 2-45**). There are two types of nucleic acids: **deoxyribonucleic acid (DNA)** and **ribonucleic acid (RNA).** Both play central roles in directing the production of proteins in living organisms, and by doing so play a central role in determining all of the inherited characteristics of an individual. In both types of nucleic acids, the molecule has the same type of backbone: a sugar molecule attached to a phosphate group attached to another sugar, then another phosphate, and so on. Attached to each sugar is one of the nitrogen-containing molecules called DNA **bases** (so named because of their chemical structure). A 10-unit nucleic acid therefore would have 10 bases, one attached to each sugar within the sugar-phosphate-sugar-phosphate backbone. But the base attached to each sugar is not always the same. It can be one of several different bases. For this reason, a nucleic acid

**FIGURE 2-45 The molecules that carry genetic information.** The structure of nucleic acids.

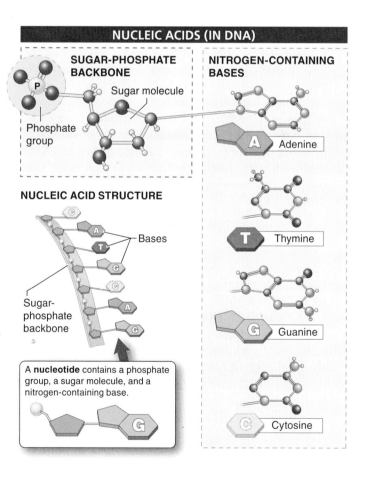

**NUCLEIC ACIDS (IN DNA)**

**SUGAR-PHOSPHATE BACKBONE**

Sugar molecule

Phosphate group

**NITROGEN-CONTAINING BASES**

A  Adenine

T  Thymine

G  Guanine

C  Cytosine

**NUCLEIC ACID STRUCTURE**

Bases

Sugar-phosphate backbone

A **nucleotide** contains a phosphate group, a sugar molecule, and a nitrogen-containing base.

is often described by the sequence of bases attached to the sugar-phosphate-sugar-phosphate backbone.

Nucleic acids are able to store information by varying which base is attached at each position in the molecule's backbone. At each position in a molecule of DNA, for example, the base can be any one of four possible bases: adenine (A), thymine (T), guanine (G), or cytosine (C). Just as the meaning of a sentence is determined by which letters are strung together, the information in a molecule of DNA is determined by its sequence of bases. One molecule may have the sequence adenine, adenine, adenine, guanine, cytosine, thymine, guanine—abbreviated as AAAGCTG. Another molecule

may have the sequence CGATTACCCGAT. Because the information differs in each case, so, too, does the protein for which the sequence codes, as we'll see.

# 2·20 DNA holds the genetic information to build an organism.

A molecule of DNA has two strands, each a sugar-phosphate-sugar-phosphate backbone with a base sticking out from each sugar molecule. The two strands wrap around each other, each turning in a spiral. Although each strand has its own sugar-phosphate-sugar-phosphate backbone and sequence of bases, the two strands are connected by the bases protruding from them.

You can picture a molecule of DNA as a ladder. The two sugar-phosphate-sugar-phosphate backbones are like the long vertical sides of the ladder that give it height. A base sticking out represents a rung on the ladder. Or, more accurately, half a rung. The bases protruding from each strand meet in the center and bind to each other (via hydrogen bonds), holding the ladder together. DNA differs from a ladder slightly, in that it has a gradual twist. The two spiraling strands together are said to form a **double helix** (**FIGURE 2-46**).

The two intertwining spirals fit together because only two combinations of bases pair up together. The base A always pairs with T, and C always pairs with G. Consequently, if the base sequence of one of the spirals is CCCCTTAGGAACC, the base sequence of the other must be GGGGAATCCTTGG. That is why researchers working on the Human Genome Project describe only one sequence of nucleotides when presenting a DNA sequence—even though that DNA is double-stranded in our bodies. With that one sequence, we can infer the identity and order of the bases in the complementary sequence, and thus we know the exact structure of the nucleic acid.

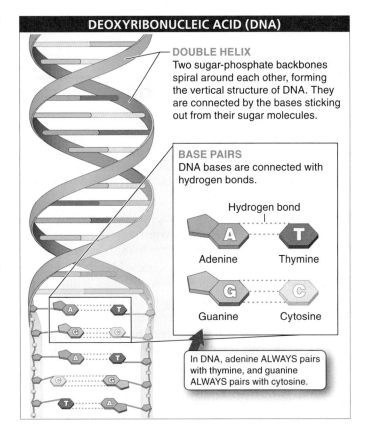

**DEOXYRIBONUCLEIC ACID (DNA)**

**DOUBLE HELIX**
Two sugar-phosphate backbones spiral around each other, forming the vertical structure of DNA. They are connected by the bases sticking out from their sugar molecules.

**BASE PAIRS**
DNA bases are connected with hydrogen bonds.

Hydrogen bond

A — T
Adenine — Thymine

G — C
Guanine — Cytosine

In DNA, adenine ALWAYS pairs with thymine, and guanine ALWAYS pairs with cytosine.

**FIGURE 2-46 A gradually twisting ladder.** The rungs of the DNA ladder are nucleotide base pairs, and the sides of the ladder are made up of two sugar-phosphate backbones.

The sequences of nucleotide bases containing the information about how to produce a particular protein have anywhere from a hundred to several thousand bases. In a human, all of the DNA in a cell, containing all of the instructions for every protein that a human must produce, contains about three billion base pairs. Almost all of this DNA is in the cell's nucleus.

# 2·21 RNA is a universal translator, reading DNA and directing protein production.

The process of building a protein from a DNA sequence is not a direct one. Rather, it incorporates a middleman, RNA, that is also a nucleic acid (**FIGURE 2-47**). Segments of the DNA are read off, directing the production of short strips of RNA that contain the information, taken from the DNA, about the amino acid sequence in a protein. The RNA moves to another part of the cell and then directs the piecing together of amino acids into a three-dimensional protein. We explore this in greater detail in Chapter 5.

RNA differs from DNA in three important ways. First, the sugar molecule of the sugar-phosphate-sugar-phosphate backbone differs slightly, containing an extra atom of oxygen. Second, RNA is single stranded. The sugar-phosphate-sugar-phosphate backbone is still there, as are the bases that protrude from each sugar. The bases, however, do not bind with anything else and do not form base pairs with another RNA strand. And third, while RNA has the bases A, G, and C, it replaces the thymine with a similar base called uracil (U).

Whether we're looking at the nucleotides that make up RNA and DNA or the lipids used to build sex hormones and cell membranes, we see a recurring theme in the construction of biological macromolecules: from relatively simple sets of building blocks linked together, infinitely complex molecules can be formed. Complex webs of one simple sugar, bonded together as glycogen, for instance, provide fuel for organisms. Similarly, sequences of amino acids of 20 different types, joined together, specify the structure of all the proteins found in every species on earth.

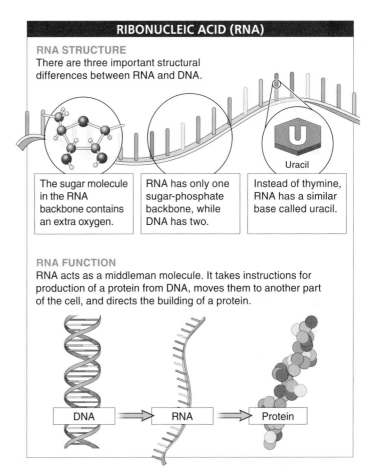

**RIBONUCLEIC ACID (RNA)**

**RNA STRUCTURE**
There are three important structural differences between RNA and DNA.

| The sugar molecule in the RNA backbone contains an extra oxygen. | RNA has only one sugar-phosphate backbone, while DNA has two. | Instead of thymine, RNA has a similar base called uracil. |

Uracil

**RNA FUNCTION**
RNA acts as a middleman molecule. It takes instructions for production of a protein from DNA, moves them to another part of the cell, and directs the building of a protein.

DNA → RNA → Protein

**FIGURE 2-47 The middleman between DNA and protein.** The structure of RNA.

## MELT-IN-YOUR-MOUTH CHOCOLATE MAY NOT BE SUCH A SWEET IDEA.

Food chemists have figured out how to make chocolate that melts in your mouth. Is that a good thing?

**Q: WHY ARE SOME FATS "LIQUIDY," LIKE OIL, AND OTHERS SOLID?** The less saturated a fat is, the more "liquidy" it is at room temperature. Most animal fats are saturated and are solid at room temperature. Best example: butter. Most plant fats are polyunsaturated and are liquid at room temperature. Best example: vegetable oils.

**Q: CAN OILS BE MADE MORE SOLID?** It's possible to increase the saturation of plant fats. Just heat them up and pass hydrogen bubbles through the liquid. In creating partially hydrogenated plant oils, this process reduces the number of carbon-carbon double bonds and makes the oil more solid. (It's easy, it's cheaper than just using butter, and it increases foods' shelf-life.)

**Q: DOES THAT IMPROVE THEIR TASTE?** By precisely controlling the level of saturation in plant fats, it is possible to create foods that are solid but have such a low melting point that they quickly melt on contact with the warmth of your mouth. This seems great, but . . .

**Q: IS THERE A DOWNSIDE?** The saturation of vegetable fats creates trans fats, due to the position taken by the newly added hydrogen atoms in the molecule. Trans fats increase levels of LDL ("bad") cholesterol and decrease HDL ("good") cholesterol, narrowing blood vessel walls and increasing the risk of heart disease and strokes.

**CONCLUSION:** Partially hydrogenated vegetable oils can give food a perfect texture and a pleasing feel in your mouth. But the creaminess comes with a high cost when it comes to your health.

# Big Ideas in ... Chemistry

## 2·1–2·3 Atoms form molecules through bonding.

- An atom is the smallest unit into which material can be divided without losing its essential properties.
- Atoms are made up of protons and neutrons in the nucleus and electrons that circle, far away, around that nucleus.
- Molecules are atoms linked together.

## 2·4–2·6 Water has features that enable it to support all life.

- Water molecules easily form hydrogen bonds, giving water great cohesiveness and the ability to resist temperature changes, and making it a versatile solvent.
- The pH of a fluid is a measure of how acidic or basic the solution is.

## 2·7–2·10 Carbohydrates are fuel for living machines.

- Carbohydrates, made up of carbon, oxygen, and hydrogen, are the primary fuel for all cellular machinery and also form much of the structure of cells.
- The simplest carbohydrates, including glucose, are called monosaccharides or simple sugars.
- Multiple simple sugars sometimes link together into more complex carbohydrates, including starch, which is the primary form of energy storage in plants, and glycogen, which is a primary form of energy storage in animals.
- Some carbohydrates are not digestible by humans.

## 2·11–2·13 Lipids store energy for a rainy day.

- Lipids are macromolecules—made up primarily of carbon, hydrogen, and oxygen—that are insoluble in water.
- Lipids are important for energy storage, hormones, and membrane structure.
- The breakdown of dietary fats releases more energy per gram than other macromolecules.

## 2·14–2·18 Proteins are versatile molecules that serve as building blocks.

- Cells and tissues are primarily built from proteins, sequences of amino acids that fold into complex, three-dimensional shapes.
- The atoms, especially nitrogen, present in the plant and animal proteins that an organism eats are essential to the organism's growth and repair.
- The amino acid sequence of a protein determines how it folds into a three-dimensional shape. This shape determines many of the protein's features, such as which molecules it will interact with.
- Enzymes are proteins that help initiate and speed up chemical reactions. They aren't permanently altered in the process.

## 2·19–2·21 Nucleic acids store information on how to build and run a body.

- The nucleic acids DNA and RNA are macromolecules that store information by having unique sequences of units called nucleotides. Both play central roles in directing protein production in organisms.
- The sequence of nucleotide bases contains the information about how to produce a particular protein.
- RNA acts like a universal translator of the genetic code into proteins. It reads DNA sequences and directs the production of a sequence of amino acids.

See the To-Go Guide (packaged with every new copy of this textbook) for an illustrated summary of the chapter.

# Check Your Knowledge

## GRAPHIC CONTENT

*Thinking critically about visual displays of data.*

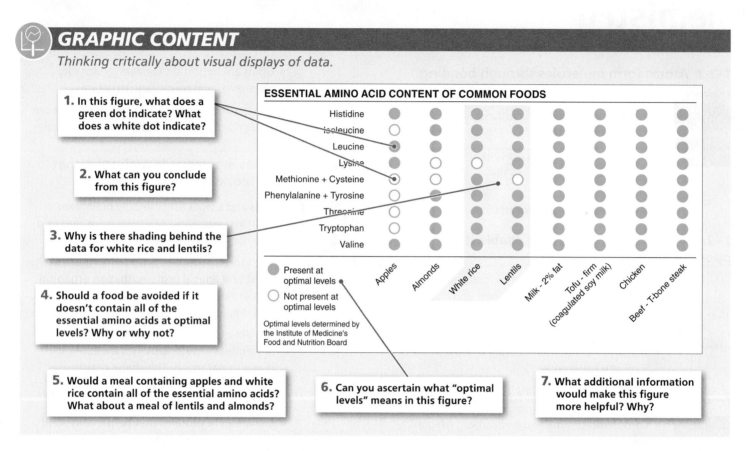

**1. In this figure, what does a green dot indicate? What does a white dot indicate?**

**2. What can you conclude from this figure?**

**3. Why is there shading behind the data for white rice and lentils?**

**4. Should a food be avoided if it doesn't contain all of the essential amino acids at optimal levels? Why or why not?**

**5. Would a meal containing apples and white rice contain all of the essential amino acids? What about a meal of lentils and almonds?**

**6. Can you ascertain what "optimal levels" means in this figure?**

**7. What additional information would make this figure more helpful? Why?**

See Answers at the back of the book.

# Key Terms in Chemistry

# Multiple-Choice Questions

**1.** The atomic number of carbon is 6. Its nucleus must contain:

a) 6 neutrons and 6 protons.
b) 3 protons and 3 neutrons.
c) 6 neutrons and no electrons.
d) 6 protons and no electrons.
e) 6 protons and 6 electrons.

100 HARD
**41**
0 EASY

**2.** The second orbital shell of an atom can hold _____ electrons.

a) 2
b) 3
c) 4
d) 6
e) 8

100 HARD
**20**
0 EASY

**3.** A covalent bond is formed when:

a) two nonpolar molecules associate with each other in a polar environment.
b) a positively charged particle is attracted to a negatively charged particle.
c) one atom gives up electrons to another atom.
d) two atoms share electrons.
e) two polar molecules associate with each other in a nonpolar environment.

100 HARD
**16**
0 EASY

**4.** Which of the following phenomena is most likely due to the high cohesiveness of water?

a) Lakes and rivers freeze from the top down, not the bottom up.
b) The fishing spider can walk across the surface of liquid water.
c) Adding salt to snow makes it melt.
d) The temperature of Santa Monica Bay, off the coast of Los Angeles, fluctuates less than the air temperature throughout the year.
e) All of the above are due to the cohesiveness of water.

100 HARD
**51**
0 EASY

**5.** Water can absorb and store a large amount of heat while increasing only a few degrees in temperature. Why?

a) The heat must first be used to break the hydrogen bonds rather than raise the temperature.
b) The heat must first be used to break the ionic bonds rather than raise the temperature.
c) The heat must first be used to break the covalent bonds rather than raise the temperature.
d) An increase in temperature causes an increase in adhesion of the water.
e) An increase in temperature causes an increase in cohesion of the water.

100 HARD
**22**
0 EASY

**6.** A chemical compound that releases H+ into a solution is called:

a) a proton.
b) a base.
c) an acid.
d) a hydroxide ion.
e) a hydrogen ion.

100 HARD
**31**
0 EASY

**7.** Which of the following foods is not a significant source of complex carbohydrates?

a) fresh fruit
b) rice
c) pasta
d) oatmeal
e) All of the above are significant sources of complex carbohydrates.

100 HARD
**31**
0 EASY

**8.** Sucrose (table sugar) and lactose (the sugar found in milk) are examples of:

a) naturally occurring enzymes.
b) simple sugars.
c) monosaccharides.
d) disaccharides.
e) complex carbohydrates.

100 HARD
**42**
0 EASY

**9.** Which of the following statements about starch is incorrect?

a) Starch is the primary form of energy storage in plants.

b) Starch consists of a hundred or more glucose molecules joined together in a line.

c) Starch tastes sweet because it is made from glucose.

d) Starch is a polysaccharide.

e) All of the above statements about starch are correct.

**10.** Which of the following statements about fiber is incorrect?

a) Dietary fiber reduces the risk of colon cancer.

b) Fiber in the diet slows the passage of food through the intestines.

c) Humans are unable to extract any caloric value from fiber.

d) The cellulose of celery stalks and lettuce leaves is fiber.

e) Fiber passing through the digestive system scrapes the wall of the digestive tract, stimulating mucus secretion and aiding in the digestion of other molecules.

**11.** In an unusually detailed dissection of your dinner, you isolate an unknown fatty acid. It is a liquid at room temperature (that is, has a low melting point) and contains carbon-carbon double bonds. What are you most likely to be eating?

a) a plant

b) a cow

c) a pig

d) a chicken

e) a lamb

**12.** An unsaturated fatty acid is one in which:

a) carbon-carbon double bonds are present in the hydrocarbon chain.

b) the hydrocarbon chain has an odd number of carbons.

c) the hydrocarbon chain has an even number of carbons.

d) no carbon-carbon double bonds are present in the hydrocarbon chain.

e) not all of the carbons in the hydrocarbon chain are bonded to hydrogen atoms.

**13.** Which statement about phospholipids is incorrect?

a) Because their phosphate groups repel each other, they are used as organisms' chief form of short-term energy.

b) They are hydrophobic at one end.

c) They are hydrophilic at one end.

d) They are a major constituent of cell membranes.

e) They contain glycerol linked to fatty acids.

**14.** Proteins are an essential component of a healthy diet for humans (and other animals). Their most common purpose is to serve as:

a) raw material for growth.

b) fuel for running the body.

c) organic precursors for enzyme construction.

d) long-term energy storage.

e) inorganic precursors for enzyme construction.

**15.** Dietary proteins:

a) are considered "complete" only if they contain the 8 essential amino acids required by humans.

b) consist of all 20 amino acids required in the human body.

c) are considered "complete" only if they contain the 12 non-essential amino acids required by humans.

d) are nutritionally identical, since all are broken down into their constituent amino acids in the digestive system.

e) can be obtained from animal sources but not plant sources.

**16.** The primary structure of proteins is often described as amino acids connected like beads on a string. In this same vein, which of the following images best describes a protein's quaternary structure?

a) threads in a cloth

b) needles in a haystack

c) rungs on a ladder

d) links on a chain

e) coils in a spring

**17.** Which of the following statements about enzymes is incorrect?

a) Enzymes can initiate chemical reactions.
b) Enzymes speed up chemical reactions.
c) Enzymes often induce conformational changes in the substrates that they bind.
d) Enzymes contain an active site for binding of particular substrates.
e) Enzymes undergo a permanent change during the reactions they promote.

100 HARD

0 EASY

**18.** Which of the following nucleotide bases are present in equal amounts in DNA?

a) adenine and cytosine
b) thymine and guanine
c) adenine and guanine
d) thymine and cytosine
e) adenine and thymine

100 HARD

22

0 EASY

**19.** Which type of macromolecule contains an organism's genetic information?

a) polysaccharide
b) monosaccharide
c) fatty acid
d) DNA
e) phospholipid

100 HARD

6 ASY

**20.** All of the following are features of RNA except:

a) a sugar-phosphate-sugar-phosphate backbone.
b) a coiled double-stranded structure.
c) a different type of sugar than is found in DNA.
d) an ability to direct protein production.
e) a different type of base than is found in DNA.

100 HARD

46

0 EASY

# Short-Answer Questions

**1.** An atom of sodium has an atomic number of 11 and an atom of chlorine has an atomic number of 17.

a) Indicate the number of electrons that are present in the outermost electron shell of each atom.
b) Predict what type of bond would form between these atoms, and explain why.

**2.** Explain how hydrogen bonding plays a part in each of the unique properties of water and, in essence, allows for life on earth.

**3.** There are many cleaning products on the market that advertise their ability to get rid of stubborn, oily dirt. Why is water alone not sufficient for removing oily dirt?

**4.** Some people who are watching what they eat may opt for a salad when they are dining out. A dinner salad may include lettuce, celery, strawberries, walnuts, chicken, and a salad dressing of oil and vinegar. Explain which category of major macromolecules each of these individual salad components falls into. Then explain the general purpose or function of each major macromolecule.

See Answers at the back of the book.
For additional study questions, go to www.prep-u.com.

## ABOUT THE CHAPTER OPENING PHOTO

Hydrogen bonding between molecules gives water cohesiveness. This causes the water to bead up when on surfaces that water molecules are less attracted to, such as the waxy leaves of many plants.

The smallest part of you

# Cells

# 3·1–3·3 What is a cell?

*Human cell, packed with organelles.*

# 3·1
## All organisms are made of cells.

Where do we begin if we want to understand how organisms work? Given their complexity, this task can be daunting. In the case of organisms, whether we are studying a creature as small as a flea or as large as an elephant or giant sequoia, they can be broken down into smaller units that are more easily studied and understood (**FIGURE 3-1**). The most basic unit of any organism is the **cell,** the smallest unit of life that can function independently and perform all the necessary functions of life, including reproducing itself. Understanding cell structure and function is the basis for

Fungi

Elephants

Head louse

**FIGURE 3-1 What do these diverse organisms have in common?** Cells.

our understanding of how complex organisms are organized.

The term "cell" was first used in the mid-1600s by Robert Hooke, an English scientist also known for his contributions to philosophy, physics, and architecture. When he was made Curator of Experiments for the Royal Society of London, Hooke suddenly had access to many of the first microscopes available, and he began to examine everything he could get his hands on. Because Hooke thought the close-up views of a very thin piece of cork resembled a mass of small, empty rooms, he named these compartments *cellulae,* Latin for "small rooms."

Today, we know that a cell is a three-dimensional structure, like a fluid-filled balloon, in which many of the essential chemical reactions of life take place (such as the breakdown of carbohydrates for energy, the modification of cholesterol to create testosterone and estrogen, and the translation of the genetic code into protein production). Generally, these reactions involve transporting raw materials and fuel into the cell and exporting finished materials and waste products out of the cell. In addition, nearly all cells contain DNA (deoxyribonucleic acid), a molecule that contains the information that directs the chemical reactions in a cell, the formation of various cellular products within the cell, and the cell's ability to reproduce itself. (A few types of cells, including mammalian red blood cells and cells called "sieve tube elements" that form part of the plant circulatory system, lose their nuclei after they are created and are unable to divide.) We explore all of these features of cell functioning in this and the next three chapters.

To see a cell, you don't have to work in a lab or use a microscope. Just open your refrigerator. Chances are you've got a dozen or so visible cells in there: eggs. Although most cells are too small to see with the naked eye, there are a few exceptions, including hens' eggs from the supermarket. As long as they are unfertilized, which most store-bought hens' eggs are, each egg tends to contain just one cell. The ostrich egg, weighing about three pounds, contains the largest of all animal cells. (It should be noted, however, that by the time the ostrich lays a fertilized egg, the embryo inside has already gone through multiple divisions.)

In addition to being among the largest cells around, eggs are also the most valuable. Almas caviar, eggs from the beluga sturgeon, sells for nearly $700 per ounce. This value is exceeded only by that of human eggs, which currently fetch as much as $25,000 for a dozen or so eggs on the open market (**FIGURE 3-2**). (Human sperm cells command only about a penny per 20,000 cells!)

Hummingbird egg: 0.02 ounce
Elephant bird (extinct) egg: two gallons

Beluga sturgeon eggs: $700 per ounce

TEM 400×

Human eggs: Thousands of dollars per egg

*Eggs are the largest and most expensive cells in the world!*

**FIGURE 3-2 Not all cells are tiny.** And some cells are extremely valuable!

Most cells are much smaller than hens' eggs and ostrich eggs. Consider that, at this very moment, there are probably more than seven billion bacteria in your mouth—even if you just brushed your teeth! This is more than the number of people on earth. It is possible to squeeze so many bacteria in there because most cells (not just bacteria) are really, really tiny—so tiny that 2,000 red blood cells, lined up end to end, would just extend across a dime.

After sufficient improvements were made to early microscopes in the 19th century, the central role of the cell in biology could be understood. By the 1830s, scientists realized that all plants and animals were made entirely from cells. Subsequent studies revealed that every cell seemed to arise from the division of another cell. You, for example, are made up of at least 60 trillion cells, all of which came from just one cell: the single fertilized egg produced when an egg cell from your mother was fertilized by a sperm cell from your father.

The facts that (1) all living organisms are made up of one or more cells and (2) all cells arise from other, preexisting cells are the foundations of **cell theory,** one of the unifying theories in biology, and one that is universally accepted by all biologists. As we see in Chapter 10, the origin of life on earth was a one-time deviation from cell theory: the first cells on earth probably originated from free-floating molecules in the oceans early in the earth's history (about 3.5 billion years ago). Since that time, however, all cells and thus all life have been produced as a continuous line of cells, originating from these initial cells.

In this chapter, we investigate the two different kinds of cells that make up all of the organisms on earth, the processes by which cells control how materials move into and out of the cell, and how cells communicate with each other. We also explore some of the important structures found in many cells and the specialized roles these structures play in a variety of cellular functions. Along the way, we learn about some of the health consequences that occur when cells malfunction.

---

**TAKE-HOME MESSAGE 3·1**

The most basic unit of any organism is the cell, the smallest unit of life that can function independently and perform all of the necessary functions of life, including reproducing itself. All living organisms are made up of one or more cells, and all cells arise from other, preexisting cells.

---

# 3·2 Prokaryotic cells are structurally simple, but there are many types of them.

Although there are millions of diverse species on earth, billions of unique organisms alive at any time, and trillions of different cells in many of those organisms, every cell on earth falls into one of two basic categories.

A **eukaryotic cell** (from the Greek for "good" and "kernel") has a central control structure called a nucleus, which contains the cell's DNA. Organisms composed of eukaryotic cells are called **eukaryotes.**

A **prokaryotic cell** (from the Greek for "before" and "kernel") does not have a nucleus; its DNA simply resides in the cytoplasm. An organism consisting of a prokaryotic cell is called a **prokaryote.**

We begin our study of the cell by exploring the prokaryotes. Prokaryotes were the first cells on earth, making their appearance about 3.5 billion years ago, and, for a long time (1.5 billion years), they had the planet to themselves.

All prokaryotes are one-celled organisms and are invisible to the naked eye. Prokaryotes have four basic structural features (**FIGURE 3-3**).

1. A **plasma membrane** encompasses the cell (and sometimes is simply called the "cell membrane"). For this reason, anything inside the plasma membrane is referred to as "intracellular," and everything outside the plasma membrane is "extracellular."

2. The **cytoplasm** is the jelly-like fluid that fills the inside of the cell.

3. **Ribosomes** are little granular bodies where proteins are made; thousands of them are scattered throughout the cytoplasm.

4. Each prokaryote has one or more circular loops or linear strands of DNA.

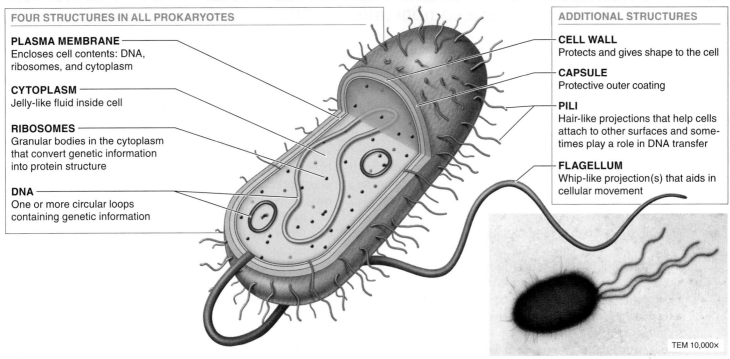

**FOUR STRUCTURES IN ALL PROKARYOTES**

**PLASMA MEMBRANE**
Encloses cell contents: DNA, ribosomes, and cytoplasm

**CYTOPLASM**
Jelly-like fluid inside cell

**RIBOSOMES**
Granular bodies in the cytoplasm that convert genetic information into protein structure

**DNA**
One or more circular loops containing genetic information

**ADDITIONAL STRUCTURES**

**CELL WALL**
Protects and gives shape to the cell

**CAPSULE**
Protective outer coating

**PILI**
Hair-like projections that help cells attach to other surfaces and sometimes play a role in DNA transfer

**FLAGELLUM**
Whip-like projection(s) that aids in cellular movement

TEM 10,000×

**FIGURE 3-3 Structural features of a prokaryote.**

In addition to the characteristics common to all prokaryotes, some prokaryotes have additional, unique features. Many have a rigid **cell wall,** for example, that protects and gives shape to the cell. Some have a slimy, sugary capsule as their outermost layer. This sticky outer coat provides protection and enhances the prokaryotes' ability to anchor themselves in place when necessary.

Many prokaryotes have a **flagellum** (*pl.* **flagella**), a long, thin, whip-like projection that rotates like a propeller and moves the cell through the medium in which it lives. Other appendages include **pili** (*sing.* **pilus**), much thinner, hair-like projections that help prokaryotes attach to surfaces and can serve as "tubes" through which they exchange DNA.

You may think you've never encountered a prokaryote, but you are already familiar with the largest group of prokaryotes: the bacteria. All bacteria are prokaryotes, from those such as *Escherichia coli* (*E. coli*), which live in your intestine and helps your body make some essential vitamins, to those responsible for illness, such as *Streptococcus pyogenes,* which causes strep

throat. Another recently discovered group of organisms, the archaea, are also prokaryotes.

From our human perspective, it is easy to underestimate the prokaryotes. Although prokaryotes are smaller, evolutionarily older, and structurally more simple than eukaryotes, they are fantastically diverse metabolically (i.e., in the way they break down and build up molecules). Among many other innovations, bacteria can fuel their activities in the presence or absence of oxygen, using almost any energy source on earth, depending on the type of bacteria—from the sulfur in deep-sea hydrothermal vents to hydrogen to the sun.

**TAKE-HOME MESSAGE 3·2**

Every cell on earth is either a eukaryotic or a prokaryotic cell. Prokaryotes, which have no nucleus, were the first cells on earth. They are all single-celled organisms. Prokaryotes include the bacteria and archaea and, as a group, are characterized by tremendous metabolic diversity.

# 3·3 Eukaryotic cells have compartments with specialized functions.

Eukaryotes showed up about 1.5 billion years after prokaryotes, and in the 2 billion years that they have been on earth, they have evolved into some of the most dramatic and interesting creatures, such as platypuses, dolphins, giant sequoias, and the Venus flytrap. Not all eukaryotes are multicellular, however. Many fungi are unicellular, and there is a huge group of eukaryotes, called the Protista (or protists), nearly all of which are single-celled organisms visible only with a microscope. Nonetheless, because all prokaryotes are single-celled and thus invisible to the naked eye, every organism that we see around us is a eukaryotic organism, including all plants and animals (**FIGURE 3-4**).

Eukaryotic cells can be distinguished from prokaryotic cells by their size and their numerous structural features. Chief among the distinguishing features of eukaryotic cells is the presence of a **nucleus,** a membrane-enclosed structure that contains linear strands of DNA. In addition to a nucleus, eukaryotic cells usually contain in their cytoplasm several other specialized structures. Many of these structures, called **organelles,** are enclosed separately within their own lipid membranes. Eukaryotic cells are also about 10 times larger than prokaryotes. All of these physical differences make it easy to distinguish eukaryotes from prokaryotes under a microscope (**FIGURE 3-5**).

Organelles enable eukaryotic cells to do many things that a prokaryotic cell cannot do. Most important, the physical separation of compartments means that there can be distinct areas within the same cell in which different chemical reactions can occur simultaneously. In the mostly non-compartmentalized interior of a prokaryotic cell, random molecular movements quickly blend the chemicals throughout the cell, reducing the ease with which different reactions can occur simultaneously.

**FIGURE 3-6** illustrates a generalized animal cell and a generalized plant cell. Because they share a common, eukaryotic ancestor, they have much in common. Both can have a plasma membrane, nucleus, cytoskeleton, and a host of organelles, including rough and smooth endoplasmic membranes, Golgi apparatus, and mitochondria. Animal cells have centrioles, which are not present in most plant cells. Plant cells have a rigid cell wall (as do fungi and many protists) and chloroplasts (also found in some protists). Plants also have a vacuole, a large central chamber (only occasionally found in animal cells). We explore each of these animal and plant organelles in detail later in this chapter.

When you compare a complex eukaryotic cell with the structurally simple prokaryotic cell, it's hard not to wonder how eukaryotic cells came about. How did they get filled up with so many organelles? We can't go back 2 billion years to watch the initial evolution of eukaryotic cells, but there is considerable evidence for some of what occurred. In particular, a theory called **endosymbiosis** provides the best explanation for the presence of two organelles in eukaryotes: chloroplasts in plants and algae, and mitochondria in plants and animals. Chloroplasts help plants and algae convert sunlight into a more usable form of energy. Mitochondria help plants and animals harness the energy stored in food molecules. (Chapter 4, on energy, explains the details of both of these processes.)

Venus flytrap

Golden-headed lion tamarin

Yellow-fuzz cone slime mold

**FIGURE 3-4 Diversity of the eukaryotes.** Every organism that we can see without magnification is a eukaryotic organism.

## EUKARYOTIC vs. PROKARYOTIC CELLS

Nucleus

Other organelles

TEM 6,000×

TEM 10,000×

**TYPICAL EUKARYOTIC CELL FEATURES**
• DNA contained in nucleus.
• Larger than prokaryotes—usually at least 10 times bigger.
• Cytoplasm contains specialized structures called organelles.

**TYPICAL PROKARYOTIC CELL FEATURES**
• No nucleus—DNA is in the cytoplasm.
• Internal structures mostly not organized into compartments.
• Much smaller than eukaryotes.

**FIGURE 3-5** Comparison of eukaryotic and prokaryotic cells.

According to the theory of endosymbiosis, two different types of prokaryotes may have set up close partnerships with each other. For example, some small prokaryotes capable of performing photosynthesis (the process by which plant cells capture light energy from the sun and transform it into the chemical energy stored in food molecules) may have come to live inside a larger

**Q**

*Humans, deep down, may be part bacteria. How can that be?*

"host" prokaryote. The photosynthetic "boarder" may have made some of the energy from its photosynthesis available to the host.

After a long while, the two cells may have become more and more dependent on each other, until neither cell could live without the other (they became "symbiotic"), and

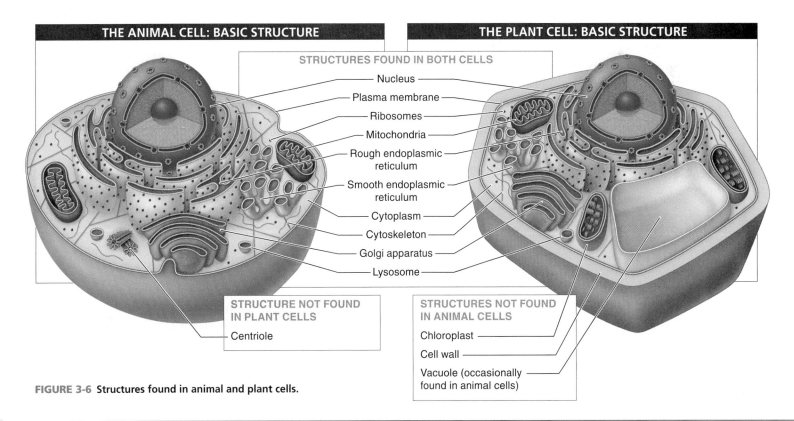

**FIGURE 3-6** Structures found in animal and plant cells.

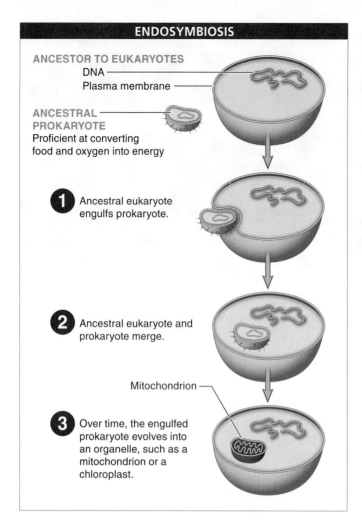

## ENDOSYMBIOSIS

**ANCESTOR TO EUKARYOTES**
DNA
Plasma membrane

**ANCESTRAL PROKARYOTE**
Proficient at converting food and oxygen into energy

**1** Ancestral eukaryote engulfs prokaryote.

**2** Ancestral eukaryote and prokaryote merge.

Mitochondrion

**3** Over time, the engulfed prokaryote evolves into an organelle, such as a mitochondrion or a chloroplast.

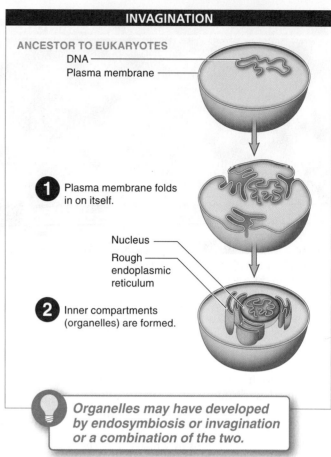

## INVAGINATION

**ANCESTOR TO EUKARYOTES**
DNA
Plasma membrane

**1** Plasma membrane folds in on itself.

Nucleus
Rough endoplasmic reticulum

**2** Inner compartments (organelles) are formed.

*Organelles may have developed by endosymbiosis or invagination or a combination of the two.*

**FIGURE 3-7 How did eukaryotic cells become so structurally complex?** Two theories.

they became a single, more complex organism. Eventually, the photosynthetic prokaryote evolved into a **chloroplast,** the organelle in plant cells in which photosynthesis occurs. A similar scenario might explain how a prokaryote unusually efficient at converting food and oxygen into easily usable energy took up residence inside another, host prokaryote and evolved into a **mitochondrion,** the organelle in plant and animal cells that converts the energy stored in food into a form usable by the cell (**FIGURE 3-7**).

The idea of endosymbiosis is supported by several observations.

1. Chloroplasts and mitcochondria are similar in size to prokaryotic cells and divide by splitting (fission), just like prokaryotes.

2. Chloroplasts and mitochondria have internal structures called ribosomes, similar to those found in bacteria, that allow them to synthesize some of their own proteins; this ability is not found in other organelles, which rely on proteins made by cytoplasmic ribosomes.

3. Chloroplasts and mitochondria have small amounts of circular DNA, similar to the circular DNA in prokaryotes and in contrast to the linear DNA strands found in a eukaryote's nucleus.

4. Analysis of chloroplast and mitochondrial DNA has revealed that it is highly related to bacterial DNA, much more closely than it is related to eurkaryotic DNA.

The best current theory about the origin of orther organelles in eukaryotes is a process called **invagination.** The idea is that the plasma membrane around the cell may have folded in on itself to form the inner compartments, which subsequently became modified and specialized (see Figure 3-7).

### TAKE-HOME MESSAGE 3·3

**Eukaryotes are single-celled or multicellular organisms consisting of cells with a nucleus that contains linear strands of genetic material. The cells also commonly have organelles throughout their cytoplasm; these organelles may have originated evolutionarily through endosymbiosis or invagination, or both.**

## 3·4–3·7 Cell membranes are gatekeepers.

*Like gatekeepers, cell membranes control the movement of material into and out of the cell.*

# 3·4 Every cell is bordered by a plasma membrane.

Just as skin covers our bodies, every cell of every living thing on earth, whether a prokaryote or a eukaryote, is enclosed by a plasma membrane, a two-layered membrane that holds the contents of a cell in place and regulates what enters and leaves the cell. Plasma membranes are thin (a stack of a thousand would be only as thick as a single hair) and flexible, and in photos or diagrams the membranes often resemble simple plastic bags, holding the cell contents in place. This image is a gross oversimplification, however. Membranes are indeed thin and flexible, but they are far from simple: a close look at a plasma membrane will reveal that its surface is filled with pores, outcroppings, channels, and complex molecules floating around within the two layers of the membrane itself (**FIGURE 3-8**).

Plasma membranes are such complex structures because they perform several critical functions beyond simply holding a cell's interior contents. Cells, after all, are perpetually interacting with their external environment. They take in food and nutrients and dispose of waste products. Cells take in water. They build and export molecules needed elsewhere in the body. They mediate communications and adhesion with the external environment and other cells. And, like the booths or checkpoints at a country's borders that control the flow of people into and out of the country, plasma membranes serve as gatekeepers that control the flow of molecules into and out of the cell.

The foundation of all plasma membranes is a layer of lipid molecules all packed together. These are a special type of lipid, called **phospholipids,** which, as you may recall from Section 2-13, have what appear to be a head and two long tails. The head consists of a molecule of **glycerol** linked to a molecule containing phosphorus (**FIGURE 3-9**). This head region is said to be **polar,** because the electrons are not shared equally among the atoms, leading to regions of partial positive and partial negative charge. As you may recall from Chapter 2, water is also a polar molecule and, for this reason, other polar molecules mix easily with water. Molecules that can mix with water are described as **hydrophilic** ("water-loving") molecules. The two tails of the phospholipid are long chains of carbon and hydrogen atoms. Because the electrons in these bonds are shared equally, the carbon-hydrogen chains are **nonpolar.** And because they are nonpolar, these tails do not mix with water and are said to be **hydrophobic** ("water-fearing"). The chemical structure of phospholipids gives them a sort of split personality: their hydrophilic head region mixes easily with water, while their hydrophobic tail region does not mix with water.

The split personality of phospholipids makes them good membrane material. Once a large number of phospholipids are packed together with all of their heads facing one way and their tails the other, we have a sheet with one side that is hydrophilic and one that is hydrophobic. In a cell's plasma membrane, two of these sheets of phospholipids are arranged

Plasma membranes are made up of two layers that are filled with a variety of pores, molecules, and channels.

Plasma membrane

TEM 100,000×

Inside cell 1

Inside cell 2

**FUNCTIONS**
• Holds contents of cell in place
• Takes in food and nutrients
• Aids in building and exporting molecules
• Allows interactions with the environment and neighboring cells

*Plasma membranes are the "gatekeepers" of the cell.*

**FIGURE 3-8 More than just an outer layer.** The plasma membrane performs several critical functions beyond simply enclosing a cell's interior contents.

so that the hydrophobic tails are all in contact with one another and the hydrophilic heads are in contact with the watery solution outside and inside the cell (see Figure 3-9). This arrangement gives us another way to describe the structure of the plasma membrane: as a **phospholipid bilayer.**

The phospholipids are not locked in place in the plasma membrane; they just float around their side of the bilayer. They cannot pop out of the membrane, or flop from one side to the other, because their hydrophobic tails always line up away from any watery solution. Just as similarly charged sides of two magnets push away from each other, so do the

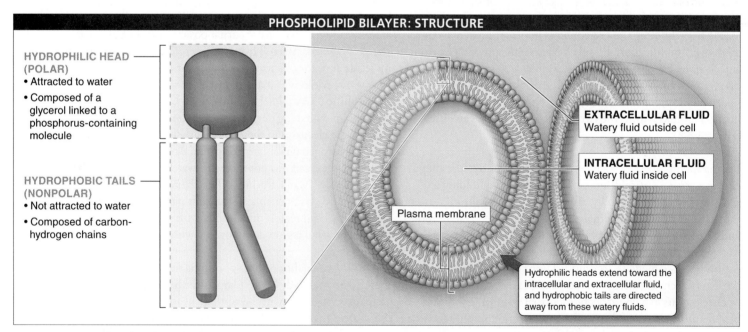

**PHOSPHOLIPID BILAYER: STRUCTURE**

**HYDROPHILIC HEAD (POLAR)**
• Attracted to water
• Composed of a glycerol linked to a phosphorus-containing molecule

**HYDROPHOBIC TAILS (NONPOLAR)**
• Not attracted to water
• Composed of carbon-hydrogen chains

**EXTRACELLULAR FLUID**
Watery fluid outside cell

**INTRACELLULAR FLUID**
Watery fluid inside cell

Plasma membrane

Hydrophilic heads extend toward the intracellular and extracellular fluid, and hydrophobic tails are directed away from these watery fluids.

**FIGURE 3-9 Good membrane material.** The phospholipid bilayer of the plasma membrane prevents fluid from leaking out of the cell.

hydrophobic tails in the center of the membrane push away from and avoid coming into contact with water molecules. Because the center part of the bilayer membrane is made up of hydrophobic lipids, the solution on one side of the membrane cannot leak across into the solution on the other side. In this way, the plasma membrane forms a boundary around the cell's contents.

# 3·5 Molecules embedded in the plasma membrane help it perform its functions.

While the plasma membrane's phospholipid bilayer construction forms the cell's basic boundary with its environment, there is much more to a plasma membrane. Embedded within or attached to the phospholipid bilayer are different types of protein, carbohydrate, and lipid molecules (**FIGURE 3-10**), which serve many different roles.

The proteins found in the plasma membrane enable it to carry out most of its gatekeeping functions. For every 50–100 phospholipids in the membrane, there is one protein molecule. Some of these proteins, called **transmembrane proteins,** penetrate right through the lipid bilayer, from one side to the other. Others, called **surface proteins** or peripheral proteins, reside primarily on the inner or outer surface of the membrane.

What determines whether a protein resides on the surface or extends through the bilayer? Its tertiary structure. Remember from Chapter 2 that all the amino acids that make up each protein have side chains that differ from one another chemically. Some of these side chains are hydrophobic, others are hydrophilic, and as a protein is assembled into its final shape, these side chains can cause parts of the protein to be attracted to hydrophobic or hydrophilic regions. Because a transmembrane protein has both hydrophobic and hydrophilic regions, part of the protein can be positioned in the hydrophobic region in the center of the membrane while the other parts reside in the hydrophilic regions. Peripheral membrane proteins, on the other hand, have an entirely hydrophilic structure and reside on the membrane surface, bound only to the head regions of the phospholipids. As a consequence, they can be positioned on either the outer or the inner side of the membrane.

Once membrane proteins are in place, the hydrophobic and hydrophilic forces keep them properly oriented. Because all of the components of the plasma membrane are held in the membrane in this manner, they can float around without ever popping out.

There are four primary types of membrane proteins, each of which performs a different function (**FIGURE 3-11**).

**1. Receptor proteins** are surface or transmembrane proteins that bind to chemicals in the cell's external environment and, by doing so, regulate certain processes within the cell. Cells in the heart, for example, have receptor proteins that bind to adrenaline, a chemical released into the bloodstream in times of extreme stress or fright. When adrenaline binds to these heart cells, the cells increase the heart's rate of contraction to pump blood through the body more quickly. You have experienced this reaction if you've ever been startled and felt your heart start to pound.

**MOLECULES WITHIN THE PLASMA MEMBRANE**

Extracellular fluid

Hydrophilic region
Hydrophobic region

Carbohydrates

Plasma membrane

Transmembrane protein

Lipids

Surface proteins

Intracellular fluid

Hydrophobic and hydrophilic forces determine the orientation of proteins in the plasma membrane.

**FIGURE 3-10 Protein, carbohydrate, and lipid molecules are embedded in the plasma membrane.**

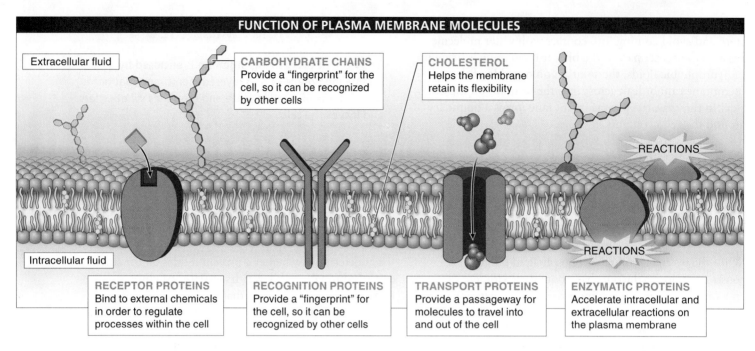

## FUNCTION OF PLASMA MEMBRANE MOLECULES

Extracellular fluid

**CARBOHYDRATE CHAINS**
Provide a "fingerprint" for the cell, so it can be recognized by other cells

**CHOLESTEROL**
Helps the membrane retain its flexibility

REACTIONS

Intracellular fluid

REACTIONS

**RECEPTOR PROTEINS**
Bind to external chemicals in order to regulate processes within the cell

**RECOGNITION PROTEINS**
Provide a "fingerprint" for the cell, so it can be recognized by other cells

**TRANSPORT PROTEINS**
Provide a passageway for molecules to travel into and out of the cell

**ENZYMATIC PROTEINS**
Accelerate intracellular and extracellular reactions on the plasma membrane

**FIGURE 3-11 Plasma membrane molecules serve diverse roles.**

**2. Recognition proteins** are surface or transmembrane proteins that give each cell a "fingerprint" that makes it possible for the body's immune system (which fights off infections) to distinguish the cells that belong inside your body from those that are invaders and need to be attacked. (Note that carbohydrates also play a role in recognition.) Recognition proteins also can help cells bind to or adhere to other cells or molecules.

**3. Transport proteins** are transmembrane proteins that help polar or charged substances pass through the plasma membrane. Transport proteins come in a variety of shapes and sizes, making it possible for a wide variety of molecules to be transported.

**4. Enzymatic proteins** are surface or transmembrane proteins (enzymes) that accelerate chemical reactions on the plasma membrane's surface. (A variety of enzymatic proteins exist, with some accelerating reactions on the inside of the plasma membrane and others accelerating reactions on the outside of the plasma membrane.)

In addition to the various kinds of membrane proteins, two other types of molecules—carbohydrate chains and cholesterol—can be incorporated in a cell's plasma membrane.

Short, branched carbohydrate chains that are attached to proteins or to phospholipid heads on the outside of the cell membrane serve as part of a membrane's fingerprint, along with recognition proteins. This fingerprint allows the cell to be recognized by other cells, such as those of the immune system.

The plasma membrane also can contain the lipid **cholesterol,** which helps the membrane maintain its flexibility. Cholesterol prevents the membrane from becoming too fluid or floppy at moderate temperatures and acts as a sort of antifreeze, preventing the membrane from becoming too rigid at freezing temperatures. The membranes of some cells are about 25% cholesterol; other plasma membranes, such as those of most bacteria and plants, have no cholesterol at all.

As we've seen, the plasma membrane is made up from several different types of molecules, like a mosaic, and many of those molecules float around, held in a proper orientation by hydrophobic and hydrophilic forces, but not always anchored in place. For these reasons, the plasma membrane is often described as a **fluid mosaic.**

### TAKE-HOME MESSAGE 3·5

**The plasma membrane is a fluid mosaic of proteins, lipids, and carbohydrates. Proteins found in the plasma membrane enable it to carry out most of its gatekeeping functions. The proteins act as receptors, help molecules gain entry into and exit from the cell, and catalyze reactions on the inner and outer cell surfaces. In conjunction with carbohydrates, some plasma membrane proteins identify the cell to other cells. And, in addition to the phospholipids that make up most of the plasma membrane, cholesterol is an important lipid in some membranes, influencing fluidity.**

# 3·6    Faulty membranes can cause disease.

With all the complexity of plasma membranes, it's not surprising that there are many ways in which they can malfunction. One disease that results from an improperly functioning membrane is cystic fibrosis, the single most common fatal inherited disease in the United States. At any given time, about 30,000 people in the United States have cystic fibrosis.

Cystic fibrosis occurs when an individual inherits from both parents incorrect genetic instructions for producing one type of transmembrane protein that allows chloride ions to get into and out of cells. This transport protein occurs primarily in the membranes of cells in the lungs and digestive tract.

There are more than a thousand different ways in which these genetic instructions can be defective, but the result is the same: malfunction of chloride passageways in a cell's membrane that causes gradual accumulation of chloride ions within cells. In nearly all cases of cystic fibrosis, two primary effects occur: an improper salt balance in the cells and a buildup of thick, sticky mucus—particularly in the lungs. Normal mucus helps to protect the lungs by trapping dust and bacteria. This mucus is then moved out of the lungs (helped along by coughing). The mucus produced by

someone with cystic fibrosis, however, is too thick and sticky to be moved out of the lungs, so it collects there, where it impairs lung function and increases the risk of bacterial infection. Because of the improper cellular salt balance, one way to test for cystic fibrosis is to measure the concentration of salt in the sweat—abnormally high concentrations indicate that the person may have the disease.

Although many high-tech treatments have been promised for the sufferers of cystic fibrosis and a great deal of research is being done on this disease, one of the most common treatments is decidedly low-tech. Parents help their children with cystic fibrosis clear the mucus out of their lungs by holding them on a steep slant, almost upside down, and vigorously patting or thumping their chest and back. This shakes loose the mucus in their lungs and moves it to a place where they can cough it up (**FIGURE 3-12**). With careful treatment, the life expectancy of someone with cystic fibrosis can be 35–40 years or longer.

While faulty membranes can cause a disorder as serious as cystic fibrosis, pharmaceutical tinkering with cell membranes can alter cellular function in ways that have beneficial effects. For example, a group of drugs called

---

**TREATING CYSTIC FIBROSIS**

The thick and sticky mucus produced by someone with cystic fibrosis collects in the lungs, impairing lung function and increasing the risk of bacterial infection.

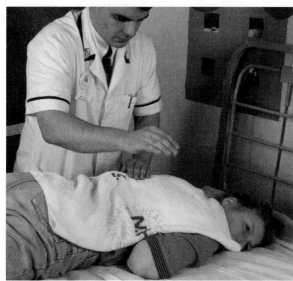

"Thumping" on the chest and back can loosen the mucus.

The vest, by inflating and deflating rapidly, can have a similar effect in the course of a 20-minute session.

**FIGURE 3-12 Moving mucus manually, or through an inhalation vest.**

Kidney    Adrenal gland

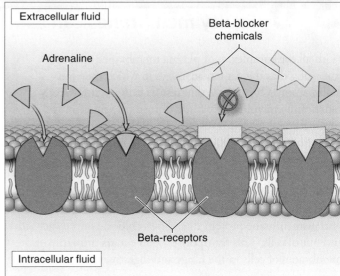

Extracellular fluid

Adrenaline

Beta-blocker chemicals

Beta-receptors

Intracellular fluid

**1** In stressful situations, the adrenal glands pump out adrenaline.

**2** Adrenaline binds with beta-receptors on cells, causing a faster heartbeat and increased blood pressure.

**3** Beta-blocker chemicals bind to receptors and prevent adrenaline from binding to the cell.

**FIGURE 3-13 Drugs can alter cell membrane function.** The fight-or-flight chemical adrenaline cannot enter the cell when beta-blocking medications bind to the cell's beta-receptors.

> *By binding to adrenaline receptors, beta-blockers reduce anxiety symptoms.*

"beta-blockers" is extremely effective at reducing anxiety. This effect was discovered almost accidentally—the drugs were actually developed as a treatment for high blood pressure.

**Q** Why do "beta-blockers" reduce anxiety?

Here's how beta-blockers work. Many cells in your body, particularly the cells of the heart, have receptor proteins on their plasma membranes that can bind to adrenaline, which helps your body cope with stressful situations. These receptor proteins are called "beta-adrenergic receptors." In stressful situations, your adrenal glands pump out adrenaline (**FIGURE 3-13**). On reaching cells in your heart (among other locations in your body) and binding to beta-receptors, the adrenaline promptly causes your heart to beat faster and more forcefully, increasing your blood pressure in the process. This reaction is fine in a short-term fight-or-flight situation, but it is not healthy over the long run because the increased pressure can damage blood vessels. Depending on its severity, this reaction can also be problematic if you are giving

a presentation or taking a test, or if you're in any other anxiety-producing situation.

When you take a beta-blocker pill, the pill dissolves and the chemicals travel throughout your body until they encounter the beta-receptors. They bind to the receptors, hold on, and block the adrenaline from doing its job. This outcome slows your heart rate, causes a reduction in blood pressure, and can bring great relief to those suffering from the sweating and trembling associated with anxiety.

### TAKE-HOME MESSAGE 3·6

Normal cell functioning can be disrupted when cell membranes—particularly the proteins embedded in them—do not function properly. Such malfunctions can cause health problems, such as cystic fibrosis. But disruption of normal cell membrane function can also have beneficial, therapeutic effects, such as in the treatment of high blood pressure and anxiety.

# 3·7 Membrane surfaces have a "fingerprint" that identifies the cell.

As mentioned earlier, every cell in your body has a "fingerprint" made from a variety of molecules on the outside-facing surface of the cell membrane. Some of these membrane molecules vary, depending on the specific function of the cell. Others are common to all of your cells and convey to your immune system: "I belong here." Cells with an improper fingerprint are recognized as foreign and are attacked by your body's defenses.

Throughout our evolutionary history, this system has been tremendously valuable in helping our bodies fight infection. In some cases, however, this vigilance is a problem, like a car alarm that goes off even when you don't want it to. Suppose you receive a liver (or any other organ) transplant. Even if the donor is a close relative, the molecular fingerprint on the cells of the donated liver is not identical to your own. Consequently, your body sees the new organ as a foreign object and puts up a fight against it (**FIGURE 3-14**). Because your body will naturally try to reject the new organ, doctors must administer drugs that suppress your immune system. Immune suppression helps you tolerate the new liver, but, as you can imagine, it leaves you without some of the defenses to fight off other foreign invaders, such as bacteria that may cause infection. The presence or absence of certain molecular markers on plasma membranes is also responsible for a person's blood type and can lead to problems with simple blood transfusions.

**Q** **Why is it extremely unlikely that a person will catch HIV from casual contact—such as shaking hands—with an infected individual?**

The AIDS-causing virus, HIV, uses the molecular markers on plasma membranes to infect an individual's cells. These same molecular markers are also the reason that it is extremely unlikely that you can catch an HIV infection from casual contact with an infected individual, such as shaking his or her hand. The specific molecular markers involved in infection by HIV belong to a group of identifying markers called "clusters of differentiation." Abbreviated as "CD markers" and having names such as CD1, CD2, and CD3, these marker molecules are proteins embedded in the plasma membrane that enable a cell

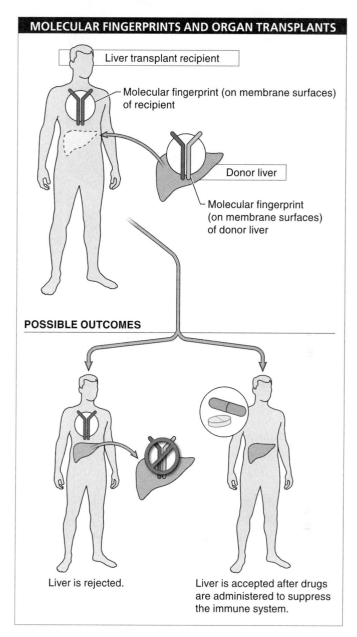

**MOLECULAR FINGERPRINTS AND ORGAN TRANSPLANTS**

Liver transplant recipient

Molecular fingerprint (on membrane surfaces) of recipient

Donor liver

Molecular fingerprint (on membrane surfaces) of donor liver

**POSSIBLE OUTCOMES**

Liver is rejected.

Liver is accepted after drugs are administered to suppress the immune system.

**FIGURE 3-14 Mismatched molecular fingerprints can cause difficulty in organ transplantation.**

to bind to outside molecules and, sometimes, transport them into the cell.

One CD marker, called the CD4 marker, is found only on cells deep within the body and in the bloodstream, such as

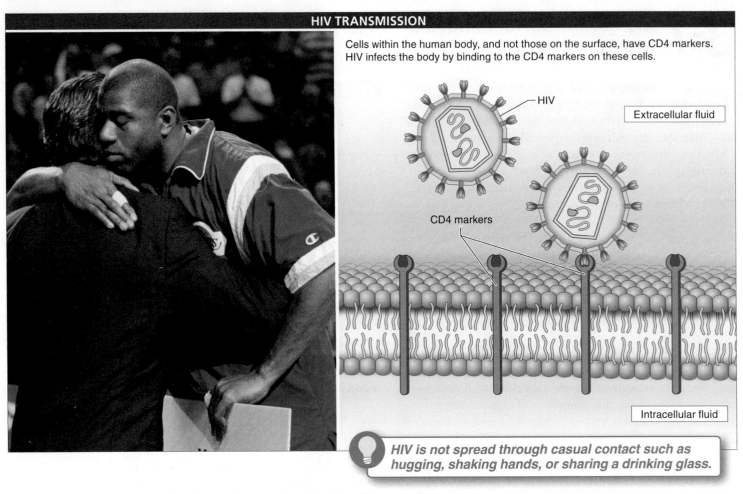

Cells within the human body, and not those on the surface, have CD4 markers. HIV infects the body by binding to the CD4 markers on these cells.

HIV

Extracellular fluid

CD4 markers

Intracellular fluid

*HIV is not spread through casual contact such as hugging, shaking hands, or sharing a drinking glass.*

**FIGURE 3-15 HIV requires CD4 markers—not found on skin cells—to infect the body.**

immune system cells and some nerve cells. It is the CD4 marker, in conjunction with another receptor, that is targeted by HIV. If the virus can find a cell with a CD4 marker, it can infect you, and because the CD4 markers *never* occur on the surface of your skin cells, casual contact such as touching is very unlikely to transmit the virus (**FIGURE 3-15**). Even if millions of HIV particles are present on a person's hands, they just can't gain access into any of the other person's surface cells.

How does HIV get transmitted? Far and away, the most common methods of transmission involve the transfer of blood, semen, vaginal fluid, or breast milk from one individual to another. Particles of the virus as well as cells infected by the virus are present in these fluids. Consequently,

the chief routes of transmission are from an infected mother to her child in breast milk, from an infected mother to her baby at birth, the use of contaminated needles, and unprotected sexual intercourse. Because an open sore or cut might expose some of the cells in your bloodstream to the outside world, however, it is not impossible for casual transmission of HIV to occur.

## TAKE-HOME MESSAGE 3·7

**Every cell in your body has a "fingerprint" made from a variety of molecules on the outside-facing surface of the cell membrane. This molecular fingerprint is key to the function of your immune system.**

# 3·8–3·11 Molecules move across membranes in several ways.

*Passage of proteins between cells can be tracked (junctions between these human cancer cells are stained red and green).*

## 3·8 Passive transport is the spontaneous diffusion of molecules across a membrane.

To function properly, cells must take in food and/or other necessary materials and must move out metabolic waste and molecules produced for use elsewhere in the body. In some cases, this movement of molecules requires energy and is called active transport. (We cover active transport later in this chapter.) In other cases, the molecular movement occurs spontaneously, without the input of energy, and is called **passive transport.** There are two types of passive transport: diffusion and osmosis. (Osmosis is discussed in Section 3-9.)

**Diffusion** is passive transport in which a particle, called a **solute,** is dissolved in a gas or liquid (a **solvent**) and moves from an area of high solute concentration to an area of lower concentration (**FIGURE 3-16**). In the absence of other forces, molecules of a substance will always tend to move from where they are more concentrated to where they are less concentrated. This movement occurs because molecules move randomly and are equally likely to move in any direction. And when certain molecules are highly concentrated, they keep

bumping into each other and eventually end up evenly distributed. For this reason, we say that molecules tend to *move down their concentration gradient.* For a simple illustration, drop a tiny bit of food coloring into a bowl of water and wait for a few minutes. The molecules of dye are initially clustered together in a very high concentration. Gradually, they disperse down their concentration gradient until the color is equally spread throughout the bowl.

In cells, molecules such as oxygen ($O_2$) and carbon dioxide ($CO_2$) that are small and carry no charge can pass directly through the phospholipid bilayer of the membrane without the assistance of any other molecules, in a process called **simple diffusion.** Each time you take a breath, for example, there is a high concentration of $O_2$ molecules in the air you pull into your lungs. That oxygen diffuses across the plasma membranes of the lung cells and into your bloodstream, where red blood cells pick it up and deliver it to parts of your body where it is needed. Similarly, because $CO_2$ in your bloodstream is at a higher concentration than in the air in

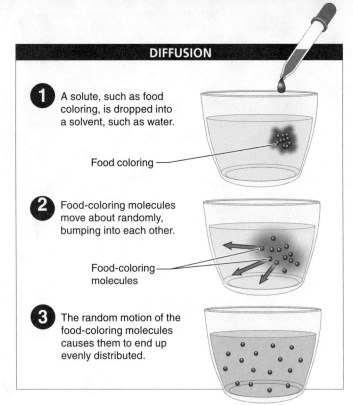

**DIFFUSION**

**1** A solute, such as food coloring, is dropped into a solvent, such as water.

Food coloring

**2** Food-coloring molecules move about randomly, bumping into each other.

Food-coloring molecules

**3** The random motion of the food-coloring molecules causes them to end up evenly distributed.

**FIGURE 3-16 Diffusion: a form of passive transport that results in an even distribution of molecules.**

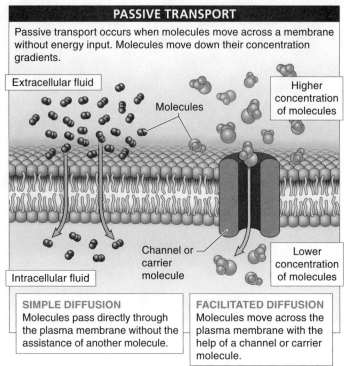

**PASSIVE TRANSPORT**

Passive transport occurs when molecules move across a membrane without energy input. Molecules move down their concentration gradients.

Extracellular fluid

Molecules

Higher concentration of molecules

Channel or carrier molecule

Intracellular fluid

Lower concentration of molecules

**SIMPLE DIFFUSION**
Molecules pass directly through the plasma membrane without the assistance of another molecule.

**FACILITATED DIFFUSION**
Molecules move across the plasma membrane with the help of a channel or carrier molecule.

**FIGURE 3-17 Simple and facilitated diffusion: no energy required.**

your lungs, it diffuses from your blood into your lungs and is released to the atmosphere when you exhale (**FIGURE 3-17**).

Most molecules, however, can't get through plasma membranes on their own. They may be electrically charged (polar) and, hence, are repelled by the hydrophobic middle region of the phospholipid bilayer. Or the molecules may be too big to squeeze through the membrane. Nonetheless, when there is a concentration gradient from one side of the membrane to the other, these molecules may still be able to diffuse across the membrane, down their concentration gradient, with the help of a channel or carrier molecule—one of the transport proteins, which we discussed in Section 3-5. Often, this transport protein spans the membrane and functions like a revolving door. When spontaneous diffusion across a plasma membrane requires a transport protein, it is called **facilitated diffusion** (see Figure 3-17).

Defects in transport proteins can reduce facilitated diffusion or even bring it to a complete stop, with serious health consequences. Many genetic diseases are the result of inheriting incorrect genetic instructions for building transport proteins. In the disease cystinuria, incorrect genetic instructions result in a malformed transport protein in the plasma membrane. When structured and functioning properly, this transport protein facilitates the diffusion of some amino acids (including cystine, from which the disease gets its name) out of the

kidneys into the urine. When the proteins are malformed, they cannot facilitate this diffusion and these amino acids build up in the kidneys, forming painful and dangerous kidney stones.

Diffusion across membranes doesn't occur just in animals—we see it in all organisms on earth. As we'll see in Chapter 4, one of the most important biological processes on earth is the diffusion of $CO_2$ from the atmosphere (an area of relatively high concentration) into the leaf cells of plants (areas of relatively low concentration), where it can be attached to other molecules, forming sugars through photosynthesis. At the same time, $O_2$ diffuses out of the leaves and into the atmosphere.

**TAKE-HOME MESSAGE 3·8**

For proper functioning, cells must acquire food molecules and/or other necessary materials from outside the cell. Similarly, metabolic waste molecules and molecules produced for use elsewhere in the body must move out of the cell. In passive transport—which includes simple and facilitated diffusion and osmosis—the molecular movement occurs spontaneously, without the input of energy. This generally occurs as molecules move down their concentration gradient.

# 3·9 Osmosis is the passive diffusion of water across a membrane.

Just as solute molecules will passively diffuse down their concentration gradients, water molecules will also move from areas of high concentration to areas of low concentration to equalize the concentration of water inside and outside the cell. The diffusion of water across a membrane is a special type of passive transport called **osmosis** (**FIGURE 3-18**). Just as solute molecules may diffuse across a plasma membrane, molecules of water also move across the membrane, equalizing the water concentration inside and outside the cell. (Although some water can pass through the lipid bilayer, the hydrophobic region severely limits this flow. Most of the rapid movement of water in and out of cells occurs through "water channels," called *aquaporins,* that are transmembrane proteins with hydrophilic channels through which the water molecules pass in single file.)

Osmosis can have some dramatic effects on cells. It turns out that many molecules just can't move across a cell membrane. They may be too big or too charged, or there may not be any carrier molecules in the membrane that can move them. Let's imagine, for instance, that there is a high concentration of a large molecule inside a cell and that it can't pass through the cell membrane. Although the molecule can't move out of the cell and down its concentration gradient, water can move into the cell down *its* concentration gradient. As water diffuses into the cell, the cell will get larger.

When a cell is in a fluid environment (referred to as a solution), the amount of dissolved substances (solutes) in that solution may be (1) equal to, (2) less than, or (3) greater than the concentration of dissolved substances in the cell. This relationship between the concentrations of solutes inside the cell and solutes outside the cell is referred to as **tonicity** (see Figure 3-18).

Let's review how water diffuses across the cell membrane in each of these different tonicity situations.

**1.** If the concentration of solutes outside the cell is *equal* to the concentration inside the cell, the outside solution is **isotonic.** Although water still moves between the solution and the cell, since it moves at the same rate in both directions, there is no net change in the amount of water inside versus outside the cell.

**2.** If the concentration of solutes outside the cell is *lower* than the concentration inside the cell, the outside solution is **hypotonic.** In this situation, if the cell membrane is not permeable to the solutes, more water will move into the cell than out of it, and the cell will swell.

**3.** If the concentration of solutes outside the cell is *higher* than the concentration inside the cell, the outside solution is **hypertonic.** In this situation, if the cell membrane is not permeable to the solutes, more water will move out of the cell than into it, and the cell will shrivel.

## OSMOSIS

Osmosis is a type of passive transport by which water diffuses across a membrane, in order to equalize the concentration of water inside and outside the cell. The direction of osmosis is determined by the total amount of solutes on either side of the membrane.

PLANT CELL    ANIMAL CELL (RED BLOOD CELL)

**ISOTONIC SOLUTION**
• Solute concentrations are balanced.
• Water movement is balanced.

Extracellular fluid

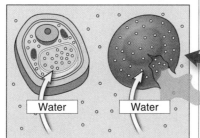

**HYPOTONIC SOLUTION**
• Solute concentrations are lower in the extracellular fluid.
• Water diffuses into cells.

Unlike plant cells, animal cells may explode in hypotonic solutions because they don't have a cell wall to limit cellular expansion.

**HYPERTONIC SOLUTION**
• Solute concentrations are higher in the extracellular fluid.
• Water diffuses out of cells.

 *Water will always move toward a region having a greater concentration of solutes.*

**FIGURE 3-18 Osmosis overview.**

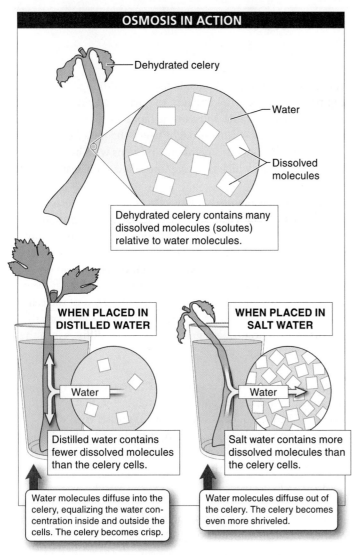

Dehydrated celery

Water

Dissolved molecules

Dehydrated celery contains many dissolved molecules (solutes) relative to water molecules.

**WHEN PLACED IN DISTILLED WATER**

**WHEN PLACED IN SALT WATER**

Water

Water

Distilled water contains fewer dissolved molecules than the celery cells.

Salt water contains more dissolved molecules than the celery cells.

Water molecules diffuse into the celery, equalizing the water concentration inside and outside the cells. The celery becomes crisp.

Water molecules diffuse out of the celery. The celery becomes even more shriveled.

**FIGURE 3-19 Osmosis in your kitchen.**

The key to understanding osmosis is to remember that water will always move toward a hypertonic solution.

You can see osmosis in action in your own kitchen (**FIGURE 3-19**). Take a stalk of celery and leave it on the counter for a couple of hours. As water evaporates from the cells of the celery stalk, it will shrink and become limp. You can make it crisp again by placing it in a solution of distilled water. Why distilled water? Because it contains very few dissolved molecules—it is a hypotonic solution. The cells in the celery stalk, on the other hand, contain many dissolved molecules, such as salt. Because those dissolved molecules can't easily pass

across the plasma membrane of the celery stalk cells, the water molecules move into the cells (moving from the pool of distilled water where they are at super-high concentration to the interior of the celery cells where water is scarcer). Would the celery regain its crispness if you placed it in concentrated salt water? No, because in the hypertonic salt solution there are more dissolved solute molecules (and fewer water molecules) outside the celery cells, so what little water remains in the celery cells actually moves, by osmosis, out of the celery, causing it to shrivel even more.

A more practical use of osmosis is seen with the laxative Milk of Magnesia. This product contains magnesium salts, which are poorly absorbed from the digestive tract. As a consequence, after a dose of the laxative, water moves by osmosis from the surrounding cells into the intestines. The water softens the feces in the intestines and increases the fecal volume, thereby relieving constipation.

**Q** How do laxatives relieve constipation?

It's important to note that the direction of osmosis is determined only by a difference in the *total concentration* of all the molecules dissolved in the water: it does not matter what the solutes are, only how many molecules of solutes there are. To determine which way the water molecules will move, you need to determine the total amount of "dissolved stuff" on either side of the membrane. The water will move toward the side with more stuff dissolved in it. For this reason, even small increases in the salinity of lakes can have disastrous consequences for the organisms living there, from fish to bacteria. Conversely, putting animal cells—such as red blood cells—in distilled water will cause them to explode, because water will diffuse into the cell (which contains more solutes) and the cell will swell and burst. This will not generally happen to plant cells because (as we see later in this chapter) the plant cell plasma membrane is surrounded by a rigid cell wall that limits the amount of cell expansion possible when water moves in by osmosis.

**TAKE-HOME MESSAGE 3·9**

**The diffusion of water across a membrane is a special type of passive transport called osmosis. Water molecules move across the membrane until the concentration of water inside and outside the cell is equalized.**

# 3·10 In active transport, cells use energy to move small molecules into and out of the cell.

Molecules can't always move spontaneously and effortlessly in and out of cells. Sometimes their transport needs energy, in which case the process is called **active transport.** Such energy expenditures may be necessary if the molecules or ions to be moved are being moved *against* their concentration gradient. In all active transport, proteins embedded in the membrane act like motorized revolving doors, pushing molecules into the cell regardless of the concentration of those molecules on either side of the membrane. There are two distinct types of active transport, primary and secondary, which differ only in the source of the fuel that keeps the revolving doors spinning.

The process of digestion provides an example of **primary active transport,** the type of active transport that occurs when energy from ATP is used to fuel the transport of molecules. To help break down the food into more digestible bits, the cells lining your stomach create an acidic environment by pumping large numbers of $H^+$ ions (also called protons) into the stomach contents, against their concentration gradient (**FIGURE 3-20**). (That is, there are more $H^+$ ions in the stomach contents than there are inside the cells lining the stomach.) All of this $H^+$ pumping increases your ability to digest the food but comes at a great energetic cost—in the form of usage of the high-energy molecule ATP—because the protons would not normally flow into a region against their concentration gradient. (We explore ATP in more detail in Chapter 4.)

Many transport proteins use an indirect method of fueling their activities rather than using energy released directly from ATP. In the process of **secondary active transport,** the transport protein simultaneously moves one molecule against its concentration gradient while letting another flow down its concentration gradient. Although no ATP is used directly in this process, at some other time and in some other location, energy from ATP was used to pump one of the types of molecule involved in the secondary transport process against its concentration gradient. The process is a bit like using energy to pump water up to the top of a high water tower. Later, the water can be allowed to run out of the tower over a water wheel, which can, in turn, power a process such as grinding wheat into flour. Our bodies frequently use the energy from one reaction that occurs spontaneously to fuel another reaction that requires energy.

## TAKE-HOME MESSAGE 3·10

In active transport, moving molecules across a membrane requires energy. Active transport is necessary if the molecules to be moved are very large or if they are being moved against their concentration gradient. Proteins embedded in the plasma membrane act like motorized revolving doors to actively transport (pump) the molecules.

## ACTIVE TRANSPORT

Active transport occurs when the movement of molecules into and out of a cell requires the input of energy. For example, in response to eating, the cells lining your stomach use ATP to pump large numbers of $H^+$ ions into the stomach.

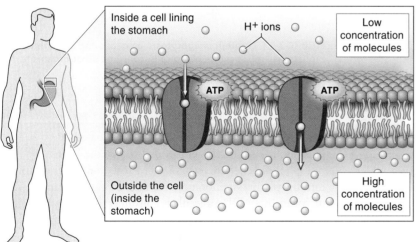

Inside a cell lining the stomach

$H^+$ ions

Low concentration of molecules

ATP    ATP

Outside the cell (inside the stomach)

High concentration of molecules

**FIGURE 3-20** With an input of energy, molecules can be moved against concentration gradients.

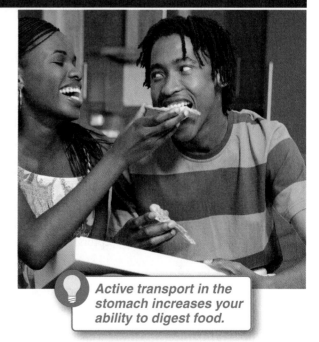

*Active transport in the stomach increases your ability to digest food.*

# Endocytosis and exocytosis are used for bulk transport of particles.

Many substances are just too big to get into a cell by passive or active transport. To absorb large particles, such as bacterial invaders, cells engulf them with their plasma membrane in a process called **endocytosis.** Cells also have another method besides passive and active transport for moving molecules out of the cell. Cells that manufacture molecules (such as digestive enzymes) for use elsewhere in the body must get those molecules out of the cell, and they often use the process of **exocytosis** to do this.

There are three types of endocytosis: phagocytosis, pinocytosis, and receptor-mediated endocytosis. All three involve the basic process of the plasma membrane oozing around an object that is outside the cell, surrounding it, forming a little pocket called a *vesicle,* and then pinching off the vesicle so that it is inside the cell but separated from the rest of the cell contents.

**Phagocytosis and pinocytosis** The process by which relatively large particles are engulfed by cells is called **phagocytosis** (**FIGURE 3-21**). Amoebas and other unicellular protists, as well as white blood cells, use phagocytosis to consume entire organisms either as food or as their way of defending against pathogens (disease-causing organisms or substances). Whereas "phagocytosis" comes from the Greek for "eat" and "container," the term **pinocytosis** comes from the Greek for "drink" and "container" and describes the process of cells taking in dissolved particles and liquids. The two processes are largely the same, except that the vesicles formed during pinocytosis are generally much smaller than those formed during phagocytosis.

**Receptor-mediated endocytosis** The third type of endocytosis, **receptor-mediated endocytosis,** is much more specific than either phagocytosis or pinocytosis. In this process, receptor molecules on the surface of a cell sit waiting until the one type of molecule they recognize bumps into them. For one receptor it might be insulin, for another it might be cholesterol. Many receptors of the same type are often clustered together in a cell's plasma membrane. When the appropriate molecule binds to each of the receptor proteins, the membrane begins to fold inward, first forming a little pit and then completely engulfing the molecules, which are still attached to their receptors.

One of the most important examples of receptor-mediated endocytosis involves cholesterol (**FIGURE 3-22**). Most cholesterol that circulates in the bloodstream is in the form of

## PHAGOCYTOSIS

Phagocytosis is a type of endocytosis by which cells engulf large particles.

TEM 4,700×

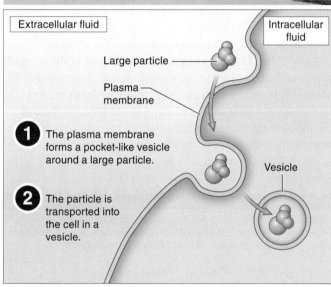

Extracellular fluid

Intracellular fluid

Large particle

Plasma membrane

**1** The plasma membrane forms a pocket-like vesicle around a large particle.

**2** The particle is transported into the cell in a vesicle.

Vesicle

**FIGURE 3-21 Through phagocytosis, amoebas and other unicellular protists, as well as white blood cells, consume other organisms for food or for defense.**

particles called low-density lipoproteins, or LDL. Each molecule of LDL is a cholesterol globule coated by phospholipids. Proteins embedded within the LDL's phospholipid coat are recognized by receptor proteins built into the plasma membranes of liver cells. Once bound to the receptors, the LDL molecule is consumed by the cell by endocytosis. Inside the cell, the cholesterol is broken down and used to make a variety of other useful molecules, such as the hormones estrogen and testosterone.

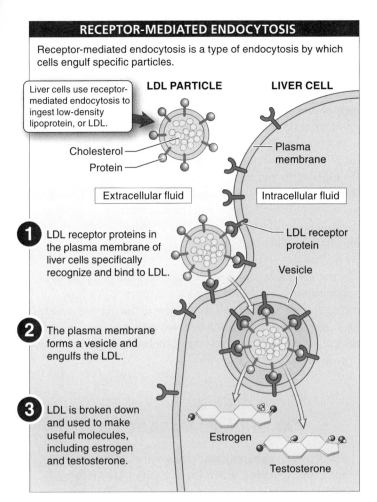

## RECEPTOR-MEDIATED ENDOCYTOSIS

Receptor-mediated endocytosis is a type of endocytosis by which cells engulf specific particles.

Liver cells use receptor-mediated endocytosis to ingest low-density lipoprotein, or LDL.

**LDL PARTICLE**

**LIVER CELL**

Cholesterol
Protein

Plasma membrane

Extracellular fluid

Intracellular fluid

**1** LDL receptor proteins in the plasma membrane of liver cells specifically recognize and bind to LDL.

LDL receptor protein

Vesicle

**2** The plasma membrane forms a vesicle and engulfs the LDL.

**3** LDL is broken down and used to make useful molecules, including estrogen and testosterone.

Estrogen

Testosterone

**FIGURE 3-22 Receptor proteins aid in endocytosis.** This process is important in the ability of your liver to remove cholesterol from your bloodstream.

Circulating cholesterol often builds up on the walls of arteries, reducing the area (that is, the diameter of the blood vessel) available for blood flow and causing the artery to harden. Because too much circulating cholesterol in LDL molecules can lead to cardiovascular disease and death (**FIGURE 3-23**), it's good to remove as much cholesterol from your bloodstream as quickly as possible. Individuals lucky enough to have large numbers of LDL receptors on the plasma membranes of their liver cells have a significantly lower risk of cardiovascular disease.

**Q** Faulty cell membranes are a primary cause of cardiovascular disease. What modification to these membranes might be an effective treatment?

Conversely, individuals who consume food laden with too much cholesterol (such as egg yolks, cheese, and sausages), or those who have the misfortune of inheriting genes that code for faulty liver cell membranes that have only a few LDL receptors, can have a heavy load of circulating blood cholesterol. This outcome can result in the early onset of cardiovascular disease. In the extreme case where an

### HEART DISEASE

Healthy artery

Partially blocked artery

Blocked artery

**FIGURE 3-23 Comparison of a healthy artery and arteries choked by the buildup of cholesterol.**

 *When circulating cholesterol levels exceed the LDL receptors' capacity for binding and removing the lipid, blood flow can become reduced or blocked and the risk of cardiovascular disease increases.*

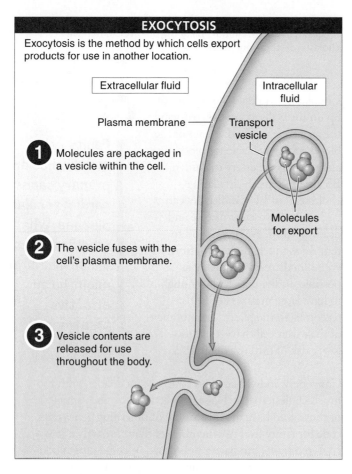

## EXOCYTOSIS

Exocytosis is the method by which cells export products for use in another location.

| Extracellular fluid |   | Intracellular fluid |

Plasma membrane

Transport vesicle

**1** Molecules are packaged in a vesicle within the cell.

Molecules for export

**2** The vesicle fuses with the cell's plasma membrane.

**3** Vesicle contents are released for use throughout the body.

**FIGURE 3-24 Exocytosis moves molecules out of the cell.** This is how insulin is exported from the cells where it is synthesized.

individual is born with no LDL receptors at all (a disorder called familial hypercholesterolemia), circulating cholesterol accumulates in the arteries so rapidly that cardiovascular disease begins to develop even before puberty, and death from a heart attack can occur before the age of 30.

**Exocytosis** Cells in the pancreas produce a chemical called insulin that moves throughout the circulatory system, informing body cells that there is glucose in the bloodstream that ought to be taken in and utilized for energy. The insulin molecule is much too large to pass out through the plasma membranes of the cells in which it is manufactured. As a result, after molecules of insulin are produced, they are coated with a phospholipid membrane to form a vesicle. The insulin-carrying vesicle then moves through the cytoplasm to the inner surface of the cell's plasma membrane. Once there, the phospholipid membrane surrounding the insulin and the phospholipid membrane of the cell fuse together, dumping the contents of the vesicle out of the cell and into the bloodstream (**FIGURE 3-24**).

The movement of molecules out of cells, called exocytosis, occurs throughout the body and is not restricted to large molecules. In the brain and other parts of the nervous system, for example, communication between cells occurs as one cell releases large numbers of very small molecules, called neurotransmitters, by exocytosis.

### TAKE-HOME MESSAGE 3·11

When materials cannot get into a cell via diffusion or a pump (for example, when the molecules are too big), cells can engulf the molecules or particles with their plasma membrane in a process called endocytosis. Similarly, molecules can be moved out of a cell by exocytosis. In both processes, the plasma membrane moves to surround the molecules or particles and forms a little vesicle that is pinched off inside the cell (endocytosis) or fuses with the plasma membrane and dumps its contents outside the cell (exocytosis).

# 3·12 Cells are connected and communicate with each other.

*Cells reach out and connect; no telephone lines required.*

## 3·12 Connections between cells hold them in place and enable them to communicate with each other.

So far in this chapter we have examined the cell as a free-living and independent entity. The majority of cells in any multicellular organism, however, are connected to other cells. There is a variety of methods by which cells adhere to other cells, involving numerous types of protein and glycoprotein adhesion molecules. We examine three primary types of connections between animal cells: (1) tight junctions, (2) desmosomes, and (3) gap junctions (**FIGURE 3-25**).

**Tight junctions** form continuous, water-tight seals around cells and also anchor cells in place. Much like the caulking around a tub or sink that keeps water from leaking into the surrounding walls, tight junctions prevent fluid flow between cells. Tight junctions are particularly important in the small intestine, where digestion occurs. Cells lining the small intestine absorb nutrients from the watery fluid moving through your gut. If the fluid—and the resident bacteria—inside the intestine were to leak between the cells and into your body cavity, you would not be able to extract sufficient energy and nutrients from your food, and the

bacteria would make you sick. The tight junctions instead force fluid to pass *into* the cells that line the intestine, where the nutrients can be utilized.

**Desmosomes** are like spot welds or rivets that fasten cells together into strong sheets. They occur at irregular intervals and function like fastened Velcro: they hold cells together but are not water-tight, allowing fluid to pass around them. Desmosomes and other similar junctions are found in much of the tissue that lines the cavities of animal bodies. They also are found in muscle tissue, holding fibers together. Genetic disorders that reduce cells' ability to form desmosome proteins or lead to destruction of desmosomes by the immune system result in the formation of blisters, where layers of skin separate from each other.

Finally, **gap junctions** are pores surrounded by special proteins that form open channels between two cells (see Figure 3-25). Functioning like secret passageways, these junctions are large enough for salts, sugars, amino acids, and the chemicals that carry electrical signals to pass through,

**TIGHT JUNCTIONS**
Form a water-tight seal between cells, like caulking around a tub

TEM 38,000×

**DESMOSOMES**
Act like Velcro and fasten cells together

TEM 70,000×

**GAP JUNCTIONS**
Act like secret passageways and allow materials to pass between cells

**FIGURE 3-25 Cell connections: tight junctions, desmosomes, and gap junctions.**

but are too small for the passage of organelles or very large molecules such as proteins and nucleic acids. Gap junctions are an important mechanism for cell-to-cell communication. In the heart, for example, the electrical signal telling muscle cells to contract is passed from cell to cell through gap junctions. Gap junctions are also important in allowing a cell to recognize that it has bumped up against another cell; chemicals flowing from one cell to the next can signal the body to stop producing cells of a particular type.

**Q**

Is a breakdown of cell-cell communication related to cancer?

Compared with normal cells, cancer cells have fewer gap junctions. And research suggests that the resulting reduction in intercellular communication among cancer cells may be important in the formation of

masses of cells, called tumors, and their spread throughout the body. (Interestingly, treatments that cause an increase in the number of gap junctions tend to reduce tumor growth.)

---

### TAKE-HOME MESSAGE 3·12

In multicellular organisms, most cells are connected to other cells. The connections can form a water-tight seal between the cells (tight junctions), can hold sheets of cells together while allowing fluid to pass between neighboring cells (desmosomes), or can function like secret passageways, allowing the movement of cytoplasm, molecules, and other signals between cells (gap junctions).

# 3·13–3·21 Nine important landmarks distinguish eukaryotic cells.

*Cell of the voodoo lily. How many different organelles can you spot in this plant cell?*

## 3·13 The nucleus is the cell's genetic control center.

As we have seen, all cells are surrounded by a complex plasma membrane that actively regulates what materials get into and out of a cell. Now it's time to look at the cellular contents that are surrounded by the plasma membrane in eukaryotic cells.

The nucleus is the largest and most prominent organelle in most eukaryotic cells. In fact, the nucleus is generally larger than any prokaryotic cell. If a cell were the size of a large lecture hall or movie theater, the nucleus would be the size of a big-rig 18-wheeler truck parked in the front rows. The nucleus has two primary functions, both related to the fact that most of the cell's DNA resides in the nucleus. First, the nucleus is the genetic control center, directing most cellular activities by controlling which molecules are produced, and in what quantity. Second, the nucleus is the storehouse for hereditary information (**FIGURE 3-26**).

Three important structural components stand out in the nucleus (**FIGURE 3-27**). First is the **nuclear membrane,** sometimes called the nuclear envelope, which surrounds the nucleus and separates it from the cytoplasm. Unlike most

**FIGURE 3-26 The nucleus holds the genetic information that makes it possible to build organisms.** That's why identical twins, carrying the same genetic information, look so similar.

**NUCLEUS**

**FUNCTIONS**
- Acts as the genetic control center of the cell
- Stores hereditary information

**NUCLEOLUS**
Area of the nucleus, where ribosomal subunits are assembled

**CHROMATIN/ CHROMOSOMES**
Thin fibers of DNA, which carry all hereditary information

**NUCLEAR MEMBRANE**
Two bilayers, covered in pores, that surround the nucleus

Pore

TEM 43,000×

**FIGURE 3-27 The nucleus: the cell's genetic control center.**

plasma membranes, however, the nuclear membrane consists of *two* bilayers, one on top of the other, much like the double-bagging of groceries at the market. The nuclear membrane is not a sack, though. It is perforated, covered with tiny pores made from multiple proteins embedded in the phospholipid membranes and spanning both bilayers. These pores enable large molecules to pass from the nucleus to the cytoplasm and from the cytoplasm to the nucleus.

The second prominent structure in the nucleus is the **chromatin,** a mass of long, thin fibers consisting of DNA with some proteins attached to it that keep it from getting impossibly tangled. Most of the time, as the DNA directs cellular activities, the chromatin resembles a plate of spaghetti. When it's time for cell division (a process described in detail in Chapter 6), the chromatin coils up and the threads become shorter and thicker until they become visible as chromosomes, the compacted, linear DNA molecules on which all of the organism's hereditary information is carried.

A third structure in the nucleus is the **nucleolus,** an area near the center of the nucleus where subunits of the ribosomes, a critical part of the cellular machinery, are assembled. Ribosomes are like little factories in which the information stored in the DNA is used to construct proteins, including enzymes and the proteins that make up tissues such as the bark of trees or the bone of vertebrates. The ribosomes are built in the nucleolus but pass through the nuclear pores and into the cytoplasm before starting their protein-production work.

**TAKE-HOME MESSAGE 3·13**

The nucleus is usually the largest and most prominent organelle in the eukaryotic cell. It directs most cellular activities by controlling which molecules are produced and in what quantity. The nucleus is also the storehouse for all hereditary information.

# 3·14 Cytoplasm and the cytoskeleton form the cell's internal environment, provide its physical support, and can generate movement.

If you imagine yourself inside a cell the size of a big lecture hall, it might come as a surprise that you can barely see that big rig of a nucleus parked in the front rows. Visibility is almost zero, not only because the room is filled with jelly-like cytoplasm, but also because there is a dense web of thick and thin, straight and branched, ropes, strings, and scaffolding, running every which way throughout the room.

This inner scaffolding of the cell, which is made from proteins, is the **cytoskeleton** (FIGURE 3-28). It has three chief purposes. First, it gives animal cells shape and support— making red blood cells look like little round doughnuts (without the hole in the middle) and giving neurons their very long, thread-like appearance. Plant cells are shaped primarily by their cell wall (a structure we discuss later in the chapter), but they also have a cytoskeleton. Second, the cytoskeleton controls the intracellular traffic flow, serving as a series of

tracks on which a variety of organelles and molecules are guided across and around the inside of the cell. And third, because the elaborate scaffolding of the cytoskeleton is dynamic and can generate force, it gives all cells some ability to control their movement.

Three types of protein fibers make up the cytoskeleton. **Microtubules,** the thickest, are linear polymers of a protein and look like rigid, hollow tubes. Like intracellular conveyor belts, they are the tracks to which molecules and organelles within the cell may attach and be moved along. Microtubules also help to pull chromosomes apart during cell division. Continuously built, disassembled, and rebuilt, microtubules rarely last more than about 10 minutes in a cell. **Intermediate filaments,** a second type of cytoskeleton fiber, are durable, rope-like systems of numerous different overlapping proteins. They give cells great strength. **Microfilaments** are the

---

## CYTOSKELETON

**FUNCTIONS**
- Acts as the inner scaffolding of the cell
- Provides shape and support
- Controls intracellular traffic flow
- Enables movement

Structural network of intermediate filaments

3,500×

---

### THREE TYPES OF PROTEIN FIBERS IN THE CYTOSKELETON

**MICROTUBULES**
- Thick, hollow tubes
- The tracks to which molecules and organelles within the cell may attach and be moved along

**INTERMEDIATE FILAMENTS**
- Durable, rope-like systems of numerous overlapping proteins
- Give cells great strength

**MICROFILAMENTS**
- Long, solid rod-like fibers
- Help with cell contraction and cell division

**FIGURE 3-28  The cytoskeleton: the cell's inner scaffolding.**

**FIGURE 3-29** Cilia and flagella assist the cell with movement.

With gentle beating, cilia can move fluid past cells, whereas flagella, with whip-like motions, can move the cells themselves.

thinnest elements in the cytoskeleton. Long, solid, rod-like fibers, microfilaments help generate forces, including those important in cell contraction and cell division.

A couple of microtubule-based structures are sometimes present in cells and can help to move the cell through its environment (or in stationary cells, can help move the cell's environment past the cell). **Cilia** (*sing.* **cilium**) are short projections that often occur in large numbers on a single cell (**FIGURE 3-29**). Cilia beat swiftly, often in unison and in ways that resemble blades of grass in a field, blowing in the wind. Cilia can move fluid along and past a cell. This movement can accomplish many important tasks, including sweeping the airways to our lungs to clear them of debris (such as dust) in the air we breathe.

**Flagella** are much longer than cilia. They occur in many prokaryotes and single-celled eukaryotes, and many algae and plants have cells with one or more flagella. But in animals, cell

types with one or more flagella are very rare. One of these cell types, however, has a critical role in every animal species: sperm cells. With a flagellum for a tail, sperm are among the most mobile of all animal cells. Some spermicidal birth control methods prevent conception by disabling the flagellum and immobilizing the sperm cells.

### TAKE-HOME MESSAGE 3·14

The inner scaffolding of the cell, which is made from proteins, is the cytoskeleton. Consisting of three types of protein fibers—microtubules, intermediate filaments, and microfilaments—the cytoskeleton gives animal cells their shape and support, gives cells some ability to control their movement, and serves as a series of tracks on which organelles and molecules are guided across and around the inside of the cell.

## 3·15 Mitochondria are the cell's energy converters.

Cars have it easy: we generally put only a single type of fuel in them, and it is exactly the same fuel every time. With human bodies it's a different story. During some meals we put in meat and potatoes. Other times, fruit or bread or vegetables, popcorn and gummy bears, pizza and ice cream. Yet no matter what we eat, we expect our body to utilize the energy in these various foods to power all the reactions that make it possible

for us to breathe, move, and think. The mitochondria are the organelles that make this possible (**FIGURE 3-30**).

**Mitochondria** (*sing.* **mitochondrion**) are the cell's all-purpose energy converters, and they are present in nearly all plant cells, animal cells, and every other eukaryotic cell. Although we consume a variety of foods, our mitochondria allow us to

### FUNCTIONS
- Act as all-purpose energy converters
- Harvest energy to be used for cellular functions

DNA
Matrix
Outer membrane
Inner membrane
Intermembrane space

TEM 73,000×

*Cells such as liver cells, which use a lot of energy, can have up to 2,500 mitochondria!*

**FIGURE 3-30** Mitochondria: the cell's all-purpose energy converters.

convert the energy contained in the chemical bonds of the carbohydrates, fats, and proteins in the food into carbon dioxide, water, and ATP—the molecule that is the energy source all cells use to fuel all their functions and activities. (ATP and how it works are described in detail in Chapter 4.) Because this energy conversion requires a significant amount of oxygen, organisms' mitochondria consume most of the oxygen used by each cell. In humans, for example, our mitochondria consume as much as 80% of the oxygen we breathe. They give a significant return on this investment by producing about 90% of the energy our cells need to function.

As with so many aspects of our bodies, form follows function. Cells that are not very metabolically active, such as some fat storage cells in humans, have very few mitochondria. Other cells, such as muscle, liver, and sperm cells in animals and fast-growing root cells in plants—all of which have large energy requirements—are packed densely with mitochondria (**FIGURE 3-31**). Some of these cells have as many as 2,500 mitochondria!

To visualize the structure of a mitochondrion, imagine a plastic sandwich bag. Now take another, bigger plastic bag and stuff it inside the sandwich bag. That's the structure of mitochondria: there is a smooth outer membrane and a scrunched-up inner membrane. This construction forms two separate compartments within the mitochondrion: a region

### NUMBER OF MITOCHONDRIA IN VARIOUS CELLS

Liver cell — ~2,500

Skeletal muscle cell — ~1,200

White blood cell in lung — ~700

Dermal cell (just under the skin) — ~200

White adipose cell (fat storage) — ~100

Red blood cell — 0

**Number of mitochondria per cell**

*Cells with high energy needs have many times more mitochondria than cells with low energy needs.*

**GRAPHIC CONTENT!**
*Thinking critically about visual displays of data.*
*Turn to p.124 for a closer inspection of this figure.*

**FIGURE 3-31** How does the number of mitochondria vary in different types of cells?

outside the inner plastic bag (called the **intermembrane space**) and another region, called the **matrix,** inside the inner plastic bag. This bag-within-a-bag structure has important implications for energy conversion. And having a heavily folded inner membrane that is much larger than the outer membrane provides a huge amount of surface area on which to conduct chemical reactions. (We discuss the details of mitochondrial energy conversions and the role of mitochondrial structure in Chapter 4.)

As we learned in our earlier discussion of endosymbiosis, mitochondria may very well have existed, billions of years ago, as separate single-celled bacteria-like organisms. They are similar to bacteria in size and shape, and may have originated when symbiotic bacteria took up permanent residence within other cells. Perhaps the strongest evidence for this is that mitochondria have their own DNA (see Figure 3-30). Mixed in among the approximately 3,000 proteins in each mitochondrion, there are anywhere from 2 to 10 copies of its own little ring-shaped DNA. This DNA carries the instructions for making 13 important mitochondrial proteins necessary for metabolism and energy production.

**Q**

**We all have more DNA from one parent than the other. Who is the bigger contributor: mom or dad? Why?**

We're always taught that our mothers and fathers contribute equally to our genetic composition, but this isn't quite true. The mitochondria in every one of your cells (and the DNA that comes with them) come from the mitochondria that were initially present in your mother's egg, which, when fertilized by your father's sperm, developed into you. In other words: all of your mitochondria are descended from your mother's

mitochondria. The tiny sperm contributes DNA, but no cytoplasm and, hence, no mitochondria. Consequently, mitochondrial DNA is something that we inherit exclusively from our mothers. This is true not only in humans but in most multicellular eukaryotes. As the fertilized egg develops into a two-celled, then four-celled, embryo, the mitochondria split themselves by a process called fission, the same process of division and DNA duplication used by bacteria—so there are always a sufficient number of mitochondria for the newly produced cells. This similarity between mitochondria and bacteria is another characteristic that supports the theory that mitochondria were originally symbiotic bacteria.

Given the central role of mitochondria in converting the energy in food molecules into a form that is usable by cells, you won't be surprised to learn that mitochondrial malfunctions can have serious consequences. Recent research has focused on possible links between defective mitochondria and diseases characterized by fatigue and muscle pain. It seems, for instance, that many cases of "exercise intolerance"—extreme fatigue or cramps after only slight exertion—may be related to defective mitochondrial DNA.

---

**TAKE-HOME MESSAGE 3·15**

In mitochondria, which are found in nearly all eukaryotic cells, the energy contained in the chemical bonds of carbohydrate, fat, and protein molecules is converted into carbon dioxide, water, and ATP, the energy source for all cellular functions and activities. Mitochondria may have their evolutionary origins as symbiotic bacteria living inside other cells.

---

# 3·16 Lysosomes are the cell's garbage disposals.

Garbage. What does a cell do with all the garbage it generates? Mitochondria wear out after about 10 days of intensive activity, for starters. And white blood cells constantly track down and consume bacterial invaders, which they then have to dispose of. Similarly, the thousands of ongoing reactions of cellular metabolism produce many waste macromolecules that cells must digest and recycle. Many eukaryotic cells deal with this garbage by maintaining hundreds of versatile floating "garbage disposals" called lysosomes (**FIGURE 3-32**).

**Lysosomes** are round, membrane-enclosed, acid-filled vesicles that dispose of garbage. They are filled with about 50 different digestive enzymes and a super-acidic fluid, a corrosive broth so powerful that if a lysosome were to burst, it would almost immediately kill the cell by rapidly digesting all of its component parts. The selection of enzymes in the lysosome represents a broad spectrum of chemicals designed for dismantling macromolecules that are no longer needed by the cell or are generated as by-products of cellular metabolism.

**FUNCTION**
- Act as floating garbage disposals for cells, digesting and recycling cellular waste products and consumed material

Membrane

Digestive enzymes and acid

Partially digested organelle

TEM 59,500×

**FIGURE 3-32 Lysosomes: digestion and recycling of the cell's waste products.**

Some of the enzymes break down lipids, others carbohydrates, others proteins, and still others nucleic acids. Consequently, when a cell consumes a particle of food or even an invading bacterium via phagocytosis, the cell directs the material to lysosomes for dismantling. And, ever the efficient system, the cell releases most of the component parts of molecules that are digested, such as amino acids, back into its cytoplasm, where they can be reused by the cell as raw materials.

Some immune system cells tend to have particularly large numbers of lysosomes, most likely because of their great need for disposing of the by-products of disease-causing bacteria.

**Q** **Why is Tay-Sachs disease like a strike by trash collectors?**

With 50 different enzymes necessary for lysosomes to carry out their metabolic salvaging act, malfunctions sometimes occur. A common genetic disorder called Tay-Sachs disease is the result of just such a genetic mishap. In Tay-Sachs disease, an individual inherits an inability to produce a critical lipid-digesting enzyme. Even though the lysosomes cannot digest certain lipids, the cells continue to send lipids to the lysosomes, where they accumulate, undigested. This leads to a sort of lysosome constipation. The lysosome swells until it bursts and digests the whole cell or until it chokes the cell to death. Within the first few years of life, this process occurs in large numbers of cells and eventually leads to the child's death.

Although it has long been debated among biologists, it does seem that plant cells also contain lysosomes, compartments with similar digestive broths that carry out the same digestive processes as in animals.

---

**TAKE-HOME MESSAGE 3·16**

**Lysosomes are round, membrane-enclosed, acid-filled organelles that function as a cell's garbage disposals. They are filled with about 50 different digestive enzymes and enable a cell to dismantle macromolecules, including disease-causing bacteria.**

# 3·17 The endoplasmic reticulum is the site where cells build proteins and disarm toxins.

The information for how to construct the molecules essential to a cell's smooth functioning and survival is stored in the DNA found in the cell's nucleus. The energy used to construct these molecules and to run cellular functions comes primarily from the mitochondria. The actual production and modification of biological molecules, however, occurs in a system of organelles called the **endomembrane system** (**FIGURE 3-33**). This mass of interrelated membranes spreads out from and surrounds the nucleus, forming chambers within the cell that contain their own mixtures of chemicals. The endomembrane system takes up as much as one-fifth of the cell's volume and is responsible for many of the fundamental functions of the cell.

Lipids are produced within these membrane-enclosed compartments, products for export to other parts of the body are modified and packaged here, polypeptide chains are assembled into functional proteins here, and many of the toxic chemicals that find their way into our bodies—from recreational drugs to antibiotics—are broken down and neutralized in the endomembrane system.

Because the process of protein production follows a somewhat linear path, we explore the endomembrane system sequentially, beginning just outside the nucleus with the endoplasmic reticulum.

**Rough endoplasmic reticulum** Perhaps the organelle with the most cumbersome name, the **rough endoplasmic reticulum,** or **rough ER** ("endoplasmic reticulum" is derived from the Greek for "within" and "anything molded" and the Latin for "small net"), is a large series of interconnected, flattened sacs that look like a stack of pancakes. These sacs are connected directly to the nuclear envelope. In most eukaryotic cells, the rough ER almost completely surrounds the nucleus (**FIGURE 3-34**). It is called "rough" because its surface is studded with little bumps. These bumps are ribosomes, the cell's protein-making machines, and cells with high rates of protein production generally have large numbers of ribosomes. We cover the details of ribosome structure and protein production in Chapter 5.

The primary function of the rough ER is to fold and package proteins that will be shipped to other locations in the endomembrane system, the cell surface, or outside the cell. Poisonous frogs, for example, package their poison in the rough ER of the cells where it is produced before transporting it to the poison glands on their skin. Proteins that are used within the cell itself are generally produced on free-floating ribosomes in the cytoplasm.

**Smooth endoplasmic reticulum** As its name advertises, the **smooth endoplasmic reticulum** is part of the

## OVERVIEW OF THE ENDOMEMBRANE SYSTEM

**FUNCTIONS**
• Produces and modifies molecules to be exported to other parts of the organism
• Breaks down toxic chemicals and cellular by-products

Rough endoplasmic reticulum

Smooth endoplasmic reticulum

Golgi apparatus

**FIGURE 3-33** The endomembrane system: the rough endoplasmic reticulum, smooth endoplasmic reticulum, and Golgi apparatus.

**FUNCTION**
• Modifies proteins that will be shipped to other locations in the endomembrane system, the cell surface, or outside the cell

Ribosomes

Rough ER is covered in ribosomes. Ribosomes are protein-making machines.

TEM 56,500×

**FIGURE 3-34** The rough endoplasmic reticulum is studded with ribosomes.

endomembrane network that is smooth, because there are no ribosomes bound to it (**FIGURE 3-35**). Although it is connected to the rough ER, it is farther from the nucleus and, besides lacking ribosomes, differs slightly in appearance. Whereas the rough ER looks like stacks of pancakes, the smooth ER sometimes looks like a collection of branched tubes.

The smooth surface gives us the first hint that smooth ER has a different job than rough ER. Because ribosomes are absolutely essential for protein production, we can guess that, with few of them, the smooth ER is not extensively involved in folding or packaging proteins. It isn't. Instead, it primarily takes part in the synthesis of lipids such as fatty acids,

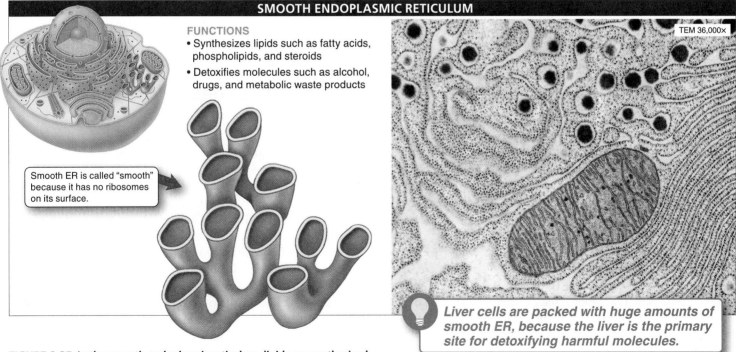

**SMOOTH ENDOPLASMIC RETICULUM**

**FUNCTIONS**
• Synthesizes lipids such as fatty acids, phospholipids, and steroids
• Detoxifies molecules such as alcohol, drugs, and metabolic waste products

TEM 36,000×

Smooth ER is called "smooth" because it has no ribosomes on its surface.

Liver cells are packed with huge amounts of smooth ER, because the liver is the primary site for detoxifying harmful molecules.

**FIGURE 3-35** In the smooth endoplasmic reticulum, lipids are synthesized and alcohol, antibiotics, and other drugs are detoxified.

phospholipids, and steroids, as well as carbohydrates. Exactly which lipids are produced varies throughout the organism and across plant and animal species. Inside the smooth ER of mammalian ovaries and testes, for example, the hormones estrogen and testosterone are produced. Inside the smooth ER of liver and fat cells, other lipids are produced. As occurs for proteins in the rough ER, following their production these lipids are packaged in transport vesicles and then sent to other parts of the cell or to the plasma membrane for export.

Another critical responsibility of the smooth ER—particularly the smooth ER in human liver cells—is to help protect us from the many dangerous molecules that get into our bodies. Alcohol, antibiotics, barbiturates, amphetamines, or other stimulants that we may consume, along with many toxic metabolic waste products formed in our bodies, are made less harmful by the detoxifying enzymes in the smooth ER. And just as a bodybuilder's muscles get bigger and bigger in response to weightlifting, our smooth ER proliferates in cells that are exposed to large amounts of particular drugs. This proliferation can be beneficial because it increases the capacity for detoxification. But as the cells become more and more efficient at detoxification, our tolerance to the very drugs the smooth ER is trying to destroy can increase.

As we see over and over in living organisms, form follows function. Not surprisingly, then, just as cells with high rates of protein production have large numbers of ribosomes, we find huge amounts of smooth ER in liver cells because they are the

*How can long-term use of one drug increase your resistance to another, different drug that you have never encountered?*

primary sites of molecular detoxification. Other cells that are packed with ER (both rough and smooth) include plasma cells in the blood that produce immune system proteins for export and pancreas cells that secrete large amounts of digestive enzymes.

**Q** Chronic exposure to many drugs (from antibiotics to heroin) can induce a proliferation of smooth ER, particularly in the liver, and the smooth ER's associated detoxification enzymes. This proliferation in turn increases tolerance to the drugs, necessitating higher doses to achieve the same effect. Moreover, the increased detoxification capacity of the cells often enables them to better detoxify other compounds, even if the person has never been exposed to these chemicals. This increased detoxification capacity can lead to problems. An individual who has been exposed to large amounts of certain drugs, for example, may end up responding less well to antibiotics.

---

### TAKE-HOME MESSAGE 3·17

**The production and modification of biological molecules in eukaryotic cells occurs in a system of organelles called the endomembrane system, which includes, among other organelles, the rough and smooth endoplasmic reticulum. In rough ER, proteins that will be shipped elsewhere in the body are folded and packaged. In the smooth ER, lipids and carbohydrates are synthesized and alcohol, antibiotics, and other drugs are detoxified.**

---

## 3·18 The Golgi apparatus processes products for delivery throughout the body.

Moving farther outward from the nucleus, we encounter another organelle within the endomembrane system: the Golgi (GOHL-jee) apparatus (**FIGURE 3-36**). The **Golgi apparatus** processes molecules synthesized in the cell—primarily proteins and lipids—and packages those that are destined for use elsewhere in the body. The Golgi apparatus is also a site of carbohydrate synthesis, including the complex polysaccharides found in many plasma membranes. The Golgi apparatus, which is not connected to the endoplasmic reticulum, is a flattened stack of membranes (each of which is called a Golgi body) that are *not* interconnected.

After transport vesicles bud from the endoplasmic reticulum, they move through the cytoplasm until they

reach the Golgi apparatus. Here, the vesicles fuse with the Golgi apparatus membrane and dump their contents into a Golgi body. There are about four successive Golgi body chambers to be visited, and the molecules get passed from one to the next. In each Golgi body, enzymes make slight modifications to the molecules (such as the addition or removal of phosphate groups or sugars). The processing that occurs in the Golgi apparatus often involves tagging molecules (much like adding a postal address or tracking number) to direct them to some other part of the organism. After they are processed, the molecules bud off from the Golgi apparatus in a vesicle, which then moves into the cytoplasm. If the molecules are destined for delivery and use elsewhere in the body, the transport vesicle eventually fuses

**FUNCTION**
• Processes and packages proteins, lipids, and other molecules for export to other locations in or outside of the cell

Transport vesicles

TEM 62,500×

**FIGURE 3-36** Golgi apparatus: processing of molecules synthesized in the cell and packaging of those molecules destined for use elsewhere in the body.

with the cell's plasma membrane and dumps the molecules into the bloodstream via exocytosis.

**FIGURE 3-37** shows how the various parts of the endomembrane system work to produce, modify, and package molecules within a cell.

> ### TAKE-HOME MESSAGE 3·18
>
> The Golgi apparatus—another organelle within the endomembrane system—processes molecules synthesized in the cell and packages those that are destined for use elsewhere in the body.

## THE ENDOMEMBRANE SYSTEM AT WORK

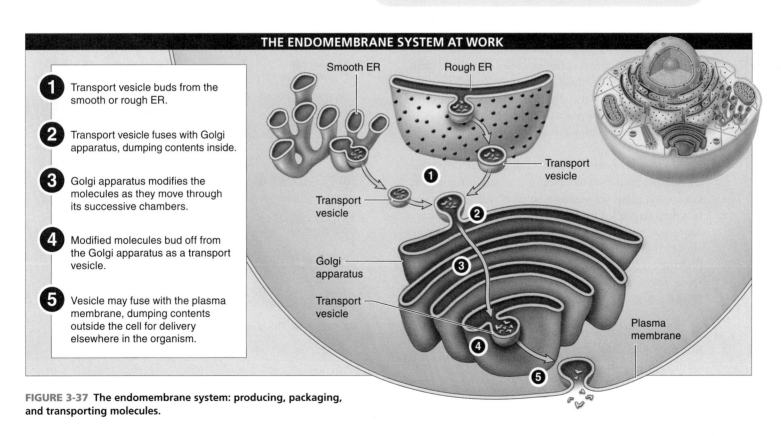

**1** Transport vesicle buds from the smooth or rough ER.

**2** Transport vesicle fuses with Golgi apparatus, dumping contents inside.

**3** Golgi apparatus modifies the molecules as they move through its successive chambers.

**4** Modified molecules bud off from the Golgi apparatus as a transport vesicle.

**5** Vesicle may fuse with the plasma membrane, dumping contents outside the cell for delivery elsewhere in the organism.

Smooth ER
Rough ER
Transport vesicle
Transport vesicle
Golgi apparatus
Transport vesicle
Plasma membrane

**FIGURE 3-37** The endomembrane system: producing, packaging, and transporting molecules.

# 3·19 The cell wall provides additional protection and support for plant cells.

Until this point, we've discussed organelles and structures that are common to both plants and animals. Now we're going to look at several structures that are not found in all eukaryotic cells. Because plants are stuck in the ground, unable to move, they have several special needs beyond those of animals. In particular, they can't outrun or outmaneuver their competitors to get more food. Nor can they outrun organisms that want to eat them. If they want more sunshine, they have few options beyond simply growing larger or taller. And if they want to resist plant-eating animals, they must simply reduce their edibility.

One structure that helps plants achieve both of these goals is the cell wall, which surrounds the plasma membrane. The cell wall is made largely from long fibers of a polysaccharide called cellulose. Note that although animal cells do not have cell walls, some non-plants such as archaea, bacteria, protists, and fungi also have cell walls; their cell walls are made of different polysaccharides rather than cellulose.

In plants, cell wall production can be a multistage process. Initially, when the cell is still growing, a primary cell wall is laid down, along with some glue that helps adjacent cells adhere to each other. Sometime later in life, a secondary cell wall is laid down. This secondary cell wall usually contains a complex molecule called lignin, which gives the wall much of its strength and rigidity. Oddly, this strength remains even after the plant cell dies—hence wood's great value to humans as a construction material, among other things.

The cell wall is nearly 100 times thicker than the plasma membrane, and the tremendous structural strength it confers on plant cells enables some plants to grow several hundred feet tall. Cell walls also help to make plants more water-tight—an important feature in organisms that cannot reduce water loss through evaporation by moving out of the hot sun—and provide some protection from insects and other animals that might eat them (**FIGURE 3-38**).

Surprisingly, despite its great strength, the cell wall does not completely seal off plant cells from one another. Rather, it is porous, allowing water and solutes to reach

## PLANT CELL WALL

**FUNCTIONS**
- Provides the cell with structural strength
- Gives the cell increased water resistance
- Provides some protection from insects and other animals that might eat plant parts

Plasmodesmata allow water and other molecules to pass between adjacent cells.

Cell 1

Cell 2

Cell 3

Primary cell wall  Secondary cell wall  Plasmodesma  Plasma membrane

TEM 440×

**FIGURE 3-38** The plant cell wall: providing strength.

the plasma membrane. Additionally, the cells in most plants have anywhere from 1,000 to 100,000 microscopic tube-like channels, called **plasmodesmata** (*sing.* **plasmodesma**) (see Figure 3-38), connecting the cells to each other and enabling communication and transport between them. In fact, because so many of the cells are connected to one another, sharing cytoplasm and other molecules, some biologists have wondered whether we should consider a plant as just one big cell. How would you respond to that idea?

# 3·20  Vacuoles are multipurpose storage sacs for cells.

If you look at a mature plant cell through a microscope, one organelle, the **central vacuole,** usually stands out more than all the others because it is so huge and appears empty (**FIGURE 3-39**). Although it may look like an empty sac, the central vacuole is anything but. Surrounded by a membrane, filled with fluid, and occupying from 50% to 90% of a plant cell's interior space, the central vacuole can play an important role in five different areas of plant life. (Vacuoles are also found in some other eukaryotes, including some protists, fungi, and animals, but they tend to be particularly prominent in plant cells.)

1. **Nutrient storage**—the vacuole stores hundreds of dissolved substances, including amino acids, sugars, and ions.

2. **Waste management**—the vacuole retains waste products and degrades them with digestive enzymes, much like the lysosome in animal cells.

3. **Predator deterrence**—the poisonous, nasty-tasting materials that accumulate inside the vacuoles of some plants make a powerful deterrent to animals that might try to eat parts of the plant.

4. **Sexual reproduction**—the vacuole may contain pigments that give some flowers their red, blue, purple, or other colors, enabling them to attract birds and insects that help the plant reproduce by transferring pollen.

5. **Physical support**—high concentrations of dissolved substances in the vacuole can cause water to rush into the cells through the process of osmosis. The increased

**VACUOLE**

FUNCTIONS
- Stores nutrients
- Retains and degrades waste products
- Accumulates poisonous materials
- Contains pigments, enabling plants to attract birds and insects that help the plant reproduce
- Provides physical support

Vacuole

TEM 5,000×

**FIGURE 3-39  The vacuole: multipurpose storage.**

fluid pressure inside the vacuole can cause the cell to enlarge a bit and push out the cell wall. This process is responsible for the pressure (called **turgor pressure**) that allows stems, flowers, and other plant parts to stand upright. The ability of non-woody plants to stand upright is due primarily to turgor pressure. Wilting is the result of a loss of turgor pressure.

# 3·21 Chloroplasts are the plant cell's solar power plant.

It would be hard to choose the "most important organelle" in a cell, but if we had to, the chloroplast would be a top contender. The chloroplast, an organelle found in all plants and eukaryotic algae, is the site of photosynthesis—the conversion of light energy into the chemical energy of food molecules, with oxygen as a by-product. Because all photosynthesis in plants and algae takes place in chloroplasts, these organelles are directly or indirectly responsible for everything we eat and for the oxygen we breathe. Life on earth would be vastly different without the chloroplast (**FIGURE 3-40**).

In a green leaf, each cell has about 40–50 chloroplasts. This means there are about 500,000 chloroplasts per square millimeter of leaf surface (that's more than 200 million chloroplasts in an area the size of a small postage stamp). Chloroplasts are oval, somewhat flattened objects. On the outside, the chloroplast is encircled by two distinct layers of membranes (similar to the structure of mitochondria). The fluid in the inside compartment, called the **stroma,** contains some DNA and much protein-making machinery.

With a simple light microscope, it is possible to see little spots of green within the chloroplasts. Up close, these spots look like stacks of pancakes. Each stack consists of numerous, interconnected little flattened sacs called **thylakoids,** and it is on the membranes of the thylakoids that the light-collecting for photosynthesis occurs. (In Chapter 4, we discuss the details of how chloroplasts convert the energy in sunlight into the chemical energy stored in sugar molecules.)

A peculiar feature of chloroplasts, mentioned in Section 3-3, is that they resemble photosynthetic bacteria, particularly with their circular DNA (which contains many of the genes essential for photosynthesis). Also, the dual

**CHLOROPLAST**

FUNCTION
• Site of photosynthesis—the conversion of light energy into chemical energy

Light is collected for photosynthesis on the membranes of the thylakoids within the chloroplasts.

Thylakoid

DNA

Stroma

TEM 16,000×

**FIGURE 3-40 The chloroplast: location of photosynthesis.**

## REVIEW OF CELL STRUCTURES

| STRUCTURE | ANIMALS | PLANTS | FUNCTION |
|-----------|---------|--------|----------|
| Nucleus | ✓ | ✓ | Directs cellular activity and stores hereditary information |
| Cytoskeleton | ✓ | ✓ | Provides structural shape and support and enables cellular movement |
| Mitochondrion | ✓ | ✓ | Harvests energy for cellular functions |
| Lysosome | ✓ | ✓ | Digests and recycles cellular waste products and consumed material |
| Rough ER | ✓ | ✓ | Modifies proteins that will be shipped elsewhere in the organism |
| Smooth ER | ✓ | ✓ | Synthesizes lipids and detoxifies molecules |
| Golgi apparatus | ✓ | ✓ | Processes and packages proteins, lipids, and other molecules |
| Cell wall | ⊘ | ✓ | Provides structural strength, protection, and increased resistance to water loss |
| Vacuole | Sometimes | ✓ | Stores nutrients, degrades waste products, provides pigments and structural support |
| Chloroplast | ⊘ | ✓ | Performs photosynthesis |

**FIGURE 3-41** Review of major cellular structures and their functions.

outer membrane of the chloroplast is consistent with the idea that, long ago, a cell engulfed a photosynthetic bacterium, enveloping it with its plasma membrane in endocytosis. These features have given rise to the belief that chloroplasts might originally have been bacteria that were engulfed by a predatory cell. According to the endosymbiosis theory (see Section 3-3), the bacteria remained alive and, rather than becoming a meal, became the cell's meal ticket, providing food for the cell in exchange for protection.

**FIGURE 3-41** summarizes the major parts of a cell, in animals and plants, and their functions.

### TAKE-HOME MESSAGE 3·21

**The chloroplast is the organelle in plants and algae that is the site of photosynthesis—the conversion of light energy into chemical energy, with oxygen as a by-product. Chloroplasts may originally have been bacteria that were engulfed by a predatory cell by endosymbiosis.**

## DRINKING TOO MUCH WATER CAN BE DANGEROUS!

People have long understood the risks of becoming dehydrated, particularly during activities such as marathon running in which a great deal of fluid can be lost through sweating. Only recently, however, has it been recognized that too much water—called "water intoxication"—can also be dangerous. In 2007, a 28-year-old woman competing in a water-drinking contest died just a few hours after consuming about two gallons of water without urinating. Similarly, the death of a runner during the 2002 Boston marathon was attributed to water intoxication.

**Q: WHEN PEOPLE SWEAT, WHAT DO THEY LOSE?** Sweat consists of water and some dissolved solids, primarily salt. As the sweat evaporates, the body is cooled.

**Q: HOW DO PEOPLE RESPOND TO NAUSEA AND DIZZINESS, THE WARNING SIGNS OF DEHYDRATION?** By drinking water. Because the water they drink lacks salt and other solutes, this can lead to sodium imbalances in the body, particularly in the bloodstream and the fluid around cells (extracellular fluid).

**Q: WHAT IS OSMOSIS?** Osmosis is the diffusion of water across a membrane toward areas where there is a greater concentration of solutes and, consequently, a lower concentration of water molecules.

**Q: IF YOU DRINK LOTS OF WATER (TWO GALLONS OR MORE)—WHICH HAS FEW OR NO SOLUTES—WHERE IS IT GOING TO MOVE ONCE IT GETS INTO YOUR BLOODSTREAM AND EXTRACELLULAR FLUID?** It moves by osmosis into your cells, where there is a higher concentration of solutes (primarily sodium) and a lower concentration of water molecules.

**Q: WHAT HAPPENS TO YOUR CELLS IF THEY ABSORB TOO MUCH WATER?** They swell, sometimes to the point of rupturing. Such swelling and rupturing in the enclosed brain cavity can cause vomiting and confusion. Disastrously, these symptoms are often mistaken for the symptoms of dehydration and so the consumption of water is recommended, making the situation worse. Further swelling can lead to seizures, coma, and even death. (The reverse situation occurs when people drink too much seawater. Their bloodstream and extracellular fluid gain in solutes, particularly salt, so water flows out of cells by osmosis. Cell dehydration and death can occur—usually from shrinkage of brain cells.)

**WHAT SHOULD YOU DO?** Many marathon organizers have been reducing the number of water stops they offer runners during the race and testing symptomatic runners' sodium levels in medical tents along the course. Consuming salt tablets and salty snacks helps keep blood sodium levels in a healthy range. Also, when possible, runners should weigh themselves during a race to make sure they are only replacing the fluid they have lost through sweating and are not gaining weight.

# Big Ideas in ... Cells

## 3·1–3·3 What is a cell?

- The cell is the smallest unit of life that can perform all of the necessary activities of life, including reproducing itself.

- All living organisms are made up of one or more cells, and all cells arise from other, preexisting cells.

- Every cell on earth is either a eukaryote or a prokaryote.

- Prokaryotes, which have no nucleus and are single-celled, were the first cells on earth.

- Eukaryotes are single-celled or multicellular organisms whose cells have a nucleus that contains linear strands of genetic material.

## 3·4–3·7 Cell membranes are gatekeepers.

- Every cell of every living organism is enclosed by a plasma membrane, a two-layered membrane that holds the contents of a cell in place and regulates what enters and leaves the cell.

- The plasma membrane is a fluid mosaic of proteins, lipids, and carbohydrates.

- Every cell in your body has a "fingerprint," made from a variety of molecules on the outside-facing surface of the cell membrane, that is key to the functioning of your immune system.

## 3·8–3·11 Molecules move across membranes in several ways.

- Cells must import food molecules and other necessary materials from outside the cell and export metabolic waste and molecules produced for use elsewhere in the body.

- In passive transport, the molecular movement occurs spontaneously, without the input of energy.

- The diffusion of water across a membrane is a special type of passive transport called osmosis.

- Active transport is necessary if the molecules to be moved are very large or if they are being moved against their concentration gradient, and this transport requires energy.

- Cells can also engulf particles with their plasma membrane in a process called endocytosis. Similarly, molecules can be moved out of a cell by exocytosis.

## 3·12 Cells are connected and communicate with each other.

- In multicellular organisms, most cells are connected to other cells.

- The connections can form a water-tight seal between the cells (tight junctions), can hold sheets of cells together while allowing fluid to pass between the cells (desmosomes), or can function like secret passageways between cells (gap junctions).

## 3·13–3·21 Nine important landmarks distinguish eukaryotic cells.

- Specific structures in eukaryotic cells perform specific life-sustaining functions.

- The nucleus directs all protein production and is the storehouse for all hereditary information.

- The cytoskeleton gives animal cells shape and support.

- In mitochondria, energy from macromolecules is converted into usable energy in the form of ATP.

- Lysosomes dismantle waste and bacterial pathogens.

- The production and modification of molecules, including drug detoxification, occurs in the rough and smooth endoplasmic reticulum; and the Golgi apparatus processes molecules synthesized in the cell and packages those destined for use elsewhere in the body.

- In plants, the cell wall confers structural strength, protection from predators, and resistance to water loss. Vacuoles function as storage spaces, and chloroplasts are the site of photosynthesis.

- In plants, plasmodesmata connect cells and enable communication and transport between them.

See the To-Go Guide (packaged with every new copy of this textbook) for an illustrated summary of the chapter.

# Check Your Knowledge

## GRAPHIC CONTENT

*Thinking critically about visual displays of data.*

**1.** What are the variables in this graph?

**2.** What additional information would make this figure more helpful? Why?

**3.** What can you conclude from this figure?

**4.** Is "number of mitochondria per cell" the best measure of a cell's "energy-generating capacity"? Can you think of a reason why this might not be a perfect measure? (Hint: muscle cells can be much, much larger than liver cells.)

**5.** Based on these data, can you make any of your own predictions?

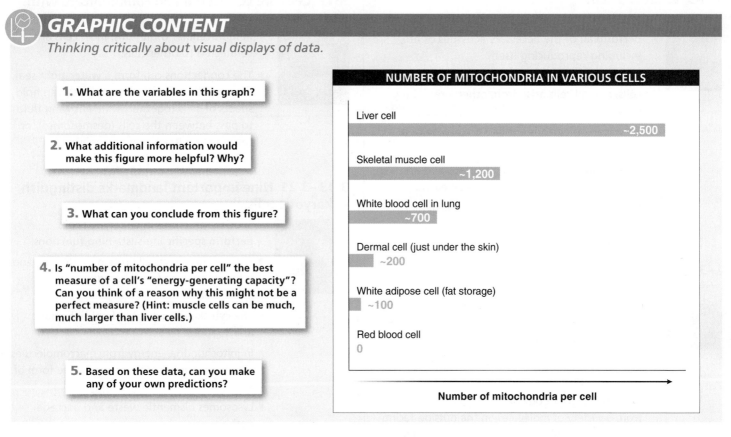

**NUMBER OF MITOCHONDRIA IN VARIOUS CELLS**

Liver cell ~2,500
Skeletal muscle cell ~1,200
White blood cell in lung ~700
Dermal cell (just under the skin) ~200
White adipose cell (fat storage) ~100
Red blood cell 0

Number of mitochondria per cell

**See Answers at the back of the book.**

## Key Terms in Cells

active transport, p. 101
cell, p. 82
cell theory, p. 84
cell wall, p. 85
chloroplast, p. 88
cholesterol, p. 92
chromatin, p. 108
cilium (*pl.* cilia), p. 110
cytoplasm, p. 84
cytoskeleton, p. 109
desmosome, p. 105
diffusion, p. 97
endocytosis, p. 102
endomembrane system, p. 114
endosymbiosis theory, p. 86
enzymatic protein, p. 92

eukaryote, p. 84
eukaryotic cell, p. 84
exocytosis, p. 102
facilitated diffusion, p. 98
flagellum (*pl.* flagella), pp. 85, 110
fluid mosaic, p. 92
gap junction, p. 105
glycerol, p. 89
Golgi apparatus, p. 116
hydrophilic, p. 89
hydrophobic, p. 89
hypertonic, p. 99
hypotonic, p. 99
intermediate filaments, p. 109
intermembrane space, p. 112

invagination, p. 88
isotonic, p. 99
lysosome, p. 112
matrix, p. 112
microfilaments, p. 109
microtubules, p. 109
mitochondrion (*pl.* mitochondria), pp. 88, 110
nonpolar, p. 89
nuclear membrane, p. 107
nucleolus, p. 108
nucleus, p. 86
organelle, p. 86
osmosis, p. 99
passive transport, p. 97
phagocytosis, p. 102
phospholipid, p. 89

phospholipid bilayer, p. 90
pilus (*pl.* pili), p. 85
pinocytosis, p. 102
plasma membrane, p. 84
plasmodesma (*pl.* plasmodesmata), p. 119
polar, p. 89
primary active transport, p. 101
prokaryote, p. 84
prokaryotic cell, p. 84
receptor protein, p. 91
receptor-mediated endocytosis, p. 102
recognition protein, p. 92
ribosome, p. 84

rough endoplasmic reticulum, p. 114
secondary active transport, p. 101
simple diffusion, p. 97
smooth endoplasmic reticulum, p. 114
solute, p. 97
solvent, p. 97
stroma, p. 120
surface protein, p. 91
thylakoid, p. 120
tight junction, p. 105
tonicity, p. 99
transmembrane protein, p. 91
transport protein, p. 92
turgor pressure, p. 120
vacuole (central), p. 119

# Multiple-Choice Questions

**1.** Which of the following statements about the cell theory is correct?

  **a)** All living organisms are made up of one or more cells.

  **b)** All cells arise from other, preexisting cells.

  **c)** All eukaryotic cells contain symbiotic prokaryotes.

  **d)** All prokaryotic cells contain symbiotic eukaryotes.

  **e)** Both a) and b) are correct.

100 HARD

 11 EASY

**2.** Which of the following statements about prokaryotes is incorrect?

  **a)** Prokaryotes appeared on earth before eukaryotes.

  **b)** Prokaryotes have circular pieces of DNA within their nuclei.

  **c)** Prokaryotes contain cytoplasm.

  **d)** Prokaryotes contain ribosomes.

  **e)** Some prokaryotes can conduct photosynthesis.

100 HARD

51

0 EASY

**3.** Which of the following facts supports the claim that mitochondria developed from bacteria that, long ago, were incorporated into eukaryotic cells by the process of phagocytosis?

  **a)** Mitochondria have flagella for motion.

  **b)** Mitochondria have proteins for the synthesis of ATP.

  **c)** Mitochondria are the "powerhouses" of the cell.

  **d)** Mitochondria are small and easily transported across cell membranes.

  **e)** Mitochondria have their own DNA.

100 HARD

37

0 EASY

**4.** Hydrophobic molecules can pass freely through the plasma membrane, but ions and polar molecules are impeded by the hydrophobic core. For this reason, plasma membranes can be considered:

  **a)** partially permeable.

  **b)** impermeable.

  **c)** hydrophobic.

  **d)** hydrophilic.

  **e)** None of these terms properly describe plasma membranes.

100 HARD

15 EASY

**5.** Drugs called beta-blockers do all of the following except:

  **a)** reduce high blood pressure.

  **b)** block signaling through adrenaline receptors.

  **c)** reduce outward symptoms of anxiety.

  **d)** bind to the cytoplasmic side of a receptor protein.

  **e)** reduce the effects of adrenaline on the heart.

100 HARD

44

0 EASY

**6.** Cellular "fingerprints":

  **a)** are exposed on the cytoplasmic side of the membrane.

  **b)** are made from cholesterol.

  **c)** are "erased" by the HIV virus.

  **d)** can help the immune system distinguish "self" from "non-self."

  **e)** All of the above are correct.

100 HARD

42

0 EASY

**7.** The movement of molecules across a membrane from an area of high concentration to one of low concentration is best described as:

  **a)** active transport.

  **b)** inactivated transport.

  **c)** passive transport.

  **d)** channel-mediated diffusion.

  **e)** electron transport.

100 HARD

31

0 EASY

8. The transport of water across a membrane from a solution of lower solute concentration to a solution of higher solute concentration is best described as:

a) osmosis.
b) facilitated diffusion.
c) receptor-mediated transport.
d) active transport.
e) general diffusion.

9. In an experiment, you measure the concentration of a polar molecule inside and outside a cell. You find that the concentration is high and gradually increasing inside the cell. What is your best hypothesis for the process you are observing?

a) facilitated diffusion
b) passive transport
c) simple diffusion
d) active transport
e) endocytosis

10. Which of the following allow the passage of small molecules between animal cells?

a) nucleoli
b) tight junctions
c) desmosomes
d) gap junctions
e) black holes

11. The largest structure in a eukaryotic cell is the _____ and it is surrounded by _____ membrane(s).

a) nucleus; one
b) nucleus; two
c) Golgi apparatus; one
d) mitochondrion; two
e) mitochondrion; one

12. The cytoskeleton:

a) is a viscous fluid found in all cells.
b) fills a cell's nucleus but not the other organelles.
c) gives an animal cell shape and support, but cannot control movement.
d) helps to coordinate intracellular movement of organelles and molecules.
e) All of the above are correct.

13. Which of the following statements about mitochondria is correct?

a) Mitochondria are found in both eukaryotes and prokaryotes.
b) There tend to be more mitochondria in fat cells than in liver cells.
c) Most plant cells contain mitochondria.
d) Mitochondria may have originated evolutionarily as photosynthetic bacteria.
e) All of the above are correct.

14. Which of the following organelles is not present in animal cells?

a) lysosome
b) Golgi apparatus
c) rough endoplasmic reticulum
d) mitochondrion
e) chloroplast

15. Given that a cell's structure reflects its function, what function would you predict for a cell with a large Golgi apparatus?

a) movement
b) secretion of digestive enzymes
c) transport of chemical signals
d) rapid replication of genetic material and coordination of cell division
e) attachment to bone tissue

16. Cell walls:

a) only occur in plant cells.
b) are not completely solid, having many small pores.
c) confer less structural support than the plasma membrane.
d) dissolve when a plant dies.
e) are made primarily from phospholipids.

17. Which one of the following organelles is not found in both plant and animal cells?

a) nucleus
b) rough endoplasmic reticulum
c) mitochondrion
d) smooth endoplasmic reticulum
e) central vacuole

**18.** In plant cells, chloroplasts:

a) serve the same purpose that mitochondria serve in animal cells.

b) are the site of conversion of light energy into chemical energy.

c) play an important role in the breakdown of plant toxins.

d) have their own linear strands of DNA.

e) Both b) and d) are correct.

100 HARD

40

0 EASY

# Short-Answer Question

**1.** Name and note the function of the organelles shown in the diagram.

**See Answers at the back of the book.**
**For additional study questions, go to www.prep-u.com.**

ORGANELLE:
FUNCTION:

ORGANELLE:
FUNCTION:

ORGANELLE:
FUNCTION:

ORGANELLE:
FUNCTION:

ORGANELLE:
FUNCTION:

ORGANELLE:
FUNCTION:

ORGANELLE:
FUNCTION:

ORGANELLE:
FUNCTION:

ORGANELLE:
FUNCTION:

## ABOUT THE CHAPTER OPENING PHOTO

Color-enhanced electron micrograph of a human liver cell (hepatocyte). Hepatocytes are responsible for protein synthesis and storage; the metabolism of carbohydrate, protein, and lipid; detoxification; and the initiation of the formation and secretion of bile.

From the sun to you in
just two steps

# Energy

## 4·1–4·4 Energy flows from the sun and through all life on earth

*Sunlight filters through the northern California redwoods.*

## 4·1 Cars that run on french fry oil? Organisms and machines need energy to work.

Imagine that you are on a long road trip. Your car's fuel gauge is nearing empty, so you pull off the highway. Instead of driving into a gas station, however, you head to the back of a fast-food restaurant and fill the fuel tank with used cooking grease. You head back to the highway, ready to drive several hundred miles before needing another pit stop.

Fast food for your car? Yes! The idea isn't as far-fetched as it sounds. In fact, on the roads of America today, many vehicles run on **biofuels,** fuels produced from plant and animal products (**FIGURE 4-1**). Most vehicles, however, run on **fossil fuels** such as gasoline. These fuels (which also include oil, natural gas, and coal) are produced from the decayed remains of plants and animals modified over millions of years by heat, pressure, and bacteria.

It turns out that biofuels, fossil fuels, and the food fuels that supply energy to most living organisms are chemically similar. This fact is not surprising because energy from the

> **Q** Humans can get energy from food. Can machines?

sun is the source of the energy stored in the chemical bonds between the atoms in all these fuels. Let's investigate how fuels provide energy.

When we burn gasoline, long chains of carbon and hydrogen atoms separate, releasing carbon dioxide ($CO_2$), water, and a lot of energy that was stored in the chemical bonds holding each gasoline molecule together. (An interesting fact: the energy released in burning one gallon of gas is equivalent to the caloric content of 15 large cheese pizzas.) In an automobile engine, some of this released energy is harnessed to push pistons, spin a crankshaft, turn wheels, and move the car.

Animal fats and the oils in many plants—such as those used to cook french fries—share an important chemical feature with gasoline. Like gasoline, these fats and oils contain chains of carbon and hydrogen atoms bound together, and just as with gasoline, breaking these bonds releases large amounts of energy (and water and $CO_2$). If this released energy can be

**FIGURE 4-1 Biofuel technology.** Biofuels are chemically similar to fossil fuels.

captured efficiently, it, too, can be used to push pistons and turn car wheels.

Cars that run on biofuels are more than just a technological trick. The production of biofuels requires only the plant or animal source, sunlight, air, water, and a relatively short amount of time—a few months or years, depending on the source. On the other hand, the production of fossil fuels such as coal or crude oil requires plant and animal remains and millions of years. This difference gives biofuels an important advantage over fossil fuels: they are a renewable resource. For this reason, biofuels point the way toward a future of reduced dependence on fossil fuels, the supplies of which are dwindling and the combustion of which has many harmful consequences, such as increasing global warming and releasing cancer-causing particles into the atmosphere. Are we at the point yet where all our cars can run on biofuels? Not quite: there are several significant drawbacks to the increased use of biofuels, chief among them the destruction of forests, wetlands, and other ecologically important habitats from the increased use of land to grow these fuels. This is why the search for better fuels continues.

In this chapter we explore how plant, animal, and other living "machines" run on energy stored in chemical bonds. Just as the forward motion of a car is fueled by breaking the chemical bonds in gasoline and harnessing the released energy, the activities of living organisms are fueled by breaking the chemical bonds in food molecules and harnessing the released energy.

All life depends on capturing energy from the sun and converting it into a form that living organisms can use. This energy capture and conversion occurs in two important

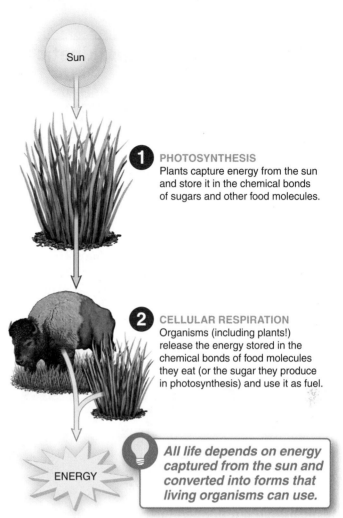

**①  PHOTOSYNTHESIS**
Plants capture energy from the sun and store it in the chemical bonds of sugars and other food molecules.

**②  CELLULAR RESPIRATION**
Organisms (including plants!) release the energy stored in the chemical bonds of food molecules they eat (or the sugar they produce in photosynthesis) and use it as fuel.

ENERGY

*All life depends on energy captured from the sun and converted into forms that living organisms can use.*

**FIGURE 4-2 Photosynthesis and cellular respiration.** The sun to you in just two steps!

processes that mirror each other: (1) **photosynthesis,** the process by which plants capture energy from the sun and store it in the chemical bonds of sugars and other food molecules they make, and (2) **cellular respiration,** the process by which all living organisms release the energy stored in the chemical bonds of food molecules and use it to fuel their lives (**FIGURE 4-2**). The sun to you in just two steps!

## TAKE-HOME MESSAGE 4·1

**The sun is the source of the energy that powers all living organisms and other "machines." The energy from sunlight is stored in the chemical bonds of molecules. When these bonds are broken, energy is released regardless of whether the bond is in a molecule of food, a fossil fuel, or a biofuel such as the oil in which french fries are cooked.**

THE NATURE OF ENERGY    PHOTOSYNTHESIS    CELLULAR RESPIRATION    ALTERNATIVE PATHWAYS

# Energy has two forms: kinetic and potential.

"Batteries not included." For a child, those are pretty depressing words. We know that many of the toys and electronic gadgets that make our lives fun or useful (or both) need energy—usually in the form of batteries. Generating ringtones, lights, and movement requires energy. The same is true for humans, plants, and all other living organisms: they need energy for their activities, from moving to reproducing to thinking.

**Energy** is the capacity to do work. And work is anything that involves moving matter against an opposing force. The energy encountered in the study of living things is of two types: kinetic and potential. **Kinetic energy** is the energy of moving objects. Legs pushing bike pedals and birds flapping wings are examples of kinetic energy (**FIGURE 4-3**). Heat, which results from lots of molecules moving rapidly, is another form of

kinetic energy. Because it comes from the movement of high-energy particles, light is also a form of kinetic energy—probably the most important form of kinetic energy on earth. (When we look at photosynthesis later in this chapter, we explore how sunlight is harnessed for producing food molecules.)

An object does not have to be moving to have the capacity to do work; it may have **potential energy,** which is stored energy or the capacity to do work that results from an object's location or position. Water behind a dam, for example, has potential energy. If a hole in the dam is opened, the water can flow through, and perhaps spin a waterwheel or turbine. A concentration gradient, which we discussed in Chapter 3, also has potential energy: if the molecules in an area of high concentration move toward an area of lower concentration, the potential energy of the gradient is converted to the kinetic

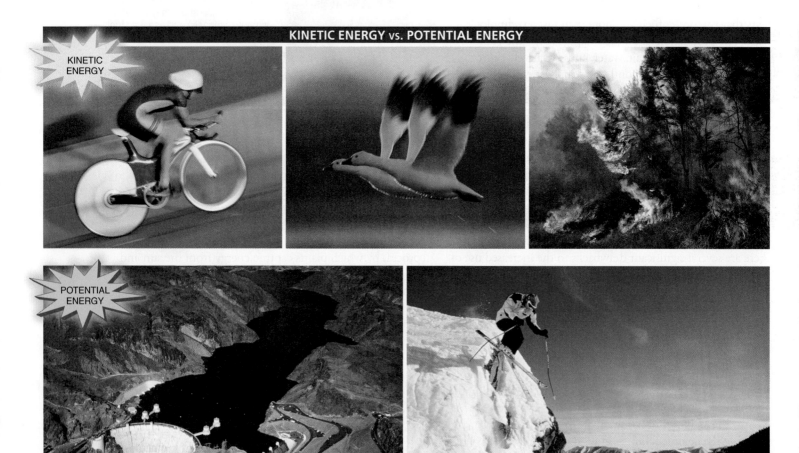

**KINETIC ENERGY vs. POTENTIAL ENERGY**

KINETIC ENERGY

POTENTIAL ENERGY

**FIGURE 4-3 Two forms of energy.** Kinetic energy is the energy of motion; potential energy is energy stored in an object, such as water trapped behind a dam, or a skier poised at the top of a hill.

**CHEMICAL ENERGY**

Glucose
$C_6H_{12}O_6$

Chemical energy is a form of potential energy stored in chemical bonds.

H — CHEMICAL ENERGY — C

*Food is a form of chemical energy!*

**FIGURE 4-4  The energy of chemical bonds.** The chemical bonds in food molecules are a form of potential energy.

energy of molecular movement, and this kinetic energy can do work. **Chemical energy,** the storage of energy in chemical bonds, is also a type of potential energy.

Because potential energy doesn't involve movement, it is a less obvious form of energy than kinetic energy. An apple has potential energy, as does any other type of food (**FIGURE 4-4**). Why? Because the chemical energy stored in the chemical bonds making up the food can be broken and the energy released during cellular respiration, enabling you to run, play, and work. We explore cellular respiration, the energy-releasing breakdown of molecules, later in this chapter. But first we need to know more about the nature of energy.

**TAKE-HOME MESSAGE 4·2**

Energy, the capacity to do work, comes in two forms. Kinetic energy is the energy of moving objects, while potential energy, such as chemical energy, is *stored* energy or the capacity to do work that results from the position or location of an object.

# 4·3 As energy is captured and converted, the amount of energy available to do work decreases.

Every minute of every day—even on cloudy ones—the sun shines brightly, releasing tremendous amounts of energy. Organisms on earth cannot capture every bit of energy released by the sun; indeed, most plants capture only a tiny fraction of the available energy. What happens to the rest? This unused energy does not simply disappear. Accountants would love to monitor the flow of energy because, as in a good accounting ledger, all of the energy numbers add up perfectly. All energy from the sun can be accounted for: some (probably less than 1%) is captured and transformed into usable chemical energy by organisms through photosynthesis (**FIGURE 4-5**). The rest of the energy from the sun is reflected back into space (probably about 30%) or absorbed by land, the oceans, and the atmosphere (about 70%) and mostly transformed into heat. Heat is not easily harnessed to do work, however, and is therefore a much less useful form of energy than the energy transformed into chemical energy in plants (and stored as carbohydrates). The same accounting also exists on a smaller level. If you eat a bowl of rice, some portion of the chemical energy stored in the bonds of the molecules that make up the rice grains is transformed into usable energy that can fuel your

cells' activities. All the rest is transformed into heat and is ultimately lost into the atmosphere.

The fact that energy can change form but never disappear is an important feature of energy in the universe, whether we are looking at the sun and the earth or a human and her rice bowl. Just as energy can never disappear or be destroyed, energy can never be created. All the energy now present in the universe has been here since the universe began, and everything that has happened since then has occurred by the transformation of one form of energy into another. In all our eating and growing, driving and sleeping, we are simply transforming energy. The study of the transformation of energy from one type to another, such as from potential energy to kinetic energy, is called **thermodynamics,** and the **first law of thermodynamics** states that energy can never be created or destroyed. It can only change from one form to another.

Because plants capture less than 1% of the sun's energy, it might seem like they are particularly inefficient. But we humans are also rather inefficient at extracting the chemical

Light energy from the sun → Energy transformed into heat / Chemical energy stored in plants

Chemical energy stored in muscles and liver → Energy transformed into heat / Kinetic energy of forward motion

**FIGURE 4-5 As energy is converted to do work, some energy is released as heat.**

energy of plants when we eat them. These inefficiencies occur because every time energy is converted from one form to another, some of the energy is converted to heat. When a human converts the chemical energy in a plate of spaghetti into the kinetic energy of running a marathon, or when a car transforms the chemical energy of gasoline into the kinetic energy of forward motion, some energy is converted to heat, the least usable form of kinetic energy. In automobiles, for example, about three-quarters of the energy in gasoline is lost as heat (**FIGURE 4-6**). Thus, for every $10 spent on gas, only about $2.50 goes toward moving the car!

**FIGURE 4-6 Inefficient conversion.** Much of the energy used to fuel a combustion engine is converted to heat rather than forward motion.

The **second law of thermodynamics** states that every conversion of energy is not perfectly efficient and invariably includes the transformation of some energy into heat. Although heat is certainly a form of energy, it is almost completely useless to living organisms for fueling their cellular activity because it is not easily harnessed to do work. Put another way, the second law of thermodynamics tells us that although the quantity of energy in the universe is not changing, its quality is. Little by little, the amount of energy that is available to do work decreases. Now that we understand that organisms on earth cannot capture every single bit of energy released by the sun—and that energy conversions are inefficient—we can look at the chief energy currency of the cell: ATP.

### TAKE-HOME MESSAGE 4·3

Energy is neither created nor destroyed but can change forms. Each conversion of energy is inefficient, and some of the usable energy is converted to less useful heat energy.

# 4·4 ATP molecules are like free-floating rechargeable batteries in all living cells.

Much of the work that cells do requires energy. But even though light energy from the sun carries energy, as do molecules of sugar, fat, and protein, none of this energy can be used directly to fuel chemical reactions in organisms' cells. First it must be captured in the bonds of a molecule called **adenosine triphosphate (ATP),** a free-floating molecule found in cells that acts like a rechargeable battery which temporarily stores energy that can then be used for cellular work in plants, animals, bacteria, and all the other organisms on earth. The use of ATP solves an important timing and coordination problem for living cells: a supply of ATP guarantees that the energy required for energy-consuming reactions will be available when it's needed.

ATP is a simple molecule with three components (**FIGURE 4-7**). At the center of the ATP molecule are two of these components: a small sugar molecule attached to a molecule called adenine. But it is the third component that makes ATP so effective in carrying and storing energy for a short time: attached to the sugar and adenine is a chain of

**ADENOSINE TRIPHOSPHATE (ATP) AND ADENOSINE DIPHOSPHATE (ADP)**

Adenine

Phosphate groups

High-energy bonds

Ribose (sugar)

Separate phosphate group

$P_i$

ENERGY

Adenine

Phosphate groups

Ribose (sugar)

ATP — Symbol for ATP used in this book

The green halo represents ATP's potential energy.

ADP — Symbol for ADP used in this book

**FIGURE 4-7 The structure of ATP and ADP.** When ATP ejects one of its phosphate groups, energy is released as ATP becomes ADP.

An input of energy from the breakdown of food attaches ADP to P$_i$.

Energy is released as a phosphate group is ejected from ATP.

**ATP can be used and recycled hundreds of thousands of times!**

**FIGURE 4-8 ATP is like a rechargeable battery.**

three negatively charged phosphate groups (hence the "tri" in "triphosphate"). Because the bonds between these three phosphate groups must hold the groups together in the face of the three electrical charges that all repel one another, each of these bonds contains a large amount of energy and is stressed and unstable. The instability of these high-energy bonds makes the three phosphate groups like a tightly coiled spring or a twig that is bent almost to the point of breaking. With the slightest push, one of the phosphate groups will pop off, displaced by water. And in the process, a little burst of energy is released that the cell can use.

It is precisely because each molecule of ATP is always on the brink of ejecting one of its phosphate groups that ATP is such an effective energy source inside a cell. As long as plenty of ATP molecules are around, they can energize the chemical reactions that make it possible for the cell to carry out reactions that require work, such as building muscle, repairing wounds, or growing roots. Each time a cell expends one of its ATP molecules to pay for an energetically expensive reaction, a phosphate is broken off and energy is

released. What is left is a molecule with two phosphates, called ADP (adenosine *di*phosphate), and a separate phosphate group (labeled P$_i$).

An organism can then use ADP, a free-floating phosphate, and an input of kinetic energy to rebuild its ATP stocks (**FIGURE 4-8**). The kinetic energy is converted to potential energy when the free phosphate group attaches to the ADP molecule and makes ATP. In this manner, ATP functions like a rechargeable battery. Where does the input of energy for recharging ATP come from? When we discuss photosynthesis, we'll see that plants, algae, and some bacteria directly use light energy from the sun to make ATP from ADP and free-floating phosphate groups. Animals use the energy contained in the bonds of their food molecules. In either case, the energy is used to re-create the unstable bond in the triphosphate chain. When energy is needed, the organism can again release it by breaking the bond holding the phosphate group to the rest of the molecule. Our bodies recycle ATP molecules in this way tens of thousands of times a day.

Here's the ATP story in a nutshell. Breaking down a molecule of sugar—in a glass of orange juice, for example—leads to a miniature burst of energy in your body. The energy from the mini-explosion is put to work building the unstable high-energy bonds that attach phosphate groups to ADP molecules, creating new molecules of ATP. Later—perhaps only a fraction of a second later—when an energy-consuming reaction is needed, your cells can release the energy stored in the new ATP molecules.

---

### TAKE-HOME MESSAGE 4·4

**Cells temporarily store energy in the bonds of ATP molecules. This potential energy can be converted to kinetic energy and used to fuel life-sustaining chemical reactions. At other times, inputs of kinetic energy are converted to the potential energy of the energy-rich but unstable bonds in the ATP molecule.**

## 4·5–4·11
## Photosynthesis uses energy from sunlight to make food.

*All plants are able to capture light energy and convert it to organic matter. Shown here: King protea (Protea cynaroides), the national flower of South Africa.*

# 4·5  Where does plant matter come from? Photosynthesis: the big picture.

Watching a plant grow over the course of a few years can seem like watching a miracle, or at least a very subtle magic trick. Of course it's neither, but the process is nonetheless amazing. Consider that in five years a tree can increase its weight by 150 pounds (68 kg) (**FIGURE 4-9**). Where does that 150 pounds of new tree come from?

**Q** When humans grow, the new tissue comes from food we eat. When plants grow, where does the new tissue come from?

Our first guess might be the soil. Could that be it? It's easy enough to weigh the soil in a pot when first planting a tree, and weigh it again 5 years later. After 5 years, though, we find that the soil in our planter has lost less than a pound, nowhere near enough to explain the massive increase in the amount of plant material. Perhaps the new growth comes from the water? Wrong again. Although the older and much larger tree holds more water in its many cells, the water provided to the plant does not come close to accounting for the increase in the dry weight of the plant.

The amazing truth is that most of the new material comes from an invisible gas in the air. In the process of photosynthesis, plants capture carbon dioxide gas ($CO_2$) from the air. And using energy they get from sunlight, along with

water and small amounts of chemicals usually found in soil, they produce solid, visible (and often tasty) sugars and other organic molecules that are used to make plant structures such as leaves, roots, stems, flowers, fruits, and seeds. In the process, the plants give off oxygen ($O_2$), a by-product that happens to

5 LB.  —— 5 years ——→  155 LB.

**FIGURE 4-9  When plants grow, where does the new tissue come from?** From the dirt? From thin air?

THE NATURE OF ENERGY    **PHOTOSYNTHESIS**    CELLULAR RESPIRATION    ALTERNATIVE PATHWAYS

**137**

Euglena                  Cyanobacterium              Dinoflagellates              Kelp

**FIGURE 4-10 Plants aren't the only photosynthesizers.** Some bacteria and other unicellular organisms, along with kelp and other multicellular algae, are capable of photosynthesis.

**FIGURE 4-11 Photosynthesis: the big picture.**

be necessary for much of the life on earth—including all animal life!

Although plants are generally the most visible organisms that can capture light energy and convert it to organic matter, they are not the only organisms capable of photosynthesis. Some bacteria and many other unicellular organisms, along with kelp and other multicellular algae, are also capable of using the energy in sunlight to produce organic materials (**FIGURE 4-10**).

There are three inputs to the process of photosynthesis (**FIGURE 4-11**): light energy (from the sun), carbon dioxide (from the atmosphere), and water (from the ground). From these three inputs, the plant produces sugar and oxygen. As we'll see, photosynthesis is best understood as two separate events: a "photo" segment, during which light is captured, and a "synthesis" segment, during which sugar is built. In the "photo" reactions, light energy is captured and temporarily saved in energy-storage molecules. During this process, water molecules split and produce oxygen. In the "synthesis" reactions, the energy in the energy-storage molecules is used to assemble sugar molecules from carbon dioxide from the air.

## TAKE-HOME MESSAGE 4·5

Through photosynthesis, plants use water, the energy of sunlight, and carbon dioxide gas from the air to produce sugars and other organic materials. In the process, photosynthesizing organisms also produce oxygen, which makes all animal life possible.

# 4·6 Photosynthesis takes place in the chloroplasts.

If a plant part is green, then you know it is photosynthetic. Leaves are green because the cells near the surface are packed full of **chloroplasts,** light-harvesting organelles, which make it possible for the plant to use the energy from sunlight to make sugars (their food) and other plant tissue (much of which animals use for food) (**FIGURE 4-12**). Other plant parts, such as stems, may also contain chloroplasts (in which case they, too, are capable of photosynthesis), but most chloroplasts are located within the cells in a plant's leaves.

Let's take a closer look at chloroplasts (**FIGURE 4-13**). The sac-shaped organelle is filled with a fluid called the **stroma.** Floating in the stroma is an elaborate system of interconnected membranous structures called **thylakoids,** which often look like stacks of pancakes. Once inside the chloroplast, you can be in one of two places: in the stroma or inside the thylakoids. The conversion of light energy to chemical energy—the "photo" part of photosynthesis—occurs inside the thylakoids. The production of sugars—which are made in the "synthesis" part of photosynthesis—occurs within the stroma.

**CHLOROPLAST**

**THYLAKOID**
Location of "photo" reactions, where light energy is converted into chemical energy.

**STROMA**
Location of "synthesis" reactions, where chemical energy from the "photo" reactions is used to synthesize sugars.

Chlorophyll

**FIGURE 4-13 Chloroplast structure.** The chloroplast is where photosynthesis takes place in a plant.

We examine both the "photo" and the "synthesis" processes in greater detail later in this chapter. First, however, we examine the nature of light energy and of **chlorophyll,** the special molecule found in chloroplasts that makes the capture of light energy possible.

## TAKE-HOME MESSAGE 4·6

In plants, photosynthesis occurs in chloroplasts, green organelles packed in cells near the plants' surfaces.

Top edge of leaf

Photosynthetic cells packed with chloroplasts

Bottom edge of leaf

**FIGURE 4-12 Photosynthesis factories.** Cells near the leaf's surface are packed with photosynthetic chloroplasts.

# 4·7 Light energy travels in waves: plant pigments absorb specific wavelengths.

You can't eat sunlight. That's because sunlight is *light* energy rather than the chemical energy found in the bonds of food molecules. Photosynthesis is powered by **light energy,** a type of kinetic energy made up of little energy packets called **photons,** which are organized into waves. Photons can do work as they bombard surfaces such as your face (heating it) or a leaf (enabling it to build sugar from carbon dioxide and water).

Photons have various amounts of energy, and the length of the wave in which they travel corresponds to the amount of energy carried by the photon. The shorter the wavelength, the more energy the light carries. Within a ray of light, there are super-high-energy packets of photons (those with short wavelengths), relatively low-energy packets (those with longer wavelengths), and everything in between. This range, which is called the **electromagnetic spectrum,** extends from extremely short high-energy gamma rays and X rays, with wavelengths as short as 1 nanometer (a human hair is about 50,000 nanometers in diameter), to very long low-energy radio waves, with wavelengths as long as a mile (**FIGURE 4-14**).

Just as we can't hear some super-high-pitched frequencies of sound (even though many dogs can), there are some wavelengths of light that are too short or too long for us to see. The light that we can see, visible light, spans all the colors of the rainbow. Humans (and some other animals) can see colors because our eyes contain light-absorbing molecules called **pigments.** These pigments absorb wavelengths of light within the visible range. The energy in these light waves excites electrons in the pigments, stimulating nerves in our eyes, which then transmit electrical signals to our brains. We perceive different wavelengths within the visible spectrum as different colors. The pigments in the human eye absorb many different wavelengths pretty well: that's why we can see so many colors. When plants use sunlight's energy to make sugar during photosynthesis, they also use the visible portion of the electromagnetic spectrum. Unlike the pigments in our eyes, however, plant pigments (the energy-capturing parts of a plant) absorb and use only a portion of visible light wavelengths.

Chlorophyll is the pigment molecule in plants that absorbs light energy from the sun. Chlorophyll molecules are embedded in the thylakoid membrane of chloroplasts, which are found primarily in plants' leaves. Just as light energy excites electrons in the pigments responsible for color vision in humans, electrons in a plant's chlorophyll can become excited by certain wavelengths of light and can capture a bit of this light energy.

Plants produce several different light-absorbing pigments (**FIGURE 4-15**). The primary photosynthetic pigment, called **chlorophyll *a,*** absorbs red and blue-violet wavelengths of light. Every other wavelength generally travels through or bounces off this pigment. Because chlorophyll *a* cannot efficiently absorb green light and instead reflects those wavelengths, we perceive the reflected light waves as green, and so the pigment and the leaves that contain it appear green. Another pigment, **chlorophyll *b,*** is similar in structure but

**ELECTROMAGNETIC SPECTRUM**

Sunlight

Radio waves | Infrared | UV light | X rays | Gamma rays

1,000 m
Longer wavelength
Lower energy

1 nm
Shorter wavelength
Higher energy

Visible light

740 nm                    400 nm

**FIGURE 4-14 A spectrum of energy.** A ray of light emits high-energy photons, low-energy photons, and everything in between. Plants use only a fraction of the light's available energy.

Plants produce several different light-absorbing pigments. Each photosynthetic pigment absorbs and reflects specific wavelengths.

Seasonal differences in the amount of pigment molecules present in leaves lead to the leaves changing color.

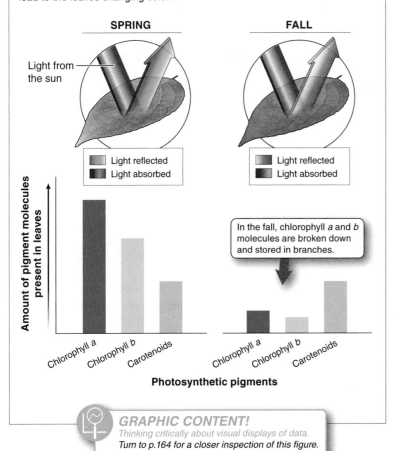

In the fall, chlorophyll *a* and *b* molecules are broken down and stored in branches.

**Photosynthetic pigments**

**GRAPHIC CONTENT!**
*Thinking critically about visual displays of data.*
*Turn to p.164 for a closer inspection of this figure.*

**FIGURE 4-15 Plant pigments.** Each photosynthetic pigment absorbs and reflects specific wavelengths.

absorbs blue and red-orange wavelengths. Chlorophyll *b* reflects yellow-green wavelengths. Some related pigments called **carotenoids** absorb blue-violet and blue-green wavelengths and reflect yellow, orange, and red wavelengths.

In the late summer, cooler temperatures cause some trees to prepare for the winter by shutting down chlorophyll production and reducing photosynthesis rates, much like an animal's hibernation. Gradually, the chlorophyll *a* and *b* molecules in the leaves are broken down and their chemical components are stored in the branches. As the amounts of chlorophyll *a* and *b* in the leaves decrease relative to the remaining carotenoids, the striking colors of the fall foliage are revealed (see Figure 4-15). During the rest of the year, chlorophylls *a* and *b* are so abundant in leaves that green masks the colors of the other pigments.

Why do the leaves of some trees turn beautiful colors each fall?

**TAKE-HOME MESSAGE 4·7**

Photosynthesis is powered by light energy, a type of kinetic energy made of energy packets called photons. Photons hit chlorophyll and other light-absorbing molecules near the green surfaces of plants. These molecules capture some of the light energy and harness it to build sugar from carbon dioxide and water.

# 4·8 Photons cause electrons in chlorophyll to enter an excited state.

An organism can use energy from the sun only if it can convert the light energy of the sun into the chemical energy in the bonds between atoms. The most important molecule in this conversion is the pigment chlorophyll (**FIGURE 4-16**).

## ENERGY MOVEMENT THROUGH CHLOROPHYLL

**1** Light energy bumps an electron in the chlorophyll molecule to a higher, excited energy level.

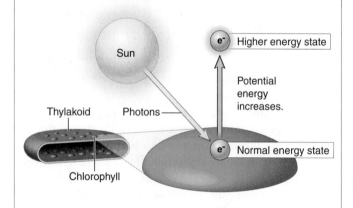

**2** The excited electron generally has one of two different fates:

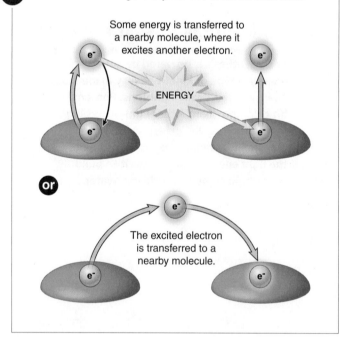

**FIGURE 4-16 Capturing light energy as excited electrons.** Chlorophyll electrons are excited to a higher energy state by light energy.

When chlorophyll is hit by photons of certain wavelengths, the light energy bumps an electron ($e^-$) in the chlorophyll molecule to a higher energy level, an *excited* state. Upon absorbing the photon, the electron briefly gains energy, and the potential energy in the chlorophyll molecule increases.

An electron in a photosynthetic pigment that is excited to a higher energy state generally has one of two fates. (1) The electron returns to its resting, unexcited state, releasing energy in the process, some of which may bump electrons in a nearby molecule to a higher energy state (while the rest of the energy is dissipated as heat). Or (2) the excited electron itself is passed to another molecule.

The passing of electrons from molecule to molecule is one of the chief ways that energy moves through cells. Many molecules carry or accept electrons during cellular activities. All that is required is that the acceptor must have a greater attraction for electrons than does the molecule from which it receives them. This receiver molecule, in turn, hands off electrons to another acceptor with an even greater attraction for them. A molecule that gains electrons always carries greater energy than it did before receiving the electron(s). For this reason, the passing of electrons from one molecule to another can be viewed as a passing of potential energy. In this way, energy moves through cells.

This transfer of electrons is one of the first steps of photosynthesis, the process that enables a plant to harness light energy from the sun and convert it to the more readily usable chemical energy. As we see later in this chapter, the dismantling of food molecules such as glucose to generate energy is also a story of breaking and rearranging chemical bonds as electrons pass from one atom or molecule to another.

Because particles in the atmosphere can block light from the sun and reduce the excitation of electrons in chlorophyll molecules, photosynthesis depends on a relatively clean atmosphere. Any reduction in the available sunlight can have serious effects on plants. Scientists believe that if a large meteor hit the earth—as one did when the dinosaurs were

**Q** *Suppose a large meteor hit the earth. How could smoke and soot in the atmosphere wipe out life far beyond the area of direct impact?*

wiped out 65 million years ago—smoke, soot, and dust in the atmosphere could block sunlight to such an extent that plants in the region, or possibly all of the plants on earth, could not conduct photosynthesis at high enough levels to survive. And when plants die off, all of the animals and other species that rely on them for energy die as well. All life on earth is completely dependent on the continued excitation of electrons by sunlight.

# 4·9 Photosynthesis in detail: the energy of sunlight is captured as chemical energy.

Photosynthesis is a complex process, but our understanding can be greatly aided by remembering one phrase: FOLLOW THE ELECTRONS. In the passages that follow, if you feel that you are losing focus or getting lost, just remember to follow the electrons. Where are they coming from? What are they passing through? Where are they going? And what will happen to them when they get there?

In the first part of photosynthesis, the "photo" part, sunlight hits a plant and, in a three-step process, some of the energy in this sunlight is captured and stored in ATP and in another molecule (NADPH) that stores energy by accepting high-energy electrons (**FIGURE 4-17**).

The energy-capturing process occurs in two **photosystems.** These photosystems are arrangements of light-catching pigments (including chlorophyll) within chloroplasts that capture energy from sunlight and transform it—first into the energy of excited electrons and ultimately into ATP and high-energy electron carriers. After these transformations, the captured energy is ready to be used to make sugar molecules in the "synthesis" part of photosynthesis. We begin following electrons as sunlight hits a leaf and the chloroplasts in it (**FIGURE 4-18**).

The chloroplasts in a leaf convert light energy from the sun into chemical energy. This conversion begins in a collection of chlorophyll molecules and other light-absorbing pigments arranged within a chloroplast's thylakoid membrane. As these pigments absorb photons from the sunlight that hits the leaves, electrons in the pigments become excited and then return to their resting state. As the electrons return to their resting state, energy (but not the electrons) is transferred to neighboring pigment molecules. This process continues until the transferred energy from many pigment molecules excites the electrons in a

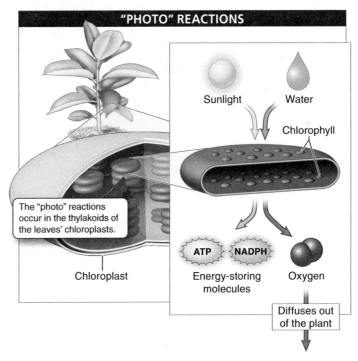

**FIGURE 4-17 Overview of the "photo" reactions.** Light energy is captured in the "photo" portion of photosynthesis. That energy is later used to power the building of sugar molecules.

chlorophyll *a* molecule at the center of the photosystem (see Figure 4-18). This is where the electron journey begins.

Each energy burst boosts chlorophyll's electrons to an excited state, but the chlorophyll *a* molecule at the center of the photosystem is special, differing from the other pigment molecules in one key feature. When its electrons are boosted to an excited state they do not return to their resting, unexcited state. Instead, the special chlorophyll *a* continually

**1** Light energy excites electrons in a pigment molecule (such as a chlorophyll molecule). The energy from the excited electrons is transferred to nearby pigment molecules.

**2** When transferred energy excites electrons in the chlorophyll *a* molecule, the primary electron acceptor grabs them and sends them to the electron transport chain.

**3** To replace electrons sent to the electron transport chain, water molecules are split and oxygen and hydrogen are released as by-products.

Chloroplast    Thylakoid

AREA OF DETAIL

Sun

e⁻

Primary electron acceptor

e⁻

To electron transport chain

Pigment molecule

Chlorophyll *a* molecule

Water

Oxygen released into the atmosphere

H⁺ H⁺ H⁺ H⁺

*The oxygen released in the photo reactions happens to be necessary for much of the life on earth—including all animal life!*

**FIGURE 4-18 The water-splitting photosystem.** Splitting water provides electrons for photosynthesis.

loses its excited electrons to a nearby molecule, called the **primary electron acceptor,** which acts like an electron vacuum.

As electrons keep getting taken away from the special chlorophyll *a* molecule, the electrons must be replaced. The replacement electrons come from water. As long as photosynthesis is occurring, a constant supply of replacement electrons is required. Molecules of water inside the thylakoid, near the special chlorophyll *a* molecule in the thylakoid membrane, are continuously split. This split—in which four photons of light split two molecules of water into four electrons, four protons (H⁺), and a molecule of $O_2$—provides the electrons necessary to replenish chlorophyll *a*'s electron supply. A convenient and life-sustaining by-product of the splitting of water in photosynthesis is the oxygen that is released from the cell, a by-product essential for much of the life on earth, including all animal life.

Once the primary electron acceptor gets hold of the high-energy electrons from chlorophyll *a,* it passes them along like hot potatoes to another molecule, which passes them to another, which passes them to yet another, in what is called an **electron transport chain** (**FIGURE 4-19**). At each step in the electron transport chain's sequence of electron handoffs, the electrons fall to a lower energy state and a little bit of energy is released. These bits of energy are harnessed to power

**Q**

Why must plants get water for photosynthesis to occur?

pumps in the thylakoid membrane that move protons (which are also referred to as hydrogen ions or H⁺ ions) from the stroma to the inside of the thylakoid. The pumps pack the protons inside the thylakoid sac at higher and higher concentrations. Think of a pump pushing water into an elevated tank, creating a store of potential energy that can gush out of the tank with great force and kinetic energy. Similarly, the protons eventually rush out of the thylakoid sacs with great force—and the force of the protons moving down their concentration gradient is harnessed to build energy-storing ATP molecules, one of the two products of the "photo" portion of photosynthesis.

Recall that the energy-capturing and energy-transforming processes of the "photo" reactions occur in two photosystems (arrangements of chlorophyll and other light-catching molecules). The electron transport chain physically links the first photosystem to the second. As the traveling electrons continue their journey, they fill electron vacancies in the reaction center of the second photosystem, right next to the first photosystem (**FIGURE 4-20**). Like the first photosystem, the second photosystem also has numerous pigments that harness photons from the sun and pass the light energy to another special chlorophyll *a* molecule. The special chlorophyll *a* molecule at the center of this second photosystem has electron vacancies because, as in the first photosystem, when electrons in the special chlorophyll *a* molecule are boosted to an excited

## THE PHOTOSYNTHETIC ELECTRON TRANSPORT CHAIN

**1** Electrons move through the electron transport chain, releasing a little energy and falling to a lower energy state.

**2** The released energy powers proton pumps that move hydrogen ions from the stroma and pack them inside the thylakoid.

**3** Protons rush out of the thylakoid with great kinetic energy, which can be used to build ATP.

Electron passed from the primary electron acceptor

Stroma

Proton pumps

Thylakoid

To NADPH-producing photosystem

Chloroplast

Thylakoid

AREA OF DETAIL

**FIGURE 4-19 Harnessing the potential of high-energy electrons.** As electrons are passed from the primary electron acceptor to a chain of molecules embedded within the thylakoid membrane, called the electron transport chain, the released energy is used to create a proton gradient.

state, they are whisked away from the chlorophyll molecule by another primary electron acceptor. This electron acceptor then passes the electrons to a second electron transport chain.

At the end of this second electron transport chain, the electrons are passed to a molecule called NADP$^+$, creating **NADPH**, a high-energy electron carrier. NADPH is the

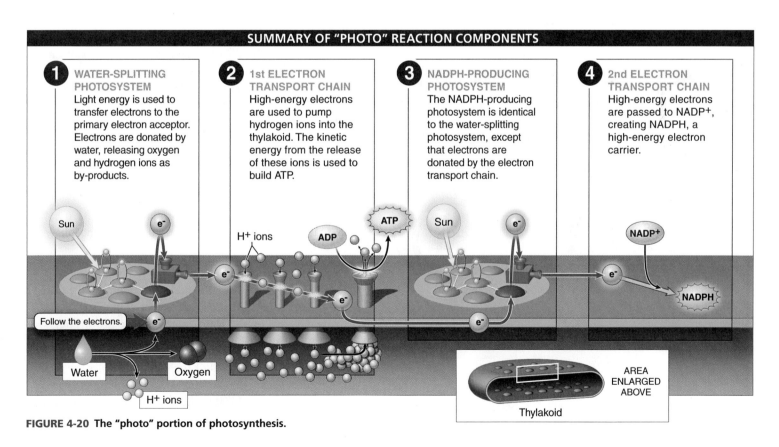

## SUMMARY OF "PHOTO" REACTION COMPONENTS

**1** WATER-SPLITTING PHOTOSYSTEM
Light energy is used to transfer electrons to the primary electron acceptor. Electrons are donated by water, releasing oxygen and hydrogen ions as by-products.

**2** 1st ELECTRON TRANSPORT CHAIN
High-energy electrons are used to pump hydrogen ions into the thylakoid. The kinetic energy from the release of these ions is used to build ATP.

**3** NADPH-PRODUCING PHOTOSYSTEM
The NADPH-producing photosystem is identical to the water-splitting photosystem, except that electrons are donated by the electron transport chain.

**4** 2nd ELECTRON TRANSPORT CHAIN
High-energy electrons are passed to NADP$^+$, creating NADPH, a high-energy electron carrier.

Sun

Follow the electrons.

Water · Oxygen · H$^+$ ions

H$^+$ ions · ADP · ATP

Sun

NADP$^+$ · NADPH

Thylakoid · AREA ENLARGED ABOVE

**FIGURE 4-20 The "photo" portion of photosynthesis.**

second important product of the "photo" portion of photosynthesis.

With the electrons' passage through the second photosystem and arrival in NADPH, we now have the final products of the "photo" part of photosynthesis (which are also called the light-dependent reactions): we've captured light energy from the sun and converted it to the chemical energy of ATP and the high-energy electron carrier NADPH (see Figure 4-20). But we haven't made any food yet. In the next section, we cover the "synthesis" part of photosynthesis and see how plants use the energy in ATP and NADPH to produce sugar from carbon dioxide.

### TAKE-HOME MESSAGE 4·9

There are two parts to photosynthesis. The first is the "photo" part, in which light energy is transformed into chemical energy, while splitting water molecules and producing oxygen. Sunlight's energy is first captured when an electron in chlorophyll is excited. As this electron is passed from one molecule to another, energy is released at each transfer, some of which is used to build the energy-storage molecules ATP and NADPH.

# 4·10 Photosynthesis in detail: the captured energy of sunlight is used to make food.

The "synthesis" part of photosynthesis takes place in a series of chemical reactions called the **Calvin cycle.** All the Calvin cycle reactions occur in the stroma of the leaves' chloroplasts, outside the thylakoids. Plants carry out these reactions using the energy stored in the ATP and NADPH molecules that are built in the "photo" portion of photosynthesis. This dependency links the light-gathering ("photo") reactions with the sugar-building ("synthesis") reactions (**FIGURE 4-21**).

If there is any part of photosynthesis that appears magical, it is the Calvin cycle. Just as a magician seems to make a rabbit appear from thin air, the Calvin cycle takes invisible molecules of $CO_2$ from the air and uses them to assemble visible—even edible—molecules of sugar. The processes in the Calvin cycle occur in three steps (**FIGURE 4-22**).

**1. Fixation.** First, using an enzyme called **rubisco,** plants pluck carbon from the air, where it occurs in the form of carbon dioxide, and then attach, or "fix," it to a visible organic (carbon-containing) molecule within the chloroplast. Not surprisingly, given its role as the critical chemical that enables plants to build food molecules, rubisco is the most abundant protein on earth.

**2. Sugar creation.** The newly built molecule is chemically modified: a phosphate from ATP is added, and the molecule receives some high-energy electrons from NADPH. This product of the Calvin cycle is a small sugar called glyceraldehyde 3-phosphate (G3P).

Some of the G3P molecules are combined to make the six-carbon sugars glucose and fructose. These sugars can be used as fuel by the plant, enabling it to grow. They can also be used as fuel by animals that eat the plant.

**3. Regeneration.** Not all of the G3P molecules are used to produce sugars. In the third and final phase of the Calvin cycle, some G3P molecules are used to regenerate

"SYNTHESIS" REACTIONS

The "synthesis" reactions occur in the stroma of the leaves' chloroplasts.

Chloroplast

ATP    NADPH

Energy-storing molecules

Carbon dioxide

CALVIN CYCLE

Sugar

**FIGURE 4-21 Overview of the "synthesis" reactions of photosynthesis.**

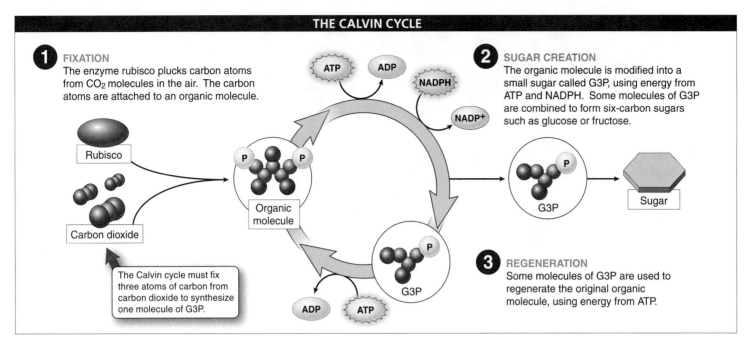

**THE CALVIN CYCLE**

**1 FIXATION**
The enzyme rubisco plucks carbon atoms from $CO_2$ molecules in the air. The carbon atoms are attached to an organic molecule.

Rubisco

Carbon dioxide

ATP    ADP

NADPH

NADP+

Organic molecule

The Calvin cycle must fix three atoms of carbon from carbon dioxide to synthesize one molecule of G3P.

**2 SUGAR CREATION**
The organic molecule is modified into a small sugar called G3P, using energy from ATP and NADPH. Some molecules of G3P are combined to form six-carbon sugars such as glucose or fructose.

G3P

Sugar

**3 REGENERATION**
Some molecules of G3P are used to regenerate the original organic molecule, using energy from ATP.

G3P

ADP    ATP

**FIGURE 4-22 Sugar synthesis.** In the Calvin cycle—the "synthesis" portion of photosynthesis—sugars are produced.

the original molecule in the chloroplast to which the carbon from $CO_2$ is attached. This regeneration process requires energy from ATP produced in the "photo" reactions of photosynthesis. With this regeneration, the Calvin cycle can continue to fix carbon and produce molecules of G3P.

Ultimately, to synthesize one molecule of G3P, the Calvin cycle must fix three atoms of carbon from carbon dioxide to the initial organic molecule; this process consumes nine molecules of ATP and six molecules of NADPH generated in the "photo" reactions of photosynthesis.

**TAKE-HOME MESSAGE 4·10**

The second part, or "synthesis" part, of photosynthesis is the Calvin cycle, which occurs in the stroma of chloroplasts. During this phase, carbon from $CO_2$ in the atmosphere is attached (fixed) to molecules in the chloroplasts, sugars are built, and molecules are regenerated to be used again in the Calvin cycle. The fixation, building, and regeneration processes consume energy from ATP and NADPH (the products of the "photo" part of photosynthesis).

# 4·11 The battle against world hunger can use plants adapted to water scarcity.

Sudan. Ethiopia. India. Somalia. Many of the world's regions with the highest rates of starvation also are places with the hottest, driest climates. This is not a coincidence. These climate conditions present difficult challenges for sustaining agriculture (**FIGURE 4-23**), and in the absence of stable crop yields, food production is unpredictable and the risk of starvation high. But evolutionary adaptations in some plants enable them to thrive in hot, dry conditions. Recent technological advances in agriculture use these innovative

evolutionary solutions to battle the problem of world hunger. In this section, we discuss some adaptations that allow plants to thrive when water is scarce. We also look at how humans use these adaptations to grow food in the dry, inhospitable climates where starvation rates are highest.

When it gets too hot and dry, animals can seek coolness in the shade. Plants, however, are anchored in place and do not have this option. Consequently, plants in hot, dry climates can lose significant amounts of water through evaporation. Evaporation is a problem for plants because water is essential to photosynthesis, growth, and the transport of nutrients. Without water, plants cannot live long.

One method of combating water loss through evaporation is for plants to close their **stomata** (*sing.* **stoma**), small pores usually on the underside of leaves (**FIGURE 4-24**). These openings are the primary sites for gas exchange in plants: carbon dioxide for photosynthesis enters through these openings and oxygen generated as a by-product in photosynthesis exits through them. When open, the stomata also allow water to evaporate from the plant. Closing their

stomata solves one problem for plants (too much water evaporation) but creates another: with the stomata shut, oxygen from the "photo" reactions of photosynthesis cannot be released from the chloroplasts, and carbon dioxide cannot enter. If there are no carbon dioxide molecules for sugar production, the Calvin cycle tries to fix carbon but instead finds only oxygen. Plant growth comes to a standstill and crops fail.

In some plants, including corn and sugarcane, a process has evolved that minimizes water loss but still enables the plants to make sugar when the weather is hot and dry. In the process called **C4 photosynthesis,** these plants add an extra set of

Stoma open

Stoma closed

**FIGURE 4-24 Plant stomata.** Carbon dioxide enters a plant through stomata, but water can be lost through the same openings.

## C3 PHOTOSYNTHESIS

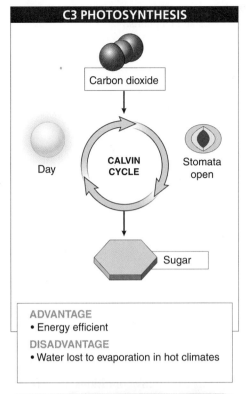

Carbon dioxide

Day

**CALVIN CYCLE**

Stomata open

Sugar

**ADVANTAGE**
• Energy efficient

**DISADVANTAGE**
• Water lost to evaporation in hot climates

Common ivy

## C4 PHOTOSYNTHESIS

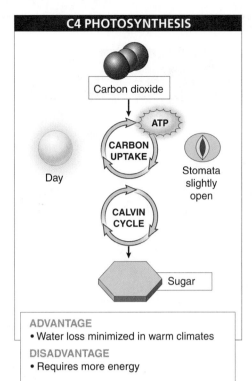

Carbon dioxide

ATP

**CARBON UPTAKE**

Day

Stomata slightly open

**CALVIN CYCLE**

Sugar

**ADVANTAGE**
• Water loss minimized in warm climates

**DISADVANTAGE**
• Requires more energy

Corn

## CAM PHOTOSYNTHESIS

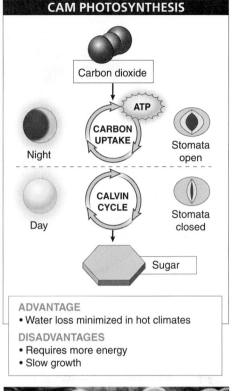

Carbon dioxide

ATP

**CARBON UPTAKE**

Night

Stomata open

Day

**CALVIN CYCLE**

Stomata closed

Sugar

**ADVANTAGE**
• Water loss minimized in hot climates

**DISADVANTAGES**
• Requires more energy
• Slow growth

Golden Barrel Cactus

**FIGURE 4-25 C3, C4, and CAM photosynthesis.**

steps to the usual process of photosynthesis (**FIGURE 4-25**). In these steps, the plants produce an enzyme that functions like the ultimate "$CO_2$-sticky tape." This enzyme has a tremendously strong attraction for carbon dioxide; it can find and bind carbon even when $CO_2$ concentration is very low. (In contrast, rubisco, the usual enzyme that plants use to pluck carbon from the atmosphere, functions poorly when $CO_2$ is scarce.) As a consequence, stomata can be opened just a tiny bit, and let in just a little $CO_2$. Having the stomata closed also reduces evaporation and conserves water for the plant.

**Q** How might global warming be bad for agriculture?

This seems like such a good solution that we would expect all plants to use it. There is a catch, though. The extra steps in C4 photosynthesis require the plant to expend additional energy. Specifically, every time the plant generates a molecule of the "$CO_2$-sticky tape" enzyme, it uses one molecule of ATP. It is acceptable to pay this energy cost only when the climate is so hot and dry that the plant would otherwise have to close its stomata and completely shut down all sugar production. If the climate is mild, however, plants conducting the more energetically expensive C4 photosynthesis would be out-competed by the

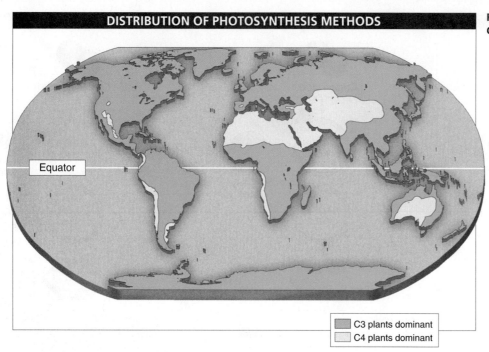

FIGURE 4-26 Global distribution of C3 and C4 plants.

DISTRIBUTION OF PHOTOSYNTHESIS METHODS

Equator

C3 plants dominant
C4 plants dominant

more efficient plants conducting standard photosynthesis (called C3 photosynthesis; see Figure 4-25). Not surprisingly, we see few C4 plants in the temperate regions of the world and little C4 photosynthesis among photosynthetic organisms living in the oceans. In hot, dry regions, however, C4 plants are dominant and displace the C3 plants wherever both occur (**FIGURE 4-26**). With global warming, many scientists expect to see a gradual expansion of the ranges over which C4 plants occur and believe that non-C4 plants will be pushed farther and farther away from the equator.

A third and similar method of carbon fixation, called **CAM** (for "crassulacean acid metabolism"), is also found in hot, dry areas. In this method, used by many cacti, pineapples, and other fleshy, juicy plants, the plants close their stomata during hot, dry days. At night, they open the stomata and let $CO_2$ into the leaves, where it binds temporarily to a holding molecule. During the day, when a carbon source is needed to make sugars in the Calvin cycle, the $CO_2$ is gradually released from the holding molecule, enabling photosynthesis to proceed while keeping the stomata closed to reduce water loss (see Figure 4-25). A disadvantage of CAM photosynthesis is that by completely closing their stomata during the day, CAM plants significantly reduce the total amount of $CO_2$ they can take in. As a consequence, they have much slower growth rates and cannot compete

well with non-CAM plants under any conditions other than extreme dryness.

C4 and CAM photosynthesis originally evolved because they made it possible for plants to grow better in the world's hot and dry regions. Researchers are now experimenting with these adaptations as a way to fight world hunger. They have introduced into rice several genes from corn that code for the C4 photosynthesis enzymes. Once in the rice, these genes increase the rice plant's ability to photosynthesize, leading to higher growth rates and food yields. Whether the addition of C4 photosynthesis enzymes will make it possible to grow new crops on a large scale in previously inhospitable environments is not certain. Early results suggest, however, that this is a promising approach.

## TAKE-HOME MESSAGE 4·11

C4 and CAM photosynthesis are evolutionary adaptations at the biochemical level that, although more energetically expensive than regular (C3) photosynthesis, allow plants in hot, dry climates to close their stomata and conserve water without shutting down photosynthesis.

# 4·12–4·15 Cellular respiration converts food molecules into ATP, a universal source of energy for living organisms.

*Watermelon powered! We extract energy from all the foods we eat.*

## 4·12 How do living organisms fuel their actions? Cellular respiration: the big picture.

Food is fuel. And all the activities of life—growing, moving, reproducing—require fuel. Plants, most algae, and some bacteria obtain their fuel directly from the energy of sunlight, which they harness through photosynthesis. Less self-sufficient organisms, such as humans, alligators, and insects, must extract the energy they need from the food they eat. This energy comes from photosynthetic organisms either directly (from eating plants) or indirectly (from eating animals that eat plants) (**FIGURE 4-27**).

All living organisms extract energy from the chemical bonds of molecules (which can be considered "food") through a process

**FIGURE 4-27 Living organisms require fuel (in one form or another).**

Oxygen from atmosphere

Sugar and other energy-rich food molecules

**CELLULAR RESPIRATION**

In the cells of plants and animals, high-energy bonds of food molecules are broken down, releasing energy.

Carbon dioxide released to atmosphere

ATP

Water

| INPUT | | OUTPUT | | |
| --- | --- | --- | --- | --- |
| Oxygen + Sugar | → | Carbon dioxide + Water + | ATP Energy |

**FIGURE 4-28 Cellular respiration: the big picture.**

called cellular respiration (**FIGURE 4-28**). This process is a bit like photosynthesis in reverse. In photosynthesis, the energy of the sun is captured and used to build molecules of sugars, such as glucose. In cellular respiration, plants and animals break down the chemical bonds of sugar and other energy-rich food molecules (such as fats and proteins) to release the energy that went into creating them. (Don't confuse cellular respiration with the act of breathing, which is also called *respiration*.) As energy is released, cells capture and store it in the bonds of ATP molecules. This plentiful, readily available stored energy can then be tapped as needed to fuel the work of the life-sustaining activities and processes of all living organisms.

In humans and other animals, cellular respiration starts after we eat food, digest it, absorb the nutrient molecules into the bloodstream, and deliver them to the cells of our bodies. At this point, our cells begin to extract some of the energy stored in the bonds of the food molecules. We focus here on the breakdown of glucose, but later in this chapter we'll see that the process is similar for the breakdown of fats or lipids. Ultimately, when a food molecule has been completely processed, the cell has used the food molecule's stored energy (along with oxygen) to create a large number of high-energy-storing ATP molecules (which supply energy to power the cell's activities), water, and carbon dioxide (which is exhaled into the atmosphere).

**TAKE-HOME MESSAGE 4·12**

Living organisms extract energy through a process called cellular respiration, in which the high-energy bonds of sugar and other energy-rich molecules are broken, releasing the energy that went into creating them. The cell captures the food molecules' stored energy in the bonds of ATP molecules. This process requires fuel molecules and oxygen and it yields ATP molecules, water, and carbon dioxide.

# 4·13 The first step of cellular respiration: glycolysis is the universal energy-releasing pathway.

To generate energy, fuels such as glucose and other carbohydrates as well as proteins and fats are broken down in three steps: (1) glycolysis, (2) the Krebs cycle, and (3) the electron transport chain. **Glycolysis** means the splitting (*lysis*) of sugar (*glyco-*) and it is the first step that all organisms on the planet take in breaking down food molecules; for many single-celled organisms, this one step is sufficient to provide all of the energy they need (**FIGURE 4-29**).

As **FIGURE 4-30** illustrates, glycolysis is a sequence of chemical reactions (there are 10 in all) through which glucose is broken down, resulting in two molecules of a substance called **pyruvate.** Glycolysis has two distinct phases: an "uphill" preparatory phase and a "downhill" payoff phase.

Just as you sometimes have to spend money to make money, before any energy can be extracted from glucose, some energy

FIGURE 4-29 Plants, animals, bacteria, and all other organisms use glycolysis to break down fuels.

Every living organism, large or small, extracts energy through glycolysis!

must be added to the molecule. This addition occurs during the "uphill" phase. The additional energy (which comes from ATP) destabilizes the glucose molecule, making it ripe for chemical breakdown. Once the glucose can be broken down chemically, the energy stored in its bonds can be harnessed as the bonds are broken.

Three of the 10 steps in glycolysis yield energy. In two of these three, as bonds from the sugar are broken, and other lower-energy bonds are formed, the energy released is quickly harnessed by the attachment of phosphate groups to molecules of ADP to create energy-rich ATP molecules. In the third energy-yielding step of glycolysis, electrons originally from the glucose are transferred to NAD⁺ to become the high-energy electron carrier NADH. Later (in an electron transport chain in the mitochondria) this energy will be converted to ATP. The net result of glycolysis is that each glucose molecule is broken down into two molecules of pyruvate. During this breakdown, some of the released energy is captured in the production of energy-rich ATP molecules and molecules of the high-energy electron carrier NADH. Two molecules of water are also produced during glycolysis.

In the absence of oxygen and in many yeasts and bacteria, glycolysis is the only game in town for fueling activity. Because single-celled organisms have much lower energy needs, they can function solely on the yields of glycolysis. For many organisms (including humans), however, glycolysis is a springboard to further energy extraction. The additional energy payoffs come from the Krebs cycle and the electron transport chain.

**GLYCOLYSIS**

**1** PREPARATORY PHASE    **2** PAYOFF PHASE

ATP (2)

Unstable molecule prepared to be broken down

ATP (4)

NADH (2)

Glucose

Gycolysis takes place in the cell's cytoplasm.

Water

Pyruvate (2)

ENERGY SPENT

−2  ATP

ENERGY ACQUIRED

+4  ATP   +2  NADH

FIGURE 4-30 Glycolysis up close.

### TAKE-HOME MESSAGE 4·13

Glycolysis is the initial phase in the process by which all living organisms harness energy from food molecules. Glycolysis occurs in a cell's cytoplasm and uses the energy released from breaking chemical bonds in food molecules to produce high-energy molecules, ATP and NADH.

# 4·14 The second step of cellular respiration: the Krebs cycle extracts energy from sugar.

Cells could stop extracting energy when glycolysis ends, but they rarely do so because that would be like leaving most of your meal on your plate. In glycolysis, only a small fraction of the energy stored in sugar molecules is recovered and converted to ATP and NADH. Cells get much more of an "energy bang" for their "food buck" in the steps following glycolysis, which occur in the mitochondria. This is why mitochondria are considered ATP "factories." In the mitochondria, the molecules produced from the breakdown of glucose during glycolysis are broken down further, during two steps that are dramatically more efficient at capturing energy: the Krebs cycle (the subject of this section) and the electron transport chain (the subject of the next section). In breaking down the products of glycolysis, the **Krebs cycle** produces some additional molecules of ATP and, more importantly, captures a huge amount of chemical energy by producing high-energy electron carriers.

**Q** Aerobic training can cause our bodies to produce more mitochondria in cells. Why is this beneficial?

Before the Krebs cycle can begin, however, the end products of glycolysis—two molecules of pyruvate for every molecule of glucose used—must be modified. First, the pyruvate molecules move from the cytoplasm into the mitochondria, where they undergo three quick modifications that prepare them to be broken down in the Krebs cycle (**FIGURE 4-31**).

**Modification 1.** Each pyruvate molecule passes a pair of its high-energy electrons (and a proton) to the electron-carrier molecule NAD⁺, building two molecules of NADH.

**Modification 2.** Next, a carbon atom and two oxygen atoms are removed from each pyruvate molecule and released as carbon dioxide. The $CO_2$ molecules diffuse out of the cell and, eventually, out of the organism. In humans, for example, these $CO_2$ molecules pass into the bloodstream and are transported to the lungs, from which they are eventually exhaled.

**Modification 3.** In the final step in the preparation for the Krebs cycle, a giant compound known as coenzyme A attaches itself to the remains of each pyruvate molecule, producing two molecules called acetyl-CoA. Each acetyl-CoA molecule is now ready to enter the Krebs cycle.

**ACETYL-CoA PRODUCTION**

Molecular model of pyruvate

1 As each pyruvate is broken down, a pair of electrons (and a proton) are passed to NAD⁺, producing NADH.

2 A carbon and two oxygen atoms are released as carbon dioxide.

3 Coenzyme A attaches itself to the remaining molecule, creating acetyl-CoA.

Modifications of pyruvate take place in the cell's mitochondria.

to Krebs cycle

**FIGURE 4-31 Preparation of pyruvate.** In the mitochondria, pyruvate must be modified before it can be broken down in the Krebs cycle.

There are eight separate steps in the Krebs cycle, but our emphasis is on its three general outcomes (**FIGURE 4-32**).

**Outcome 1.** *A new molecule is formed.* Acetyl-CoA adds its two-carbon acetyl group to a molecule of the starting material of the Krebs cycle, a four-carbon chemical called oxaloacetate, creating a six-carbon molecule.

**Outcome 2.** *High-energy electron carriers (NADH) are made and carbon dioxide is exhaled.* The six-carbon molecule then gives electrons to NAD⁺ to make the high-energy electron carrier NADH. (Don't forget that the main purpose of the Krebs cycle is the capture of energy.) The six-carbon molecule releases two carbon atoms along with four oxygen atoms to form two carbon dioxide molecules. In mammals (including humans), this $CO_2$ is carried by the bloodstream to the lungs, from where it is exhaled into the atmosphere.

**Outcome 3.** *The starting material of the Krebs cycle is re-formed, ATP is generated, and more high-energy electron carriers are formed.* After the $CO_2$ is released, the four-

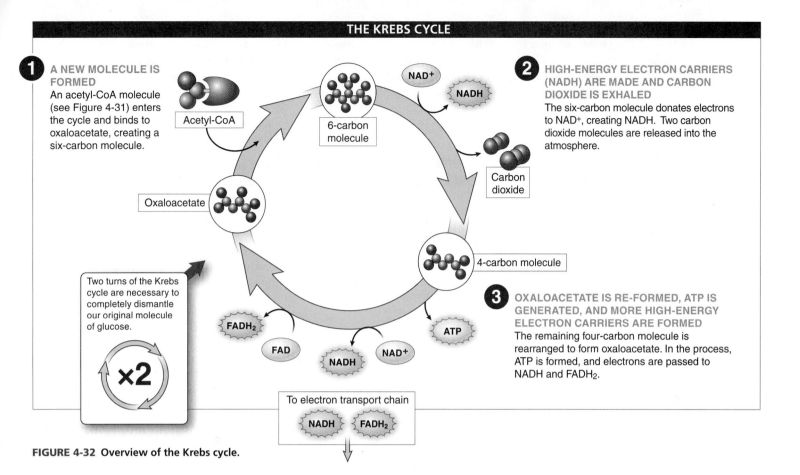

**1** **A NEW MOLECULE IS FORMED**
An acetyl-CoA molecule (see Figure 4-31) enters the cycle and binds to oxaloacetate, creating a six-carbon molecule.

Acetyl-CoA

6-carbon molecule

NAD+

NADH

**2** **HIGH-ENERGY ELECTRON CARRIERS (NADH) ARE MADE AND CARBON DIOXIDE IS EXHALED**
The six-carbon molecule donates electrons to NAD+, creating NADH. Two carbon dioxide molecules are released into the atmosphere.

Carbon dioxide

Oxaloacetate

4-carbon molecule

Two turns of the Krebs cycle are necessary to completely dismantle our original molecule of glucose.

×2

**3** **OXALOACETATE IS RE-FORMED, ATP IS GENERATED, AND MORE HIGH-ENERGY ELECTRON CARRIERS ARE FORMED**
The remaining four-carbon molecule is rearranged to form oxaloacetate. In the process, ATP is formed, and electrons are passed to NADH and FADH2.

FADH2

FAD

NADH

NAD+

ATP

To electron transport chain

NADH    FADH2

**FIGURE 4-32 Overview of the Krebs cycle.**

carbon molecule that remains from the original pyruvate-oxaloacetate molecule formed in Outcome 1 is modified and rearranged to once again form oxaloacetate, the starting material of the Krebs cycle. In the process of this reorganization, one ATP molecule is generated and more electrons are passed to one familiar high-energy electron carrier, NADH, and a new one, FADH₂. The formation of these high-energy electron carriers increases the energy yield of the Krebs cycle. One oxaloacetate is re-formed, and the cycle is ready to break down the second molecule of acetyl-CoA. Two turns of the cycle are necessary to completely dismantle our original molecule of glucose.

Now that we have seen the Krebs cycle in its entirety, let's trace the path of the original six carbons in the original glucose molecule. In a sense, the carbon atoms that were first plucked from the atmosphere to make sugar during photosynthesis have been exhaled back into the atmosphere as six molecules of carbon dioxide.

1. **Glycolysis:** the six-carbon starting point. Glucose is broken down into two molecules of pyruvate. No carbons are removed.

2. **Preparation for the Krebs cycle:** two carbons are released. Two pyruvate molecules are modified to enter the Krebs cycle, and they each lose a carbon atom in the form of two molecules of carbon dioxide.

3. **Krebs cycle:** the last four carbons are released. A total of four carbon atoms enter the Krebs cycle in the form of acetyl-CoA. For each turn of the Krebs cycle, two molecules of carbon dioxide are released. So the two final carbons are released into the atmosphere during the second turn of the wheel. Poof! The six carbon atoms that were originally present in our single molecule of glucose are no longer present.

So we've come full circle. In photosynthesis, carbon atoms from the air were used to build sugar molecules, which had energy stored in the bonds between their carbon, hydrogen, and oxygen atoms. In cellular respiration, the energy previously stored in the bonds of the sugar is converted to molecules of ATP, NADH, and FADH₂. Carbon atoms from sugar are exhaled back into the air as CO₂, and water is produced.

What happens to the high-energy electron carriers, NADH and FADH$_2$? They eventually give up their high-energy electrons to the electron transport chain. The energy released as those electrons pass through the transport chain is captured in the bonds of more ATP molecules. We explore that process in the next section.

**Q**

Why might the malfunctioning of mitochondria play a role in lethargy or fatigue?

As we explore the important roles of the Krebs cycle and electron transport chain in generating usable energy for organisms, it's important to note that mitochondrial malfunctions have serious consequences for health. More than a hundred genetic mitochondrial disorders have been identified, all of which can lead to energy shortage, including muscle weakness, fatigue, and muscle pain. And, as noted in Section 3-15, it appears that many cases of extreme fatigue or cramps that occur after only slight exertion may be related to inherited mutations in the mitochondrial DNA. Damage to our cellular powerhouses, not surprisingly, can create a personal energy crisis.

---

### TAKE-HOME MESSAGE 4·14

A huge amount of additional energy can be harvested by cells after glycolysis. First, the end product of glycolysis, pyruvate, is chemically modified. Then, in the Krebs cycle, the modified pyruvate is broken down step by step. This breakdown releases carbon into the atmosphere (as $CO_2$) as bonds are broken, and captures some of the released energy in two ATP molecules and numerous high-energy electron carriers.

---

# 4·15 The third step of cellular respiration: ATP is built in the electron transport chain.

How do we finally get a big payoff of usable energy from our glucose molecule? Glycolysis and the Krebs cycle produce a few molecules of ATP for each molecule of glucose broken down, but it is the energy held in the high-energy electron carriers NADH and FADH$_2$ formed in these processes that ultimately generates the largest amount of usable energy as ATP. In fact, almost 90% of the energy payoff from a molecule of glucose is harvested in the final step of cellular respiration when the electrons from NADH and FADH$_2$ move along an electron transport chain. This process, like the Krebs cycle, takes place in the mitochondria.

In a manner similar to that seen in the chloroplast during photosynthesis, mitochondria convert kinetic energy (from electrons) into potential energy (a concentration gradient of protons). Two structural features of mitochondria are essential to their impressive ability to harness energy from molecules.

**Feature 1.** Mitochondria have a "bag-within-a-bag" structure that makes it possible for a concentration gradient to form between the regions inside and outside the "inner bag" and makes it possible to harness the potential energy in the bonds of NADH and FADH$_2$ molecules to produce ATP (**FIGURE 4-33**).

Material inside the mitochondrion can lie in one of two spaces: (1) the intermembrane space, which is outside the inner bag, or (2) the **mitochondrial matrix,** which is inside

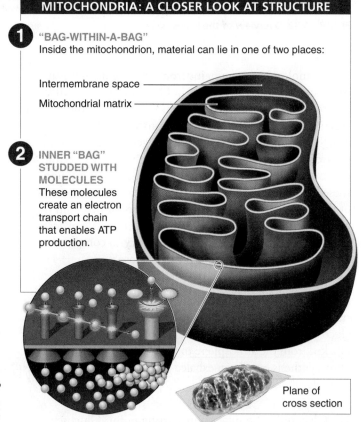

### MITOCHONDRIA: A CLOSER LOOK AT STRUCTURE

**1** "BAG-WITHIN-A-BAG"
Inside the mitochondrion, material can lie in one of two places:

Intermembrane space

Mitochondrial matrix

**2** INNER "BAG" STUDDED WITH MOLECULES
These molecules create an electron transport chain that enables ATP production.

Plane of cross section

**FIGURE 4-33** "A bag-within-a-bag." The structure of mitochondria makes possible their impressive ability to harness energy from food molecules.

# THE MITOCHONDRIAL ELECTRON TRANSPORT CHAIN

High-energy electrons are passed from the carriers NADH and FADH₂ to a series of molecules embedded in the inner mitochondrial membrane called the electron transport chain.

**1** At each step in the electron transport chain's sequence of handoffs, the electrons fall to a lower energy state, releasing a little bit of energy.

**2** At the end of the chain, the lower-energy electrons are handed off to oxygen, which then combines with free H⁺ ions to form water.

**3** The energy is used to power proton pumps, which pack hydrogen ions from the mitochondrial matrix into the intermembrane space.

**4** The protons rush back into the mitochondrial matrix with great kinetic energy, which can be used to build ATP.

**FIGURE 4-34** **The big energy payoff.** Most of the energy harvested during cellular respiration is generated by the electron transport chain in the mitochondria.

the inner bag. With two distinct regions separated by a membrane, the mitochondrion can create higher concentrations of molecules in one area or the other. And because a concentration gradient is a form of potential energy—molecules move from the high-concentration area to the low-concentration area the way water rushes down a hill—once a gradient is created, the energy released as the gradient equalizes itself can be used to do work. In the electron transport chain, this energy is used to build the energy-rich molecule ATP.

**Feature 2.** The inner bag of the mitochondrion is studded with molecules, mostly electron carriers, which are sequentially arranged as a "chain." This arrangement makes it possible for the molecules to hand off electrons in an orderly sequence.

Now let's explore how these features of mitochondria make it possible to harness energy from high-energy electron carriers (**FIGURE 4-34**).

Step 1 of the electron transport chain begins with NADH and FADH₂ in the mitochondrial matrix (inside the inner bag) moving to the membrane. There, the high-energy electrons they carry are transferred to molecules embedded within the membrane.

**Q** Over-the-counter NADH pills provide energy to sufferers of chronic fatigue syndrome. Why might this be?

After they donate their electrons, the molecules that remain, NAD⁺ and FAD, are recycled back to the Krebs cycle.

The membrane-embedded molecules pass the electrons to the next carrier, which passes the electrons to the next, and so on. At each handoff, a bit of energy is released. Thus, as electrons move from one carrier to another through the electron transport chain, they lose energy at each handoff.

At the end of the chain (step 2), the lower-energy electrons are handed off to oxygen, which then combines with free H⁺ ions in the mitochondrial fluid to form water.

As shown in step 3 of Figure 4-34, most of the energy released at each handoff from one electron carrier to another in the electron transport chain is used to pump protons (H⁺ ions) from the mitochondrial matrix across the membrane and into the intermembrane space. As more and more protons are pumped across the membrane and packed into the intermembrane space, a concentration gradient is created. This gradient represents a significant source of potential energy.

If this description seems familiar, it is. In chloroplasts, during photosynthesis, great numbers of protons are pumped from the stroma outside the thylakoid sacs to the inside

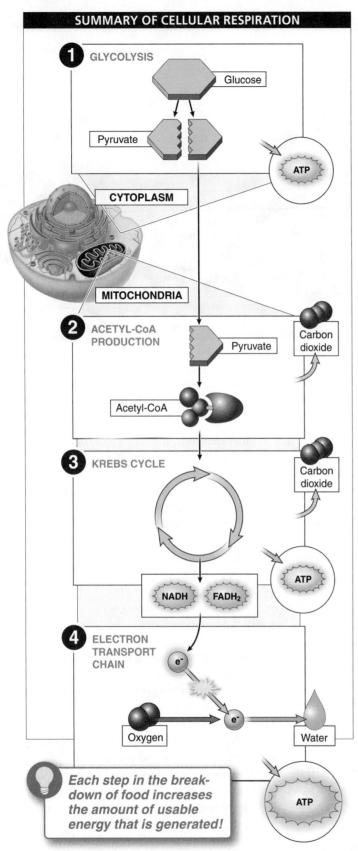

## SUMMARY OF CELLULAR RESPIRATION

**1** GLYCOLYSIS

Glucose

Pyruvate

ATP

**CYTOPLASM**

**MITOCHONDRIA**

**2** ACETYL-CoA PRODUCTION

Pyruvate

Carbon dioxide

Acetyl-CoA

**3** KREBS CYCLE

Carbon dioxide

ATP

NADH    FADH₂

**4** ELECTRON TRANSPORT CHAIN

e⁻

Oxygen    e⁻    Water

ATP

*Each step in the breakdown of food increases the amount of usable energy that is generated!*

**FIGURE 4-35 The steps of cellular respiration: from glucose to usable energy.**

of the thylakoids. We likened this potential energy to the potential energy of water in an elevated tower, which can be released with great force. Similarly, in step 4 of the mitochondrial electron transport chain, the protons pumped into the intermembrane space rush back to the inner mitochondrial matrix through channels in the inner mitochondrial membrane. And as the protons pass through, the force of their flow fuels the attachment of free-floating phosphate groups to ADP to produce ATP.

In the end, the number of ATP molecules generated from the complete dismantling of one molecule of glucose is about 36, most of which are produced with the energy harnessed from high-energy electron carriers as they pass their electrons down the electron transport chain (**FIGURE 4-35**).

Given the central role of the electron transport chain in the generation of usable energy from the breakdown of food molecules, any interference in its functioning has dire consequences. And in fact, murder by cyanide poisoning, an old tradition in detective stories, is just such an interference. When cyanide gets into the mitochondria, it binds to a molecule in the electron transport chain, preventing it from accepting electrons. This halts the transfer of electrons and the pumping of protons across the mitochondrial membrane. As a consequence, the production of ATP that would occur when protons rush back across the membrane down their concentration gradient ceases. Halting the production of ATP removes a cell's energy source, starving it very quickly. For this reason, cyanide poisoning can cause death within minutes.

**Q** Cyanide blocks the passage of electrons to oxygen in the electron transport chain. Why does this make it a toxic poison?

### TAKE-HOME MESSAGE 4·15

**The largest energy payoff of cellular respiration comes as electrons from NADH and FADH₂ produced during glycolysis and the Krebs cycle move along the electron transport chain. The electrons are passed from one carrier to another and energy is released, pumping protons into the mitochondrial intermembrane space. As the protons rush back into the mitochondrial matrix, the force of their flow fuels the production of large amounts of ATP.**

# 4·16–4·17 There are alternative pathways to energy acquisition.

*Unlike cars, living organisms (such as this cheetah) use multiple types of fuel.*

## 4·16 Beer, wine, and spirits are by-products of cellular metabolism in the absence of oxygen.

Every beer brewery, the entire wine industry, and all distilleries of whiskey, vodka, tequila, and other alcoholic beverages owe their existence to microscopic yeast cells scrambling to break down their food for energy under stressful conditions. To better understand how yeast metabolism produces alcohol, it helps to investigate what happens when humans and other animals try to metabolize energy from sugar molecules under some stressful conditions.

If you run or swim as fast as you can, you soon feel a burning sensation in your muscles. Why? Your muscle cells are becoming very acidic. This acid buildup occurs when we demand of our bodies bursts of energy beyond that which they can sustain (**FIGURE 4-36**). (The next-day muscle soreness is not due to acid buildup, which goes away in a matter of minutes or hours; this soreness is caused by damage to the muscle fibers. They break down a bit before growing stronger.)

With rapid, strenuous exertion, our bodies soon fall behind in delivering oxygen from the lungs to the bloodstream to the cells and finally to the mitochondria. Oxygen deficiency then limits the rate at which the mitochondria can break down fuel and produce ATP (**FIGURE 4-37**). This slowdown occurs because the electron transport chain requires oxygen as the final acceptor of all the electrons generated during glycolysis and the Krebs cycle. If oxygen is in short supply, the electrons from NADH (and FADH$_2$) have nowhere to go.

Consequently, the regeneration of NAD$^+$ (and FAD) in the electron transport chain is halted, leaving no recipient for the high-energy electrons harvested from the breakdown of glucose and pyruvate, and the whole process of cellular respiration can grind to a stop. Organisms don't let this interruption last long, though; most have a backup method for breaking down sugar.

**FIGURE 4-36 Energy production without oxygen.** Organisms have a backup method for breaking down sugar when oxygen is not present.

Among animals, there is an acceptor for the NADH electrons in the absence of oxygen: pyruvate, the end product of glycolysis. When pyruvate accepts the electrons, it forms lactic acid (see Figure 4-37). Once the NADH gives up its electrons, NAD⁺ is regenerated, and glycolysis can continue. But as lactic acid builds up, it causes a burning feeling in our muscles. It's not ideal, but to escape a predator or to exercise strenuously, the two ATP molecules generated from each glucose molecule during glycolysis—which can serve as an immediate energy source—are better than nothing.

Like humans, yeast normally use oxygen during their breakdown of food. And like humans, they have a backup method when oxygen is not available. But yeast make use of a different electron acceptor, and the resulting reaction leads to the production of all drinking alcohol.

After glycolysis in these single-celled organisms, pyruvate is usually converted to a molecule called *acetaldehyde,* generating bubbles of $CO_2$ (which, when yeast is used in baking, allows bread to rise). In the absence of oxygen, this molecule accepts the electrons released from NADH, allowing glycolysis to resume. Acetaldehyde's acceptance of NADH's electrons results in the production of **ethanol,** the molecule that gives beer, wine, and spirits their kick. Ethanol can also be used as a fuel source because it can be combusted in much the same way as the biofuels generated from animal fats and plant oils described earlier in this chapter.

**Fermentation** is the process by which cell respiration occurs in the absence of oxygen. It occurs when, following glycolysis, alternative molecules are used as electron acceptors. Interestingly, although ethanol is always the alcohol produced by fermentation, the flavor of the output of fermentation depends on the type of sugar metabolized by the yeast. Fruits, vegetables, and grains all give different results. If the sugar comes from grapes, wine is produced. If the sugar comes

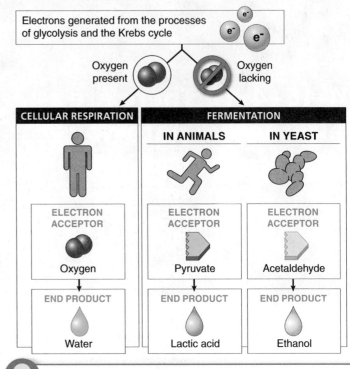

**Exertion without enough oxygen leads to burning cramps in animals, but to alcohol in yeast!**

FIGURE 4-37 **Energy production with and without oxygen**.

from a germinating barley plant, beer is produced. Potatoes, on the other hand, are the sugar source usually used to produce vodka.

Because yeast prefer the more efficient process of aerobic respiration, they produce alcohol only in the absence of oxygen. Fermentation tanks used in producing wine, beer, and other spirits are built to keep oxygen out so that yeast cells are forced to use their backup respiratory pathway of fermentation (**FIGURE 4-38**).

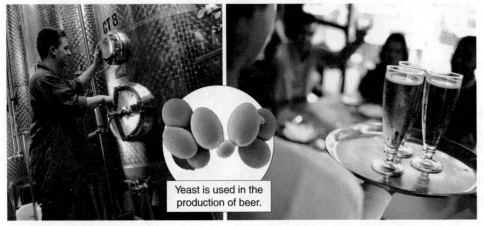

Yeast is used in the production of beer.

FIGURE 4-38 **Just a by-product of metabolism under stressful conditions.** Beer, wine, and spirits are by-products of cellular metabolism in the absence of oxygen.

## TAKE-HOME MESSAGE 4·16

Oxygen deficiency limits the breakdown of fuel because the electron transport chain requires oxygen as the final acceptor of electrons during the chemical reactions of glycolysis and the Krebs cycle. When oxygen is unavailable, yeast resort to fermentation, in which they use a different electron acceptor, pyruvate, and in the process generate ethanol, the alcohol in beer, wine, and spirits.

# 4·17 Eating a complete diet: cells can run on protein and fat as well as on glucose.

Most automobiles can run on only one type of fuel, gasoline. If you run out of fuel and cannot get to a gas station, you are out of luck. In this chapter, we have examined the steps by which plants, animals, and other organisms use glucose as fuel. But living organisms are more flexible than autos when it comes to fuel sources. They also have more complex needs because their fuel must also provide raw materials for growth.

**Q**

**How can someone on a low-carb diet fuel their body?**

Evolution has built humans and other organisms with the metabolic machinery that allows them to extract energy and other valuable chemicals from proteins, fats, and a variety of carbohydrates (**FIGURE 4-39**). For that reason, we are able to consume and efficiently utilize meals comprising various combinations of molecules.

**Sugars.** In the case of dietary carbohydrates, many are polysaccharides—multiple simple sugars linked together—rather than solely the simple sugar glucose. Before they can be broken down by cellular respiration, the polysaccharides must first be separated by enzymes into glucose or related simple sugars.

**Lipids.** Dietary lipids are broken down into their two constituent parts: a glycerol molecule and fatty acids. The glycerol is chemically modified into one of the molecules produced during glycolysis. It enters the glycolysis process at that step and is broken down to yield energy. The fatty acids, meanwhile, are chemically modified into acetyl-CoA, at which point they enter the Krebs cycle.

**Proteins.** Proteins are chains of amino acids. Upon consumption the chains are broken down chemically, after which each amino acid is broken down into (1) an amino group that may be used in the production of tissue or excreted in the urine, and (2) a carbon compound that is converted to one of the intermediate compounds in glycolysis or the Krebs cycle, allowing the energy stored in its chemical bonds to be harnessed.

In the end, humans (indeed, all animals) are able to harvest energy from a variety of food sources beyond simple glucose. Whether a meal contains carbohydrates, lipids, proteins, or

**ENERGY FROM FATS, CARBOHYDRATES, AND PROTEINS**

FIGURE 4-39 **"More flexible than autos."** Animals are able to harvest energy from proteins, carbohydrates, and lipids.

some combination thereof, the nutrients are chemically modified in some preliminary steps and then fed into one of the intermediate steps in glycolysis or the Krebs cycle to furnish usable energy for the organism (see Figure 4-39).

---

**TAKE-HOME MESSAGE 4·17**

Humans and other organisms have metabolic machinery that allows them to extract energy and other valuable chemicals from proteins, fats, and carbohydrates in addition to the simple sugar glucose.

---

## IF YOU FEED AND PROTECT YOUR FLOWERS IN A VASE, THEY'LL LAST LONGER.

Like animals, plants require fuel for the cellular activities that keep them alive.

**Q: HOW DO PLANTS GET THE ENERGY THEY NEED TO STAY ALIVE?** Plants use photosynthesis to harness light energy, converting it to sugar molecules that serve as their food.

**Q: CAN PLANTS STILL PHOTOSYNTHESIZE ONCE THEY'RE IN A VASE IN YOUR HOUSE?** Yes, but humans don't make it easy. Once cut, plants generally cannot produce sufficient sugar through photosynthesis. Light levels in houses tend to be a bit too low, and the loss of many or most plant leaves reduces the number of chloroplasts in which photosynthesis can occur. So they're starving to death.

**Q: CAN YOU SLOW THEIR DEMISE?** Yes. Plants are able to take up sugar in the vase water and use it as an energy source for cell activities. So adding a bit of sugar to the vase is like putting fuel in their tank.

**Q: IS IT THAT EASY?** No. Putting sugar—a molecule with lots of energy stored within its chemical bonds—in the vase is like offering a free lunch. And many, many organisms are looking for a free lunch. Unfortunately, when you add sugar, bacteria on the flower stems can grow rapidly, blocking the water-conducting tubes in the stems. This slows the flow not just of sugar, but of water as well.

**Q: WHAT SHOULD YOU DO?** With the addition to vase water of both sugar and an antibacterial chemical such as chlorine bleach, you can feed and protect your cut flowers, significantly increasing their longevity.

**CONCLUSION:** Here's a useful summary from Martha Stewart (www.marthastewart.com): "Flowers last longer if their stems are cut under water. Use a sharp knife, and cut the stem on a slant so it can absorb as much water as possible. Next, place [the flowers] in a flask filled with 2 inches of warm water (no hotter than 110 degrees); this expedites the flow of water into the flower. Add either cut-flower food or a teaspoon of sugar and a couple drops of liquid bleach to a vase filled with cool water up to the point where the flower foliage will hit. After 5 minutes, transfer [the flowers] to the vase."

# Big Ideas in ... Energy

## 4·1–4·4 Energy flows from the sun and through all life on earth.

- Energy is neither created nor destroyed but can change form.

- Energy from the sun fuels all life on earth as it is converted to different forms. Each conversion is inefficient, and some of the usable energy is converted to less useful heat energy.

- Cells temporarily store energy in the bonds of ATP molecules. This potential energy can be converted to kinetic energy and used to fuel life-sustaining chemical reactions.

## 4·5–4·11 Photosynthesis uses energy from sunlight to make food.

- In the "photo" part of photosynthesis, chloroplasts transform light energy into the chemical energy of ATP and NADPH, while splitting water molecules and producing oxygen.

- In the Calvin cycle, the "synthesis" part of photosynthesis, carbon from $CO_2$ in the atmosphere is attached to molecules in chloroplasts to build sugars. This production of sugars consumes ATP and NADPH generated in the "photo" part of photosynthesis.

- C4 and CAM photosynthesis are evolutionary adaptations at the biochemical level that, although having competitive drawbacks, allow plants in hot, dry climates to close their stomata and conserve water without shutting down photosynthesis.

## 4·12–4·15 Cellular respiration converts food molecules into ATP, a universal source of energy for living organisms.

- In cellular respiration, glucose and oxygen are converted to carbon dioxide, water, and energy that is stored in the high-energy bonds of ATP molecules.

- In the electron transport chain, high-energy electrons from NADH and $FADH_2$ pass from carrier to carrier, releasing energy with each step. This energy fuels the creation of a proton concentration gradient. The energy released as the protons move back down their concentration gradient is used to produce ATP.

- Overall, the cellular respiration of one glucose molecule yields about 36 molecules of ATP.

## 4·16–4·17 There are alternative pathways to energy acquisition.

- Organisms are built with multiple systems for acquiring energy. They can generate ATP even when oxygen is lacking or when organic molecules other than glucose, such as lipids, proteins, and polysaccharides, are consumed.

See the To-Go Guide (packaged with every new copy of this textbook) for an illustrated summary of the chapter.

# Check Your Knowledge

*Thinking critically about visual displays of data.*

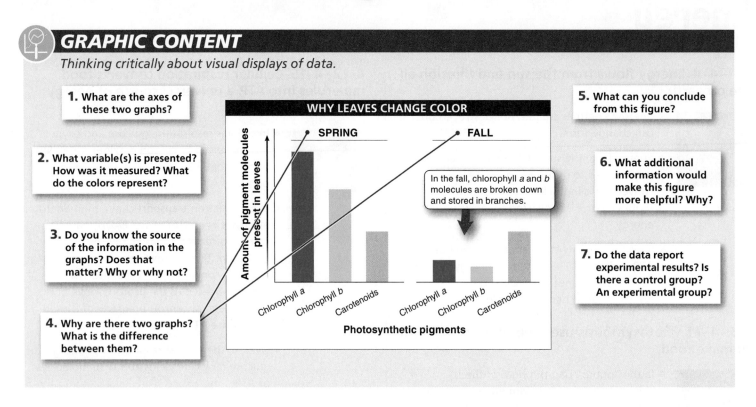

1. What are the axes of these two graphs?

2. What variable(s) is presented? How was it measured? What do the colors represent?

3. Do you know the source of the information in the graphs? Does that matter? Why or why not?

4. Why are there two graphs? What is the difference between them?

5. What can you conclude from this figure?

6. What additional information would make this figure more helpful? Why?

7. Do the data report experimental results? Is there a control group? An experimental group?

**WHY LEAVES CHANGE COLOR**

SPRING    FALL

Amount of pigment molecules present in leaves

In the fall, chlorophyll *a* and *b* molecules are broken down and stored in branches.

Chlorophyll *a*   Chlorophyll *b*   Carotenoids   Chlorophyll *a*   Chlorophyll *b*   Carotenoids

**Photosynthetic pigments**

**See Answers at the back of the book.**

## Key Terms in Energy

adenosine triphosphate (ATP), p. 135
biofuel, p. 130
C4 photosynthesis, p. 148
Calvin cycle, p. 146
CAM (crassulacean acid metabolism), p. 150
carotenoid, p. 141
cellular respiration, p. 131

chemical energy, p. 133
chlorophyll, p. 139
chlorophyll *a*, p. 140
chlorophyll *b*, p. 139
chloroplast, p. 140
electromagnetic spectrum, p. 140
electron transport chain, p. 144
energy, p. 132

ethanol, p. 160
fermentation, p. 160
first law of thermodynamics, p. 133
fossil fuel, p. 130
glycolysis, p. 152
kinetic energy, p. 132
Krebs cycle, p. 154
light energy, p. 140

mitochondrial matrix, p. 156
NADPH, p. 145
photon, p. 140
photosynthesis, p. 131
photosystem, p. 143
pigment, p. 140
potential energy, p. 132
primary electron acceptor, p. 144

pyruvate, p. 152
rubisco, p. 146
second law of thermodynamics, p. 135
stoma (*pl.* stomata), p. 148
stroma, p. 139
thermodynamics, p. 133
thylakoid, p. 139

# Multiple-Choice Questions

1. **Animal fats and plant oils are sometimes used as sources of fuel for automobile engines. How is energy harvested from these molecules?**

   a) They contain long chains of hydrogen and carbon that, when broken, release the energy stored in the bonds linking the atoms together.

   b) They contain hydrogen and carbon tails linked by disulfide bridges that, when broken, release chemical energy.

   c) They contain multiple phosphate groups that each release energy when "liberated" from the molecule chemically.

   d) They contain long hydrophobic regions that, when mixed with water, generate explosive resistances.

   e) They contain long carbon tails, and each atom has unpaired electrons that are released on exposure to extreme heat and pressure.

   100 HARD
   33
   0 EASY

2. **A cyclist rides her bike up a very steep hill. Pick from the flowing choices the statement that properly describes this activity in energetic terms.**

   a) Potential energy in food is converted to kinetic energy as the cyclist's muscles push her up the hill.

   b) Kinetic energy is highest when the cyclist is at the crest of the hill.

   c) The cyclist produces the most potential energy as she cruises down the hill's steep slope.

   d) Potential energy is greatest when the cyclist is at the top of the hill.

   e) Both a) and d) are correct.

   100 HARD

   39
   0 EASY

3. **Every time a source of energy is converted from one form to another:**

   a) the potential energy of the system increases.

   b) heat is required.

   c) the second law of thermodynamics is violated.

   d) the total amount of energy in the universe is reduced by a tiny amount.

   e) some of the energy is converted to heat, which is not a very usable form of kinetic energy.

   100 HARD
   33
   0 EASY

4. **In your body, when energy is released from the breakdown of a molecule such as glucose:**

   a) adenosine monophosphate is created.

   b) adenosine diphosphate is created.

   c) some of the energy may be harnessed by building the unstable high-energy bonds that attach phosphate groups to ADP molecules.

   d) molecules of ATP are required to capture and absorb the heat generated by the reaction.

   e) all of the energy is lost as heat.

   100 HARD

   46
   0 EASY

5. **A green plant can carry out photosynthesis if given nothing more than:**

   a) water, light, and carbon dioxide.

   b) water, light, and oxygen.

   c) carbon dioxide.

   d) oxygen.

   e) oxygen and carbon dioxide.

   100 HARD

   39
   0 EASY

**6. The actual production of sugars during photosynthesis takes place:**

a) within the outer membrane of the chloroplast.

b) within the stroma, inside the thylakoids of the chloroplast.

c) within the stroma, outside the thylakoid, but still inside the chloroplast.

d) just outside the chloroplast, within the mitochondria.

e) within the thylakoid membrane of the chloroplast.

100 HARD
57
0 EASY

**7. The leaves of plants can be thought of as "eating" sunlight. From an energetic perspective, this makes sense because:**

a) light energy, like chemical energy released when the bonds of food molecules are broken, is a type of kinetic energy.

b) both light energy and food energy can be converted to kinetic energy without the loss of heat.

c) the carbon-oxygen bonds within a photon of light release energy when broken by the enzymes in chloroplasts.

d) the carbon-hydrogen bonds within a photon of light release energy when broken by the enzymes in chloroplasts.

e) photons are linked together by hydrogen bonds, which release energy when striking the surface of a leaf.

100 HARD
51
0 EASY

**8. A molecule of chlorophyll increases in potential energy:**

a) when it binds to a photon.

b) when a photon strikes it, boosting electrons to a higher-energy excited state.

c) when it loses an electron.

d) only in the presence of oxygen.

e) None of the above. The potential energy of a molecule cannot change.

100 HARD
24
0 EASY

**9. Photosynthesizing plants rely on water:**

a) to provide the protons necessary to produce chlorophyll.

b) to concentrate the beams of light hitting a leaf, focusing them on the reaction center.

c) to replenish oxygen molecules that are lost during photosynthesis.

d) to replace electrons that are excited by light energy and passed from molecule to molecule down an electron transport chain.

e) to serve as a high-energy electron carrier.

100 HARD
29
0 EASY

**10. During photosynthesis, which step is most responsible for a plant's acquisition of new organic material?**

a) the "building" of NADPH during the Calvin cycle

b) the excitation of chlorophyll molecules by photons of light

c) the "plucking" of carbon atoms from the air and fixing of them to organic molecules within the chloroplast

d) the loss of water through evaporation

e) the production of ATP during the light reactions

100 HARD
50
0 EASY

**11. During C4 photosynthesis:**

a) plants use less ATP when producing sugar.

b) plants are able to continue producing sugars even when they must close their stomata to reduce water loss on hot days.

c) plants are able to generate water molecules to cool their leaves.

d) plants are able to reduce water loss by producing more rubisco.

e) plants are able to produce sugars without any input of carbon dioxide.

100 HARD
36
0 EASY

**12. Cellular respiration is the process by which:**

a) oxygen is used to transport chemical energy throughout the body.

b) oxygen is produced during metabolic activity.

c) light energy is converted to kinetic energy.

d) ATP molecules are converted to water and sugar.

e) energy from the chemical bonds of food molecules is captured by an organism.

100 HARD
47
0 EASY

**13. During cellular respiration, most of the energy contained within the bonds of food molecules is captured in:**

a) the conversion of the kinetic energy of food to the potential energy of ATP.

b) the Krebs cycle and electron transport chain.

c) digestion.

d) glycolysis.

e) None of the above. Energy is lost, not gained, during cellular respiration.

100 HARD

59
0 EASY

**14.** Which of the following energy-generating processes is the only one that occurs in all living organisms?

a) the Krebs cycle
b) glycolysis
c) combustion
d) photosynthesis
e) None of the above. There are no energy-generating processes that occur in all living organisms.

 100 HARD
 31
0 EASY

**15.** During the Krebs cycle:

a) the products of glycolysis are further broken down, generating additional ATP and the high-energy electron carrier NADH.
b) high-energy electron carriers pass their energy to molecules of sugar, which store them as potential energy.
c) the products of glycolysis are further broken down, generating additional ATP and ethanol.
d) cellular respiration can continue even in the absence of oxygen.
e) the products of glycolysis are converted to acetyl-CoA.

 100 HARD
 37
0 EASY

**16.** Mitochondria have a "bag within a bag" structure. This structure is necessary in order to:

a) create a concentration gradient of certain molecules between the regions inside and outside the "inner bag," a form of potential energy.
b) keep molecules of ADP in close proximity to molecules of the ATP-synthesizing enzyme.
c) allow light-reactive accessory pigments to be embedded within the membranes.
d) segregate the most toxic digestive enzymes from molecules of ATP and $NADP^+$.
e) None of the above.

100 HARD
44
0 EASY

**17.** All alcoholic beverages are produced as the result of:

a) cellular respiration by bacteria that occurs in the absence of oxygen.
b) cellular respiration by bacteria that occurs in the absence of free electrons.
c) cellular respiration by yeast that occurs in the absence of free electrons.
d) cellular respiration that occurs in the absence of sugar.
e) cellular respiration by yeast that occurs in the absence of oxygen.

 100 HARD
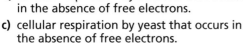 18
0 EASY

**18.** In harvesting the chemical energy of the molecules in food:

a) all macromolecules must first be converted to glucose.
b) all macromolecules must first be converted to their hydrocarbon chains.
c) all macromolecules must first be converted to glucose or another simple sugar.
d) organisms can use sugars, lipids, and proteins.
e) all macromolecules must first be converted to proteins or free-form amino acids.

 100 HARD
 56
0 EASY

## Short-Answer Questions

**1.** A plant and a mouse are placed in a sealed container with a watering system. For a period of time, both will survive. Explain what the two organisms can provide for each other in order to allow them, temporarily, to survive.

**2.** It is possible to measure the rate of the "photo" reactions in plants by measuring the levels of $O_2$ produced by chloroplasts. If bicarbonate solution, a source of $CO_2$, is provided to the plants, the rate of the "photo" reactions increases.

a) Explain why measuring $O_2$ levels provides an indication of the rate of the "photo" reactions.
b) Being as specific as possible, describe how the plants use the $CO_2$ provided by the bicarbonate solution.
c) Propose an explanation of how increasing the $CO_2$ source for the plants affects the rate of the "photo" reactions.

**3.** Some plants grow much more quickly than others. For a plant that is growing at a rapid rate, propose a hypothesis on whether cellular respiration or photosynthesis is occurring at a faster rate. Justify your hypothesis.

See Answers at the back of the book.
For additional study questions, go to www.prep-u.com.

### ABOUT THE CHAPTER OPENING PHOTO

Tulips in Skagit County, Washington, bask in the sun.

What is the genetic code,
and how is it harnessed?

# 5

# DNA, Gene Expression, and Biotechnology

Visual representation of the human genome. Each color represents a different nucleotide.

# 5·1–5·5 DNA: what is it, and what does it do?

## 5·1 "The DNA 200"—knowledge about DNA is increasing justice in the world.

In 1981, Julius Ruffin, a 27-year-old maintenance worker at Eastern Virginia Medical School, got onto an elevator and lost more than 20 years of his life. Several weeks earlier, a nursing student at the school had been raped by an attacker who broke into her apartment. When Ruffin got on the elevator, the student thought she recognized him as her attacker and called the police, who immediately arrested him.

**Q** **What is the most common reason that DNA analyses overturn incorrect criminal convictions?**

Ruffin's girlfriend testified that he was with her at the time of the attack, but on the basis of the victim's eyewitness testimony, a jury found Ruffin guilty and sentenced him to life in prison. Ruffin maintained his innocence—to no avail, until 2003. At that time, the state's Division of Forensic Science performed a DNA analysis on a swab of evidence that remained from the investigation. The analysis revealed that Ruffin was not the attacker. Rather, the analysis showed that the DNA perfectly matched that of a man who, in 2003, was already in a Virginia prison, serving time for rape. Ruffin was freed, but only after having served more than two decades in prison (**FIGURE 5-1**).

Julius Ruffin's case is tragic, but it is not unusual. He is one of a group of 200 people in the United States who have been freed from prison as a result of DNA analyses. Called "the DNA 200," these unjustly imprisoned people spent an average of 12 years behind bars. Eighty percent had been convicted of

**FIGURE 5-1 Vindicated by DNA evidence.** Julius Ruffin was released from prison after two decades of wrongful incarceration.

sexual assault; 28% had been convicted of murder. In more than three-quarters of the cases, inaccurate eyewitness testimony played an important role in the guilty verdict. (Recall, from Chapter 1, the experiments that revealed the unreliability of eyewitness identification.)

In this chapter, we take a close look at DNA, the molecule responsible for Julius Ruffin's exoneration and the deferred justice served to the DNA 200. All living organisms—people, plants, animals, bacteria, and otherwise—carry DNA in almost every cell in their body (with just a few exceptions). Like a social security number, every person's DNA is unique (with the exception of identical twins). In addition to being contained in our living cells, our DNA exists in what we leave behind. It is in our saliva, hair, blood, and even the dead skin cells that fall from our bodies. This is why DNA can serve as an individual identifier; it is a trail we all leave that is increasingly being used to ensure greater justice in our society.

The importance of DNA goes far beyond its function as an individual identifier, however. The information carried within this molecule, which is organized into individual units called genes, is among the most important of all biological knowledge. It controls the operation of cells and carries a record of the evolutionary history of lineages of cells and organisms. And, as witnessed by the following excerpts, it is often in the news.

> Selfish dictators may owe their behaviour partly to their genes, according to a study that claims to have found a genetic link to ruthlessness.
> —*Nature*, April 2008

> *Too Many One-Night Stands? Blame Your Genes . . .* according to a new study, it may be fair to say that while you jolly well could help cheating, your particular genes did make things more difficult.
> —*Time* magazine, December 2010

In fact, it's nearly impossible to open a newspaper or watch a news report these days without being informed that yet another complex human characteristic or trait has been linked to a newly discovered gene (**FIGURE 5-2**). But what does that actually mean? What is this molecule that apparently exerts influence over everything from our fidelity to our ruthlessness?

**FIGURE 5-2 Genetics issues are in the news.**

We explore these issues and more in this chapter, beginning with a look at the structure of DNA itself and how it contains the information for producing organisms. In later sections of the chapter, we learn how modern manipulations of DNA are having far-reaching implications not just for human health but for agriculture as well.

---

### TAKE-HOME MESSAGE 5·1

DNA is a molecule that all living organisms carry in almost every cell in their body. Because every person's DNA is unique and because we leave a trail of DNA behind us as we go about our lives, DNA can serve as an individual identifier. This is the basis of the DNA analysis that is increasingly being used to ensure greater justice in our society, often by establishing the innocence of individuals wrongly convicted of crimes.

# 5·2 The DNA molecule contains instructions for the development and functioning of all living organisms.

Beginning in the 1900s and continuing through the early 1950s, a series of experiments revealed two important features of DNA. First, molecules of DNA are passed down from parent to offspring. Second, the instructions on how to create a body and control its growth, development, and behavior are encoded in the DNA molecule.

Given that DNA must be able to hold the instructions for how to produce every possible type of structure in every living organism, scientists were in a frenzy to learn all they could about it. There was a spirited race to determine the chemical structure of DNA and to understand how the molecule was assembled and shaped so that it could hold and transmit so much information.

Many formidable scientists rose to the challenge of determining the structure of DNA. American Linus Pauling had already won a Nobel prize in chemistry for his work on elucidating the structure of molecules when he began investigating the structure of DNA. Simultaneously, Maurice Wilkins and Rosalind Franklin in England devoted their research to this task as well, and produced X-ray pictures of DNA that were critical to decoding its shape. But it was Englishman Francis Crick and American James

Watson, working in Cambridge, England, who happened to put all the pieces together and deduce the exact structure of DNA (**FIGURE 5-3**). As we'll see later in this chapter, their discovery was more than just a description of a molecule. As soon as they figured out DNA's structure, the answers to several other thorny problems in biology—such as how DNA might be able to duplicate itself—became apparent. We explore that process later, in the next chapter, but first we need to examine the structure of DNA in more detail.

**DNA (deoxyribonucleic acid)** is a **nucleic acid,** a macromolecule that stores information. It consists of individual units called **nucleotides,** which have three components: a molecule of sugar, a phosphate group (containing four oxygen atoms bound to a phosphorus atom), and a nitrogen-containing molecule called a **base.** The physical structure of DNA is frequently described as a "double helix." But what exactly is a double helix? Picture a long ladder twisted around like a spiral staircase and you'll have a good idea of what a DNA molecule looks like (**FIGURE 5-4**). The molecule has two distinct strands, like the vertical sides of a ladder. These are the "backbones" of the DNA molecule, and each is made from two alternating

**FIGURE 5-3 Watson and Crick.** James Watson (left) and Francis Crick (right) figured out the exact structure of DNA. The first sketch of the double helix, shown here, was rendered by Crick's wife, Odile.

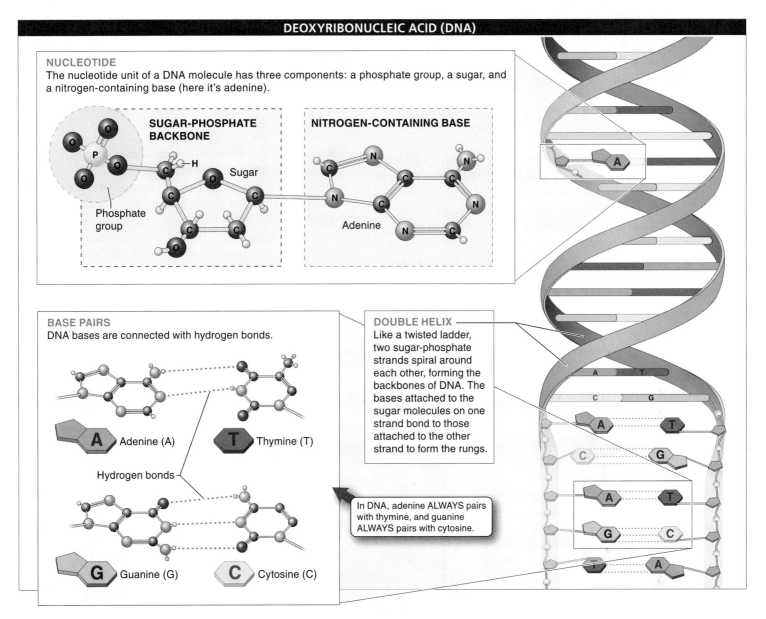

**NUCLEOTIDE**
The nucleotide unit of a DNA molecule has three components: a phosphate group, a sugar, and a nitrogen-containing base (here it's adenine).

SUGAR-PHOSPHATE BACKBONE

NITROGEN-CONTAINING BASE

Sugar

Phosphate group

Adenine

**BASE PAIRS**
DNA bases are connected with hydrogen bonds.

A  Adenine (A)

T  Thymine (T)

Hydrogen bonds

G  Guanine (G)

C  Cytosine (C)

**DOUBLE HELIX**
Like a twisted ladder, two sugar-phosphate strands spiral around each other, forming the backbones of DNA. The bases attached to the sugar molecules on one strand bond to those attached to the other strand to form the rungs.

In DNA, adenine ALWAYS pairs with thymine, and guanine ALWAYS pairs with cytosine.

**FIGURE 5-4  Overview of the structure of DNA.**

molecules: a sugar, then a phosphate group, then a sugar, then a phosphate group, and so on. The sugar is always deoxyribose and the phosphate group is always the same, too. It is the shapes of the molecules in the backbone that cause the DNA "ladder" to twist.

The alternating sugars and phosphates hold everything in place, but they play only a supporting role. The rungs of the ladder are where things get interesting. Attached to each sugar, and protruding inward like half of a rung on the ladder, is one of the four nitrogen-containing bases. Each sugar always has a base attached to it; the four bases are adenine, thymine,

cytosine, and guanine. When discussing DNA, these bases are usually referred to by their first letter: A, T, C, and G.

Both backbones of the ladder have a base protruding from each sugar. The base on one side of the ladder binds, via hydrogen bonds, to a base on the other side, and together these **base pairs** form the rungs of the ladder. They don't just pair up at random, though. Every time a C protrudes from one side, it forms hydrogen bonds with a G on the other side (and vice versa: a G always bonds to a C). Similarly, every time a T protrudes from one side, it forms hydrogen bonds with an A on the other side (and vice versa). For this reason, each DNA molecule always has the same number of Gs and Cs, and the same number of As and Ts. Because of these base-pairing rules, it also is true that if we know the base sequence for one of the strands in a DNA molecule, we know the sequence in the other. For this reason, a DNA sequence is described by writing the sequence of bases on only one of the strands.

If a human DNA molecule were really a spiraled ladder, it would be a very, very long one. One molecule of DNA can have as many as 200 million base pairs, or rungs. This length would make it difficult to squeeze the molecule into the cell. But the rungs are small, and the twisting of the molecule—twists upon twists—shortens it considerably: think about how, if you repeatedly twist your shoelace around and around, it becomes shorter and shorter. That's

what happens with DNA. Let's now investigate how DNA's structure—the rungs of the ladder, in particular—enables it to carry information.

## 5·3 Genes are sections of DNA that contain instructions for making proteins.

**Q** **Why is DNA considered the universal code for all life on earth?**

One of DNA's most amazing features is that it embodies the instructions for building the cells and structures for almost every single living organism on earth (**FIGURE 5-5**). Thus, DNA is like a universal language, the letters of which are the bases A, T, C, and G. (Note that the sugar-phosphate backbone only serves to hold the bases in sequence, like the binding of a book. It does not convey genetic information.)

We've seen how the *structure* of DNA is like a spiral staircase. Another analogy may help you understand the *information-containing* aspect of DNA. You can think of an organism's DNA as a cookbook. Just as a cookbook contains detailed instructions on how to make a variety of foods (such as french toast, macaroni and cheese, or chocolate chip cookies), an organism's DNA carries the detailed instructions to build an organism and keep it running. And just as a book can be viewed as a

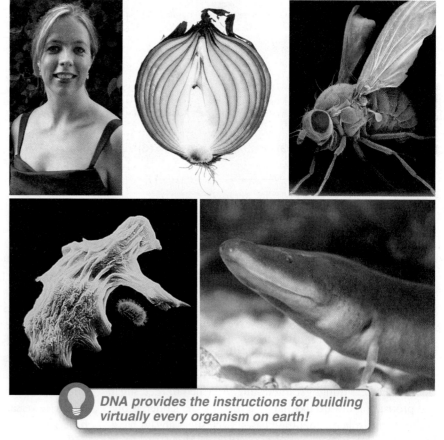

DNA provides the instructions for building virtually every organism on earth!

**FIGURE 5-5 DNA is the universal code for all life on earth.**

sequence of letters, with the book's meaning determined by which letters are strung together and in what order, a molecule of DNA can be viewed as a sequence of bases. Letters don't have much meaning on their own, of course, but when they are put together into words and sentences, their order holds a great deal of information. Similarly, the sequence in which bases appear in a molecule of an organism's DNA makes up a **code** that holds the detailed instructions for the building of the organism—chiefly in the form of instructions for the amino acid sequences of polypeptides. Upon processing and folding, these polypeptides become the numerous and varied protein molecules that make up an organism—whether it is a one-celled amoeba, a giant oak tree, or a biology student.

The full set of DNA present in an individual organism is called its **genome** (FIGURE 5-6). In prokaryotes, including all bacteria, the information is contained within circular pieces of DNA. In eukaryotes, including humans, this information is laid out in long linear strands of DNA in the nucleus. Rather than being one super-long DNA strand, eukaryotic DNA exists as many smaller, more manageable pieces, called **chromosomes.** Humans, for example, have three billion base pairs, divided into 23 unique pieces of DNA (and we have two copies of each piece, one from our mother and one from our father, for a total of 46 chromosomes, and six billion base pairs in every cell).

Many mysteries about the organization of our DNA remain. The number of chromosomes, for example, varies from species to species, but is not related to how simple or complex an organism is. Corn has 10 unique chromosomes, yet fruit flies have only 4. Dogs and chickens have 39 different chromosomes, while goldfish put most other animal species to shame with 50 different chromosomes. As in humans, individuals in each of these species inherit one copy of each chromosome from each parent, so the total number of chromosomes in a cell is doubled—a dog, for example, has 78 chromosomes in each cell. Chromosomes vary in length, too. In humans, the longest chromosome has more than 200 million base pairs and the shortest has fewer than 50 million.

Within the long sequences of bases are relatively short sequences, on average about 3,000 bases long, that are the actual genes. "Gene" may seem like an impossibly nebulous concept, because the word is often used casually in the media as if it were some magical, irresistible, and mysterious force that controls our bodies and behavior. Beyond these vague descriptions, though, the word has a literal meaning. A **gene** is a sequence of bases (strictly, as we've noted, base pairs) in a DNA molecule that carries the information necessary for

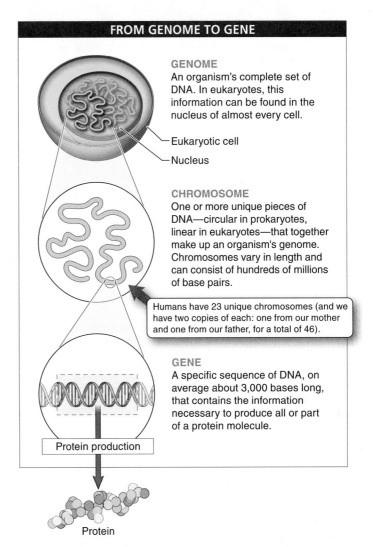

**FROM GENOME TO GENE**

**GENOME**
An organism's complete set of DNA. In eukaryotes, this information can be found in the nucleus of almost every cell.

— Eukaryotic cell
— Nucleus

**CHROMOSOME**
One or more unique pieces of DNA—circular in prokaryotes, linear in eukaryotes—that together make up an organism's genome. Chromosomes vary in length and can consist of hundreds of millions of base pairs.

Humans have 23 unique chromosomes (and we have two copies of each: one from our mother and one from our father, for a total of 46).

**GENE**
A specific sequence of DNA, on average about 3,000 bases long, that contains the information necessary to produce all or part of a protein molecule.

Protein production

Protein

FIGURE 5-6 **The genome unpacked.**

producing a functional product, usually a polypeptide or RNA molecule. Nothing more, nothing less.

Remember that a DNA molecule is like a ladder in which half of each rung is any one of the four bases—A, T, G, or C. Strung together, a segment might be read as "AAAGGCTAGGC . . ." continuing on for another 3,000 or so bases. Returning to our cookbook analogy, just as a particular sequence of letters in a cookbook may be read and understood as the directions for baking chocolate chip cookies, the sequence of bases in DNA also carries information. Perhaps the sequence spells out part of the instructions for producing a red blood cell, for constructing the keratin that will form part of a curly strand of hair, or for assembling a chemical that alters your brain chemistry so that you exhibit a mood disorder or suicidal behavior.

**ALLELES**

Within a species, different individuals may have alternative versions of a gene for a given protein. These instructions can result in different versions of the same trait.

**ALLELES**
Different versions of a gene that code for the same trait

**ALLELE 1** or **ALLELE 2** or **ALLELE 3**

Protein production | Protein production | Protein production

The protein, influenced by the environment and in some cases other genes, then produces the trait.

**FIGURE 5-7 "Different versions of the same thing."** Alleles are alternative versions of a gene.

Each gene is the instruction set for producing one particular molecule, usually a protein. For example, there is a gene in silk moths that codes for fibroin, the chief component of silk. And, there is a gene in humans that codes for triglyceride lipase, an enzyme that breaks down dietary fat.

Within a species, individuals sometimes have slightly different instruction sets for a given protein, and these instructions can result in a different version of the same characteristic. These alternative versions of a gene that code for the same feature are called **alleles** (**FIGURE 5-7**)—and function like alternative recipes for chocolate chip cookies. Any single characteristic or feature of an organism is referred to as a **trait.** A simple hypothetical example will clarify the meaning of these terms. The color of a daisy's petals is a trait. The instructions for producing this trait are found in a gene that controls petal color. This gene may have many different alleles; one allele may specify the trait of red petals, another may specify white petals, and yet another may specify yellow petals. Similarly, one allele for eye color in fruit flies may carry the instructions for producing a red eye, while another, slightly different allele may have instructions for brown eyes. (Ultimately, though, the trait may be influenced not just by the genes an individual carries but by the way those genes interact with the environment, too.)

**TAKE-HOME MESSAGE 5·3**

DNA is a universal language that provides the instructions for building all the structures in all living organisms. The full set of DNA that an organism carries is called its genome. In prokaryotes, the DNA occurs in circular pieces. In eukaryotes, the genome is divided among smaller, linear strands of DNA. An organism's DNA pieces are generally called chromosomes. A gene is a sequence of bases in a DNA molecule that carries the information necessary for producing a functional product, usually a polypeptide or RNA molecule.

# 5·4 Not all DNA contains instructions for making proteins.

It is debatable whether humans are the most complex species on the planet, but surely we must be more complex than an onion. "Complexity" is somewhat subjective and can be assessed in a variety of ways, such as by counting the number of different cell types in the organism. But if we measured complexity simply as the amount of DNA an organism has, we'd have to say an onion is more complex—it has more than five times as much DNA as a human (**FIGURE 5-8**)! We

**Q** *An onion has five times as much DNA as a human. Why doesn't that make onions more complex than humans?*

don't fare any better when compared with some other seemingly simple organisms, either. The salamander species *Amphiuma means,* for example, has about 25 times as much DNA as we do, and one species of amoeba—a single-celled organism—has almost 200 times as much!

Comparing the amount of DNA present in various species, in terms of both numbers of chromosomes and numbers of base pairs, reveals a paradox: there does not seem to be any

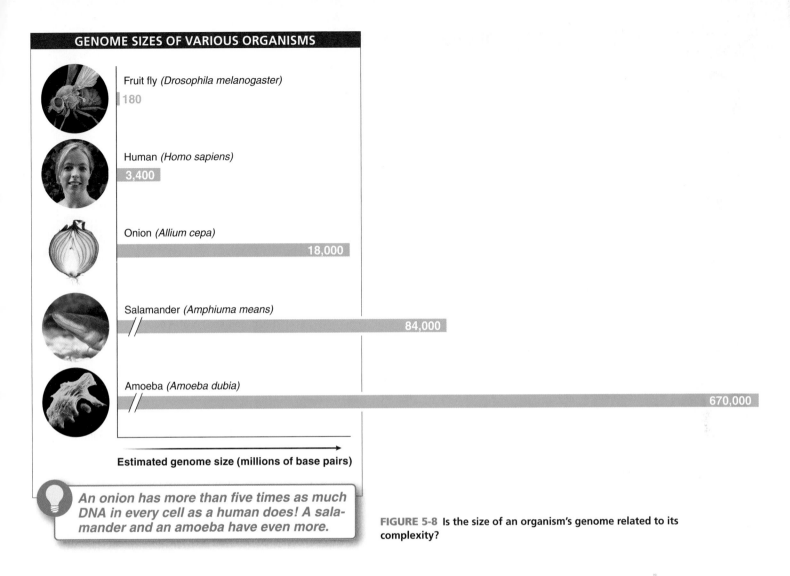

## GENOME SIZES OF VARIOUS ORGANISMS

Fruit fly (*Drosophila melanogaster*)
180

Human (*Homo sapiens*)
3,400

Onion (*Allium cepa*)
18,000

Salamander (*Amphiuma means*)
84,000

Amoeba (*Amoeba dubia*)
670,000

Estimated genome size (millions of base pairs)

*An onion has more than five times as much DNA in every cell as a human does! A salamander and an amoeba have even more.*

**FIGURE 5-8 Is the size of an organism's genome related to its complexity?**

relationship between the size of an organism's genome and the organism's complexity.

The description earlier in this chapter about what DNA is and how genes code for proteins is logical and tidy, but it doesn't completely explain what we observe in cells. In humans, for example, genes make up only about 2% of the DNA (**FIGURE 5-9**). In many species, the proportion of the DNA that consists of genes is even smaller. In almost all eukaryotic species, the amount of DNA present far exceeds the amount necessary to code for all of the proteins in the organism. The fact is, a huge proportion of base sequences in DNA do not code for anything and have no obvious purpose. When it was first observed, some biologists even referred to this non-coding DNA as "junk DNA."

In what types of organisms do we find the most "junk DNA"? Bacteria and viruses tend to have very little non-coding DNA; genes make up 90% or more of their DNA.

It is in the eukaryotes (with the exception of yeasts) that we see an explosion in the amount of non-coding DNA (**FIGURE 5-10**). Non-coding regions of DNA often take the form of sequences that are repeated, sometimes thousands (or even hundreds of thousands) of times; often these repeats exist because some DNA sequences can make copies of themselves and move throughout the genome. Occasionally, the non-coding DNA consists of gene fragments, duplicate versions of genes, and pseudogenes (sequences that evolved from actual genes but accumulated mutations that made them lose their protein-coding ability). About 25% of the non-coding regions occur *within* genes—in which case they are called **introns.** About 75% of the non-coding regions occur *between genes* (see Figure 5-10).

In the end, the presence of this non-coding DNA is still not completely understood. Recent evidence has revealed, however, that some of the non-coding DNA actually does encode extremely short RNA molecules (~20 nucleotides

## PERCENTAGES OF CODING DNA FOUND IN VARIOUS ORGANISMS

Human (*Homo sapiens*)
2%

Fruit fly (*Drosophila melanogaster*)
19%

Round worm (*Caenorhabditis elegans*)
25%

Arabidopsis (*Arabidopsis thaliana*)
28%

E. coli (*Escherichia coli*)
90%

**Percentage of DNA that codes for proteins**

**FIGURE 5-9 "Junk DNA"?** The proportion of DNA that codes for proteins or RNA varies greatly among species. Of the species listed, which has the least "junk DNA"?

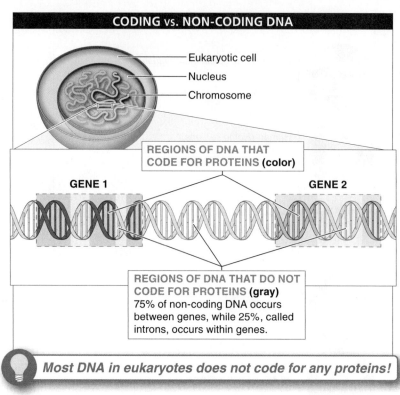

## CODING vs. NON-CODING DNA

- Eukaryotic cell
- Nucleus
- Chromosome

GENE 1          GENE 2

**REGIONS OF DNA THAT CODE FOR PROTEINS (color)**

**REGIONS OF DNA THAT DO NOT CODE FOR PROTEINS (gray)**
75% of non-coding DNA occurs between genes, while 25%, called introns, occurs within genes.

*Most DNA in eukaryotes does not code for any proteins!*

**FIGURE 5-10 Non-coding regions of DNA.** These regions are found both between and within genes.

long) that function in gene regulation; they act as a "switch" that regulates when genes are turned on or off, or as a "volume knob" that influences the rate at which gene products are produced. Non-coding DNA may also serve as a reservoir of potentially useful sequences. In any case, the label "junk DNA" is a bad description.

### TAKE-HOME MESSAGE 5·4

Only a small fraction of the DNA in eukaryotic species is in genes that code for proteins; the function of the rest is still poorly understood, although at least some of it plays important roles in the cell, such as gene regulation.

# 5·5 How do genes work? An overview.

Just as having a recipe for chocolate chip cookies is not the same thing as having the actual cookies, having a wealth of hereditary information about how to build muscle cells or leaf cells is not sufficient to produce an organism. Think about it: an organism's every cell contains all of the information needed to manufacture every protein in its body. This means that the skin cells on your arm contain the genes for producing proteins found only in liver cells or red blood cells or muscle

tissue—but they don't produce them. Having the instructions is not the same as having the products.

The genes in strands of DNA are a storehouse of information, an instruction book, but they are only one part of the process by which an organism is built. If the genes that an organism carries for a particular trait—its **genotype**—are like a recipe in a cookbook, the physical manifestation of the instructions—

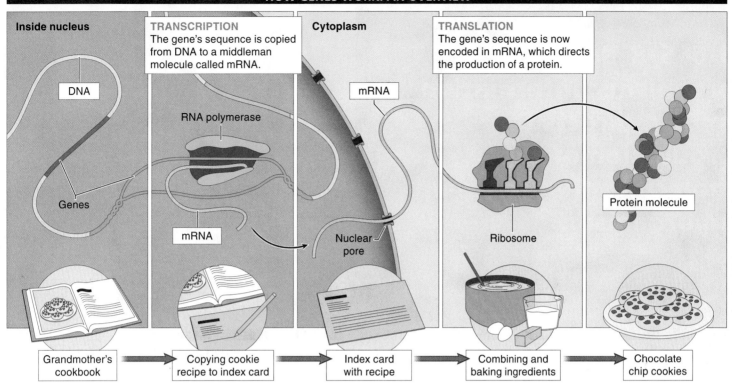

**Inside nucleus**

DNA

Genes

**TRANSCRIPTION**
The gene's sequence is copied from DNA to a middleman molecule called mRNA.

RNA polymerase

mRNA

**Cytoplasm**

mRNA

Nuclear pore

**TRANSLATION**
The gene's sequence is now encoded in mRNA, which directs the production of a protein.

Protein molecule

Ribosome

Grandmother's cookbook → Copying cookie recipe to index card → Index card with recipe → Combining and baking ingredients → Chocolate chip cookies

**FIGURE 5-11 Overview of the steps from gene to protein.**

the organism's **phenotype**—is the cookie, or any of the other foods described by the recipes. And, just as you have to assemble the ingredients, mix them, then bake the dough to get a cookie, there are several steps in the production of the molecules, tissues, and even behaviors that make up a phenotype.

How does a gene (a sequence of bases in a section of DNA) affect a flower's color or the shape of a nose or the texture of a dog's fur (the phenotype)? The process occurs in two main steps: **transcription,** in which a copy of a gene's base sequence is made, and **translation,** in which that copy is used to direct the production of a polypeptide, which then, in response to a variety of factors, including the cellular environment, folds into a protein.

**FIGURE 5-11** presents an overview of the processes of transcription and translation. In transcription, which in eukaryotes occurs in the nucleus, the gene's base sequence, or

code, is copied into a middleman molecule called **messenger RNA (mRNA).** (Because prokaryotes don't have a nucleus, transcription occurs in the cytoplasm.) This is like copying the information for the chocolate chip cookie recipe out of the cookbook and onto an index card. The mRNA then moves out of the nucleus into the cytoplasm, where translation allows the messages encoded in the mRNA to be used to build proteins.

## TAKE-HOME MESSAGE 5·5

The genes in strands of DNA are a storehouse of information, an instruction book. The process by which this information is used to build an organism occurs in two main steps: transcription, in which a copy of a gene's base sequence is made, and translation, in which that copy is used to direct the production of a polypeptide.

DNA microarrays measure the expression levels of many genes simultaneously.

# 5·6

## In transcription, the information coded in DNA is copied into mRNA.

If DNA is like a cookbook filled with recipes, transcription and translation are like cooking. In cooking, you use information about how to make chocolate chip cookies to actually produce the cookies. In an organism, the information about putting together proteins is used to build the proteins the organism needs to function. In this section, we examine transcription, the first step in the two-step process by which DNA regulates a cell's activity and its synthesis of proteins (see Figure 5-11). In transcription, a copy of one specific gene within the DNA is made. Continuing our cookbook analogy, transcription is like copying a single recipe from the cookbook onto an index card. It happens in four steps (**FIGURE 5-12**).

**Step 1. Recognize and bind.** To start the transcription process, the enzyme RNA polymerase, recognizes a **promoter site,** a part of the DNA molecule that indicates the start of a gene and, in effect, tells the RNA polymerase, "Start here." RNA polymerase binds to the DNA molecule at the promoter site and unwinds it just a bit, so that only one strand of the DNA can be read.

**Step 2. Transcribe.** As the DNA strand is processed through the RNA polymerase, the RNA polymerase builds a copy—

called a "transcript"—of the gene from the DNA molecule, just as (to use another analogy) a court reporter transcribes and creates a record of everything that is said in a courtroom. This copy is called messenger RNA (mRNA) because, once this copy of the gene is created, it can move elsewhere in the cell and its message can be translated into a protein. Throughout transcription, DNA is unwound ahead of the RNA polymerase so that a single strand of the DNA can be read, and it is rewound after the polymerase passes.

The mRNA transcript is constructed from four different nucleotides, like DNA nucleotides but with a different sugar. Each pairs up with an exposed base on the now unwound and separated DNA strand, following these rules:

If the DNA strand has a thymine (T), an adenine (A) is added to the mRNA.

If the DNA strand has an adenine (A), a uracil (U) is added to the mRNA.

If the DNA strand has a guanine (G), a cytosine (C) is added to the mRNA.

If the DNA strand has a cytosine (C), a guanine (G) is added to the mRNA.

**FIGURE 5-12 Transcription: copying the base sequence of a gene.** The first step in a two-step process by which DNA regulates a cell's activity and synthesis of proteins. (Shown here is eukaryotic transcription.)

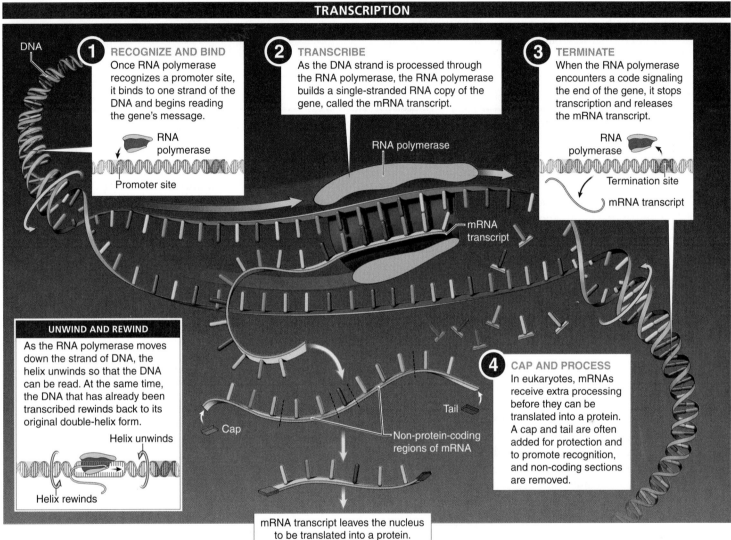

## TRANSCRIPTION

**1 RECOGNIZE AND BIND**
Once RNA polymerase recognizes a promoter site, it binds to one strand of the DNA and begins reading the gene's message.

RNA polymerase

Promoter site

**2 TRANSCRIBE**
As the DNA strand is processed through the RNA polymerase, the RNA polymerase builds a single-stranded RNA copy of the gene, called the mRNA transcript.

RNA polymerase

mRNA transcript

**3 TERMINATE**
When the RNA polymerase encounters a code signaling the end of the gene, it stops transcription and releases the mRNA transcript.

RNA polymerase

Termination site

mRNA transcript

DNA

**UNWIND AND REWIND**
As the RNA polymerase moves down the strand of DNA, the helix unwinds so that the DNA can be read. At the same time, the DNA that has already been transcribed rewinds back to its original double-helix form.

Helix unwinds

Helix rewinds

Cap

Non-protein-coding regions of mRNA

Tail

**4 CAP AND PROCESS**
In eukaryotes, mRNAs receive extra processing before they can be translated into a protein. A cap and tail are often added for protection and to promote recognition, and non-coding sections are removed.

mRNA transcript leaves the nucleus to be translated into a protein.

---

Because our court reporter (RNA polymerase) transcribes a specific sequence of DNA letters (the gene), the mRNA transcript carries this specific bit of the DNA's information. And because it is separate from the DNA, the mRNA transcript can move throughout the cell, to the places where the information is needed, while leaving the original information within the DNA.

**Step 3. Terminate.** When the RNA polymerase encounters a sequence of bases on the DNA at the end of the gene (called a termination sequence), it stops creating the transcript and detaches from the DNA molecule. After termination, the mRNA molecule is released as a free-floating, single-strand copy of the gene.

**Step 4. Capping and editing.** In prokaryotic cells, once an mRNA transcript is produced and begins to separate from the DNA it is ready to be translated into a protein (it doesn't have a nuclear membrane to cross). In eukaryotes, mRNAs receive extra processing before they can be translated into a protein.

First, a cap and a tail may be added at the beginning and end of the transcript. Like the front and back covers of a book, these serve to protect the mRNA from damage and help the protein-making machinery recognize the mRNA. Second, because (as we saw earlier in the chapter) there may be some non-coding bits of the DNA that are transcribed into mRNA, those sections—the introns—are snipped out. Once the mRNA transcript has been edited, it is ready to leave the nucleus for the cytoplasm, where it will be translated into a polypeptide that will fold into a protein.

## TAKE-HOME MESSAGE 5·6

Transcription is the first step in the two-step process by which DNA directs the production of proteins. In transcription (which occurs in the nucleus in eukaryotic cells), a single copy of one specific gene in the DNA is made in the form of a molecule of mRNA. When the mRNA copy of a gene is completed, it moves to the cytoplasm, where it can be translated into a polypeptide.

# 5·7 In translation, the mRNA copy of the information from DNA is used to build functional molecules.

Once the mRNA molecule moves out of the cell's nucleus and into the cytoplasm, the translation process begins. In translation, the information carried by the mRNA is read, and ingredients present in the cytoplasm are used to produce a protein (see Figure 5-11). The process of translation is like the combining and baking of the ingredients listed in our chocolate chip cookie recipe to produce a cookie.

Several ingredients must be present in the cytoplasm for translation to occur. First, there must be large numbers of free amino acids floating around. Recall from Chapter 2 that amino acids are the raw materials for building proteins and an essential component of our diet. Second, there must be **ribosomal subunits,** the protein-production factories where amino acids are linked together in the proper order to produce the protein. And there must also be molecules that can read the mRNA code and translate that message from a sequence of bases into a protein.

To read an encrypted message, you need to know the secret code by which to translate each of the encrypted characters into a readable character in the message. Similarly, to understand which sequence of amino acids in a polypeptide is specified by a sequence of nucleotides in a DNA molecule, you need to know a secret code of sorts.

A special type of RNA molecules holds the secret code. These molecules, called

**transfer RNA (tRNA),** interpret the mRNA code, translating the language of DNA—coded in the linear sequence of bases—into the language of proteins, coded in the linear sequences of amino acids.

Picture a two-sided molecule (**FIGURE 5-13**). On one side of the two-sided tRNA molecule is an attachment site consisting of a three-base sequence that matches up with a three-base sequence on the mRNA transcript. This matchup enables the

**TRANSFER RNA (tRNA)**

Transfer RNA (tRNA) molecules translate the mRNA code by linking specific bases on the mRNA with specific amino acids that will be used to build a protein.

Amino acid — thr

tRNA —

Attachment site —

Attachment sites consist of a three-base sequence that matches up with a three-base sequence on the mRNA transcript. Each three-base sequence in mRNA—called a codon—matches with a tRNA that carries one particular amino acid.

mRNA

Codon

**FIGURE 5-13 Each transfer RNA attaches a particular amino acid to the mRNA.**

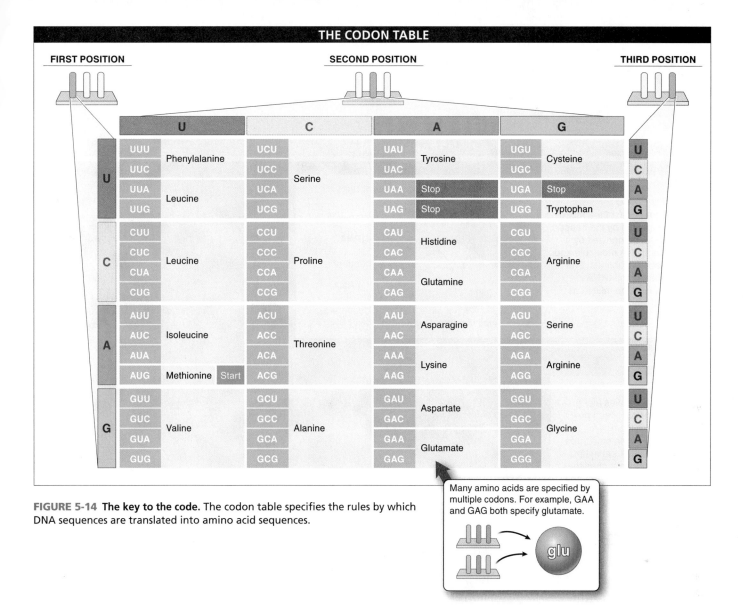

**FIGURE 5-14 The key to the code.** The codon table specifies the rules by which DNA sequences are translated into amino acid sequences.

Many amino acids are specified by multiple codons. For example, GAA and GAG both specify glutamate.

tRNA molecule to attach to the mRNA. Attached to the other side of the tRNA molecule is an amino acid. Each three-base sequence in mRNA—called a **codon**—matches with a tRNA molecule that carries a particular amino acid. The codon ACG, for example, is recognized by the tRNA molecule that carries the amino acid threonine. And the codon CAG is recognized by the tRNA molecule that carries glutamine. For every possible codon, there is one type of tRNA molecule that will recognize and bind to the mRNA at that point, and it will always carry the same amino acid.

The codon table (**FIGURE 5-14** ), also called the "genetic code," describes which amino acid is specified by each codon. With four possible bases in each of the three positions of a codon, 64 different codons are possible. Sixty-one of these codons specify amino acids and 3 are "stop" codons, indicating the

end of translation. Because there are only 20 amino acids, many amino acids are specified by multiple codons. UGU and UGC, for example, both specify cysteine. With just a few small exceptions, the genetic code is the same in every organism on earth.

The translation of an mRNA molecule into a sequence of amino acids (that will then fold into the complex three-dimensional shape of a protein) occurs in three steps (**FIGURE 5-15**).

**Step 1. Recognize and initiate protein building.** Translation begins in the cell's cytoplasm when the subunits of a ribosome, essentially a two-piece protein-building factory, recognize and assemble around a "start sequence" on the mRNA transcript. This start sequence is always the sequence A, U, and G (AUG). As the ribosomal subunits assemble

# TRANSLATION

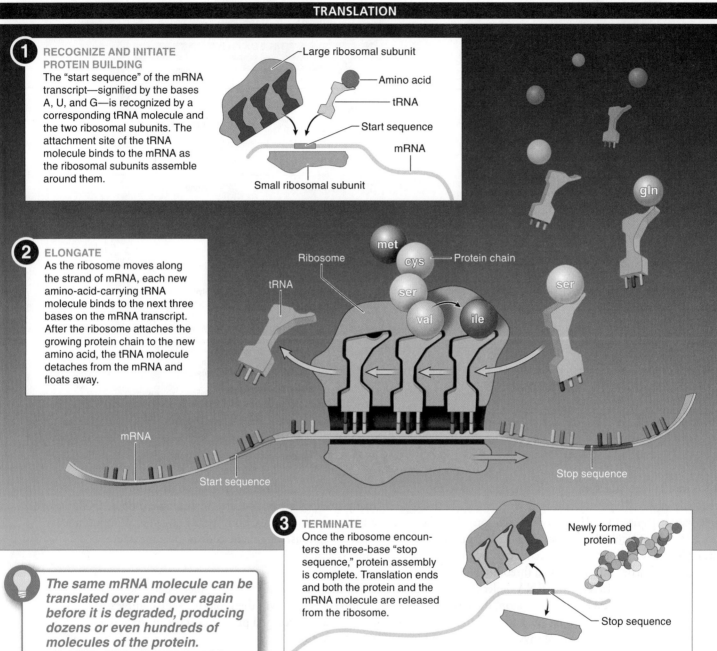

**1 RECOGNIZE AND INITIATE PROTEIN BUILDING**
The "start sequence" of the mRNA transcript—signified by the bases A, U, and G—is recognized by a corresponding tRNA molecule and the two ribosomal subunits. The attachment site of the tRNA molecule binds to the mRNA as the ribosomal subunits assemble around them.

- Large ribosomal subunit
- Amino acid
- tRNA
- Start sequence
- mRNA
- Small ribosomal subunit

**2 ELONGATE**
As the ribosome moves along the strand of mRNA, each new amino-acid-carrying tRNA molecule binds to the next three bases on the mRNA transcript. After the ribosome attaches the growing protein chain to the new amino acid, the tRNA molecule detaches from the mRNA and floats away.

- Ribosome
- Protein chain
- tRNA
- mRNA
- Start sequence
- Stop sequence

**3 TERMINATE**
Once the ribosome encounters the three-base "stop sequence," protein assembly is complete. Translation ends and both the protein and the mRNA molecule are released from the ribosome.

- Newly formed protein
- Stop sequence

*The same mRNA molecule can be translated over and over again before it is degraded, producing dozens or even hundreds of molecules of the protein.*

**FIGURE 5-15 Translation: reading a sequence of nucleotides and producing protein.** The second step in a two-step process by which DNA regulates a cell's activity and synthesis of proteins.

themselves into a ribosome, one side of a particular tRNA molecule also recognizes the start sequence on the mRNA and binds to it. This initiator tRNA has the amino acid methionine (met) bound to its other side. Thus, methionine is always the first amino acid in any protein that is produced. (Occasionally, in eukaryotes, this initial methionine is "edited out" later in the protein-building process.)

**Step 2. Elongate.** After the mRNA start sequence (AUG), the next three bases on the mRNA specify which amino acid–carrying tRNA molecule should bind to the mRNA. If the next three bases on the mRNA transcript are GUU, for example, a tRNA molecule that recognizes this sequence, and carrying its particular amino acid (valine, or val), will attach to the mRNA at that point. The ribosome then facilitates the connection of this second amino acid to the first. After the amino acid carried by a tRNA molecule is attached to the preceding amino acid, the tRNA molecule detaches from the mRNA and floats away.

As this process continues, the amino acid chain grows. The next three bases on the mRNA specify the next amino acid to be added to the first two. And the three bases after that specify the fourth amino acid, and so on. This process of progressively linking together the amino acids specified by an mRNA strand is called **protein synthesis,** because all proteins are chains of amino acids, like beads on a string. Precision in this process is essential. Any skipping or backing up or misreading as the bases of the mRNA are used to specify the amino acid sequence could alter the amino acid sequence (and possibly the normal functioning) of the protein specified by the mRNA.

The mRNA continues to be "threaded" through the ribosome, with the ribosome moving down the mRNA strand reading and translating its message in little three-base chunks. Each three-base sequence specifies the next amino acid, lengthening the growing amino acid strand.

**Step 3. Terminate.** Eventually, the ribosome arrives at the three-base sequence on the mRNA that signals the end of translation. Once the ribosome encounters this stop sequence, the assembly of the amino acid sequence is complete. Translation ends, and the amino acid strand and mRNA molecule are released from the ribosome. As it is being produced, the amino acid sequence folds and bends, based on the chemical features in the amino acid side chains (as we saw in Chapter 2). Through the folding and bending, the protein acquires its three-dimensional structure. The completed protein—such as a membrane protein or insulin or a digestive enzyme—may be used within the cell or packaged for delivery via the bloodstream to somewhere else in the body where it is needed.

Following the completion of translation, the mRNA strand may remain in the cytoplasm to serve as the template for producing another molecule of the same protein. In bacteria, an mRNA strand may last from a few seconds to more than an hour; in mammals, an mRNA may last several days. Depending on how long it lasts, the same mRNA strand may be translated hundreds of times. Eventually, it is broken down by enzymes in the cytoplasm.

---

**TAKE-HOME MESSAGE 5·7**

Translation is the second step in the two-step process by which information carried in DNA directs the synthesis of proteins. In translation, the information from a gene that has been encoded in the nucleotide sequence of an mRNA is read, and ingredients present in the cell's cytoplasm are used to produce a protein.

---

# 5·8 Genes are regulated in several ways.

Why do some people get sick more than others? Why is one person taller or shorter than another? More generally, what makes us look and act the way we do? We say that our health, growth, and development are the result of interactions between our genes and our environment. But how do our genes and our environment actually interact?

It matters which genes a person has, but carrying a gene is not the only thing that matters when it comes to **gene** expression—the production of the protein that the gene's sequence codes for. There is also the issue of **gene regulation,** the question of whether a gene is turned on—producing its protein product—or turned off. And this, commonly, hinges on a person's environment.

A powerful tool in the study of gene regulation is the microarray, a small chip that looks like a microscope slide that can be used to monitor the expression levels of thousands of genes simultaneously. Taking cells from one part of the body, researchers examine gene expression in that tissue. Microarrays are particularly useful in exploring how gene expression differs in response to an illness, or the treatment of an illness, or in response to aging.

For example, in a five-year medical study exploring the health consequences of loneliness, microarray analyses of gene activity in immune system cells revealed that 209 genes were expressed differently in two groups of participants. One group included individuals who were experiencing chronic social isolation and loneliness, while individuals in the other group were not.

Individuals in the "lonely" group had impaired transcription (decreased activity) of some genes associated with an effective stress response. They also had increased activity in genes associated with inflammatory processes that are implicated in a variety of diseases (FIGURE 5-16). Related studies have shown a similarly reduced immune function among stressed-out individuals—including one study of students during a stressful exam period!

**Controlling gene expression** Cells have a wide variety of mechanisms by which they can control when individual genes are expressed. One of the main ways expression is

*Lonely, isolated individuals have very different patterns of immune system gene expression.*

**FIGURE 5-16 Social isolation.** Very different patterns of immune-system gene expression occur in lonely, isolated individuals.

controlled is by transcription factors, proteins that bind to specific regulatory sites on the DNA. These sites are often located in front of genes. The regulation may be "positive

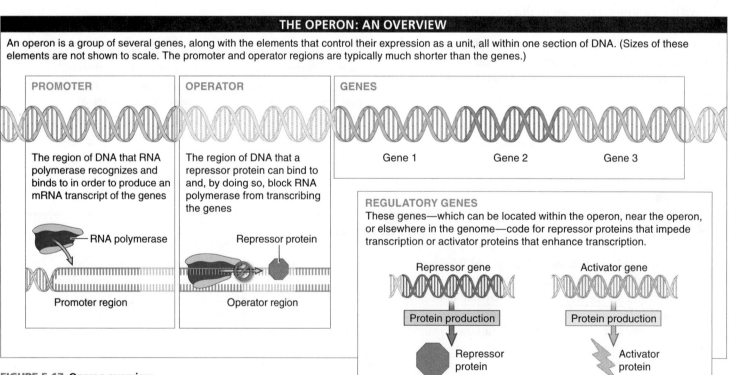

## THE OPERON: AN OVERVIEW

An operon is a group of several genes, along with the elements that control their expression as a unit, all within one section of DNA. (Sizes of these elements are not shown to scale. The promoter and operator regions are typically much shorter than the genes.)

**PROMOTER**

The region of DNA that RNA polymerase recognizes and binds to in order to produce an mRNA transcript of the genes

RNA polymerase

Promoter region

**OPERATOR**

The region of DNA that a repressor protein can bind to and, by doing so, block RNA polymerase from transcribing the genes

Repressor protein

Operator region

**GENES**

Gene 1    Gene 2    Gene 3

**REGULATORY GENES**

These genes—which can be located within the operon, near the operon, or elsewhere in the genome—code for repressor proteins that impede transcription or activator proteins that enhance transcription.

Repressor gene

Protein production

Repressor protein

Activator gene

Protein production

Activator protein

**FIGURE 5-17 Operon overview.**

control," in which binding of the regulatory protein initiates or speeds up gene expression, or "negative control," in which the protein slows or blocks gene expression.

Regulatory proteins may come from nearby cells. During the initial development of an organism, for example, gene regulation is extremely important in directing cells to differentiate into one type of tissue—perhaps muscle, rather than bone—and proteins produced by one cell commonly influence gene expression in neighboring cells.

## Prokaryotic gene control and the lac operon

Regulation may also be influenced by the nutrients an organism takes in. One well-studied example concerns the case of how the bacterium *E. coli* adjusts its diet depending on which sugars are available in its environment. More specifically, although its preference is for glucose, in the absence of glucose, it will utilize lactose as an energy source. (*E. coli* cells, many of which live in your intestine, are likely to encounter lactose, which is the chief sugar in milk.)

The absorption and digestion of lactose, however, requires transcription of three genes, and bacteria do not waste their energy transcribing these genes, translating the proteins, folding them, and sending them to the correct destinations unless lactose is actually present. That regulation is accomplished with one set of regulatory sequences called an **operon,** which is a group of several genes and the elements that control their expression as a unit, all within one section of the cell's DNA (**FIGURE 5-17**).

**Promoter.** For a gene to be transcribed, RNA polymerase recognizes and binds to the promoter region, the specific sequence of nucleotides in the DNA that signals the beginning of the gene.

**Operator.** A molecule called a repressor protein can bind to the operator, the regulatory portion of the DNA, and by doing so it blocks RNA polymerase from transcribing the genes necessary for lactose metabolism.

**Regulatory gene.** The regulatory gene codes for the repressor protein that, when bound to the operator region, does the blocking of RNA polymerase's binding to the promoter site.

In the absence of glucose (more on this below), here's how this operon (called the "lac operon") functions. If lactose is not present: (1) The regulatory gene produces the repressor protein. (2) The repressor protein binds to the operator region. And (3) RNA polymerase is blocked from transcribing the lactose metabolism genes (**FIGURE 5-18**).

### THE LAC OPERON IN ACTION

In the bacterium *E. coli*, glucose is the preferred food source, but it can utilize lactose as a backup. The lac operon makes it possible for the bacteria to not waste energy transcribing the three genes necessary for lactose absorption and digestion unless glucose is absent and lactose is actually present. Shown here is how on/off switching of the lac operon occurs **in the absence of glucose**.

#### IF LACTOSE IS NOT PRESENT: OPERON OFF

A repressor protein binds to the operator, preventing RNA polymerase from binding to the promoter and transcribing the genes necessary for lactose metabolism.

RNA polymerase
Repressor protein

Promoter   Operator   Gene 1   Gene 2   Gene 3

#### IF LACTOSE IS PRESENT: OPERON ON

Lactose (slightly modified) binds to the repressor protein, preventing it from binding to the operator. RNA polymerase can then bind to the promoter and transcribe the genes necessary for lactose metabolism.

Repressor protein
Lactose
RNA polymerase

Promoter   Operator   Gene 1   Gene 2   Gene 3
mRNA transcript

**FIGURE 5-18 Operon in action.** In *E. coli,* environmental conditions control the expression of the genes necessary for the metabolism of lactose.

If, on the other hand, lactose *is* present: (1) The regulatory gene produces the repressor protein. (2) Lactose binds to the repressor protein, altering its shape so that it can no longer bind to the operator. And (3) with no repressor bound to the DNA, RNA polymerase binds to the promoter and transcribes the genes necessary for lactose metabolism.

Note that the lac operon can only be switched on in the absence of glucose. Without glucose, a regulatory protein called an *activator* helps the DNA unwind in the vicinity of the lac operon, enabling RNA polymerase to bind and

## MECHANISMS OF REGULATING GENE EXPRESSION

### TRANSCRIPTION REGULATION

**POSITIVE CONTROL**
- Activators can initiate or speed up gene expression
- Enhancer sequences can speed RNA polymerase binding and gene transcription

**NEGATIVE CONTROL**
- Repressors can block or slow down gene expression
- Chemicals can bind to DNA and block gene transcription

### POST-TRANSCRIPTION REGULATION

- mRNA processing, transport, and enzymatic breakdown can be blocked or accelerated
- mRNA translation and protein processing rates can be altered

**FIGURE 5-19 Regulating gene expression.** Cells can increase or decrease gene expression in multiple ways.

**Eukaryotic gene control**   There are many other ways that genes can be regulated, besides operons. Many eukaryotes, for example, have enhancer sequences that increase transcription rates (positive control). Often far from the gene they regulate, these sequences, after binding a regulatory protein, bind "transcription factors" associated with RNA polymerase and enhance the rate at which the RNA polymerase binds to a promoter and transcribes a gene.

Additional mechanisms of gene control include the binding of chemicals to DNA that can block genes from being transcribed (negative control) and mRNA regulation. Messenger RNA can be regulated by altering mRNA processing and transport, or by altering the rate at which mRNA is broken down in the cytoplasm. Translation of mRNAs can be regulated, too, as can the processing of proteins following translation (**FIGURE 5-19**).

### TAKE-HOME MESSAGE 5·8

**Environmental signals influence the turning of genes on and off. By binding to DNA, regulatory proteins can block or facilitate the binding of RNA polymerase and subsequent transcription of genes. Regulation of gene expression also can occur in a variety of other ways that enhance or impede transcription, or alter mRNA's longevity and rate of degradation, or influence translation or protein processing.**

transcribe genes. When glucose is present, however, activator protein cannot bind to the DNA, and this prevents RNA polymerase from binding to and transcribing the lactose metabolism genes.

# 5·9–5·10 Damage to the genetic code has a variety of causes and effects.

*Damage to the genetic code can interfere with normal development.*

## 5·9 What causes a mutation, and what are its effects?

Through the two-step process of transcription and translation, an organism converts the information held in its genes into the proteins necessary for life. But the process is only as good as the organism's underlying genetic information. Sometimes something occurs to alter the sequence of bases in an organism's DNA. Such an alteration is called a **mutation,** and it can lead to changes in the structure and function of the proteins produced. Mutations can have a range of effects. Sometimes they result in a serious, even deadly, problem for an organism. Sometimes they have little or no detrimental effect. And occasionally—but very rarely—they may even turn out to be beneficial to the organism.

As an example of how mutations can affect organisms, consider the case of breast cancer in humans. When two human genes, called BRCA1 and BRCA2, are functioning properly, they help to prevent breast cancer by helping to repair DNA damage, preventing cells from accumulating the changes that lead to cancer. If the DNA sequence of either of these genes is altered through mutation and the gene's normal function is lost, the person carrying the gene has a significantly increased risk of developing breast cancer. (Because a variety of other factors, including environmental variables, are involved in development of cancer, it's impossible to know for certain whether these individuals will develop breast cancer.) Currently, more than 200 different changes in the DNA sequences of these genes have been detected, each of which results in an increased risk of developing breast cancer.

Given the havoc they can cause for an organism, it's not surprising that mutations have a bad reputation. After all, because they can change the protein produced, they can disrupt normal processes and harm the individual (**FIGURE 5-20**). But there are a couple of reasons why mutations' bad reputation may not be fully deserved. First, it turns out that many—perhaps even *most*—mutations are neutral, having neither a positive nor a negative effect on an organism's phenotype. This may be the case when a mutation occurs in a non-coding region of DNA. It can also occur when a change in DNA within a gene doesn't alter the function of the protein produced. Based on a recent study, researchers estimate that the rate of germ line mutations is approximately $10^{-8}$ per base pair per generation.

A second reason that mutations' bad reputation may be undeserved is the paradoxical fact that mutations are essential to evolution. Those mutations that don't kill an organism, or reduce its ability to survive and reproduce, can be beneficial. Every genetic feature in every organism was, initially, the result of a mutation. (In Chapter 8, we explore the relationship between mutation and evolution.) Ultimately, most mutations you inherit from your parents will have no effect. And all of you are carrying mutations that you will never know about!

It's important to note that mutations can occur in an organism's gamete-producing cells (that is, cells that produce sperm or eggs) as well as in its non-sex cells (such as skin cells

Normal fruit fly

Mutant fruit fly

FIGURE 5-20 **Wreaking havoc.** Mutations can change the protein produced by the altered gene, with disastrous consequences.

or cells in the lungs). Mutations in non-sex cells can have bad health consequences for the person carrying them. Many forms of cancer, such as lung cancer and skin cancer, result from such mutations. On the upside, non-sex-cell mutations are not passed on to your children. Mutations in the sex cells (gametes), on the other hand, do not have any adverse health effects on the person carrying them, but these mutations can be passed on to offspring, with terrible effects such as the induction of a miscarriage or the occurrence of birth defects. Individuals inheriting certain mutations from a parent—because the mutation occurred in the parent's sex cells or because the parent inherited the mutation from his or her parent—can be at increased risk for certain diseases such as breast cancer or cardiovascular disease.

Mutations generally fall into two types: point mutations and chromosomal aberrations (**FIGURE 5-21**). In point mutations, one base pair is changed. In chromosomal aberrations, entire sections of a chromosome are altered.

**Point mutations** occur when one base pair in the DNA is substituted for another, or when a base pair is inserted or deleted. Insertions and deletions can be much more harmful than substitutions, because the amino acid sequence of a protein is thrown off. If a single base is added or removed, the three-base groupings in an mRNA get thrown off, and the entire sequence of amino acids stipulated "downstream" from that point will be wrong—the reading frame is shifted. It's almost like putting your hands on a computer keyboard, but offset by one key to the left or right, and then typing what should be a normal sentence. It comes out as gibberish.

*Why do dentists put a heavy apron over you when they X-ray your teeth?*

**Chromosomal aberrations** are changes to the overall organization of the genes on a chromosome. Chromosomal aberrations are like the manipulation of large chunks of text when you are editing a term paper. The aberrations can involve the complete deletion of an entire section of DNA, the relocation of a gene from one part of a chromosome to elsewhere on the same chromosome or even to a different chromosome, or the duplication of a gene, with the new copy inserted elsewhere on the chromosome or on a different chromosome. Whatever the type of aberration, a gene's expression—the production of the protein that its sequence codes for—can be altered, as well as the expression of the genes around it.

Given the potentially hazardous health consequences of mutations, it is advisable to minimize their occurrence. Can this be done? Yes and no. There are three chief causes of mutation and, although one of them is beyond our control, the risk of occurrence of the other two can be significantly reduced (**FIGURE 5-22**).

**1. Spontaneous mutations.** Some mutations arise by accident as long strands of DNA are duplicating themselves—at the rate of more than a thousand bases a minute in humans—when cells are dividing (you'll read more details on this process in Chapter 6). Most errors are corrected by DNA repair enzymes, but some still slip through and there's not much we can do about them.

**2. Radiation-induced mutations.** Ionizing radiation, such as X rays, is radiation with enough energy to disrupt atomic structure—even break apart chromosomes—by removing tightly bound electrons. But even non-ionizing lower-energy radiation (which is not able to remove electrons) can damage

FIGURE 5-21 **Point mutations and chromosomal aberrations.**

*In point mutations, one nucleotide is changed, whereas in chromosomal aberrations, entire sections of a chromosome are altered.*

DNA. Ultraviolet (UV) rays from the sun, for example, can be absorbed by certain bases in DNA and cause them to rearrange bonds. This can prevent them from pairing correctly with the complementary DNA strand and can transform a cell into a cancer cell. This is why long-term sun exposure can contribute to the development of skin cancer. Because ionizing rays cannot pass through lead, the lead apron a doctor or dentist puts over you when you get an X ray protects your body from the ionizing radiation.

Another source of dangerous radiation is found in the core of nuclear power plants, where radioactive atoms are used and produced in energy-generating reactions. The high energy of the radioactivity that fuels the production of usable energy can pass through your body and disrupt your DNA, causing point mutations and chromosomal aberrations. With the proper

**Q**

Why is it dangerous to be near the core of a nuclear power plant?

**FIGURE 5-22 Gambling with mutation-inducing activities.** You can increase or decrease your risk of mutations with your behavior: smoking, tanning, and radiation (but notice the protective lead apron).

safety precautions, however, nuclear power plant workers can minimize their exposure to harmful radiation.

**3. Chemical-induced mutations.** Many chemicals, such as those found in cigarette smoke and in the exhaust from internal combustion engines, can also react with the atoms in DNA molecules and induce mutations.

In the next section, we examine how even tiny changes in the sequence of bases in DNA can lead to errors in protein production and profound health problems.

> **TAKE-HOME MESSAGE 5·9**
>
> Mutations are alterations in a single base or changes in large segments of DNA that include several genes or more. They are rare, but when they do occur, they may disrupt normal functioning of the body (although many mutations are neutral). Extremely rarely, mutations may have a beneficial effect. Mutations play an important role in evolution.

# 5·10  Faulty genes, coding for faulty enzymes, can lead to sickness.

Isabella joins her friends in sipping wine during a dinner party. As the meal progresses, her companions become tipsy. Their conversations turn racy, their moods relaxed. They refill their glasses, reveling in a little buzz. Not so for Isabella. Before her first glass is empty, she experiences a "fast-flush" response: her face turns crimson, her heart begins to race, and her head starts to pound. Worse still, she soon feels the need to vomit.

How can people respond so differently to alcohol? It comes down to a difference in a single base pair in their DNA, a single difference that can influence dramatically a person's behavior, digestion, respiration, and general ability to function. The base-pair change leads to the production of a non-functional enzyme, and the lack of a functional version of this enzyme leads to physical illness. Let's look at the details.

When we consume alcohol, our bodies start a two-step process to convert the alcohol molecules from their

intoxicating form into innocuous molecules. Each of the two steps is made possible by a specific enzyme, whose assembly instructions are, for most people, coded in their DNA.

"Fast-flushers" such as Isabella complete the first step of breaking down alcohol, but cannot fully complete the breakdown because they carry defective genetic instructions for making aldehyde dehydrogenase, the enzyme that makes possible the second step of the process. A poisonous substance subsequently accumulates, and the symptoms of the fast-flush reaction are due to this substance's toxic effects in the body.

Approximately half of the people living in Asia carry a non-functional form of the gene for aldehyde dehydrogenase, a mutation that may confer a greater benefit than harm. In

**Q** Why do many Asians have unpleasant experiences associated with alcohol consumption?

**FROM MUTATION TO ILLNESS**

**1** A mutated gene codes for a non-functioning protein, commonly an enzyme.

DNA

mRNA

Functioning enzyme should look like this:

Non-functioning enzyme

**2** The non-functioning enzyme can't catalyze the reaction as it normally would.

**3** The molecule it would have reacted with accumulates.

**4** The accumulating chemical causes sickness (fast-flushing, for example) or death (Tay-Sachs disease, for example).

**FIGURE 5-23 Faulty enzymes can interfere with metabolism.**

a study of 1,300 alcoholics in Japan, not a single one was a fast-flusher, even though half of all Japanese people are fast-flushers. The minor change in the genetic code that makes alcohol consumption an unpleasant experience may be responsible for the lower incidence of alcoholism among Japanese and other Asian people.

In many other cases—perhaps in the majority of genetic diseases—the link between a particular defective DNA sequence and physical illness is equally direct. Recall from Chapter 3 the case of Tay-Sachs disease. In Tay-Sachs disease, an individual inherits genes with a mutation that causes an inability to produce a critical lipid-digesting enzyme in their lysosomes, the cellular garbage disposals. Because these organelles cannot digest certain lipids, the lipids accumulate, undigested. The lysosomes swell until they eventually choke the cell to death. This occurs in numerous cells in the first few years of life, and ultimately leads to the child's death.

Although the details differ from case to case, the overall picture is the same for many, if not most, inherited diseases. The pathway from mutation to illness includes just four short steps (**FIGURE 5-23**).

1. A mutated gene codes for a non-functioning protein, commonly an enzyme.

2. The non-functioning enzyme can't catalyze a particular reaction as it normally would, bringing the reaction to a halt.

3. The molecule with which the enzyme would have reacted (to convert it to another substance) accumulates, just as half-made products would pile up on a blocked assembly line.

4. The accumulating chemical causes sickness and/or death.

The fact that many genetic diseases involve illnesses brought about by faulty enzymes suggests some strategies for treatment. These include administering medications that contain the normal-functioning version of the enzyme. For instance, lactose-intolerant individuals can consume the enzyme lactase, which for a short while gives them the ability to digest lactose. Alternatively, lactose-intolerant individuals can reduce their consumption of lactose-containing foods to keep the chemical from accumulating, thus reducing the problems that come from lactose overabundance.

**TAKE-HOME MESSAGE 5·10**

**Most genetic diseases result from mutations that cause a gene to produce a non-functioning enzyme, which in turn blocks the functioning of a metabolic pathway.**

# 5·11–5·13
## Biotechnology is producing improvements in agriculture.

*A scientist works with transgenic plants in a greenhouse.*

# 5·11
## What is biotechnology?

Understanding a phenomenon (such as how DNA is transcribed and translated) is nice, but controlling it can take the satisfaction to an entirely new level. This may explain why there is so much excitement surrounding the field of **biotechnology,** in which organisms, cells, and their molecules are modified to achieve practical benefits. The modern emphasis in biotechnology is on **genetic engineering,** the manipulation of organisms' genetic material by adding, deleting, or transplanting genes from one organism to another.

In the remainder of this chapter, we explore the two primary areas in which biotechnology is applied: agriculture and human health. We begin by introducing the most important tools and techniques used in biotechnology.

How would you create a plant resistant to being eaten by insects? Or a colony of bacteria that can produce human insulin? Although there are many different uses of biotechnology, it employs a surprisingly small number of processes and tools that enable researchers to do the following (**FIGURE 5-24**).

1. **Chop** up the DNA from a donor organism that exhibits the trait of interest.

**FIGURE 5-24 Five important tools and techniques of most biotechnology procedures.**

### FIVE TOOLS OF BIOTECHNOLOGY

**CHOP** up DNA from a donor species that exhibits a trait of interest.

**AMPLIFY** small samples of DNA into more useful quantities.

**INSERT** pieces of DNA into bacterial cells or viruses.

**GROW** separate colonies of bacteria or viruses, each containing some donor DNA.

**IDENTIFY** colonies of bacteria or viruses that have DNA for a trait of interest.

*Not all of the tools are used in all biotechnology applications—some only utilize one or a few of these techniques.*

2. **Amplify** the small amount of DNA into larger quantities.

3. **Insert** pieces of DNA into bacterial cells or viruses.

4. **Grow** separate colonies of the bacteria or viruses, each of which contains a different inserted piece of donor DNA.

5. **Identify** colonies of bacteria or viruses that have DNA for a trait of interest (and propagate them).

To be sure, it's easier said than done, but these tools capture the essence of most modern biotechnology. Not all of the techniques are used in all biotech applications; nonetheless, they all are mainstays of today's biotechnology world. Let's explore each technique in a bit more detail.

**Tool 1. Chopping up DNA from a donor organism.** To begin, researchers select an organism that has a desirable trait. For example, they might want to produce human growth hormone in large amounts. Their first step would be to obtain a sample of human DNA and cut it into smaller pieces. Cutting DNA into small pieces requires the use of **restriction enzymes** (**FIGURE 5-25**).

Restriction enzymes have a single function: when they encounter DNA, they cut it into small pieces. These enzymes evolved to protect bacteria from attack by viruses. Upon encountering DNA from an invading virus, a restriction enzyme recognizes and binds to a particular sequence of four to eight bases on the invader's DNA and cuts there, thus making it impossible for the virus to reproduce within the bacterial cell. Since all DNA has the same structure, these enzymes can cut DNA from any source as long as the specific four-eight base sequence is present (bacteria protect their own DNA by modifying it, or it would be cut up also). Dozens of different restriction enzymes exist, each of which cuts DNA at a different location.

**Tool 2. Amplifying DNA pieces into larger quantities.** In many situations, only a small amount of DNA is available for analysis or for some other biotechnology use. The **polymerase chain reaction (PCR)** is a laboratory technique that allows a tiny piece of DNA (perhaps one recovered from a crime scene) to be duplicated repeatedly (**FIGURE 5-26**).

The process involves heating up the DNA of interest, causing the two strands to separate. Then, in the presence of enzymes and plenty of free nucleotides, the DNA is cooled. As it cools, an enzyme matches free nucleotides

**CHOP: RESTRICTION ENZYMES**

**1** The gene of interest is located on a section of DNA from the donor species.

Gene of interest

Source DNA

**2** Restriction enzymes that target a particular base-pair sequence on either side of the gene are introduced.

Restriction enzymes

In this example, the restriction enzyme recognizes the sequence ATCGAT and cuts between the first A and T.

**3** The restriction enzymes bind to their target base-pair sequence and cut the strand of DNA.

**4** The gene of interest has now been separated from the donor's DNA.

**FIGURE 5-25 Restriction enzymes are used to isolate a gene of interest.**

with their complementary bases on each of the single strands, and covalently links these nucleotides together to form a new single strand across from each original. The result is two complete double-stranded copies of the DNA of interest. This process of heating and cooling can be repeated again and again until there are billions of identical copies of the target sequence.

**Tool 3. Inserting foreign DNA into the target organism.** In the human growth hormone example mentioned above,

**1** A solution containing an isolated segment of DNA is heated, separating the double-stranded DNA into two single strands.

Heat

**2** The enzyme DNA polymerase is added along with a large number of free nucleotides, and the solution containing the segments is cooled.

Cool

DNA polymerase

Free nucleotides

**3** DNA polymerase adds complementary bases to each single strand.

**4** The result is two identical copies of the original segment of DNA.

*This process can be repeated again and again until there are billions of identical copies of the target sequence.*

**FIGURE 5-26 The polymerase chain reaction can duplicate a small strand of DNA into billions of copies.**

the researchers might want to transfer the human growth hormone gene into the bacterium *E. coli,* creating **transgenic organisms.**

To create a transgenic organism (that is, an organism with DNA inserted from a different species), researchers must physically deliver the DNA from a donor species into the recipient organism. This delivery often is accomplished using **plasmids,** circular pieces of DNA that can be

incorporated into a bacterium's genome (**FIGURE 5-27**). Genes on the plasmid can then be expressed in the bacterial cell and are replicated whenever the cell divides, so both of the new cells contain the plasmid. In some cases, for the delivery process, genes are incorporated into viruses instead of plasmids. The viruses can then be used to infect organisms and transfer the genes of interest into those organisms.

**Tool 4. Growing bacterial colonies that carry the DNA of interest: cloning.** Once a piece of foreign DNA has been transferred to a bacterial cell, every time the bacterium divides, it creates a **clone,** a genetically identical cell that contains that inserted DNA. The term **cloning** describes

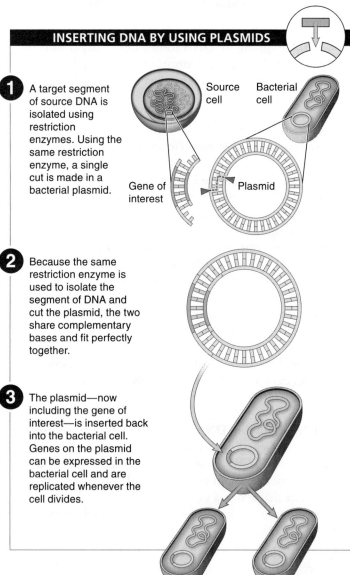

**1** A target segment of source DNA is isolated using restriction enzymes. Using the same restriction enzyme, a single cut is made in a bacterial plasmid.

Source cell

Bacterial cell

Gene of interest

Plasmid

**2** Because the same restriction enzyme is used to isolate the segment of DNA and cut the plasmid, the two share complementary bases and fit perfectly together.

**3** The plasmid—now including the gene of interest—is inserted back into the bacterial cell. Genes on the plasmid can be expressed in the bacterial cell and are replicated whenever the cell divides.

**FIGURE 5-27 A gene chauffeur.** Plasmids can transfer DNA from one species to another.

the production of genetically identical cells, organisms, or DNA molecules, a process that occurs each time a bacterium divides. With numerous rounds of cell division, it is possible to produce a huge number of clones, all of which transcribe and translate the gene of interest.

In a typical experiment, a large amount of DNA may be chopped up with restriction enzymes, incorporated into plasmids, and introduced into bacterial cells. The bacteria are then allowed to divide repeatedly, with each producing a clone of the foreign DNA fragment they carry. Together, all of the different bacterial cells containing all of the

different fragments of the original DNA are called a **clone library** or a **gene library** (FIGURE 5-28).

**Tool 5. Identifying bacterial colonies that have received the gene of interest (and propagating them).** Chopping up human DNA and inserting the pieces (each of which carries different genes) into the genomes of bacteria, which divide repeatedly, leads to a large population of bacterial cells, with many carrying useful genetic information. But the information is in no particular order, much like a bookstore in which all the books are uncatalogued and in complete disarray. How can a researcher interested in working with bacterial cells that contain just the one human gene capable of producing human growth hormone identify and separate those bacteria from the other bacteria in the population?

Researchers have developed a way to make the bacteria of interest identify themselves (FIGURE 5-29). First, a chemical

## GROW: CREATING A GENE LIBRARY

❶ To create a gene library, a large amount of DNA is chopped up using restriction enzymes. ❷ Each piece is inserted into a plasmid that is then introduced into a bacterial cell. ❸ The bacteria are allowed to divide repeatedly, each producing a clone of the foreign DNA fragment it carries. ❹ Together, all of the different bacterial cells contain all of the different fragments of the original DNA.

❶ Gene A    Gene B    Gene C

❷ Plasmid A    Plasmid B    Plasmid C

❸ Bacterium A    Bacterium B    Bacterium C

❹ GENE LIBRARY

**FIGURE 5-28 DNA collection.** A gene library (or clone library) is a collection of cloned DNA fragments.

## IDENTIFYING A GENE BY USING A DNA PROBE

GENE LIBRARY

❶ To locate a gene of interest among the millions of clones in a gene library, the bacteria are washed with a chemical that breaks down the DNA, making it single-stranded.

Gene of interest

❷ A radioactive probe is then washed over the single-stranded DNA. The probe is a short length of DNA that contains a sequence of bases complementary to the gene of interest.

DNA probe

❸ The DNA probe binds to the complementary base sequence found in the gene of interest and glows with radioactivity, allowing the gene to be easily identified.

G T A C T
C A T G A

T A G C T

A T C C G

**FIGURE 5-29 Finding the needle in a haystack.** A DNA probe is used to locate the desired clone in a gene library.

is added to the entire population of bacterial cells, separating the double-stranded DNA into single strands. Next, a short sequence of single-stranded DNA is washed over the bacteria. Called a **DNA probe,** this DNA contains part of the sequence of the gene of interest and has also been modified so that it is radioactive. Bacteria with the gene of interest bind to this probe and glow with radioactivity. These cells can then be separated out and grown in large numbers—for example, vats of *E. coli* that produce human growth hormone. (Note that because the "correct" folding of a protein depends not just on the amino acid sequence specified by a gene but also on the appropriate cellular conditions, genes transplanted from one species to another don't always produce proteins with the exact same shape. Such folding problems can lead to proteins that do not function as expected.)

We turn next to how these tools and techniques are used to develop products. Keep in mind, however, that the field is still in its infancy.

# 5·12 Biotechnology can improve food nutrition and make farming more efficient and eco-friendly.

Your breakfast cereal is probably fortified with vitamins and minerals. And for snacking you may eat protein bars that have as much protein as a full chicken breast. It shouldn't come as a surprise, then, that farmers have begun using biotechnology to improve on the natural levels of vitamins, minerals, and other nutrients in the fruits, vegetables, and livestock they produce.

For thousands of years, humans have been practicing a relatively crude and slow form of **genetic engineering**—the manipulation of a species' genome in ways that do not normally occur in nature. In its simplest form, genetic engineering is the careful selection of the plants or animals used as the breeders for a crop or animal population. Through this process, farmers and ranchers have produced meatier turkeys, seedless watermelons, and big, juicy corn kernels (**FIGURE 5-30**). But what used to take many generations of breeding can now be accomplished in a fraction of the time, using **recombinant DNA technology,** the combination of DNA from two or more sources into a product.

In crop plants, for example, the process begins with the identification of a new characteristic, such as larger size or faster ripening time, that farmers would like in a particular crop. Traditionally, breeders would then search for an organism within the same species that had the desirable trait, breed it with their crop organisms, and hope that the offspring would express the trait in the desired way. With recombinant DNA technology, the desired trait can come from *any* species, so the pool of organisms from which the trait can be taken becomes much larger. Organisms produced with recombinant DNA technology are referred to as genetically modified organisms, or GMOs.

Although the rewards are potentially huge, in practice the process of creating transgenic species that are more nutritious or have other desirable traits turns out to be difficult.

Ancestral corn

Modern corn

**FIGURE 5-30 Ancestral corn and modern corn—selected for larger, juicier kernels—compared.**

**Q** Nonetheless, the results so far hint at a fruitful marriage of agriculture and technology.

**How might a genetically modified plant help 500 million malnourished people?**

### Nutrient-rich "golden rice"

Almost 10% of the world's population suffers from vitamin A deficiency, which causes blindness in a quarter-million children each year and a host of other illnesses in people of all ages. These nutritional problems are especially severe in southern Asia and sub-Saharan Africa, where rice is a staple of most diets. Addressing this global health issue, researchers have developed what may be the model for solving problems with biotechnology. It involves the creation of a new crop called "golden rice."

Mammals generally make vitamin A from beta-carotene, a substance found in abundance in most plants (it's what makes carrots orange), but not in the edible part of rice grains. Researchers set out to change this by inserting into the rice genome, from other species, three genes that code for the enzymes used in the production of beta-carotene: two genes from the daffodil plant and one bacterial gene (**FIGURE 5-31**). It's clear that the transplanted genes are working because the normally white rice grain takes on a golden color from the accumulated beta-carotene. And since golden rice was first developed in 1999, new lines have been produced that use only two transplanted genes, yet produce almost 25 times the vitamin A found in the original strains. Field tests of golden rice are still under way, but it is viewed as one of the most promising applications yet of biotechnology.

> "Italians come to ruin most generally in three ways, women, gambling, and farming. My family chose the slowest one."
> —POPE JOHN XXIII

While the development of golden rice demonstrates that biotechnology, in the near future, may help us produce food that is more nutritious, biotechnology has already had a much more profound impact than this on agricultural practices in the United States. It is not a stretch to say that we are in the midst of a revolution—a green revolution—and that few people are aware of it. This revolution is the extent to which biotechnology has reduced the costs, both environmental and financial, of producing the plants and animals we eat.

Currently, more than 170 million acres worldwide are planted with genetically modified crops, most containing built-in insecticides and herbicide

**GOLDEN RICE**

Daffodil

Bacteria

Genes in the pathway for beta-carotene production are introduced into the white rice genome.

White rice

Beta-carotene

Golden rice

💡 *The addition of beta-carotene-producing genes to white rice has increased its vitamin A content almost 25-fold.*

**FIGURE 5-31 The potential to prevent blindness in 250,000 people each year.** Engineering rice to prevent blindness by increasing its vitamin A content.

resistance, representing more than a forty-fold increase over the past 10 years. The financial benefits to farmers—at least in the short run—are so great that more and more of them are embracing the genetically modified crops.

**Q**

**What genetically modified foods do most people in the United States consume (usually without knowing it)?**

The numbers are surprising: 86% of all corn grown in the United States is genetically modified. Ninety-three percent of all cotton grown is genetically modified. And 93% of all soybeans grown are genetically modified (**FIGURE 5-32**). Two factors explain much of the extensive adoption of genetically modified plants in U.S. agriculture. (1) Many plants have had insecticides engineered into them, which can reduce the amounts of insecticides used

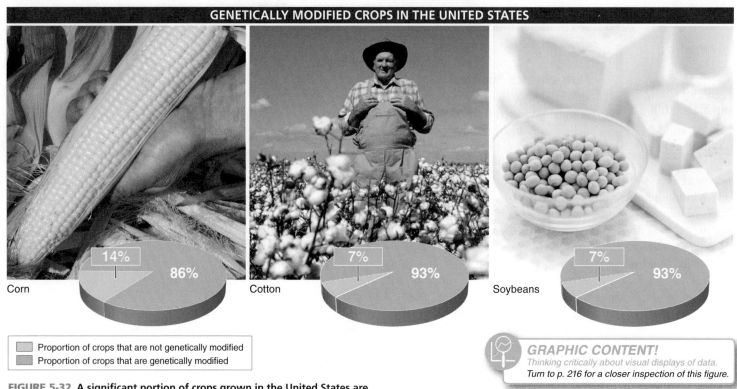

Corn 14% 86%

Cotton 7% 93%

Soybeans 7% 93%

☐ Proportion of crops that are not genetically modified
☐ Proportion of crops that are genetically modified

*GRAPHIC CONTENT!*
*Thinking critically about visual displays of data.*
*Turn to p. 216 for a closer inspection of this figure.*

**FIGURE 5-32 A significant portion of crops grown in the United States are genetically modified.**

in agriculture. (2) Many plants also have herbicide-resistance genes engineered into them. Such herbicide-resistant plants as well as insect-resistant plants can reduce the amount of plowing required around crops to remove weeds. As a consequence, then, the use of genetically modified plants can reduce both the costs of producing food and the loss of topsoil to erosion.

Here are some of the biggest successes in the application of recombinant DNA technology to agriculture.

**Insect and herbicide resistance** Insect pests have a field day on agricultural crops (**FIGURE 5-33**). Crops planted at high densities and nurtured with ample water and fertilizer represent a huge potential food resource for insects. Every year, about 40 million tons of corn are unmarketable as a consequence of insect damage. Increasingly, however, farmers have been enjoying greater success in their battles against insect pests.

Farmers owe much of this success to soil-dwelling bacteria of the species *Bacillus*

*thuringiensis* (in brief, Bt). These bacteria produce spores containing crystals that are poisonous to insects but harmless to the crop plants and to people. In insects, within an hour of

**FIGURE 5-33 Competing with us for food.** Insects can ravage crops.

Bacterium (*Bacillus thuringiensis*)

Bt crystal gene

Plant cell

Corn genome

**1** Corn plant destroyed by butterfly larvae (caterpillars).

**2** Bacterial gene coding for Bt crystals, which are poisonous to the caterpillars, is inserted directly into the corn plant's DNA.

**3** Bt crystals—toxic to caterpillars—are now produced by the corn plant itself and present in all cells, reducing the amount of pesticides the farmer must use.

**FIGURE 5-34 Help from bacteria in growing disease-resistant corn.** Corn engineered to contain spores of the bacterium *Bacillus thuringiensis* (Bt) kills insect pests but does not harm humans.

ingestion, Bt crystals cause pores to develop throughout the insects' digestive systems, paralyzing their gut and making them unable to feed. Within a few days, the insects die from a combination of tissue damage and starvation.

**Q** How can genetically modified plants lead to reduced pesticide use by farmers?

Beginning in 1961, the toxic Bt crystals were included in the pesticides sprayed on crop plants. Then in 1995, the gene coding for the production of the Bt crystals was inserted directly into the DNA of many different crop plants, including corn, cotton, and potatoes. As a consequence, farmers no longer need to apply huge amounts of Bt-containing pesticides (**FIGURE 5-34**); instead, the plants themselves produce insect-killing Bt crystals. There is no evidence that humans are harmed by the Bt crystals, even when they are exposed to very high levels.

Bacteria have come to the aid of farmers in their fight against pests in another way, too. In the 1990s, researchers discovered a bacterial gene that gives the bacteria resistance to herbicides, and they introduced the gene into crop plants. Integration of this gene into the plants' DNA gives them resistance to herbicides, allowing farmers to kill weeds with herbicides but leave the crop plants unharmed, greatly increasing yields (**FIGURE 5-35**).

**Faster growth and bigger bodies** Agriculture includes the cultivation not just of plants but also of animals. And for

**FIGURE 5-35 Crop duster.** Herbicides like the one applied by this crop duster must kill weeds while leaving the crop unharmed.

**FIGURE 5-36 Bigger salmon.** The larger salmon carries a growth hormone gene that keeps it growing year-round rather than in the summer only.

the first time, the U.S. Food and Drug Administration is close to approving for human consumption a genetically modified animal. The animal in question is a transgenic Atlantic salmon that carries a growth hormone gene from another species (Chinook salmon), along with a region of DNA from a third species (ocean pout) that acts as an "on" switch, facilitating transcription of the growth hormone gene. The transgenic fish, which is reported by its creators to taste the same as regular Atlantic salmon, grows much more quickly and reaches market size within 18 months rather than the usual three years (**FIGURE 5-36**).

The FDA has reported that the transgenic salmon "is as safe to eat as food from other Atlantic salmon." Numerous fisheries experts, food safety experts, environmental groups,

and consumer groups, however, continue to express concerns about a wide variety of safety and environmental issues and the process by which the safety and environmental impacts of the transgenic species have been evaluated. Health concerns include the possibility that consuming the salmon will cause increased rates of allergic reactions, as well as unknown effects that may stem from potentially higher levels of hormones present in the fish.

Most troubling to environmental groups is the risk that the larger, faster-growing transgenic fishes will escape from their enclosed breeding facilities and back into their natural habitat—something that many experts believe is inevitable. If this occurs, environmentalists fear that the fish might harm wild salmon populations, many of which are listed as endangered, because the transgenic salmon can consume more resources and may grow too large to be consumed by its natural predators. It is unclear what the outcome would be.

---

### TAKE-HOME MESSAGE 5·12

**Biotechnology has led to important improvements in agriculture by using transgenic plants and animals to produce more nutritious food. Even more significant is the extent to which biotechnology has reduced the environmental and financial costs of producing food, through the creation of herbicide-resistant and insect-resistant crops. The ecological and health risks of such widespread use of transgenic species are not fully understood and are potentially great.**

---

# 5·13 Fears and risks: are genetically modified foods safe?

Chickens without feathers look ridiculous (**FIGURE 5-37**). But such a genetically modified breed was developed with a valuable purpose in mind: "naked" birds are easier and less expensive to prepare for market, benefiting farmers by lowering their costs and benefiting consumers by lowering prices. These chickens, however, have turned out to be unusually vulnerable to mosquito attacks, parasites, and disease, and ultra-sensitive to sunlight. They also have difficulty mating, because the males are unable to flap their wings. Researchers currently are working to address these problems.

Naked chickens teach us an important lesson about genetically modified plants and animals. Although the new breed of

featherless chickens was produced by relatively low-tech genetic engineering—the traditional animal husbandry method of crossbreeding two different types of chickens—rather than by more modern recombinant DNA technology, the new breed ended up having not just the desired trait of no feathers but also some unintended and undesirable traits. Now, as more genetically modified foods are created using modern methods of recombinant DNA technology, the same risks of also creating unintended and potentially harmful traits must be weighed. For these and other reasons—some legitimate and rational, others irrational—many people have concerns about the production and consumption of genetically modified foods (**FIGURE 5-38**). These concerns are outlined in the following discussion.

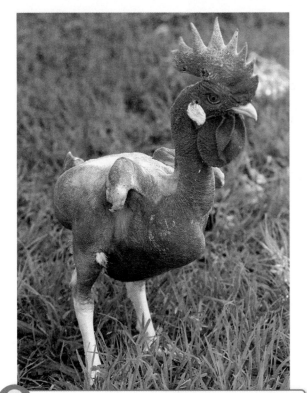

Featherless birds are cheaper for farmers and consumers. But there are unintended consequences, including vulnerability to mosquitoes and other parasites.

**FIGURE 5-37** **"Naked" birds.** The breeding of featherless chickens has benefited farmers but led to some unanticipated consequences.

*Organisms that we want to kill may become invincible.* Pesticide-resistant canola plants were cultivated in Canada, making it possible for farmers to apply herbicides freely to kill the weeds but not the canola crop. But the pesticide-resistant canola plants accidentally spread to neighboring farms and grew out of control, because traditional herbicides could not kill them. Similarly, there is the possibility that insect pests will develop resistance to the Bt produced by genetically modified crops, which will also make these pests resistant to Bt pesticides applied to crops that are not genetically modified.

*Organisms that we don't want to kill may be killed inadvertently.* Monarch butterflies feed on milkweed plants. Recent research has demonstrated that if pollen from plants genetically modified to contain the insect-killing Bt genes accidentally lands on milkweed plants and is consumed by monarch butterflies, the butterflies can be killed, which may significantly reduce their populations. Although such an incident has not occurred outside experimental fields, it illustrates a risk that may be hard to control.

*Genetically modified crops are not tested or regulated adequately.* It is impossible to really know whether a new technology has been tested adequately. Still, scientists and lawmakers have been working toward an organized and responsible set of policies designed to ensure that sufficient safety testing is done. For example, laboratory procedures for working with recombinant DNA have been established, and researchers have developed techniques that make it impossible for most genetically engineered organisms to survive outside the specific conditions for which they are developed.

As an example of the degree of testing of genetically engineered foods, the Monsanto Company has had its strain of herbicide-resistant soybeans evaluated and approved by 31 different regulatory agencies in 17 different countries, including, in the United States, the Department of Agriculture, the Food and Drug Administration, and the Environmental Protection Agency. In a recent report on genetically modified animals, however, an expert committee of the U.S. National Academy of Sciences warned that GMOs still pose risks that the government is unable to evaluate. Technology is moving so fast that it is difficult to even know what the new risks might be.

> ❝And he gave it for his opinion, 'that whoever could make two ears of corn, or two blades of grass, to grow upon a spot of ground where only one grew before, would deserve better of mankind, and do more essential service to his country, than the whole race of politicians put together.'❞
>
> —JONATHAN SWIFT, *Gulliver's Travels,* 1726

**FIGURE 5-38** **Consumer fears.** Protesters voice opposition to the use of genetically modified organisms (GMOs).

*Eating genetically modified foods is dangerous.* In the 1990s, a gene from Brazil nuts was used to improve the nutritional content of soybeans. The genetically modified soybeans had better nutritional content, but they also acquired some allergy-causing chemicals previously present in the Brazil nuts but not in soybeans. This outcome illustrates the risk that some unwanted features might be passed from species to species in the creation of transgenic organisms. In this case, all of the genetically modified soybeans were destroyed and this research program was suspended. To date, no evidence has appeared to suggest that consumption of any genetically modified foods is dangerous.

*Loss of genetic diversity among crop plants is risky.* As increasing numbers of farmers stop using non-genetically modified crops in favor of one or a few genetically modified strains of crops, the genetic diversity of the crops declines. This can make them more vulnerable to environmental changes or pests. The Irish Potato Famine is an example of the value of genetic diversity in crops. In the mid-1800s, much of the population of Ireland depended on a diet of potatoes. Because most of the potato crops had been propagated from cuttings from the same plant, they were all genetically the same. When the crops were infected by a rot-causing mold, all of the potato plants were susceptible and most were wiped out, causing a famine responsible for the deaths of more than a million people.

*Hidden costs may reduce the financial advantages of genetically modified crops.* When seed companies create genetically modified seeds with crop traits desirable to farmers, the companies also engineer sterility into the seeds. As a consequence, the farmers must purchase new seeds for each generation of their crops. Such increases in long-term costs and dependency on seed companies must be factored in by the farmers.

Another argument sometimes made in opposition to genetically modified foods is that such organisms are not "natural" and, for that reason, must be harmful. This is one argument that is flawed and should not be a cause for concern. Smallpox, HIV, poison ivy, and cyanide, after all, are natural. The smallpox vaccine, on the other hand, is unnatural. Innumerable other valuable technological developments are equally unnatural. There simply is no value in knowing whether something is natural or unnatural when evaluating whether or not it is good and desirable.

In the end, we must compare the risks of producing genetically modified foods with the benefits. The cost-benefit analyses will have to include the potential to reduce food costs and the ability to reduce environmental degradation by agriculture. For example, with genetically modified, pest-resistant crops, farmworkers will greatly benefit from spending less time applying pesticides. These benefits of reduced pesticide exposure will be significantly greater for workers in the less-developed countries, where safety regulations for pesticide use are more frequently ignored. Unfortunately, the difficulties in establishing the risks of genetically modified foods—as well as the ecological magnitude of those risks—ensures that this important issue must be evaluated extremely closely and regulated carefully.

## TAKE-HOME MESSAGE 5·13

**More and more genetically modified foods are being created using modern methods of recombinant DNA technology. Numerous legitimate fears among the public remain, however, about the potentially catastrophic risks of these foods, given that their development relies on such new technology, and about the long-term financial advantages they offer.**

5·14–5·17
Biotechnology
has the potential
for improving human
health (and criminal
justice).

*With the cloning of sheep, a new era in biotechnology began.*

## 5·14 The treatment of diseases and production of medicines are improved with biotechnology.

You can't always get what you want. In the best of all worlds, biotechnology would *prevent* humans from ever getting debilitating diseases. Next best would be to *cure* diseases once and for all. But these noble goals are not always possible, so biotechnology often is directed at the more practical goal of *treating* diseases, usually by producing medicines more efficiently and more effectively than they can be produced with traditional methods. Biotechnology has had some notable successes in achieving this goal. The treatment of diabetes is one such success story.

Diabetes is a chronic disease in which the body cannot produce insulin, a chemical that allows cells to take up and break down sugar from the blood. Complications from diabetes can include vascular disease, kidney damage, and nerve damage. As recently as 1980, if you were one of the approximately 15 million Americans with diabetes, each day you would treat the disease by injecting yourself with insulin extracted from the pancreas of cattle or pigs that had been killed for meat. For most diabetics, the insulin injections kept the disease under control. But the traditional process of collecting insulin this way was difficult and costly.

Everything changed in 1982 when a 29-year-old entrepreneur, Bob Swanson, joined scientist Herbert Boyer to transform the

**Q** Why do some bacteria produce human insulin?

potential of recombinant DNA technology. In doing so, they started the biotech revolution. Working with the scientist Stanley Cohen, Swanson and Boyer used restriction enzymes to snip out the human DNA sequence that codes for the production of insulin. They then inserted this sequence into the bacterium *E. coli,* creating a transgenic organism. After cloning the new, transgenic bacteria, the team was able to grow vats and vats of the bacterial cells, all of which churned out human insulin (**FIGURE 5-39**). The drug could be produced efficiently in huge quantities and made available for patients with diabetes. This was the first genetically engineered drug approved by the Food and Drug Administration and it continues to help millions of people every day.

Perhaps even more significant than providing a better source of insulin, Swanson, Boyer, and Cohen's application of biotechnology revealed a generalized process for genetic engineering. It instantly opened the door to a more effective method of producing many different medicines to treat diseases. Today, more than 1,500 companies work in the recombinant DNA technology industry, and their products generate more than $40 billion in revenues each year.

Several important achievements followed the development of insulin-producing bacteria. Here are just two examples.

**FIGURE 5-39 Life-saving insulin.** Human insulin is engineered through recombinant DNA technology.

**FIGURE 5-40 Bulking up with a little (illegal) help.** Sylvester Stallone—age 61 in this photo—used human growth hormone (along with strength training) to develop larger muscles for a film role.

**1. Human growth hormone (HGH).** Produced by the pituitary gland, human growth hormone has dramatic effects throughout the body. It stimulates protein synthesis, increases the utilization of body fat for energy to fuel metabolism, and stimulates the growth of virtually every part of the body (**FIGURE 5-40**). Insufficient growth hormone production, usually due to pituitary malfunctioning, leads to dwarfism.

When treated with supplemental HGH, individuals with dwarfism experience additional growth. Until 1994, however, HGH treatment was prohibitively expensive because the growth hormone could only be produced by extracting and purifying it from the pituitary glands of human cadavers.

*What is "blood doping"? How does it improve some athletes' performance?*

Through the creation of transgenic bacteria, using a technique similar to that used in the creation of insulin-producing bacteria, HGH can now be produced in virtually unlimited supplies and made available to more people who need it.

The availability of HGH, which can increase strength and endurance, may be irresistibly tempting to some people (who don't need it for medical reasons)—even at $7,500 for a month's supply. Recent sporting scandals suggest that the illegal use of HGH occurs frequently among elite swimmers, cyclists, and other athletes.

**2. Erythropoietin.** Produced primarily by the kidneys, erythropoietin (also known as EPO) is a hormone that regulates the production of red blood cells. Numerous clinical conditions (nutritional deficiencies and lung disease, among others) and treatments (such as chemotherapy) can lead to anemia, a lower than normal number of red blood cells, which reduces an individual's ability to transport oxygen to tissues and cells. This lack of sufficient oxygen, in turn, can cause a variety of symptoms, including weakness, fatigue, and shortness of breath.

First cloned in 1985, recombinant human erythropoietin (rhu-EPO) is now produced in large amounts in hamster ovaries. It is used to treat many forms of anemia. Worldwide sales of EPO are in the billions of dollars.

EPO has been at the center of several "blood doping" scandals in professional cycling. This hormone increases the oxygen-carrying capacity of the blood, so some otherwise healthy athletes have used EPO to improve their athletic performance. It can be very dangerous, though. By increasing the number of red blood cells, the blood can become much thicker and this can increase the risk of heart attack.

Beyond these and other medicines currently produced by transgenic organisms, plans are under way to create a variety of other useful products—including potatoes that produce antibodies enabling a more effective response to illnesses—for treating disease. In the next section we examine the strategies for preventing genetic diseases, and the much less successful attempts to cure diseases through biotechnology.

> **TAKE-HOME MESSAGE 5·14**
>
> Biotechnology has led to some notable successes in treating diseases, usually by producing medicines more efficiently and effectively than they can be produced with traditional methods.

# 5·15 Gene therapy: biotechnology can help diagnose and prevent diseases, but has had limited success in curing them.

Would you want to know? Once, this was just a hypothetical question: if you carried a gene that meant you were likely to develop a particular disease later in your life, would you want to know about it? Or another question: would you want to know if there's a good chance that your future children will be born with a genetic disease? Now, for better or for worse, these are becoming real-life questions that we all must address. And there is more at stake than simply peace of mind. As biotechnology develops the tools to identify some of the genetic time-bombs that many of us carry, it also carries the danger that such information may become the basis for greater discrimination than we have ever known.

Intervening to prevent diseases through biotechnology focuses on answering certain questions posed at three different points in time.

**1. Is a given set of parents likely to produce a baby with a genetic disease?** Many genetic diseases occur only if an individual inherits two copies of the disease-causing gene, one from each parent. This is true for Tay-Sachs disease, cystic fibrosis, and sickle-cell anemia, among others. Individuals with only a single copy of the disease-causing gene never fully show the disease, but they may pass on the disease gene to their children. Consequently, two healthy parents (that is, having no disease symptoms) may produce a child with the disease. In these cases, it can be beneficial for the parents to be screened to determine whether they carry a disease-causing copy of the gene. Such screening, combined with genetic counseling and testing of embryos following fertilization, can reduce the incidence of the disease dramatically. Since screening began in 1969, the incidence of Tay-Sachs disease, for example, has fallen by more than 75% (**FIGURE 5-41**).

*Since screening began in 1969, the incidence of Tay-Sachs disease has been reduced by more than 75%!*

**FIGURE 5-41 Genetic screening can determine the presence of the Tay-Sachs gene.**

**2. Will a baby be born with a genetic disease?** Once fertilization has occurred, it is possible to test an embryo or developing fetus for numerous genetic problems. Prenatal genetic screening can detect cystic fibrosis, sickle-cell anemia, Down syndrome, and a rapidly growing list of other disorders.

To screen the fetus, doctors must examine some of the fetal cells and/or the amniotic fluid (which surrounds the fetus in the uterus and carries many chemicals produced by the developing embryo). Cells and fluid are usually collected by amniocentesis or chorionic villus sampling (CVS), techniques that we explore in detail in Chapter 6.

**3. Is an individual likely to develop a genetic disease later in life?** DNA technology can also be used to detect disease-causing genes in individuals who are currently healthy but are at increased risk of developing an illness later. Early detection of many diseases, such as breast cancer, prostate cancer, and skin cancer, greatly enhances the ability to treat the disease and reduce the risk of more severe illness or death.

These potential benefits of genetic technology come with significant potential costs. People who have a gene that puts them at increased risk of developing a particular disease, for example, might be discriminated against, even though they are not currently sick and may never suffer from the particular disease. Although a federal Genetic Information Nondiscrimination Act was signed into law in 2008, the law does not cover life insurance, disability insurance, and long-term care insurance. Insurance companies have already denied such coverage on discovering that an individual carries a gene that puts him or her at increased risk of disease. Additionally, parents who discover that their developing fetus will develop a painful, debilitating, or fatal disease soon after birth are confronted with the difficult question of how to proceed.

When it comes to curing a disease by using biotechnology, there is good news and bad news. The good news is that, in the 1990s, a handful of humans with a usually fatal genetic disease called severe combined immunodeficiency disease (SCID) were completely cured through the application of biotechnology. The bad news is that it has not been possible to apply these promising techniques to other diseases.

It's not for a lack of trying. There have been more than 500 other clinical trials for **gene therapies** designed to treat or cure a variety of diseases by inserting a functional gene into an individual's cells to replace a defective version of the gene. But no clear successes. Not one.

Let's examine the case of SCID, which has served as a model for gene therapy. SCID is a condition in which a baby is born with an immune system unable to properly produce a type of white blood cell. This leaves the infant vulnerable to most infections and usually leads to death within the first year of life (**FIGURE 5-42**). In gene therapy for

**FIGURE 5-42 Protected by a bubble.** This child with severe combined immunodeficiency lived 12 years in a protective bubble at Baylor Medical Center.

SCID, researchers removed from an affected baby's bone marrow some **stem cells,** cells that have the ability to develop into any type of cell in the body. In bone marrow, they normally produce white blood cells, but in individuals with SCID, a malfunctioning gene disrupts normal white blood cell production.

Next, in a test tube, the bone marrow stem cells were infected with a transgenic virus carrying the functioning gene. Ideally, the virus inserted the good gene into the DNA of the stem cells, which were then injected back into the baby's bone marrow. There, the cells could produce normal white blood cells, permanently curing the disease. Although this strategy worked to cure several cases of SCID, treatment has been suspended indefinitely following the recent deaths of two patients from illness related to their treatment.

**Q** *Why has gene therapy had such a poor record of success in curing diseases?*

Difficulties with gene therapy have been encountered in several different areas, usually related to the organism used to transfer the normal-functioning gene into the cells of a person with a genetic disease. They include:

1. Difficulty getting the working gene into the specific cells where it is needed.

2. Difficulty getting the working gene into enough cells and at the right rate to produce a physiological effect.

3. Difficulty arising from the transfer organism getting into unintended cells.

4. Difficulty regulating gene expression.

Beyond these technical problems, for most diseases the malfunctioning gene has not been identified or the disease is caused by more than one malfunctioning gene. And finally, it is important to keep in mind that gene therapy targets cells in the body other than sperm and eggs. Consequently, while a disease might, in theory, be cured in an individual, he or she can still pass on the disease-causing gene(s) to offspring. It's not clear what the future holds for gene therapy, but a great deal of research is still in progress.

**TAKE-HOME MESSAGE 5·15**

Biotechnology tools have been developed to identify whether a set of parents is likely to produce a baby with a genetic disease, whether a baby is likely to be born with a genetic disease, and whether an individual carries disease-causing genes that may have their effect later in life. These tools can help reduce suffering and the incidence of diseases, but come with significant potential costs. Gene therapy has had a poor record of success in curing human diseases, primarily because of technical difficulties in transferring normal-functioning genes into the cells of a person with a genetic disease.

# 5·16 Cloning—ranging from genes to organs to individuals—offers both promise and perils.

Cloning. Perhaps no scientific word more readily conjures horrifying images of the intersection of curiosity and scientific achievement. But is fear the appropriate emotion to feel about this burgeoning technology? Perhaps not.

For starters, let's clarify what the word means. "Cloning" actually refers to a variety of different techniques. To be sure, cloning can refer to the creation of new individuals that have exactly the same genome as the donor individual—a process called "whole organism cloning." That is, a clone is like an identical twin, except that it may differ in age by years or even decades. It is also possible to clone tissues (such as skin) and entire organs from an individual's cells. And, as we saw in Section 5-11, it is possible to clone genes.

Cloning took center stage in the public imagination in 1997, when Ian Wilmut, a British scientist, and his colleagues first reported that they had cloned a sheep—which they named Dolly. Their research was based on ideas that went back to 1938, when Hans Spemann first proposed the experiment of removing the nucleus from an unfertilized egg and replacing it with the nucleus from the cell of a different individual. Although the process used by Wilmut and his research group was difficult and inefficient, it was surprisingly simple in concept (**FIGURE 5-43**). They removed a cell from the mammary gland of a grown sheep, put its nucleus into another sheep's egg from which the nucleus had been removed, induced the egg to divide, and

*Are there any medical justifications for cloning?*

transplanted it into the uterus of a surrogate mother sheep. Out of 272 tries, they achieved just one success. But that was enough to show that the cloning of an adult animal was possible.

Shortly after news of Dolly's birth, teams set about cloning a variety of other species, including mice, cows, pigs, and cats (**FIGURE 5-44**). Not all of this work was driven by simple curiosity. For farmers, cloning could have real value. It can take a long time to produce animals with desirable traits from an agricultural perspective—such as increased milk production in cows. And with each successive generation of breeding, it can be difficult to maintain these traits in the population. But through the process of transgenic techniques and whole animal cloning, large numbers of valuable animals with such traits can be produced and maintained.

**Q** Medical researchers, too, see much to gain from cloning. In particular, transgenic animals containing human genes—such as the hamsters producing rhu-EPO, discussed earlier—can be very valuable. But can a human be cloned? At this point, it is almost certain that the cloning of a human will be possible. Many people wonder, though, whether such an endeavor should be pursued. There is near unanimity among scientists that human cloning to produce children should not be attempted. Some of the reasons cited relate to problems of safety for the mother and the child, legal and philosophical issues relating to the inability of

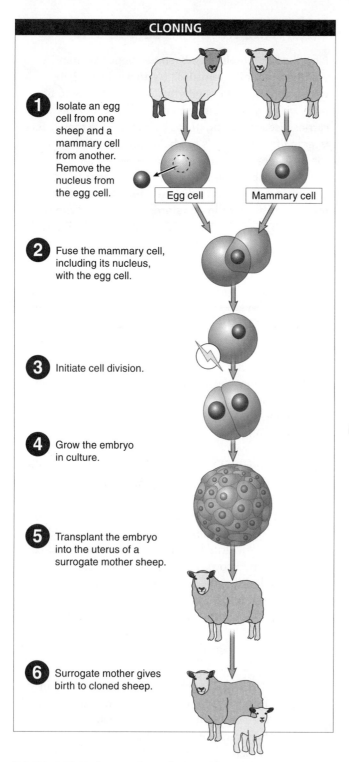

## CLONING

1. Isolate an egg cell from one sheep and a mammary cell from another. Remove the nucleus from the egg cell.

   Egg cell    Mammary cell

2. Fuse the mammary cell, including its nucleus, with the egg cell.

3. Initiate cell division.

4. Grow the embryo in culture.

5. Transplant the embryo into the uterus of a surrogate mother sheep.

6. Surrogate mother gives birth to cloned sheep.

**FIGURE 5-43 No longer science fiction.** The steps used in the cloning of Dolly the sheep.

**FIGURE 5-44 Genetically identical cloned animals.** The cloning of animals can maintain desirable traits from generation to generation.

cloned individuals to give consent, problems of the exploitation of women, and concerns regarding identity and individuality. Governments are struggling to develop wise regulations for this new world.

### TAKE-HOME MESSAGE 5·16

Cloning of individuals has potential benefits in agriculture and medicine, but ethical questions linger.

# 5·17 DNA is an individual identifier: the uses and abuses of DNA fingerprinting.

In another time, Colin Pitchfork, a murderer and rapist, would have walked free. But in 1987, he was captured and convicted, betrayed by his DNA, and is now serving two life sentences in prison. Pitchfork's downfall began when he raped and murdered two 15-year-old high-school girls in a small village in England in the 1980s. The police thought they had their perpetrator when a man confessed, but only to the second murder. He denied any involvement with the first murder, which perplexed the police because the details of the two crimes strongly suggested that the same person committed both.

At the time, British biologist Alec Jeffreys made the important discovery that there are small pieces of DNA in every person's chromosomes that are tremendously variable in their base sequences. In much the same way that each person has a driver's license number or social security number that differs from everyone else's, these DNA fragments are so variable that it is extremely unlikely that two people would have identical sequences at these locations. Thus, a comparison between these regions in a person's DNA sample and in DNA-containing evidence left at a crime scene would enable police to determine whether the evidence sample came from that person.

Jeffreys analyzed DNA left by the murderer-rapist on the two victims and found that it did indeed come from a single person, and that person was *not* the man who had confessed to one of the crimes. The original suspect was released and has the distinction of being the first person cleared of a crime through DNA fingerprinting. To track down the criminal, police then requested blood samples from all men in the area of the crimes who were between 18 and 35 years old—a practice that many viewed as an invasion of privacy. The police collected and analyzed more than 5,000 blood samples. This procedure led them to Colin Pitchfork, whose DNA matched perfectly the DNA left on both victims, and ultimately was the evidence responsible for his conviction (**FIGURE 5-45**). (He almost slipped through, having persuaded a friend to give a blood sample in his name. But when the friend was overheard telling the story in a pub, police tracked down Pitchfork to get a blood sample.)

DNA fingerprinting is now used extensively in forensic investigations, in much the same way that regular fingerprints have been used for the past 100 years. But traditional fingerprinting is limited in its usefulness for

**FIGURE 5-45 Betrayed by his DNA.** Colin Pitchfork, the first criminal brought to justice with DNA fingerprinting.

many crimes because, often, no fingerprints are left behind. DNA samples, though, more frequently are left behind, usually in the form of semen, blood, hair, skin, or other tissue. As a consequence, this technology has been directly responsible for bringing thousands of criminals to justice and, as we saw at the beginning of the chapter, for establishing the innocence of more than 200 people wrongly convicted of murder and other capital crimes. Let's examine how DNA fingerprinting is done, why it is such a powerful forensic tool, and why it is not foolproof.

The DNA from different humans is almost completely identical. More than 99.9% of the DNA sequences of two individuals are the same, because we're all of the same species and thus share a common evolutionary history. Even so, in a genome of three billion base pairs, one-tenth of a percent difference still translates to about three million base-pair differences. These differences are responsible for the fact that all individuals have their own unique genome. (The lone exception? Identical twins, whose DNA is exactly the same.) Thus, when we are trying to evaluate whether the DNA from a crime scene matches that from a suspect, the analysis focuses on the parts of our DNA that differ. There are thousands of these highly variable regions in the human genome.

Among the thousands of variable regions in the human genome, one particular type is used for the determination of a person's genetic fingerprint. These regions are called **STRs** (for **short tandem repeats**) and are characterized by having a short sequence (commonly four or five nucleotides) that repeats over and over a dozen or more times.

An individual—we'll call her Individual A—has two copies of each chromosome, one from her mother and one from her father. At one STR location (on chromosome 3, for example), the number of times the sequence repeats is likely to differ on the maternal and the paternal copies of that chromosome in Individual A. The sequence may repeat 14 times on the maternal copy of chromosome 3 and it may repeat only 3 times on the paternal copy that she carries. In such a case, Individual A is said to have two different alleles for this STR region: 14 and 3. In contrast, in Individual B, for the same STR region on chromosome 3, the sequence may repeat 5 and 11 times (**FIGURE 5-46**).

For an STR region within the human genome, there typically are about 10 different alleles that occur within a population, and each is shared by about 10% of all individuals in a population. So the likelihood that two individuals carry the same two alleles is about 1 in 100.

**SHORT TANDEM REPEATS (STRs)**

Short tandem repeats, or STRs, are sequences of DNA (commonly 4 or 5 nucleotides) that repeat over and over again. They occur in some of the most highly variable regions of an individual's DNA.

INDIVIDUAL A

STR locus

Chromosome from mother — Number of repeats: 14

Chromosome from father — 3

Short tandem repeat (STR)

14 / 3
Individual A's alleles for this STR locus

INDIVIDUAL B

Number of repeats

Chromosome from mother — 5

Chromosome from father — 11

For a given region, it's the same sequence that repeats, but the number of times it repeats differs from individual to individual (and usually differs on the maternal and paternal copies of a chromosome that an individual carries).

5 / 11
Individual B's alleles for this STR locus

**FIGURE 5-46 Biotechnology in forensics.** Forensic scientists can use highly variable regions of DNA to genetically link a person to DNA-containing evidence left at a crime scene.

This is unlikely, but given enough people, many are likely to carry the same alleles. If two different STR regions are analyzed, the likelihood that two individuals have the same four alleles is then $1/10 \times 1/10 \times 1/10 \times 1/10$, or 1/10,000. This is much rarer, but still would lead to multiple individuals within the same large city carrying the same four alleles. The real power of DNA fingerprinting comes from simultaneously determining the alleles an individual carries (that is, their genotype) not simply at one or two STR locations but at 13 *different* STR locations. This is the number used by the FBI in constructing DNA fingerprints in the United States, and it makes the probability that two individuals would have exactly the same genotype extremely low (see Figure 5-46).

A DNA fingerprint is created by determining which alleles an individual carries for 13 different STR loci.

**1** For each of the 13 STR loci used to construct an individual's DNA fingerprint, the DNA fragment containing each STR region is amplified using PCR, resulting in huge numbers of those fragments. The amplified DNA fragments differ in size, depending on how many times the repeating unit of that STR is repeated.

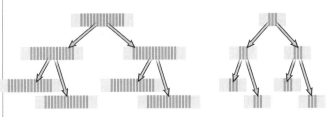

**2** The amplified DNA fragments are separated by size using electrophoresis. In this process, DNA fragments are poured into a solution and an electrical charge is applied. Because DNA is a negatively charged molecule, the pieces of DNA move toward the positively charged electrode. Smaller pieces—those with fewer repeats—move more quickly than larger pieces.

Individual A's DNA sample

Direction of DNA strand movement

Electrical charge

**3** Computer software is used to analyze the results. The number of repeats within an STR region (indicating an individual's genotype) is determined by comparing the length of the fragments containing that STR region to DNA fragments of known lengths.

FIGURE 5-47 **Your unique identifier: a DNA fingerprint.** DNA fingerprints are being used to match a suspect's DNA to DNA found at the scene of a crime. (The DNA fingerprint of the crime scene evidence matches that from Individual B.)

For each STR locus analyzed, an individual's genotype is determined by using PCR to amplify that region, then measuring the length of the STR region using electrophoresis. The length of the region can then be used to determine the number of times that the STR is repeated. For a single STR region, an individual's genotype is two numbers, reflecting the number of STR repeats in the copies inherited from the mother and from the father. And a person's full DNA fingerprint is a string of 26 numbers that includes the two numbers for each of 13 STRs.

**Q** What is a DNA fingerprint?

In court, a suspect's genotype might be compared with the DNA fingerprint obtained from evidence found at the crime scene. DNA samples from different people will produce different 26-number "fingerprints," whereas different samples of DNA from one person will have exactly the same genotype (**FIGURE 5-47**).

In the end, despite universally accepted methods, DNA fingerprinting still is not foolproof. Numerous incidences of human error—accidental as well as intentional—have been documented, from mislabeled test tubes to tissue from a suspect being added to evidence from a crime scene. So we should not blindly draw conclusions solely from this one type of evidence. Nonetheless, DNA fingerprinting is an increasingly valuable tool for law enforcement. The FBI, for example, has reported that nearly one-third of their suspects are cleared immediately by DNA testing, and that many more criminals, because of DNA fingerprinting, now plead guilty to the crimes they have committed.

**TAKE-HOME MESSAGE 5·17**

Comparisons of highly variable DNA regions can be used to identify tissue specimens and determine the individual from whom they came.

## MIXING ASPIRIN AND ALCOHOL CAN LEAD TO METABOLIC INTERFERENCE AND UNEXPECTED INEBRIATION.

Ethanol is the form of alcohol found in cocktails, beer, and wine. When you consume alcohol, the first step in your body's metabolic breakdown of the ethanol is carried out by the enzyme alcohol dehydrogenase.

**Q: WHAT HAPPENS TO ETHANOL MOLECULES THAT ARE *NOT* BROKEN DOWN?** Any ethanol molecules in the bloodstream that are not broken down by alcohol dehydrogenase make their way to the brain. Once there, they cause you to feel a bit inebriated—or drunk, if the amount of ethanol is sufficiently large. Aspirin has the unintended side effect of disabling alcohol dehydrogenase, thus interfering with its normal functioning.

**Q: WHAT HAPPENS IF SOMEONE TAKES TWO ASPIRINS AND THEN DRINKS SEVERAL ALCOHOLIC BEVERAGES?** By blocking the activity of alcohol dehydrogenase, aspirin interferes with a person's ability to break down ethanol. As a consequence, the ethanol remains in the body longer and has a more pronounced effect on the brain, producing greater inebriation (26% greater, according to one study).

**Q: WHAT CAN YOU CONCLUDE?** Medications, even over-the-counter products, can have unexpected (and unintended) physiological consequences, particularly when mixed with alcohol consumption. Exercise great caution when taking them.

# Big Ideas in ... DNA, Gene Expression, and Biotechnology

## 5·1–5·5 DNA: what is it and what does it do?

- DNA, the genetic material for all living organisms, is a double-helical molecule consisting of two "backbone" chains of sugar and phosphate molecules, with the chains linked by pairs of nucleotide bases.

- Genes are sequences of bases in DNA that contain the code for the construction of protein molecules. Only a small fraction of the DNA in eukaryotes consists of protein-coding genes.

## 5·6–5·8 Building organisms: information in DNA directs the production of the molecules that make up an organism.

- Information in the genes is transcribed into the sequences of mRNA molecules. These mRNA sequences are then translated as they direct the construction of proteins.

- Regulation of gene expression can occur in ways that enhance or impede transcription, or alter the longevity and degradation of mRNA, or influence translation or protein processing.

## 5·9–5·10 Damage to the genetic code has a variety of causes and effects.

- Mutations are rare alterations in a single base or changes in large segments of DNA that include several genes.

- Mutations in a gene may disrupt normal physiological functioning.

- Mutations play an important role in evolution.

- Most genetic diseases result from a mutation that causes a gene to produce a non-functioning enzyme, resulting in a blocked metabolic pathway.

## 5·11–5·13 Biotechnology is producing improvements in agriculture.

- Biotechnology is the use of technology to modify organisms, cells, and their molecules to achieve practical benefits.

- With modern molecular methods, researchers can cut and copy DNA and deliver it to new organisms, not necessarily of the same species.

- Biotechnology has led to important advances in agriculture by using transgenic plants and animals to produce more nutritious food. Biotechnology has also reduced the environmental and financial costs of producing food through the creation of herbicide- and insect-resistant crops.

- Some legitimate concerns among the public remain as to the safety of genetically modified foods.

## 5·14–5·17 Biotechnology has the potential for improving human health (and criminal justice).

- Biotechnology has had some successes in treating diseases, usually by producing medicines more efficiently and effectively than with traditional methods.

- Gene therapy has so far had a poor record of success in curing human diseases.

- With biotechnology, many tools have been developed that enable us to identify individuals carrying disease-causing genes. These tools can help reduce the symptoms and incidence of diseases, but come with significant potential costs, particularly the risk of discrimination.

- Advances in biotechnology have led to automated methods for analyzing DNA sequences. Comparisons of highly variable DNA regions have forensic value in identifying tissue specimens and determining the individual from whom they came.

- Cloning of individuals has potential benefits in agriculture and medicine, but ethical questions linger.

See the To-Go Guide (packaged with every new copy of this textbook) for an illustrated summary of the chapter.

# Check Your Knowledge

*Thinking critically about visual displays of data.*

1. What proportion of corn grown in the United States is genetically modified?

2. From the figures, can you determine whether there is more genetically modified corn or genetically modified cotton produced in the United States? How do you do this? And if it is not possible, why isn't it?

3. What is the "take-home message" from these figures? Does it influence your thoughts on genetically modified crops? Why or why not?

4. Why are data given for proportions of genetically modified crops rather than absolute amounts? Does this alter your interpretation of the graph?

**GENETICALLY MODIFIED CROPS IN THE UNITED STATES**

14%
86%
Corn

7%
93%
Cotton

7%
93%
Soybeans

Proportion of crops that are not genetically modified
Proportion of crops that are genetically modified

5. Worldwide, the percent of corn grown that is genetically modified is just 26%, while the proportion of soybeans that is genetically modified is 77% and cotton is 49%. How would this information add value to the figure?

6. The proportion of corn grown in the United States that is genetically modified has changed since 2001 as follows:

| Year | Proportion |
|------|-----------|
| 2001 | 26% |
| 2002 | 34% |
| 2003 | 40% |
| 2004 | 47% |
| 2005 | 52% |
| 2006 | 61% |
| 2007 | 73% |
| 2008 | 80% |
| 2009 | 85% |
| 2010 | 86% (shown in the graph) |

Show two different ways that you could display these data. What is the clearest conclusion someone would draw from the figures?

See Answers at the back of the book.

## Key Terms in DNA, Gene Expression, and Biotechnology

# Multiple-Choice Questions

**1.** A person's DNA is carried in:

   **a)** muscle cells.

   **b)** hair.

   **c)** saliva.

   **d)** skin cells.

   **e)** All of the above contain a person's DNA.

100 HARD

28

0 EASY

**2.** Which of the following are always the same in every DNA molecule?

   **a)** the sugar

   **b)** the base

   **c)** the phosphate group

   **d)** Only a) and b) are always the same.

   **e)** Only a) and c) are always the same.

100 HARD

37

0 EASY

**3.** The full set of an individual organism's DNA is called its:

   **a)** complement.

   **b)** genome.

   **c)** nucleosome.

   **d)** nucleotide.

   **e)** chromosome.

100 HARD

21

0 EASY

**4.** In humans, genes make up _____ of the DNA.

   **a)** about 75%

   **b)** 100%

   **c)** less than 5%

   **d)** about 10%

   **e)** about 50%

100 HARD

33

0 EASY

**5.** Genotype is to phenotype as:

   **a)** cookie is to oven.

   **b)** cookie is to recipe.

   **c)** cookbook is to cookie.

   **d)** recipe is to cookie.

   **e)** oven is to cookie.

100 HARD

26

0 EASY

**6.** To start the transcription process, a large molecule, _____, recognizes a _____.

   **a)** RNA polymerase; messenger RNA

   **b)** DNA polymerase; termination site

   **c)** DNA polymerase; promoter site

   **d)** RNA polymerase; promoter site

   **e)** DNA polymerase; messenger RNA.

100 HARD

39

0 EASY

**7.** During transcription, at the point at which the DNA strand being copied has an adenine, a(n) _____ is added to the _____.

   **a)** thymine; tRNA

   **b)** cytosine; DNA

   **c)** uracil; tRNA

   **d)** adenine; mRNA

   **e)** uracil; mRNA.

100 HARD

46

0 EASY

**8.** There are different _____ molecules for each of the 20 different amino acids that are used in building proteins.

   **a)** ribosomal subunit

   **b)** tRNA

   **c)** mRNA

   **d)** DNA

   **e)** elongation

100 HARD

60

0 EASY

**9.** Deletions and substitutions are two types of point mutations. Which type is more likely to cause mistranslations of proteins?

a) Substitutions, because they shift the reading frame and cause downstream amino acids to be changed.

b) Substitutions, because one protein is substituted for another protein.

c) Deletions, because they shift the reading frame and cause downstream amino acids to be changed.

d) Deletions, because one protein is deleted.

e) None of the above is correct.

100 HARD
40
0 EASY

**10.** Which of the following statements about the metabolism of ethanol (which is present in alcoholic beverages) is incorrect?

a) Individuals who produce non-functioning aldehyde dehydrogenase exhibit "fast-flushing."

b) The process requires two enzymes: alcohol dehydrogenase and isopropyl dehydrogenase.

c) Individuals who are "fast-flushers" are less likely to become alcoholics.

d) Aspirin interferes with the action of alcohol dehydrogenase.

e) All of the above are correct.

100 HARD
69
0 EASY

**11.** The polymerase chain reaction (PCR):

a) makes it possible to create huge numbers of copies of tiny pieces of DNA.

b) enables researchers to determine the sequence of a complementary strand of DNA when they have only single-stranded DNA.

c) utilizes RNA polymerase to build strands of DNA.

d) can create messenger RNA molecules from small pieces of DNA.

e) All of the above are correct.

100 HARD
60
0 EASY

**12.** Which of the following is not a difficulty that medicine has encountered in its attempts to cure human diseases through gene therapy?

a) The transfer organism—usually a virus—may get into unintended cells and cause disease.

b) It is difficult to get the working gene into the specific cells where it is needed.

c) It is difficult to get the working gene into enough cells at the right rate to have a physiological effect.

d) For many diseases, a malfunctioning gene has not been identified.

e) All of the above are difficulties encountered in attempts to cure human diseases through gene therapy.

100 HARD
31
0 EASY

**13.** Golden rice:

a) grows without a husk, thereby reducing the processing required before it can be consumed.

b) can make vitamin A without beta-carotene.

c) could help prevent blindness in the 250,000 children who lose their sight each year because of vitamin A deficiency.

d) supplies more vitamin A in one serving than an individual needs in a full week.

e) is one of the most recent developments in organic farming.

100 HARD
30
0 EASY

**14.** Which of the following statements about Bt crystals is correct?

a) They are produced by soil-dwelling bacteria of the species Bacillus thuringiensis.

b) The gene coding for the production of Bt crystals has been genetically engineered into the genome of dairy cows, increasing their milk production sixfold.

c) They are produced by the polymerase chain reaction (PCR).

d) They are produced by most weedy species of plants.

e) All of the above are correct.

100 HARD
58
0 EASY

**15.** Highly repetitive sequences of DNA:

a) are characteristic of genes that code for biochemical traits rather than structural traits.

b) are used in biotechnology when creating a clone.

c) are produced when a mutation occurs in a non-sex cell.

d) can be used to produce a DNA fingerprint.

e) are produced when a mutation occurs in a sperm-producing or egg-producing cell.

100 HARD
60
0 EASY

## Short-Answer Questions

**1.** After sequencing a molecule of DNA, you discover that 20% of the bases are cytosine. What percentage of the bases would you expect to be thymine? Explain why.

**2.** A cell is in the process of transcription. The single strand of DNA that is being copied has the sequence ATC GAC GGA TCC. Indicate the mRNA strand that will be synthesized from this strand, listing the correct sequence of bases. Explain why transcription is important in the process of producing protein.

**3.** What is being "translated" in translation? What is the role of transfer RNA molecules in this step?

4. The genetic disorder known as sickle-cell disease results in an altered form of the protein hemoglobin, which ultimately can affect the structure of red blood cells. A portion of the sequence of the normal hemoglobin allele is:

CTG ACT CCT GAG GAG AAG TCT

Individuals with the sickle-cell disorder have the following sequence:

CTG ACT CCT GTG GAG AAG TCT

Identify the type of mutation and explain how it would result in an altered (in this case non-functional) protein.

5. Describe three major applications of biotechnology and their impact on human health.

See Answers at the back of the book.
For additional study questions, go to www.prep-u.com.

Continuity and variety

# 6

# Chromosomes and Cell Division

# 6·1–6·5 There are different types of cell division.

*Cell division by fission in* Staphylococcus aureus, *a disease-causing bacterium having resistance to multiple antibiotics.*

# 6·1

## Immortal cells can spell trouble: cell division in sickness and in health.

Once you are fully grown, do you have just one set of cells that live as long as you do? The answer is *no:* your cells are continually dying off, and the ones that remain divide and replace the cells you've lost, in an ongoing process. But how long can this last? Can it go on forever? And does a cell even know how old it is?

Actually, a cell does have a measure of how old it is. Just as a car comes with an odometer, which keeps track of how far the car has been driven, animal cells have a sort of counter that keeps track of how many times a cell has divided. This "counter" is a section of non-coding, repetitive DNA, called a **telomere,** that serves as a protective cap and is located at each tip of every chromosome, right next to the genes that direct the processes that keep the organism alive (**FIGURE 6-1**). Every time a cell divides, making an exact copy of itself, its

> « Cancer cells are those which have forgotten how to die. »
> — HAROLD PINTER, playwright, from the poem "Cancer Cells," 2002

DNA divides as well. However, each time the DNA divides, the telomere at each end of every chromosome gets a bit shorter. Eventually, after some critical number of cell divisions, the telomeres can become so short that additional cell divisions cause the loss of functional, essential DNA, and that means almost certain death for the cell.

At birth, the telomeres in most human cells are long enough to support about 50 cell divisions. Occasionally, however, individuals are born with telomeres that are much shorter than normal. In these people, the normal functioning of many genes is disrupted, and their cells and tissues begin to appear aged very soon after birth (**FIGURE 6-2**). As a consequence, children born with this disorder rarely live beyond the age of 13.

Just as someone might think it's a good idea to disconnect the car's odometer, you might think that disconnecting the cellular odometer by continually rebuilding the telomere would be an effective strategy for allowing a cell and its descendants to function for a longer time than normal. Such a line of cells

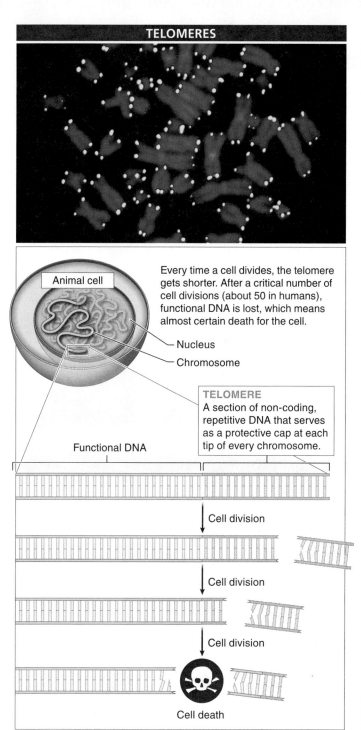

**TELOMERES**

Every time a cell divides, the telomere gets shorter. After a critical number of cell divisions (about 50 in humans), functional DNA is lost, which means almost certain death for the cell.

Animal cell

— Nucleus
— Chromosome

**TELOMERE**
A section of non-coding, repetitive DNA that serves as a protective cap at each tip of every chromosome.

Functional DNA

Cell division

Cell division

Cell division

Cell death

**FIGURE 6-1 A cellular "odometer."** Telomeres limit cells to a fixed number of divisions. The photo shows chromosomes (in blue) with the telomeres highlighted in yellow.

would never die—so you'd think that a constantly rebuilding telomere would be like a fountain of youth. It's not. We know this because there are some cells with a sort of perpetually resetting odometer. These cells rebuild their telomeres after each cell division, restoring the chromosomes' protective caps. Unfortunately, most cells that rebuild their telomeres with

**FIGURE 6-2 Just a child.** Individuals with the genetic condition known as Hutchinson-Gilford progeria syndrome are born with shorter-than-normal telomeres and appear to age at a rapid rate.

each cell division present a big problem: they don't know when to stop dividing. There is a word for such cells: cancer. Because cancer is runaway cell division, discovering a cure for cancer will necessarily involve a deep understanding of cell division.

In this chapter, we investigate the processes that enable cells to divide and create new cells. Prokaryotes exhibit one method of cell division (called binary fission), and it serves all of their cell division needs. Eukaryotes exhibit two methods of cell division, mitosis and meiosis, each of which has a specific purpose in a eukaryote's life cycle. We'll explore normal cell division and discuss what happens when cell division does not proceed in the normal way.

**TAKE-HOME MESSAGE 6·1**

Cell division is an ongoing process in most organisms and their tissues; disruptions to normal cell division can have serious consequences. In eukaryotic cells, a protective section of DNA called the telomere, at each end of every chromosome, plays a role in keeping track of cell division, getting shorter every time the cell divides. If telomeres become too short, additional cell divisions cause the loss of essential DNA and cell death. Cells that rebuild their telomeres with each division can become cancerous.

# 6·2 Some chromosomes are circular, others are linear.

As a method for storing genetic information, DNA has complete market saturation. All life on earth uses it. This is pretty remarkable considering the tremendous diversity of life that exists on our planet—from single-celled bacteria to multicellular plants and animals. One way in which different organisms' DNA varies is in how it is organized into chromosomes.

In most bacteria and archaea, the genetic information is carried in a single, circular chromosome, a strand of DNA that is attached at one site to the cell membrane (**FIGURE 6-3**). Eukaryotes, which have much more DNA than do bacteria, organize it into free-floating linear chromosomes within the nucleus. As we saw in Chapter 5, a eukaryotic organism may have as few as 2 chromosomes or as many as 1,000. For plants and animals, the number is usually between 10 and 50.

The most important part of a eukaryotic chromosome may be the DNA molecule, which carries information on how to carry out the processes needed to support the life of the organism. But eukaryotic chromosomes (and some prokaryotic chromosomes) are made of more than just DNA. The long, linear strand is wrapped around proteins called **histones,** which keep the DNA from getting tangled and enable it to be tightly and efficiently packed in an orderly manner inside the cell (see Figure 6-3). In the next sections we examine how prokaryotic and eukaryotic chromosomes behave during cell division.

## TAKE-HOME MESSAGE 6·2

In most bacteria and archaea, the genetic information is carried in a single, circular chromosome, a strand of DNA that is attached at one site to the cell membrane. Eukaryotes have much more DNA than do bacteria and organize it into free-floating linear chromosomes within the nucleus.

**PROKARYOTIC AND EUKARYOTIC CHROMOSOMES**

**PROKARYOTIC CELLS**
Prokaryotic cells have a single circular chromosome attached to the cell membrane.

Chromosome
Attachment site

**EUKARYOTIC CELLS**
Eukaryotic cells contain free-floating linear chromosomes within a nucleus.

Chromosome
Nucleus
Histones
DNA

**FIGURE 6-3 Chromosomes compared.** Prokaryotic cells have a single circular chromosome attached to the cell membrane. Eukaryotic cells contain free-floating linear chromosomes within a nucleus.

# 6·3 Prokaryotes divide by binary fission.

When it is time for bacteria and archaea to reproduce, they use a method called **binary fission,** which means "division in two" (**FIGURE 6-4**). This process begins with **replication,** the method by which a cell creates an exact duplicate of each chromosome. Replication begins as the double-stranded DNA molecule unwinds from its coiled-up configuration. Once the strands are uncoiled, they split apart like a zipper. As the double-stranded molecule unzips, free-floating nucleotide bases attach to the growing DNA backbone, with bases exposed on each of the two separated, single-stranded circular molecules of DNA, matching A to T and G to C, thus creating two identical double-stranded DNA molecules. The newly created circular chromosomes of DNA attach to the inside of the plasma membrane at different spots from each other. The original cell, called the **parent cell,** then pinches in until it divides into two new cells, called **daughter cells.** Each of the daughter cells has an exact and precise two-stranded copy of the original two-stranded circular chromosome.

In some prokaryotes, such as the bacteria called *E. coli* that live in our digestive system, the complete process of binary fission can occur very quickly—often in as little as 20 minutes. Binary fission is considered **asexual reproduction** because the daughter cells inherit their DNA from a single parent cell and thus are genetically identical to the parent cell. In **sexual reproduction,** a combination of DNA from two separate individuals is passed on to offspring, resulting in offspring that are genetically different both from one another and from their parents. We discuss sexual reproduction in more detail later in this chapter.

## TAKE-HOME MESSAGE 6·3

Bacteria divide by a type of asexual reproduction called binary fission: first, the circular chromosome duplicates itself, then the parent cell splits into two new, genetically identical daughter cells.

FIGURE 6-4 **Dividing in two.** Asexually reproducing prokaryotes reproduce rapidly through binary fission.

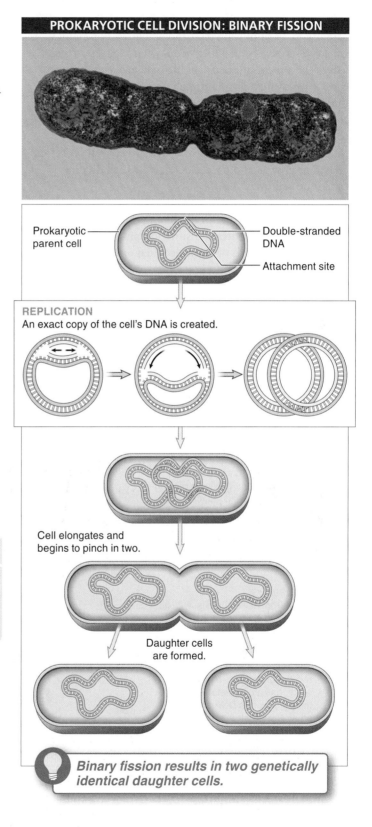

**PROKARYOTIC CELL DIVISION: BINARY FISSION**

Prokaryotic parent cell

Double-stranded DNA

Attachment site

**REPLICATION**
An exact copy of the cell's DNA is created.

Cell elongates and begins to pinch in two.

Daughter cells are formed.

*Binary fission results in two genetically identical daughter cells.*

# 6·4 There is a time for everything in the eukaryotic cell cycle.

In life, we often go through phases that are defined by the primary focus of our interests and activities. For many years we go to school. For a while we may tend to our career. Then, for a spell, our personal life may take center stage. Most eukaryotic cells go through phases, too. They spend long periods of time occupied with activities relating solely to their growth and then may suspend those activities as they segue into a period devoted exclusively to reproducing themselves. This alternation of activities between processes related to growth and processes related to cell division is called the **cell cycle** (FIGURE 6-5).

Before we go any farther in discussing the cell cycle, we need to note an important distinction among the cells of the body. All the cells of a multicellular eukaryotic organism can be divided into two types: **somatic cells** are the cells forming the body of the organism; **reproductive cells** are the sex cells, or **gametes.** The processes of cell division by which somatic cells and reproductive cells are produced differ from each other. In this section, we focus on cell division as it occurs in somatic cells. Later in the chapter, we examine cell division that leads to the production of sex cells.

The cell cycle describes the series of phases in somatic cell division. There are two main phases in the cell cycle: **interphase,** during which the cell grows and prepares to divide, and the **mitotic phase** (or **M phase**), during which first the nucleus and genetic material within the cell divide, then the rest of the cellular contents divide. Interphase is further divided into three distinct subphases, described below. A eukaryotic somatic cell moves through the phases in precise order and, at any given moment, the cell is at some point in this cycle.

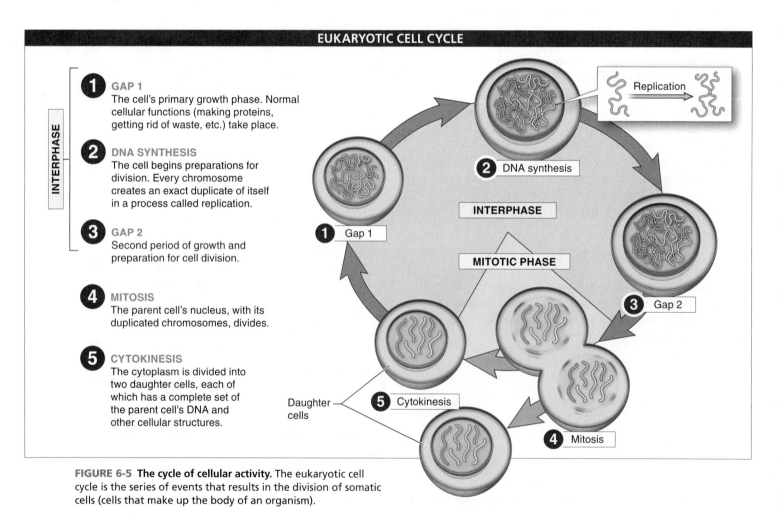

**EUKARYOTIC CELL CYCLE**

**INTERPHASE**

**1 GAP 1**
The cell's primary growth phase. Normal cellular functions (making proteins, getting rid of waste, etc.) take place.

**2 DNA SYNTHESIS**
The cell begins preparations for division. Every chromosome creates an exact duplicate of itself in a process called replication.

**3 GAP 2**
Second period of growth and preparation for cell division.

**4 MITOSIS**
The parent cell's nucleus, with its duplicated chromosomes, divides.

**5 CYTOKINESIS**
The cytoplasm is divided into two daughter cells, each of which has a complete set of the parent cell's DNA and other cellular structures.

Replication

**2** DNA synthesis

**INTERPHASE**

**1** Gap 1

**MITOTIC PHASE**

**3** Gap 2

Daughter cells

**5** Cytokinesis

**4** Mitosis

**FIGURE 6-5 The cycle of cellular activity.** The eukaryotic cell cycle is the series of events that results in the division of somatic cells (cells that make up the body of an organism).

**Interphase** The three subphases of interphase can be summarized as follows.

**Gap 1.** During this period, a cell grows and performs all cellular functions (making proteins, getting rid of waste, and so on). Cells that divide infrequently, such as most neurons and heart muscle cells, spend most of their time in the Gap 1 phase.

**DNA synthesis** During this phase, the cell begins to prepare for cell division, first by creating an exact duplicate of each chromosome by replication. Before replication, each chromosome is a long, linear piece of genetic material. After replication, each chromosome has become a pair of identical long, linear pieces, held together near the center; the region where the two pieces are in contact is called the **centromere.** (The centromere also serves as the point to which spindle fibers can attach.)

**Gap 2.** In this phase, the cell continues to grow and prepare for division. This phase differs from Gap 1, though, because the genetic material has now been duplicated. Gap 2 is usually much shorter than Gap 1.

**Mitotic phase** This period begins with **mitosis,** a process in which the parent cell's nucleus, including its chromosomes, duplicates. Mitosis is generally followed by **cytokinesis,** during which the cytoplasm is divided into two daughter cells, each of which has a complete set of the parent cell's DNA and other cellular structures. (Note that each daughter cell becomes a parent cell in the next cell cycle.) The mitotic phase is usually the shortest period in the eukaryotic cell cycle.

The transition from one phase of the cell cycle to the next is triggered by specific events. For instance, a cell moves from the Gap 1 phase (denoted $G_1$) to DNA synthesis only if environmental conditions outside the cell are favorable for cell division. Until conditions are favorable, the cell remains in $G_1$. Likewise, a cell begins mitosis only when the genetic material has been successfully duplicated. Later in this chapter we'll see that errors in this orderly, step-by-step process can lead to out-of-control cell division and cancer.

> ### TAKE-HOME MESSAGE 6·4
>
> **Eukaryotic somatic cells alternate in a cycle between cell division and other cell activities. The cell division portion of the cycle is called the mitotic phase. The remainder of the cell cycle, called interphase, consists of two gap phases (during which cell growth and other metabolic activities occur) separated by a DNA synthesis phase during which the genetic material is replicated.**

# 6·5 Cell division is preceded by replication.

In one of the great understatements in the scientific literature, James Watson and Francis Crick wrote the following sentence in their paper describing the structure of DNA: "It has not escaped our notice that the specific pairing we have postulated immediately suggests a possible copying mechanism for the genetic material." The researchers' ability to explain how DNA copied itself was a critical feature of their description of DNA's structure. After all, every single time any cell in any organism's body divides, that cell's DNA must first duplicate itself so that each of the two new daughter cells has all of the genetic material of the original parent cell. The process of DNA duplication, as we've seen, is called replication.

What is it about DNA's structure that makes duplication so straightforward? Watson and Crick were referring to the feature of DNA called **complementarity,** the characteristic that in the double-stranded DNA molecule, the base on one strand always has the same pairing partner (called the **complementary base**) on the other strand: A pairs with T (and vice versa), and G pairs with C (and vice versa).

With this consistent pattern of pairing, when the DNA molecule separates into two strands, it is possible to perfectly reconstruct all the information on the missing half because one strand carries all the information needed to construct its complementary strand. Just before cells divide, the DNA molecule unwinds and "unzips," and each half of the unzipped molecule serves as a template on which the missing half is reconstructed. At the end of the process of reconstructing the missing halves, there are two DNA molecules—each identical to the original DNA molecule— one for each of the two new cells.

So, the process of DNA replication occurs in two steps: unwinding and rebuilding (**FIGURE 6-6**).

1. **Unwinding.** Replication begins when the coiled, double-stranded DNA molecule unwinds and separates into two strands, like a zipper unzipping.

2. **Rebuilding.** In the rebuilding process, each of the single strands becomes a double strand as the

appropriate complementary base pairs with the exposed base, and an enzyme links the nucleotides of the new strand together. If an A is exposed, a T binds to it; if a C is exposed, a G binds to it.

The end result of replication is two double-stranded DNA molecules, each identical to the cell's original double-stranded DNA molecule. DNA replication occurs in all types of cells, somatic and reproductive, prior to cell division.

A variety of mutations (such as mismatched bases and added or deleted segments) can occur during replication, but many of these errors are caught and repaired during or after replication. If an error remains, however, the sequences in a replicated DNA molecule (including the genes) can be different from those in the parent molecule. A changed sequence may then produce a different mRNA sequence, which may in turn produce a different amino acid sequence. As a result, the changed sequence may lead to the production of a different protein. In this way, new genes can enter a population and be acted on by evolution. We explore the relationship between mutation and evolution in more detail in Chapter 8.

**Q** *Errors sometimes occur when DNA duplicates itself. Why might that be a good thing?*

**DNA REPLICATION**

**1 UNWINDING**
The coiled, double-stranded DNA molecule unwinds and separates into two strands.

**2 REBUILDING**
Enzymes connect a nucleotide with the appropriate base to the growing new strand, as the base bonds with the exposed, complementary base.

Complementary nucleotide bases attached

**3** Each of the single strands becomes a double strand, each identical to the cell's original double-stranded DNA molecule.

Adenine
Cytosine
Guanine
Thymine

**FIGURE 6-6  How DNA replicates: unwinding and rebuilding.**

---

**TAKE-HOME MESSAGE 6·5**

Every time a cell divides, that cell's DNA must first duplicate itself so that each of the two new daughter cells has all the genetic material of the original parent cell. The process of DNA duplication is called replication. Errors in replication can lead to mutations, which are changes in the DNA sequence.

## 6·6–6·9 Mitosis replaces worn-out old cells with fresh new duplicates.

*For most of the cell cycle, chromosomes resemble a plate of spaghetti more than the familiar condensed, replicated "X" seen in photos.*

# 6·6 Most cells are not immortal: mitosis generates replacements.

Look around your dorm room. Dust is everywhere. What is it? It is primarily dead skin cells. In fact, you (and your roommates) slough off millions of dead skin cells each day—yet your skin doesn't disappear. Why not? Because your body replaces the sloughed-off cells. When the cells wear out, your body

**Q**

What is dust? Why is it your fault?

creates replacements through the process of mitosis.

Mitosis has just one purpose: to enable existing cells to generate new, genetically identical cells. There are two different reasons for this need (**FIGURE 6-7**).

**THE NEED FOR NEW CELLS: GROWTH AND REPLACEMENT**

The tip of the root is a site of particularly rapid cell division in most plants.

The act of shaving damages thousands of cells, which must be replaced.

**FIGURE 6-7 Reasons for mitosis.** Mitosis is important in organism growth and the replacement of cells.

**1. Growth.** During development, organisms get bigger. Growth happens in part through the creation of new cells. In fact, if you want to see cell division in action, one surefire place to find it is at the tip of a plant root: at a growth rate of about half an inch per day, the root is one of the fastest-growing parts of a plant.

**2. Replacement.** Cells must be replaced when they die. The wear and tear that come from living can physically damage cells. The daily act of shaving, for example, damages thousands of cells on a man's face. It's nothing to worry about, though. Microscopic views of human skin reveal several distinct layers, with the outermost layers—the layers under assault during shaving—made up primarily of dead cells. These cells help protect us from infection and also reduce the rate at which the underlying living cells dry out. The living cells that exist just below the layers of dead cells are produced at a high rate by mitosis.

**FIGURE 6-8 The red blood cells in your body carry oxygen to your organs and tissues and are replaced about every 120 days.**

Some other cells that must be replaced actually die on purpose, in a planned process of cell suicide called **apoptosis** (A-pop-TOE-siss). This seemingly counterproductive strategy is employed in parts of the body where the cells are likely to accumulate significant genetic damage over time and are therefore at high risk of becoming cancer cells (a process described later in this chapter). Cells targeted for apoptosis include many of the cells lining the digestive tract as well as those in the liver, two locations where cells are almost constantly in contact with harmful substances.

Every day, a huge number of cells in an individual must be replaced by mitosis. In humans this number is in the billions. Nearly all the somatic cells of the body—that is, all cells except sperm- and egg-producing cells—undergo mitosis. There are a few notable exceptions. Brain cells and heart muscle cells, in particular, do not seem to divide, or, if they do divide, they do so at very, very slow rates. (We don't know why this is so.)

The rate at which mitosis occurs in animals varies dramatically for different types of cells. The most rapid cell division takes place in the blood and in the cells lining various tissues and organs. The average red blood cell, for example, is in circulation only for about two to four months and then must be replaced (**FIGURE 6-8**). The cells lining the intestines are replaced about every three weeks. Hair follicles, too, are among the most rapidly dividing cells.

### TAKE-HOME MESSAGE 6·6

Mitosis enables existing cells to generate new, genetically identical cells. This makes it possible for organisms to grow and to replace cells that die.

# 6·7 Overview: mitosis leads to duplicate cells.

For mitosis to begin, the parent cell replicates its DNA, creating a duplicate copy of each chromosome. Once this task is completed, the remainder of mitosis can take place, in which the chromosomes are separated into identical sets in two separate nuclei, and then the cell can divide into two duplicate cells, the daughter cells (**FIGURE 6-9**).

**FIGURE 6-9** **These two human embryonic stem cells have just completed mitosis.**

Mitosis occurs in just four steps. It begins with the condensing (or compacting) of the chromosomes, all of which are uncondensed and tangled during interphase, a bit like a big plate of spaghetti. Next, all the duplicated, condensed chromosomes move to the center of the cell. Each chromosome is then pulled apart from its duplicate. In the final step, new membranes form around each complete set of chromosomes to create the new nuclei, after which the cytoplasm and the rest of the cell are divided into two cells that pinch apart. Where once there was one parent cell, now there are two daughter cells (**FIGURE 6-10**).

## TAKE-HOME MESSAGE 6·7

**Mitosis is the process by which cells duplicate themselves. Mitosis follows chromosome replication and leads to the production of two daughter cells from one parent cell.**

**MITOSIS CREATES DUPLICATE CELLS**

Parent cell

**INTERPHASE**
The genetic material is replicated during the DNA synthesis portion of interphase.

**MITOSIS**

**1** Chromosomes condense.

Spindle

**2** Chromosomes line up in the middle of the cell.

**3** Each chromosome is pulled apart from its duplicate.

**4** New nuclear membranes form around each complete set of chromosomes and the cell divides.

**FIGURE 6-10** **A simplified introduction to mitosis.**

# 6·8 The details: mitosis is a four-step process.

Let's look at the process of mitosis in a bit more detail, keeping in mind that the ultimate result is the production of two cells with identical chromosomes.

### Interphase: in preparation for mitosis, the chromosomes replicate
Processes essential to cell division take place even before the mitotic phase of the cell cycle begins. During the DNA synthesis part of interphase, every chromosome replicates itself. Recall from Section 6-4 that prior to replication, each chromosome is just a long, linear piece of genetic material; after replication, each chromosome is a *pair* of identical long, linear pieces, held together at the centromere.

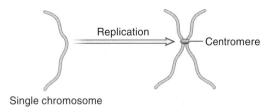

Single chromosome → Replication → Centromere

**Mitosis** The actual process of cell division occurs in four steps (**FIGURE 6-11**).

**1. Prophase: the long, linear chromosomes that have replicated condense.** When you look at a cell through a microscope, you generally won't see any chromosomes because the cells you're looking at are in some part of their interphase. During all of the cell cycle except mitosis, chromosomes are uncoiled and spread out in a diffuse way, like a mass of spaghetti. Because they are so stretched out, they are not dense enough to be visible. Mitosis begins with prophase, when the chromosomes in the cell's nucleus become more and more tightly coiled. As they condense, they become thick enough to be seen through a light microscope.

At this point, in prophase, each chromosome looks like the letter X. This appearance is misleading; chromosomes are not actually X-shaped. They are *linear*. Each X consists of two identical linear DNA molecules—a chromosome and its

**INTERPHASE**

Nucleus

Nuclear membrane
Centromere
Replicated chromosome
Parent cell

**INTERPHASE**
• Chromosomes replicate in preparation for mitosis.

**MITOSIS**

Spindle fibers

Nuclear membrane
Sister chromatids
Centriole
Spindle fibers

**1 PROPHASE**
• Nuclear membrane breaks down.
• Sister chromatids (replicated chromosomes) condense.
• Spindle forms.

**2 METAPHASE**
• Sister chromatids line up at the center of the cell.

**FIGURE 6-11 Mitosis: cell duplication, step by step.**

**Q** Animal chromosomes are linear. Why do they look like the letter X in pictures?

identical, replicated copy—joined at the centromere. (Each of the identical DNA molecules is called a **chromatid;** together, the two are called **sister chromatids.**) The reason for the X shape in most photos of chromosomes is that the only time the chromosome is coiled tightly and thus thick enough to be seen (and photographed) is during cell division.

across the cell between its two ends, or poles—where the threads connect at each pole to a structure called the **centriole.** These threads (known as **spindle fibers**) pull the sister chromatids to the middle of the cell and will eventually be used to pull the chromatids apart as cell division proceeds.

Centromere

Sister chromatids

Because a lot of room is required for the duplicated chromosomes to separate, the membrane around the nucleus is dismantled and disappears near the end of this first stage of mitosis. At the same time, a structure called the **spindle** is assembled. The spindle, which is part of the cell's cytoskeleton, can be thought of as a group of parallel threads stretching

**2. Metaphase: the chromosomes congregate at the cell center.** After condensing, the pairs of chromatids seem to move aimlessly around the cell, but eventually they line up at the cell's center, pulled by spindle fibers attached to the centromere of each pair. Metaphase can be considered the half-time of mitosis. At the end of metaphase, all the chromatid pairs are lined up in an orderly fashion, straddling the center in a "single-file" congregation that is called the metaphase plate. All the chromatids are at their most condensed during this part of mitosis.

**3. Anaphase: the chromatids separate and move in opposite directions.** In anaphase, the fibers attached to the centromeres begin pulling each chromatid in the sister chromatid pairs toward opposite poles of the cell. From each pair of sister chromatids, one strand of DNA is pulled in one direction and the other, identical strand is pulled in the opposite direction. At the end of anaphase, one full

**CYTOKINESIS**

Chromosomes

Spindle fiber

Nuclear membrane

Daughter cells

**3 ANAPHASE**
• The sister chromatid pairs are pulled apart by the spindle fibers. One full set of chromosomes goes to one side of the cell and another identical set goes to the other.

**4 TELOPHASE**
• The chromosomes begin to uncoil as the nuclear membrane reassembles around them.
• The cell begins to pinch in two.

**CYTOKINESIS**
• Cytoplasm is divided into the two daughter cells.

set of chromosomes is at one end of the cell and another, identical full set is at the other end. These chromosome sets will eventually reside in the nucleus of each of the two new daughter cells that result from the cell division.

**4. Telophase: new nuclear membranes form around the two complete chromosome sets.** With two full, identical sets of chromosomes collected at either end of the cell, the cell is prepared to divide into two genetically identical cells. In this last step, called telophase, the chromosomes begin to uncoil and fade from view, nuclear membranes reassemble, and the cell begins to pinch into two.

The process of mitosis, the division of a cell's nucleus into two nuclei, each of which includes an identical set of chromosomes, is generally followed by cytokinesis, during which the cell's cytoplasm is also divided into approximately equal parts, with some of the organelles going to each of the two new cells. When cytokinesis is complete, the two new daughter cells, each with an identical nucleus containing identical genetic material, enter interphase and begin the business of being cells.

Usually, mitosis occurs without errors and only when needed. In rare cases, however, something goes wrong that can cause cell division to proceed unchecked. In such cases, cancer can arise.

### TAKE-HOME MESSAGE 6·8

**The ultimate result of mitosis and cytokinesis is the production of two genetically identical cells.**

# 6·9  Cell division out of control means cancer.

Too much of a good thing can be bad. This is especially true when cell division runs amok, a situation that can lead to cancer. **Cancer** is defined as unrestrained cell growth and division that can damage adjacent tissues. Some cancers can metastasize, or spread to other locations in the body. Cancer can cause serious health problems and is the second leading cause of death in the United States, responsible for more than 20% of all deaths. Only heart disease causes more deaths.

Cancer occurs when some disruption of the DNA in a normal cell interferes with the cell's ability to regulate cell division. DNA disruption can be caused by chemicals that mutate DNA or by sources of high energy such as X rays, the sun, or nuclear radiation. Cancer can even be caused by some viruses. However it begins, once a cell loses control over its cell cycle, cell division can proceed unrestrained (**FIGURE 6-12**).

Cancer cells have several features that distinguish them from normal cells; the three most significant differences are:

**1. Cancer cells lose their "contact inhibition."** Most normal cells divide until they bump up against other cells or collections of cells (tissues). At that point, they stop dividing. Cancer cells, however, ignore the signal that they are at high density and continue to divide.

**2. Cancer cells can divide indefinitely.** As we saw in Section 6-1, a normal cell can divide approximately 50 times. After that point, the cell may continue living but it loses the ability to divide. Cancer cells, on the other hand, never lose their ability to divide and continue to do so indefinitely, even in the presence of conditions that would normally halt the cell cycle before cell division. (Cancer cells can divide indefinitely because they are able to rebuild their telomeres following each cell division.)

**3. Cancer cells have reduced "stickiness."** Cells are normally held together by adhesion molecules, proteins within cell membranes. And cancer cells, too, usually group together, forming a tumor. But the membranes of cancer cells tend to have reduced adhesiveness, causing them to stick to each other less than do non-cancerous cells.

Tumors caused by excessive cell growth and division are of two very different types: benign and malignant. Benign tumors, such as many moles, are just masses of normal cells that do not spread. They can usually be removed safely without any lasting consequences. Malignant tumors, on the other hand, are the result of unrestrained growth of cancerous cells. They grow continuously and shed cells (**FIGURE 6-13**). The shedding of cancer cells from malignant tumors is how cancer spreads, a process called metastasis (meh-TASS-tuh-siss) In this process, cancer cells separate from a tumor and invade the circulatory system, which spreads them to different parts of the body where they can cause the growth of additional tumors.

## CANCER CELLS

Cancer cells have several features that distinguish them from normal cells.

### CANCER CELLS HAVE NO CONTACT INHIBITION

Normal cells divide until they bump up against other cells or tissues, at which point they stop dividing.

Cancer cells, however, ignore the signal and continue to divide, piling on top of one another.

Normal cells

Cancer cells

### CANCER CELLS DIVIDE INDEFINITELY

Normal somatic cells can divide a limited number of times. Cancer cells never lose their ability to divide, and continue to do so indefinitely.

Normal cell

Cancer cell

Eventually, normal cells are no longer able to divide.

Cancer cells continue to divide indefinitely.

### CANCER CELLS HAVE REDUCED "STICKINESS"

The membranes of cancer cells tend to have reduced adhesiveness, causing them to stick to each other less than do non-cancerous cells.

Normal cells

Cancer cells

**FIGURE 6-12 Problem cells.** Cancer occurs when there is a disruption in cells' ability to regulate cell division and a reduced adhesion among those cells.

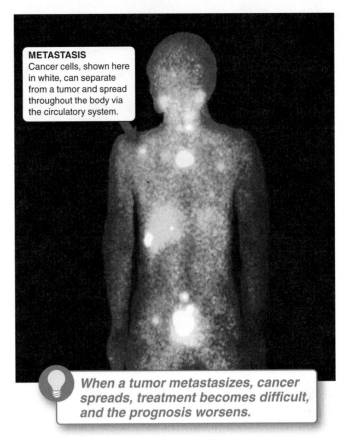

**METASTASIS**
Cancer cells, shown here in white, can separate from a tumor and spread throughout the body via the circulatory system.

*When a tumor metastasizes, cancer spreads, treatment becomes difficult, and the prognosis worsens.*

**FIGURE 6-13 Multiple tumors.** Metastasis spreads cancer cells throughout the body.

How does cancer actually cause death? Somewhat surprisingly, it's not because of some chemical or genetic property of the cancer cells themselves. It's simpler than that. As a tumor gets larger, it takes up more and more space, and presses against neighboring cells and tissues (**FIGURE 6-14**). Eventually, the tumor may block other cells and tissues from carrying out their normal functions and even kill them. This cell dysfunction or cell death can have disastrous consequences when the affected normal tissue controls processes critical to life, such as breathing, heart function, or the detoxification processes in the liver.

**Q** What is cancer? How does it usually cause death?

To treat cancer, the rapidly dividing cells must be removed surgically or killed, or their division at least slowed down. Currently, the killing and slowing down are done in two ways: by chemotherapy and by radiation. In chemotherapy, drugs that interfere with cell division are administered,

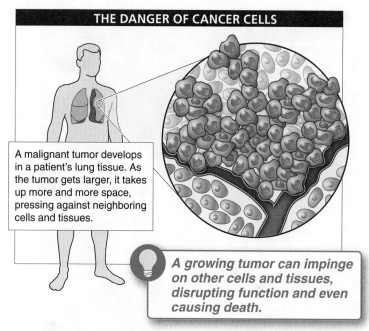

**THE DANGER OF CANCER CELLS**

A malignant tumor develops in a patient's lung tissue. As the tumor gets larger, it takes up more and more space, pressing against neighboring cells and tissues.

*A growing tumor can impinge on other cells and tissues, disrupting function and even causing death.*

**FIGURE 6-14 Cancer's effects.** Cancer cells harm the body by crowding and disrupting normal cells.

**FIGURE 6-15 Cancer treatments.** The treatments can be as painful and difficult as the disease.

slowing down the growth of tumors. Because these drugs interfere with rapidly dividing cells throughout the body (not just the rapidly dividing cancer cells), they can have very unpleasant side effects. In particular, chemotherapy drugs disrupt normal systems that rely on the rapid and constant production of new cells. For instance, chemotherapy often causes extreme fatigue and shortness of breath because it reduces the rate at which red blood cells are produced, thus limiting the amount of oxygen that can be transported throughout the body. By interfering with the division of bone marrow stem cells, chemotherapy also reduces the production of platelets and white blood cells and thus increases bruising and bleeding, and increases susceptibility to infection. Another location of rapidly dividing cells commonly affected by chemotherapy and radiation is the hair follicle. Many people lose their hair when undergoing chemotherapy (it usually grows back when treatments stop).

Like chemotherapy, radiation works by disrupting cell division. Unlike the drugs used in chemotherapy, however, which circulate throughout the entire body, radiation therapy directs high-energy radiation only at the part of the body where a tumor is located. Like chemotherapy, the radiation process is not perfect and nearby tissue is often harmed as well. The significant negative effects of chemotherapy and radiation treatment on normal tissue and the suffering this leads to have caused some patients to comment that the treatment for cancer can initially feel as bad as the disease itself—especially because patients may be asymptomatic at the time of treatment (**FIGURE 6-15**).

Is there hope for a complete cure soon? Not yet. Nonetheless, many potentially successful therapies for cancer treatment and prevention are on the horizon. Extensive research is being conducted on the mechanisms by which genes controlling the cell cycle are damaged in cancer cells, for example, and how such damage might be prevented or reversed.

In addition, there is a rapidly growing field of research into cancer-inhibiting chemicals that have evolved in many plant species. One such chemical is resveratrol, which is found in grapes and peanuts. Many of these naturally occurring chemicals may be less harsh than the drugs currently used in chemotherapy. We are also becoming more conscious of the role that the environment plays in increasing cancer risk; our greater awareness is leading to changes in industrial processes and, on the individual level, to dietary or lifestyle changes. Advances in early screening, particularly in identifying cancer prior to metastasis, continue to have significant benefits as well.

**TAKE-HOME MESSAGE 6·9**

Cancer is unrestrained cell growth and cell division, which lead to large masses of cells that may cause serious health problems. Treatment focuses on killing or slowing down the fast-growing and dividing cells, usually using chemotherapy and/or radiation.

# 6·10–6·14 Meiosis generates sperm and eggs and a great deal of variation.

*Human gametes: sperm attempting to fertilize an egg.*

## 6·10 Overview: sexual reproduction requires special cells made by meiosis.

There are two ways in which an organism can reproduce. Many organisms, including bacteria, fungi, and even some plants and animals, undergo asexual reproduction, in which a single parent produces identical offspring. Other organisms, including most animals and plants, undergo sexual reproduction, in which offspring are produced by the fusion of two reproductive cells in the process of **fertilization.** And some species, particularly among plants, can use both methods.

Sexually reproducing organisms, humans included, have a problem to solve when it comes to cell division. When offspring are created, they carry the genetic material from two individuals. But think about the difficulties this presents. If reproductive cells were produced through mitosis, both parents would contribute a full set of genes—that is, 23 pairs of chromosomes in humans—to create a new individual, and the new offspring would inherit 46 pairs of chromosomes in all. And when that individual reproduced, if he or she contributed 46 pairs of chromosomes and his or her mate also contributed 46 pairs, their offspring would have 92 pairs of chromosomes. Where would it end? The genome would double in size every generation. That wouldn't work at all. At

the very least, within a few generations the cells would be so overloaded with chromosomes that they would explode.

In sexually reproducing organisms, a way has evolved to avoid such chromosome overload. The solution is **meiosis,** a process that enables organisms to make special reproductive cells, the gametes, which have only half as many chromosomes as the rest of the cells in the organism's body (the somatic cells). In other words, in anticipation of combining one individual's genome with another's during fertilization, meiosis reduces each individual's genome by half in the gametes. In humans, for example, each gamete cell has only one set of 23 chromosomes, rather than two sets. In genetics, the term **diploid** refers to cells that have two copies of each chromosome (in humans that is two sets of 23 chromosomes, for 46 chromosomes in total), and the term **haploid** refers to cells that have one copy of each chromosome. Thus, somatic cells are diploid, and gametes, the cells produced in meiosis, are haploid. At fertilization, two haploid cells, each with one set of 23 chromosomes, merge and create a new individual with the proper diploid human genome of 46 chromosomes. And when the time comes to reproduce, this new individual, through meiosis, also will produce haploid gametes that have only a single set of 23 chromosomes. With sexual reproduction, then, diploid organisms produce haploid gametes that fuse at fertilization to restore the diploid state (**FIGURE 6-16**). In this way, meiosis maintains a stable genome size in a species.

Although there are some variations on this pattern of alternation between the haploid and diploid states, in most cases, multicellular animals produce simple haploid cells for reproduction. And after two of those gametes come together to form a diploid fertilized egg, multiple cell divisions via mitosis again produce a diploid, multicellular animal.

Meiosis achieves more than just a reduction in the amount of genetic material in gametes. As a diploid individual, you have two copies (i.e., two alleles) of every gene in every somatic cell: one from your mother and one from your father. When making haploid cells from cells that are diploid, an individual creates cells that have one allele for each trait rather than two. Which of these two alleles is included in each gamete cannot be predicted.

Each egg or sperm is produced with a varied combination of maternal and paternal alleles so that all of the gametes an individual makes carry a unique set of alleles. As a consequence, each offspring inherits a slightly different assemblage of each parent's alleles. All offspring resemble their parents, yet none resembles them in exactly the same way. This variation among offspring can be seen among the five Jackson brothers in the photo that opens this chapter (and among the Kennedy brothers in **FIGURE 6-17**).

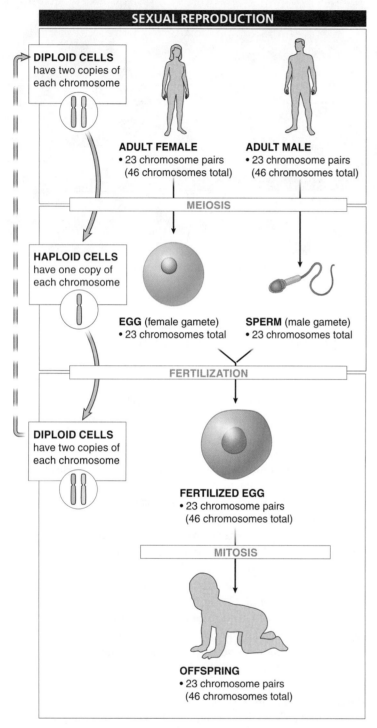

**FIGURE 6-16 Sexual reproduction.** Diploid organisms produce haploid gametes that fuse at fertilization and return to the diploid state.

**FIGURE 6-17 One mother, one father: varied offspring.** Variation generated in the production of sperm and eggs causes each sibling to carry a unique set of alleles.

In all, then, meiosis has two important outcomes:

1. It reduces the amount of genetic material in gametes.

2. It produces gametes that differ from one another with respect to the combinations of alleles they carry.

In the next sections, we examine the exact steps by which these outcomes occur and investigate the consequences of all the variation that meiosis and sexual reproduction can generate.

# 6·11  Sperm and egg are produced by meiosis: the details, step by step.

Mitosis is an all-purpose process for cell division. It occurs all over the body, all the time. Meiosis, on the other hand, is a special-purpose process. It takes place in just a single place: the **gonads** (the ovaries and testes in sexually reproducing animals). And it takes place for just a single reason: the production of gametes (reproductive cells). To better understand meiosis, let's examine in detail the steps by which a diploid cell creates haploid cells, each of which is different from the other.

Meiosis starts with a diploid cell—not just any diploid cell, but one of the specialized diploid cells, found in the gonads, that is capable of undergoing meiosis. Thus, for humans, meiosis starts with a cell that has 46 chromosomes: a maternal copy and a paternal copy of each of the 23 human chromosomes. Together, the maternal and paternal copies of a chromosome are called a **homologous pair,** or **homologues.** This means that there are 23 homologous pairs in a diploid human cell (**FIGURE 6-18**).

Before meiosis can occur (and just as we saw with mitosis), each of these 46 chromosomes is duplicated during the cell's interphase. So now, instead of having just a maternal and paternal copy of, say, chromosome 1, we have two copies of the maternal copy of chromosome 1, bound together as sister chromatids, and two copies of the paternal copy of chromosome 1, bound together as sister chromatids. This means that when meiosis begins, we have 92 (2 × 46) strands of DNA, after replication.

Unlike mitosis, which has only one cell division, cells undergoing meiosis divide twice. In the first division, the homologues separate. In other words, for each of the 23 chromosome pairs, the maternal sister chromatid pairs and the paternal sister chromatid pairs separate into two new cells. In the second division, each of the two new cells

**HOMOLOGUES AND SISTER CHROMATIDS**

Homologues are the maternal and paternal copies of a chromosome. A sister chromatid is one of the two identical copies of a chromosome created during replication. The two sister chromatids are held together at a centromere.

FIGURE 6-18 **Chromosome vocabulary: homologues and sister chromatids.**

divides again, and this time the sister chromatids separate from each other into two newer cells. At the end of meiosis, there are four new cells, each of which has 23 strands of DNA—that is, 23 chromosomes (**FIGURE 6-19**). Note that none of these four cells will undergo any further cell division—they do not become parent cells for a next cycle of cell division. And remember, only specialized types of diploid cells located in the gonads can undergo meiosis.

**Interphase: in preparation for meiosis, the chromosomes replicate** Before meiosis begins, preparations for cell division occur that are similar to those taking place in the interphase of mitosis in the somatic cell cycle. In particular, every chromosome creates an exact duplicate of itself by replication. The chromosomes that were each a single, long, linear piece of genetic material become a pair of identical long, linear pieces, held together at the centromere.

## MEIOSIS: REDUCING THE GENOME BY HALF

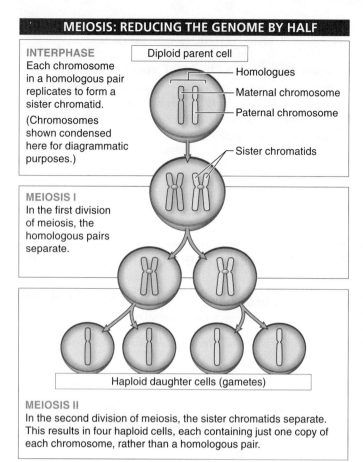

**INTERPHASE**
Each chromosome in a homologous pair replicates to form a sister chromatid.
(Chromosomes shown condensed here for diagrammatic purposes.)

Diploid parent cell

— Homologues
— Maternal chromosome
— Paternal chromosome

— Sister chromatids

**MEIOSIS I**
In the first division of meiosis, the homologous pairs separate.

Haploid daughter cells (gametes)

**MEIOSIS II**
In the second division of meiosis, the sister chromatids separate. This results in four haploid cells, each containing just one copy of each chromosome, rather than a homologous pair.

**FIGURE 6-19 Meiosis reduces the genome by half in anticipation of combining it with another genome.**

Meiosis begins following this replication step during interphase; it proceeds through eight steps (**FIGURE 6-20**).

### Meiosis division I: the homologues separate

The first meiotic division takes place in four stages.

**1. Prophase I: chromosomes condense and crossing over occurs.** This is by far the most complex of all the phases of meiosis. It begins with all of the replicated genetic material condensing.

As the sister chromatids become shorter and thicker, the homologous chromosomes come together. That is, the maternal and paternal sets of sister chromatids for chromosome 1 come together so that there are *four* versions of this chromosome lined up together. The maternal and paternal sets of chromosome 2, 3, 4, and so on, also line up. Under a microscope, the two homologous pairs of sister chromatids appear as pairs of X's lying on top of each other.

At this point, the sister chromatids that are next to each other do something that makes every sperm or egg cell genetically

**INTERPHASE**

**MEIOSIS DIVISION I: HOMOLOGUES SEPARATE**

Centromere
Replicated chromosome

Homologues

Piece of maternal chromatid now on paternal chromatid following crossing over

Spindle fiber

Nuclear membrane

Spindle

**INTERPHASE**
• Chromosomes (uncondensed in this phase) replicate in preparation for meiosis.

**1** **PROPHASE I**
• Replicated chromosomes condense.
• Spindle is formed.
• Homologous pairs of sister chromatids come together and cross over.
• Nuclear membrane disintegrates.

**2** **METAPHASE I**
• Homologues move toward the center of the cell (at this point, referred to as the metaphase plate) and line up.

**3** **ANAPHASE I**
• Homologues separate and are pulled to opposite poles. Sister chromatids going to each side are a mix of maternal and paternal genetic material.

**FIGURE 6-20 Meiosis: generating reproductive cells, step by step.**

unique: they swap little segments of DNA. In other words, when this is happening inside of you, some of the genes that you inherited from your mother get swapped onto the strand of DNA you inherited from your father, and the corresponding bit from your father is inserted into the DNA strand from your mother. This event is called **recombination** or **crossing over,** and it can take place at several spots (up to dozens) on each chromatid. As a result of crossing over, every sister chromatid ends up having a unique mixture of the genetic material that you received from your two parents. Note that crossing over occurs only during the production of gametes, in meiosis. It does not occur during mitosis. Following crossing over, the nuclear membrane disintegrates.

We explore crossing over in more detail in Section 6-13. The remaining steps of meiosis are relatively straightforward.

## 2. Metaphase I: all chromosomes line up along the center of the cell.

After crossing over, each pair of homologous chromosomes (that is, the pairs of X's lying on top of each other) moves to the center of the cell, pulled by the spindle fibers to form the arrangement called the metaphase plate. Remember that each pair of homologous chromosomes includes the maternal and paternal versions of the chromosome (with crossed-over segments) *and* the replicated copy of each—four strands in all.

## 3. Anaphase I: homologues are pulled to either side of the cell.

This phase is the beginning of the first cell division that occurs during meiosis. In anaphase, the homologues are pulled apart toward opposite poles of the cell. One of the homologues (consisting of two sister chromatids) goes to one pole, the other to the opposite pole. At this point, something else occurs that, along with crossing over, contributes to making all the products of meiosis genetically unique. The maternal and paternal sister chromatid pairs are pulled to the ends of the cell in a random fashion. As a result, the pairs of sister chromatids gathered at each pole are a mix of maternal and paternal sister chromatids.

Imagine all the different combinations that can occur in a species with 23 pairs of chromosomes. For chromosome pair 1, perhaps the maternal homologue goes to the "top" pole and the paternal to the "bottom" pole. And for pair 2, perhaps the maternal homologue also goes to the top and the paternal to the bottom. But maybe for pair 3 it is the paternal homologue that goes to the top and the maternal homologue to the bottom. For each of the 23 pairs, you cannot predict which will go to the top and which will go to the bottom. There is a *huge* number of possible combinations. (Can you figure out how many? Here's a hint: if there were 2 chromosome pairs, the answer would be 4; if there were 3 chromosome pairs, the answer would be 8.)

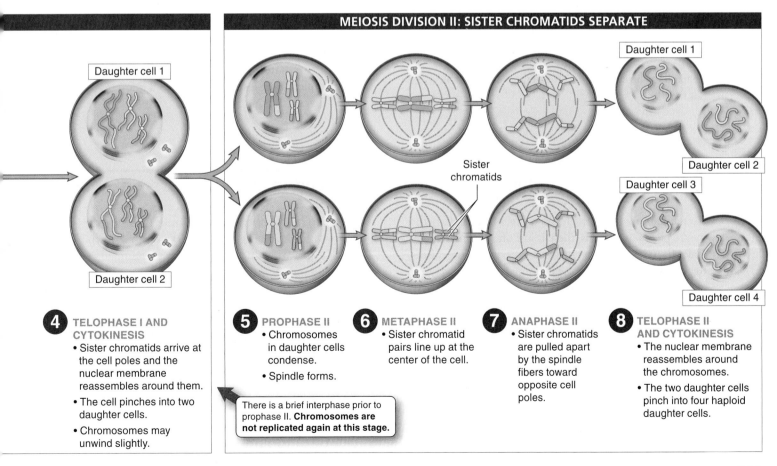

**MEIOSIS DIVISION II: SISTER CHROMATIDS SEPARATE**

Daughter cell 1

Daughter cell 2

Daughter cell 1

Daughter cell 2

Daughter cell 3

Daughter cell 4

Sister chromatids

**4 TELOPHASE I AND CYTOKINESIS**
- Sister chromatids arrive at the cell poles and the nuclear membrane reassembles around them.
- The cell pinches into two daughter cells.
- Chromosomes may unwind slightly.

There is a brief interphase prior to prophase II. **Chromosomes are not replicated again at this stage.**

**5 PROPHASE II**
- Chromosomes in daughter cells condense.
- Spindle forms.

**6 METAPHASE II**
- Sister chromatid pairs line up at the center of the cell.

**7 ANAPHASE II**
- Sister chromatids are pulled apart by the spindle fibers toward opposite cell poles.

**8 TELOPHASE II AND CYTOKINESIS**
- The nuclear membrane reassembles around the chromosomes.
- The two daughter cells pinch into four haploid daughter cells.

**4. Telophase I and cytokinesis: nuclear membranes reassemble around sets of sister chromatid pairs, and two daughter cells form.** After the pairs of chromatids arrive at the two poles of the cell, nuclear membranes re-form, then cytokinesis occurs: the cytoplasm divides, and the cell membrane pinches the cell into two daughter cells. Each daughter cell has its own nucleus that contains the genetic material—two sister chromatids for each of the 23 chromosomes in humans.

### Meiosis division II: separating the sister chromatids

There is a brief interphase after the first division of meiosis. In some organisms, the DNA molecules (now in the form of chromatid pairs) briefly uncoil and fade from view. In others, the second part of meiosis begins immediately. *It is important to note that in the brief interphase before prophase II, there is no replication of any of the chromosomes.* The second part of meiosis, like the first division of meiosis, is a four-phase process.

**5. Prophase II: chromosomes re-condense.** The second division of meiosis begins with prophase II. In this phase, the genetic material in each of the two daughter cells once again coils tightly, making the pairs of chromatids visible under the microscope. (Unlike prophase I, no crossing over occurs during prophase II.)

**6. Metaphase II: sister chromatid pairs line up at the center of the cell.** In each of the two daughter cells, the sister chromatid pairs (each pair appearing as an X) move to the center of the cell, pulled by spindle fibers attached to the centromere, where the sister chromatids are held together. The congregation of all the genetic material in the center of each daughter cell is visible as a flat metaphase plate.

**7. Anaphase II: sister chromatids are pulled to opposite sides of the cell.** This phase starts with 46 pieces of DNA in each cell created by meiosis I: that is, 23 sister chromatid pairs. During this anaphase, the fibers attached to the centromere begin pulling each chromatid in the 23 sister chromatid pairs toward opposite poles of the daughter cell. When anaphase II is finished, each of what will become the four daughter cells (you could think of them as granddaughter cells!) has one single copy of each of the 23 chromosomes.

**8. Telophase II and cytokinesis: nuclear membranes reassemble and the two daughter cells pinch into four haploid gametes.** After the sister chromatids for all 23 chromosomes have been pulled to opposite poles, the cytoplasm then divides. The cell membrane pinches each cell into two new daughter cells, nuclear membranes begin to re-form, and the process comes to a close.

In humans, the outcome of one diploid cell undergoing meiosis is the creation of four haploid daughter cells, each with just one set of 23 individual chromosomes. These chromosomes contain a combination of traits from the individual's diploid set of chromosomes.

> ### TAKE-HOME MESSAGE 6·11
>
> Meiosis occurs only in gamete-producing cells. It is preceded by DNA replication and consists of two rounds of cellular division, one in which homologous pairs of sister chromatids separate and a second in which the sister chromatids separate. The final product of meiosis in a diploid organism is four haploid gametes.

# 6·12 Male and female gametes are produced in slightly different ways.

No matter how small or large they are, no matter whether they're plants or animals, there is just one way to distinguish males from females in all sexually reproducing organisms. Regardless of the species, the defining feature is always the same (and if you're thinking of anything visible to the naked eye, you're wrong). When there are two sexes—as there are in nearly every sexually reproducing plant and animal species—the females are the sex that produces the larger gamete, and the males produce the smaller, more motile, gamete (**FIGURE 6-21**). Because all gametes are produced through meiosis, it is a slight variation in how meiosis works in males and females that leads to this difference.

The size difference between the male gamete and the female gamete (sperm and eggs in animals) all comes down to the fact that sperm cells have very little cytoplasm, while eggs have a huge amount. During the production of sperm, the two divisions occur just as described in Section 6-11, resulting in four evenly sized cells that become sperm. During the production of eggs, things are a bit different. The first division occurs just as described in Section 6-11, except that in telophase I, as the cell divides, instead of forming two cells of the same size, the division is lopsided. The genetic material is evenly divided, but nearly all of the cytoplasm goes to one of the cells and almost none goes to the other. The smaller cell is

The definition of male vs. female is always the same: females produce the larger gamete, males produce the smaller, more motile gamete.

FIGURE 6-21 **Egg and sperm.** In sexually reproducing organisms, the female produces the larger gamete.

called a **polar body** and degrades almost immediately in most animal cells (although it sometimes goes through with a second division, after which, in most animal cells, both of those polar bodies degrade; we see in Chapter 18 that polar bodies have some function in producing nutritive tissue for the developing embryo in flowering plants). Then, in the second meiotic division of the larger cell, there is again an unequal division of cytoplasm.

As in the first division, one of the new cells gets nearly all of the cytoplasm and the other gets almost none, forming another polar body. The net result of meiosis in the production of eggs is one large egg with lots of cytoplasm and two or three small polar bodies with very little cytoplasm that degrade and never function as gametes (**FIGURE 6-22**).

Ultimately, whether eggs or sperm are produced, each gamete ends up with just one copy of each chromosome. That way, the fertilized egg that results from the fusion of sperm and egg carries two complete sets of chromosomes, so the developing individual will be diploid. The extra cytoplasm carried by the egg contains a large supply of nutrients and other chemical resources to help with the initial development of the organism following fertilization.

## TAKE-HOME MESSAGE 6·12

In species with two sexes—including nearly every sexually reproducing plant and animal species—females are the sex that produces the larger gamete, and males produce the smaller gamete. Whether it is male or female gametes that are being produced, each gamete ends up with just one copy of each chromosome.

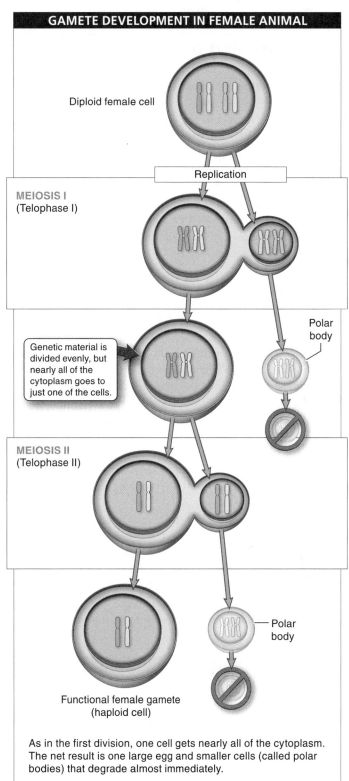

**GAMETE DEVELOPMENT IN FEMALE ANIMAL**

Diploid female cell

Replication

MEIOSIS I
(Telophase I)

Polar body

Genetic material is divided evenly, but nearly all of the cytoplasm goes to just one of the cells.

MEIOSIS II
(Telophase II)

Polar body

Functional female gamete
(haploid cell)

As in the first division, one cell gets nearly all of the cytoplasm. The net result is one large egg and smaller cells (called polar bodies) that degrade almost immediately.

FIGURE 6-22 **Unequal distribution of cytoplasm results in one large egg.**

# 6·13 Crossing over and meiosis are important sources of variation.

Genetically speaking, there are two ways to create unique individuals. The obvious way is for an organism to carry an allele that is not present in any other individuals. Alternatively—and equally successful in creating uniqueness—an individual can carry a *collection* of alleles, no single one of which is unique, that has never before occurred in another individual. Both types of novelty introduce important variation into a population of organisms. The process of crossing over (**FIGURE 6-23**), which occurs during prophase I in meiosis, creates a significant amount of the second type of variation.

As we saw in Section 6-11, during the first prophase of meiosis, the sister chromatids of homologous chromosomes all come together. That previous sentence is a mouthful! Let's review exactly what it means. Let's look at just one of your 23 homologous pairs of chromosomes, the homologous pair of chromosome 15. When we refer to the "homologous pair" of chromosome 15, remember that this pair includes two copies of chromosome 15: one copy from your mother (which you inherited from the egg that was fertilized to create you) and one copy from your father (which you inherited from the sperm that fertilized the egg). Each chromosome in the pair carries the same genes, but because they came from different people, they don't necessarily have the same alleles.

Once the sister chromatids of the homologous chromosome pairs line up (so that there are now four chromatids in two pairs lying very close together), regions that are close together can swap segments. A piece of one of the maternal chromatids—perhaps including the first 100 genes on the strand of DNA—may swap places with the same segment in a paternal chromatid. Elsewhere, a stretch of 20 genes in the

middle may be swapped from the other maternal chromatid with one of the paternal chromatids. Every time a swap of DNA segments occurs, an identical amount of genetic material is exchanged, so all four chromatids still contain the complete set of genes that make up the chromosome. The *combination* of alleles on each chromatid, though, is now different.

Suppose there are genes relating to eye color, hair color, and height on a particular chromosome. After crossing over, the linear strands still have instructions for all of those traits. But whereas one chromatid previously may have had instructions for brown eyes, brown hair, and short height, it may now have some differences—perhaps brown eyes, blond hair, and tall height. All of the alleles from your parents are still carried on one DNA molecule or another. But the *combination of traits* that are linked together on a single chromatid is new. And when a gamete, let's say it's an egg, carrying a new combination of alleles is fertilized by a sperm, the developing individual will carry a completely novel set of alleles. And so without creating new versions of any traits (such as yellow eyes or purple hair), crossing over still creates gametes with collections of alleles that may never have existed together. In Chapter 8, we'll see that this variation is tremendously important for evolution.

---

**TAKE-HOME MESSAGE 6·13**

Although it doesn't create new versions of any traits, crossing over during the first prophase of meiosis creates gametes with combinations of traits that may never have existed before; this variation is important for evolution.

---

CROSSING OVER

HOMOLOGOUS CHROMOSOMES

Sister chromatids

Maternal copy    Paternal copy

Crossing over between the sister chromatids of the homologous chromosomes

Homologous chromosomes after the exchange of genetic information

Each of these 4 chromatids gets packaged into a haploid gamete (sperm or egg cell).

Chromatids with recombined DNA

**FIGURE 6-23 Swapping DNA.** Crossing over creates new combinations of alleles on each chromatid.

*Crossing over doesn't create new alleles but it does create new combinations of alleles on a chromatid.*

# 6·14

## What are the costs and benefits of sexual reproduction?

As we've seen, there are two fundamentally different ways by which cells and organisms can reproduce. On the one hand there is mitosis and asexual reproduction via binary fission; on the other hand, meiosis and sexual reproduction. Is one method better than the other? It depends. In fact, the more appropriate question is, what are the advantages and disadvantages of each method and under what conditions do the benefits outweigh the costs?

### What are the advantages of sexual reproduction?

Sexual reproduction leads to offspring that are genetically different from one another and from either parent, through three different routes (**FIGURE 6-24**).

1. **Combining alleles from two parents at fertilization.** First and foremost, with sexual reproduction, a new individual comes from the fusion of gametes from two different individuals. Each of these parents comes with his or her own unique set of genetic material.

2. **Crossing over during the production of gametes.** As we saw, crossing over during prophase I of meiosis causes every chromosome in a gamete to carry a mixture of an individual's maternal and paternal genetic material.

3. **Shuffling and reassortment of homologues during meiosis.** Recall that as the homologues for each chromosome are pulled to opposite poles of the cell during the first division of meiosis, maternal and paternal homologues are randomly pulled to each pole. This means that there is an extremely large number of different combinations of maternal and paternal homologues that could end up in each gamete.

**Q** Bacteria reproduce asexually, whereas most plants and animals reproduce sexually. Which is the better method?

The variability among the offspring produced by sexual reproduction enables populations of organisms to cope better with changes in their environment. After all, if the environment is gradually changing from one generation to the next, individuals producing many genetically different offspring increase the likelihood that one of their offspring will carry a set of genes particularly suited to the new environment. Over time, populations of sexually reproducing organisms can quickly adapt to changing environments. It's like buying different lottery tickets—the more tickets you buy, the more likely it is that one of them will be a winner.

**SOURCES OF GENETIC VARIATION**

There are multiple reasons why offspring are genetically different from their parents and one another.

**ALLELES COME FROM TWO PARENTS**
Each parent donates his or her own set of genetic material.

**CROSSING OVER**
Crossing over during meiosis produces a mixture of maternal and paternal genetic material on each chromatid.

**REASSORTMENT OF HOMOLOGUES**
The homologues and sister chromatids distributed to each daughter cell during meiosis are a random mix of maternal and paternal genetic material.

**FIGURE 6-24 Crossing over and meiosis: creating many different combinations of alleles.** As a result of crossing over and meiosis, each individual developing from a separate fertilized egg is genetically unique.

### What are the disadvantages of sexual reproduction?

In the yellow dung fly (*Scathophaga stercoraria*), males sometimes wrestle each other for mating access to a female. The female awaits the outcome of the battle in a pile of dung. Occasionally, females drown in the dung pile as they wait. Males, too, can be at risk during reproduction. This is seen most dramatically in species such as praying mantises and black widow spiders in which the female will attempt to eat the male (and often succeed in doing so) during or after mating, gaining not just gametes but a nutritious meal as well (**FIGURE 6-25**)!

Dangers associated with mating are just one of the downsides to sexual reproduction. There are several others. First, when an individual reproduces, only half of its offsprings' alleles will come from that organism. The other half will come from the other parent. With asexual reproduction, an individual produces nearly identical offspring, so there is a very efficient transfer of genetic information from one generation to the next. Also, with sexual reproduction it takes time and energy to find a partner. This is energy that asexual organisms can devote to additional reproduction. Moreover, as we see with the dung flies, sex can be a risky proposition because organisms make themselves vulnerable to predation, disease, and other calamities during mating and reproduction. And finally, as we'll see later in this chapter, the complex cellular division required for sexual reproduction offers opportunities for mistakes, sometimes leading to chromosomal disorders.

*Asexual reproduction can be fast! Some bacteria can divide every 20 minutes.*

**FIGURE 6-26 Pluses and minuses.** This asexually reproducing bacterial colony can reproduce quickly and efficiently, but because each new cell's genome is an exact duplicate of the parent genome, there is no genetic variety to enable adaptation should the environment change.

beneficial for organisms to produce offspring as similar to themselves as possible.

### What are the disadvantages of asexual reproduction?
The downside to asexual reproduction is that the more closely an offspring's genome resembles its parent's, the less likely it is that the offspring will be suited to the environment when it changes.

In the end, we still see large numbers of species using asexual reproduction and large numbers, including the vast majority of animals and plants, that reproduce sexually. It seems that conditions favoring each occur in the great diversity of habitats in the world. That we see both sexual and asexual reproduction also highlights the recurring theme in biology that there often is more than one way to solve a problem.

**FIGURE 6-25 Sexual reproduction has its risks.** Female black widow spiders attempt to eat the male during or after mating.

### What are the advantages of asexual reproduction?
With asexual reproduction, the advantages and disadvantages are more or less reversed. Because asexual reproduction involves only a single individual, it can be fast and easy. Some bacteria can divide, forming a new generation, every 20 minutes (**FIGURE 6-26**). And for organisms in isolated habitats or when establishing new populations, asexual reproduction can be advantageous as well. Asexual reproduction is efficient, too. Offspring carry all of the genes that their parent carried— they are genetically identical. If the environment is stable, it is

---

**TAKE-HOME MESSAGE 6·14**

There are two fundamentally different ways that cells and organisms can reproduce: (1) mitosis and asexual reproduction via binary fission, and (2) meiosis and sexual reproduction. Asexual reproduction can be fast and efficient, but leads to genetically identical offspring that carry all of the genes that their parent carried, which could be disadvantageous in a changing environment. Sexual reproduction leads to offspring that are genetically different from one another and from either parent, but takes more time and energy and can be risky.

# 6·15–6·16 There are sex differences in the chromosomes.

*XX and XY: only one of the 23 chromosome pairs determines sex in humans.*

# 6·15

## How is sex determined in humans?

**Q**

**Which parent determines a baby's sex? Why?**

In humans, the sex of a baby is determined by its father. The sequence of events involved in sex determination is instigated by one special pair of chromosomes, the sex chromosomes, which carry information that directs a growing fetus to develop as a male or as a female.

Let's take a closer look at the human sex chromosomes. We noted that there are 23 pairs of chromosomes in every somatic cell. These can be divided into two different types: 1 pair of sex chromosomes and 22 pairs of non-sex chromosomes. The human sex chromosomes are called the **X and Y chromosomes** (FIGURE 6-27).

How do the X and Y chromosomes differ from the other chromosomes? As we described earlier, all of the genetic information is stored on the chromosomes in all the cells of an organism's body. But most of this information is not sex specific—that is, if you are building an eye or a neuron or a skin cell or a digestive enzyme, it doesn't matter whether it is for a male or for a female; the instructions are the same for both sexes. Some genetic information, however, instructs the body to develop into one sex or the other. That information is found on the sex chromosomes.

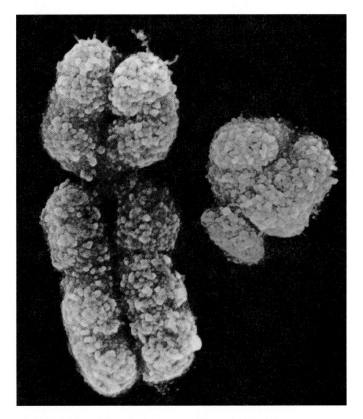

**FIGURE 6-27** X and Y: the human sex chromosomes. (The X chromosome is on the left.)

An individual has two copies of all the non-sex chromosomes. One copy is inherited from the mother, one from the father. Individuals also have two copies of the sex chromosomes, but not always two copies of the same kind. Males have one copy of the X chromosome and one copy of the Y chromosome. Females, on the other hand, don't have a Y chromosome but instead have two copies of the X chromosome.

So how does the father determine the sex of the baby? During meiosis in females, the gametes that are produced carry only one copy of each chromosome. This is true for the sex chromosomes, too. So from the two copies of the X chromosome carried by females, half of the gametes end up with a copy of one of those X chromosomes while the rest of the gametes inherit the other X chromosome. Thus, every egg has an X for its one sex chromosome. During meiosis in males, the sperm that are produced also carry one copy of each chromosome, including the sex chromosomes, but in this case, half of the sperm produced inherit the X chromosome and the other half inherit the Y chromosome. At fertilization, an egg bearing a single X chromosome (and one copy of all the non-sex chromosomes) is fertilized by a sperm bearing one copy of all the non-sex chromosomes and *either* an X chromosome or a Y chromosome. When the sperm carries an X, the baby will have two X chromosomes and will develop as a female. When the sperm carries a Y, the baby will have an X and a Y and so will develop as a male (**FIGURE 6-28**).

**We know that no genetic information on the Y chromosome is essential for producing a normally functioning human. Why?**

Given the distribution of X and Y chromosomes into males and females, it must be true that no essential genetic information is carried on the Y chromosome. Why? Because females don't have a Y chromosome in any of their cells, yet they are able to develop and live normal, healthy lives. For this reason, we know that nothing on the Y chromosome is absolutely necessary for the development of a normally functioning human.

Physically, the X and Y chromosomes look very different from each other. The X chromosome is relatively large and carries a great deal of genetic information relating to a large number of non-sex-related traits. The Y chromosome is tiny and carries genetic information about only a very small number of traits. The genetic instructions on the Y chromosome instruct the fetal gonads to develop as testes rather than ovaries. Once this is done, very little additional genetic input from the Y chromosome is necessary. Instead, the hormones produced by the testes or ovaries generally direct the rest of the body to develop as a male or female.

## HOW SEX IS DETERMINED IN HUMANS

Individuals have two copies of the sex chromosomes in every cell.

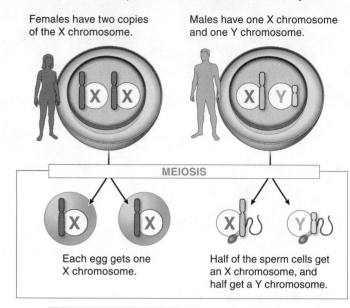

Females have two copies of the X chromosome.

Males have one X chromosome and one Y chromosome.

MEIOSIS

Each egg gets one X chromosome.

Half of the sperm cells get an X chromosome, and half get a Y chromosome.

FERTILIZATION

When the sperm carries an X, the baby will have two X chromosomes and develop as female.

When the sperm carries a Y, the baby will have an X and a Y and develop as male.

**FIGURE 6-28 Sex determination in humans.** The sex of offspring is determined by the father.

In the next sections we investigate some of the ramifications of males having two different sex chromosomes and look at some of the systems for sex determination that have evolved in other groups of organisms.

### TAKE-HOME MESSAGE 6·15

In humans, the sex chromosomes carry information that directs a growing fetus to develop as either a male (if a Y chromosome is present) or a female (if no Y chromosome is present). Sex determination depends on the sex chromosome inherited from the father.

# 6·16 The sex of offspring is determined in a variety of ways in non-human species.

For something as fundamental to a species as sex determination, you might imagine that one method evolved and all species use it. The world is more diverse than that. In most plants there aren't even distinct males and females. In corn, for example, every individual produces both male and female gametes. All earthworms and garden snails are also capable of producing both male and female gametes. Such organisms are called **hermaphrodites,** because male and female gametes are produced by a single individual. But even among the species with separate males and females there are several different methods of sex determination. The human method, in which having both an X and a Y chromosome makes you male and having two X chromosomes makes you female, is common among eukaryotes and is seen in all mammals. But at least three other methods are used by organisms for determining sex (**FIGURE 6-29**).

**Birds**  In birds (as well as in some fish and butterfly species), the mother determines the sex in much the same way that fathers determine sex in humans. Females have one copy of two different sex chromosomes, called the Z and the W chromosomes, while males have only one type, carrying two copies of the Z chromosome. Consequently, the sex of bird offspring is determined by the mother rather than the father.

**Ants, bees, and wasps**  In these insects, sex is determined by the number of chromosome sets an individual possesses. Males are haploid, having a *single* set of chromosomes, and females are diploid, carrying two sets of chromosomes. In this method, females produce haploid eggs by meiosis. They mate with males and store the sperm in a sac. As each egg is produced, the female can fertilize it with sperm she has stored, in which case the offspring will be diploid and female. Alternatively, she can lay the unfertilized egg, which develops into a haploid, male individual. Just think, in these species, males don't have a father, yet they do have a grandfather!

**Turtles**  In some species, sex determination is controlled by the environment rather than by the number or types of chromosomes an organism has. In most turtles, for example, offspring's sex is determined by the temperature at which the eggs are kept. Eggs kept relatively hot during incubation become females, while eggs incubated at cooler temperatures become males. The sex of some lizards and crocodiles is also determined this way.

## TAKE-HOME MESSAGE 6·16

A variety of methods are used for sex determination across the world of plant and animal species. These include the presence or absence of sex chromosomes, the number of chromosome sets, and environmental factors such as incubation temperature.

### SEX DETERMINATION IN VARIOUS OTHER SPECIES

**BIRDS**
In birds, the mother determines the sex in a way that is similar to how fathers determine sex in humans.

FEMALE        MALE

**ANTS, BEES, AND WASPS**
In ants, bees, and wasps, sex is determined by the number of chromosome sets an individual possesses.

FEMALE        MALE

**TURTLES**
In most turtles, sex is determined by the temperature at which the eggs are incubated.

FEMALE        MALE

**FIGURE 6-29 Other methods of sex determination.** In some species, sex is determined by the type of chromosomes, the number of chromosomes, or incubation temperature of the eggs.

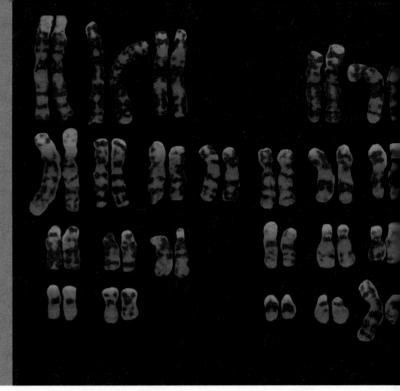

Karyotype displaying the chromosome pairs of a normal human female.

## 6·17 Down syndrome can be detected before birth: karyotypes reveal an individual's entire chromosome set.

Reproduction becomes riskier for women as they become older. Increasingly, their gametes contain incorrect numbers of chromosomes or chromosomes that have been damaged (**FIGURE 6-30**). These problems can lead to effects on the offspring that range from minor to fatal.

To find out whether their offspring may have a disorder associated with incorrect numbers of chromosomes, parents can request a quick test for some common genetic problems even before their baby is born. This information is available from an analysis of an individual's **karyotype,** a visual display of a complete set of chromosomes. A karyotype can be made for adults or children, but it is most commonly done for a fetus. A karyotype is a useful diagnostic tool because it can be prepared very early in the fetus's development to assess whether there is an abnormality in the number of chromosomes or in their structure. And because it shows all of the chromosomes, even the sex chromosomes, it also reveals the sex of the fetus.

Preparing a karyotype takes five steps. (1) First it is necessary to obtain some cells from the individual. (2) The cells are then encouraged to divide by culturing them in a test tube with nutrients. (3) After a few days to two weeks of cell

**DOWN SYNDROME AND MOTHER'S AGE**

*Incidence of Down syndrome per 1,000 live births* (y-axis: 0, 5, 10, 15, 20, 25)

*Age of mother* (x-axis: 20, 25, 30, 35, 40, 45)

**GRAPHIC CONTENT!**
*Thinking critically about visual displays of data.*
*Turn to p. 256 for a closer inspection of this figure.*

**FIGURE 6-30 Reproduction risks and maternal age.** Down syndrome incidence increases as the age of the mother increases.

division, the cells are treated with a chemical that stops them exactly midway through cell division—a time when the chromosomes are coiled thickly and more visible than usual. (4) The cells are then placed on a microscope slide and a

**FIGURE 6-31 A human karyotype.** A visual display of a complete set of chromosomes.

stain is added that binds to the chromosomes, making them more easily visible. (5) And finally, the chromosomes are arranged by size and shape and displayed on a monitor or in a photograph (**FIGURE 6-31**).

The first step—obtaining cells to prepare the karyotype—is relatively easy in an adult or child. Usually cells are collected from a small blood sample. Collecting cells from a fetus, on the other hand, poses a special problem. Two different methods for collecting cells have been developed.

**1. Amniocentesis.** This procedure can be done approximately three to four months into a pregnancy (**FIGURE 6-32**). In one quick motion (and without the use

**FIGURE 6-32 Amniocentesis.** Fluid surrounding a fetus (containing some of its cells) is extracted and analyzed to determine the genetic composition of the developing fetus.

**FIGURE 6-33 Trisomy 21.** An extra copy of chromosome 21 causes Down syndrome.

of anesthetic), a 3- to 4-inch needle is pushed through the abdomen, through the amniotic sac, and into the amniotic fluid that surrounds and protects the fetus. Using ultrasound for guidance, the doctor aims for a small pocket of fluid as far as possible from the fetus and withdraws about 2 tablespoons of fluid. This fluid contains many cells from the fetus, which can then be used for karyotype analysis. Any chromosomal abnormalities in the fetus will be present in these cells.

**2. Chorionic villus sampling (CVS).** In this procedure, rather than sampling cells from the amniotic fluid, a small bit of tissue is removed from the **placenta,** the temporary organ that allows the transfer of gases, nutrients, and waste products between a mother and fetus. A needle is inserted either through the abdomen or through the vagina and cervix, again using ultrasound for guidance. Then a small piece of the finger-like projections from the placenta is removed via the syringe. Because the fetus and placenta both develop from the same fertilized egg, their cells have the same genetic composition. The chief advantage of CVS over amniocentesis is that it can be done several weeks earlier in the pregnancy, usually between the 10th and 12th weeks.

The resulting karyotype, whether from amniocentesis or CVS, will reveal whether the fetus carries an extra copy of any of the chromosomes or lacks a copy of one or more chromosomes. Of all the chromosomal disorders detected by karyotyping, Down syndrome is the most commonly observed. Named after John Langdon Down, the doctor who first described it, in 1866, the syndrome is revealed by the presence of an extra

copy of chromosome 21 (**FIGURE 6-33**). (For this reason, the condition that causes Down syndrome is also called "trisomy 21.") Striking about 1 of every 1,000 children born, Down syndrome is characterized by a suite of physical and mental characteristics that includes learning disabilities, a flat facial profile, heart defects, and increased susceptibility to respiratory difficulties.

Down syndrome and other disorders caused by a missing chromosome or an extra copy of a chromosome are a consequence of **nondisjunction,** the unequal distribution of chromosomes during meiosis. Nondisjunction can occur at two different points in meiosis: homologues can fail to separate during meiosis I, or sister chromatids can fail to separate during meiosis II (**FIGURE 6-34**). In both cases, nondisjunction results in an egg or sperm with zero or two copies of a chromosome rather than a single copy. Any of the chromosomes can fail to separate during cell division, but the ramifications of trisomy (having an extra copy of one chromosome in every cell) are greater for chromosomes with larger numbers of genes. When trisomy occurs for chromosomes with greater numbers of genes, the likelihood that the developing embryo will survive to birth is reduced. Consequently, we tend to see cases of trisomy that involve only the chromosomes with the fewest genes, such as chromosomes 13, 15, 18, 21, and 22. In fact, observations show that trisomy 1 is never seen (all such fertilized eggs die before implantation in the uterus), trisomy 13 occurs in 1 in 20,000 newborns (and most die soon after birth), while trisomy 21, as we've seen, occurs in 1 in 1,000 newborns (many of whom live long lives).

## NONDISJUNCTION

Nondisjunction is the unequal distribution of chromosomes during meiosis. The resulting gametes have zero or two copies of a chromosome, rather than a single copy.

METAPHASE I — Homologues

ANAPHASE I — Homologues do not separate.

METAPHASE II

RESULTING GAMETES

Both gametes are missing a chromosome. | Both gametes have an extra chromosome.

**FIGURE 6-34 An error during meiosis produces gametes with too many or too few chromosomes.**

As we mentioned at the beginning of this section, as women become older there are increased problems associated with reproduction. Older women, for example, have more babies with Down syndrome. Why does the risk of having a baby with Down syndrome or some other disorder that results from trisomy increase with increasing age for women but significantly less so for men? As women age, their gametes tend to have more errors. The problem is that those eggs began meiosis near the time the woman was born, and they may not complete it until she is 40 or more years old. During those decades, the cells may accumulate defects that interfere with normal cell division. In men, the cells that undergo meiosis are relatively young because new sperm-producing cells are produced every couple of weeks after puberty.

Whereas lacking a non-sex chromosome or having an extra chromosome usually has serious consequences for health, we'll see in the next section that lacking or having an extra X or Y chromosome has consequences that are much less severe.

**Q** Why do older women have more babies with Down syndrome?

### TAKE-HOME MESSAGE 6·17

A karyotype is a visual display of a complete set of chromosomes. It is a useful diagnostic tool because it can be prepared early in a fetus's development to assess whether there is an abnormality in the number of chromosomes or in their structure, such as in Down syndrome. Down syndrome is caused by having an extra copy of chromosome 21.

# Life is possible with too many or too few sex chromosomes.

It is usually fatal to have one too many or one too few of the non-sex chromosomes. When it comes to the sex chromosomes, however, the situation is less extreme. Many individuals are born lacking one of the sex chromosomes or with an additional X or Y chromosome, and they usually survive. We can get a glimpse into the role of the sex chromosomes by looking at the physical and mental consequences of having an abnormal number of sex chromosomes. In each of the cases below, the condition is caused by nondisjunction of the sex chromosomes during the production of sperm or eggs (**FIGURE 6-35**).

**Turner syndrome: X** Approximately 1 in 5,000 females carry only one X chromosome (and no Y chromosome), exhibiting a condition called Turner syndrome, denoted as X_ (or sometimes XO). This is the only condition in humans in which a person can survive without one of a pair of chromosomes. Instead of having 46 chromosomes in every cell, these individuals have only 45. As common as Turner syndrome is, in 98% of the fertilized eggs in which this condition occurs, the egg is spontaneously aborted long before a fetus can come to term.

There are both physical and mental consequences of the absence of a second sex chromosome.

- Women with Turner syndrome are usually relatively short, averaging 4 feet 8 inches in height.

- They develop a web of skin between the neck and shoulders.

- The ovaries never fully mature, so the women are almost always sterile.

- The breasts and other secondary sex characteristics develop incompletely.

- Intelligence is usually normal, but some learning difficulties are common.

**Klinefelter syndrome: XXY** An individual who has two X chromosomes and a Y chromosome, a condition known as Klinefelter syndrome, develops as male. This is because, as we saw earlier, the Y chromosome carries genetic

**Q** *If a person has two X chromosomes but also has a Y chromosome, is the individual male or female?*

instructions that cause fetal gonads to develop as testes; if these instructions are absent, the fetal gonads develop as ovaries. The extra X chromosome, however, does cause Klinefelter males to be somewhat feminized, although this effect can be reduced through treatments such as testosterone supplementation. Approximately 1 in 1,000 males have the genotype XXY, making this one of the most common genetic abnormalities in humans.

## TOO MANY OR TOO FEW SEX CHROMOSOMES

| GAMETE GENOTYPES | OFFSPRING GENOTYPE | SYNDROME CHARACTERISTICS |
|---|---|---|
| | | **TURNER SYNDROME (FEMALE)**<br>• Short height<br>• Web of skin between neck and shoulders<br>• Underdeveloped ovaries; often sterile<br>• Some learning difficulties |
| | | **KLINEFELTER SYNDROME (MALE)**<br>• Underdeveloped testes<br>• Lower testosterone levels; usually infertile<br>• Development of some female features<br>• Long limbs and slightly taller than average |
| | | **XYY MALE**<br>• Taller than average<br>• Moderate to severe acne<br>• Intelligence may be slightly lower than average<br>• Sometimes called "super males" |
| | | **XXX FEMALE**<br>• May be sterile<br>• No obvious physical or mental problems<br>• Sometimes called "metafemales" |

**FIGURE 6-35 Characteristics of individuals with too many or too few sex chromosomes.**

Klinefelter syndrome has some physical and mental consequences:

- Men with this syndrome have testes, but they are smaller than average. Because the testes are small, levels of testosterone are low and the men are almost always infertile.

- They develop some female features, including reduced facial and chest hair and some breast development.

- They have long limbs and are slightly taller than average (about 6 feet on average).

- They learn to speak at a later age than average and tend to have language impairments.

- Some individuals have further additional X chromosomes, exhibiting the karyotypes XXXY or even XXXXY. These males also exhibit Klinefelter syndrome but more frequently have mental retardation.

A hermaphrodite is an individual with functioning male and female reproductive organs capable of producing both male and female gametes. Often it is mistakenly assumed that a person with Klinefelter syndrome must be a hermaphrodite because he has both two X chromosomes (which would usually make an individual a female) and an X and a Y chromosome (usually making the individual a male). Although hermaphroditism is common among invertebrates and occurs in some fish and other vertebrates, contrary to urban legends, human hermaphrodites do not exist. Some men with Klinefelter syndrome may have some features of the opposite sex; however, they do not produce female gametes and so are not hermaphrodites.

**Q** Do human hermaphrodites exist?

**XYY males** There is no official name for the condition in which an individual has one X chromosome and two Y chromosomes, although such individuals are sometimes referred to as "super males." This chromosomal abnormality occurs in approximately 1 in 1,000 males (and in about 1 in 325 males who are 6 feet or taller). There are no distinguishing features at birth to indicate that an individual carries an extra Y chromosome, and the vast majority live their lives normally without knowing they have an extra chromosome in every cell.

Several consequences of the XYY condition have been well documented:

- XYY males are relatively tall (about 6 feet 2 inches on average).

- They tend to have moderate to severe acne.

- Their intelligence usually falls within the same range as that of XY males, although the average may be slightly lower.

Because, in the United States, about 3 to 5 times as many XYY males are found in prisons as in the population as a whole, there has been tremendous controversy over whether the extra Y chromosome predisposes individuals to criminal behavior. These observations were not part of randomized, controlled, double-blind experiments, however, so they are open to numerous alternative interpretations.

In one study, for example, researchers identified all male Danish citizens born in Copenhagen between 1944 and 1947, and obtained sex chromosome determinations for the 4,139 of these men who were taller than 6' 0.4" (184 cm). From this sample, they found that the level of intellectual functioning (as indicated by scores on an army selection test taken by the men) was significantly lower among the XYY males as compared to XY males. Based on this finding, they suggested that the increased representation of XYY males in prison may be more likely due to the fact that males of lower intelligence are more likely to be apprehended, regardless of which chromosomes they carry. The controversy remains unresolved, and the vast majority of XYY males are not in prison.

Although an error during meiosis in either the father or the mother can cause Klinefelter or Turner syndrome, it is only an error in meiosis in a male that can give rise to an XYY child. Why? Because only males carry Y chromosomes. Consequently, an XYY child must receive both of his Y chromosomes from his father. This means that the father's sperm cell must have contained two copies of the Y, a result of nondisjunction during meiosis in the father.

**XXX females** Sometimes called "metafemales," individuals with three X chromosomes occur at a frequency of about 1 in 1,000 women. Very few studies of this condition have been completed, although initial observations suggest that some XXX females are sterile but otherwise have no obvious physical or mental problems.

---

**TAKE-HOME MESSAGE 6·18**

Although it is usually fatal to have one too many or one too few of the non-sex chromosomes, individuals born with only a single sex chromosome that is an X, or with an additional X or Y chromosome, usually survive—though often with physical and/or mental problems.

## CAN YOU SELECT THE SEX OF YOUR BABY?
### (Would you want to?)

Some fertility clinics now promise that for as little as $200 they can help a couple choose the sex of their baby.

**Q: GIVEN OUR KNOWLEDGE OF THE BEHAVIOR OF THE X AND Y CHROMOSOMES, IS IT MORE LIKELY THAT THESE SEX-SELECTION TECHNIQUES INVOLVE MANIPULATIONS OF SPERM OR OF EGGS? WHY?** Because all eggs have X chromosomes, the X chromosome doesn't play a role in determining the sex of a baby. Instead, the fertility clinics must somehow sort and separate sperm cells based on whether they carry an X or a Y chromosome (men produce approximately equal numbers of each).

**Q: WHAT FEATURE OF SPERM MUST BE USED IN SEPARATING THEM?** This process must be based on a way of distinguishing between sperm carrying an X chromosome and sperm carrying a Y chromosome. After the two types of sperm are distinguished, they must then be separated. Sperm with the desired sex chromosome are then used to fertilize the woman's egg (which may be done within her body, using artificial insemination, or in a Petri dish, after which a fertilized embryo is transferred into the woman's reproductive tract).

**Q: WHAT TECHNIQUES MIGHT MAKE IT POSSIBLE TO SEPARATE SPERM WITH AN X CHROMOSOME FROM THOSE WITH A Y CHROMOSOME?** One method involves determining, by weighing sperm, which cells have more DNA. The heavier sperm must be carrying the X rather than the Y chromosome because the X is so much larger. Another method is to add a fluorescent dye to sperm that temporarily attaches to DNA. Because sperm with an X chromosome have more DNA, more of the dye attaches to them and they are more fluorescent. A machine then sorts the sperm one by one, and at ovulation, insemination is performed using the sperm with the desired sex chromosome.

**Q: DOES IT WORK?** The technique is still fairly new, but initial results suggest that the procedures have a success rate of 70% to 90%.

**CONCERNS FOR YOU TO PONDER.** What are the ethical concerns raised by selecting the sex of one's baby? Are there some circumstances in which such a selection would be more acceptable than in others? Would it be acceptable to select for a baby with one eye color versus another? Why would that process be more difficult than selecting for sex?

# Big Ideas in ...
# Chromosomes and Cell Division

## 6·1–6·5 There are different types of cell division.

- In eukaryotic cells, a protective section of DNA called the telomere, at each end of every chromosome, plays a role in keeping track of cell division, getting shorter every time the cell divides. If telomeres become too short, additional cell divisions cause the loss of essential DNA and cell death. Cells that rebuild their telomeres with each division can become cancerous.

- In most prokaryotes, genetic information is carried in a single, circular DNA strand (chromosome). Eukaryotes have much more DNA, organized into linear chromosomes in the nucleus.

- Bacteria divide by a type of asexual reproduction called binary fission.

- In the cell cycle, eukaryotic somatic cells alternate between division, called the mitotic phase, and other cell activities, called interphase, which consists of two gap phases separated by a DNA synthesis phase.

- Every time a cell divides, its DNA must first replicate so that each of the two new cells has all the genetic material of the parent cell.

## 6·6–6·9 Mitosis replaces worn-out old cells with fresh new duplicates.

- In mitosis, cells generate new, genetically identical cells, enabling organisms to grow and replace cells.

- Mitosis occurs in four steps, following replication of the chromosomes, to produce two genetically identical daughter cells from one parent cell.

- Cancer is unrestrained cell growth and division, leading to large masses of cells (tumors). Treatment focuses on killing or slowing down cell growth and division.

## 6·10–6·14 Meiosis generates sperm and eggs and a great deal of variation.

- Meiosis occurs only in gamete-producing cells of sexually reproducing organisms.

- Gametes have only half as much genetic material as the parent cell, and differ from one another in the allele combinations they carry.

- Meiosis is preceded by DNA replication and consists of two rounds of cell division: separation of homologous pairs of sister chromatids, then separation of the sister chromatids. The final product in a diploid organism are haploid gametes.

- In species with two sexes, the female is the sex producing the larger gamete, while males produce the smaller gamete.

- Crossing over during the first prophase of meiosis creates gametes with new collections of alleles which may never have existed together; this variation is important for evolution.

- Cells and organisms can reproduce by mitosis and asexual reproduction via binary fission or by meiosis and sexual reproduction.

## 6·15–6·16 There are sex differences in the chromosomes.

- In humans, the sex chromosomes carry information that directs a fetus to develop as male (if a Y chromosome is present) or female (no Y present). The sex is determined by the sex chromosome inherited from the father.

- Several methods for sex determination occur in plant and animal species.

## 6·17–6·18 Deviations from the normal chromosome number lead to problems.

- A karyotype is a display of a complete set of chromosomes. It can be prepared very early in fetal development to detect abnormalities in the chromosome number or structure.

- Having one too many or one too few non-sex chromosomes is usually fatal, but individuals born with a single sex chromosome that is an X, or with an additional X or Y, usually survive, but often with physical and/or mental problems.

See the To-Go Guide (packaged with every new copy of this textbook) for an illustrated summary of the chapter.

# Check Your Knowledge

## GRAPHIC CONTENT

*Thinking critically about visual displays of data.*

**1.** According to this figure, what is the probability (i.e., 1 out of 800, or 1 out of 256, etc.) that a baby born to a 35-year-old woman will have Down syndrome?

**2.** Is a woman of age 45 who gives birth more likely than not to have a baby with Down syndrome? What is the exact probability that her baby will not have Down syndrome?

**3.** From this figure, can you conclude that most babies with Down syndrome are born to women older than 35?

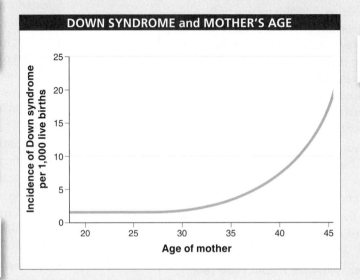

**DOWN SYNDROME and MOTHER'S AGE**

**4.** Why does the graph not show data for women older than 45?

**5.** Among women aged 35 and higher (but not among women younger than 35), the incidence of fetuses produced that have an extra copy of chromosome 21 may be higher than the graph indicates. Why?

See Answers at the back of the book.

## Key Terms in Chromosomes and Cell Division

anaphase, p. 233
apoptosis, p. 230
asexual reproduction, p. 225
binary fission, p. 225
cancer, p. 234
cell cycle, p. 226
centriole, p. 233
centromere, p. 227
chromatid, p. 233

complementarity, p. 227
complementary base, p. 227
crossing over, p. 241
cytokinesis, p. 227
daughter cell, p. 225
diploid, p. 237
fertilization, p. 237
gamete, p. 226
gonad, p. 239

haploid, p. 237
hermaphrodite, p. 249
histone, p. 224
homologous pair (homologues), p. 239
interphase, p. 226
karyotype, p. 250
meiosis, p. 237
metaphase, p. 233

mitosis, p. 227
mitotic phase (M phase), p. 226
nondisjunction, p. 252
parent cell, p. 225
placenta, p. 252
polar body, p. 243
prophase, p. 232
recombination, p. 241
replication, p. 225

reproductive cell, p. 226
sexual reproduction, p. 225
sister chromatids, p. 233
somatic cell, p. 226
spindle, p. 233
spindle fiber, p. 233
telomere, p. 222
telophase, p. 234
X and Y chromosomes, p. 247

# Multiple-Choice Questions

1. Which of the following statements about telomeres is incorrect?

   a) They function like a counter, keeping track of how many times a cell has divided.

   b) At birth, they are long enough to permit approximately 50 cell divisions in most cells.

   c) They are slightly shorter in prokaryotic cells than in eukaryotic cells.

   d) They function like a protective cap on chromosomes.

   e) They contain no critical genes.

   100 HARD
   **62**
   0 EASY

2. Prokaryotic cells can divide via:

   100 HARD

   a) mitosis.

   b) binary fission.

   c) meiosis.

   **31**

   d) both mitosis and binary fission.

   0 EASY

   e) None of the above.

3. In multicellular organisms, cells that undergo mitotic division but not meiotic division are called _____ cells.

   a) somosis

   100 HARD

   b) skin

   c) interphase

   d) somatic

   **22**

   e) germ

   0 EASY

4. The _____ marks the break in the cell cycle between the end of mitosis and beginning of the DNA synthesis stage.

   a) mitotic phase

   100 HARD

   b) synthesis phase

   **64**

   c) Gap 2 phase

   d) Gap 1 phase

   e) None of the above.

   0 EASY

5. Mitosis results in:

   a) daughter cells with twice as much genetic material as the parent cell and a unique collection of alleles.

   100 HARD

   b) eight daughter cells.

   c) daughter cells with the same number and composition of chromosomes.

   **45**

   d) gametes.

   0 EASY

   e) daughter cells with the same number of chromosomes but different combinations of alleles.

6. Using a light microscope, it is easiest to see chromosomes:

   a) during mitosis and meiosis, because the condensed chromosomes are thicker and therefore more prominent.

   b) during interphase, when they are concentrated in the nucleus.

   100 HARD

   c) in the mitochondria, because the chromosomes are circular.

   **35**

   d) during asexual reproduction.

   0 EASY

   e) during interphase, because they are uncoiled and have a more linear structure.

7. The division of the cytoplasm during cell division is referred to as:

   a) cytoplasm splicing.

   100 HARD

   b) cytokinesis.

   c) vegetative growth.

   d) cytodivision.

   e) hybridization.

   0 **10** EASY

**8. Which of the following statements about tumors is incorrect?**

a) Benign tumors pose less of a health risk than do malignant tumors.

b) Malignant tumors shed cancer cells that can spread to other parts of the body.

c) Tumors are caused by excessive cell growth and division.

d) Cancer cells from malignant tumors can travel to other parts of the body in a process called metastasis.

e) They contain cells with abnormally high contact inhibition.

100 HARD
48
0 EASY

**9. During meiosis but not during mitosis:**

a) haploid gametes are produced that are all identical in their allelic composition.

b) division of the cytoplasm occurs.

c) chromosomes line up in the center of the cell during metaphase.

d) genetic variation among the daughter cells is increased.

e) two identical daughter cells are produced.

100 HARD
52
0 EASY

**10. Sister chromatids are:**

a) the result of crossing over.

b) identical molecules of DNA resulting from replication.

c) homologous chromosomes.

d) produced in meiosis but not in mitosis.

e) single-stranded.

100 HARD
39
0 EASY

**11. Which of the following events does not occur during prophase I of meiosis?**

a) Spindle fibers form.

b) Homologous pairs of chromosomes align at the metaphase plate.

c) Homologues come very close together.

d) Crossing over occurs between sister chromatids.

e) Chromosomes condense.

100 HARD
43
0 EASY

**12. Which of the following is the best way to distinguish male from female?**

a) Males are larger.

b) Males are more brightly colored.

c) Males produce motile gametes.

d) Males are more aggressive.

e) All of the above statements are correct.

100 HARD
32
0 EASY

**13. A potential disadvantage of asexual reproduction is that:**

a) it increases the time required to find a mate.

b) it allows perpetuation of a population even when the members are isolated.

c) it allows population size to increase rapidly.

d) it produces genetically uniform populations.

e) it requires additional rounds of meiosis, which is energetically much more costly than mitosis.

100 HARD
35
0 EASY

**14. Which of the following is not a method of chromosomal sex determination that occurs in nature?**

a) In birds, the mother's sex chromosomes determine the sex of the offspring.

b) In sea turtles, eggs laid in hot sand become females and eggs laid in cooler sand become males.

c) In humans, the presence of a Y chromosome makes an individual male, even if he also possesses two X chromosomes.

d) In bees, the eggs that the queen allows to be fertilized become females and the eggs she doesn't fertilize become males.

e) All of the above are naturally occurring examples of sex determination.

100 HARD
38
0 EASY

**15. A karyotype reveals:**

a) the shape of the spindle.

b) 23 pairs of chromosomes.

c) the number, shapes, and sizes of chromosomes in an individual cell.

d) the sex chromosomes but not the non-sex chromosomes.

e) the non-sex chromosomes but not the sex chromosomes.

100 HARD
20
0 EASY

**16. What type of chromosomal abnormality leads to the phenotype known as Down syndrome?**

a) trisomy 21

b) monosomy 21

c) monosomy 22

d) trisomy 22

e) disomy 21

100 HARD
20
0 EASY

**17. Nondisjunction:**

a) is the unequal division of the genetic material during cell division.

b) occurs during meiosis but not mitosis.

c) occurs in males but not females among mammals, and in females but not males among birds.

d) leads to a missing chromosome or extra chromosome.

e) Both a) and d) are correct.

100 HARD
36
0 EASY

# Multiple-Choice Questions

1. **Which of the following statements about telomeres is incorrect?**

   a) They function like a counter, keeping track of how many times a cell has divided.

   b) At birth, they are long enough to permit approximately 50 cell divisions in most cells.

   c) They are slightly shorter in prokaryotic cells than in eukaryotic cells.

   d) They function like a protective cap on chromosomes.

   e) They contain no critical genes.

   100 HARD
   **62**
   0 EASY

2. **Prokaryotic cells can divide via:**

   a) mitosis.

   b) binary fission.

   c) meiosis.

   d) both mitosis and binary fission.

   e) None of the above.

   100 HARD
   **31**
   0 EASY

3. **In multicellular organisms, cells that undergo mitotic division but not meiotic division are called _____ cells.**

   a) somosis

   b) skin

   c) interphase

   d) somatic

   e) germ

   100 HARD
   **22**
   0 EASY

4. **The _____ marks the break in the cell cycle between the end of mitosis and beginning of the DNA synthesis stage.**

   a) mitotic phase

   b) synthesis phase

   c) Gap 2 phase

   d) Gap 1 phase

   e) None of the above.

   100 HARD
   **64**
   0 EASY

5. **Mitosis results in:**

   a) daughter cells with twice as much genetic material as the parent cell and a unique collection of alleles.

   b) eight daughter cells.

   c) daughter cells with the same number and composition of chromosomes.

   d) gametes.

   e) daughter cells with the same number of chromosomes but different combinations of alleles.

   100 HARD

   **45**
   0 EASY

6. **Using a light microscope, it is easiest to see chromosomes:**

   a) during mitosis and meiosis, because the condensed chromosomes are thicker and therefore more prominent.

   b) during interphase, when they are concentrated in the nucleus.

   c) in the mitochondria, because the chromosomes are circular.

   d) during asexual reproduction.

   e) during interphase, because they are uncoiled and have a more linear structure.

   100 HARD
   **35**
   0 EASY

7. **The division of the cytoplasm during cell division is referred to as:**

   a) cytoplasm splicing.

   b) cytokinesis.

   c) vegetative growth.

   d) cytodivision.

   e) hybridization.

   100 HARD

   0 **10** EASY

**8. Which of the following statements about tumors is incorrect?**

a) Benign tumors pose less of a health risk than do malignant tumors.

b) Malignant tumors shed cancer cells that can spread to other parts of the body.

c) Tumors are caused by excessive cell growth and division.

d) Cancer cells from malignant tumors can travel to other parts of the body in a process called metastasis.

e) They contain cells with abnormally high contact inhibition.

**9. During meiosis but not during mitosis:**

a) haploid gametes are produced that are all identical in their allelic composition.

b) division of the cytoplasm occurs.

c) chromosomes line up in the center of the cell during metaphase.

d) genetic variation among the daughter cells is increased.

e) two identical daughter cells are produced.

**10. Sister chromatids are:**

a) the result of crossing over.

b) identical molecules of DNA resulting from replication.

c) homologous chromosomes.

d) produced in meiosis but not in mitosis.

e) single-stranded.

**11. Which of the following events does not occur during prophase I of meiosis?**

a) Spindle fibers form.

b) Homologous pairs of chromosomes align at the metaphase plate.

c) Homologues come very close together.

d) Crossing over occurs between sister chromatids.

e) Chromosomes condense.

**12. Which of the following is the best way to distinguish male from female?**

a) Males are larger.

b) Males are more brightly colored.

c) Males produce motile gametes.

d) Males are more aggressive.

e) All of the above statements are correct.

**13. A potential disadvantage of asexual reproduction is that:**

a) it increases the time required to find a mate.

b) it allows perpetuation of a population even when the members are isolated.

c) it allows population size to increase rapidly.

d) it produces genetically uniform populations.

e) it requires additional rounds of meiosis, which is energetically much more costly than mitosis.

**14. Which of the following is not a method of chromosomal sex determination that occurs in nature?**

a) In birds, the mother's sex chromosomes determine the sex of the offspring.

b) In sea turtles, eggs laid in hot sand become females and eggs laid in cooler sand become males.

c) In humans, the presence of a Y chromosome makes an individual male, even if he also possesses two X chromosomes.

d) In bees, the eggs that the queen allows to be fertilized become females and the eggs she doesn't fertilize become males.

e) All of the above are naturally occurring examples of sex determination.

**15. A karyotype reveals:**

a) the shape of the spindle.

b) 23 pairs of chromosomes.

c) the number, shapes, and sizes of chromosomes in an individual cell.

d) the sex chromosomes but not the non-sex chromosomes.

e) the non-sex chromosomes but not the sex chromosomes.

**16. What type of chromosomal abnormality leads to the phenotype known as Down syndrome?**

a) trisomy 21

b) monosomy 21

c) monosomy 22

d) trisomy 22

e) disomy 21

**17. Nondisjunction:**

a) is the unequal division of the genetic material during cell division.

b) occurs during meiosis but not mitosis.

c) occurs in males but not females among mammals, and in females but not males among birds.

d) leads to a missing chromosome or extra chromosome.

e) Both a) and d) are correct.

**18.** Of all the varying types of sex chromosome abnormalities, one possibility—that of an individual having one Y but no X—has never been reported. Which of the following statements is the most likely reason that medical researchers have yet to report a person having a karyotype with one Y and no X?

a) There is no meiotic event that could produce an egg that does not have an X chromosome.

b) Individuals are probably born with this karyotype; however, because the phenotypic abnormalities caused are so slight, they live out their lives normally and are unaware of this genetic anomaly.

c) The Y chromosome needs the X chromosome to be present in the cell in order for its genes to be expressed.

d) The X chromosome contains genes that are unrelated to sex determination and are essential for life.

e) Our current technology for karyotyping would not reveal such an abnormality.

100 | HARD
65
0 | EASY

# Short-Answer Questions

1. Seventy to ninety percent of the genetic material in a sperm or egg cell made in your body could be inherited from your mother. Why isn't there always a 50/50 split (half from your mother and half from your father)? Is the reverse situation (more genetic material from your father than from your mother) possible?

2. Healthy individuals may have just one sex chromosome, as long as it is an X chromosome. Why can't a person survive with a Y chromosome and no X chromosome?

3. Some organisms can switch between sexual and asexual reproduction. Under what conditions would you expect them to reproduce sexually? What conditions would be more conducive to asexual reproduction? How would you test your hypotheses under laboratory conditions?

See Answers at the back of the book.
For additional study questions, go to www.prep-u.com.

## ABOUT THE CHAPTER OPENING PHOTO

Sibling similarities—such as those seen among the Jackson brothers (shown here performing in 1973)—illustrate the phenotypic consequences of genetic similarity; but the subtle differences also reveal variability among the offspring in sexually reproducing species. The phenotypic variation stems from a variety of sources, including crossing over, the shuffling and reassortment of homologues during meiosis, and the contributions from two parents.

**Family resemblance:
how traits are inherited**

# 7

# Genes and Inheritance

# 7·1–7·5 Why do offspring resemble their parents?

*Mirror image? Not quite, but a canine parent and happy puppy reveal parent-offspring similarity.*

## 7·1 Family resemblance: your father and mother each contribute to your genetic makeup.

**Q** *How can a single bad gene make you smell like a rotten fish?*

Can a gene be cruel? Of course not. But if one could, consider this candidate: in humans, there is a gene for an enzyme called FMO3 (flavin-containing mono-oxygenase-3), which breaks down a chemical in our bodies that smells like rotting fish. Some unfortunate individuals inherit a defective FMO3 gene and can't break down the noxious chemical. Instead, their urine, sweat, and breath excrete it, causing them to smell like rotting fish. Worst of all, because the odor comes from within their bodies, they cannot wash it away no matter how hard they try. Called "fish odor syndrome," this disorder often causes those afflicted by it to suffer ridicule, social isolation, and depression. (Currently, there is no cure for fish odor syndrome, although there are a few ways of reducing the odor such as reducing consumption of eggs, fish, and some meats that contain certain chemicals, including sulfur.)

For individuals born with this malady, beyond their own suffering there looms a scary question: "Will I pass this condition on to my children?" They might also wonder how

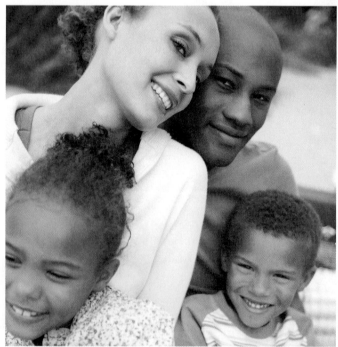

**FIGURE 7-1 Where do we begin and our parents end?** Family resemblances reveal that many traits are inherited.

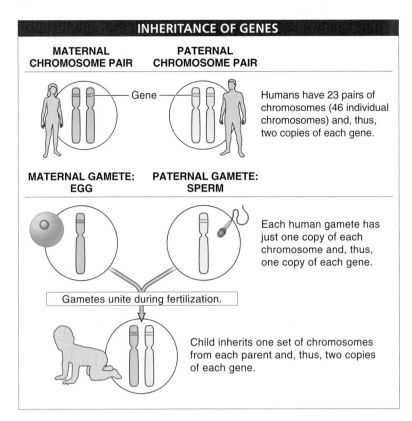

## INHERITANCE OF GENES

**MATERNAL CHROMOSOME PAIR**    **PATERNAL CHROMOSOME PAIR**

Gene

Humans have 23 pairs of chromosomes (46 individual chromosomes) and, thus, two copies of each gene.

**MATERNAL GAMETE: EGG**    **PATERNAL GAMETE: SPERM**

Each human gamete has just one copy of each chromosome and, thus, one copy of each gene.

Gametes unite during fertilization.

Child inherits one set of chromosomes from each parent and, thus, two copies of each gene.

**FIGURE 7-2 How we inherit our genes.** Each human offspring inherits one maternal set of 23 chromosomes and one paternal set.

they came to have the disorder, particularly if neither of their parents suffered from it.

Fortunately, most of us do not have to worry about fish odor syndrome, but we do wonder about many other inherited traits that we may or may not pass down to our children. Whom do I blame for my straight hair (or curly hair or big feet)? Hearing a new baby described as having "his father's nose" and "his grandmother's lips" could make someone think that we're all cobbled together from a bunch of used parts. Where do we begin and our parents end (**FIGURE 7-1**)?

In many cases, the answers to questions about heredity are simple. Fish odor syndrome, for example, is just such a case. Recall from Chapter 6 that sexually reproducing organisms generally carry two copies of every chromosome in every cell (except their sex cells), because they inherit one copy of each chromosome from each parent. Humans, for example, have 23 pairs of chromosomes (46 individual chromosomes). At each location on the two chromosomes of a pair is the same gene: one copy from the mother and one from the father (**FIGURE 7-2**). As we noted in Chapter 5, each of the two copies of the gene is called an **allele.** The gene for FMO3 is on chromosome 1. There is a normal version—or allele—of the gene for FMO3, which most people carry, and there is a

rare, defective version that is responsible for fish odor syndrome. As long as a person has at least one normal version of the FMO3 gene, he or she will produce enough of the enzyme to break down the fishy chemical. But if a person inherits two copies of the defective version of the fish odor gene, one from each parent, that person will inherit the disorder (**FIGURE 7-3**).

When it comes to having children, there is a bright side, at least. Although individuals with fish odor syndrome carry two defective copies of the allele and will pass on one set of the bad instructions in every sperm or egg cell that they make, their children won't necessarily inherit the disease. As long as the other parent supplies a normal version of the FMO3 gene, the child will not have fish odor syndrome. The unaffected children, though, will carry a silent copy of the fish odor allele, and if, when they have children, it comes together with a defective FMO3 allele from another person, the fish odor trait will be expressed. In this way, some alleles can exist in a population without always revealing themselves.

## FISH ODOR SYNDROME

**MATERNAL CHROMOSOME PAIR**    **PATERNAL CHROMOSOME PAIR**

Normal version of the FMO3 gene

Defective version of the FMO3 gene, responsible for fish odor syndrome

**MATERNAL GAMETES: EGGS**    **PATERNAL GAMETES: SPERM**

Gametes unite during fertilization.

If the child inherits two copies of the defective version of the FMO3 gene, the child will develop fish odor syndrome.

**FIGURE 7-3 Unlucky catch.** A baby inheriting two copies of the defective FMO3 gene will develop fish odor syndrome.

In this chapter, we explore how heredity works. Inheritance follows some simple rules that allow us to make sense of the patterns of family resemblance we see, and even to predict accurately how many offspring are likely to inherit particular traits—or, put another way, predict the likelihood that an offspring will inherit a trait. We'll also examine why the behavior of some traits is easy to predict while many other traits have less straightforward patterns of inheritance, yet still can be studied experimentally and yield predictable patterns.

# 7·2 Some traits are controlled by a single gene.

Like father, like son. No one will ever seek a paternity test to prove that Michael Douglas is the son of Kirk Douglas. All it takes is a quick look at the legendary dimpled chins shared by the two actors and we are sure that the men are related. We see all around us that offspring resemble their parents more than they resemble other random individuals in the population—a consequence of the passing of characteristics from parents to offspring through their genes. This is **heredity.**

Observing heredity is easy. Elucidating how it works is not. For thousands of years before the mechanisms of heredity were discovered and understood, plant and animal breeders recognized that there is a connection from parents to offspring across generations. In ancient Greece, for example, the poet Homer extolled the tremendous benefits to society that came from the skillful breeding of horses. The awareness of beneficial breeding practices enabled farmers to systematically create strains of crops, livestock, and even pets with desirable traits (**FIGURE 7-4**). From flocks of sheep and herds of reindeer to litters of puppies and fields of wheat, the similarity among relatives is clear. Once breeders recognized the existence of heredity, they began selecting individual plants or animals with the desired traits to breed with each other, in the hope that their offspring would also have those desirable traits. Since then, it has been possible to create a rich world of sweeter corn, loyal dogs, docile livestock, beautiful flowers, and more. And these breeders did it all without ever really understanding how heredity works.

Practical successes were many, but people didn't understand exactly how the outcomes were achieved. They knew the result of selective breeding, but why and how it worked, or didn't always work, was a mystery. Patterns of similarity among related individuals are impossible to make sense of without an underlying understanding of *how* heredity works. This understanding is the province of the field of genetics.

Mechanisms of heredity are like a puzzle, and they become a bit easier to grasp when we focus on aspects of family resemblances that are commonplace. Virtually everyone

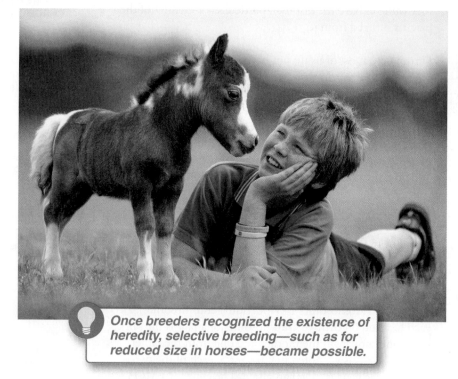

*Once breeders recognized the existence of heredity, selective breeding—such as for reduced size in horses—became possible.*

**FIGURE 7-4  Honey, I shrunk the horse!** Selective breeding was used to produce this tiny horse.

## EXAMPLES OF SINGLE-GENE TRAITS

Cleft chin

Non-cleft chin

Unattached earlobes

Attached earlobes

Widow's peak

Straight hairline

Tongue-roller

Non-tongue-roller

**FIGURE 7-5 Chins, ears, hairlines, and tongue-rolling: single-gene traits.** Some traits are determined by the instructions a person carries on one gene.

with a cleft chin, for instance, has a similarly dimpled parent. Everyone who has earlobes that hang from their ears rather than being attached directly to their head has at least one parent with the same feature. Everyone who has a widow's peak has a parent with the same hairline. Everyone who is farsighted has a similarly farsighted parent. The list goes on and on. In fact, there are more than 9,000 human traits that exhibit these straightforward patterns of inheritance. Traits that are determined by the instructions a person carries on one gene are called **single-gene traits** (**FIGURE 7-5**).

> ❝HEREDITY
>
> I AM the family face;
> Flesh perishes, I live on,
> Projecting trait and trace
> Through time to times anon,
> And leaping from place to place
> Over oblivion.
>
> The years-heired feature that can
> In curve and voice and eye
> Despise the human span
> Of durance—that is I;
> The eternal thing in man,
> That heeds no call to die.❞
>
> —THOMAS HARDY, in *Moments of Vision and Miscellaneous Verses*, 1917

It is important to note here that most human characteristics are influenced by multiple genes as well as by the environment. However, because the mechanism by which single-gene traits pass from parent to child is the easiest pattern of inheritance to decipher, we first explore this process. Then we'll expand our model of heritability to account for the passing on of more complex traits.

### TAKE-HOME MESSAGE 7·2

**Many human traits are determined by instructions a person carries on a single gene, and these traits exhibit straightforward patterns of inheritance.**

## Mendel learned about heredity by conducting experiments.

What do parents "give" their offspring that confers similarity? In other words, how is a physical entity, such as a cleft chin, inherited? How is such a trait passed from parent to offspring? It wasn't until the mid-1800s that any real headway was made on answering these questions. At that time, Gregor Mendel, a monk living in what is now the Czech Republic, began some studies that not only shed light on these questions but practically answered them completely. Mendel understood the essence of the genetics puzzle and set out to piece it all together.

When Mendel turned to these questions about heredity, there were no obvious answers. One idea put forth suggested that perhaps an entire, pre-made human—albeit a very, very tiny one—was contained in every sperm cell (**FIGURE 7-6**). That idea had been popular since the late 1600s. It doesn't take much imagination to see why it ran into difficulty, though, because children resemble both their mothers and their fathers, not just their fathers. Another popular idea was that offspring reflect a simple blending of their two parents' traits via the blood. But this couldn't explain how brown-eyed parents could give birth to blue-eyed children. Or why one tall and one short parent sometimes produced a tall

**Q**

Individuals who share common ancestry are called "blood relatives," yet they don't actually share any blood. How might the phrase "blood relatives" be a reflection of early conceptions of inheritance?

child, rather than always producing children of intermediate height.

Sometimes scientific breakthroughs are made because an innovative new technique is invented or a lucky observation is made by chance. Neither played a role in Mendel's success. He didn't do anything radically new, but simply applied the tried and true methodical experimentation and scientific thinking (as described in Chapter 1) that had been a boon to chemists and physicists. In particular, three features of Mendel's research were critical to its success (**FIGURE 7-7**).

**1.** He chose a good organism to study: the garden pea. It's not that Mendel had a particular fondness for vegetables. In fact, his goals were to understand inheritance in all organisms, not just plants. Pea plants were simply the right tool. Cats and dogs or even mice wouldn't have served his purposes very well, because they would have been too hard to take care of in the large numbers— thousands and thousands of individuals— that Mendel required. Humans, too, would have made a terrible study organism. We take too long to breed (and won't produce offspring on command). Pea plants, on the other hand, have a lot going for them as experimental organisms. They are relatively easy to fertilize manually by "pollen dusting." It is easy to collect dozens or even hundreds of offspring from a single **cross,** the process in which male pollen (carrying the sperm) is used to fertilize female eggs. And, on top of all this, pea plants are fast enough breeders that Mendel could conduct experiments that lasted for multiple generations.

**2.** Mendel chose to focus on easily categorized traits. For instance, shape: all peas of the variety that Mendel studied are either round or wrinkled in shape, with nothing in between. In addition, all peas are either yellow or green in color, never any intermediate shade. In all, Mendel looked at seven different traits of the pea plants but, for each trait, only two variants ever appeared, so he and his research assistants could easily observe and unambiguously identify them.

**3.** Mendel began his studies by first repeatedly breeding together similar plants until he had many distinct populations, each of which was unvarying for a particular trait. He described these plants as **true-breeding** for that trait because they always produced offspring with the same

*The mistaken idea that a tiny, pre-made human existed in every sperm cell was introduced in the 1600s. This theory remained popular through the 1800s.*

**FIGURE 7-6 The "pre-made human."** Tiny humans in every sperm cell? Why would children have any resemblance to their mothers?

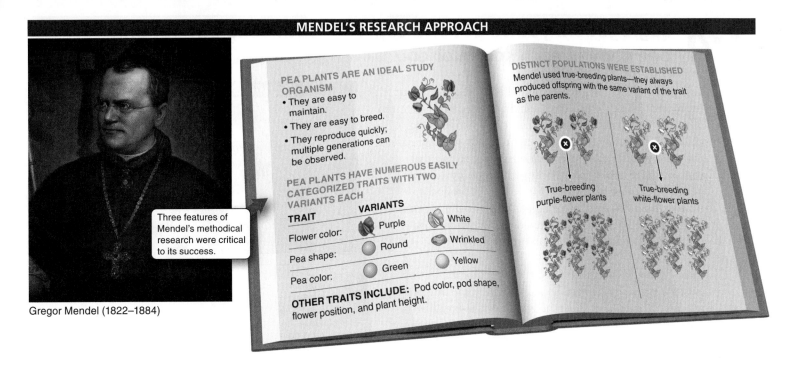

**FIGURE 7-7 Mendel's notebook.** Careful planning and a well-selected study organism contributed to the success of Mendel's experiments.

variant of the trait as the parents. True-breeding round-pea plants always produced plants with round peas when they were crossed together. True-breeding purple-flowered pea plants always produced purple-flowered offspring, while true-breeding white-flowered pea plants always produced white-flowered offspring. It took a lot of prep work to establish these populations, but once Mendel had them, he was in a position to set up all of the different crosses that enabled him to piece together the genetics puzzle.

Once he had obtained his groups of true-breeding plants, Mendel began a straightforward process of experimentation. He crossed plants with different traits—for example, he crossed plants with green peas and plants with yellow peas—and observed large numbers of their offspring, counting the number of plants that produced green peas and the number of plants that produced yellow peas. (This cross, written as "green × yellow cross," always resulted in plants that produced only yellow peas.) Mendel devised a hypothesis that would explain his observations and generate predictions about the outcomes of further crosses. Then he conducted those crosses to see whether the predictions generated by his hypothesis were borne out.

There was a simple elegance to Mendel's work: his predictions were always in the form of simple questions that could be answered by his experiments, such as, "If I cross a plant that produces wrinkled peas with a plant that produces smooth peas, will the offspring plants produce wrinkled or smooth peas?" His crosses would come out either one way or the other. Everything was black and white. This process of performing well thought-out, rigorous experiments was a radical innovation for biology. In the end, Mendel's patience, perseverance, and cleverness paid off. We explore his results in the next section and see that they allowed Mendel to clearly and definitively describe how the traits he studied were inherited.

## TAKE-HOME MESSAGE 7·3

In the mid-1800s, Gregor Mendel conducted studies that help us understand heredity. He focused on easily observed and categorized traits in garden peas and applied methodical experimentation and rigorous hypothesis testing to determine how traits are inherited.

# 7·4 Segregation: you've got two copies of each gene but put only one copy in each sperm or egg.

One odd and recurring result spurred Mendel to figure out the mechanism by which traits could be passed from parent to offspring. Just as brown-eyed parents can have blue-eyed children, sometimes traits that weren't present in either parent pea plant would show up in their offspring. Crosses that entailed the fertilization of plants with purple flowers by pollen from other plants with purple flowers produced *mostly* purple-flowered offspring, for instance, but *sometimes* they produced plants with white flowers. How was that possible? Where did the whiteness come from?

Here's where Mendel's meticulous and methodical experiments paid off. First, he started with some true-breeding white-flowered plants. Then he got some true-breeding purple-flowered plants. He wondered: which color wins out when a white-flowered plant is crossed with a purple-flowered plant? The answer was definitive: purple wins (**FIGURE 7-8**). All of the offspring from these crosses were purple, every time. For this reason, Mendel called the purple-flower trait **dominant,** and he considered the white-flower trait to be the **recessive** trait. In general, a dominant trait masks the effect of a recessive trait when an individual carries both the dominant and the recessive versions of the instructions for the trait.

Things got a bit more interesting when Mendel took the purple-flowered plants that came from the cross between true-breeding purple- and white-flowered plants and bred these purple offspring with each other. He found that these mixed-parentage plants were no longer true-breeding. Occasionally, they would produce white-flowered offspring. (To be exact, of the 929 offspring plants Mendel examined, 705 had purple flowers, and 224 had white flowers.) Apparently, the directions for building white flowers—last seen in one of their grandparents—were still lurking inside their purple-flowered parent plants. The existence of traits that could disappear for a generation and then show up again was quite perplexing. Mendel devised a simple hypothesis to explain these observations (**FIGURE 7-9**).

Mendel's hypothesis for explaining this pattern of inheritance incorporated three ideas that then helped him (and now help us) make predictions about crosses he hadn't yet done.

**1.** Rather than passing on the trait itself, **each parent puts into every sperm or egg it makes a single set of instructions for building the trait.** Today, we call that instruction set a *gene.*

**2. Offspring receive two copies of the instructions for any trait.** Often, both sets of instructions are identical

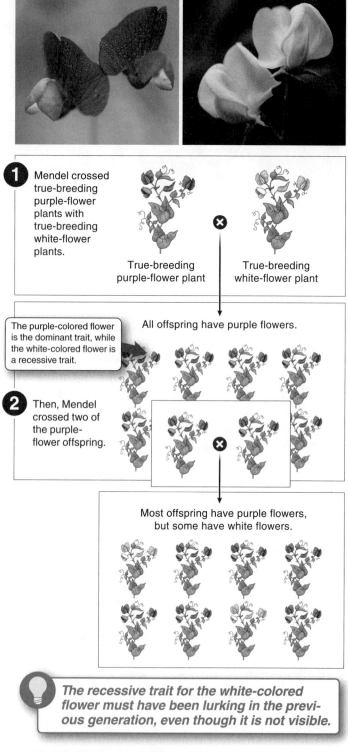

**DOMINANT AND RECESSIVE TRAITS**

**1** Mendel crossed true-breeding purple-flower plants with true-breeding white-flower plants.

True-breeding purple-flower plant ✕ True-breeding white-flower plant

The purple-colored flower is the dominant trait, while the white-colored flower is a recessive trait.

All offspring have purple flowers.

**2** Then, Mendel crossed two of the purple-flower offspring.

Most offspring have purple flowers, but some have white flowers.

💡 *The recessive trait for the white-colored flower must have been lurking in the previous generation, even though it is not visible.*

**FIGURE 7-8 White or purple?** By careful and repeated crosses among pea plants, Mendel determined that there were "dominant" and "recessive" traits.

## MENDEL'S LAW OF SEGREGATION

According to Mendel's law of segregation, only one of the two alleles for a gene is put into a gamete. At fertilization, offspring receive from each parent one allele for each gene.

Heterozygous pea plant          Heterozygous pea plant

Two different alleles (white, purple) for the same gene (flower color)

**MEIOSIS** Each gamete gets one copy of each gene.

**FERTILIZATION** Each fertilized egg gets two copies of each gene.

Homozygous recessive     Heterozygous     Heterozygous     Homozygous dominant

**FIGURE 7-9 Segregation of alleles in meiosis.** Organisms have two copies of each gene but place only one copy into each gamete during the process of meiosis (as described in Chapter 6).

and the offspring produce the traits according to those instructions. Other times, though, each parent contributes a slightly different set of instructions—that is, a different allele—for the trait. So, for example, in pea plants there are two alleles for flower color: a purple-flower allele and a white-flower allele. Each is a different allele, but because both specify instructions for producing color in flowers,

they are both genes for flower color. Same gene (flower-color gene), different alleles (purple and white).

**3. The trait observed in an individual depends on the *two* copies of the gene it inherits from its parents.** When an individual inherits the same two alleles for this gene, the individual's genotype for that gene is said to be **homozygous** and the individual shows the trait specified by the instructions embodied in those alleles. When an individual inherits a different allele from each parent, the individual's genotype for that gene is said to be **heterozygous.** Dominant and recessive alleles are defined by their action when they are in the heterozygous state: if one of the alleles is fully apparent in the individual while the other is not, the first is called the dominant allele and is said to "mask" the effect of the other allele, which is called the recessive allele.

*"There were no questions."*
—Entry in meeting notes following Mendel's first public presentation of his ideas on how heredity works. (*Tragically, no one in the audience of scientists had any idea what he was talking about. Not until about 40 years later did the world understand his discoveries.*)

Later, when an offspring grows up and gets ready to reproduce, it only gets to contribute one of its two copies of the gene to its own offspring. The other parent will contribute the other allele. So, when sperm and eggs are made, each sex cell gets only one copy of the gene—as opposed to the two copies present in every other cell in the body. For a male who is heterozygous, for example, it means that half of the sperm he produces will have one of the alleles and half will have the other. The idea that, of the two copies of each gene everyone carries, only one of the two alleles gets put into each gamete is significant and important enough that it is called **Mendel's law of segregation.**

### TAKE-HOME MESSAGE 7·4

**Each parent puts a single set of instructions for building a particular trait into every sperm or egg it makes. This instruction set is called a gene. The trait observed in an individual depends on the two copies (alleles) of the gene it inherits from its parents.**

# 7·5 Observing an individual's phenotype is not sufficient for determining its genotype.

Things are not always as they appear. Take skin coloration, for example. Humans and many other mammals have a gene that contains the information for producing melanin, one of the chemicals responsible for giving our skin its coloring (**FIGURE 7-10**). This gene is one of many that influence skin color. Unfortunately, there is also a defective, non-functioning version of the melanin gene that is passed along through some families. An individual who inherits two copies of the defective version of the gene cannot produce pigment and has a condition known as albinism, a disorder characterized by little or no pigment in the eyes, hair, and skin. But it is impossible to tell whether a normally pigmented individual carries one of these defective alleles just by looking at his or her appearance—we would need to get a genetic analysis done to discover this information.

The inability to deduce an individual's genetic makeup through simple observation is a general problem in genetics: physical appearances don't always exactly reflect the underlying genes. A normally pigmented individual may have two copies of the pigment-producing allele or may have only one copy. In either case, the individual will look the same. The outward appearance of an individual is called its **phenotype.** A phenotype includes features visible to the naked eye such as flashy coloration, height, or the presence of antlers.

A phenotype also includes less easily visible characteristics such as the chemicals an individual produces to clot blood or digest lactose. An individual's phenotype even includes the behaviors it exhibits.

Underlying the phenotype is the **genotype.** This is an organism's genetic composition. We usually speak of an individual's genotype in reference to a particular trait. For example, an individual's genotype might be described as "homozygous for the recessive allele for albinism" (e.g., two recessive alleles of the gene or two dominant alleles of the gene.) Another individual's genotype for the melanin gene might be described as "heterozygous" (e.g., a recessive allele and a dominant allele.) Occasionally, the word "genotype" is also used as a way of referring to *all* of the genes an individual carries.

When an organism exhibits a recessive trait, such as albinism, we know with certainty what its genotype is for the melanin gene. When it shows the dominant trait, on the other hand, it's impossible to distinguish whether the organism carries two copies of the dominant allele (homozygous dominant) or carries one copy of the dominant allele and one of the recessive (heterozygous). Because it's not possible to discern the genotypes of two individuals with the same phenotype just by looking at them, much of genetic analysis must make use of clever experiments and careful record-keeping to determine organisms' genotypes. Let's explore a useful method of analyzing the relationships between genotypes and phenotypes.

How do we analyze and predict the outcome of crosses? First, we must assign symbols to represent the different variants of a gene. Generally we use an upper-case letter for the dominant allele and lower-case letter for the recessive allele. In the case of the albino giraffe, we represent the albino individual as *aa,* because it must carry two copies of the recessive allele, *a*. A giraffe that is pigmented must have the genotype *AA* or *Aa*. If we don't know which of the two possible genotypes the pigmented individual has, we can write *A_*, where _ is a placeholder for the unknown second allele, whose identity we aren't certain of.

We can trace the possible outcomes of a cross between two individuals using a handy tool called the **Punnett square.** In **FIGURE 7-11** we illustrate the

PHENOTYPE: Little or no pigment in the eyes, hair, and skin
GENOTYPE: Homozygous for the recessive allele for albinism

**FIGURE 7-10 An albino giraffe stands out from its peers.** Carrying two non-functioning versions of the gene that produces the pigment melanin causes albinism.

A Punnett square is a useful tool for determining the possible outcomes of a cross between two individuals.

**FIGURE 7-11 Predicting the outcome of crosses.** A Punnett square helps in determining the likelihood of inheriting albinism.

Cross 1

MOTHER
albino
homozygous
**aa**

GAMETES
a          a

FATHER
pigmented
homozygous
**AA**

GAMETES
A

A

Aa        Aa

Aa        Aa

| OFFSPRING | GENOTYPE | PHENOTYPE |
|---|---|---|
| | All heterozygous **Aa** | All pigmented |

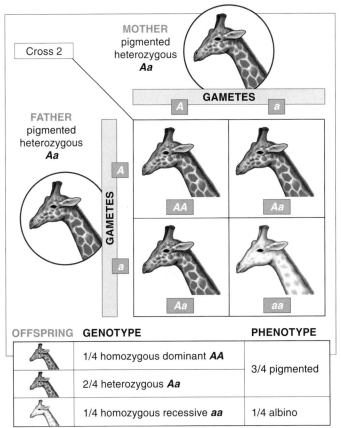

Cross 2

MOTHER
pigmented
heterozygous
**Aa**

GAMETES
A          a

FATHER
pigmented
heterozygous
**Aa**

GAMETES
A

a

AA        Aa

Aa        aa

| OFFSPRING | GENOTYPE | PHENOTYPE |
|---|---|---|
| | 1/4 homozygous dominant **AA** | 3/4 pigmented |
| | 2/4 heterozygous **Aa** | |
| | 1/4 homozygous recessive **aa** | 1/4 albino |

cross between a true-breeding pigmented individual, *AA*, and an albino, *aa*. Along the top of the square we list, individually, the two alleles that one of the parents produces, and along the left side of the square we list the two alleles that the other parent produces. We split up an individual's two alleles in this way because, although the individual carries two alleles, only one of the alleles is contained in each sperm or egg cell that it produces.

In the four cells of the Punnett square, we enter the genotypes of all the possible offspring resulting from our cross—a cell contains the allele given at the head of the column, which is one parent's contribution, and the allele at the left of the row, which is the other parent's contribution. The two gametes that come together at fertilization produce the genotype of the offspring. In Cross 1 illustrated in Figure 7-11, every possible offspring would be heterozygous and would be normally pigmented, because it receives a dominant allele from the pigmented parent and a recessive allele from the albino parent.

In the bottom half of Figure 7-11 (Cross 2), we trace the cross between two heterozygous individuals. Note that each parent produces two kinds of gametes, one with the dominant allele and one with the recessive allele. This cross has three possible outcomes: one-quarter of the time the offspring will be homozygous dominant (*AA*), one-quarter of the time the offspring will be homozygous recessive (*aa*), and the remaining half of the time the offspring will be heterozygous (*Aa*). Phenotypically, three-quarters of the offspring will be normally pigmented (*AA* or *Aa*) and one-quarter will be albino (*aa*).

## TAKE-HOME MESSAGE 7·5

It is not always possible to determine an individual's genetic makeup, known as its genotype, by observation of the organism's outward appearance, known as its phenotype. For a particular trait, an individual may carry a recessive allele whose phenotypic effect is masked by the presence of a dominant allele. Much genetic analysis therefore makes use of clever experiments and careful record-keeping, often using Punnett squares, to determine organisms' genotypes.

FAMILY
RESEMBLANCE

GENETICS
AND CHANCE

TRANSLATION
OF GENOTYPES

GENE
LINKAGE

273

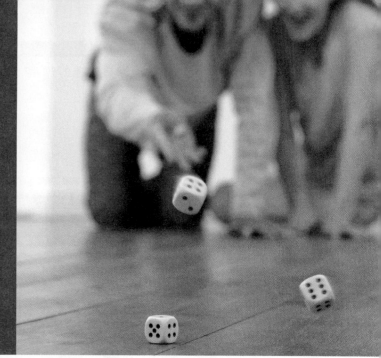

# 7·6–7·8 Probability and chance play central roles in genetics.

*Because of the role of chance in genetics, we cannot always predict the exact outcome of a genetic cross.*

## 7·6 Chance is important in genetics.

Sometimes genetics is a bit like gambling. Even with perfect information, it can still be impossible to know the genetic outcome with certainty. It's like flipping a coin: you can know every last detail about the coin, but you still can't know whether the coin will land on heads or tails. The best you can do is define the probability of each possible outcome.

The rules of probability (the same ones that govern coin tosses and the rolling of dice) have a central role in genetics, for two reasons. The first is a consequence of segregation, the process Mendel described, in which each gamete an individual produces receives only one of the two copies of each gene that the individual carries in most of its other cells. As a result, it is equally likely that the haploid gamete—the sperm or egg—will include one allele or the other allele of the two that the individual carries. It is impossible to know which allele it will be. The second reason probability plays a central role in genetics is that fertilization, too, is a chance event. All of the sperm or eggs produced by an individual are different from one another, and any one of those gametes may be the gamete involved in fertilization. Thus, knowing everything about the alleles a parent carries is not enough to be able to determine with certainty which alleles his or her offspring will carry.

Let's explore how we can make predictions in games of chance, as well as in matings, based on probabilities. As with the probability of getting heads in a coin flip, any gamete produced by an individual heterozygous for a trait has a 50% probability of carrying the dominant allele and a 50% probability of carrying the recessive allele. If an individual is homozygous for a trait, 100% of his or her gametes will carry that allele. Let's use these probabilities in an example.

If a male is heterozygous for albinism (*Aa*) and a female is albino (*aa*), what is the probability that their child will be homozygous for albinism (*aa*)? To get the *aa* outcome, two events must occur. First, the father's gamete must carry the recessive allele (*a*), and second, the mother's gamete must carry the recessive allele (*a*). In this case, the probability of a homozygous recessive offspring is 0.5 (the probability that the father's gamete carries *a*) times 1.0 (the probability that the mother's gamete carries *a*), for a probability of 0.5. This is a general rule when determining the likelihood of a complex event occurring: if you know the probability of each component that must occur, you multiply all the probabilities together to get the overall probability of that complex event occurring (**FIGURE 7-12**).

Consider another example, involving Tay-Sachs disease. As described in Chapter 3, Tay-Sachs is caused by malfunctioning lysosomes that do not digest cellular waste properly. It leads to death in early childhood. Tay-Sachs occurs if a child inherits two recessive alleles for the

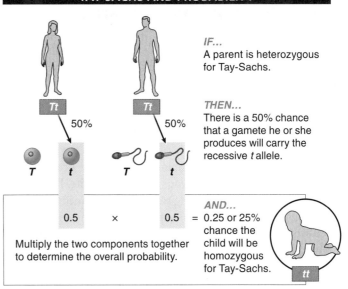

IF...
The mother is albino, and the father is heterozygous.

THEN...
There is a 100% chance that the mother's egg will carry the recessive *a* allele and a 50% chance that a sperm will carry the recessive *a* allele.

AND...
1.0  ×  0.5  = 0.5 or 50% chance the offspring will be albino.

Multiply the two components together to determine the overall probability.

**FIGURE 7-12 Using probability to determine the chance of inheriting albinism.**

IF...
A parent is heterozygous for Tay-Sachs.

THEN...
There is a 50% chance that a gamete he or she produces will carry the recessive *t* allele.

0.5  ×  0.5  = 0.25 or 25% chance the child will be homozygous for Tay-Sachs.

Multiply the two components together to determine the overall probability.

**FIGURE 7-13 The chance of inheriting Tay-Sachs disease.** The probability that a child will inherit Tay-Sachs disease from two heterozygous parents is 25%.

Tay-Sachs gene. If each parent is heterozygous for the Tay-Sachs gene, what is the probability that their child will have Tay-Sachs disease?

To solve this, break down the event of the child having Tay-Sachs into two separate events: first, it must inherit a recessive allele, *t*, from its heterozygous father. Because half the father's sperm will have the dominant *T* allele and half will have the *t* allele, the probability of the child inheriting the *t* allele is 0.5. Second, the child must also inherit a recessive allele from its heterozygous mother. Because half of the mother's eggs will have the *T* allele, while the other half have the *t* allele, the probability of the child inheriting the *t* allele from its mother is also 0.5. So the overall probability of the child having Tay-Sachs disease is 0.5 × 0.5 = 0.25, or 1 in 4 (**FIGURE 7-13**). Of course, if the couple has only one child, we can't predict *with certainty* whether the child will have Tay-Sachs. The mathematical probability is, however,

that if the couple had an infinite number of children, we could be certain that one-fourth of them would have Tay-Sachs disease. This fact may not be much help for heterozygous couples trying to decide whether to have a child or not.

**TAKE-HOME MESSAGE 7·6**

Probability plays a central role in genetics. In segregation, each gamete that an individual produces receives only one of the two copies of each gene the individual carries in its other cells, but it is impossible to know which allele goes into the gamete. Chance plays a role in fertilization, too: all of the sperm or eggs produced by an individual are different from one another, and any one of those gametes may be the gamete involved in fertilization.

# 7·7 A test-cross enables us to figure out which alleles an individual carries.

How can you see something that is invisible to the naked eye? When it comes to genetics, this is a practical issue that has been causing problems for as long as humans have been breeding plants and animals for food or other purposes. Genes are too small to be seen, and so determining an individual's genotype requires indirect methods. Suppose you are in charge of the alligators at a zoo. Some of your alligators come from a population in which white, albino alligators have occasionally occurred, although none of your alligators is white. Because white alligators—those having two recessive pigmentation alleles, *mm* ("m" for melanin)—are popular with zoo visitors, you would like to produce some at your zoo through a mating program.

The problem is that you cannot be certain of the genotype of your alligators. They might be homozygous dominant, *MM,* or they might be heterozygous, *Mm.* In either case, their phenotype is normal coloration. How can you figure out which of these two possibilities is the genotype of a particular alligator? This is a challenge to animal breeders, but not an insurmountable one. Genes may be invisible, but their identity can be revealed by a simple tool called the **test-cross.**

In the test-cross, you take an individual exhibiting a dominant trait but whose genotype is unknown. You cross (i.e., mate) that individual with an individual that is homozygous recessive and look at the phenotypes of their offspring. In the case of your breeding program for albino alligators, you could borrow an albino alligator from another zoo and breed your unknown-genotype alligator (genotype: *M_*) with that albino alligator (genotype: *mm*). There are two possible outcomes, and they will reveal the genotype of your unknown-genotype alligator (**FIGURE 7-14**). If your alligator is homozygous dominant (*MM*), it will contribute a dominant allele, *M,* to every offspring. Even though the albino alligator will contribute the recessive allele, *m,* to all its offspring, all the offspring of this cross will be heterozygous, *Mm,* and none of them will be albino.

If, on the other hand, your unknown-genotype alligator is heterozygous, *Mm,* half of the time it will contribute a recessive allele, *m,* to the offspring. In every one of those cases, the offspring will be homozygous recessive and thus albino.

**FIGURE 7-14 A test-cross can reveal an unknown genotype.** In this test-cross, a homozygous white female alligator is bred with a normally colored male of unknown genotype. The color of their offspring will help identify whether the male is homozygous dominant or heterozygous.

TEST-CROSS: WHITE ALLIGATORS

Test-cross

MOTHER
white
homozygous
*mm*

FATHER
pigmented
unknown genotype
*M_*
(*MM* or *Mm*)

GAMETES

*m*    *m*

GAMETES

*M*

*Mm*    *Mm*

*M or m*

*Mm* or *Mm*

*mm* or *mm*

Unknown allele could be *M* or *m*.

OFFSPRING (if unknown genotype is *MM*)

| | GENOTYPE | PHENOTYPE |
|---|---|---|
| | All heterozygous *Mm* | All pigmented |

OFFSPRING (if unknown genotype is *Mm*)

| | GENOTYPE | PHENOTYPE |
|---|---|---|
| | 2/4 heterozygous *Mm* | 2/4 pigmented |
| | 2/4 homozygous recessive *mm* | 2/4 white |

So the cleverness of the test-cross is that when you cross your unknown-genotype organism with an individual showing the recessive trait (and so having the known genotype of *mm*), the offspring will reveal the previously unknown makeup of the parent. For example, if your alligator sometimes produces offspring with the recessive trait (and if it does, it will do so half the time, on average), its genotype must be heterozygous. If it never produces offspring with the recessive trait, it must be homozygous for the dominant allele. To be confident in concluding that the unknown-genotype alligator has the *MM* genotype, though, you'd have to observe many, many offspring. After all, even if its genotype is *Mm*, quite a few offspring in a row might be normally pigmented, with the genotype *Mm*. Eventually, however, a heterozygous individual is likely to produce an offspring with the homozygous recessive genotype, *mm*, and the albino phenotype.

> ### TAKE-HOME MESSAGE 7·7
>
> In a test-cross, an individual that exhibits a dominant trait but has an unknown genotype is mated with an individual that is homozygous recessive. The phenotypes of the offspring reveal whether the unknown-genotype individual is homozygous dominant (all of the offspring exhibit the dominant trait) or heterozygous (half of the offspring show the dominant trait and half show the recessive trait).

# 7·8 We use pedigrees to decipher and predict the inheritance patterns of genes.

People want to know things about the future, such as: what is the likelihood that I will have a child with a particular genetic disease, say hemophilia? Or what is my own risk of developing a genetic disease, such as Huntington's disease, later in my life? Geneticists who study these and other diseases want to know other things: how is a particular disease or trait inherited? Is it recessive or dominant? Is it carried on the sex chromosomes or on one of the other chromosomes? A **pedigree** is a type of family tree that can help answer these questions.

In a pedigree, information is gathered from as many related individuals as possible across multiple generations (**FIGURE 7-15**). Starting from the bottom, each row in the chart represents a generation, listing all of the children in their order of birth, their sex, and whether or not they have a particular trait. Working up the pedigree, the children's parents are indicated and, above them, their parents' parents, for as far back as data are available. Squares represent males and circles represent females, and these shapes are shaded to indicate that an individual exhibits the trait of interest. Sometimes the genotype (as much of it as is known) is also listed for each individual.

By analyzing which individuals manifest the trait and which do not, it may be possible to deduce a pattern of inheritance for the trait—or, at least, rule out certain patterns. For example, for dominant traits, all affected individuals must have at least one parent who exhibits the trait. In contrast, an individual can exhibit a recessive trait even if both parents are unaffected (i.e., do not exhibit the

trait). In this case, the individual's parents must be heterozygous for that trait, each carrying one dominant and one recessive allele. Similarly, it is sometimes possible to determine whether a trait is carried on the sex chromosomes (X or Y, which we discuss in Section 7-13) or on one of the non-sex chromosomes (also called "autosomes"). Traits that are controlled by genes on the sex chromosomes are called **sex-linked traits.** Recessive sex-linked traits, for example, appear more frequently in males than in females, whereas dominant sex-linked traits appear more frequently in

---

**PEDIGREE**

A pedigree is a useful tool to document a trait of interest across multiple generations of family members.

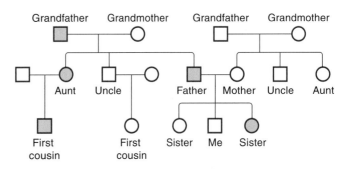

○ Female exhibiting trait of interest   ■ Male exhibiting trait of interest
○ Female not exhibiting trait   □ Male not exhibiting trait

**FIGURE 7-15 Family tree.** A pedigree maps the occurrence of a trait in a family.

Anury is a condition in dogs and other animals in which they are born without a tail. The condition is inherited as a recessive trait.

**IF...**
A dog has no tail, but both parents do (such as individual "3").

**THEN...**
We know with certainty each parent's genotype. (Can you figure out the genotypes of 1 and 2?)

**PEDIGREE**

- ⬤ Female with anury
- ◯ Female not exhibiting anury
- ⬛ Male with anury
- ☐ Male not exhibiting anury

This puppy definitely inherits a recessive allele from its father. It will have anury if (1) its mother has a recessive allele **and** (2) she passes that allele to the puppy. What is that probability?

**FIGURE 7-16 Pedigree puzzle: tracing the occurrence of anury in a family of dogs.**

females. These patterns may become obvious only on inspection of a large pedigree.

An example of how pedigrees can help determine how traits are inherited is given in **FIGURE 7-16**. Anury is a condition seen in dogs and some other animals in which the animal has no tail. The pedigree reveals that anury is inherited as a recessive trait, because unaffected parents can have offspring with the disorder.

Can you determine the genotype of the individual labeled "1" in Figure 7-16? Why must that female be heterozygous (*Aa*) for anury? (By the way, an individual that carries one allele for a recessive trait, and so does not exhibit the trait but can have offspring that do, is referred to as a **carrier** of the trait.) Must her mate have the same genotype?

On the same pedigree, note that two individuals in the second generation have a puppy (indicated by "?"). What is the probability that this puppy has anury? Examine the pedigree carefully and see whether you can come up with the answer; the next paragraph will guide you through if you get stuck.

In the cross producing the mystery puppy, the father has anury and so must be homozygous recessive; he will definitely pass on one *a* to the son. The mother's genotype can be *AA*, *Aa*, or *aA*, given that both of her parents are heterozygous. Consequently, she has a ⅔ (2 in 3) probability of being a carrier of the *a* allele. If she is, then there is a ½ (50%) chance that she will pass it on to her offspring, and he will have anury. The probability, therefore, is ⅔ × ½ = ⅓, or a 1 in 3 chance that the puppy will have anury.

As we'll discuss in later sections of this chapter, some traits may not show complete dominance and many traits are also influenced by the environment, so it is not always completely obvious what a trait's mode of inheritance is. In such cases, the more individuals we can include in the pedigree, the more accurate the analysis. Some human pedigrees contain thousands of individuals and stretch back six or more generations. With some other species it is possible to analyze tens of thousands of individuals per generation for a dozen or more generations. This is why plants and small insects (among other organisms) are excellent for studying inheritance patterns.

**Q** *Why do breeders value "pedigreed" horses and dogs so much?*

Dog and horse breeders often speak of "pedigreed" animals, a feature that adds tremendous value to the animals. This is because, with knowledge of an animal's family tree, or complete lineage, it is much less likely that any genetic surprises will occur as the animal develops and reproduces.

Once we have an idea about how a trait is inherited, it is possible to make informed predictions about a couple's risk of having a child with a particular disorder. This can be useful in conjunction with prenatal testing and treatment. It also enables the identification of individuals who are carriers of a recessive trait.

---

**TAKE-HOME MESSAGE 7·8**

Pedigrees help scientists, doctors, animal and plant breeders, and prospective parents determine the genes that individuals carry and the likelihood that the offspring of two individuals will exhibit a given trait.

---

# The translation of genotypes into phenotypes is not a black box.

*Phenotype diversity has multiple sources and is all around us.*

## 7·9 Incomplete dominance and codominance: the effects of both alleles in a genotype can show up in the phenotype.

As Mendel saw it, the world of genetics was straightforward and simple. We should be so lucky. Unfortunately, the world in which each trait is coded for by a single gene with two alleles—one completely dominant and one recessive—and with no environmental effects at all doesn't quite capture the complexity of the world beyond Mendel's pea plants. So, here and in the next sections, we build up a more complex model of how genes influence the building of bodies. We begin with the observation that, sometimes, the phenotype of heterozygous individuals differs from that of either of the homozygotes, and instead reflects the influence of both alleles rather than a clearly dominant allele.

One situation in which complete dominance is not observed is called **incomplete dominance,** in which the phenotype of a heterozygote is intermediate between the phenotypes of the two homozygotes. In humans, there is a plasma membrane receptor that allows cells (chiefly those in the liver) to remove cholesterol from the bloodstream (see Section 3-11), and the gene that codes for this receptor exhibits incomplete dominance. Individuals that carry two recessive alleles for this gene produce few (or even no) LDL receptors, and cardiovascular disease develops early, at a very young age. Individuals carrying one recessive allele and one dominant allele (that codes for normal-functioning LDL receptors) have significantly reduced levels of circulating cholesterol and a corresponding lower risk of cardiovascular disease. Individuals carrying two copies of the dominant allele produce twice as many LDL receptors as heterozygotes and experience significantly lower levels of circulating cholesterol and a much lower risk of cardiovascular disease.

An example of incomplete dominance that can be easily observed is the flower color of snapdragons (**FIGURE 7-17**). We can obtain true-breeding (homozygous) lines of snapdragons with red flowers, and true-breeding (homozygous) lines that produce only white flowers. When plants from these two populations are crossed, we would expect—if one allele were dominant over the other—either all red or all white flowers. Instead, such crosses always produce plants with pink flowers. Interestingly, when we cross two plants with pink flowers, we get ¼ red-flowered plants, ½ pink-flowered plants, and ¼ white-flowered plants.

How can we interpret this cross? It seems that the plants with white flowers have the genotype $C^W C^W$ and produce no pigment. At the other extreme, the plants with red flowers have the genotype $C^R C^R$ and produce a great deal of pigment. The letter "$C$" refers to the fact that the gene codes for **c**olor, and the superscript "$W$" or "$R$" refers to an allele producing no pigment (**w**hite) or **r**ed pigment. We use these designations for the genotypes because it isn't clear that either white or red is dominant over the other, and so neither should be upper-case or lower-case. The pink flowers receive one of the pigment-producing $C^R$ alleles and one of the no-pigment-producing $C^W$ alleles, and so produce an intermediate amount of pigment. Ultimately, the intensity of pigmentation just

Incomplete dominance occurs when a heterozygote exhibits an intermediate phenotype between the two homozygotes.

Because neither allele is dominant, we can't use upper case and lower case for the alleles. Instead, the superscript *W* represents the allele that produces the white flower, and *R* represents the red flower allele.

**Cross 1**

MOTHER
white-flowered
homozygous
$C^W C^W$

FATHER
red-flowered
homozygous
$C^R C^R$

| OFFSPRING | GENOTYPE | PHENOTYPE |
|---|---|---|
| | All heterozygous $C^W C^R$ | All pink flowers |

**Cross 2**

MOTHER
pink-flowered
heterozygous
$C^W C^R$

FATHER
pink-flowered
heterozygous
$C^W C^R$

| OFFSPRING | GENOTYPE | PHENOTYPE |
|---|---|---|
| | 1/4 homozygous $C^R C^R$ | 1/4 red flowers |
| | 2/4 heterozygous $C^W C^R$ | 2/4 pink flowers |
| | 1/4 homozygous $C^W C^W$ | 1/4 white flowers |

**FIGURE 7-17 Pink snapdragons demonstrate incomplete dominance.** When true-breeding white and red snapdragons are crossed, offspring have pink flowers.

## CODOMINANCE

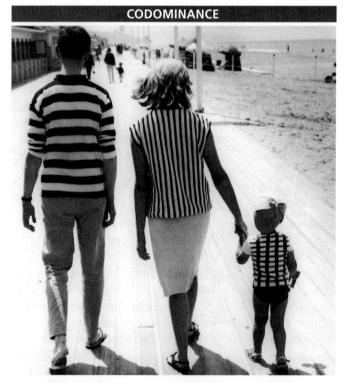

**FIGURE 7-18 With codominance, a heterozygous individual shows features of both alleles.**

depends on the amount of pigment chemical that is made by the flower-color gene.

A second situation in which complete dominance is not observed is called **codominance,** in which the heterozygote displays characteristics of both homozygotes, playfully represented in **FIGURE 7-18** (although "shirt phenotype," of course, has no genetic basis). In codominance, neither allele masks the effect of the other. An actual example of codominance occurs with feather color in chickens. When white chickens are crossed with black chickens, the offspring all have both white and black feathers.

### TAKE-HOME MESSAGE 7·9

Sometimes the effects of both alleles in a heterozygous genotype are evident in the phenotype. With incomplete dominance, the phenotype of a heterozygote appears to be an intermediate blend of the phenotypes of the two homozygotes. With codominance, a heterozygote has a phenotype that exhibits characteristics of both homozygotes.

# 7·10 What's your blood type? Some genes have more than two alleles.

What is your blood type? It can be O, A, B, or AB. Each of these different blood types indicates something about the physical characteristics of your red blood cells and has implications for blood transfusions—both giving and receiving blood. The blood groups are also interesting from a genetic perspective, because they illustrate a case of **multiple allelism,** in which a single gene has more than two alleles. Each individual still carries only two alleles—one from the mother and one from the father. But if you survey all of the alleles present in the *population,* you will find more than just two alleles.

Inheritance of the ABO blood groups provides the simplest example of multiple allelism, because there are only three alleles. We can call these alleles *A, B,* and *O.* The *A* and *B* alleles are both completely dominant to *O,* so individuals are considered to have blood type A whether they have the genotype *AA* or *AO* (**FIGURE 7-19**). Similarly, an individual with the genotype *BB* or *BO* is considered to have blood type B. If you carry two copies of the *O* allele, you have blood type O. The *A* and *B* alleles are codominant with each other, so the genotype *AB* gives rise to blood type AB. Consequently, with these three alleles in the population, individuals can be one of four different blood types: A, B, AB, or O.

What are the phenotypes of these alleles? An individual's blood-type alleles carry instructions that direct construction of a specific set of chemicals, called *antigens,* that protrude from every red blood cell. Antigens are proteins that jut from the surface of a cell and can "turn on" its defenses against foreign invaders. The *A* allele directs the production of A antigens all over the surface of red blood cells. Similarly, the *B* allele directs the production of B antigens on all red blood cells.

**MULTIPLE ALLELISM: BLOOD TYPE**

Multiple allelism occurs when there are three or more alleles for a gene within a population. An individual still inherits only two alleles—one from each parent.

Gene that determines blood type

Three alleles possible:
- **A** (dominant to **O** and codominant with **B**)
- **B** (dominant to **O** and codominant with **A**)
- **O** (recessive to **A** and **B**)

Type A / Type B / Type AB / Type O

Red blood cells have 6 different genotypes (**AA**, **AO**, **BB**, **BO**, **AB**, and **OO**). These genotypes result in 4 different phenotypes (type A, type B, type AB, and type O).

**FIGURE 7-19 Multiple allelism.** There are three different alleles—*A, B,* and *O*—for blood type in humans.

The *O* allele does not code for either type of antigen. This means that individuals with blood type AB have red blood cells with both A and B antigens, while individuals with blood type O have red blood cells with a "smoother" surface because they produce no A or B antigens (**FIGURE 7-20**).

**BLOOD TYPE, ANTIGENS, AND ANTIBODIES**

**BLOOD TYPE**

**ANTIGENS**
Chemicals on the surface of some cells. They act as signposts that tell the immune system whether the cell belongs in the body.

**TYPE A** — A antigens
**TYPE B** — B antigens
**TYPE AB** — A and B antigens
**TYPE O** — Neither A nor B antigens

**ANTIBODIES**
Immune system molecules that attack cells with foreign antigens.

B antibodies / A antibodies / Neither A nor B antibodies / A and B antibodies

Individuals produce antibodies to the antigens they don't have on their cells.

**FIGURE 7-20 Friend or foe?** Antigens are signposts that tell the immune system whether or not a cell belongs in the body.

Antigens on red blood cells play a role in the body's disease-fighting immune system. Antigens are like signposts, telling the immune system whether a cell belongs in the body or not. If a red blood cell with the wrong antigens enters your bloodstream, your immune system recognizes it as a foreign invader and destroys it. Such an attack is initiated by molecules in the bloodstream called *antibodies,* which attack only foreign antigens. Individuals with only A antigens on their red blood cells produce antibodies that attack B antigens. If they encounter a red blood cell with B antigens, they attack it. Such an immune response can lead to destruction of red blood cells, low blood pressure, and even death. Under normal circumstances, antibodies do not encounter a red blood cell with foreign antigens. Such an event may occur, however, if red blood cells with foreign antigens are accidentally injected into the person's bloodstream in a transfusion.

Individuals with only B antigens on their red blood cells produce antibodies that attack A antigens. Individuals with blood type O, who have neither A nor B antigens on their red

**Q** Why are people with type O blood considered "universal donors"? Why are those with type AB considered "universal acceptors"?

blood cells, produce antibodies that attack both A and B antigens. Individuals with blood type AB don't produce either type of antibody (or else they would have antibodies that attacked their own blood cells). From this information, we can deduce which blood types can be used in transfusions. Individuals with blood type O are universal donors because their red blood cells have no antigens and so do not trigger a reaction from either type of antibody. And individuals with blood type AB are universal recipients because they do not produce antibodies to either the A or B antigen. This is shown in **FIGURE 7-21**.

Another marker on the surface of red blood cells is the Rh blood group marker. This marker results from the effects of a single gene with just two alleles. Individuals' red blood cells carry the Rh cell surface marker if they have one or two copies of the dominant Rh marker allele, and they are said to be "Rh-positive." This "positive" (or "+") is then noted in their blood type, in addition to their ABO blood type, as in "O-positive" or "B-negative." Individuals who have two copies of the recessive allele for this gene do not have any

**THE SCIENCE BEHIND BLOOD DONATION**

| BLOOD TYPE | CAN DONATE TO | CAN RECEIVE FROM |
|---|---|---|
| **TYPE A** • Red blood cells have A antigens. • Individual produces antibodies that attack B antigens. | Type A    Type AB | Type A    Type O |
| **TYPE B** • Red blood cells have B antigens. • Individual produces antibodies that attack A antigens. | Type B    Type AB | Type B    Type O |
| **TYPE AB** • Red blood cells have A and B antigens. • Individual produces neither A nor B antibodies. • Individual is universal recipient. | Type AB | ANYONE! Type A  Type B  Type AB  Type O |
| **TYPE O** • Red blood cells have neither A nor B antigens • Individual produces antibodies that attack A and B antigens • Individual is universal donor. | ANYONE! Type A  Type B  Type AB  Type O | Type O |

Individuals with type O blood are universal donors. Individuals with type AB are universal recipients.

**FIGURE 7-21 Mapping blood compatibility.** Individuals mount an immune response to blood cells if they produce an antibody that attacks an antigen present on the donated red blood cells.

Rh markers, and they are described as "negative," as in "O-negative" or "A-negative." If individuals who are Rh-negative are exposed to Rh-positive blood, their immune system attacks the Rh antigens as foreign invaders—an immune response that can vary from mild, and passing unnoticed, to severe, which can lead to death. Thus, the Rh marker further restricts whose blood a person can receive in a transfusion.

Beyond the ABO and the Rh marker groups, there are many, many genes with multiple alleles—a dozen or even more alleles in some cases. In fact, one gene for eye color in fruit flies has more than 1,000 different alleles!

**TAKE-HOME MESSAGE 7·10**

In multiple allelism, a single gene has more than two alleles. Each individual still carries only two alleles, but in the population, more than just two alleles exist. This is the case for the ABO blood groups in humans.

# 7·11 Multigene traits: how are continuously varying traits such as height influenced by genes?

When babies are born, the parents are often curious about how tall their child will grow to be. Old wives' tales suggest a couple of ways for predicting height. If the baby is a boy, they say to add 5 inches to the mother's height and average that with the father's height. If it is a girl, subtract 5 inches from the father's height and average that with the mother's height. Alternatively, the lore says, just take the child's height at two years of age and double it.

While these methods can be surprisingly accurate at making predictions, they don't really help us understand the underlying reason *why* height can be predicted so well. The reason these methods work is that genes play a strong role in influencing height, and so offspring do resemble their parents in measurable ways. But unlike the simple case of Mendel's pea plants, where a single gene with two alleles determines the height of the plant, for humans and most animals, adult height is influenced by many different genes. Such a trait is said to be **polygenic.**

Recent research has identified two genes that help determine height in humans. Individuals with the "tall" allele for both genes are taller than those with the tall allele for only one. Individuals with one tall allele, in turn, are taller than people who do not have a tall allele for either gene. The term **additive effects** describes what happens when the effects of alleles from multiple genes all contribute to the ultimate phenotype. One of these height genes is on chromosome 15 and codes for an enzyme that converts testosterone to estrogen. This enzyme influences height because estrogen helps bones fuse at their ends and thus stop growing. In all, the two height genes identified in humans account for only about 2 inches of a person's height, but they exemplify how multiple genes can influence a single characteristic. The variety of heights from very short to very tall, with every height in between, reflects the fact that height is a trait influenced by contributions from multiple genes as well as the environment (**FIGURE 7-22**).

Many other physical traits are influenced by multiple genes, including eye color in humans. Long believed to be controlled by a single gene with a dominant brown-eye allele and a recessive blue-eye allele, it now seems that eye color is the result of the interactions between at least two genes and

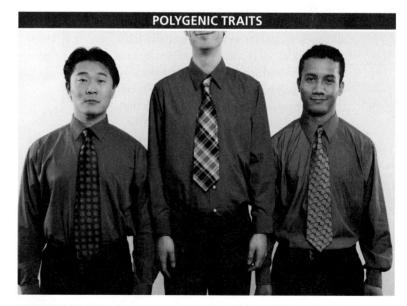

**POLYGENIC TRAITS**

**FIGURE 7-22 From many genes, one trait.** Height and skin color are multigene traits.

possibly more—a situation that makes more sense, given the significant continuous variation in eye color seen among adults.

Many behavioral traits are influenced by multiple genes. The developmental disorder known as autism, for example, seems to be the result of alterations in numerous—perhaps as many as 10 or even 20—different genes. Individuals with autism have difficulty interacting with others, particularly in making emotional connections. Autistic individuals also tend to have narrowly focused and repetitive interests. Autism is notoriously difficult to study, though, because its symptoms are varied, as is their intensity. It turns out that these variations in the disorder may result from the different combinations of the many genes involved in autism, much as different alleles for several height genes may work together to produce a variety of heights.

Interestingly, the genes responsible for autism may also influence some desirable characteristics. Behaviors that cause problems when manifest in the extreme may turn out to be beneficial when less extreme, such as unusual abilities of

perception, analytical skills, and focus. This idea—called the "geek theory of autism"—was first suspected when an upsurge in cases of autism in children was noticed in areas with large numbers of high-tech workers, including Silicon Valley in California and near Cambridge, England, in the heart of the United Kingdom's high-technology industry. Subsequently, researchers have published findings showing an over-representation of autistic children among parents working in the fields of engineering, physics, computer science, and math (although the researchers point out that autistic children are born to parents across all professions and socio-economic backgrounds). Perhaps alleles for genes that contribute to making individuals good at computer programming and solving complex technology problems can produce autism when an individual carries too many of them.

---

**TAKE-HOME MESSAGE 7·11**

Many traits, including continuously varying traits such as height, eye color, and skin color, are influenced by multiple genes.

---

# 7·12  Sometimes one gene influences multiple traits.

Just as multiple genes can influence one trait, some individual genes can influence multiple, unrelated traits, a phenomenon called **pleiotropy.** In fact, this may be true of nearly all genes. Consider sickle-cell anemia (also called sickle-cell disease), a potentially fatal condition in which individuals produce defective red blood cells that change their shape, becoming sickle-shaped, when they lose the oxygen they carry. The defective blood cells cannot effectively transport oxygen to tissues, and they accumulate in blood vessels, causing extreme pain. Individuals with sickle-cell disease suffer shortness of breath and numerous other problems that lead to a significantly reduced lifespan. The gene responsible for sickle-cell disease encodes the molecule hemoglobin, the oxygen-carrying molecule in red blood cells: $Hb^A$ is the allele for normal hemoglobin, and $Hb^S$ is the abnormal, "sickle-cell" allele. Anemia occurs in individuals homozygous for the sickle-cell allele, $Hb^S Hb^S$, but not in individuals carrying at least one copy of the normal allele, $Hb^A$ (although heterozygous individuals produce both

**Q** What is the benefit of "almost" having sickle-cell disease?

normal and sickling red blood cells, just not enough to cause sickle-cell anemia).

This hemoglobin gene is pleiotropic because individuals homozygous or heterozygous for the sickle-cell allele, $Hb^S Hb^S$ or $Hb^A Hb^S$, are resistant to the parasite that causes malaria. The resistance is due to the fact that the malarial parasite—which lives in red blood cells—cannot survive well in cells that carry the defective version of the hemoglobin gene. And because even individuals who are heterozygous for the sickle-cell allele have a significant number of sickling red blood cells, their bloodstream is just not a hospitable environment for the malarial parasite (**FIGURE 7-23**). Consequently, this one gene influences two traits: both the building of hemoglobin and the susceptibility to malaria.

Another example of pleiotropy is the SRY gene. Named for "**s**ex-determining **r**egion on the **Y** chromosome," this gene causes fetal gonads to develop as testes shortly after fertilization. Following the gonads' secretion of testosterone, a

Pleiotropy occurs when one gene influences multiple, different traits.

*Hb^A^Hb^A* **HOMOZYGOTE**
• Does not have sickle-cell disease
• Is susceptible to malaria

*Hb^S^Hb^A* **HETEROZYGOTE**
• Does not have sickle-cell disease
• Is immune to malaria

*Hb^S^Hb^S* **HOMOZYGOTE**
• Has sickle-cell disease
• Is immune to malaria

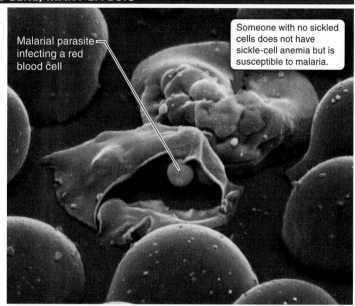

Malarial parasite infecting a red blood cell

Someone with no sickled cells does not have sickle-cell anemia but is susceptible to malaria.

**FIGURE 7-23 From one gene, multiple traits.** The allele for sickle-cell disease is pleiotropic: it causes red blood cells to form an unusual, sickled shape, and it also provides resistance to malaria.

cascade of other developmental changes then occurs, such as development of the internal and external male reproductive structures. Ultimately, the SRY gene is responsible for numerous behavioral characteristics as well. These characteristics are described further in Chapter 9.

> **TAKE-HOME MESSAGE 7·12**
>
> In pleiotropy, one gene influences multiple, unrelated traits. Most, if not all, genes may be pleiotropic.

# 7·13 Why are more men than women color-blind? Sex-linked traits differ in their patterns of expression in males and females.

The patterns of inheritance of most traits do not differ between males and females. When a gene is on an autosome (one of the non-sex chromosomes), both males and females inherit two copies of the gene, one from their mother and one from their father. The likelihood that an individual inherits one particular genotype versus another does not differ between males and females.

Traits coded for by the sex chromosomes, on the other hand, have different patterns of expression in males and females. One of the most easily observed examples of this phenomenon is red-green color-blindness. On the X chromosome in humans, there is a gene that carries the instructions for producing light-sensitive proteins in the eye that, when produced normally, respond differently to red wavelengths and green wavelengths of light. These proteins make it possible to distinguish between the colors red and

green. As long as an individual has at least one functioning copy of the gene carrying the instructions for production of these light-sensitive proteins, he or she produces sufficient amounts of the protein to have normal color vision.

There is a rare allele for this gene, however, that produces a non-functioning version of the protein. Having some of this non-functioning protein is not a problem as long as the person also carries another, normal version of the gene and produces some of the functioning protein. But this poses a problem for males, because they have only one X chromosome. In contrast, females inherit two X chromosomes.

Here's the problem: men get only one chance to inherit the normal version of the gene that codes for red-green color vision. It must be on the X chromosome they inherit from

A sex-linked trait is carried on the X chromosome. Women carry two copies of the X chromosome, while men carry an X chromosome and a Y chromosome.

Gene with instructions for light-sensitive proteins within the eye

Two alleles possible:

*R* (produces functioning light-sensitive proteins)

*r* (produces defective light-sensitive proteins)

**TO BE COLOR-BLIND...**

Male must inherit color-blindness allele (*r*) from his mother.

Female must inherit color-blindness allele (*r*) from both parents.

**TO HAVE NORMAL VISION...**

Male must inherit normal color-vision allele (*R*) from his mother.

Female can inherit normal color-vision allele (*R*) from either her mother or her father.

**FIGURE 7-24** Sex-linked traits such as color-blindness are not expressed equally among males and females.

their mother. Women get two chances. Although a woman may inherit the defective allele from one parent, she can still inherit the normal allele from the other parent. As long as she inherits the normal gene from one parent, she will have normal color vision. (If she inherits the defective allele from both parents, she will be red-green color-blind.) As we would predict, then, the frequency of red-green color-blindness is significantly greater in males than in females (**FIGURE 7-24**). Approximately 7% to 10% of men exhibit red-green color-blindness, while fewer than 1% of women are red-green color-blind.

Although males exhibit sex-linked recessive traits more frequently than do women, the situation is reversed for sex-linked dominant traits. In these cases, because females have two chances to inherit the allele that causes the trait, they are more likely to have the allele and thus exhibit the trait than are males, who have only one chance to inherit the allele.

**Q** If a man is color-blind, did he inherit this condition from his mother, his father, or both parents?

---

**TAKE-HOME MESSAGE 7·13**

The patterns of inheritance of most traits do not differ between males and females. However, when a trait is coded for by a gene on a sex chromosome, such as color vision on the X chromosome, the pattern of expression differs for males and females.

---

# 7·14 Environmental effects: identical twins are not identical.

It is a very serious warning, in boldfaced capital letters: "**PHENYLKETONURICS: CONTAINS PHENYLALANINE.**" But it's not next to a skull and crossbones on a glass bottle in a chemistry lab, it's on cans of diet soda intended for human consumption (**FIGURE 7-25**). Most of us either don't notice the warning or ignore it. Still, it's always there, and to some people it's a matter of life and death. What does it mean?

**Q** Drinking diet soda can be deadly if you carry a single bad gene. What gene is it and why is it so deadly?

At the most fundamental level, this warning is on products that contain phenylalanine because an organism's phenotype is a product of its genes in combination with its environment. In this case, specifically, the warning is for people who have a particular genotype that, in the presence of phenylalanine, can be deadly.

Phenylalanine is one of the amino acids. Sometimes our bodies use phenylalanine directly to

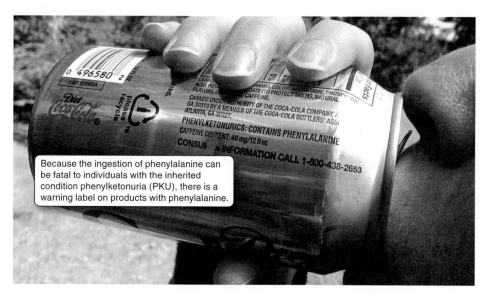

Because the ingestion of phenylalanine can be fatal to individuals with the inherited condition phenylketonuria (PKU), there is a warning label on products with phenylalanine.

**FIGURE 7-25 Change the environment, "cure" the disease.** The ingestion of phenylalanine in diet soda can be fatal to individuals with the inherited condition phenylketonuria (PKU).

build proteins, adding it to a growing amino acid chain. At other times, phenylalanine is chemically converted into another amino acid, called tyrosine. The body then uses tyrosine to build proteins and in a variety of other ways.

The problem is this: at birth, some people carry two copies of a mutant version of the gene that is supposed to produce the enzyme that converts phenylalanine into tyrosine. The mutant gene produces a malfunctioning enzyme, and none of the body's phenylalanine is converted into tyrosine. Perhaps this doesn't seem as though it should be a big problem, but it is. Little by little, as these individuals consume phenylalanine, it builds up in their bodies because none of it is converted into tyrosine. Eventually, so much accumulates that it reaches toxic levels and poisons the brain, leading to mental retardation and death. The disease is called phenylketonuria, or PKU.

Here's where the warning label comes in. By limiting the amount of phenylalanine in their diet (from diet soda and other sources), individuals with the two mutant alleles for processing phenylalanine can avoid the toxic buildup of the amino acid in their brain. In essence, they modify their environment (their diet) so that it contains little or no phenylalanine. And in that environment, the PKU mutant alleles are harmless. This example highlights the fact that genes by themselves do not "code" for physical characteristics. Rather, genes interact with the environment to produce physical characteristics. Unless you have information about

**Q** Could you create a temporarily spotted Siamese cat with an ice pack? Why?

both the genes *and* the environment, it is not usually possible to know what the phenotype will be.

A non-life-threatening illustration of the interaction of genes with the environment can be seen at a pet shop. Siamese cats (as well as Himalayan rabbits) carry genes that produce dark pigmentation. These genes interact strongly with the environment and are heat-sensitive. Dark pigment is produced only in relatively cold areas of the animal's body, while warm areas remain very light in color. This is why the fur on the coldest parts of the body—the ears, paws, tail, and tip of the face—becomes the darkest, while the fur on the rest of the body remains cream-colored or white (**FIGURE 7-26**). For Siamese cats living in cold climates and that spend a lot of time outside, it's interesting to notice that they become significantly darker in color during the winter months. Those that lounge indoors all winter remain lighter in color.

There are thousands of other cases in which genes' interactions with the environment influence their ultimate effects in the body. The very fact that identical twins—who inherit exactly the same set of alleles at the time of fertilization—don't die on the same day at the same moment and from the same cause, for example, reveals that environmental variation influences the expression of genes. The scope of environmental influences ranges from traits with large and obvious environmental effects, such as body weight and its relationship

Some pigment genes produce dark pigment only under cold conditions—such as on the tail, nose, ears, and feet of these animals.

**FIGURE 7-26 Heat-sensitive fur color.** Some pigment-producing genes produce the dark pigment of fur only under cold conditions. That's why these animals have darker patches of fur on their extremities.

with caloric intake, to traits such as eye color that are barely influenced by environmental effects, to traits with complex and subtle interactions with the environment, such as intelligence or personality.

Because of the role that environmental factors play in influencing phenotypes, DNA is not like a blueprint for a house. There is nearly always significant interaction between the genotype and the environment that influences the exact phenotype produced. The use of this metaphor is problematic to the extent that it suggests the phenotype is determined *solely* by the genotype. If that were the case, there would be no reason to invest in better schools, physical fitness regimens, nutritional monitoring, self-help efforts, or any other process by which individuals or societies try to improve people's lives (i.e., alter phenotypes) by enriching their environment.

**Q** James Watson, the co-discoverer of DNA, once wrote that when we completed the Human Genome Project, we would have "the complete genetic blueprint of man." Why might that be a poor metaphor?

**TAKE-HOME MESSAGE 7·14**

Genotypes are not like blueprints that specify phenotypes. Phenotypes are generally a product of the genotype in combination with the environment.

## 7·15–7·16 Some genes are linked together.

*Different genes influence red hair and freckles, so why are they often inherited together?*

# 7·15 Most traits are passed on as independent features: Mendel's law of independent assortment.

Sometimes you can be right about something for the wrong reason. This happened to Gregor Mendel. He didn't know that genes were carried on chromosomes. He believed that the units of heredity were just free-floating entities within cells. Given this perspective, it made sense to him that the inheritance pattern of one trait wouldn't influence the inheritance of any other trait. He believed that all genes behaved independently.

It helps to consider an example. Earlier in this chapter we saw that the characteristic of having a dimpled chin is caused by a single dominant gene. Imagine that you had a true-breeding population of people with dimpled chins (remember, "true-breeding" for a trait means that all offspring always manifest the trait; in the case of dimples, all have the genotype *DD*). Now suppose that an individual in this population had children with someone from a population of true-breeding, non-dimple-chinned people (everyone has the genotype *dd*). In this case, all of their children would have dimpled chins, but they would be heterozygous (*Dd*), getting a dominant allele from their dimple-chinned parent and a recessive allele from the other parent. If two heterozygotes had children together, though, they would produce three-quarters dimpled and one-

quarter non-dimpled children, with genotypes in the ratio of ¼ *DD*, ½ *Dd*, and ¼ *dd*. That is just what Mendel observed for traits in pea plants.

But what if we concurrently observed another characteristic of these people? Suppose the original true-breeding population of dimpled-chin people (*DD*) was also true-breeding for albinism (all *aa*), a disorder caused by carrying two recessive alleles for a single gene. And suppose that individuals in the normal-chinned population (*dd*) were also true-breeding for normal pigmentation (all *AA*). The question is, do the alleles an individual inherits for the dimpled-chin trait influence which alleles that individual inherits for pigmentation? And the answer is that, in most cases, they do not (**FIGURE 7-27**). Rather, the first cross of a dimple-chinned albino with a normal-chinned, normally pigmented individual would result in a child heterozygous for both traits, expressing each of the dominant traits. Phenotypically, this and all other offspring of this cross would be dimple-chinned and normally pigmented. In a mating between two doubly heterozygous individuals, three-quarters of their children would have the dominant trait and one-quarter would have the recessive trait, regardless of which trait you are tallying. In other words,

Mendel's law of independent assortment states that one trait does not influence the inheritance of another trait.

**IF...**
Parents are both homozygous for both traits. (But each parent is homozygous for a different allele than the other parent.)

**MOTHER**
albino
homozygous **aa**
dimpled chin
homozygous **DD**

**GAMETES**
All are **aD**

**FATHER**
pigmented
homozygous **AA**
non-dimpled chin
homozygous **dd**

GAMETES
All are **Ad**

All are **Aa Dd**

**THEN...**
All offspring are heterozygous for both traits (i.e., "doubly heterozygous"): **Aa Dd** (and exhibit the dominant trait in their phenotype).

| OFFSPRING | GENOTYPE | PHENOTYPE |
|---|---|---|
| 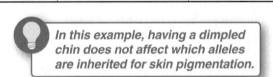 | All heterozygous **Aa** | All pigmented |
| | All heterozygous **Dd** | All dimpled chin |

*In this example, having a dimpled chin does not affect which alleles are inherited for skin pigmentation.*

**IF...**
Parents are both heterozygous for both traits (i.e., "doubly heterozygous"). Four different types of gametes are produced by each: **AD**, **Ad**, **aD**, and **ad**.

**MOTHER**
pigmented
heterozygous **Aa**
dimpled chin
heterozygous **Dd**

**GAMETES**

| | AD | Ad | aD | ad |
|---|---|---|---|---|
| **AD** | AA DD | AA Dd | Aa DD | Aa Dd |
| **Ad** | AA Dd | AA dd | Aa Dd | Aa dd |
| **aD** | Aa DD | Aa Dd | aa DD | aa Dd |
| **ad** | Aa Dd | Aa dd | aa Dd | aa dd |

**FATHER**
pigmented
heterozygous **Aa**
dimpled chin
heterozygous **Dd**

**THEN...**
The genotype proportions for **A/a** are still ¼ **AA**, ½ **Aa**, and ¼ **aa**. And the genotype proportions for **D/d** are still ¼ **DD**, ½ **Dd**, and ¼ **dd**.

| OFFSPRING | GENOTYPE | PHENOTYPE |
|---|---|---|
| | 1/4 homozygous dominant **AA** | 3/4 pigmented |
| | 2/4 heterozygous **Aa** | |
| | 1/4 homozygous recessive **aa** | 1/4 albino |
| | 1/4 homozygous dominant **DD** | 3/4 dimpled chin |
| | 2/4 heterozygous **Dd** | |
| | 1/4 homozygous recessive **dd** | 1/4 non-dimpled chin |

**FIGURE 7-27 Independent assortment of genes.**

neither trait influences the inheritance pattern for the other trait; all traits are inherited independently of each other. This is known as **Mendel's law of independent assortment.**

In the next section we'll see that, despite Mendel's correct understanding that separate traits are inherited independently, his belief that this happened because all genes just float freely around in the cell was not correct. The genes, as we now know, are carried on chromosomes. And this fact leads

occasionally to some situations in which independent assortment does *not* occur.

**TAKE-HOME MESSAGE 7·15**

Genes tend to behave independently, such that the inheritance pattern of one trait doesn't usually influence the inheritance of any other trait.

# 7·16 Red hair and freckles: genes on the same chromosome are sometimes inherited together.

Most redheads have pale skin and freckles. This simple observation is problematic for the law of independent assortment as Mendel imagined it. After all, in his law, he asserted that the inheritance of one trait does not influence the inheritance of another. But clearly that doesn't happen in this case. Having one of the traits, such as red hair, seems to influence the presence of another trait, pale skin (**FIGURE 7-28**). Strictly speaking, Mendel's second law is not true for *every* pair of traits. Sometimes the alleles for two genes are inherited together and expressed almost as a package.

**Q** *Why do most redheads have pale skin?*

In the case of humans, for example, there are about 23,000 genes in our genome. Yet we have only 23 unique chromosomes (two copies of each). Thus, genes influencing different traits must be on the same chromosome, maybe even right next to each other. When they are close together, we say that they are **linked genes.** One 2008 study, for example, demonstrated a link between human genes that influence hair color and skin pigmentation (including freckles).

Why are linked genes inherited together? To answer this, we must revisit the behavior of chromosomes during the production of gametes, discussed in Chapter 6. When you made that sperm or egg by meiosis, only one of the two copies of each of your chromosomes ended up in the gamete. It may have been the one from your mother or it may have been the one from your father. In either case, all of the alleles that were on the chromosome from that one parent were passed on as a group to the child that resulted from the fertilization involving that gamete. This process continues generation after generation. The linked alleles never get split up unless, during meiosis, recombination occurs between them, moving one or more to

**FIGURE 7-28 Violating the law of independent assortment.** Red hair and freckles are often inherited as a package deal.

the other chromosome in the pair so that they now become linked with the alleles on that chromosome (**FIGURE 7-29**).

When alleles are linked closely on the same chromosome, Mendel's second law doesn't hold true. It is very surprising—and was fortunate for Mendel—that of the seven pea plant traits he chose for his studies, none of them were close together on the same chromosome. For this reason, they all behaved as if they weren't linked.

---

**TAKE-HOME MESSAGE 7·16**

Sometimes, having one trait influences the presence of another trait. This is because the alleles for two genes are inherited and expressed almost as a package deal when the genes are located close together on the same chromosome.

---

## LINKED GENES

If this gamete is fertilized, the linked genes will be passed on to the offspring together as a group.

**HOMOLOGOUS CHROMOSOMES**

Linked genes

Maternal copy    Paternal copy

When crossing over occurs, linked genes usually stay together.

Homologous chromosomes after the exchange of genetic information

Each of the four chromatids gets packaged into a gamete.

*If genes are far apart on a chromosome or are on different chromosomes, the inheritance of one gene does not influence the inheritance of another.*

**FIGURE 7-29 Gene linkage.** Alleles that are closely linked on the same chromosome will be passed on to offspring in one bundle.

## CAN A GENE NUDGE US TOWARD NOVELTY-SEEKING (AND SPICY FOODS)?
### Novelty-seeking and a gene that influences it.

On human chromosome 11 there is a gene called DRD4. This gene carries the instructions for building a protein embedded in the membranes of many of your brain cells. The protein is a receptor for the brain chemical dopamine, which is active in several areas of the brain called "pleasure centers" and influences initiative and motivation.

Although all humans have some form of the DRD4 gene, there are several different alleles that a person might carry (though only two per person, of course). One aspect of the gene that varies is a short sequence near the beginning of the gene, called "promoter polymorphism –521C/T," that alters the activity of the brain's pleasure centers. There are two different alleles, *C* and *T*, and it appears that having one or two copies of the *T* allele leads to reduced transcription of the dopamine receptor gene, while having the *CC* genotype is associated with greater gene activity.

**Q: CAN A SINGLE GENE INFLUENCE YOUR PERSONALITY?** In 2000, researchers reported that individuals carrying two copies of the *C* allele for the DRD4 gene, as opposed to zero or one copy of the *C* allele, were more likely to exhibit certain personality traits, particularly novelty-seeking. The subjects in the study responded to a commonly used questionnaire, called the temperament and character inventory. On the questionnaire, subjects ranked how well certain statements described them, such as "I have sometimes done things just for kicks or thrills" or "I think things through before coming to a decision." The subjects were then evaluated for seven dimensions of personality traits. When it came to their score for "novelty-seeking," there was a small but statistically significant difference in how the subjects scored, depending on their genotype for this DRD4 gene:

**GRAPHIC CONTENT!**
*Thinking critically about visual displays of data.*
*Turn to p. 294 for a closer inspection of this figure.*

By 2008, 11 studies had been published on the relationship between the *CC* genotype and novelty-seeking. In a "meta-study" of all the published findings, researchers concluded that, taken together, these studies demonstrated a significant association between novelty-seeking and the –521*C/T* genotype. They summarized the results graphically:

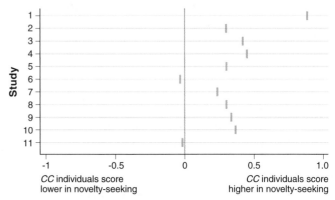

**Novelty-seeking score of *CC* individuals compared with *CT* and *TT* individuals**

Researchers have subsequently reported a variety of behavioral differences among people with higher novelty-seeking scores, including increased propensities for engaging in high-risk sports, alcohol and drug consumption, and attraction to vocations with a higher degree of risk. These people have even been reported to show a preference for spicy foods!

**Q: DO YOU WANT YOUR EMPLOYERS OR INSURANCE COMPANIES TO KNOW WHICH ALLELES YOU CARRY FOR THE DRD4 GENE?** Information about your genetic predisposition for diseases such as breast cancer or colon cancer can put you at risk for genetic discrimination. Might a person be at risk for discrimination based on whether his or her genotype is *CC* or *CT* or *TT* for this one variant of the DRD4 gene? Should it be taken into account that the personality differences influenced by DRD4 are small, that they seem to be affected by numerous other genes, and that there is tremendous variation in personality traits that is unrelated to the DRD4 gene? Would you want to know what your genotype is? Why?

# Big Ideas in ... Genes and Inheritance

## 7·1–7·5 Why do offspring resemble their parents?

- Offspring resemble their parents because they inherit genes—instruction sets for biochemical, physical, and behavioral traits, some of which are responsible for diseases—from their parents.

- Gregor Mendel conducted studies on garden peas that help us understand how traits are inherited, by applying methodical experimentation and rigorous hypothesis testing and by focusing on easily observed and categorized traits in the pea plants.

- Each parent puts into every sperm or egg it makes a single set of instructions for building a particular trait. The trait observed in an individual depends on the two copies of the gene that it inherits from its parents.

- It is not always possible to determine an individual's genotype through observation of its phenotype. Much genetic analysis therefore makes use of clever experiments and careful record-keeping to determine organisms' genotypes.

## 7·6–7·8 Probability and chance play central roles in genetics.

- Probability plays a central role in genetics. In segregation, each gamete that an individual produces receives only one of the two copies of each gene the individual carries in its other cells; it is impossible to know which allele this will be.

- Chance plays a role in fertilization, too: all of the sperm or eggs produced by an individual are different from one another, and any one of those gametes may be the gamete involved in fertilization.

- Test-crosses and pedigrees help scientists, doctors, plant and animal breeders, and prospective parents determine which genes individuals carry and the likelihood that the offspring of two individuals will exhibit a given trait.

## 7·9–7·14 The translation of genotypes into phenotypes is not a black box.

- Sometimes the effects of both alleles in a heterozygous genotype are seen in the phenotype.

- Many traits are influenced by multiple genes, and many genes may influence multiple unrelated traits.

- When a trait is coded for by a gene on the sex chromosomes, the pattern of expression differs for males and females.

- Genotypes are not like blueprints that specify phenotypes. Rather, phenotypes are a product of the genotype in combination with the environment.

## 7·15–7·16 Some genes are linked together.

- Many genes tend to behave independently, such that the inheritance pattern of one trait doesn't usually influence the inheritance of any other trait.

- When genes are located close together on the same chromosome, the alleles for the genes are inherited and expressed almost as a package deal.

See the To-Go Guide (packaged with every new copy of this textbook) for an illustrated summary of the chapter.

# Check Your Knowledge

## GRAPHIC CONTENT

*Thinking critically about visual displays of data.*

**1.** What variable was measured? How was it measured? Is there a better way that it could have been measured?

**2.** What does the *CC* genotype refer to? How does it differ from the *CT* or *TT* genotypes?

**3.** What information is conveyed by the light blue bars? Why is this information helpful?

**4.** The figure is from a "meta-study." What does each data point represent?

**5.** The 11 studies shown all evaluate the same thing, yet their results are not identical. Why might that be?

**6.** What conclusion can you draw from this figure? How certain are you about that conclusion?

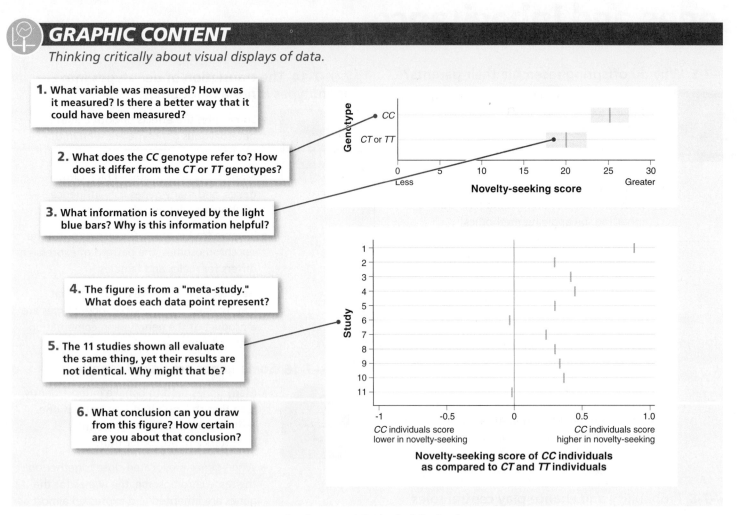

See Answers at the back of the book.

# Key Terms in Genes and Inheritance

# Multiple-Choice Questions

1. **Most genes come in alternative forms called:**

   a) alleles.

   b) heterozygotes.

   c) gametes.

   d) chromosomes.

   e) homozygotes.

2. **Traits that are determined by a single gene:**

   a) occur in single-celled organisms, but not in humans.

   b) are common in humans.

   c) must occur on the X chromosome, because males have only a single X and so the gene must be able to function in the absence of its homologous allele.

   d) include eye color and skin color.

   e) can have only two alleles.

3. **Pea plants were well suited for Mendel's breeding experiments for all of the following reasons except:**

   a) Peas exhibit variations in a number of observable characteristics, such as flower color and seed shape.

   b) Mendel and his staff could control the pollination between different pea plants.

   c) It is easy to obtain large numbers of offspring from any given cross.

   d) Many of the characteristics that vary in pea plants are not linked closely on the same chromosome.

   e) Peas have a particularly long generation time.

4. **The law of segregation states that:**

   a) the transmission of genetic diseases within families is always recessive.

   b) an allele on one chromosome will always segregate from an allele on a different chromosome.

   c) gametes cannot be separate and equal.

   d) the number of chromosomes in a cell is always divisible by 2.

   e) each of two alleles for a given trait segregate into different gametes.

5. **In pea plants, purple flower color is dominant to white flower color. If two pea plants that are true-breeding for purple flowers are crossed, in the offspring:**

   a) all of the flowers will be purple.

   b) three-quarters of the flowers will be purple and one-quarter will be white.

   c) half of the flowers will be purple and one-quarter will be white.

   d) one-quarter of the flowers will be purple and three-quarters will be white.

   e) all of the flowers will be white.

6. **Pea flowers may be purple (*P*) or white (*p*). Pea seeds may be round (*R*) or wrinkled (*r*). What proportion of the offspring from a cross between purple-flowered, round-seeded individuals (heterozygous for both traits) will have both white flowers and round seeds?**

   a) $\frac{1}{16}$

   b) $\frac{1}{2}$

   c) $\frac{3}{16}$

   d) $\frac{3}{4}$

   e) $\frac{9}{16}$

**7. The test-cross:**

a) makes it possible to determine the genotype of an individual of unknown genotype that exhibits the dominant version of a trait.

b) is a cross between an individual whose genotype for a trait is not known and an individual homozygous recessive for the trait.

100 HARD

43

c) sometimes requires the production of multiple offspring to reveal the genotype of an individual whose genotype is unknown (but who exhibits the dominant phenotype).

0 EASY

d) Only a) and b) are correct.

e) Choices a), b), and c) are correct.

**8. Which of the following statements is correct regarding pedigree analysis?**

a) Darkened squares or circles always represent individuals with the trait being traced.

100 HARD

b) White squares or circles always represent heterozygous individuals.

c) Horizontal lines connect siblings.

33

d) The length of the vertical lines is dependent on the relatedness between two individuals.

0 EASY

e) Squares represent females, and circles represent males.

**9. All of the offspring of a black hen and a white rooster are gray. The simplest explanation for this pattern of inheritance is:**

a) multiple alleles.

100 HARD

b) codominance.

c) incomplete dominance.

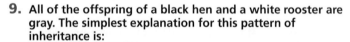
44

d) incomplete heterozygosity.

e) sex linkage.

0 EASY

**10. A woman with type B blood and a man with type A blood could have children with which of the following phenotypes?**

a) AB only

100 HARD

b) AB or O only

c) A, B, or O only

d) A or B only

27

e) A, B, AB, or O

0 EASY

**11. Which of the following traits shows a polygenic method of inheritance?**

a) flower color in snapdragons

100 HARD

b) blood type in humans

c) seed color in peas

44

d) sickle-cell disease in humans

e) skin color in humans

0 EASY

**12. The impact of a single gene on more than one characteristic is called:**

100 HARD

a) incomplete dominance.

b) environmentalism.

c) balanced polymorphism.

31

d) pleiotropy.

e) codominance.

0 EASY

**13. A rare, X-linked dominant condition in humans, congenital generalized hypertrichosis, is marked by excessive hair growth all over a person's body. Which of the following statements about this condition is incorrect?**

a) The son of a woman with this disease has just slightly more than a 50% chance of having this condition.

b) The daughter of a man with this condition has a 100% chance of having this condition.

100 HARD

c) The daughter of a woman with this disease has just slightly more than a 50% chance of having this condition.

69

d) The son of a woman with this disorder has a 100% chance of having this condition.

0 EASY

e) Males with this condition almost always have a more severe phenotype, because women usually have at least one normal allele that dilutes the effect of the mutated allele.

**14. Individuals carrying two non-functioning alleles for the gene producing the enzyme that converts phenylalanine into tyrosine:**

a) will develop phenylketonuria (PKU).

b) may or may not develop phenylketonuria, because the environment (diet, in this case) also plays a role in determining their phenotype.

100 HARD

c) will develop both PKU and Tay-Sachs disease, because the genes causing the two traits are on the same chromosome.

58

d) will develop both PKU and Tay-Sachs disease, because one pleiotropic gene influences both traits.

0 EASY

e) do not need to worry about their phenylalanine consumption, because PKU depends on both genes and the environment.

**15. Because of Mendel's law of independent assortment:**

a) individuals with red hair are more likely to have freckles.

100 HARD

b) skin color and hair texture tend to be inherited together.

c) we can deduce that genes cannot exist as free-floating entities within a cell but must be carried on chromosomes.

51

d) the alleles coding for one trait do not usually influence the inheritance pattern for another trait.

0 EASY

e) Both a) and b) are correct.

**16.** Thousands (or even tens of thousands) of different traits make up an individual. For this reason:

a) in a species with 23 different chromosomes, some traits must be coded for by genes on the same chromosome.

b) the environment must influence more than half of our traits.

c) all (or nearly all) genes must be pleiotropic.

d) knowing an individual's phenotype is not sufficient for determining the person's genotype.

e) All of the above are correct.

100 HARD
76
0 EASY

## Short-Answer Questions

**1.** Two healthy parents have two children, both of whom have a degenerative muscular disorder called SMA (spinal muscular atrophy).

a) Explain how SMA must be inherited and what the genotypes of the parents must be.

b) The parents wish to have a third child and are contemplating adopting versus having a biological child. They seek the guidance of a genetic counselor and ask what the probability of having another child with SMA would be. What does the genetic counselor tell them? Explain.

**2.** Mackenzie's favorite flower is the hydrangea. She plants a large hydrangea bush in her yard and enjoys the big purple blossoms. Her sister, Anna, would like some hydrangeas for her yard as well, so they transplant a small portion of the large bush to Anna's yard. Anna is surprised when, the first time the plant in her yard blooms, the blossoms are pink instead of purple. If the genotype of this plant has not changed, what might have influenced a change in the phenotype?

**3.** Your friend explains to you that her father has hemophilia, a genetic disorder that affects the process of blood clotting, but her mother does not.

a) Knowing that the inheritance pattern for this disorder is sex-linked, explain to your friend why it is more common for males to inherit a sex-linked trait.

b) Using a Punnett square, indicate the probability that a brother of your friend would have hemophilia, assuming that her mother carries two functional copies of the gene involved in this sex-linked recessive disorder. Explain your answer.

**See Answers at the back of the book.**
**For additional study questions, go to www.prep-u.com.**

Darwin's dangerous idea

# 8

# Evolution and Natural Selection

# 8·1 Evolution is an ongoing process.

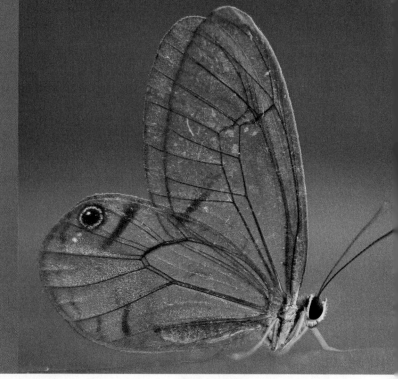

*What big eyes you have! The glasswing butterfly,* Cithaerias aurorina, *from Peru.*

## 8·1 We can see evolution occurring right before our eyes.

What's the longest that you've ever gone without food? Twenty-four hours? Thirty-six hours? If you've ever gone hungry that long, you probably felt as if you were going to die of starvation, but humans can survive days, even weeks, without food. In 1981, for example, 27-year-old Bobby Sands, a political prisoner in Northern Ireland, went on a hunger strike. Forsaking all food and consuming only water, he gradually deteriorated and ultimately died—after 66 days without food.

Each hour without food is even more dangerous when you're tiny. How long do you think a fruit fly can last without food? The answer is just under a day—20 hours, give or take a few (**FIGURE 8-1**). Fruit flies can't live long without food because their tiny bodies don't hold very large caloric reserves. But could you breed fruit flies that could live longer than 20 hours, on average? Yes.

First, start with a population of 5,000 ordinary fruit flies. In biology, the term **population** specifically means a group of organisms of the same species living in a particular geographic

FIGURE 8-1 **How long can a fruit fly survive without food?**

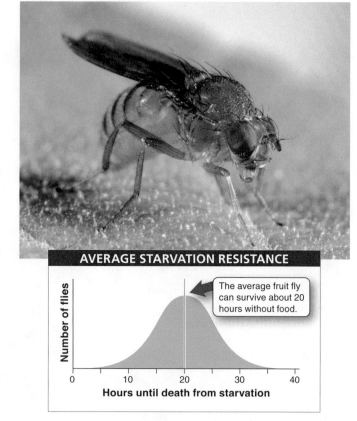

**AVERAGE STARVATION RESISTANCE**

The average fruit fly can survive about 20 hours without food.

Number of flies

Hours until death from starvation

region. The region can be a small area such as a test tube or a large area such as a lake. In your experiment, your population occupies a cage in your laboratory.

From these 5,000 fruit flies, you choose only the 20% of flies that can survive the longest without food:

1. Remove the food from the cage with the population of 5,000 flies.

2. Wait until 80% of the flies starve to death, then put a container of food into the cage.

3. After the surviving flies eat, they'll have the energy to reproduce. When they do, collect and transfer the eggs to a new cage.

You can now start a new generation populated only by the offspring of those fruit flies able to survive without food for the longest amount of time. When these eggs hatch, do you think the flies in this new generation will live longer than 20 hours without food? Indeed, these flies do show increased resistance to starvation. The average fly in the new generation can live for about 23 hours without food. And again, some of the new flies survive for more than 23 hours, some for less. The longest-surviving flies tend to be those that pack on the fat when the food is available. The new starvation-resistance curve for this second generation is similar to that of the original flies, but shifted to the right a bit so that the highest part of the curve occurs at 23 hours, indicating that most flies of the new generation survive for an average of 23 hours without food (**FIGURE 8-2**).

What if you kept repeating these three steps? Start each new generation using eggs only from the fruit flies in the top 20% of starvation resistance. After five generations, would the average starvation-resistance time for a fly in your latest

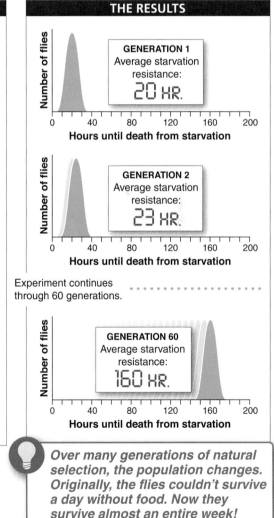

FIGURE 8-2 **Evolution in action: increasing fruit fly resistance to starvation.**

population be even higher? In fact, it would. With each new generation, you would see a slight increase, and with the fifth generation, the population's average survival would be about 32 hours.

One benefit to conducting research on fruit flies is that they have very short lives—they reach maturity at about two weeks and live for one month, on average. Such an experiment can be continued for many generations. After 60 generations of allowing the flies that are best at surviving without food to reproduce, how has the population changed? Amazingly, the *average* fly in the resulting population can survive for more than 160 hours without food (see Figure 8-2). In 60 generations, the flies have gone from a starvation resistance of less than a day to one of nearly a week! At the point where the food is removed, the flies in generation 60 are noticeably fatter than the flies in generation 0.

What happened? In a word: **evolution.** That is, there was a genetic change in the population of fruit flies living in the cage. Every fly in the generation 60 population, even the fly with the worst starvation resistance, is still more than seven times better at resisting starvation than the best fly in the original population. Later in this chapter we'll discuss the mechanisms by which evolution can occur, but, in short, in this situation it is the consequence of certain individual organisms in a population being born with characteristics that enable them to survive better and reproduce more than the offspring of other individuals in the population. In this experiment, the 20% of fruit flies that were the most starvation resistant had a huge reproductive advantage over less starvation-resistant flies because they were the *only* flies in the population that survived to reproduce.

This experiment answers a question that is sometimes perceived as complex or controversial: does evolution occur? The answer is an unambiguous *yes.* We can watch it happen in the lab whenever we want. Recall from our discussion of the scientific method (see Chapter 1) that for an experiment's results to be valid, they must be reproducible. To be certain of these results, researchers carried out the starvation-resistance experiment described above five separate times. The results were the same every time.

In this experiment we changed starvation resistance in fruit flies, but what if we used dogs instead of flies and, instead of allowing the most starvation-resistant individuals to reproduce, we allowed only the smallest dogs to reproduce? Would the average body size of individual dogs decrease over time? Yes. What if the experiment were done in a natural habitat, in the absence of human intervention, on rabbits, and only the fastest rabbits escaped death from predation by foxes? Would the average running speed in rabbits increase over time? Yes. We know these answers because each of these experiments has already been done. An unfortunate example of evolution in action is the evolution of resistance to antibiotics in numerous illness-causing bacteria, including some bacteria responsible for pneumonia and tuberculosis.

In this chapter, we examine how evolution occurs and the types of changes it can cause in a population. We also review the five primary lines of evidence indicating that evolution and natural selection are processes that help us clarify all other ideas and facts in biology. Let's begin our investigation with a look at how the idea of evolution by natural selection was developed. This knowledge will help us better understand why evolution by natural selection is regarded as one of the most important ideas in human history, why it is sometimes considered a dangerous idea, and why it generates emotional debate.

## TAKE-HOME MESSAGE 8·1

**The characteristics of individuals in a population can change over time. We can observe such change in nature and can even cause it to occur.**

# 8·2–8·5 Darwin journeyed to a new idea.

The *Galápagos tortoise*, Geochelone elephantopus.

## 8·2 Before Darwin, many people believed that all species had been created separately and were unchanging.

Charles Darwin grew up in an orderly world. In the Victorian British society he inhabited, many beliefs about humans and our place in the world had changed little over the previous two centuries. Biblical explanations were sufficient for most natural phenomena: the earth was thought by most people to be about 6,000 years old. With the occasional exception of a flood or earthquake or volcanic eruption, the earth was believed to be mostly unchanging. People recognized that organisms existed in groups called species. (In Chapter 10, we discuss in more detail what a species is; for now, we'll just say that individual organisms in a given species can interbreed with each other but not with members of another species.) People in Darwin's society also generally believed that all species, including humans, had been created at the same time and that, once created, they never changed and never died out.

Before he left this world, Darwin had thrown into question long- and dearly held beliefs about the natural world, and he forever changed our perspective on the origins of humans and our relationship to all other species. He didn't smash his society's worldview to pieces all at once, though, and he didn't do it by himself (**FIGURE 8-3**).

In the 1700s and 1800s, scientific thought was advancing at a rapid pace. In 1778, the respected French naturalist

Georges Buffon shook things up by suggesting that the earth must be about 75,000 years old. He arrived at this age by estimating that 75,000 years was the minimum time required for the planet to cool from a molten state. In the 1790s, Georges Cuvier began to explore the bottoms of coal and slate mines and found fossil remains that had no obvious similarity to any living species. Many interpreted the Bible in such a way that made Cuvier's discoveries unthinkable, believing that biblical accounts did not allow for species to be wiped out. Cuvier's publications documented giant fossils (including the Irish elk, the mastodon, and the giant ground sloth) that bore no resemblance to any currently living animals (**FIGURE 8-4**). And although Cuvier was not a proponent of the idea of evolution, the fossils he discovered allowed only one explanation: extinction was a fact. Troubling as this observation was for the prevailing Western worldview, it was only the beginning.

Not only was it starting to seem that species could disappear from the face of the earth, but several scientists, including Darwin's own grandfather, Erasmus Darwin, began to suggest that living species might change over time. And in the early 1800s, the biologist Jean-Baptiste Lamarck put forth a popular idea about the mechanisms by which this change might occur. Lamarck's idea was wrong—he thought that

EVOLUTION
IS ONGOING
DARWIN'S
JOURNEY
FOUR MECHANISMS
NATURAL SELECTION
EVIDENCE
FOR EVOLUTION
303

**GEORGES BUFFON**
(1707–1788)
Suggested that the earth was much older than previously believed.

**GEORGES CUVIER**
(1769–1832)
By documenting fossil discoveries, showed that extinction had occurred.

**JEAN-BAPTISTE LAMARCK**
(1744–1829)
Suggested that living species might change over time.

**CHARLES LYELL**
(1797–1875)
Argued that geological forces had gradually shaped the earth and continue to do so.

The young Charles Darwin

**FIGURE 8-3 Scientists who shaped Darwin's thinking.**

change came about chiefly through the use or disuse of features—but the increased willingness among scientists to question previously sacred "truths" contributed to an atmosphere of unfettered scientific thought in which it was possible to challenge convention.

Perhaps the revolutionary ideas that most inspired Darwin were those of the geologist Charles Lyell. In his 1830 book *Principles of Geology,* Lyell argued that geological forces had shaped the earth and were continuing to do so, producing mountains and valleys, cliffs and canyons, through gradual but relentless change. This idea that the physical features of the earth were constantly changing would most closely parallel

Darwin's idea that the living species of the earth, too, were gradually—but constantly—changing.

## TAKE-HOME MESSAGE 8·2

In the 18th and 19th centuries, scientists began to overturn many commonly held beliefs in the Western world, including that the earth was only about 6,000 years old and that all species had been created separately and were unchanging. These gradual changes in scientists' beliefs helped shape Charles Darwin's thinking.

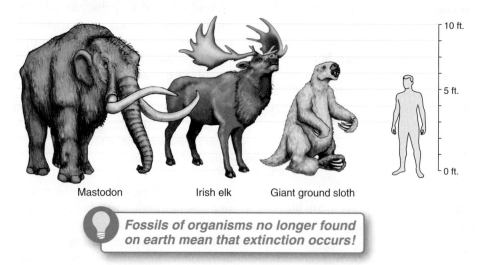

Mastodon    Irish elk    Giant ground sloth

10 ft.

5 ft.

0 ft.

*Fossils of organisms no longer found on earth mean that extinction occurs!*

**FIGURE 8-4 Extinction occurs.** Deep in coal mines, Cuvier discovered the fossilized remains of very large animals no longer found on earth.

# 8·3 A job on a 'round-the-world survey ship allowed Darwin to indulge and advance his love of nature.

Charles Darwin was born into a wealthy family in England in 1809, and his early life was unremarkable. Charles was never at the top of his class and professed to hate schoolwork. Nonetheless, at 16, he went to the University of Edinburgh to study medicine, following his father's footsteps. He was bored, though, and left at the end of his second year, when the prospect of watching gruesome surgeries (in the days before anesthesia) was more than he could bear.

Upon the abrupt ending of Charles's medical career, his father urged him to pursue the ministry and sent him to study theology at Cambridge University. Although he never felt great inspiration in his theology studies, Darwin was in heaven at Cambridge, where he could pursue his real love, the study of nature.

Shortly after graduation in 1831, Darwin—with significant help from his botany professor—landed his dream job, a position as a "gentleman companion" for the captain of the HMS *Beagle,* on a five-year, 'round-the-world surveying expedition (**FIGURE 8-5**). This job came as a huge relief to the young Darwin, who was not really interested in becoming a minister.

Once on the *Beagle,* Darwin found that he didn't actually like sea travel. He was seasick for much of his time on board, writing to his cousin: "I hate every wave of the ocean with a fervor which you . . . can never understand." To avoid nausea, Charles spent as much time as possible on shore. It was there that he found that fieldwork was his true calling. At each stop, he would eagerly investigate the new worlds he found. In Brazil, he was enthralled by tropical forests. In Patagonia, near the southernmost part of South America, he explored beaches and cliffs, finding spectacular fossils from huge extinct mammals. Elsewhere, he explored coral reefs and barnacles, always packing up specimens to bring back to museums in England and recording his observations for later use. He was like a schoolboy on permanent summer vacation.

**DARWIN'S 'ROUND-THE-WORLD VOYAGE**

London, England

Atlantic Ocean

Galápagos Islands

Pacific Ocean

Indian Ocean

Galápagos Islands

**FIGURE 8-5** **Like a schoolboy on permanent summer vacation.** In 1831, Darwin set out on the HMS *Beagle* on a five-year surveying expedition that took him around the globe.

## TAKE-HOME MESSAGE 8·3

After initially training in medicine and theology, Charles Darwin was able to focus on studying the natural world when, in 1831, he got a job on a ship conducting a five-year, 'round-the-world survey.

# 8·4 Observing geographic similarities and differences among fossils and living plants and animals, Darwin developed a theory of evolution.

The only book Darwin brought with him on the *Beagle* was Lyell's *Principles of Geology,* which he read again and again. Intrigued by the book's premise that the earth has been, and continues to be, constantly changing, Darwin's mind was ripe for fresh ideas—particularly as he began observing seemingly inexplicable things. How could he explain marine fossils high in the Andes, hundreds of miles from the nearest ocean? The idea that the earth was an ever-changing planet was seeping into Darwin's mind. That idea would serve him well when, a few years into its journey, the *Beagle* stopped at the Galápagos Islands, off the northwest coast of South America.

This group of volcanic islands was home to many unusual species, from giant tortoises to extremely docile lizards, which made so little effort to run away that Darwin had to avoid stepping on them. Darwin was particularly intrigued by the wide variety of birds, especially the finches, which seemed dramatically more variable than those he had seen in other locations.

Darwin noticed two important and unexpected patterns on his voyage that would be central to his discovery of a mechanism for evolution. The first involved the finches he collected and donated to the Zoological Society of London. Darwin had assumed that they were the equivalent of tall and short, curly-haired and straight-haired people. That is, he thought that all the finches were of the same species but with different physical characteristics or **traits,** such as body size, beak shape, or feather color. The staff of the Zoological Society, however, could see from the birds' physical differences that the birds were not reproductively compatible and that there were 13 unique species—a different species for every one of the Galápagos Islands that Darwin had visited. Moreover, although the birds were different species, they all resembled very closely the single species of finch living on the closest mainland, in Ecuador (**FIGURE 8-6**).

This resemblance seemed a suspicious coincidence to Darwin. Perhaps the island finches resembled the mainland species because they used to be part of the same mainland population.

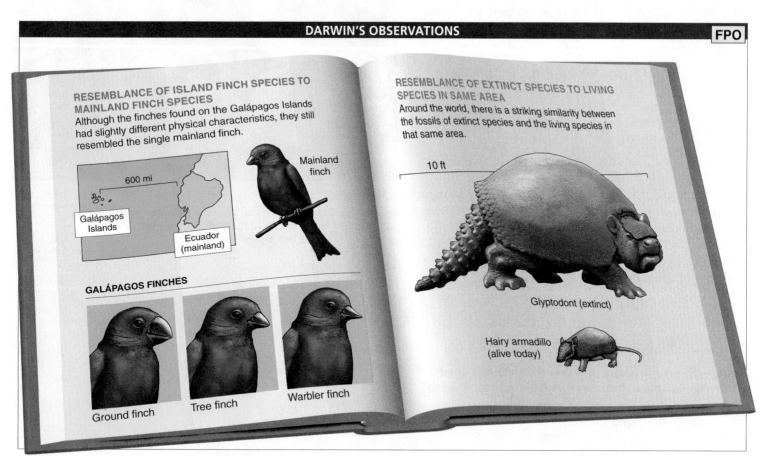

**DARWIN'S OBSERVATIONS**

FPO

**RESEMBLANCE OF ISLAND FINCH SPECIES TO MAINLAND FINCH SPECIES**
Although the finches found on the Galápagos Islands had slightly different physical characteristics, they still resembled the single mainland finch.

600 mi

Galápagos Islands

Ecuador (mainland)

Mainland finch

**GALÁPAGOS FINCHES**

Ground finch

Tree finch

Warbler finch

**RESEMBLANCE OF EXTINCT SPECIES TO LIVING SPECIES IN SAME AREA**
Around the world, there is a striking similarity between the fossils of extinct species and the living species in that same area.

10 ft

Glyptodont (extinct)

Hairy armadillo (alive today)

**FIGURE 8-6 Darwin observed unexpected patterns.**

Over time they may have separated and diverged from the original population and gradually formed new—but similar—species. Darwin's logic was reasonable, but his idea flew in the face of all the scientific thinking of the day.

The second important but unexpected pattern Darwin noted was that, throughout his voyage, at every location there was a striking similarity between the fossils of extinct species and the living species in that same area. In Argentina, for instance, he found some giant fossils from a group of organisms called "glyptodonts." The extinct glyptodonts looked just like armadillos, a species that still flourished in the same area. Or rather, they looked like armadillos on steroids: the average armadillo today is about the size of a house cat and weighs 10 pounds, while the glyptodonts were giants at 10 feet long and more than 4,000 pounds (see Figure 8-6).

If glyptodonts had lived in South America in the past, and armadillos were currently living there, why was only one of the species still alive? And why were the glyptodont fossils found only in the same places that modern armadillos lived? Darwin deduced that glyptodonts resembled armadillos because they were their ancestors. Again, it was a logical deduction, but it contradicted the scientific dogma of the day that species were unchanging and extinction did not occur.

The wheels were turning in Darwin's head, but he didn't have a great "Eureka!" moment until several years after his voyage, when he was back in England. He was reading "for amusement" *Essay on the Principle of Population,* by the economist Thomas Malthus. Malthus prophesied doom and gloom for populations, including humans, based on his belief that populations had the potential to grow much faster than food supplies could. Darwin speculated that, rather than the future holding certain catastrophe for all, maybe the best

individuals would "win" in the ensuing struggle for existence, and the worst would "lose." If so, he suddenly saw that "favourable variations would tend to be preserved, and unfavourable ones to be destroyed."

In 1842, Darwin finally prepared a draft of his ideas in a 35-page paper, written in pencil, and fleshed it out over the next couple of years. He knew that his idea was important—so important that he wrote a letter to his wife instructing her, in the case of his sudden death (an odd request, given that he was only 35 years old), to give his "sketch" to a competent person, along with about $1,000 and all of his books, so that this person could complete it for him.

He also had an inkling that his ideas would rock the world. In a letter to a close friend, he wrote: "At last gleams of light have come, and I am almost convinced (quite contrary to the opinion I started with) that species are not (it is like confessing to murder) immutable." But "confessing to murder" was apparently more than Darwin was ready for. Inexplicably, he put his sketch into a drawer, where it remained for 14 years.

## TAKE-HOME MESSAGE 8·4

**Darwin noted unexpected patterns among fossils he discovered and living organisms he observed while on the voyage of the *Beagle*. Fossils resembled but were not identical to the living organisms in the same area where they were found. And finch species on each of the Galápagos Islands differed from each other in small but significant ways. These observations helped Darwin develop his theory of how species might change over time.**

# 8·5 In 1859, after decades of mulling and procrastinating, Darwin published his thoughts on natural selection.

In a letter to Darwin, Alfred Russel Wallace (FIGURE 8-7), a young British biologist in the throes of malaria-induced mania in Malaysia, laid out a clear description of the process of evolution by natural selection, after having read Malthus's book. He asked Darwin to "publish it if you think it is worthy." Although he may have been inspired by his fever to put his thoughts in writing, they were careful, precise, and resulted from many years of thought and exposure to the writings of many of the same scientists who influenced Darwin. Crushed at having been scooped, Darwin wrote that "all my originality will be smashed." He had numerous friends

among the most prominent scientists of the day, however, and they arranged for a joint presentation of Wallace's and Darwin's work to the Linnaean Society of London. As a result, both Darwin and Wallace are credited for the first description of evolution by natural selection.

Darwin then sprang into action, rapidly putting together his thoughts and observations and completing, 16 months later, a full book. In 1859, he published *The Origin of Species* (its full title is *On the Origin of Species by Means of Natural Selection, or the Preservation of Favoured Races in the Struggle for Life*).

**FIGURE 8-7** **Alfred Russel Wallace.** Darwin and Wallace independently identified the process of evolution by natural selection.

**FIGURE 8-8** **Reconsidering the world.** Commonly held ideas before and after Darwin.

The book was an instant hit, selling out on its first day, provoking public discussion and debate and ultimately causing a wholesale change in the scientific understanding of evolution. Where once the worldview was of a young earth, populated by unchanging species all created at one time, with no additions or extinctions, now there was a new dynamic view of life on earth: species could and did change over time, and as some species split into new species, others became extinct (**FIGURE 8-8**).

Darwin's theory has proved to be among the most important and enduring contributions in all of science. It has stimulated an unprecedented diversity of theoretical and applied research, and it has withstood repeated experimental and observational testing. With the background of Darwin's elegant idea in hand, we can now examine its details.

## TAKE-HOME MESSAGE 8·5

After putting off publishing his thoughts on natural selection for more than 15 years, Darwin did so only after Alfred Russel Wallace independently came up with the same idea. The two men published a joint presentation on their ideas in 1858, and Darwin published a much more detailed treatment in *The Origin of Species* in 1859, sparking wide debate and discussion of natural selection.

Barn owl swoops to seize a mouse.

# 8·6–8·11 Four mechanisms can give rise to evolution.

# 8·6 Evolution occurs when the allele frequencies in a population change.

Suppose you were put in charge of a large population of tigers in a zoo. Almost all of them are orange and brown with black stripes. Occasionally, though, unusual all-white tigers are born. This white phenotype is the result of a rare pair of alleles that suppresses the tiger's production of most fur pigment (**FIGURE 8-9**). (Recall from Section 7-1 that alleles are variants of a gene, and that most mammals inherit one allele from

their mother and one from their father for each gene.) Because visitors flock to zoos to see the white tigers, you want to increase the proportion of your tiger population that is white. How would you do so?

One possibility is that you could alter the population by trying to produce more animals with the white phenotype.

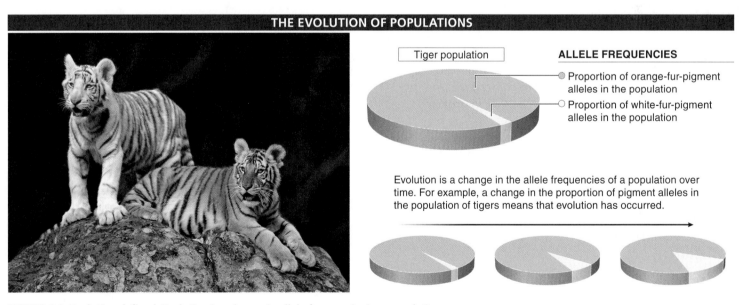

**THE EVOLUTION OF POPULATIONS**

Tiger population

**ALLELE FREQUENCIES**

- Proportion of orange-fur-pigment alleles in the population
- Proportion of white-fur-pigment alleles in the population

Evolution is a change in the allele frequencies of a population over time. For example, a change in the proportion of pigment alleles in the population of tigers means that evolution has occurred.

FIGURE 8-9 **Evolution defined.** Evolution is a change in allele frequencies in a population.

In this case, your best strategy would be to try to breed the white tigers with each other. Over time, this would lead to more and more white tigers. And as the generations go by (this will take a while, though, since the generation time for tigers is about eight years), your population will include a higher proportion of white tigers. When this happens, you will have witnessed evolution, a change in the allele frequencies of the population.

Another way you might increase the proportion of white tigers in your population would be to acquire some white tigers from another zoo. By directly adding white tigers to your population, the population would include a higher proportion of white tigers (you might even trade some of your orange tigers, which would further increase the proportion of your population that is white). By adding white tigers or removing orange tigers, you will again have witnessed evolution in your population.

As this example illustrates, evolution doesn't involve changing the genetics or physical features of *individuals*. Individuals do not evolve. Rather, you change the proportions of the alleles in the *population*. Within a population, for a particular allele type, the proportion of all the alleles that are of this type is that allele's frequency, like its "market share." In the tiger example, the white-fur alleles initially had little market share, perhaps as small as 1%. Over time, though, they came to make up a larger and larger proportion of the total fur-pigment alleles in the population. And as this happened, evolution occurred.

Any time an allele's market share changes, evolution is occurring. Perhaps the most dramatic way we see this is in acts of predation. When one organism kills another, the dead animal's alleles will no longer be passed on in the population. And if certain alleles make it more likely that one animal will be killed by another—maybe they alter its coloration, or speed, or some other defense mechanism—those alleles are likely to lose market share. The converse is equally true: evolution occurs when alleles increase in frequency.

Darwin demonstrated that natural selection could be an efficient mechanism of evolution and a powerful force in adapting populations to their environments. Evolution and natural selection, however, are not the same thing. Natural selection is one way that evolution can occur, but it is not the only agent of evolutionary change. It is one of four (**FIGURE 8-10**). They are:

## THE MECHANISMS OF EVOLUTION

**MUTATION**
An alteration of the base-pair sequence in the DNA of an individual's gamete-producing cells that changes an allele's frequency.

**GENETIC DRIFT**
A random change in allele frequencies, unrelated to any allele's influence on reproductive success.

**MIGRATION**
A change in allele frequencies caused by individuals moving into or out of a population.

**NATURAL SELECTION**
A change in allele frequencies that occurs when individuals with one version of a heritable trait have greater reproductive success than individuals with a different version of the trait.

**FIGURE 8-10  The four ways in which evolution can occur.**

1.  Mutation

2.  Genetic drift

3.  Migration

4.  Natural selection

Keeping in mind that evolution is genetic change in a population, we'll now explore each of these four agents that are capable of causing such genetic changes.

### TAKE-HOME MESSAGE 8·6

**Evolution is a change in allele frequencies within a population. It can occur by four different mechanisms: mutation, genetic drift, migration, and natural selection.**

# 8·7 Mutation—a direct change in the DNA of an individual—is the ultimate source of all genetic variation.

In describing the first of the four mechanisms of evolution, it is helpful to keep in mind our precise definition of evolution: a change in the allele frequencies found in a population. **Mutation** is an alteration of the base-pair sequence of an individual's DNA, and when this alteration occurs in the DNA that codes for a particular gene, the change in the DNA sequence may change that allele. (See Section 5-9 for a detailed discussion of the mutation process.) Say, for example, that a mutation changes one of a person's two blue-eye alleles into a brown-eye allele. If this mutation occurs in the sperm- or egg-producing cells, it can be passed on to the next generation; the offspring may carry the brown-eye allele. When this happens, the proportion of blue-eye alleles in the population is slightly reduced, and the proportion of brown-eye alleles is slightly increased. Evolution has occurred.

When considering mutation as an agent of evolution, it is important to remember that the cells of a multicellular eukaryotic organism can be divided into two types: somatic cells and reproductive cells. Somatic cells—the cells forming the body of an organism—are not passed from parent to offspring. Only reproductive cells are. For that reason, only mutations that affect reproductive cells can be inherited, and so we consider only those mutations as altering the allele frequencies within a population. The causes of mutations, however, are the same for somatic and reproductive cells.

Mutations can occur spontaneously during the complex process of cell division and, as we saw in Chapter 5, can be induced by a variety of environmental factors—including radiation, such as X rays, and some chemicals. But although these factors can induce mutations and influence the rate at which mutations occur, they do not generally influence exactly *which* mutations occur. Thus, mutations are random, and whether a mutation happens or not is unrelated to how useful or harmful it might be. Treatment of an agricultural pest with harmful chemicals, for example, might increase the number of mutations occurring in that pest population, but it does not increase the likelihood that a particular mutation is beneficial or detrimental.

There are many potential DNA-damaging threats in our world today. These include tanning beds, for example, which cause individuals to become tan by bombarding the skin with ultraviolet rays similar to those produced by the sun. These rays induce tanning just as exposure to the sun does: by stimulating pigment-producing cells to produce more melanin, a pigment that darkens skin color. But because ultraviolet rays are a form of energy, they can damage DNA, just as the sun's rays can. Not surprisingly, researchers have found that the greater an individual's exposure to ultraviolet rays, the greater the likelihood of DNA damage and, as a consequence, of skin cancer.

Another potential source of DNA-damaging energy is of concern to many more people: mobile phones. The scientific community is still divided on this question, however. Although mobile phones do emit electromagnetic radiation, that radiation is sufficiently low level that it is generally unable to break chemical bonds and therefore cause mutations. The majority of studies have not found any consistent causal relationship between mobile phone usage and any short- or medium-term health hazards.

Since 2004, however, several studies have suggested that further investigation is warranted. A study from 12 laboratories found evidence of DNA damage in cell cultures exposed to levels of radiation similar to those produced by mobile phones. Another study found that individuals who had used mobile phones for 10 years or more had an increased risk of brain tumors. A third study discovered an increased risk of brain tumors only for the side of the head on which the phone was used. Stay tuned for more information on this important question. (And, as with tanning beds, it is important to note that any cell phone–induced mutations are almost certainly *somatic* cell mutations, and so unlikely to influence evolution.)

In addition to being an agent of evolutionary change, mutation has another important role relevant to all agents of evolution, including natural selection: mutation is the ultimate source of genetic variation in a population. We saw that a mutation may lead to the conversion of one allele to another that is already found within the population, as in our blue eye and brown eye example. More importantly, though, a mutation may create a completely novel allele that codes for the production of a new protein (**FIGURE 8-11**). That is, a change in the base-pair sequence of a person's DNA may cause the production of a gene product that has never existed before in the human

**Q** Tanning beds and mobile phones expose people to radiation. Can they damage DNA?

## EVOLUTIONARY CHANGE: MUTATION

Mutation causes evolution when an alteration of the base-pair sequence in the DNA of an individual's gamete-producing cells changes an allele's frequency.

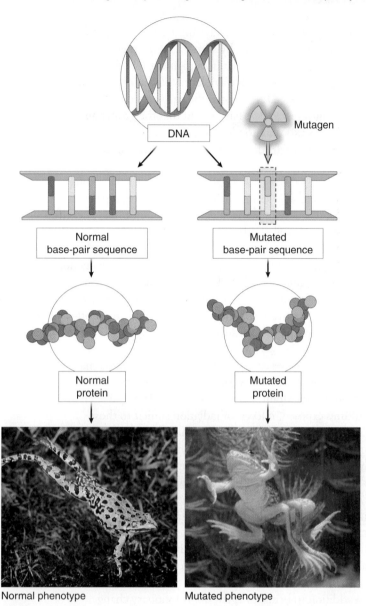

Normal phenotype

Mutated phenotype

💡 *Despite mutation's vital role in the generation of variation, mutations almost always cause early death or lower the reproductive success of an organism.*

**FIGURE 8-11 Agents of evolutionary change: mutation.**

population: instead of blue or brown eyes, the mutated gene might code for yellow or red eyes. If such a new allele occurs in the sperm- and egg-producing cells, and if it does not significantly reduce an individual's "fitness" (which we'll consider below), the new allele can be passed on to offspring and remain in the population. At some future time, the mutation might even confer higher fitness, in which case natural selection could cause it to increase in frequency. For this reason, mutation is critical to natural selection: all variation—the raw material for natural selection—must initially come from mutation.

Despite this vital role in the generation of variation, however, nearly all mutations reduce an organism's fitness by causing early death or reducing its reproductive success. Suppose you wrote a 10-page paper. If you were to randomly select one letter in the paper and change it to another letter, is the change likely to make your paper better or worse? The answer is obvious: except in *very* rare situations (can you imagine one?) the change will hurt your paper. Similarly, mutations usually change a normally functioning allele that codes for a normally functioning protein into an allele that codes for a non-functioning protein. Almost inevitably, that reduces an organism's fitness. For this reason, our bodies protect our sperm- or egg-producing DNA with a variety of built-in error-correction mechanisms. As a result, mutations are rare (on the order of 1 mutation in every 30 million base pairs each generation in humans).

### TAKE-HOME MESSAGE 8·7

**Mutation is an alteration of the base-pair sequence in an individual's DNA. If such an alteration changes an allele in an individual's gamete-producing cells, this constitutes evolution within the population. Mutations can be caused by high-energy radiation or chemicals in the environment, and also can appear spontaneously. Mutation is the only way that new alleles can be created within a population, and thus it generates the variation on which natural selection can act.**

# 8·8

## Genetic drift is a random change in allele frequencies in a population.

Along with mutation, another evolutionary agent is **genetic drift,** a random change in allele frequencies in a population. For example, imagine that there are two alleles for a particular trait, let's say a cleft chin. This is a dominant trait, so individuals with either one or two copies of the dominant allele (*CC* or *Cc*) exhibit the cleft chin (**FIGURE 8-12**). Now suppose that two heterozygous (*Cc*) people have one child. Which combination of alleles will that child receive? This is impossible to predict, because it depends completely on which sperm fertilizes which egg—the luck of the draw. If the couple's sole child inherits a recessive allele from each parent, will the *population's* allele frequencies change? Yes. The new individual's two recessive alleles (*cc*) increase the proportion of recessive alleles in the population. It is equally likely that this couple's only child will receive two dominant alleles (*CC*). In either case, because a change in allele frequencies has occurred, evolution has happened.

This impact of genetic drift is much greater in small populations than in large populations. In a large population, any one couple having a child with two recessive alleles is likely to be offset by another couple having one child with two dominant alleles. When that happens, the overall allele frequencies in the population do not change, and evolution has not occurred.

One of the most important consequences of genetic drift is that it can lead to **fixation** for one allele of a gene in a population (see Figure 8-12). Fixation occurs when an allele's frequency in a population reaches 100% (and the frequency of all other alleles of that gene becomes 0%). If this happens, there is no more variability in the population for this gene; all individuals will always produce offspring carrying only that allele (until new alleles arise through mutation). For this reason, genetic drift reduces the genetic variation in a population.

The important factor that distinguishes genetic drift from natural selection is that the change in allele frequencies is not related to the alleles' influence on reproductive success. A cleft chin does not affect an individual's ability to reproduce.

**FIGURE 8-12 Agents of evolutionary change: genetic drift.** Genetic drift has the greatest impact in small populations.

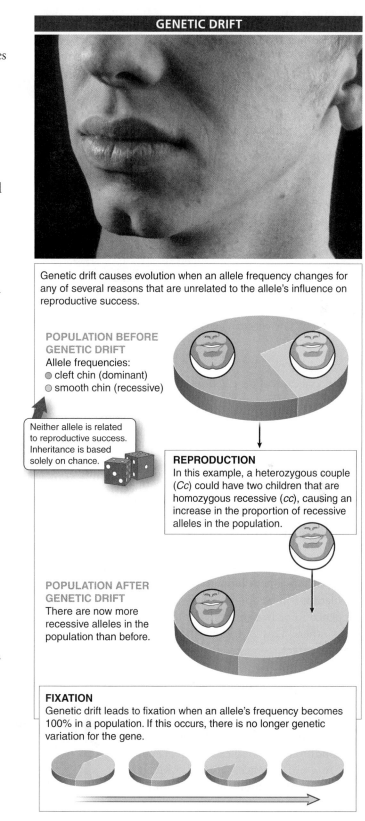

**GENETIC DRIFT**

Genetic drift causes evolution when an allele frequency changes for any of several reasons that are unrelated to the allele's influence on reproductive success.

**POPULATION BEFORE GENETIC DRIFT**
Allele frequencies:
- cleft chin (dominant)
- smooth chin (recessive)

Neither allele is related to reproductive success. Inheritance is based solely on chance.

**REPRODUCTION**
In this example, a heterozygous couple (*Cc*) could have two children that are homozygous recessive (*cc*), causing an increase in the proportion of recessive alleles in the population.

**POPULATION AFTER GENETIC DRIFT**
There are now more recessive alleles in the population than before.

**FIXATION**
Genetic drift leads to fixation when an allele's frequency becomes 100% in a population. If this occurs, there is no longer genetic variation for the gene.

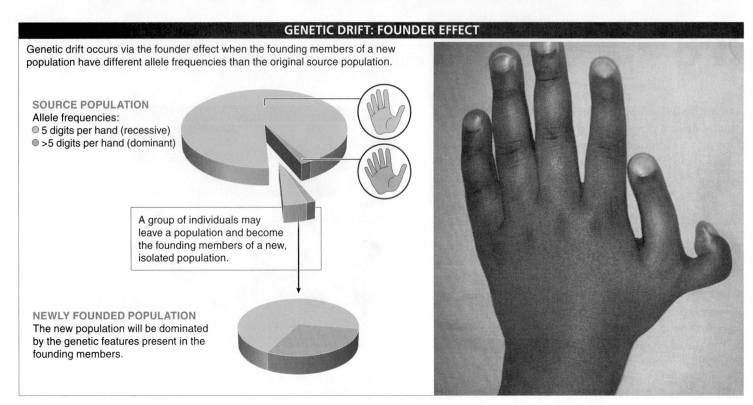

Genetic drift occurs via the founder effect when the founding members of a new population have different allele frequencies than the original source population.

**SOURCE POPULATION**
Allele frequencies:
- 5 digits per hand (recessive)
- >5 digits per hand (dominant)

A group of individuals may leave a population and become the founding members of a new, isolated population.

**NEWLY FOUNDED POPULATION**
The new population will be dominated by the genetic features present in the founding members.

**FIGURE 8-13** One way that genetic drift occurs: the founder effect.

Two special cases of genetic drift, the founder effect and population bottleneck effect, are important in the evolution of many populations.

**Founder effect** A small number of individuals may leave a population and become the founding members of a new, isolated population. The founder population may have different allele frequencies than the original, "source" population, particularly if the founders are a small group. If this new population does have different allele frequencies, evolution has occurred. Because the founding members of the new population will give rise to all subsequent individuals, the new population will be dominated by the genetic features that happened to be present in that group of founding fathers and mothers. This type of genetic drift is called the **founder effect.**

The Amish population in the United States is believed to have been established by a small number of founders, some of whom happened to carry the allele for *polydactyly*— the condition of having extra fingers and toes. As a consequence, this trait, while rare, now occurs much more frequently among the Amish than among the rest of the U.S. population (**FIGURE 8-13**).

**Q** Why are Amish people more likely than other people to have extra fingers and toes?

**Population bottleneck effect** Occasionally, a famine, disease, or rapid environmental change causes the deaths of a large proportion of individuals in a population. Because the population is quickly reduced to a small fraction of its original size, this reduction is called a **bottleneck.** If the catastrophe is equally likely to strike any member of the population, the remaining members are essentially a random, small sample of the original population. For this reason, the remaining population may not possess the same allele frequencies as the original population. Thus, the consequence of such a population bottleneck would be evolution through genetic drift (**FIGURE 8-14**).

Just such a population bottleneck occurred with cheetahs near the end of the last ice age, about 10,000 years ago. Although the cause is unknown— possibly environmental cataclysm or human hunting pressures—it appears that nearly all cheetahs died. And although the population rebounded, all

Genetic drift occurs via the bottleneck effect when famine, disease, or rapid environmental change causes the deaths of a large, random proportion of the population, and the surviving individuals have different allele frequencies than the original population.

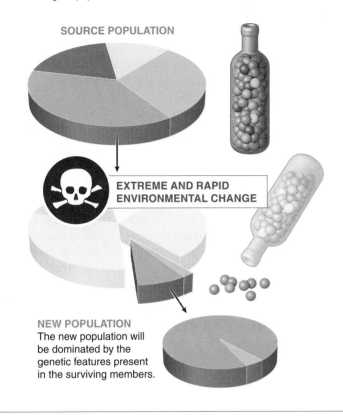

SOURCE POPULATION

EXTREME AND RAPID ENVIRONMENTAL CHANGE

NEW POPULATION
The new population will be dominated by the genetic features present in the surviving members.

FIGURE 8-14 Another way that genetic drift occurs: the bottleneck effect.

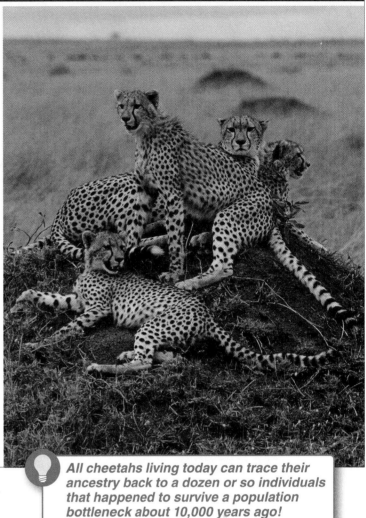

*All cheetahs living today can trace their ancestry back to a dozen or so individuals that happened to survive a population bottleneck about 10,000 years ago!*

cheetahs living today can trace their ancestry back to a dozen or so lucky individuals that survived the bottleneck. As a result of this past instance of evolution by genetic drift, there is almost no genetic variation left in the current population of cheetahs. (And, in fact, a cheetah is able to accept a skin graft from any other cheetah, much as identical twins can from each other.)

---

**TAKE-HOME MESSAGE 8·8**

Genetic drift is a random change in allele frequencies within a population, unrelated to the alleles' influence on reproductive success. Genetic drift is a significant agent of evolutionary change, primarily in small populations.

---

# 8·9

## Migration into or out of a population may change allele frequencies.

The third agent of evolutionary change is **migration.** Migration, also called **gene flow,** is the movement of some individuals of a species from one population to another (**FIGURE 8-15**). This movement from population to

population within a species distinguishes migration from the founder effect, in which individuals migrate to a new habitat, previously unpopulated by that species. If migrating individuals survive and reproduce in the new population, and

FIGURE 8-15 Agents of evolutionary change: migration (gene flow).

## MIGRATION (GENE FLOW)

Gene flow causes evolution if individuals move from one population to another, causing a change in allele frequencies in either population.

**① BEFORE MIGRATION**
Two populations of the same species exist in separate locations. In this example, they are separated by a mountain range.

Population 1                     Population 2

**② MIGRATION**
A group of individuals from Population 1 migrates over the mountain range.

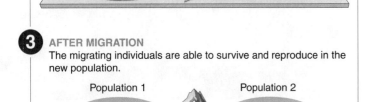

Population 1                     Population 2

**③ AFTER MIGRATION**
The migrating individuals are able to survive and reproduce in the new population.

Population 1                     Population 2

if they carry a different proportion of alleles than the individuals in their new home, then the recipient population experiences a change in allele frequencies and, consequently, experiences evolution. And because alleles are simultaneously lost from the population that the migrants left behind, that population, too, will experience a change in its allele frequencies and, thus, will evolve.

Gene flow between two populations is influenced by the mobility of the organisms and by barriers, such as mountains or rivers. And as we'll see in Chapter 16, human activities, too, can dramatically influence the migratory potential of organisms, as we sometimes transport species. This can occur intentionally—as when people purchase exotic pets or non-native species for their garden—or unintentionally—such as when marine species get trapped in ballast water tanks of cargo ships, later to be released in ports on the other side of the world when the tanks are flushed.

### TAKE-HOME MESSAGE 8·9

Migration, or gene flow, leads to a change in allele frequencies in a population as individuals move into or out of the population.

# 8·10 When three simple conditions are satisfied, evolution by natural selection is occurring.

The fourth agent of evolutionary change is **natural selection.** This is the mechanism that Darwin identified in *The Origin of Species,* in which he noted that three conditions are necessary for natural selection to occur:

1. There must be variation for the trait within a population.

2. That variation must be heritable (that is, capable of being passed from parents to offspring).

3. Individuals with one version of the trait must produce more offspring than those with a different version of the trait.

Let's examine these conditions more closely.

**Condition 1: Variation for a trait** Close your eyes and imagine a dog. What does the dog look like? If 50 people were asked this question, we would probably get 50 different dog descriptions (**FIGURE 8-16**). Some are big, some are small. Some have short hair, some long. They vary in just about every way you can imagine. Likewise, if 50 people were to imagine a human face, a similarly broad range of images would pop into their heads. Variation is all around us. Beyond making the world an interesting place to live in, variation serves another purpose: it is the raw material on which evolution feeds.

Variation is not limited to physical features such as fur color or face shape. Organisms vary in physiological and

FIGURE 8-16 The first condition for natural selection is variation for a trait.

biochemical ways, too. Some people can quickly and efficiently metabolize alcohol, for example. Others find themselves violently ill soon after sipping a glass of wine. The same goes for digesting milk. Similarly, we vary in our susceptibility to poison ivy or diseases such as malaria. Behavioral variation—from temperament to learning abilities to interpersonal skills—is dramatic and widespread, too. So impressed was Darwin with the variation he observed throughout the world, he devoted the first two chapters of *The Origin of Species* to a discussion of variation in nature and among domesticated animals. Darwin knew that variation was an essential component of evolution by natural selection and considered it the first of three necessary conditions for natural selection.

### Condition 2: Heritability

The second condition that Darwin identified as necessary for natural selection was a no more complex discovery than the first: for natural selection to act on a trait within a population, offspring must inherit the trait from their parents. Although inheritance was poorly understood in Darwin's time, it was not hard to see that, for many traits, offspring look more like their parents than like some other, random individual in the population (**FIGURE 8-17**). Animal breeders had long known that the

FIGURE 8-17 The second condition for natural selection is heritability. Goldie Hawn and daughter Kate Hudson resemble each other.

fastest horses generally give birth to the fastest horses. Farmers, too, understood that the plants with the highest productivity generally produce seeds from which highly productive plants grow. And everyone knew that children resemble their parents, from their appearance to their behavior to their temperament. It was enough to know that this similarity between offspring and parents exists—it was not necessary to understand how it occurs or to be able to quantify it. We call the transmission of traits from parents to their children through genetic information **inheritance** or **heritability.**

## Condition 3: Differential reproductive success

It would be nice to say that Darwin made a stunning and insightful discovery for the third of the three conditions necessary for natural selection to occur, but he didn't. Rather, he derived the third condition for natural selection from three fairly simple observations. First, more organisms are born than can survive. Second, organisms are continually struggling for existence. Lastly, some organisms are more likely than others to survive and reproduce. In a world of scarce, limited resources, finding food or shelter is a zero-sum game: if one organism is feasting, another is more likely to be starving.

The tiniest dog in a litter has reduced differential reproductive success. Its more robust siblings prevent access to the food it needs to grow and thrive.

**FIGURE 8-18 The third condition for natural selection is differential reproductive success.**

**EVOLUTION BY NATURAL SELECTION: A SUMMARY**

Natural selection causes evolution as individuals with one version of a heritable trait have greater reproductive success than individuals with a different version of the trait.

**1 VARIATION FOR A TRAIT**
Different versions of a trait are present within a population.

**+**

**2 HERITABILITY**
The different versions of a trait may be passed from parents to offspring.

**+**

**3 DIFFERENTIAL REPRODUCTIVE SUCCESS**
Individuals with the version of a trait most suited to reproduction in their environment generally leave more offspring than individuals with other versions of the trait.

*NATURAL SELECTION*
When these three conditions are satisfied, the population's allele frequencies change and, consequently, evolution by natural selection occurs.

**FIGURE 8-19 Agents of evolutionary change: natural selection.**

This three-part observation led Darwin to his third condition for natural selection, which goes by the name **differential reproductive success:** from all the variation existing in a population, individuals with traits most suited to reproduction in their environment generally leave more offspring than do individuals with other traits (**FIGURE 8-18**). For example, in the fruit fly food experiments, we saw that the flies that inherit the ability to pack on fat when food is available end up leaving more offspring than those inheriting a poor ability to pad their little fruit fly frames with fat deposits. The portly fruit flies have greater reproductive success than other individuals in the population.

That's it. Natural selection—certainly one of the most influential and far-reaching ideas in the history of science—occurs when three basic conditions are met (**FIGURE 8-19**):

1. Variation for a trait

2. Heritability of that trait

3. Differential reproductive success based on that trait

When these three conditions are satisfied, evolution by natural selection is occurring. It's nothing more and nothing less. Over time, the traits that lead some organisms to have greater reproductive success than others will increase in frequency in a population, while traits that reduce reproductive success will fade away.

When it comes to some traits, the reason they confer greater reproductive success is that they make the individual more attractive to the opposite sex. Such traits—including the brightly colored tail feathers of male peacocks, the large antlers of male red deer, and a variety of other "ornaments" that increase an individual's status or appeal—increase in frequency because they satisfy the three conditions for natural selection. This natural selection for mating success is called **sexual selection.**

**Q** **Most agricultural pests evolve resistance to pesticides. How does this happen?** Another way of looking at natural selection is to focus not on the winners (the individuals who are producing more offspring), but on the losers. Natural selection can be viewed as the elimination of some heritable traits from a population. If you carry a trait that makes you a slower-running rabbit, for example, you are more likely to be eaten by a fox (**FIGURE 8-20**). If running speed is a heritable trait (and it is), the next generation in a population contains fewer slow rabbits. Over time, the population is changed by natural selection. It evolves.

**1** **VARIATION FOR A TRAIT**
Running speed in rabbits can vary from one individual to the next.

Speed →

**2** **HERITABILITY**
The trait of running speed is passed on from parents to their offspring.

**3** **DIFFERENTIAL REPRODUCTIVE SUCCESS**
In a population, rabbits with slower running speeds are eaten by the fox, and their traits are not passed on to the next generation.

**FIGURE 8-20 Removing the losers.** Natural selection can be thought of as the elimination from a population of traits that confer poor reproductive success.

One of Darwin's contemporaries, Thomas Huxley, supposedly cursed himself when he first read *The Origin of Species,* saying that he couldn't believe he didn't figure it out on his own. Each of the three basic conditions is indeed simple and obvious. The brilliant deduction,though, was to put the three together and appreciate the consequences.

**TAKE-HOME MESSAGE 8·10**

Natural selection is a mechanism of evolution that occurs when there is heritable variation for a trait, and individuals with one version of the trait have greater reproductive success than do individuals with a different version of the trait. Natural selection can also be thought of as the elimination of alleles that reduce the reproductive rate of individuals carrying those alleles, relative to the reproductive rate of individuals who do not.

# 8·11
## A trait does not decrease in frequency simply because it is recessive.

Do recessive traits gradually become less common in a population? In the early 1900s, there was much discussion among biologists about this question. They wondered: "If the allele for brown eyes is dominant over the allele for blue eyes, why doesn't a population eventually become all brown-eyed?"

R. C. Punnett (originator of the Punnett square, described in Chapter 7) posed this question to G. H. Hardy, a mathematician with whom Punnett played cricket. Hardy replied the next day, and published his answer in *Science* in 1908. It turns out that Wilhelm Weinberg answered the question independently six months earlier. But because he published his findings in an obscure journal, scientists didn't appreciate his results for another 35 years.

Their result underlies all evolutionary genetics and is referred to as the Hardy-Weinberg Law. It reveals several things, including that the answer to the question of whether recessive traits become rarer in a population is *no*. A trait does not decrease in frequency simply because it is recessive. Here's how Hardy and Weinberg demonstrated it.

First, we refer to the frequency of the dominant allele in the population, *A,* as "*p*" and the frequency of the recessive allele, *a,* as "*q*." Since every allele in the population has to be either *A* or *a*, we can say that $p + q = 1$. The percentage of the two alleles must total 100%. If we know the frequency of either allele in the population, we can subtract it from 1 to calculate the frequency of the other allele.

Next, we can predict how common each genotype in the population will be (**FIGURE 8-21**). The frequency of *AA* is just the probability that an individual gets two copies of allele *A*, which is $p \times p$, or $p^2$. Applying the same math, the frequency of *aa* individuals in the population is simply $q \times q$, or $q^2$. Predicting the frequency of heterozygous individuals, *Aa*, is slightly more complicated. Because the dominant allele may come from either the mother or the father, the frequency of the *Aa* genotype is $2 \times p \times q$, or $2pq$.

Consider an example. Suppose a population of 1,000 kangaroo rats has the following phenotype and genotype frequencies: 16 are dark brown (*BB*), 222 are spotted (*Bb*), and 762 are light brown (*bb*). The trait shows incomplete dominance, with the allele for dark brown color, *B,* dominant over the allele for light brown color, *b,* and the

heterozygote having a spotted phenotype. Because every individual in the population has two alleles for the coat-color gene (one from each parent), there are twice as many alleles as members of the population. So a population of 1,000 kangaroo rats has 2,000 alleles of the gene for coat color.

Now let's look at the allele frequencies. Each *BB* individual has two copies of *B,* and each *Bb* individual has one copy:

Frequency of $B = (2 \times 16 + 222)/2{,}000 = 0.127 = p$

Frequency of $b = (2 \times 762 + 222)/2{,}000 = 0.873 = q$

Since we know the allele frequencies, *p* (0.127) and *q* (0.873), we can determine the genotype frequencies of the offspring produced in this population. According to the equations above, they should be:

Frequency of $BB = p^2 = (0.127)^2 = 0.016$

Frequency of $Bb = 2pq = 2(0.127)(0.873) = 0.222$

Frequency of $bb = q^2 = (0.873)^2 = 0.762$.

If 1,000 kangaroo rats are produced, we expect to see genotype frequencies that are the same as in the parent generation:

16 *BB*, 222 *Bb*, and 762 *bb*. And from these genotype frequencies, we expect the following allele frequencies among the offspring they produce:

Frequency of $B = (2 \times 16 + 222)/2{,}000 = 0.127 = p$

Frequency of $b = (2 \times 762 + 222)/2{,}000 = 0.873 = q$

Notice that the frequencies are unchanged. The allele frequencies 0.127 and 0.873 will always produce the same genotype frequencies—0.16, 0.222, and 0.762—which, in turn, will always have the same allele frequencies. Put another way: the recessive allele *doesn't* decrease over time! It doesn't change at all.

This is true as long as individuals are not dying off specifically because they carry the recessive allele (or because they carry the dominant), in which case the allele frequencies would be changing due to natural selection. It also holds true only if mutations aren't changing the allele frequencies, and if migration or genetic drift isn't altering the allele frequencies. In each of these exceptions, the allele

# HARDY-WEINBERG EQUILIBRIUM

If a population is in Hardy-Weinberg equilibrium, and the allele frequencies of the population are known, predictions can be made about the genotypes of the offspring that the population produces.

**PARENT GENERATION** (Population of 1,000 kangaroo rats)

|  | Dark brown kangaroo rat | Spotted kangaroo rat | Light brown kangaroo rat |
|---|---|---|---|
| Genotype | BB | Bb | bb |
| # of individuals in population | 16 | 222 | 762 |

**ALLELE FREQUENCIES**

B  $p = (2 \times 16 + 222) / 2{,}000 = \textbf{0.127}$
  Each Bb individual has one copy of B.
  Each BB individual has two copies of B.

b  $q = (2 \times 762 + 222) / 2{,}000 = \textbf{0.873}$
  Each Bb individual has one copy of b.
  Each bb individual has two copies of b.

**REPRODUCTION** (We assume random mating)

$p = 0.127$   $q = 0.873$

| | B $p = 0.127$ | b $q = 0.873$ |
|---|---|---|
| B $p = 0.127$ | $p^2 = 0.016$ | $pq = 0.111$ |
| b $q = 0.873$ | $pq = 0.111$ | $q^2 = 0.762$ |

**GENOTYPE FREQUENCIES**

BB  $p^2 = 0.016$
Bb  $2pq = 0.222$
bb  $q^2 = 0.762$

**NEXT GENERATION** (If 1,000 kangaroo rat offspring are produced)

|  | Dark brown kangaroo rat | Spotted kangaroo rat | Light brown kangaroo rat |
|---|---|---|---|
| Genotype | BB | Bb | bb |
| # of individuals in population | 16 | 222 | 762 |

 *As long as there is random mating and no evolution, the frequencies of recessive alleles and dominant alleles do not change over time.*

**FIGURE 8-21 Recessive alleles don't necessarily disappear from populations.**

frequencies would be changing through evolution. The Hardy-Weinberg conclusion also assumes random mating, in which the alleles are randomly coming together in all possible genotypes.

As long as these assumptions hold true—that is, random mating and no evolution—allele frequencies will not change over time, and the Hardy-Weinberg equations allow us to predict the genotype frequencies we should see.

What if we examine a population and find that the genotype frequencies we observe are *not* those predicted by the Hardy-Weinberg equations? There may be, for example, fewer heterozygotes than predicted. If this occurs, we say that the population is not in *Hardy-Weinberg equilibrium,* and we know that either evolution is occurring, or non-random mating, or both. If there are fewer heterozygotes than we expected, for example, it may be that more heterozygotes are dying than homozygotes, for some reason—perhaps predation. Or maybe individuals are preferentially mating with individuals having a similar phenotype. In either case, our calculations help us better understand the forces influencing the population and suggest further lines of investigation.

## TAKE-HOME MESSAGE 8·11

If we know the frequency of each allele in a population, we can predict the genotypes and phenotypes we should see in that population. If the phenotypic frequencies in a population are not those predicted from the allele frequencies, the population is not in Hardy-Weinberg equilibrium, because an assumption has been violated. Either non-random mating or evolution is occurring. But as long as the Hardy-Weinberg assumptions are not violated, recessive alleles and dominant alleles do not change their frequencies over time.

## 8·12–8·17 Through natural selection, populations of organisms can become adapted to their environments.

*The polar bear is well adapted to the snowy environment of the higher latitudes.*

# 8·12 Traits causing some individuals to have more offspring than others become more prevalent in the population.

"Survival of the fittest." This is perhaps one of the most famous but misunderstood phrases. Taken literally, it seems to be circular: those organisms that survive *must* be the fittest. If being fit is defined as the ability to survive. But we'll see that in evolution, the word *fitness* is not defined by an organism's ability to survive or its physical strength or its health. Rather, fitness has everything to do with an organism's reproductive success.

Here's an interesting side note: the phrase "survival of the fittest" was coined not by Darwin but by Herbert Spencer, an influential sociologist and philosopher. Moreover, the phrase did not appear in *The Origin of Species* when Darwin first published it. It wasn't until the fifth edition, 10 years later, that he used the phrase in describing natural selection.

Before we see how fitness affects natural selection and a population's adaptation to its environment, let's define fitness. **Fitness** is a measure of the relative amount of reproduction of an individual with a particular phenotype, compared with the reproductive output of individuals of the same species with alternative phenotypes.

Suppose there are two fruit flies. One fly carries the genes for a version of a trait that allows it to survive a long time without food. The other has the genes for a different version of the trait that allows it to survive only a short while without food. Which fly has the greater fitness? If the environment is one in which there are long periods of time without food, such as in the experiment described at the beginning of the chapter, the fly that can live a long time without food is likely to produce more offspring than the other fly, and so over the course of its life has greater fitness. The alleles carried by an individual with high fitness will increase their proportion in a population over time, and the population will evolve.

There are three important elements to an organism's fitness.

**1. An individual's fitness is measured relative to other genotypes or phenotypes in the population.** Those traits that confer on an individual the highest fitness will generally increase in frequency in a population, and their increase will always come at the expense of alternative traits that confer lower fitness.

**2. Fitness depends on the specific environment in which the organism lives.** The fitness value of having one trait versus another depends on the environment. A sand-colored mouse living in a beach habitat will be more fit than a chocolate-colored mouse. But that same sand-colored mouse will practically call out to potential predators if it lives in the darker brush away from the beach. An organism's fitness, although genetically based,

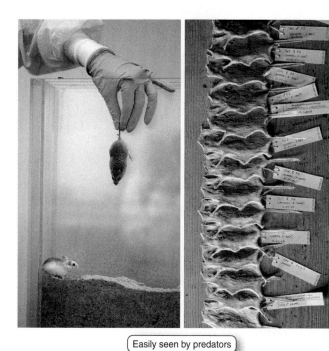

Easily seen by predators

In a lighter habitat, the allele for light-colored fur is fitter and increases in frequency:

In a darker habitat, the allele for dark-colored fur is fitter and increases in frequency:

**FIGURE 8-22** An organism's fitness depends on the environment in which it lives.

is not fixed in stone and unchanging—it can change over time and across habitats (**FIGURE 8-22**).

**3. Fitness depends on an organism's reproductive success compared with other organisms in the population.** If you carry an allele that gives you the trait of surviving for 200 years, but that also causes you to be sterile, your fitness is zero; that allele will never be passed down to future generations, and its frequency in the population will soon be zero. On the other hand, if you inherit an allele that gives you a trait that causes you to die at half the age of everyone else, but also causes you to have twice as many offspring as the average individual, your fitness is increased. It is reproductive success that is all-important in determining whether particular traits increase in frequency in a population.

"Survival of the fittest" is a misleading phrase because it is the individuals with the greatest *reproductive output* that are the most fit in any population. It becomes a more meaningful phrase if we consider it as a description of the fact that those alleles that increase an individual's fitness will "survive" in a population more than those that decrease an individual's fitness.

**Q** "Survival of the fittest" is a misnomer. Why?

---

**TAKE-HOME MESSAGE 8·12**

Fitness is a measure of the relative amount of reproduction by an individual with a particular phenotype, compared with the reproductive output of individuals with alternative phenotypes. An individual's fitness can vary, depending on the environment in which it lives.

---

# 8·13 Organisms in a population can become better matched to their environment through natural selection.

If you put a group of humans on the moon, will they flourish? If you take a shark from the ocean and put it in your swimming pool, will it survive? In both cases, the answer is *no.* Organisms are rarely successful when put into novel environments. And the stranger the new environment, the less likely it is that the transplanted organism will survive. Why is that?

As Darwin noted over and over during his travels, the organisms that possess traits that allow them to better exploit the environment in which they live will tend to produce more

offspring than the organisms with alternative traits. With passing generations, a population will be made up of more and more of these fitter organisms. And, as a consequence, organisms will tend to be increasingly well matched or adapted to their environment.

**Adaptation** refers both to the process by which organisms become better matched to their environment *and* to the specific features that make an organism more fit. Examples of adaptations abound. Bats have an extremely accurate type of hearing (called echolocation) for navigating and finding

**FIGURE 8-23 Adaptations increase fitness.** Quills are an adaptation that reduces (but doesn't eliminate) the risk of predation for porcupines.

food, even in complete darkness. Porcupine quills make porcupines almost impervious to predation (**FIGURE 8-23**). Mosquitoes produce strong chemicals that prevent blood from clotting, so that they can extract blood from other animals.

**TAKE-HOME MESSAGE 8·13**

Adaptation, which refers to the process by which organisms become better matched to their environment and to the specific traits that make an organism more fit, occurs as a result of natural selection.

# 8·14 Natural selection does not lead to perfect organisms.

In Lewis Carroll's *Through the Looking-Glass,* the Red Queen tells Alice that "in this place it takes all the running you can do, to keep in the same place." She might have been speaking about the process of evolution by natural selection. After all, if the least fit individuals are continuously weeded out of the population, we might logically conclude that, eventually, fitness will reach a maximum, and all organisms in all populations will be perfectly adapted to their environment. But this never happens. That is where the Red Queen's wisdom comes in.

Consider one of the many clearly documented cases of evolution in nature: beak size of Galápagos finches. Over the course of a multidecade study, biologists Rosemary and Peter Grant closely monitored the average size of the finches' beaks. They found that the average beak size within a population fluctuated according to the food supply. During dry years—when the finches had to eat large, hard seeds—bigger, stronger

beaks became the norm. During wet years, smaller-beaked birds were more successful, since there was a surplus of small, soft seeds.

> ❝I was taught that the human brain was the crowning glory of evolution so far, but I think it's a very poor scheme for survival. ❞
> —KURT VONNEGUT, American writer

The ever-changing "average" finch beak illustrates that adaptation does not simply march toward some optimal endpoint (**FIGURE 8-24**). Evolution in general, and natural selection specifically, does not guide organisms toward "betterness" or perfection. Natural selection is simply a process by which, in each generation, the alleles that cause organisms to have the traits that make them most fit in that environment

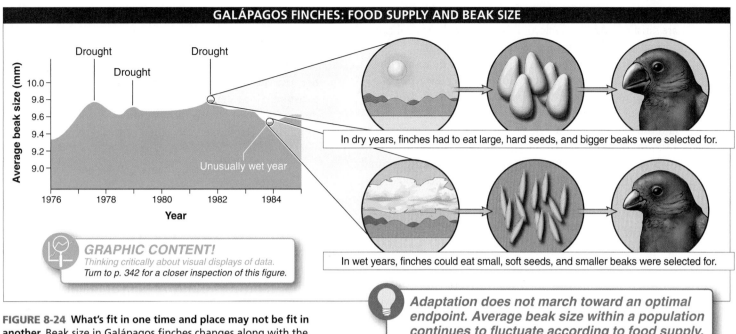

GALÁPAGOS FINCHES: FOOD SUPPLY AND BEAK SIZE

In dry years, finches had to eat large, hard seeds, and bigger beaks were selected for.

In wet years, finches could eat small, soft seeds, and smaller beaks were selected for.

*GRAPHIC CONTENT!*
*Thinking critically about visual displays of data.*
*Turn to p. 342 for a closer inspection of this figure.*

**FIGURE 8-24 What's fit in one time and place may not be fit in another.** Beak size in Galápagos finches changes along with the average rainfall.

*Adaptation does not march toward an optimal endpoint. Average beak size within a population continues to fluctuate according to food supply.*

---

tend to increase in frequency. If the environment changes, the alleles that are most favored may change, too.

**Q** In the next to last paragraph of *The Origin of Species,* Darwin wrote: "as natural selection works solely by and for the good of each being, all corporeal [physical] and mental endowments will tend to progress towards perfection." In this passage, he overlooks several different factors that prevent populations from progressing inevitably toward perfection:

**Why doesn't natural selection lead to perfect organisms?**

1. Environments change quickly. Natural selection may be too slow to adapt the organisms in a population to such a constantly moving target.

2. Variation is needed as the raw material of selection—remember, it is the first necessary condition for

natural selection to occur. If a mutation creating a new, "perfect" version of a gene never occurs, the individuals within a population will never be perfectly adapted.

3. There may be different alleles for a trait, each causing individuals to have the same fitness. In this case, each allele represents an equally fit "solution" to the environmental challenges.

**TAKE-HOME MESSAGE 8·14**

Natural selection does not lead to organisms perfectly adapted to their environment, because (1) environments can change more quickly than natural selection can adapt organisms; (2) mutation does not produce all possible alleles; (3) there is not always a single, optimum adaptation for a given environment.

---

# 8·15 Artificial selection is a special case of natural selection.

In practice, plant and animal breeders understood natural selection before Darwin did; they just didn't know that they understood it. Farmers bred crops for maximum yield, and dog, horse, and pigeon fanciers selectively bred the animals with their favorite traits to produce more and more of

the offspring with more and more exaggerated versions of those traits.

The process used by animal breeders and farmers is called **artificial selection,** but the underlying genetic process does not differ from natural selection because the three conditions are satisfied. Artificial selection is considered a special case of natural selection, however, because the differential reproductive success is being determined by humans rather than by nature. Apple growers, for example, use artificial selection to produce the wide variety available. What is important is that it is still differential reproductive success, and the results are no different. It was a stroke of genius for Darwin to recognize that the same process farmers were using to develop new and better crop varieties is occurring naturally in every population on earth, and always has been.

# 8·16 Natural selection can change the traits in a population in several ways.

Certain traits are easily categorized—blue eyes and brown eyes, white tiger fur and orange tiger fur, cleft chin and smooth chin. Other traits, such as height, are influenced by many genes and environmental factors, so that a continuous range of phenotypes occurs (**FIGURE 8-25**). Whether a given trait is influenced by one gene or by a complex interaction of many genes and the environment, it can be subject to natural selection and be changed in any of several ways.

**TRAIT CATEGORIZATION**

**SPECIFIC CATEGORY**
Some traits, such as chin shape, can be easily placed into one category or the other.

**or**

Cleft chin    Smooth chin

**CONTINUOUS RANGE**
Other traits, such as height, may fall into a wide range of values.

Height (feet)

*Traits influenced by a single gene or by multiple genes interacting with the environment can be changed in various ways by natural selection.*

**FIGURE 8-25 Some traits fall into clear categories, others range continuously.**

**FIGURE 8-26 More milk?** Cows have been selected for their ability to produce more and more milk.

**Directional selection** In **directional selection,** individuals with one extreme of the range of variation in the population have higher fitness. Milk production in cows is an example. There is a lot of variation in milk production from cow to cow. As you might expect, farmers select for breeding those cows with the highest milk production and have done so for many decades. The result of such selection is not surprising: between 1920 and 1945, average milk production increased by about 50% in the United States (**FIGURE 8-26**).

Those at the other end of the range, the cows that produce the smallest amounts of milk, have reduced fitness, since the farmers do not allow them to reproduce, limiting the number of "less milk" alleles in the next

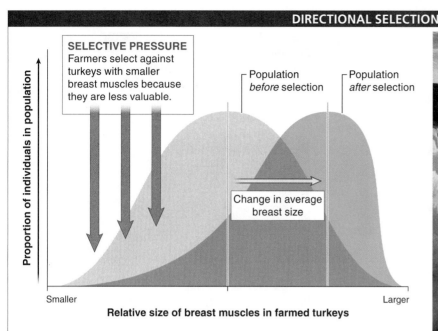

**SELECTIVE PRESSURE** Farmers select against turkeys with smaller breast muscles because they are less valuable.

Population *before* selection

Population *after* selection

Change in average breast size

Proportion of individuals in population

Smaller

Larger

**Relative size of breast muscles in farmed turkeys**

**FIGURE 8-27 Patterns of natural selection: directional selection.**

*Turkeys on poultry farms can no longer mate naturally! Their breast muscles have become so large that artificial insemination is required.*

generation. In subsequent generations, the average value for the milk-production trait increases. Note, however, that the variation for the trait in the cow population decreases a bit, too, because the alleles contributing to one of the phenotypic extremes are eventually removed from the population.

Hundreds of experiments in nature and laboratories and on farms demonstrate the power of directional selection. In fact, it is one of the marvels of laboratory selection and animal breeding that nearly any trait chosen to be exaggerated via directional selection responds dramatically—sometimes with absurd results (**FIGURE 8-27**). Turkeys, for instance, have been selected for increased size of their breast muscles, which makes them more profitable for farmers. Directional selection has been so successful that the birds' breast muscles are now so large that it has become impossible for them to mate. All turkeys on large-scale, industrial poultry farms must now reproduce through artificial insemination!

**Q**

Turkeys on poultry farms have such large breast muscles that they can't get close enough to each other to mate. How can such a trait evolve?

**Stabilizing selection** How much did you weigh at birth? Was it more than 10 pounds? More than 11? Was it close to the 22½ pounds that Christina Samane's child weighed

when born in South Africa in 1982? Or was it 5 pounds or less? Unlike directional selection, for which there is increased fitness at one extreme and reduced fitness at the other, **stabilizing selection** is said to occur when individuals with intermediate phenotypes are the most fit. The death rate among babies, for example, is lowest between 7 and 8 pounds, increasing for both lighter and heavier babies (**FIGURE 8-28**). This outcome has kept the average weight of a baby constant over many generations; but at the same time, the variation in birth weight has decreased as stabilizing selection has reduced the frequencies of genes associated with high and low birth weights.

Modern technologies, including Caesarean deliveries and premature-birth wards, allow many babies to live who would not have survived without such technology. This has the unintended consequence of reducing the selection against any alleles causing those "extreme" traits, reducing the rate of their removal from the population.

**Disruptive selection** There is a third kind of natural selection, in which individuals with extreme phenotypes experience the highest fitness, and those with intermediate phenotypes have the lowest. Although examples of this type of

**SELECTIVE PRESSURE** Babies with low birth weights are more vulnerable to complications.

Population *before* selection

Population *after* selection

**SELECTIVE PRESSURE** Babies with high birth weights are more vulnerable to complications.

Proportion of individuals in population

Birth weight of human baby (lbs.)

**FIGURE 8-28 Patterns of natural selection: stabilizing selection.**

selection, called **disruptive selection,** are rare in nature, the results are not surprising. Among some species of fish—the coho salmon, for instance—only the largest males acquire good territories. They generally enjoy relatively high reproductive success. While the intermediate-size fish regularly get run out of the good territories, some tiny males are able to sneak in and fertilize eggs before the territory owner detects their presence. Consequently, we see an increase in the frequency of small and large fish with a reduction in the frequency of medium-size fish (**FIGURE 8-29**).

## TAKE-HOME MESSAGE 8·16

Acting on multigene traits for which populations show a large range of phenotypes, natural selection can change populations in several ways. These include directional selection, in which the average value for the trait increases or decreases; stabilizing selection, in which the average value of a trait remains the same while extreme versions are selected against; and disruptive selection, in which individuals with extreme phenotypes have the highest fitness.

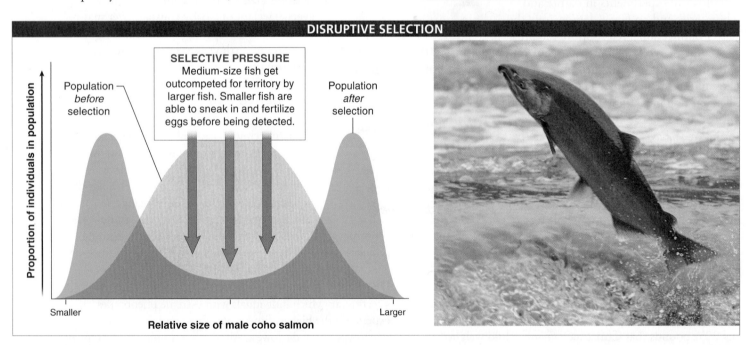

Population *before* selection

**SELECTIVE PRESSURE** Medium-size fish get outcompeted for territory by larger fish. Smaller fish are able to sneak in and fertilize eggs before being detected.

Population *after* selection

Proportion of individuals in population

Smaller                                                        Larger

Relative size of male coho salmon

**FIGURE 8-29 Patterns of natural selection: disruptive selection.**

# 8·17 Natural selection can cause the evolution of complex traits and behaviors.

We have seen that natural selection can change allele frequencies and modify the frequency with which simple traits, such as fur color or turkey-breast size, appear in a population. But what about complex traits, such as behaviors, that involve numerous physiological and neurological systems? For instance, can natural selection improve maze-running ability in rats?

The short answer is *yes*. Remember, evolution by natural selection is occurring, changing the allele frequencies for traits, whenever (1) there is variation for the trait, (2) that variation is heritable, and (3) there is differential reproductive success based on that trait. These conditions can easily be satisfied for complex traits, including behaviors.

In 1954, to address this question, William Thompson trained a group of rats to run through a maze for a food reward (**FIGURE 8-30**). There was a huge amount of variation in the rats' abilities: some rats learned much more quickly than others how to run the maze. Thompson then selectively bred the fast learners with each other and the slow learners with each other. Over several generations, he developed two separate populations: rats descended from a line of fast

maze-learners and rats descended from a line of slow maze-learners. After only six generations, the slow learners made twice as many errors as the fast learners before mastering the maze, while the fast learners were adept at solving complex mazes that would give many humans some difficulty. Fifty-five years later, it still is unclear which particular genes are responsible for maze-running behavior, yet the selection experiment still demonstrates a strong genetic component to the behavior.

> « All things must change
> To something new, to something strange. »
> —HENRY WADSWORTH LONGFELLOW

Natural selection can also produce complex traits in unexpected, roundabout ways. One vexing case involves the question of how natural selection could produce an organ as complex as a fly wing, when 1% or 2% of a wing—that is, an incomplete structure—doesn't help an insect to fly. In other words, while it is clear how natural selection can preserve and increase the frequencies of fitness-increasing traits in populations, how do these soon-to-be-useful traits increase

## NATURAL SELECTION FOR COMPLEX TRAITS

**SLOW LEARNERS**
Rats making most errors in maze are selected and bred with each other.

After 6 generations of selection, the rats selected for fast learning made only half as many errors as the slow learners.

**FAST LEARNERS**
Rats making fewest errors in maze are selected and bred with each other.

*Average number of errors committed* (y-axis: 100, 140, 180, 220, 260, 300)

*Generation* (x-axis: 1, 2, 3, 4, 5, 6)

*Natural selection can produce and alter complex traits, including behaviors.*

**FIGURE 8-30 "Not too complex . . ."** Natural selection can produce and alter complex traits, including behaviors.

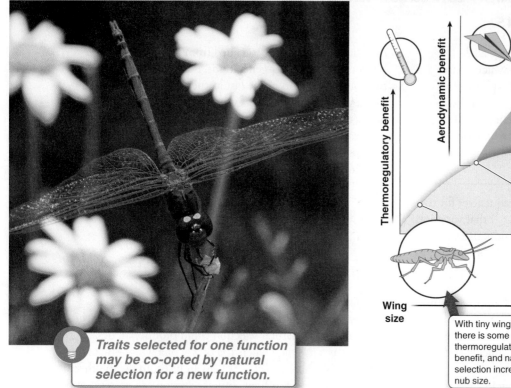

Traits selected for one function may be co-opted by natural selection for a new function.

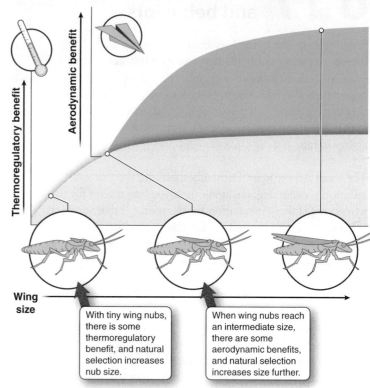

**Wing size**

With tiny wing nubs, there is some thermoregulatory benefit, and natural selection increases nub size.

When wing nubs reach an intermediate size, there are some aerodynamic benefits, and natural selection increases size further.

**FIGURE 8-31** How could natural selection produce an insect wing, when the primitive wing nub had no aerodynamic benefits?

during the early stages if they don't increase the organism's fitness (**FIGURE 8-31**)?

**How can a wing evolve if 1% of a wing doesn't help an organism fly or glide at all?**

**Q** The key to answering this question is that 1% of a wing doesn't actually need to function as a wing to increase an individual's fitness. Often, structures are enhanced or elaborated on by natural selection because they enhance fitness by serving some other purpose. In the case of little nubs or "almost-wings" on an insect, experiments using models of insects demonstrated that, as expected, the nubs confer no benefit at all when it comes to flying. (They don't even help flies keep their orientation during a "controlled fall.") The incipient wings *do* help the insects address a completely different problem, though. They allow much more efficient temperature control, so that an insect can gain heat from the environment when the insect is cold and dissipate heat when the insect is hot. Experiments on heat-control efficiency, in fact, show that as small nubs become more and more pronounced, they are more and more effective, probably conferring increased fitness

on the individual in which they occur—but only up to a point.

Eventually, the thermoregulatory benefit stops increasing, even if the nub length continues to increase. But it is right around this point that the proto-wing starts to confer some aerodynamic benefits (see Figure 8-31). Consequently, natural selection may continue to increase the length of this "almost-wing," but now the fitness increase is due to a wholly different effect. Such functional shifts explain the evolution of numerous complex structures that we see today, and these shifts may be common in the evolutionary process.

---

**TAKE-HOME MESSAGE 8·17**

Natural selection can change allele frequencies for genes involved in complex physiological processes and behaviors. Sometimes a trait that has been selected for one function is later modified to serve a completely different function.

---

# 8·18–8·22 The evidence for evolution is overwhelming.

The pygmy seahorse (Hippocampus bargibanti) is so well camouflaged by the coral it lives near that the first specimens were discovered only after coral had been collected and put in an aquarium.

# 8·18
## The fossil record documents the process of natural selection.

> It is indeed remarkable that this theory [evolution] has been progressively accepted by researchers, following a series of discoveries in various fields of knowledge. The convergence, neither sought nor fabricated, of the results of work that was conducted independently is in itself a significant argument in favor of this theory.
>
> —POPE JOHN PAUL II, 1996

In the some 150 years since Darwin first published *The Origin of Species,* thousands of studies involving his theory of evolution by natural selection have been conducted, both in the laboratory and in natural habitats. A wide range of modern methodologies has also been developed, all contributing to a much deeper understanding of the process of evolution. This ongoing accumulation of evidence overwhelmingly supports the basic premise that Darwin put forward, while filling in many of the gaps that frustrated him.

In the remainder of the chapter we review the five primary lines of evidence demonstrating the occurrence of evolution:

1. **The fossil record**—physical evidence of organisms that lived in the past

2. **Biogeography**—patterns in the geographic distribution of living organisms

3. **Comparative anatomy and embryology**—growth, development, and body structures of major groups of organisms

4. **Molecular biology**—examination of life at the level of individual molecules

5. **Laboratory and field studies**—implementation of the scientific method to observe and study evolutionary mechanisms

The first of the five lines of evidence is the fossil record. Although it has been central to much documentation of the occurrence of evolution, it is a very incomplete record. After all, the soft parts of an organism tend to decay rapidly and completely after death. And there are only a few environments (such as tree resin, tar pits, and the bottom of deep lakes) in which the forces of decomposition are so reduced that an organism's hard parts, including bones, teeth, and shells, can be preserved for thousands or even millions of years. These remains, called **fossils,** can be used to reconstruct what organisms must have looked like long ago. Such reconstructions

**FIGURE 8-32 Evidence for evolution: the fossil record.** Fossils can be used to reconstruct the appearance of organisms that lived long ago.

often provide a clear record of evolutionary change (**FIGURE 8-32**).

The use of **radiometric dating** helps in painting a clearer picture of organisms' evolutionary history by telling us the age of the rock in which a fossil is found (**FIGURE 8-33**). In Darwin's time, it was assumed that the deeper in the earth a fossil was found, the older it was. Radiometric dating goes a step further, making it possible to determine not just the *relative* age of fossils, but also their *absolute* age. This is accomplished by evaluating the amounts of certain radioactive isotopes present in the fossil-containing rocks. Radioactive isotopes in a rock begin breaking down into more stable compounds as soon as the rock is formed, and they do so at a constant rate. Nothing can alter this. By measuring the relative amounts of the radioactive isotope and its leftover decay product in the rock where a fossil is found, the age of the rock, and thus of the fossil, can be calculated.

Radiometric dating confirms that the earth is very old. Rocks more than 3.8 billion years old have been found on all of the earth's continents, with the oldest so far found in northwestern Canada. By using the radioactive isotope uranium-238, with a half-life of 4.5 billion years, researchers have determined that the earth is about 4.6 billion years old and that the earliest organisms appeared at least 3.5 billion years ago. Radiometric dating also makes it possible to put the fossil record in order. By dating all the fossils discovered in one locale, paleontologists can learn whether and how the organisms were related to each other and how groups of organisms changed over time.

Paleontologists must deal with the fact, however, that fossil formation is an exceedingly unlikely event, and when it does

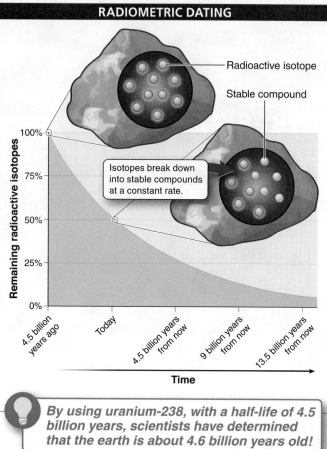

**By using uranium-238, with a half-life of 4.5 billion years, scientists have determined that the earth is about 4.6 billion years old!**

**FIGURE 8-33 How old is that fossil?** Radiometric dating using uranium-238 helps to determine the age of rocks and the fossils in them.

occur it represents only those organisms that happened to live in that particular area and also had physical structures that could leave fossils. For this reason, the fossil record is annoyingly incomplete. Entire groups of organisms have left no fossil record at all, and some others have numerous gaps. Still, fossils have been found that link all of the major groups of vertebrates.

The evolutionary history of horses is among the most well preserved in the fossil record. First appearing in North America about 55 million years ago, horses radiated around the world, with more recent fossils appearing in Eurasia and Africa. These fossils exhibit distinct adaptations to those differing environments. Later, about 1.5 million years ago, much of the horse diversity—including all North American horse species—disappeared, leaving only a single remaining genus, or group of species, called *Equus*. Because there is now only one horse genus on earth, it is tempting to imagine a simple linear path from modern horses straight back through their 55-million-year evolutionary history. But that's just not how evolution works. In reality, there have been numerous branches of horses that have split off over evolutionary time,

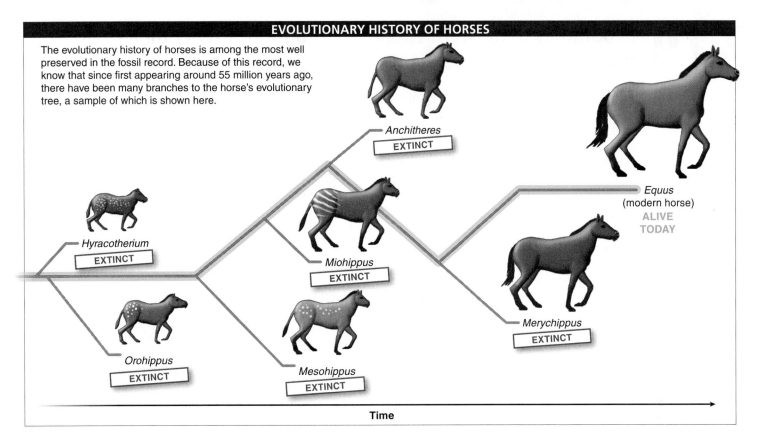

## EVOLUTIONARY HISTORY OF HORSES

The evolutionary history of horses is among the most well preserved in the fossil record. Because of this record, we know that since first appearing around 55 million years ago, there have been many branches to the horse's evolutionary tree, a sample of which is shown here.

*Anchitheres*
EXTINCT

*Hyracotherium*
EXTINCT

*Equus*
(modern horse)
ALIVE TODAY

*Miohippus*
EXTINCT

*Orohippus*
EXTINCT

*Mesohippus*
EXTINCT

*Merychippus*
EXTINCT

**Time**

**FIGURE 8-34  An evolutionary family tree.** The branching evolutionary tree of the horse.

flourished for millions of years, and only recently gone extinct. What we see living today is only a single branch of a greatly branched evolutionary tree (**FIGURE 8-34**).

The fossil record provides another valuable piece of evidence for the process of natural selection, in the form of "missing links." These are fossils that demonstrate a link between groups of species believed to have shared a common ancestor. One such "missing link" is *Tiktaalik* (**FIGURE 8-35**). First found in northern Canada and estimated to be 375 million years old, *Tiktaalik* fossils seem to represent a transitional phase between fishes and land animals. These creatures, like fishes, had gills, scales, and fins, but they also had arm-like joints in their fins and could drag their bodies across land in much the same way as a marine mammal such as a seal.

### TAKE-HOME MESSAGE 8·18

**Radiometric dating confirms that the earth is very old and allows scientists to determine the age of fossils. By analyzing fossil remains, biologists can reconstruct what organisms looked like long ago, learn how organisms were related to each other, and understand how groups of organisms evolved over time.**

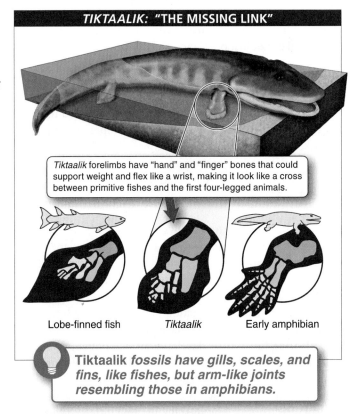

## *TIKTAALIK:* "THE MISSING LINK"

*Tiktaalik* forelimbs have "hand" and "finger" bones that could support weight and flex like a wrist, making it look like a cross between primitive fishes and the first four-legged animals.

Lobe-finned fish   *Tiktaalik*   Early amphibian

*Tiktaalik fossils have gills, scales, and fins, like fishes, but arm-like joints resembling those in amphibians.*

**FIGURE 8-35  A missing link?** *Tiktaalik* seems to be a transitional phase between fishes and land animals.

# 8·19 Geographic patterns of species distributions reflect their evolutionary histories.

The study of the distribution patterns of living organisms around the world is called **biogeography.** This is the second line of evidence that helps us see that evolution occurs and better understand the process. The patterns of biogeography noticed by Darwin and by many subsequent researchers provide strong evidence that evolutionary forces are responsible for these patterns. Species often more closely resemble other species that live less than a hundred miles away but in radically different habitats than they resemble species living thousands of miles away in nearly identical habitats. On Hawaii, for example, nearly every bird is some sort of modified honeycreeper, from seed-eating honeycreepers to curved-bill nectar-feeding honeycreepers (**FIGURE 8-36**).

When it comes to species distributions, history matters. Species were not designed from scratch to fill a particular niche. Rather, whatever arrived in a geographic region first—usually a nearby species—took up numerous different lifestyles in numerous different habitats, and the populations ultimately adapted to and evolved in each environment. In Hawaii, it seems that a finch-like descendant of the honeycreepers arrived 4–5 million years ago and rapidly evolved into a large number of diverse species. The same process occurred and continues to occur in all locales, not just on islands.

Large isolated habitats also have interesting biogeographic patterns. Australia and Madagascar are filled with unique organisms that are clearly not closely related to organisms elsewhere. In Australia, for example, marsupial species, rather than placental mammals, fill all of the usual roles. There are marsupial "wolves," marsupial "mice," marsupial "squirrels," and marsupial "anteaters" (**FIGURE 8-37**).

The marsupials of Australia physically resemble their placental counterparts for most traits, but molecular analysis shows that they are actually more closely related to one another, sharing a common marsupial ancestor. Their relatedness to each other is also revealed by similarities in their reproduction: females give birth to offspring at a relatively early stage of development, and the offspring finish their development in a pouch. The presence of marsupials in Australia does not simply mean that marsupials are better adapted than placentals to Australian habitats. When placental organisms have been transplanted to Australia they do just fine, often thriving

**BIOGEOGRAPHY: HAWAIIAN HONEYCREEPERS**

The honeycreepers of Hawaii have adapted to a wide range of habitats, yet still closely resemble a finch-like shared ancestor found nearly 2,000 miles away.

Mainland finch (probable shared ancestor)

Hawaiian Islands (enlarged relative to North America to show detail)

Finch-like ancestor arrives in Hawaii 4–5 million years ago.

NORTH AMERICA

'Akeke'e honeycreeper

**HABITAT**
Wet forests above 1,100 meters

**DIET**
Arthropods, primarily spiders and caterpillars

Maui Parrotbill honeycreeper

**HABITAT**
Northwestern slopes of Haleakalā between 1,230 and 2,370 meters

**DIET**
Insects, primarily beetle larvae and moth pupae

'I'iwi honeycreeper

**HABITAT**
Wet forests above 1,250 meters

**DIET**
Primarily nectar from a variety of flowers

**FIGURE 8-36 Evidence for evolution: biogeography.** The Hawaiian honeycreepers resemble a common ancestor from mainland North America, but all have unique features.

to the point of endangering the native species. Instead, it appears that the terrestrial placental mammals in Australia disappeared about 55 million years ago—although the reasons are not clear—giving the marsupials the chance to flourish and diversify.

**AUSTRALIAN MARSUPIALS**

Sugar glider

Numbat

Tasmanian wolf

**PLACENTAL COUNTERPARTS**

Flying squirrel

Giant anteater

Gray wolf

**FIGURE 8-37 Evidence for evolution: biogeography.** Many Australian marsupials resemble placental counterparts, though they are not closely related.

 *Though less related to each other than you are to a shrew, these marsupials and their placental counterparts have come to resemble each other as natural selection has adapted them to similar habitats.*

Biogeographic patterns such as those seen in honeycreepers and in the marsupials of Australia illustrate that evolution doesn't necessarily lead to the same "solutions" each time a particular set of environmental conditions occurs. Rather, the traits in populations that happen to be in a particular location gradually change, and those species become better adapted to the habitats they occupy.

**TAKE-HOME MESSAGE 8·19**

Observing geographic patterns of species distributions —noting similarities and differences among species living close together but in very different habitats and among species living in similar habitats but located far from one another—helps us understand the evolutionary histories of populations.

# 8·20 Comparative anatomy and embryology reveal common evolutionary origins.

If you observe any vertebrate embryo, you will see that it passes through a stage in which it has little gill pouches on the sides of the neck. It will also pass through a stage in which it has a long bony tail. This is true whether it is a human embryo or that of a turtle or a chicken or a shark. The gill pouches disappear before birth in all but the fishes. Similarly, we don't find humans with tails. Why do these features exist during an embryo's development? Such common embryological stages indicate that the organisms share a

common ancestor, from which all have been modified (**FIGURE 8-38**). Study of these developmental stages and the adult body forms of organisms provides our third line of evidence for the occurrence of evolution.

Among adult animals, several features of anatomy reveal the ghost of evolution in action. We find, for example, that many related organisms show unusual similarities that can be explained only through evolutionary relatedness. The

## HOMOLOGOUS STRUCTURES: EMBRYO STAGE

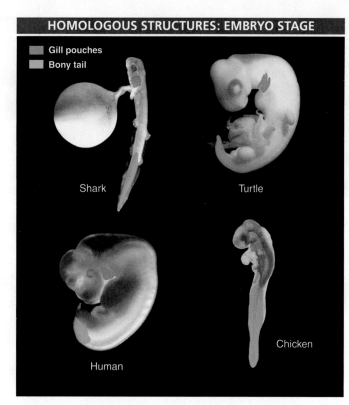

Gill pouches
Bony tail

Shark

Turtle

Human

Chicken

**FIGURE 8-38 Evidence for evolution: embryology.** Structures derived from common ancestry can be seen in embryos.

that betray the fact that they share a common ancestor. These features are called **homologous structures.**

At the extreme, homologous structures sometimes come to have little or no function at all. Such evolutionary leftovers, called **vestigial structures,** exist because they had value ancestrally. Some vestigial structures in mammals include the molars that continue to grow in vampire bats, even though these bats consume a completely liquid diet (**FIGURE 8-40**); eye sockets (with no eyes) in some populations of cave-dwelling fishes; and in whales, pelvic bones that are attached to nothing (but in nearly all other mammals serve as an important attachment point for leg bones). Even humans may have a vestigial organ—the appendix. It is greatly enlarged in our relatives the great apes, acting as host to cellulose-digesting bacteria that aid in breaking down the plants in the ape diet, but in humans it seems to serve no purpose (although recent studies suggest it provides some immune system support).

Why do all vampire bats grow teeth specialized for grinding solid food, when the bats have a completely liquid diet?

Not all organisms with adaptations that appear similar actually share recent ancestors. We see flying mammals (bats) and flying insects (locusts) (**FIGURE 8-41**). Likewise, dolphins and penguins live in similar habitats and have flippers that help them swim. In both examples, however, the analogous structures developed from different original structures. Natural selection, in a process called **convergent evolution,** may act on different starting

forelimbs of mammals are used for a variety of very different functions in bats, porpoises, horses, and humans (**FIGURE 8-39**). If each had been designed specifically for the uses necessary to that species, we would expect dramatically different designs. And yet, in each of these species, we see the same bones—modified extensively—

## HOMOLOGOUS BONE STRUCTURES

Humerus

Radius
Ulna

Phalanges
Metacarpals
Carpals

Human

Horse

Bat

Porpoise

**FIGURE 8-39 Evidence for evolution: comparative anatomy.** Homologous bone structures among some mammals.

*The similarities in the bone structure of the forelimbs of mammals demonstrate common ancestry.*

Vampire bat skull
Molars

Like their ancestors, vampire bats have molars (adapted for grinding food), even though they consume a completely liquid diet and so have no use for the molars.

FIGURE 8-40 Evidence for evolution: vestigial structures.

materials (such as a flipper or a forelimb) and modify them until they serve similar purposes—much as we saw in the marsupial and placental mammals in Figure 8-37.

### TAKE-HOME MESSAGE 8·20

Similarities in the anatomy and development of different groups of organisms and in their physical appearance can reveal common evolutionary origins.

## CONVERGENT EVOLUTION

Different starting materials come to perform the same function through convergent evolution.

FIGURE 8-41 Evidence for evolution: convergent evolution and analogous structures.

# 8·21 Molecular biology reveals that common genetic sequences link all life forms.

New technologies for deciphering and comparing the genetic code provide our fourth line of evidence that evolution occurs. From microscopic bacteria to flowering plants to insects and primates, the molecular instructions for building organisms and transmitting hereditary information are the same: four simple bases, arranged in an almost unlimited variety of sequences. All living organisms share the same genetic code.

When we examine the similarity of DNA among related individuals within a species, we find that they share a

greater proportion of their DNA than do unrelated individuals. This is not unexpected; you and your siblings got all of your DNA from the same two parents, while you and your cousins got half of your DNA from the same two grandparents. The more distantly you and another individual are related, the more your DNA differs.

We can measure the DNA similarity between two species by comparing their DNA sequences for individual genes. For example, let's look at the gene that codes for a polypeptide used to build hemoglobin. In vertebrate animals, hemoglobin is found inside red blood cells, where its function is to carry oxygen throughout the body. It is made up of two types of chains of amino acids, the alpha chain and the beta chain. In humans, the beta chain has 146 amino acids. In rhesus monkeys, this beta chain is nearly identical: of the 146 amino acids, 138 are the same as those found in human hemoglobin, and only 8 are different. In dogs, the sequence is still somewhat similar, with 114 of the same amino acids. When we look at non-mammals such as birds, we find about 101 amino acid similarities with humans. And between humans and lamprey eels (still vertebrates, but with more than 500 million years since our last common ancestor), there are 21 amino acid similarities.

The differences in the amino acid sequence of the beta hemoglobin chain (and thus the alleles that govern this structure) seem to indicate that humans have more recently shared a common ancestor with rhesus monkeys than with dogs. And that we have more recently shared an ancestor with dogs than with birds or lampreys. These findings are just as we would expect, based on estimates of evolutionary relatedness made from comparative anatomy and embryology, as well as from the fossil record. It is as if a molecular clock is ticking. The longer two species have been evolving separately, the greater the number of changes in amino acid sequences—or "ticks of the clock" (**FIGURE 8-42**).

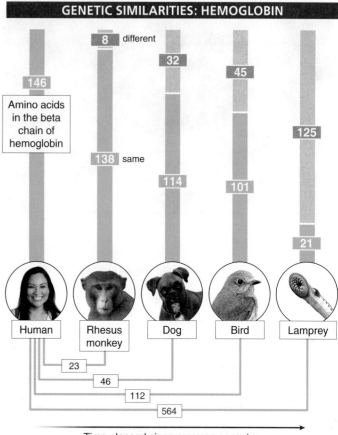

**GENETIC SIMILARITIES: HEMOGLOBIN**

Amino acids in the beta chain of hemoglobin

Time elapsed since common ancestor was last shared (millions of years)

 *The longer two species have been evolving on their own, the greater the number of genetic differences that accumulate.*

**FIGURE 8-42  An evolutionary clock?** Genetic similarities (and differences) demonstrate species relatedness.

> ### TAKE-HOME MESSAGE 8·21
>
> All living organisms share the same genetic code. The degree of similarity in the DNA of different species can reveal how closely related they are and the amount of time that has passed since they last shared a common ancestor.

# 8·22  Laboratory and field experiments enable us to watch evolution in progress.

A fifth line of evidence for the occurrence of evolution comes from multigeneration experiments and observations. People once thought that evolution was too slow a process to be observed in action. However, by choosing the right species—preferably organisms with very short life spans—and designing careful experiments, we can observe and measure evolution as it is happening.

In one clever study, researchers looked at populations of grass on golf courses—a habitat where lawn-mower blades represent a significant source of mortality. Although all the grass on golf courses was of the same species, on the putting greens it was cut very frequently, whereas on the fairways it was cut only

occasionally, and in the rough it was almost never cut at all (**FIGURE 8-43**). Over just a few years, significant changes occurred in these different grass populations. The grass plants on the greens came to be short lived, with rapid development to reproductive age and a very high seed output. For the plants in these populations, life was short and reproduction came quickly. (For those in which it did not come quickly, it did not come at all; hence their lack of representation in the population.) Plants on the fairways had slightly slower development and a reduced seed output, while those in the rough had the slowest development and the lowest seed output of all. When plants from each of the habitats were collected and grown in greenhouses under identical conditions, the dramatic differences

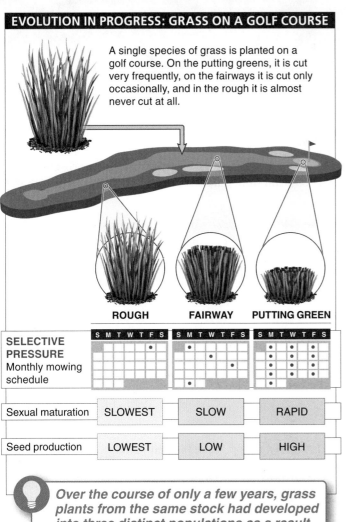

A single species of grass is planted on a golf course. On the putting greens, it is cut very frequently, on the fairways it is cut only occasionally, and in the rough it is almost never cut at all.

**ROUGH**   **FAIRWAY**   **PUTTING GREEN**

| SELECTIVE PRESSURE Monthly mowing schedule | | |
|---|---|---|

| | ROUGH | FAIRWAY | PUTTING GREEN |
|---|---|---|---|
| Sexual maturation | SLOWEST | SLOW | RAPID |
| Seed production | LOWEST | LOW | HIGH |

*Over the course of only a few years, grass plants from the same stock had developed into three distinct populations as a result of the frequency at which they were cut.*

**FIGURE 8-43 Evolution in progress: grasses.** Different mowing patterns can cause evolution in the grass on a golf course.

in growth, development, and life span remained, confirming that the frequencies of the various alleles controlling the traits of life span and reproductive output in the three populations had changed. In other words, evolution had occurred.

One particularly disturbing line of evidence for the occurrence of evolution in nature comes from the evolution of antibiotic-resistant strains of bacteria that cause illness in humans. In the 1940s, when penicillin was first used as a treatment for bacterial infections, it was uniformly effective in killing *Staphylococcus aureus*. Today, more than 90% of isolated *S. aureus* strains are resistant to penicillin (**FIGURE 8-44**). Because penicillin has become such a pervasive toxin in the environment of *Staphylococcus,* natural selection has led to an increase in the frequency of alleles that make these strains resistant to the drug. As a consequence, humans are increasingly at risk for becoming infected with *Staphylococcus* and getting diseases such as

When first used as medicine in the 1940s, penicillin was uniformly effective in killing the bacterium *Staphylococcus aureus*. Today, natural selection has led to an increase in antibiotic-resistance alleles, and humans are increasingly at risk from untreatable *Staphylococcus* infections.

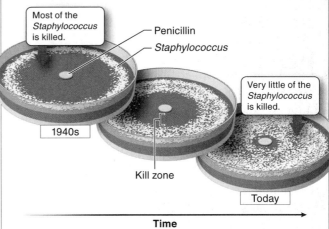

Most of the *Staphylococcus* is killed.

— Penicillin
— *Staphylococcus*

1940s

Kill zone

Very little of the *Staphylococcus* is killed.

Today

**Time**

**FIGURE 8-44 Evolution in progress: disease-causing bacteria.** Antibiotic resistance has evolved in *Staphylococcus*.

pneumonia and meningitis. In the 1960s, the antibiotics methicillin and oxacillin had nearly complete effectiveness against *Staphylococcus*. Today, though, nearly a third of staph infections are resistant to these antibiotics as well. The meaning of such unintentional natural selection "experiments" is clear and consistent: evolution is occurring all around us.

Finally, let's return to the starvation-resistant fruit flies introduced at the beginning of this chapter. They are an unambiguous demonstration that natural selection can produce dramatic changes in a population, and that it can bring about these changes very quickly. And, perhaps most important, we saw that replicating the same evolutionary process over and over again produced the same predictable results. In the laboratory, on the farm, in the doctor's office, and in deserts, streams, and forests, evolution is occurring.

Reflecting on both the process and products of evolution and natural selection, in the final paragraph of *The Origin of Species* Darwin eloquently wrote: "There is grandeur in this view of life . . . and that . . . from so simple a beginning endless forms most beautiful and most wonderful have been, and are being, evolved."

**TAKE-HOME MESSAGE 8·22**

**Replicated, controlled laboratory selection experiments and long-term field studies of natural populations allow us to watch and measure evolution as it occurs.**

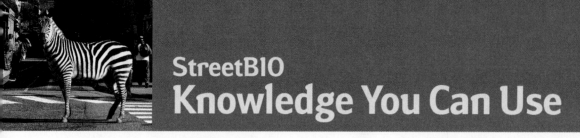
## EVOLUTION: WHAT IT IS AND WHAT IT IS NOT . . .

Misconceptions about evolution and how it occurs can arise from misunderstandings about subtleties in the process, from incomplete or erroneous accounts of evolution in schools and the media, and from purposeful attempts to interfere with the understanding of evolution.

**MISCONCEPTION: EVOLUTION IS A THEORY ABOUT THE ORIGIN OF LIFE.** *Clarification:* Although scientists are interested in (and investigating) how life began, evolutionary biology is focused on the processes that occurred (and still occur) *after* life began, including those responsible for diversification and branching of species and those adapting populations to their environments.

**MISCONCEPTION: EVOLUTION IS NOT OBSERVABLE OR TESTABLE.** *Clarification:* Evolution is both observable and testable. Long-term, controlled and replicated, laboratory experiments on many species have demonstrated evolution (see an example in Section 8-1), including speciation. Numerous studies of natural populations have also documented evolution occurring (see Section 8-14). And, as in astronomy and geology, sophisticated hypothesis testing can be done using observations in the natural world.

**MISCONCEPTION: EVOLUTION IS ONLY A THEORY AND HASN'T BEEN PROVED.** *Clarification:* This misconception stems from confusion between the colloquial definition of a theory as a "guess" or "hunch" and the scientific definition of a theory as an explanation, based on many lines of evidence, that generates and supports predictions and is tested in many ways (see Section 1-10).

**MISCONCEPTION: INDIVIDUALS EVOLVE (AND ARE "TRYING" TO ADAPT).** *Clarification:* Natural selection can lead to populations of organisms adapted to their environment (see Section 8-13). But the process does not involve "trying" to adapt—it is simply a consequence of some individuals passing on their genes at a higher rate than individuals carrying alternative versions of those genes, and it is the population rather than any one individual that changes over time (see Section 8-12).

**MISCONCEPTION: EVOLUTION IS GOAL-ORIENTED AND LEADS TO PROGRESS AND "OPTIMAL" SOLUTIONS.** *Clarification:* Alleles that make it impossible for the individuals carrying them to live in a particular environment are weeded out of the population. But this is not the same as creating perfect or optimal organisms. Rather, the traits and features of organisms that increase in frequency within a population are simply those that cause the organisms carrying them to leave more descendants than are left by individuals with alternative traits (see Section 8-10). As the environment changes, so do the traits that are most fit. Several factors can prevent populations from becoming better adapted to a particular environment (see Section 8-14).

**MISCONCEPTION: THERE ARE NO TRANSITIONAL FOSSILS; THE GAPS IN THE FOSSIL RECORD DISPROVE EVOLUTION.** *Clarification:* Because fossils are formed only under a narrow range of environmental conditions, biologists do not expect to find all transitional forms, so the gaps do not disprove evolution. Numerous transitional fossils *have* been found, including those between dinosaurs and modern birds, and those between whales and their terrestrial mammalian ancestors (see Section 8-18).

**MISCONCEPTION: THE THEORY OF EVOLUTION SAYS THAT LIFE ORIGINATED, AND EVOLUTION PROCEEDS, RANDOMLY, BY CHANCE.** *Clarification:* Although chance is one element of evolution—there is a random element to the generation of mutations, for example—many non-random factors are central to how evolution proceeds. For example, the differential reproductive success of some variants—such as bacteria with antibiotic-resistance genes—that leads to natural selection is not random (see Section 8-13).

**MISCONCEPTION: MOST BIOLOGISTS HAVE REJECTED EVOLUTION.** *Clarification:* All modern theories of science are continually tested, verified, and modified as new and deeper understandings of the natural world emerge. That is how science works. The presidents of the National Academy of Sciences, the American Association for the Advancement of Science, and the National Science Teachers Association recently wrote a joint statement, saying that "evolutionary theory has stood the test of time in serving as the most comprehensive scientific explanation for the diversity of life on Earth and it is accepted by the overwhelming majority of scientists." There have been no credible challenges to the basic principles of evolution and how it proceeds.

# Big Ideas in ... Evolution and Natural Selection

## 8·1 Evolution is an ongoing process.

- The characteristics of individuals in a population can change over time. We can observe such change in nature and can even cause such change to occur.

## 8·2–8·5 Darwin journeyed to a new idea.

- Following years of observations on worldwide patterns of the living organisms and fossils of plant and animal species, Charles Darwin developed a theory of evolution by natural selection that explained how populations of species can change over time.

## 8·6–8·11 Four mechanisms can give rise to evolution.

- Evolution—a change in the allele frequencies in a population—can occur by genetic drift, migration, mutation, or natural selection.

## 8·12–8·17 Through natural selection, populations of organisms can become adapted to their environments.

- When there is variation for a trait and the variation is heritable, and there is differential reproductive success based on that trait, evolution by natural selection is occurring.

- Fitness is an individual's relative reproductive output. The alleles that yield traits conferring the highest fitness on the individuals that carry them will increase their frequency in a population over time, but this process does not necessarily lead to "perfect" organisms.

- Artificial selection, in which humans determine which individuals will have the highest reproductive success, is a type of natural selection.

- Directional, stabilizing, and disruptive selection can alter the average value of a trait and/or the variation for a trait in a population.

- Natural selection can produce and modify complex traits, including behaviors. This modification may occur as traits selected for one purpose later confer increased fitness for a different purpose.

## 8·18–8·22 The evidence for evolution is overwhelming.

- Many lines of evidence document the occurrence of evolution and point to the central and unifying role of evolution by natural selection in helping us better understand all other ideas and facts in biology.

---

See the To-Go Guide (packaged with every new copy of this textbook) for an illustrated summary of the chapter.

# Check Your Knowledge

**1.** What does the curvy line represent?

**2.** Finch beaks don't grow or shrink much once the bird reaches maturity. What is responsible for the average beak size changing from year to year?

**3.** Three droughts and one very wet year are highlighted. Why is that information included here?

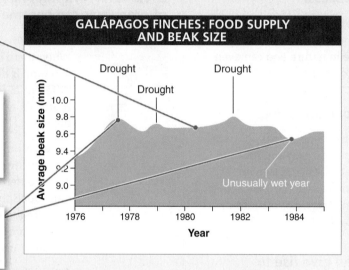

GALÁPAGOS FINCHES: FOOD SUPPLY AND BEAK SIZE

**4.** If you wanted to make predictions about finch beak size in the future, what information would be helpful?

**5.** Is there an "optimum" beak size for a finch? How does this figure help you answer that question?

**6.** What do you think is the "Take-Home Message" of this graph?

See Answers at the back of the book.

## Key Terms in Evolution and Natural Selection

artificial selection, p. 325
adaptation, p. 323
biogeography, p. 334
bottleneck effect, p. 314
convergent evolution, p. 336

differential reproductive success, p. 319
directional selection, p. 326
disruptive selection, p. 328
evolution, p. 302
fitness, p. 322

fixation, p. 313
fossil, p. 331
founder effect, p. 314
gene flow, p. 315
genetic drift, p. 313
heritability, p. 318

homologous structure, p. 336
inheritance, p. 318
migration, p. 315
mutation, p. 311
natural selection, p. 316

population, p. 300
radiometric dating, p. 332
sexual selection, p. 319
stabilizing selection, p. 327
trait, p. 306
vestigial structure, p. 336

# Multiple-Choice Questions

**1.** The average time to death from starvation in a fruit fly is about 20 hours. Selecting for increased starvation resistance in fruit flies:

a) has no effect, because starvation resistance is not a trait that influences fruit fly fitness.

b) has little effect, because ongoing mutation continuously reduces starvation resistance, counteracting any benefits from selection.

c) cannot increase their survival time, because there is no genetic variation for this trait.

d) has no effect, because starvation resistance is too complex a trait, dependent on the effects of too many genes.

e) can produce populations in which the average time to death from starvation is 160 hours.

**2.** Georges Cuvier's discovery of fossils of Irish elk and giant ground sloths:

a) supports scientific evidence for extinction.

b) was possible because of Buffon's determination that the earth was more than 6,000 years old.

c) was possible only following Darwin's publication of *The Origin of Species.*

d) was made in deep ocean trenches.

e) suggested that species are immutable.

**3.** While on the voyage of the HMS *Beagle,* Darwin:

a) wrote *The Origin of Species.*

b) nurtured his love of studying nature, exploring plant and animal diversity, and collecting fossils.

c) discovered a love for sea travel.

d) studied Malthus's book *Essay on the Principle of Population.*

e) corresponded extensively with Alfred Russel Wallace about their ideas on evolution.

**4.** Which one of the following statements best describes the difference between artificial and natural selection?

a) Natural selection is limited to physical traits and artificial selection is not.

b) Artificial selection has produced many of the most delicious food items for humans; natural selection has not.

c) Natural selection acts without the input of humans; artificial selection requires human input.

d) Charles Darwin understood natural selection but, at that time, was unaware of artificial selection.

e) Natural selection works on all species; artificial selection works only on laboratory-raised species.

**5.** Which of the following statements about Charles Darwin is **incorrect**?

a) He spent five years traveling the world, observing living organisms and collecting fossils.

b) He was under constant pressure from his father to make something of himself.

c) He dropped out of medical school.

d) He was enthusiastic about unleashing his theory of natural selection on the world as soon as he thought of it.

e) He and Alfred Russel Wallace independently came up with the theory of evolution by natural selection.

**6.** Evolution occurs:

a) only when the environment is changing.

b) only through natural selection.

c) almost entirely because of directional selection.

d) only via natural selection, genetic drift, migration, or mutation.

e) by altering physical traits but not behavioral traits.

7. **Which of the following statements about mutations is incorrect?**

a) Mutations are almost always random with respect to the needs of the organism.

b) A mutation is any change in an organism's DNA.

c) Most mutations are harmful or neutral to the organism in which they occur.

d) The origin of genetic variation is mutation.

e) All of these statements are correct.

8. **Genetic drift has the strongest effects in populations that are:**

a) small.

b) large.

c) characterized by extensive gene flow.

d) in Hardy-Weinberg equilibrium.

e) undergoing natural selection.

9. **When a group of individuals colonizes a new habitat, the event is likely to be an evolutionary event, because:**

a) members of a small population have reduced rates of mating.

b) gene flow increases.

c) mutations are more common in novel environments.

d) new environments tend to be inhospitable, reducing survival there.

e) small founding populations are rarely genetically representative of the initial population.

10. **To establish that evolution by natural selection is operating in a population, one must demonstrate variability for a trait, heritability of that trait, differential reproductive success based on that trait, and:**

a) increased complexity of the organism.

b) random mating.

c) progress.

d) continuous change in the environment.

e) nothing else.

11. **"Survival of the fittest" may be a misleading phrase to describe the process of evolution by natural selection, because:**

a) it is impossible to determine the fittest individuals in nature.

b) survival matters less to natural selection than does reproductive success.

c) natural variation in a population is generally too great to be influenced by differential survival.

d) during population bottlenecks, it is the least fit individuals that have the greatest survival.

e) reproductive success on its own does not necessarily guarantee evolution.

12. **Evolutionary adaptation:**

a) refers both to the process by which populations become better matched to their environment and to the features of an organism that make it more fit than other individuals.

b) cannot occur in environments influenced by humans.

c) is possible only when there is no mutation.

d) is responsible for the fact that porcupines are at an unusually high risk of predation.

e) occurs for physical traits but not behaviors.

13. **Adaptations shaped by natural selection:**

a) are magnified and enhanced through genetic drift.

b) are unlikely to be present in humans living in industrial societies.

c) may be out of date, having been shaped in the past under conditions that differed from those in the present.

d) represent perfect solutions to the problems posed by nature.

e) are continuously modified so that they are always matched to the environment in which a population lives.

14. **Artificial selection was used on corn to produce a single strain of corn with increased growth rates and greater resistance to a fungus. Although farmers have continued to select for these traits, the productivity of this strain is no longer increasing. This suggests that:**

a) the population size has been decreasing.

b) all or most of the natural variation for these traits has been eliminated.

c) gene migration is a major evolutionary agent in corn.

d) long-term disruptive selection may lead to speciation.

e) artificial selection is not as strong as natural selection.

15. **In a population in which a trait is exposed to stabilizing selection over time:**

a) neither the average value nor the variation for the trait changes.

b) both the average value and the variation for the trait increase.

c) the average value increases or decreases, and the variation for the trait decreases.

d) the average value for the trait stays approximately the same, and the variation for the trait decreases.

e) the average value for the trait stays approximately the same, and the variation for the trait increases.

**16.** Maze-running behavior in rats:

a) is too complex a trait to be influenced by natural selection.

b) is a heritable trait.

c) is not influenced by natural selection, because it does not occur in rats' natural environment.

d) shows no variation.

e) is influenced primarily by mutation in the laboratory.

 100 HARD — 52 — 0 EASY

**17.** A fossil is defined most broadly as:

a) the preserved pieces of hard parts (e.g., shell or bone) of extinct animals.

b) any preserved remnant or trace of an organism from the past.

c) the preserved bones of vertebrates.

d) a piece of an organism that has turned into rock.

e) the process of preservation of intact animal bodies.

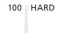 100 HARD — 41 — 0 EASY

**18.** Which of the following statements about marsupial mammals is correct?

a) They fill many niches in Australia that are occupied by placental mammals in other parts of the world.

b) They are less fit than placental mammals.

c) They have become extinct as a result of the greater fitness of placental mammals.

d) They are more closely related to each other than they are to placental mammals.

e) Both a) and d) are correct.

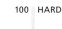 100 HARD — 46 — 0 EASY

**19.** Convergent evolution can occur only when two species:

a) have a recent common ancestor.

b) live in the same geographic area.

c) are separated by a barrier such as a new river.

d) evolve under similar selective forces.

e) are both unpalatable to predators.

 100 HARD — 54 — 0 EASY

**20.** What can be concluded about comparing differences in molecular biology between different species?

a) Extremely different species are fundamentally unrelated in any way.

b) Only DNA sequences can be used to compare relatedness between species.

c) Birds are more closely related to humans than dogs are.

d) Genetic similarities and differences demonstrate species relatedness.

e) The longer two species have been evolving on their own, the fewer the genetic differences between them.

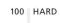 100 HARD — 47 — 0 EASY

**21.** Evolution:

a) occurs too slowly to be observed in nature.

b) can occur in the wild but not in the laboratory.

c) is responsible for the increased occurrence of antibiotic-resistant bacteria.

d) does not occur in human-occupied habitats.

e) None of these statements is correct.

 100 HARD — 48 — 0 EASY

## Short-Answer Questions

**1.** Distinguish between evolution and natural selection. (Restrict your answer to 30 words or fewer.)

**2.** What is genetic drift? Why is it a more potent agent of evolution in small populations than in large populations?

**3.** In *The Origin of Species,* Charles Darwin wrote: "We may look with some confidence to a secure future of great length. And as natural selection works solely by and for the good of each being, all corporeal and mental endowments will tend to progress towards perfection." Describe three reasons why he is wrong.

**4.** How does the increasing frequency of antibiotic-resistant strains of bacteria represent an example of the occurrence of evolution?

See Answers at the back of the book.
For additional study questions, go to www.prep-u.com.

### ABOUT THE CHAPTER OPENING PHOTO

Camels in the Sahara are adapted to life in the hot, dry desert. They have a "flexible thermostat," which allows their body temperature to drop during cool nights, storing coolness, and to increase over the course of a hot day. They have two rows of eyelashes, which protect them from blowing sand. And they have webbed feet, which help them to avoid sinking in the sand.

Communication, cooperation,
and conflict in the animal world

# 9

# Evolution and Behavior

## 9·1–9·4 Behaviors are traits that can evolve.

From taste preferences to mate preferences, behaviors can evolve just as physical features can. (Shown here, a chimpanzee forages for termites.)

# 9·1

## Behavior has adaptive value, just like other traits.

What are your favorite things to eat? Perhaps donuts or hot fudge sundaes top your list. Or maybe pizza, or cheeseburgers and french fries. One thing is almost certain about the food preferences of any human: the list is filled with calorie-rich substances. Put another way, when it comes to eating, humans show an almost complete aversion to dirt or pebbles or other substances from which no energy can be extracted (**FIGURE 9-1**).

In controlled studies, humans demonstrate a clear and consistent preference for sweet and fatty foods. Specifically, when people are presented with different mixtures, their preferences increase as the sugar concentration in the food increases—up to a point, after which adding more sugar makes the food less preferable. But the preference becomes stronger and stronger as the food's fat content increases. In short, we prefer sweet (but not too sweet) foods that are packed with as much fat as possible (**FIGURE 9-2**).

Watch birds in a field and you'll see that they, too, have preferences. A starling, for example, does not eat everything in its path. Rather, it walks through the grass, probing the ground and only occasionally picking up something to eat, such as a beetle larva. Shore crabs, too, have definite

preferences. When presented with mussels of different sizes, shore crabs prefer the mussels that provide the greatest amount of energy relative to the amount of energy a crab

**FIGURE 9-1 Mud pie for dinner?** Humans show an aversion to eating dirt (and other substances from which we cannot extract nutrition).

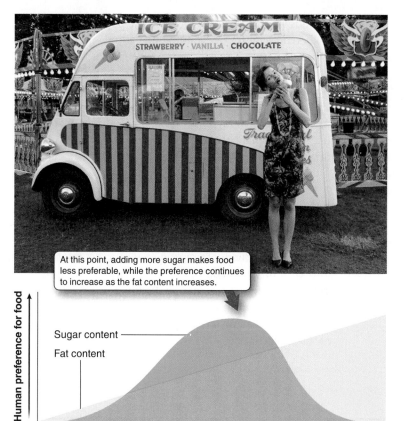

At this point, adding more sugar makes food less preferable, while the preference continues to increase as the fat content increases.

**Human preference for food**

Sugar content

Fat content

**Sugar and fat content in food**

 *In controlled experiments, humans exhibit preferences for sugary foods, high in fat. And the more fat, the stronger the preference is.*

**FIGURE 9-2 Fatty and sweet.** Across cultures, humans prefer sweet, but not too sweet, foods that are packed with fat.

offspring they can produce. As a consequence, natural selection can shape feeding behaviors just as it can bring about changes in physical structures.

Before we continue, let's define exactly what we mean by "behavior." **Behavior** encompasses any and all of the actions performed by an organism, often in response to its environment or to the actions of another organism. Feeding behavior is only one of many behaviors influenced by natural selection.

The scope of animal behavior is vast and varied. Even a brief listing of well-studied topics in animal behavior is impressively long:

- Conflict, aggression, and territoriality
- Cooperation, alliance building, and sociality
- Competing for food and predator avoidance
- Migration and navigation
- Behavioral control of body temperature
- Courtship and mate choice
- Pair bonding and fidelity
- Breeding and parental behavior
- Communication
- Learning and tool use

must expend to open the mussel. The largest mussels take too long to open and so are less preferred; the smallest mussels are easily opened but don't provide enough caloric content to make the effort worthwhile (**FIGURE 9-3**). Thus, the research shows, shore crabs prefer medium-size mussels.

Humans, starlings, shore crabs, and all other animals have taste preferences for the same reason: animals (including humans), unlike plants, cannot create their own food, so they must consume materials from which they can extract the most energy and acquire essential nutrients. If they don't, they die. We may believe that some things simply taste good while others do not, but these preferences are a direct result of the actions of molecules within our taste buds. And those molecules are produced by a large number of genes that have been shaped by natural selection.

Taste preferences directly influence the evolutionary fitness of organisms; the feeding choices of animals affect the number of

**ENERGY-EFFICIENT FEEDING BEHAVIOR HAS EVOLVED**

*The feeding behavior of shore crabs has evolved; they preferentially choose mussels that provide the most energy relative to the effort it takes to open the shell.*

**FIGURE 9-3 Efficient eater.** A shore crab feeds on a mussel.

In Chapter 8, we learned about the evolutionary origins of many *physical traits*. We did not investigate every physical trait that occurs in organisms, however. It was sufficient to look at a few examples, such as birth weight in babies, milk production in cows, and fur color in tigers, to understand the general process by which natural selection acts and how it can lead to the evolution of organisms adapted to their environments. Similarly, in this chapter, although we explore the evolution of *behavioral traits,* we won't investigate each and every behavioral trait in organisms. Rather, we'll explore three broad areas of animal behavior—cooperation and conflict, communication, and mating and parenting—focusing on why those behaviors evolved. That is, we'll be answering this question: How do these behaviors contribute to an organism's fitness, survival, and reproduction?

Our approach to the study of evolution and behavior parallels our approach to the study of the evolution of non-behavioral traits. Recall from the discussion of natural selection in the last chapter that when a heritable trait increases an individual's reproductive success relative to that of other individuals, that

trait tends to increase in frequency in the population. This is Darwin's mechanism for evolution by natural selection and, as we saw in Chapter 8, its effects are evident all around us, from the fancy ornamentation of male peacocks' feathers produced by sexual selection to the cryptic coloration that camouflages so many organisms. In this chapter, we see that *behavior* is just as much a part of an organism's phenotype as is its anatomical structures, and that behavior is produced and shaped by natural selection.

## TAKE-HOME MESSAGE 9·1

**Behavior encompasses any and all of the actions performed by an organism. When a heritable trait increases an individual's reproductive success relative to that of other individuals, that trait tends to increase in frequency in the population. Behavior is as much a part of an organism's phenotype as is an anatomical structure, and as such it can be produced and shaped by natural selection.**

# 9·2 Some behaviors are innate.

In his study of pea plants, Gregor Mendel described a single gene for plant height that caused a plant to be either tall or short. The production of a trait such as plant height,

however, is not completely genetically determined. Certain environmental conditions, such as the type of soil and the availability of water, nutrients, and sunlight, also have a role

**SOME BEHAVIORS ARE INNATE**

In geese, the sight of an egg outside the nest triggers a fixed action pattern: the goose uses a side-to-side egg-retrieval movement all the way back to the nest, even if the egg is taken away during the process.

Male stickleback during non-breeding season

Male stickleback during breeding season

During breeding season, a red belly on any other male stickleback fish triggers an aggressive response.

**FIGURE 9-4 No learning required.** Egg-retrieval behavior in geese and the aggressive behavior of male sticklebacks are fixed action patterns.

to play. Nearly all physical traits of all organisms are the products not only of genes but also of environmental conditions. When it comes to the production of behaviors, the environment also plays an important role.

The degree to which a behavior depends on the environment for expression, however, varies a great deal. At one extreme are behaviors—called **instincts** or **innate behaviors**—that don't require any environmental input to develop. Innate behaviors are present in all individuals in a population and do not vary much from one individual to another or over an individual's life span. An example of innate behavior is a **fixed action pattern.** Triggered in response to a specific signal called a **sign stimulus,** a fixed action pattern is a sequence of behaviors that requires no learning, does not vary, and, once started, runs to completion. Here are just two examples of fixed action patterns.

**Egg retrieval in geese.** When a goose spots an egg outside its nest (a sign stimulus), a fixed action pattern is triggered. The goose gets out of the nest and rolls the egg back, using a side-to-side motion, keeping the egg tucked underneath its

bill. Once started, a goose continues the behavior, even if the egg is taken away on the way to the nest (**FIGURE 9-4**).

**Aggressive displays and attacks by stickleback fish.** The bellies of male stickleback fish turn bright red when the breeding season arrives. During this time, a male stickleback reacts aggressively to the sight of a red belly (a sign stimulus) on any other male stickleback. In fact, the males become antagonistic at the sight of anything remotely resembling a red belly. One researcher even noticed that his sticklebacks performed aggressive displays every day when a red mail truck drove by his window.

### TAKE-HOME MESSAGE 9·2

The degree to which a behavior depends on the environment for expression varies a great deal. At one extreme are instincts, innate behaviors present in all individuals in a population that do not vary much from one individual to another or over an individual's life span. Innate behaviors develop without any environmental input.

# 9·3 Some behaviors must be learned (and some are learned more easily than others).

In contrast to innate behaviors are those behaviors that are influenced to a much greater degree by the individual's environment. Requiring some degree of **learning,** these behaviors are altered and modified over time in response to past experiences. There can be tremendous variation among behaviors that require learning: some come relatively easily and are learned by most individuals in a population, while other behaviors are less easily learned.

Consider a trait common to most primates, including humans: fear of snakes. In the wild, rhesus monkeys are afraid of snakes. The sight of a snake causes the monkeys to engage in fear-related responses, including making alarm calls and rapidly moving away from the snake.

Observations of monkeys in captivity, however, reveal that they aren't born with a fear of snakes. Captive Rhesus monkeys will reach over a plastic model of a snake to get a peanut. Experiments also show that monkeys learn to fear snakes if they see another monkey

*Why is it so much easier for an infant to learn a complex language than for a college student to learn biology?*

terrified at the sight of a snake (**FIGURE 9-5**). Even if they see only a videotape of another monkey expressing fear at the sight of a snake, monkeys immediately and permanently become afraid of snakes. They will no longer reach over a snake model to get a peanut, even when they are very hungry. Instead, they scream and move as far away from the artificial snake as possible. Studies on humans show an equally easily learned fear of snakes. Rather than shrieking and cowering in the corner of a cage, however, we respond with sweaty palms and an increased heart rate.

**Q** When behaviors are learned easily and by all (or nearly all) individuals, it is called **prepared learning.** In addition to the snake-fearing behavior of monkeys, examples of prepared learning abound. The acquisition of language in humans is a dramatic example. Most children don't talk until they are a year old, but by the age of three they understand most rules of sentence construction, and the average six-year-old who is a native English-speaker already has a vocabulary of about 13,000 words. These skills are impressive, particularly given that, at this age, children are

**PREPARED LEARNING**

**1** **BEFORE EXPOSURE TO FEAR OF SNAKES**
A captive monkey will reach over a plastic snake for food.

Monkey has no fear of snakes.

**2** **EXPOSURE TO FEAR OF SNAKES**
A captive monkey views another monkey expressing fear at the sight of a snake.

In real life     or     On video

**3** **AFTER EXPOSURE TO FEAR OF SNAKES**

Monkey expresses fear of the plastic snake.

*A monkey quickly learns to fear snakes if it sees another monkey express such fear.*

**FIGURE 9-5 Learning to be afraid.** The monkey can easily learn to fear snakes by observing fear in other monkeys. This is a type of prepared learning.

generally not competent in reading, writing, and fine motor coordination.

Examples of prepared learning also underscore the fact that organisms don't learn everything with equal ease. For example, researchers demonstrated how easily monkeys learned to fear snakes, but in a related experiment, they demonstrated that the monkeys were not prepared to learn everything with equal ease. The researchers altered the videotape that previously showed a monkey expressing fear on encountering a snake. On the modified tape, a monkey was shown having the same

fear reaction in response to a flower or a toy rabbit. The captive monkeys—which had never seen flowers or rabbits or snakes—did not learn to fear flowers or rabbits the way they learned to fear snakes.

These observations point to an evolutionary basis for the acquisition of certain behaviors. It seems that organisms are well-prepared to learn behaviors that have been important to their ancestors' survival and reproductive success over the course of their evolutionary history, and are less prepared to learn behaviors irrelevant to their evolutionary success. Consider, for example, that although human babies quickly and easily develop a fear of snakes, they don't easily develop a fear of guns. Given that more than 30,000 people are killed by guns in the United States each year, while fewer than three dozen people are killed by snakes, it would seem that we ought to be very afraid of guns and relatively unconcerned about snakes. But we are built in just the opposite way (**FIGURE 9-6**). The likely explanation is that evolution can be slow in producing populations that are

**Q**

**Human babies quickly and easily develop a fear of snakes. Yet they don't easily develop a fear of guns. Why?**

**FIGURE 9-6 Instincts out of date.** A human infant is born with a fear of snakes.

adapted to their environments—that is, evolutionary change cannot always keep up with a rapidly (in evolutionary terms) changing environment.

Like all genes, any genes involved in behaviors have been handed down to us from our ancestors. Snakes caused many human deaths over the course of our evolutionary history. In contrast, guns didn't kill a single person until very recently on the evolutionary time scale. Accordingly, we still fear our ancient enemy, the snake, and have no instinctual response to novel threats, regardless of how deadly.

## TAKE-HOME MESSAGE 9·3

In contrast to innate behaviors are those behaviors that are influenced more by the individual's environment, requiring some learning, and are often altered and modified over time in response to past experiences. Organisms are well-prepared to learn behaviors that have been important to the reproductive success of their ancestors, and less prepared to learn behaviors irrelevant to their evolutionary success.

# 9·4 Complex-appearing behaviors don't require complex thought in order to evolve.

Why do humans and other animals have sex? Is it because they are thinking, robot-like: "Must maximize reproductive success. Must maximize reproductive success"? Of course not. Yet they go about their lives behaving as if they were.

In actuality, natural selection produces organisms that exhibit relatively simple behaviors in response to certain situations or environmental conditions—behaviors that to us may seem complex and sophisticated. Let's look at an example. An animal that experiences pleasure when it has sex has the incentive to seek out additional opportunities to experience that pleasure. As long as reproductive success is enhanced in the process, it is not necessary (from the reproductive-success standpoint) for the animal to deliberately seek that outcome. In other words, natural selection doesn't have to produce animals consciously trying to maximize reproductive success. It only needs to produce animals that behave in a way that actually results in reproductive success, and this outcome will occur if organisms experience a pleasurable sensation by having sex. It's like an evolutionary shortcut. Behaviors that lead to a specific outcome that increases the animal's relative reproductive success will be favored by natural selection.

We can investigate experimentally whether natural selection works in this way by tricking animals. Consider the example of egg retrieval in geese. When an egg falls out of the nest, a goose retrieves it in exactly the same way every time. Is a goose able to do this because it vigilantly keeps tabs on all of

**Q** Animals don't consciously try to maximize their reproductive success, yet they behave as if they do. How does this happen?

its eggs? This seemed to be the case, until researchers decided to trick some geese. The researchers learned that putting *any* object that remotely resembles an egg near a goose's nest triggers the retrieval process. Geese will retrieve beer cans, door knobs, and a variety of other objects that will not increase the animal's reproductive success. Moreover, if multiple egg-like items are just outside the nest, the goose will retrieve the largest item first.

To test the limits of the retrieval behavior, researchers began putting larger and larger models of eggs outside the nests. When given a choice between an actual goose egg and an artificial egg the size of a basketball, the goose always tried to retrieve the giant egg. Rather than keeping tabs on their eggs, geese seem to be following a rule of thumb to retrieve any nearby egg-like objects, preferentially retrieving the larger objects first (**FIGURE 9-7**). Thus, natural selection can produce organisms that exhibit silly and clearly maladaptive (i.e., fitness-decreasing) behaviors in experiments—the goose trying to retrieve a basketball-sized egg—if, in nature, individuals' performance of these simple actions nearly always leads to fitness-increasing behaviors.

Taking an evolutionary approach to the study of the behavior of animals, including humans, is not new. Charles Darwin wrote in *On the Origin of Species:* "In the distant future I see open fields for far more important researches. Psychology will be based on a new foundation." And in the past several decades, Darwin's prediction has

NATURALLY OCCURRING FIXED ACTION PATTERN

By following simple rules that have evolved, an animal can exhibit complex-appearing behaviors.

When an egg rolls out of a nest, a goose retrieves it in exactly the same way every time.

EXPERIMENTAL FIXED ACTION PATTERN

A goose retrieves any object near its nest that remotely resembles its eggs.

A goose retrieves the largest egg-like object first instead of the actual goose eggs.

*Setting up a situation that would never occur in nature, we can reveal that an animal is following simple behavioral rules.*

**FIGURE 9-7 Programmed to retrieve.** The goose will retrieve any egg-like object outside its nest.

increasingly proved true. In a synthesis of evolutionary biology and psychology, researchers taking an approach called "evolutionary psychology" have begun to view the human brain and human behaviors, including emotions, as traits produced by natural selection, selected as a result of their positive effects on survival and reproduction. So, too, has the evolutionary approach to studying behavior begun to influence the field of economics. The 2002 Nobel prize for economics was awarded to researchers who used insights from evolutionary biology and psychology in their analyses of human decision making.

In the rest of this chapter, working from the understanding that organisms' traits, including their behaviors, have evolved by natural selection, we examine the question of why organisms behave as they do. We begin with an exploration of selfishness and cooperation in animals, investigating in particular how natural selection has produced organisms that engage in apparent acts of altruism.

## TAKE-HOME MESSAGE 9·4

If an animal's behavior in natural situations usually increases its relative reproductive success, the behavior will be favored by natural selection. The natural selection of such behaviors does not require the organism to consciously try to maximize its reproductive success.

# 9.5–9.9 Cooperation, selfishness, and altruism can be better understood with an evolutionary approach.

*Placing herself at risk, a female arctic ground squirrel alerts the colony to approaching danger.*

# 9.5

## "Kindness" can be explained.

If we look closely, we see many behaviors in the animal world that *appear* to be **altruistic behaviors**—that is, they seem to come at a cost to the individual performing them while benefiting a recipient. When discussing altruism, we define costs and benefits in terms of their contribution to an individual's fitness.

Take the case of the Australian social spider (*Diaea ergandros*). After giving birth to about 50 hungry spiderlings, the mother's body slowly liquefies into a nutritious fluid that the newborn spiders consume. Over the course of about 40 days, her offspring literally eat her alive; they ultimately kill their mother, but start their lives well-nourished. The cost to the mother is huge—she is unable to produce any more offspring—but there is a clear benefit to her many gene-bearing recipients (**FIGURE 9-8**).

Such altruistic-appearing behavior is common enough in the natural world that it puzzled Darwin. Natural selection, he believed, generally works to produce selfish behavior. The alleles that cause the individual carrying them to have the greatest reproductive success should increase their market share—that is, should increase as a proportion of all the alleles for that gene within the population—in any given generation. At the same time, alleles that cause the individual carrying

*This mother spider is actually being selfish—not altruistic—when she lets her offspring eat her alive.*

**FIGURE 9-8 Parental care to the extreme.** The female *Diaea ergandros* spider feeds her offspring with her own body.

them to help increase other individuals' reproductive success should decrease their relative frequency in a population. Put another way, natural selection should never produce altruistic behavior. Darwin worried that if the apparent instances of altruism were indeed truly altruistic, they would prove fatal to his theory.

As it turns out, Darwin's theory is safe. Virtually all of the apparent acts of altruism in the animal kingdom prove, on closer inspection, to be not truly altruistic; instead, they have evolved as a consequence of either **kin selection** or **reciprocal altruism.**

**Kindness toward close relatives: kin selection.** Kin selection can lead to the evolution of apparently altruistic behavior toward close relatives. Suppose that, for one gene, you carry allele $K$, which causes you to behave in a way that increases the fitness of a close relative of yours, while decreasing your own fitness. Allele $K$ might actually increase its frequency in the population, because you and your relatives tend to share many of the same alleles—including allele $K$. Thus the increased fitness of the relatives you help might compensate for your own reduced fitness, because allele $K$ will, overall, increase its frequency in the population when you help your relatives increase their reproductive output.

**Kindness toward unrelated individuals: reciprocal altruism.** Reciprocal altruism can lead to the evolution of apparently altruistic behavior toward unrelated individuals. Suppose that, for another gene, you carry allele $R$, which causes you to help individuals to whom you are not related. This helps to increase their fitness and decreases yours in the process. Allele $R$ might still increase its market share in the population, if the individuals whom you help become more likely to return the favor and help you in the future, increasing your fitness.

Seen in this light, both kin selection and reciprocal altruism can lead to the evolution of behaviors that are *apparently* altruistic but, in actuality, are beneficial—from an evolutionary perspective—to the individuals engaging in the behaviors.

In the next two sections, we explore kin selection and reciprocal altruism in more detail. We also investigate some of the many testable predictions about when acts of apparent kindness should occur, whom they should occur between, and how we could increase or decrease the frequency of their occurrence through a variety of modifications to the environment.

> "It is not from the benevolence of the butcher, the brewer, or the baker that we expect our dinner, but from their regard to their own interest. We address ourselves, not to their humanity, but to their self-love, and never talk to them of our own necessities, but of their advantages."
> —ADAM SMITH, *An Inquiry into the Nature and Causes of the Wealth of Nations,* 1776

Occasionally, we see individuals engaging in behavior that *is* genuinely altruistic. We'll examine how and why certain environmental situations cause individuals to behave in a way that decreases their fitness and, as such, is evolutionarily maladaptive. It is also important to note that even as genes play a central role in cooperation and conflict, humans' ability to override impulses toward selfishness is responsible for some of the rich diversity in behavior that we see around us.

---

**TAKE-HOME MESSAGE 9·5**

Many behaviors in the animal world appear altruistic. In almost all cases, the apparent acts of altruism are not truly altruistic and have evolved as a consequence of either kin selection or reciprocal altruism and, from an evolutionary perspective, are beneficial to the individual engaging in the behavior.

---

# 9·6 Apparent altruism toward relatives can evolve through kin selection.

The grasslands of the western United States are home to large colonies of Belding's ground squirrels, living in underground burrows. Because the colonies are as large as hundreds or thousands of individuals, they attract many birds of prey, which succeed in killing a squirrel in about 10% of attempts. Squirrels have a system for reducing predation risk that resembles a neighborhood watch program. When an aerial predator approaches, it is common for a squirrel standing on top of a burrow to produce a loud whistle-like alarm call, warning the colony of the impending danger. On hearing an alarm call, squirrels quickly take cover in their burrows, reducing their risk of death. Making an alarm call is a very

An individual Belding's ground squirrel will warn members in its colony that a predator is nearby. Eighty percent of squirrels making the alarm calls are older females, who have more relatives to protect in the colony.

Predator

*Making an alarm call seems like a bad idea—a calling squirrel frequently is killed in the process—but because she protects her relatives, the alarm-calling behavior is favored by natural selection.*

**FIGURE 9-9 Protecting relatives by making an alarm call.** At great risk to herself, a female Belding's ground squirrel will make an alarm call likely to save her family members from a predator.

dangerous activity: about half the time that an alarm call is made, the caller is killed by the predator (**FIGURE 9-9**).

Some squirrels see predators and make alarm calls, while other squirrels see predators and keep their mouths shut. Why would any squirrel make an alarm call? And which squirrels are most likely to engage in this altruistic-appearing behavior? It certainly isn't random: about 80% of the squirrels making alarm calls are female. Moreover, older females are five times more likely than young females to make alarm calls.

These differences reveal that alarm calling is about protecting relatives. The more kin an individual is likely to have, the more likely that individual is to sound the alarm. Because males travel long distances to live in new colonies shortly after reaching maturity, most adult males don't live near their parents, siblings, or any other relatives except their own offspring. Females, on the other hand, remain near the area where they were born and are likely to have many close relatives nearby. Older females, then, are likely to have the largest number of relatives.

The biologist W. D. Hamilton expressed the idea of "kin selection," that one individual assisting another could compensate for its own decrease in fitness if it helped a close relative in a way that increased the relative's fitness. After all,

he realized, the recipient of the aid is likely to have at least some genes in common with the altruistic individual. And the more genes they share (i.e., the more closely the individuals are related), the more likely it is that the alleles being propagated by the recipient of the altruistic-like behavior are the same alleles found in the altruistic individual.

Thus, altruistic-appearing behavior will most likely occur when the benefits to close relatives are greater than the cost to the individual performing the behavior. That is, the more closely related two individuals are, the more likely they are to act altruistically toward each other. It seems that individuals are acting selflessly/altruistically, but they are really acting in their genes' best interests.

Belding's ground squirrels probably don't know exactly how closely they are related to the squirrels around them, so it is likely that natural selection has led to an evolutionary shortcut that allows them to behave as if they did know. In a clever experiment that tested this idea, researchers trapped adult female squirrels and relocated them to distant ground squirrel colonies in which they had no close relatives. If the transplanted females were able to determine exactly how closely they were related to the squirrels around them, they would not make alarm calls. To do so would put the caller at risk without benefiting her genes. What the researchers

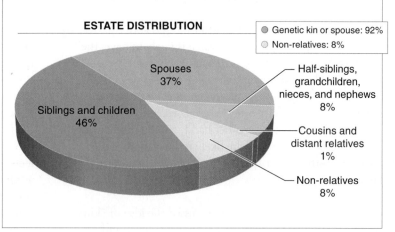

## INAPPROPRIATE ALARM CALLING IN UNNATURAL SITUATIONS

Researchers transplanted an older female Belding's ground squirrel from her original colony to a new colony, where she had no relatives. Even though this older female is not related to any individuals in the colony, she behaves as if she were and is likely to make an alarm call.

Transplanted ground squirrel (not related to any individuals in the colony)

Predator

As with the egg-retrieval experiments in geese, by tricking an animal, setting up a situation that would never occur in nature, we can reveal that the animal is following simple behavioral rules.

**FIGURE 9-10 Following simple behavioral rules.** A transplanted female Belding's ground squirrel that is not related to the squirrel colony behaves as though she is.

found, though, was that transplanted females were just as likely to sound the alarm in the colony of strangers as they were in their home territories, surrounded by close relatives (**FIGURE 9-10**). It seems that female ground squirrels have evolved to follow a simple rule that says, "If I am an older female, I behave as if I have many close relatives around me."

**Q** **Who are people most likely to bequeath money to in their wills?**

Another example of apparent altruistic behavior toward kin is seen in how people give their money and property away when they die. Kin selection theory gives rise to the prediction that people will bequeath their assets based on their degree of relatedness to others. In an analysis of 1,000 randomly chosen wills, researchers tested and supported two hypotheses: (1) that individuals would leave more of their estates to genetic kin, including their parents, children, and siblings (or spouses, who, the researchers assumed, would eventually distribute the assets to the dead person's

genetic kin), and (2) that closer relatives would be left greater amounts than more distant relatives (**FIGURE 9-11**).

Based on the idea of kin selection, it is necessary to redefine an individual's fitness. An individual's fitness is not just measured by his or her total reproductive output (this is called **direct fitness**). Fitness also includes the reproductive output that individuals bring about through their seemingly altruistic behaviors toward their close kin. This is called an individual's **indirect fitness.** The sum of an individual's indirect and direct fitness is called **inclusive fitness.**

No two individuals—with the exception of identical twins—are genetically identical. And because different individuals do not share all of the same alleles, we expect that they should experience some conflict. One disturbing example of such conflict takes place between a pregnant woman and her developing fetus.

A mother's and a fetus' interests differ when it comes to the question of how much food—doled out as nutrients in the blood flow across the placenta—the fetus ought to get. There is a point at which it is in the mother's best interests

**Q** *James Joyce wrote: "Whatever else is unsure in this stinking dunghill of a world, a mother's love is not." According to the concept of kin selection, he may not be completely right. Why?*

## KIN SELECTION: DIVIDING WEALTH

• Individuals are more likely to leave their estates to genetic kin, which includes parents, children, and siblings, and to spouses (who ultimately distribute the assets to genetic kin).

• Closer relatives receive more of the estate than distant relatives.

**ESTATE DISTRIBUTION**

- ● Genetic kin or spouse: 92%
- ○ Non-relatives: 8%

Spouses 37%

Siblings and children 46%

Half-siblings, grandchildren, nieces, and nephews 8%

Cousins and distant relatives 1%

Non-relatives 8%

**FIGURE 9-11 Reading the will.** How likely are you to inherit money from a family member?

to reduce the amount of glucose and other nutrients given to the fetus. Her genes gain if she saves some of her resources for future fetuses, to whom she will be related as closely as to the present fetus.

Now consider the fetus' "point of view." Future siblings will carry some but not all of the same genes as the fetus. Consequently, the fetus does not necessarily benefit from sacrificing nutritional intake for the sake of future siblings. In a sense, it is a sibling rivalry that starts before the sibling is even conceived! The conflict results in a physiological battle throughout pregnancy. The fetus produces chemicals that increase the diameter of the mother's blood vessels, thus increasing the amount of sugar delivered to the fetus. In response, the mother produces more insulin, a chemical that has exactly the opposite effect, causing a reduction in the amount of sugar in the bloodstream available to the fetus. In some mothers this conflict causes gestational diabetes, an inability to properly regulate her blood sugar levels—a condition that disappears as soon as the baby is born. In all pregnancies, though, the conflict escalates until the mother is producing a thousand times the normal amount of insulin.

Gestational diabetes might also be the result of conflict between the fetus's alleles that come from its father and those that come from its mother. The parents may have different interests concerning how much to invest in the current offspring versus how much to save for future offspring. (Paternally inherited genes may benefit from greater resource uptake by the fetus, while maternally inherited genes may benefit from reduced resource uptake.)

Just as close relatives are more likely to help each other than are strangers, the converse is also true: the less closely related two individuals are, the more likely they are to experience conflict. Male lions, for example, upon taking over a pride usually kill unrelated cubs (although females defend their young aggressively), causing the females to become reproductively ready sooner than they would if they continued nursing their cubs.

The idea of kin selection gives rise to another prediction about conflict among relatives and non-relatives. Specifically, it generates the prediction that children in step-families will be injured or killed more frequently than those living with their biological families. And in fact, evidence from numerous studies across many different cultures, including evaluation of 20,000 reports from the American Humane Association in the United States, supports what has been called the "Cinderella syndrome." These findings included, for example, an estimate that the risk of abuse for a preschooler is about 1 in 3,000 for a child living with two biological parents (with whom he or she shares considerable genetic relatedness) and 40 in 3,000 for a child living with a step-parent (with whom he or she has no genetic relatedness). This difference in risk for children living in a home with a step-parent versus those living with their biological parents remains even when socioeconomic factors are taken into account. Although the effect has been noted in multiple other cultures, it is important to note that in the overwhelming majority of step-families no abuse occurs.

**Q** Is a child living with one or more step-parents at greater risk of abuse than a child of the same age living with his or her biological parents?

### TAKE-HOME MESSAGE 9·6

Kin selection describes apparently altruistic behavior in which an individual that assists a genetic relative compensates for its own decrease in direct fitness by helping increase the relative's fitness and, consequently, its own inclusive fitness.

# 9·7 Apparent altruism toward unrelated individuals can evolve through reciprocal altruism.

It is ironic that studies of altruism among animals reveal that natural selection has primarily produced selfish behavior. And, as we saw in the previous section, when behavior appears altruistic, it is frequently because individuals are helping kin and, by doing so, are promoting the reproduction of the copies of the genes (i.e., the alleles) that they share with close relatives. Does any apparently altruistic behavior toward non-relatives occur in the animal world? Yes, but there's not much of it. Here we'll explore the conditions that give rise to reciprocal altruism, how it may have arisen, and why it is so common among humans but so rare among most other animal species.

We start by examining one species with well-documented altruistic-appearing behavior: the vampire bat. Vampire bats live in social groups of 8–12 mostly unrelated adults, roosting primarily in caves and hollow trees. They feed by landing on large mammals such as cattle, horses, and pigs, piercing the skin with their razor-sharp teeth, and drinking the blood that flows from the wound. A chemical in their saliva keeps the blood from clotting, and thus allows the bats to feed for a longer period of time.

Because of their small body size (about the size of your thumb) and their very high metabolic rate, vampire bats must consume almost their entire body weight in blood each night. If they go for more than about 60 hours without finding a meal, they are likely to die from starvation. Here's where the apparent altruism comes in: a bat that has not found food and is close to death will beg food from a bat that has recently eaten. In many cases, the bat that has just eaten will regurgitate some of the blood it has consumed into the mouth of the hungry bat, saving it from starvation. This act obviously has very high benefit for the recipient of the blood, but it comes at a cost to the sharing bat, which loses some of the caloric content of a meal it has just obtained (**FIGURE 9-12**).

### RECIPROCAL ALTRUISM IN VAMPIRE BATS

A vampire bat may vomit up a blood meal for an unrelated individual—saving that individual's life. Bats do this only for individuals that will return the favor, so their own survival is improved in the long run and the behavior is favored by natural selection.

**FIGURE 9-12 The gift of life.**

Kin selection is responsible for some of the blood sharing (females often regurgitate blood for their own offspring), but, in many cases, individuals give blood to unrelated individuals. How might this behavior have arisen? To answer the question, it is important to note three other features of vampire bats. First, they are able to recognize more than a hundred distinct individuals. Second, bats that receive blood donations from non-relatives reciprocate significantly more than average with the bats that have shared blood with them. And third, bats that are not familiar with each other (and do not have a history of helping each other) generally do not regurgitate for each other.

One method proposed to explain the evolution of this apparent altruism is that the bats giving blood to other bats in need are repaid the favor when they are in need of blood. In other words, the act only seems to be selfless, when in actuality it is selfish. With such reciprocal altruism, both individuals (at different encounters) give up something of *relatively* low value in exchange for getting something of great value at a later time when they need it most. In other words, they are storing goodwill in another individual, in much the same way that a person might put money in a bank for a rainy day. In both cases, individuals are protected from some of the world's uncertainties.

Reciprocal altruism has also been documented in vervet monkeys. Monkeys that would otherwise be searching for food or, perhaps, for a mate spend time grooming other monkeys instead (mostly by picking parasites from their fur). In return for their grooming behavior, the monkeys are more likely to receive assistance in response to their solicitations for aid (**FIGURE 9-13**).

Taken together, studies of primates, bats, and other mammals show that reciprocal altruism can evolve if certain conditions are met:

1. Repeated interactions among individuals, with opportunities to be both the donor and the recipient of altruistic-appearing acts

2. Benefits to the recipient that are significantly greater than the costs to the donor

3. The ability to recognize and punish cheaters, individuals that are recipients of altruistic-appearing acts but do not return the favor

In the absence of these conditions, selfishness is expected to be the norm among unrelated individuals. But in the presence of all three conditions, reciprocity masquerading as altruism is likely to occur, as it does with blood sharing among the vampire bats.

## RECIPROCAL ALTRUISM IN GROOMING BEHAVIOR

💡 *Time that could be spent finding food? Individuals that groom others are more likely to receive assistance if they ever are in danger.*

**FIGURE 9-13** You scratch my back, I'll scratch yours.

The three conditions required for the evolution of reciprocal altruism are not satisfied in many animal species, which may be why altruistic-appearing behavior among unrelated individuals is rare. But there are some exceptions. The cleaner fish, for example, is a small fish that is allowed to swim right into the mouth of a grouper, a much larger fish, where it eats the grouper's parasites. Stomach analyses reveal that groupers almost never eat the cleaner fish, opting to pass up this opportunity for a meal. And so, while the grouper is cleaned of parasites, the cleaner fish gets to eat a meal of parasites. In this case, it is easy to rule out kin selection as the reason for the altruistic-appearing behavior; the fishes are from different species and can't possibly be close relatives. Many such mutually beneficial relationships seem to develop when the environment is stressful. In other words, behaviors can evolve that help organisms live more harmoniously with one another—when doing so improves their reproductive success.

As rare as it is in other species, reciprocal altruism is very common among humans. Friendship, among other human relationships, is built on reciprocity and is almost universal.

The opportunity for friendship may be enhanced by our long life span and our ability to recognize thousands of faces and keep track of cheaters. These features are essential in individuals engaging in reciprocal altruism, because an individual becomes very vulnerable when he or she acts in an altruistic manner toward an unrelated individual. The risk is that the altruism will not be repaid, in which case the cheater enjoys greater fitness than the altruist.

**Q** Why are humans among the few species to have friendships?

While cooperators, those who repay altruistic-appearing behavior, have evolutionary advantages over loners, this advantage disappears if the cooperators are the givers all or most of the time, never, or rarely, getting anything in return. The importance of keeping track of cheaters and of identifying good potential reciprocity partners (i.e., other cooperators) may be why humans are so interested in social information and seemingly trivial gossip. As with many of the complex behaviors we have explored, humans probably use some rules of thumb when making decisions about reciprocal altruism. This might include keeping track of *all* available social information to best determine any promising individuals with whom to engage in reciprocal altruism. This includes information about health, generosity, social status, and reproduction. Researchers have even hypothesized that the evolution of maintaining an interest in social information, while valuable in most contexts, may now have maladaptive manifestations. We may find ourselves interested in and distracted by the social lives of Angelina Jolie, Brad Pitt, or other individuals whom we will most likely never meet.

**Q** Why is it easier to remember gossip than physics equations?

### TAKE-HOME MESSAGE 9·7

In reciprocal altruism, an individual engages in an altruistic-appearing act toward another individual. Although giving up something of value, the actor does so only when likely to get something of value at a later time. Reciprocal altruism occurs only if individuals have repeated interactions and can recognize and punish cheaters, conditions satisfied in humans but in few other species.

# 9·8 In an "alien" environment, behaviors produced by natural selection may no longer be adaptive.

A transplanted female squirrel makes an alarm call, putting herself in grave danger to help out dozens of squirrels that share none of her genes. A person in the United States donates money to refugees in Africa. Or someone craves high-fat food even as she knows it may increase her risk of heart disease and shorten her life.

These situations have something in common. In each, genetically influenced traits, behaviors favored by natural selection over evolutionary time, cause individuals to behave in a way that *reduces* their fitness. But how can that be? How can natural selection lead to fitness-reducing behaviors?

To better understand these situations, it is helpful to return to our discussion of fitness in Chapter 8. Recall that fitness not only is an individual's reproductive success, relative to that of other individuals in the population, but also depends on the specific environment in which the organism lives. An adult female Belding's ground squirrel that makes alarm calls generally increases her inclusive fitness by doing so, because all around her are individuals who share her genes. For this reason, such behavior has been favored by natural selection and has increased in frequency in the population.

The world in which alarm calling evolved did not include biologists with pickup trucks who trapped female squirrels, drove them long distances, and released them into colonies of unrelated individuals. So, when a female squirrel suddenly finds herself in that world, her evolved behaviors can no longer be expected to be adaptive—no more than a human could be expected to survive on the moon. And an individual's genes and all the behaviors they influence cannot change overnight. Adaptation to a new environment takes time, and the more quickly an environment changes, the more likely it is that the evolved behaviors of a population will no longer be appropriate.

Let's return to the donations to refugees on another continent. To understand how natural selection could lead to this behavior, we need to understand a bit about human evolutionary history. We know from archaeological deposits that for more than two million years, our ancestors lived as hunter-gatherers in small groups of a few hundred people, at most. Their success depended on joint efforts against predators and in killing prey; being "nice" paid off when the prospect of hunting alone or sleeping outside the camp meant almost certain death. The loners died, so we are descended from those who could work well with others.

It is only recently, the blink of an eye in evolutionary time, that humans invented agriculture, industrialization, and the means of food production and distribution. As a result of these changes, group size has increased dramatically; on a given day, you may see 10 or even 100 *times* as many people as a hunter-gatherer ancestor might have seen. Additionally, if you were a hunter-gatherer, all of the people in your group would be a regular part of your life. You would have many opportunities to help them and, in turn, to be helped by them. Reciprocity paid. And we evolved so that "altruistic" acts gave us pleasure, stimulating parts of our brain in ways that made us want to repeat such actions.

It is with this brain that you approach the issue of refugees halfway across the world, or a homeless person somewhere in the United States. It is almost certain that you will not have repeated interactions with any of those people. And you will probably never be in a position to be helped by them. From an evolutionary perspective, your action almost certainly will not increase your fitness. But in the world that humans evolved in, such altruistic-appearing behavior would most likely have been reciprocated at some future time—and thus your instincts guide you to, and reward you for, your kindness (**FIGURE 9-14**).

In addition to having small population sizes, our ancestors' hunter-gatherer world was characterized by unpredictable food sources that could not be stored for long periods of time. In response to this environment, humans evolved to have great appetites and taste preferences for calorie-laden foods. Fitness increased with food consumption and the storage of fat, helping to protect individuals during regular times of starvation. From an evolutionary perspective, today's world is truly an alien environment: most of us have easy access to unlimited amounts of food, including fat-laden fast foods. Because humans have not yet adapted to this environment of plentiful food, our instincts to consume large amounts are strong, making it difficult for most people to control their body weight.

From battles against obesity to charitable contributions to alarm calling in transplanted squirrels, when organisms of any

## MALADAPTIVE BEHAVIOR

In the small hunter-gatherer groups common for much of human evolutionary history, altruistic behaviors would probably have been reciprocated. Today, the pleasure we feel in response to such behaviors remains—even though in our world we may behave altruistically to people unable to ever return the favor, making our act of kindness maladaptive.

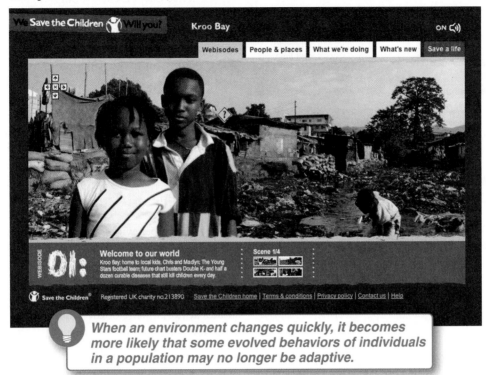

> When an environment changes quickly, it becomes more likely that some evolved behaviors of individuals in a population may no longer be adaptive.

**FIGURE 9-14** Charitable acts can give us pleasure (but in today's world can be evolutionarily maladaptive).

species find themselves in an environment that differs from the environment to which they are evolutionarily adapted, we expect (and see) behaviors that appear to be (and are) not evolutionarily adaptive. Still, understanding the process of natural selection can help us make sense of these behaviors (which in some cases can even seem nonsensical).

Can you think of other ways in which the modern industrial environment differs from the environment to which humans are adapted? How do these differences lead to situations in

which our behavior, like the behavior of the transplanted Belding's ground squirrel, may not be adaptive from an evolutionary perspective?

---

**TAKE-HOME MESSAGE 9·8**

When organisms are in environments that differ from the environment to which they are adapted, the behaviors they exhibit are not necessarily evolutionarily adaptive.

---

# 9·9 Selfish genes win out over group selection.

Kin selection and reciprocal altruism can evolve in a population of animals, as we have seen; and examples of kin selection abound in nature. As a consequence, casual observers of nature frequently see individuals acting in ways

that appear altruistic, even though these individuals are truly acting—from the perspective of evolutionary fitness— in their own selfish interests. This nearly universal selfishness raises the question of whether evolution ever

**DOES GROUP SELECTION OCCUR?**

Group selection describes the evolution of a trait that is beneficial for the species or population while decreasing the fitness of the individual exhibiting the trait.

**ALLELE FREQUENCIES**

- proportion of **selfless behavior** allele in the population ("do what's best for the group, even though it reduces your own reproductive output")

- proportion of **selfish behavior** allele in the population ("do what's best for you, even if it hurts the group")

Regardless of the initial frequencies, over time, selfish behavior alleles increase their market share relative to alleles for selfless behaviors.

**Time**

Because group selection decreases the reproductive success of individuals, it very rarely occurs.

**FIGURE 9-15 Can a "selfless" gene increase in frequency in a population?**

leads to behaviors that are good for the species or population but detrimental to the individual exhibiting the behavior, a process called **group selection.**

It might seem that evolution would favor individuals that behave in a manner that benefits the group, even if it comes at a cost to the individual's reproductive success. But this does not happen. Behaviors that reduce an individual's reproductive output (relative to that of other individuals in the population) are not likely to evolve. Natural selection favors individuals producing more offspring relative to those producing fewer offspring.

Let's look at an example. Imagine that a new allele appears in a population (perhaps through mutation) that causes the individual carrying the allele to double its reproductive output, even though this might spell doom for the species as individuals over-use their resources. An individual carrying this "selfish" allele will pass it to more offspring than an individual carrying the alternative allele, coding for the production of fewer offspring, will pass its allele—and this will lead to excessive, "selfish" consumption of the species' resources. And the selfish offspring also will pass on the selfish allele at a higher rate than the alternative allele is passed on. This scenario may lead to extinction of the species, yet it still occurs. With its market share perpetually increasing, the selfish allele eventually is favored by evolution and will predominate (**FIGURE 9-15**).

Just because a behavior (determined by an "unselfish" allele) leads to a better outcome for the group doesn't mean that natural selection will favor that behavior. In the end, natural selection generally causes increases in the alleles that benefit the *individual* carrying them, even when this comes at the expense of the group. In some special situations, it is possible for natural selection to lead to group selection. But the stringent conditions necessary for this to occur are so rarely found in nature that we almost never see it.

---

**TAKE-HOME MESSAGE 9·9**

Behaviors that are good for the species or population but detrimental to the individual exhibiting such behaviors are not generally produced in a population under natural conditions.

# 9·10–9·15 Sexual conflict can result from disparities in reproductive investment by males and females.

*"Look at me!" The red-crowned crane (Grus japonensis) extends its wings during a courtship display as potential mates look on.*

## 9·10 There are big differences in how much males and females must invest in reproduction.

As we've seen, many behaviors have evolved that influence the ways in which animals interact with each other, reducing conflict and sometimes leading to cooperation. One aspect of life that necessarily involves interaction—for sexually reproducing species—is reproduction, from courtship and mating to parental investment and the forming and breaking of ties. In this and the next few sections, we'll explore these behaviors and the factors that influence them.

How many babies can a woman produce over her lifetime? The number is probably higher than you would guess: a Russian woman had 69 children (in 27 pregnancies). But even this high number is greatly exceeded by the 888 offspring produced by one man (the Emperor of Morocco from 1672 to 1727). The large difference between the number of offspring that can be produced by females and males is less surprising. Among other species of mammals, the pattern is consistent: a male elephant seal can produce 100 offspring, while the maximum produced by a female over her lifetime is 8; a male red deer can produce 24 offspring, while the maximum produced by a female is 14. Here we examine the physical differences between males and females and how they lead to differences in sexual behavior.

The very definition of "male" and "female" hinges on a physical difference between the sexes. Recall from Chapter 6

that in species with two distinct sexes, a **female** is defined as the sex that produces the larger gamete, while a **male** produces the smaller gamete (**FIGURE 9-16**). At conception, the mother's material and energetic contribution to the offspring, the

*At conception, the mother's energetic contribution to the offspring, her reproductive investment, exceeds the father's.*

**FIGURE 9-16** Small sperm, big egg.

energy she will expend in the growth, feeding, and care of offspring—her **reproductive investment**—exceeds the father's. This is true for all animals, whether mammals, birds, insects, or sharks. (It is also true for plants.) Not only are female gametes larger, they tend to be relatively immobile and produced in smaller numbers. Male gametes, on the other hand, while smaller, are more plentiful and very motile.

This discrepancy in size and quantity between the sperm produced by a male and the eggs produced by a female sets the stage for evolutionary developments that magnify this initial difference in reproductive investment. For starters, the difference in the number of gametes that males and females can produce means that males have the potential to produce many, many more offspring than females. Put another way, a male's **total reproductive output,** the lifetime number of offspring he can produce, tends to increase as the number of females he is able to fertilize increases. A female, on the other hand, does not generally increase her reproductive success by mating with additional males beyond the first (**FIGURE 9-17**).

**Q** **Why do males usually compete for females rather than the opposite?**

Because additional matings usually lead to greater increases in reproduction (and fitness) for males, selective pressure has resulted in the evolution of some differences in male and female reproductive behavior. For males, the most effective way to maximize reproductive success often is to find and gain access to mating opportunities with additional females. For females, an effective way to maximize reproductive success often is to put more effort into parenting and less effort into mating.

Two physical differences that can exist between males and females are particularly important when it comes to reproduction. First, in species with internal fertilization, which includes most mammals, fertilization takes place in the female. The offspring also grow and develop within the female's body. The amount of energy females invest in reproduction is therefore much greater than males' investment. Females' reproductive investment also limits their reproductive output; they can be pregnant only once at a time. A second important physical difference between females and males occurs in the mammals: lactation takes place in females and not in males (with a very small number of exceptions). In these species, then, nurturing during both pregnancy and lactation can be accomplished only by the female. This difference in reproductive investment has led to the evolution of some very different reproductive behaviors between males and females.

Although the gamete size difference is consistent across all animals (i.e., the egg is always bigger and energetically more

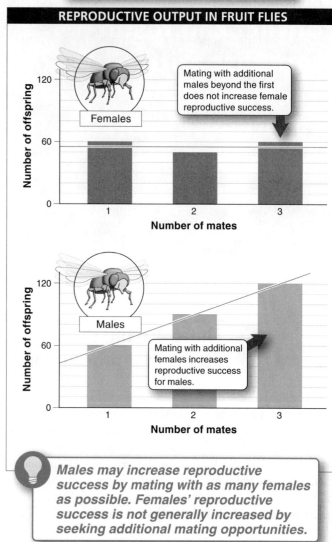

**FIGURE 9-17 Maximizing reproductive success in fruit flies.**

costly to make than the sperm), the physical differences between the sexes in the early nurturing (i.e., gestation and feeding) of offspring can vary considerably across animal species. In some cases, male and female investments become more nearly equal after fertilization. In birds, for example, before emergence of the chicks, much of the development of the fertilized egg is external: the female lays an egg, but either the male or the female can protect and incubate the developing embryo. Further, birds do not lactate. Once hatched, the chicks must be fed—a task that can be done by both parents.

Accordingly, in many bird species, the maximum lifetime reproductive output of males and females is similar. In the

The female mammal invests much energy in the production and care of offspring.

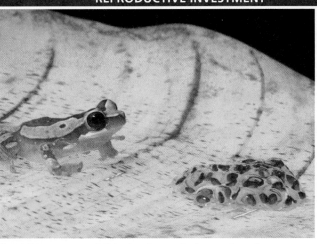

Amphibian eggs are left to develop on their own; there is very little maternal or paternal reproductive investment after fertilization.

Among most bird species, males and females have an equal reproductive investment.

**FIGURE 9-18 Male and female reproductive investment differs across species.** Female investment is greatest when offspring develop internally and are fed by lactation after birth. In animals whose offspring develop externally, males or females (or neither parent) can provide the care.

kittiwake gull, for example, the largest number of offspring produced by a male is 26, while for a female the number is 28. External fertilization in fish and amphibian species further reduces the reproductive investment of the female—she does not spend any energy as the fertilized eggs begin developing into embryos (**FIGURE 9-18**).

We have seen, then, that in species in which fertilization occurs inside the female body, there is an unequal energetic investment in reproduction, and as a result the female expends more energy in the growth and care of offspring (much more so for mammals than for birds). Another profound consequence of internal fertilization is that a male cannot be 100% certain that any offspring a female produces are his progeny. Because it is possible for a female to mate with multiple males, any of whom could be the father, male mammals and birds will always have some degree of **paternity uncertainty.**

In the next few sections, we'll continue to explore how the physical differences between males and females in reproductive investment, along with paternity uncertainty, have led to the evolution of differences in male and female reproductive behavior.

### TAKE-HOME MESSAGE 9·10

In mammals and many other animals, there are important physical differences between males and females relating to reproduction. Fertilization usually takes place in the female. Lactation takes place only in the female. And in species in which fertilization occurs inside the female, males cannot be certain that offspring are their progeny. These physical differences have led to the evolution of differences in male and female reproductive behavior.

# 9·11 Males and females are vulnerable at different stages of the reproductive exchange.

Suppose you are on your college campus and a person of the opposite sex comes up to you and says, "Hi. I have been noticing you around campus. I find you very attractive. Would you go out with me tonight?" What percentage of men would answer *yes?* And women? In a study conducted at

Florida State University in 1978 and 1982, the percentage answering *yes* was 50% for both males and females.

Now imagine the question was "Would you have sex with me tonight?" Among the men, 75% said *yes;* among the women, not a single one said *yes*. It is tempting to interpret such results as a consequence solely of the culture of Western society, characterized by significant differences in the expectations and tolerances of male and female sexual conduct. However, cultural expectations don't adequately explain the consistency of the pattern found across Western and non-Western societies: men and women differ in their approach to sexuality.

Studies of male-female differences in selectiveness about sexual partners, drawing on a wide range of human cultures—including many as far removed from Western influence as possible, such as the Trobriand Islanders of Melanesia in the South Pacific—reveal a consistent difference in men's and women's willingness to have sex. Why might this difference exist?

Humans, like nearly all mammals, are characterized by a greater initial reproductive investment by females. Women are more hesitant to take a mating opportunity, while men are much more willing. For females, the cost of a poor mating choice can have significant consequences—pregnancy and lactation, with offspring from a low-quality male or a male who does not provide any parental investment or access to valuable resources (**FIGURE 9-19**). For a male, the consequences of a poor choice are less dire—little beyond the time and energy involved in mating.

Two differences in the sexual behavior of males and females have evolved:

1.  The sex with the greater energetic investment in reproduction is more discriminating about mating.

2.  Members of the sex with the lower energetic investment in reproduction compete among themselves for access to the higher-investing sex.

A dramatic illustration of how a high reproductive investment leads to the evolution of choosiness in mating behavior comes from the insect world. When bush crickets mate, the male loses about a quarter of its body weight contributing a massive ejaculate (the equivalent of nearly 50 pounds of semen in a human), which the female then uses for energy (**FIGURE 9-20**). (It can represent up to one-tenth of her lifetime caloric intake.)

Not surprisingly, male crickets are very choosy when selecting a mate. They reject small females that would produce relatively few offspring. (Females, you won't be surprised to learn, spend a great deal of effort courting males.)

The sex with the greater reproductive investment must be choosy: a poor choice of mate could be disastrous. The sex with the lower initial reproductive investment (usually males), on the other hand, is not made vulnerable through its mating choices, because the matings are of little energetic consequence. However, there is a point at which males *are* vulnerable in the reproductive exchange.

The point of greatest vulnerability for males comes when they provide parental care to offspring. Due to paternity uncertainty, there is some chance that the male may be investing in offspring that are not his own. This has significant evolutionary costs: rather than increasing his own fitness, he is increasing the fitness of another male. Females, conversely, are less vulnerable at the point of providing parental care to offspring, because a female can be completely certain that the offspring she gives birth to are her own.

In the next two sections, we explore the evolution of reproductive behaviors that help males and females reduce their vulnerability and maximize their reproductive success. First we explore competition and courtship tactics used in

FIGURE 9-19 **The choosier sex.** Males and females differ in their attitude toward mating opportunities. For the female more than the male, the choice of the wrong mate could have expensive consequences.

*When bush crickets mate, the male loses about a quarter of his body weight contributing a massive ejaculate that the female uses for energy. This makes males very choosy.*

**FIGURE 9-20 A costly decision.** Because of the investment the male bush cricket is required to make when fertilizing a female, he chooses a mate very carefully.

cases than researchers originally predicted, offspring have different fathers than observers assumed. It seems that things are not always as they appear to observing biologists. As a consequence, much of the new research calls into question some long-held assumptions about the behavioral consequences of physical differences in reproductive investment among birds versus mammals. There are general behavioral patterns among animals, but as is becoming increasingly clear, these patterns are not universal features of their biology, and a reasonable skepticism is important when identifying and interpreting broad trends across large groups of species.

**2.** Throughout history, there have been many cases of people using observations and scientific findings to justify a wide variety of discriminatory thoughts and behaviors—for example, that if male mammals "naturally" tend to be less faithful to their mates, this reduces individual responsibility for behaviors among humans. Such thinking ignores both the tremendous variation in behavior among species and the power of cultural norms and socialization to influence and shape human behavior, while encouraging the inappropriate assumption that biology can supply meaningful insights into morality or ethical decision making.

selecting and securing a mate, then we investigate the various strategies of "mate guarding" employed by males to increase their paternity certainty.

As we explore some general patterns of reproductive behavior among males and females, it is important to keep two critical points in mind.

**1.** There is tremendous variability across species in male and female behaviors. For example, the use of DNA fingerprinting to identify the parents of bird offspring has led to some surprising revelations. In significantly more

---

**TAKE-HOME MESSAGE 9·11**

Differing patterns of investment in reproduction make males and females vulnerable at different stages of the reproductive process. This has contributed to the evolution of differences in their sexual behavior. The sex with greater energetic investment in reproduction is more discriminating about mates, and members of the sex with a lower energetic investment in reproduction compete among themselves for access to the higher-investing sex.

---

# 9·12 Tactics for getting a mate: competition and courtship can help males and females secure reproductive success.

Female choosiness (and the male-male competition it leads to) tends to increase the likelihood that a female will select only those males that have plentiful resources or relatively high-quality genes, either of which is beneficial to the female, causing her to produce more or better offspring—where "better" may mean increased disease resistance or physical traits that will be found attractive by future mates. Female choosiness is manifested by four general rules.

**1. Mate only after subjecting a male to courtship rituals.** In many bird species, including the western grebe, the female requires the male to perform an elaborate and time-consuming courtship dance before she will mate with him. For the grebe, this courtship dance involves fancy dives into water, graceful hovering, various head movements, and flamboyant twists and turns. The courtship process can go on for several days. But if the male passes the time-

**COURTSHIP RITUALS**
A female grebe requires the male to perform a courtship dance before she will mate with him.

**GIFTS UP FRONT**
A female hanging fly will not mate with a male unless he brings her a large offering of food.

**CONTROL OF VALUABLE RESOURCES**
Female yellow-bellied marmots prefer rock outcroppings that provide retreats for predator escape and for hibernation (and are controlled by dominant males).

**GOOD LOOKS**
A female peacock is attracted to a male with the most beautiful tail feathers.

**FIGURE 9-21 Four factors that influence a female's choice of mate.**

consuming audition, he can generally be counted on to stick around to see a brood through hatching and early care (**FIGURE 9-21**).

Among bower birds, a courting male builds a small thatched structure, at which he makes displays that can involve complex dance steps and dramatic poses. The female walks around the structure and inspects it. Only if she deems it satisfactory will she mate. A male that is rejected by multiple females usually tears down his bower and rebuilds it from scratch to try again to attract a female.

**2. Mate only with a male who controls valuable resources.** Territorial defense is a common means by which males compete for access to females. Among arctic ground squirrels, for example, a female chooses a mate based, in part, on the territory he defends, which is where she will reside after mating. With greater quality and quantity of resources in his territory, a male is better able to attract females, whose reproductive success can be increased if the territory is rich in resources.

> *It is a truth universally acknowledged, that a single man in possession of a good fortune must be in want of a wife.*
>
> —JANE AUSTEN, *Pride and Prejudice,* 1813

**3. Mate only with a male who contributes a large parental investment up front.** Better than a believable pledge to commit resources to future offspring is an actual, on-the-spot exchange in which a female requires a male to give her his parental investment up front, in the form of resources that will help her maximize her reproductive success. In the hanging fly, for example, a female will not mate with a male unless he brings her a big piece of food—usually a dead insect. If a male offers an acceptable food item or other token, called a **nuptial gift,** as part of his courtship behavior, she will mate with him (see Figure 9-21). The larger the food item, the longer she will mate; and the more she eats, the larger the number of eggs she will lay. After about 20 minutes of mating, though, when a male has transferred all of the sperm that he can, he is likely to break off the mating and take back whatever remains of the "gift," which he may use to try to attract another mate. Nuptial feeding is common among birds and insects.

**4. Mate only with a male that has a valuable physical attribute.** Male-male competition for the chance to mate with females can also take a more literal form: actual physical contests. Across the animal kingdom, from dung beetles to hippopotamuses, male-male contests determine the dominance rankings of males. Females then mate primarily with the highest-ranking males.

In a similar process, rather than choosing the best-fighting or largest males, females sometimes base their choice on some physical attribute, such as antler size in red deer, the bright red chest feathers of frigate birds, or the elaborate tail feathers of the male peacock (see Figure 9-21). In each case, for the female, the physical feature serves as an indicator of the relative quality of the male, possibly because the feature is correlated with the male's health.

**Why do so few women get into barroom brawls?**

**Q** ▶ With so many examples of male-male fighting as part of the courtship rituals that have evolved for attracting a mate, it is reasonable to ask: Why is it so rare for females to fight?

And why do females generally not have to advertise their health with flashy feathers or other ornamentation? The answer is that as long as females are making the greater investment in reproduction, nearly any male will mate with them. Consequently, there is nothing more to be gained by trying to outcompete other females or otherwise attract the attention of males.

Moreover, among humans, social and cultural values have powerful influences over mating behavior, complicating interpretation. We are not lumbering robots, destined to follow some genetic program. Researchers have noted some subtle manifestations of female health and fertility, including waist-to-hip ratios and patterns of facial and body symmetry. There is a rich and complex world of mating tactics, many of which we do not fully understand.

**TAKE-HOME MESSAGE 9·12**

As a consequence of male-female differences in initial reproductive investment, males tend to increase reproductive success by mating with many females and have evolved to compete among themselves to get the opportunity to mate. Females increase their reproductive success not through extra matings but rather by being choosy about selecting a mate and by caring for their offspring.

# 9·13 Tactics for keeping a mate: mate guarding can protect a male's reproductive investment.

If a male simply abandons a female after mating and searches for other mating opportunities, rather than making any investment in the potential offspring, he has no risk of investing further energy in offspring that are not his. This is one way to minimize the potential costs associated with paternity uncertainty. If you don't play, you can't lose.

This strategy, however, is not necessarily the most effective way for a male to maximize his reproductive success. If a male has a hundred or even a thousand matings, but no offspring survive, that behavior is not evolutionarily successful. Consequently, in species for which offspring's survival can be enhanced with greater parental investment—including those species not well-developed at birth, such as humans—there is an incentive for a male to provide some additional investment

in offspring, even though such behavior makes him vulnerable to paternity uncertainty.

In situations in which males provide parental care, it is common for the male to reduce his vulnerability through some form of **mate guarding.** In addition to simply reducing opportunities for a female to mate with other males (which can benefit all males), mate guarding can improve a male's reproductive success by increasing his paternity certainty and thus reducing his vulnerability when he makes investment in offspring. As long as the offspring emerge from the female's body, she can be certain that they contain her genes. In contrast, a male inhabits a "danger zone" that lasts as long as the female is fertile. If she mates with any other males during this time, the offspring she produces may not be his. If he is going to make any investments of time or energy that will benefit the offspring, he benefits by minimizing his risk in the danger zone. It is during this period that mate guarding is particularly common.

**Q** **Why do so few females guard their mates as aggressively as males do?**

Mate-guarding methods range from the simple to the macabre. If a male wants to ensure that a female does not mate with another male, why bother to stop mating at all? In many species, males take this approach to reduce their risk in the danger zone. Among house flies, even though the male has completed the transfer of sperm to the female in 10 minutes of copulation, he does not separate from her for a full hour. Moths go even farther and continue to mate for a full 24 hours. And in the extreme case of this strategy, certain frog species continue individual bouts of mating for several months (**FIGURE 9-22**). In humans, this would be equivalent to almost 10 years for a single round of intercourse.

In a slightly subtler form of mate guarding that occurs in reptiles, insects, and many mammalian species, after copulation, males block the passage of additional sperm into

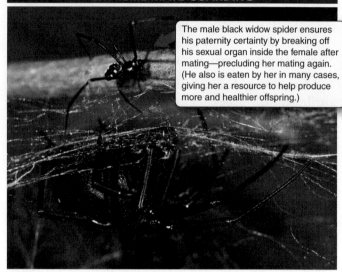

**FIGURE 9-23 A reasonable trade-off?** The male black widow spider ensures his paternity by an extreme form of mate guarding.

the female by producing a copulatory plug. Formed in the female reproductive tract from coagulated sperm and mucus, copulatory plugs can be very effective. Male garter snakes that encounter a female snake with a copulatory plug, for example, do not attempt to court or mate with her, treating her instead as if she were not available.

A much more extreme form of mate guarding occurs in the black widow spider: the male breaks off his sexual organ inside the female, preventing her from ever mating again. Interestingly, when the act is completed, the female usually kills and eats the male (**FIGURE 9-23**). In sealing his mate's reproductive tract, the male assures himself of fathering the offspring, and in consuming her mate's nutrient-filled body, the female gets resources that help her produce the offspring.

*Some frog species carry on individual bouts of mating for several months continuously. If humans mated for a similar percentage of our lives, a single round of intercourse would last almost ten years!*

**FIGURE 9-22 Preventing paternity uncertainty.** A prolonged period of mating prevents the female from accessing other males, assuring the male of reproductive success.

**TAKE-HOME MESSAGE 9·13**

Mate guarding can, in general, increase reproductive success by reducing additional mating opportunities for a partner, and can improve a male's reproductive success by increasing his paternity certainty and thus reducing his vulnerability when he makes investment in offspring.

# 9·14 Monogamy versus polygamy: mating behaviors can vary across human and animal cultures.

As we continue our tour of animal mating behavior, we turn again to the elephant seal. In December of each year, the males appear on islands off the coast of northern California, where they compete with each other for possession of the beach. Through bloody fights they establish dominance hierarchies, with the biggest males—which are 13 feet long and weigh more than 2 tons—generally winning.

In mid-January, the females arrive and are ready to mate. They congregate in large groups on just a few prime beaches. Because the females stick close together, the biggest males, who control the prime beaches, can dominate the other males in the competition for sexual access to females (**FIGURE 9-24**). In one study that observed 115 males, the 5 highest-ranking males fathered 85% of the offspring. While nearly every female will mate and produce offspring, the majority of males never get the chance to mate during the 10–20 years of their lives.

The elephant seals' mating pattern exemplifies **polygamy,** a system in which some individuals attract multiple mates while other individuals attract none. Polygamous mating systems can be subdivided into **polygyny,** in which individual males

mate with multiple females, and **polyandry,** in which individual females mate with multiple males. Polygamy can be contrasted with **monogamy,** in which most individuals mate and remain with just one other individual. Polygamy and monogamy are two types of **mating systems,** which describe the patterns of mating behavior in a species. In this section, we explore the features of environments and species that influence mating systems and survey the range of mating systems observed in nature.

As we have seen, throughout the animal world, males tend to make less parental investment than females. As a consequence, females are choosy about which males they mate with, and males compete for access to mating opportunities. Not surprisingly, multiple females often end up selecting the same male—usually a male on a territory rich in resources or a male with unusually pronounced physical features, such as antler size. Such female selection is the reason for the fancy ornamentation we see in male peacocks. Although the male peacock's tail is like a giant bull's-eye to predators, the number of eyespots on the tail is directly related to how well he can attract a mate: below 140 eyespots, he gets no mates; at 150, he gets two to three mates on average; and with 160 eyespots

## POLYGAMY IN ELEPHANT SEALS

**FIGURE 9-24 King of the beach.** The biggest, best-fighting male elephant seals control the beach and mate with as many females as possible.

In a polygamous mating system, some individuals have multiple mates, while others have few or none.

| DOMINANT MALE ELEPHANT SEAL | LOW-RANKING MALE ELEPHANT SEALS |
|---|---|

OFFSPRING

 *In an extremely polygynous mating system, such as in elephant seals, the vast majority of males have no reproductive success at all.*

he gets six or more mates (see Figure 9-21). Given their greater parental investment, virtually *all* females are able to attract a mate, regardless of fitness or status.

Mating systems are not as easy to define as the elephant seal example might lead us to believe. Three issues, in particular, complicate the task. First, there are often differences between animals' mating behavior and their bonding behavior. That is, it may seem that a male and female have formed a **pair bond**—in which they spend a high proportion of their time together, often over many years, sharing a nest or other "home" and contributing equally to parental care of offspring in what appears to be a monogamous relationship. Closer inspection (often including DNA analysis of the offspring) sometimes reveals, however, that the male and/or the female may be mating with other individuals in the population, and perhaps the mating system is better described as a variation on polygamy. A second difficulty in defining a species' mating system arises because the mating system may vary within the species. That is, some individuals may be monogamous, while others are polygamous. And the mating system may even change over the course of an individual's life. A third difficulty is that males and females often differ in their mating behavior. In the elephant seals described above, it could be said that the females are all mating monogamously, while the males are polygamous.

Within almost any large group of related animal species, there will be a variety of species with each type of mating system, so it is not usually possible to make generalizations. In a recent surprising discovery, researchers were able to alter the behavior of males of a polygynous species of meadow voles by inserting into their brains a single gene from a related but monogamous prairie vole species. With the new gene, the formerly polygynous males showed a preference for their current partner over other females, and the males' ability to form a long-term bond with their partner increased.

Examination of birds and mammals in general, however, reveals one sharp split. As we discussed in Section 9-10, the vast majority of female mammals have a greater parental investment than male mammals. In birds, females and males have a more equal parental investment.

Does the difference in parental investment patterns in birds and mammals lead to different mating systems? Yes. In mammals, polygyny is the most common mating system

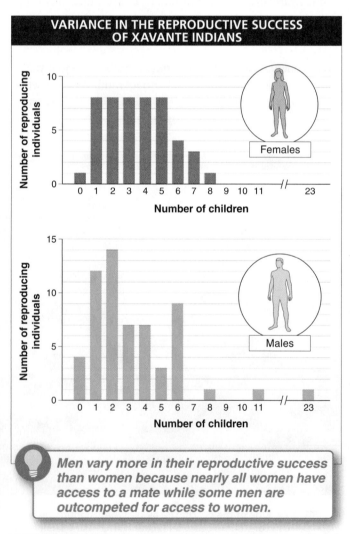

**VARIANCE IN THE REPRODUCTIVE SUCCESS OF XAVANTE INDIANS**

*Men vary more in their reproductive success than women because nearly all women have access to a mate while some men are outcompeted for access to women.*

**FIGURE 9-25** **Let's stay together.** Parental investment that is roughly equal often leads to monogamous mating behavior in birds.

**FIGURE 9-26** **Larger male variance in reproductive success.** Studies of the Xavante Indians of Brazil show that the range in number of offspring over their lifetime is much greater for men than for women.

**MONOGAMOUS BIRDS**

*Of the approximately 10,000 species of birds, more than 90% appear to be monogamous.*

across all large groups, from rodents to primates. Polygyny is a consequence of the significant female investment and lesser male investment. Males can generally benefit more by seeking additional mating opportunities, leading to male-male competition. In birds, the relatively equal parental investment by males and females has led to much less polygyny. More than 90% of the approximately 10,000 bird species we know about appear to be monogamous (although in most cases, it is serial monogamy) (FIGURE 9-25).

**Q**

**Are humans monogamous or polygamous?**

And what of humans? Across a variety of cultures, males consistently have greater variance in reproductive success than females—that is, some males have very high reproductive success and many others have little or none. FIGURE 9-26, for example, presents data from one of the first investigations of this phenomenon. It is from a study of the Xavante Indians of Brazil. The Xavante were selected because they were a pre-industrial population subsisting primarily as hunter-gatherers, with no access to reliable birth control and with almost no contact with Western cultures. The average number of children (3.6) does not differ from men and women, but some men have very large numbers (as many as 23!), and many have none. There is significantly less variability in reproductive success

from one woman to another. Interestingly, a 2010 study using data from 7,710 women and men living in the contemporary United States similarly found significantly greater variance in reproductive success among men as compared to women.

Such differences in the variance in reproductive success among males and females, which can be significantly influenced by socialization and other cultural forces, are generally associated with polygynous mating systems. However, the difference between the sexes found in the Brazilian study is quite small when compared with many other mammalian species, such as elephant seals, and is close to that seen in populations with a monogamous mating system. Humans, consequently, seem to have a mating system close to, but not completely, monogamous.

**TAKE-HOME MESSAGE 9·14**

Mating systems—monogamy, polygyny, and polyandry—describe the variation in number of mates and the reproductive success of males and females. They are influenced by the relative amounts of males' and females' parental investment.

# 9·15 Sexual dimorphism is an indicator of a population's mating behavior.

Male elephant seals are three to four times the size of female elephant seals. In contrast, the males and females of most bird species are the same size. In these species, even expert bird-watchers often cannot distinguish between the sexes, except when the female bird is carrying eggs (FIGURE 9-27). When the sexes of a species do differ in size or appearance, this is called **sexual dimorphism.** Why do species have such dramatic differences in the degree to which males and females resemble each other?

Body size is an important clue to behavior. We have seen how male elephant seals have a winner-take-all tournament for control of the beach, and as the winner, a male has access to mating opportunities. Because the largest individual will have the most offspring, there is selection for larger and larger body size. The smaller males have relatively low reproductive success. There is no selection for larger body size among female elephant seals,

**Q**

It's almost impossible to distinguish males from females in most bird species. Why does this tell us they are monogamous?

because females of any size can mate. Hence the dramatic size difference between the sexes.

In addition to body size, coloration is also a clue to behavior. Because females of some polygynous species choose the males with the brightest or flashiest coloration, rather than the largest males, male-male competition sometimes results in differences in physical appearance between the sexes rather than (or in addition to) differences in size.

What happens when there is little male-male competition for mates? Among bird species, males can provide significant investment in offspring. Eggs have to be incubated, and because many chicks emerge in a very poorly developed condition, it takes two parents to raise them. So a pair tends to stay together for a season or longer. If each female can pair up with only one male, there really isn't much

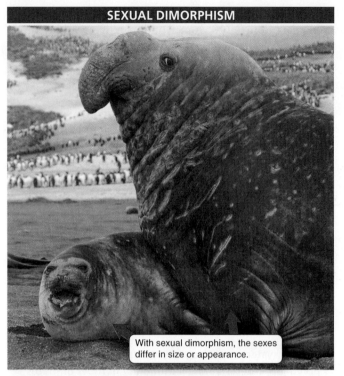

**SEXUAL DIMORPHISM**

With sexual dimorphism, the sexes differ in size or appearance.

**BEHAVIORS ASSOCIATED WITH SEXUAL DIMORPHISM**
• One parent invests more in caring for the offspring.
• Mating system tends toward polygamy.
• One sex (usually females) is choosier when selecting a mate.
• One sex (usually males) competes for access to mating opportunities with the other sex.

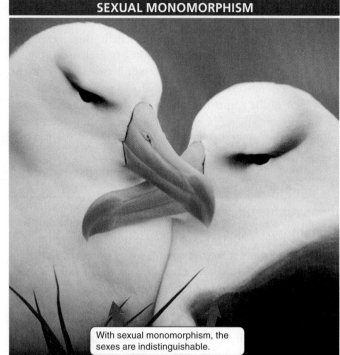

**SEXUAL MONOMORPHISM**

With sexual monomorphism, the sexes are indistinguishable.

**BEHAVIORS ASSOCIATED WITH SEXUAL MONOMORPHISM**
• Both parents invest (approximately) equally in caring for the offspring.
• Mating system tends toward monogamy.
• Both sexes are equally choosy when selecting a mate.

**FIGURE 9-27 Size says a lot.** Species with large males and small females are very likely to be polygamous. Species with similarly sized males and females are likely to be monogamous.

**Q** Men are bigger than women. What does that tell us about our evolutionary history of monogamy versus polygamy?

competition for mates. And without male-male competition, there is little selection for increased size. As a consequence, little sexual dimorphism occurs in most bird species.

In turn, we can predict a great deal about the parental practices of a species just by looking at a picture of a male and a female and examining the ratio of body sizes. If the two are dramatically different, as in elephant seals, you can be confident that the smaller sex is doing most of the care of the offspring

and that the species is more likely to be polygamous than monogamous.

**TAKE-HOME MESSAGE 9·15**

Differences in the level of competition among individuals of each sex for access to mating opportunities can lead to the evolution of male-female differences in body size and other aspects of appearance. In polygynous species, this results in larger males that are easily distinguished visually from females. In monogamous species, there are few such differences between males and females.

*Howling strengthens the social bonds among wolves in a pack.*

# 9·16
## Animal communication and language abilities evolve.

The challenge of communication is central to the evolution of animal behaviors, whether they relate to cooperation, conflict, or the many interactions inherent to selecting a mate. An animal must be able to convey information about its status and intentions to other individuals. And equally important, animals must be able to interpret and assess the information they get from other individuals so as to reduce the likelihood of being tricked or taken advantage of. In this section and the next, we explore the evolution of a variety of methods of animal communication and the ways in which communication can influence fitness and the evolution of other behaviors.

Not all animals need to open their mouths when they have something important to communicate. When the female silkworm moth is ready to mate, for example, she releases a potent chemical called bombykol into the air. As molecules of the chemical come in contact with the bushy antennae of a male silkworm moth—who might be as far as three miles away—he recognizes the mate attractant and responds by flying upwind, looking for the source of the chemical, until he zeroes in on the female and they mate (**FIGURE 9-28**).

This is just one example—a chemical one—of **communication,** an action or signal on the part of one individual that informs or alters the behavior of another individual. But a huge range of animal behaviors requires communication, including defending territory, alerting individuals to threats, announcing one's readiness for mating, establishing one's position in a dominance hierarchy, and caring for offspring.

Animals communicate with many different types of signals. Three types are most common.

1. **Chemical.** Molecules released by an individual into the environment that trigger behavioral responses in other individuals are called **pheromones.** They include the airborne mate attractant of the silkworm moth, territory markers such as those found in dog urine, and trail markers used by ants. They are even used by humans: there are chemicals that influence the length of the menstrual cycle in women and are responsible for the synchronization of menstrual cycles that occurs in women living in group housing situations, such as in dormitories and prisons.

2. **Acoustical.** Sounds that trigger behavioral responses are abundant in the natural world. These include the alarm calls of Belding's ground squirrels described earlier, the complex songs of birds, whales, frogs, and crickets vying for mates, and the territorial howling of wolves.

3. **Visual.** Individuals often display visual signals of threat, dominance, or health and vigor. Examples include the male baboon's baring of his teeth and the vivid tail feathers of the male peacock.

With increasingly complex forms of communication, increasingly complex information can be conveyed. One extreme case is the honeybee **waggle dance.** When a honeybee scout returns to the hive after successfully locating a source of food, she performs a set of maneuvers on a honeycomb that resemble

**CHEMICAL COMMUNICATION**
Pheromones released by one individual and detected by another using its antennae can trigger behavioral responses.

**AUDITORY COMMUNICATION**
Sounds, such as the howl of a coyote, are a common method of triggering behavior responses.

**VISUAL COMMUNICATION**
Organisms can convey information, such as threat or receptivity, with visual displays. Here, a balloon-fish puffs up its body in response to a predator.

**FIGURE 9-28 Chemical, auditory, and visual communication.** Many animal behaviors require communication: the ability to convey and receive information from other animals.

a figure 8, while wiggling her abdomen. Based on the specific angle at which the scout dances (relative to the sun's position in the sky) and the duration of her dance, the bees that observe the dance are able to leave the hive and quickly locate the food source.

The complexity and power of the waggle dance raises an important question: at what point does communication become **language?** Language is a very specific type of communication in which arbitrary symbols represent concepts, and a system of rules, called grammar, dictates the way the symbols can be manipulated to communicate and express ideas (**FIGURE 9-29**).

It's not always easy to identify language. Consider one type of communication in vervet monkeys. Vervet monkeys live in groups, and when an individual sees a predator coming, it makes an alarm call to warn the others. Unlike in most other alarm-calling species, the vervet monkey's alarm call differs depending on whether the predator is a hawk, a snake, or a leopard. And the response of other vervet monkeys to the call is appropriate to the type of predator approaching. The snake alarm causes individuals to stand on their toes and look down. The leopard alarm causes individuals to quickly climb the nearest tree. And the hawk alarm causes vervet monkeys to look to the sky.

Does alarm calling in vervet monkeys or the honeybee waggle dance constitute language? This is debatable. Is the form of American Sign Language taught to chimpanzees and gorillas a language (see Figure 9-29)? There are opinions on both sides.

**FIGURE 9-29 Dancing, signing, and speaking.** A variety of ways have evolved in animals to convey complex information to others.

Interestingly, the great apes have shown a capacity for creating new signs, expressing abstract ideas, and referring to things that are distant in time or space—important features of human language. One thing that is certain, however, is that human language is near the most complex end of the communication continuum.

Language influences many of the other behaviors discussed throughout this chapter. The evolution of reciprocal altruism, for example, may be influenced by language, as this makes it easier for individuals to convey their needs and resources to other individuals. Similarly, in courtship and the maintenance of reproductive relationships, language gives individuals a tool for conveying complex information relating to the resources and value they may bring to the interaction.

# 9·17 Honest signals reduce deception.

Among Natterjack toads, found in northern Europe, males produce booming calls to attract females. Because females desire large males and size determines the volume of the calls, females will push their way through murky swamps until they reach the loudest croaker. They are drawn by calls that can be heard a mile away and are often louder than the legal noise limit for a car engine (**FIGURE 9-30**).

The Natterjack's call is an **honest signal,** a signal that cannot be faked and is given when both the individual making the signal and the individual responding to it have the same interests. An honest signal is one that carries the most accurate information about an individual or situation, and animals that respond to signals of any sort have evolved to value most highly those signals that cannot be faked. A loud Natterjack's call, for example, cannot be faked by a small Natterjack toad. And so there is evolutionary pressure for animals to both produce and respond to honest signals, because these behaviors enable them to maximize their fitness by selecting the best possible mate.

**HONEST SIGNAL**

The volume of a male Natterjack toad's call impacts females' mating decisions. Loud calls can only be made by large males—smaller males can't fake it—so it is an honest signal, carrying accurate information.

**FIGURE 9-30 Loud and true.**

When one animal can increase its fitness by deceiving another, deception can also be expected to evolve. An example might be baby birds begging for food from their parent. If an allele that causes a chick to exaggerate its need for food leads to faster growth and better health for that individual, the allele is likely to increase in frequency in the population.

Communication and signaling, therefore, are features of populations that are continually evolving. We would expect this evolutionary "arms race" between honest signals and deception to lead, over time, to both increasingly unambiguous signals and ever more sophisticated patterns of deception.

In this chapter, we've seen how our understanding of the behavior of animals, including humans, has been greatly expanded by applying scientific thinking, an experimental approach, and careful consideration of the process by which the evolutionary mechanism of natural selection shapes populations. We return to an important point highlighted at the beginning of this chapter: an organism's phenotype is not limited to its physical traits but includes its behaviors as well. Consequently, behaviors respond to selective pressures and evolve.

## HOW TO WIN FRIENDS AND INFLUENCE PEOPLE

As we have seen, in the animal world, it is very, very rare to find individuals engaging in costly behaviors that benefit unrelated individuals; with few exceptions, we see little that resembles animal "friendship."

Humans are a rare exception—every day, in myriad ways, we see friendly behaviors. Just as we put money in the bank for a rainy day, we buffer ourselves from the world's uncertainties by storing goodwill in our neighbors. Cooperators have evolutionary advantages over loners.

**Q: WHY IS "KINDNESS" A RISKY BEHAVIOR EVOLUTIONARILY?** From an evolutionary perspective, we remain vulnerable. With each costly act we perform for unrelated individuals, strangers and friends alike, there is the risk that our efforts will be in vain, our energies lost. For that reason, in our evolution as reciprocal altruists, we acquired a cautiousness, a hesitancy to stick our necks out.

**Q: HOW CAN WE HELP OTHERS FEEL LESS VULNERABLE AND MORE WILLING TO COOPERATE?** Our ability to override the impulses that often push us toward selfishness is one of the hallmarks of being human. Still, there is no reason to put people in situations where they may be better off rejecting friendship. In each interaction, there is an individual on the other side who is also feeling vulnerable. By taking steps, often absurdly simple, to address the unconscious vulnerabilities that others feel, we can increase the likelihood that they choose cooperation, reciprocity, and friendship.

- Learn and use other people's names. It tells them you recognize them specifically and, consequently, that you understand exactly who it is that you "owe." Smile and make eye contact—even when attempting to change lanes in traffic. These gestures feel like the beginning of a relationship, thus stimulating favor-granting instincts.

- Embrace etiquette. Acknowledge your debts to others. Be effusive and public in your thanks for kindnesses done to you. Send thank-you cards.

- Take the first step in reciprocity. Human cooperation is so tied to reciprocal exchange that even tiny gestures of good faith can play an important role in building relationships. Whenever possible, give gifts. Even small ones. Even when they are not required. Especially when they are not required.

- Develop a good reputation. Become known for generosity, for loyalty, for remembering and acknowledging kindnesses done to you.

**CAN YOU THINK OF OTHER STEPS?**

# Big Ideas in ...
# Evolution and Behavior

## 9·1–9·4 Behaviors are traits that can evolve.

- Behavior encompasses all of an organism's actions.
- Behavior is as much a part of an organism's phenotype as are its anatomical structures and, as such, can be produced and shaped by natural selection.
- Innate behaviors are present in all individuals in a population and do not vary much from one individual to another or over an individual's life span.
- Organisms are well-prepared to learn behaviors that have been important to their reproductive success over the course of their evolutionary history and less prepared to learn behaviors irrelevant to their evolutionary success.
- It is not necessary for an organism to consciously try to maximize its reproductive success in order for a behavior to evolve.

## 9·5–9·9 Cooperation, selfishness, and altruism can be better understood with an evolutionary approach.

- Many behaviors in the animal world appear altruistic. In almost all cases, such acts are not truly altruistic and have evolved as a consequence of kin selection or reciprocal altruism, and from an evolutionary perspective are beneficial to the individual engaging in the behavior.
- Kin selection describes apparently altruistic behavior in which an individual's assistance to a genetic relative helps increase the relative's fitness and, consequently, its own inclusive fitness.
- Reciprocal altruism describes altruistic-appearing acts in which the individual performing the act is likely to get something of value at a later time.
- Behaviors that are good for the species or population but detrimental to the individual exhibiting the behavior are not generally produced under natural conditions.

## 9·10–9·15 Sexual conflict can result from disparities in reproductive investment by males and females.

- In mammals and many other animals, there are important physical differences between males and females relating to reproduction. These differences have led to the evolution of differences in male and female reproductive behaviors.
- Males tend to increase their reproductive success by mating with many females and have evolved to compete among themselves to get the opportunity to mate. Females do not increase their reproductive success through extra matings, but rather by caring for their offspring and being choosy when selecting a mate.
- Mate guarding reduces additional mating opportunities for a partner and can improve a male's reproductive success by increasing his paternity certainty, reducing his vulnerability when investing in offspring.
- Mating systems describe the variation in mate number and reproductive success of males and females.
- Differences in the level of competition among the individuals of each sex for access to mating opportunities can lead to the evolution of male-female differences in body size and other aspects of appearance.

## 9·16–9·17 Communication and the design of signals evolve.

- Methods of communication—chemical, acoustical, and visual—have evolved in animal species, enabling individuals to convey information about their condition and situation. These abilities influence fitness and the evolution of almost all other behaviors.
- Animals have evolved to rely primarily on signals that cannot easily be faked, in order to gain the maximum amount of information from them.

See the To-Go Guide (packaged with every new copy of this textbook) for an illustrated summary of the chapter.

# Check Your Knowledge

*Thinking critically about visual displays of data.*

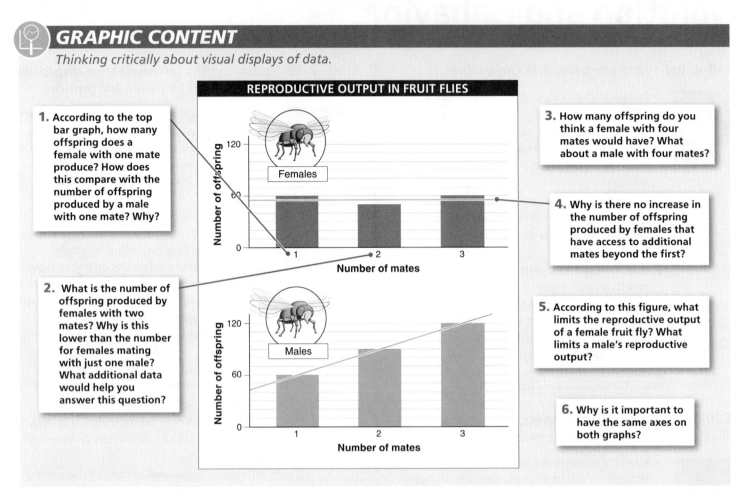

**1.** According to the top bar graph, how many offspring does a female with one mate produce? How does this compare with the number of offspring produced by a male with one mate? Why?

**2.** What is the number of offspring produced by females with two mates? Why is this lower than the number for females mating with just one male? What additional data would help you answer this question?

**3.** How many offspring do you think a female with four mates would have? What about a male with four mates?

**4.** Why is there no increase in the number of offspring produced by females that have access to additional mates beyond the first?

**5.** According to this figure, what limits the reproductive output of a female fruit fly? What limits a male's reproductive output?

**6.** Why is it important to have the same axes on both graphs?

**See Answers at the back of the book.**

# Key Terms in Evolution and Behavior

altruistic behavior, p. 355
behavior, p. 349
communication, p. 377
direct fitness, p. 358
female, p. 365
fixed action pattern,
    p. 351
group selection, p. 364

honest signal, p. 379
inclusive fitness, p. 358
indirect fitness, p. 358
innate behavior,
    p. 351
instinct, p. 351
kin selection, p. 356
language, p. 378

learning, p. 351
male, p. 365
mate guarding,
    p. 371
mating system, p. 373
monogamy, p. 373
nuptial gift, p. 370
pair bond, p. 374

paternity uncertainty,
    p. 367
pheromone, p. 377
polyandry, p. 373
polygamy, p. 373
polygyny, p. 373
prepared learning, p. 351
reciprocal altruism, p. 356

reproductive investment,
    p. 366
sexual dimorphism,
    p. 375
sign stimulus, p. 351
total reproductive output,
    p. 366
waggle dance, p. 377

# Multiple-Choice Questions

**1. Why do human taste preferences exist?**

a) They are culturally generated; consumers are influenced by suggestive advertising.

b) Human feeding choices influence energy intake and, consequently, fitness.

c) Humans can extract energy from a variety of non-food sources, so taste preferences cause us to focus on foods that are more plentiful in our environment.

d) Fats actually do taste better than sugars.

e) Vitamins and minerals give foods unusual tastes, which causes us to seek them out.

**2. From an evolutionary perspective, behavior can best be viewed as:**

a) a trait that arises by learning, not by natural selection.

b) non-heritable.

c) a trait subject to drift and mutation, but not natural selection.

d) part of the phenotype.

e) All of the above are correct.

**3. Why is it so much easier for an infant to learn a complex language than for a college student to learn biology?**

a) Language involves memorization only, without the need for understanding rules, while biology involves both.

b) Language is a feature with great evolutionary relevance for humans.

c) Learning biology was not a behavior with evolutionary relevance for humans.

d) Biology involves much more vocabulary than learning a complex language.

e) Both b) and c) are correct.

**4. Babies in the United States quickly and easily develop a fear of snakes. Yet they don't easily develop a fear of guns. Why?**

a) Humans cannot develop fears of inanimate objects.

b) Evolution can be slow in producing populations that are adapted to their environments.

c) Babies are more likely to encounter snakes than guns as they develop in the United States.

d) Fewer individuals are killed by guns than by snakes in the United States each year.

e) All of the above are correct.

**5. In Belding's ground squirrels, why are females much more likely than males to engage in altruistic behavior by sounding alarm calls?**

a) Belding's ground squirrels have a sex ratio that is biased toward females.

b) Females invest more in foraging and food storage, so they are more likely to lose their lives or their food if a predator attacks.

c) Belding's ground squirrels have a sex ratio that is biased toward males.

d) Females tend to remain in the area where they were born, so the females that call are warning their own kin.

e) Males forage alone, so their alarm calls are useless.

**6. Gestational diabetes is thought to be the consequence of:**

a) pregnant women decreasing their average daily amount of activity.

b) a mother withholding investment in future offspring in order to invest more in the current pregnancy.

c) a mother consuming too much sugar during gestation.

d) conflict between the mother and fetus with respect to how much food the fetus should be given; the mother is equally related to the fetus and to any future offspring, while the fetus is not equally related to itself and any subsequent siblings.

e) physiological constraints on the amount of insulin a mother can provide for the fetus.

**7. Vampire bats:**

a) sometimes regurgitate blood into the mouth of another bat that is close to starving, but the likelihood is a function of whether the individuals are genetically related.

b) are unusual in that they are one of the few animal species that exhibit kin selection.

c) sometimes regurgitate blood into the mouth of an unrelated bat that is close to starving.

d) exhibit reciprocal altruism but not kin selection.

e) There are no such things as vampire bats; they're found only in a Dracula novel.

100 HARD
**37**
0 EASY

**8. Altruistic behavior in animals may be a result of kin selection, a theory maintaining that:**

a) genes promote the survival of copies of themselves when behaviors by animals possessing those genes assist other animals that share those genes.

b) aggression within sexes increases the survival and reproduction of the fittest individuals.

c) companionship is advantageous to animals because, in the future, they can recognize those that have helped them and provide help to those individuals.

d) aggression between the sexes increases the survival and reproduction of the fittest individuals.

e) companionship is advantageous to animals because, in the future, they can recognize those that have helped them and request help once again.

100 HARD
**25**
0 EASY

**9. All of the following conditions are necessary for reciprocal altruism to evolve in a species except:**

a) the ability to recognize different individuals.

b) the ability to punish cheaters who do not reciprocate.

c) repeated interactions with the same individuals.

d) at least one of the sexes must not disperse so that some individuals always live near their kin.

e) None of the above is necessary for the evolution of reciprocal altruism.

100 HARD
**31**
0 EASY

**10. In a situation in which males guard eggs and care for the young without help from the female, which of the following statements would most likely be correct?**

a) Males are large and more brightly colored in order to attract the very best females.

b) Males and females are equally brightly colored, but males court females aggressively.

c) The population is monogamous with no sexual dimorphism.

d) A single male controls a harem of females to which he has exclusive reproductive access.

e) Females are more brightly colored than males and court males aggressively.

100 HARD
**44**
0 EASY

**11. In mammals, as well as many other species, males generally compete for females. The best explanation for this phenomenon is:**

a) males are more aggressive.

b) males, on average, have higher fitness.

c) females have a higher parental investment.

d) males are choosy.

e) females are better looking.

100 HARD
 **26**
0 EASY

**12. Mate guarding is a reproductive tactic that functions to:**

a) reduce paternity uncertainty.

b) increase the female's investment in offspring.

c) reduce the male's reproductive investment.

d) reduce the female's fitness.

e) increase the number of mates to which a male has access.

100 HARD
**32**
0 EASY

**13. Relative to birds, more mammalian species are:**

a) polygynous.

b) monogamous.

c) polyandrous.

d) hermaphroditic.

e) sexually monomorphic.

100 HARD
**59**
0 EASY

**14. In a species such as pigeons, in which males are almost indistinguishable in appearance from females, the most likely mating system is:**

a) monomorphism.

b) monogamy.

c) polygyny.

d) polyandry.

e) It is impossible to predict the mating system with only this information.

100 HARD
**37**
0 EASY

**15. If you find a species of fish in which males are much more brightly colored and larger than females, what might you infer about their mating system?**

a) The degree of sexual dimorphism does not give any information about the mating system.

b) They are simultaneous hermaphrodites.

c) They exhibit parallel monogamy.

d) They are serially monogamous.

e) They are polygynous.

100 HARD
**39**
0 EASY

**16. Polygynous species:**

a) usually employ external fertilization.

b) are usually sexually dimorphic, with males larger and more highly ornamented.

c) are usually sexually dimorphic, with females larger and more highly ornamented.

d) usually have males and females that are physically indistinguishable.

e) are more commonly found among birds than among mammals.

100 HARD
 **29**
0 EASY

## Short-Answer Question

1. Female moorhens are larger and more aggressive than males. They also compete among themselves for access to the smaller, fatter males. Which sex do you think provides more parental care? Explain your answer.

**See Answers at the back of the book.**
**For additional study questions, go to www.prep-u.com.**

**See Answers at the back of the book.**
**For additional study questions, go to www.prep-u.com.**

Now hear this! Two common European toads (*Bufo bufo*) communicate.

Understanding biodiversity

# 10

# The Origin and Diversification of Life on Earth

## 10·1–10·2 Life on earth most likely originated from non-living materials.

*Life on earth arose in a very different environment than we experience today. (Shown here: the formation of volcanic rock in Hawaii.)*

# 10·1

## Complex organic molecules arise in non-living environments.

In the beginning, there was nothing. Now there is something. That, in a nutshell, describes one of the most important, yet difficult to resolve, questions in science: How did life on earth begin? We open our investigation by describing a few things that we do know and exploring how they help us speculate about the things we do not yet know. First, let's give a basic definition of what we mean by "life." **Life** is defined by the ability to replicate and by the presence of some sort of metabolic activity (the chemical processes by which molecules are acquired and used and energy is transformed in controlled reactions). In the next section we expand on this definition and explore the transition from non-living to living.

Earth formed about 4.5 billion years ago from clouds of dust and gases, and as it very gradually cooled, a crust formed at the surface and condensing water formed the oceans. This probably took several hundred million years. The oldest rocks, found in Canada, are about 3.8 billion years old. And the earliest life forms appeared not long after these first rocks formed: fossilized bacteria-like cells have been found in rocks that are 3.4 billion years old.

From that initial point in earth's formation, there is now tremendous **biodiversity**—variety and variability among all genes, species, and ecosystems. In this chapter we explore how

this biodiversity might have come to be, how we name groups of organisms, and how we determine the relatedness of these groups to one another. We begin by returning to the question of the origin of life on earth.

How did these first organisms arise? Some have suggested that life may have originated elsewhere in the universe and traveled to earth, possibly on a meteor. It is hotly debated, however, whether microbes could even survive the multi-million-year trip to earth in the cold vacuum of space with no protection from ultraviolet and other forms of radiation. Experimental data have been unable to answer this question definitively. Most scientists believe, instead, that life originated on earth probably in several distinct phases.

**Phase 1: The formation of small molecules containing carbon and hydrogen** The conditions on earth around the time of the origin of life were very different from those of today. In particular, chemical analyses of old rocks reveal that no oxygen gas was present. The atmosphere included large amounts of carbon dioxide, nitrogen, methane, ammonia, hydrogen, and hydrogen sulfide. Most of these molecules were produced by volcanic eruptions. It was this environment that probably served as the cradle of life, or what Darwin called the "warm little pond" (**FIGURE 10-1**).

Earth's early atmosphere had almost no oxygen and was filled with many compounds released during volcanic eruptions.

Small organic molecules eventually formed, providing the building blocks of life.

**FIGURE 10-1 Darwin's "warm little pond."** The first life on earth tolerated an atmosphere without oxygen.

Critical to the origin of life was the formation of small molecules containing carbon and hydrogen. Because of their chemical structures, these molecules bond very easily and so have a huge variety of forms and functions, making them indispensable for the chemical processes of life. There are several plausible scenarios for how the creation of these small organic molecules may have occurred. The most likely comes from some simple but revealing four-step experiments done in 1953 by a 23-year-old student named Stanley Miller and his advisor, Harold Urey (**FIGURE 10-2**).

1. They created a model of the "warm little pond" and their best estimate of earth's early atmosphere: a flask of water with $H_2$, $CH_4$ (methane), and $NH_3$ (ammonia).

2. They subjected this mini-world to sparks, to simulate lightning.

3. They cooled the atmosphere so that any compounds formed in it would rain back down into the water.

4. They waited and then examined the contents of the water to see what happened.

They didn't have to wait long to get exciting results. Within a matter of days—not millions of years or even a few months—they discovered many organic molecules, including five different amino acids, in their primordial sea. (Using more sensitive equipment, recent re-analyses revealed that all 20 amino acids present in living organisms were produced during the course of these experiments.) This was the first demonstration that complex organic molecules

In 1953, Stanley Miller (shown) and Harold Urey developed a simple four-step experiment that demonstrated how complex organic molecules could have arisen in earth's early environment.

**1** They created a model of the "warm little pond" and atmosphere early in the earth's history.

**2** The atmosphere was subjected to sparks, to simulate lightning.

**3** The atmosphere was cooled so that any compounds in it would rain back down into the water.

**4** They examined the water, looking for organic molecules.

Electrical discharge

**2**

**1**

**3**

Cool

**4**

Heat

Within a matter of days, they discovered many organic molecules (including five different amino acids) in their primordial sea.

**FIGURE 10-2 A promising first step.** The Urey-Miller experiment generated organic molecules from hydrogen, methane, and ammonia.

could have arisen in earth's early environment. However, questions remain, such as whether the environment that Urey and Miller assumed to exist on the early earth was reasonable. Still, the experiments are a promising first step, suggesting that complex organic compounds—including amino acids, the primary constituents of proteins and an essential component of living systems—could have been produced from inorganic chemicals and lightning energy in the primitive environment of earth.

In the next section, we consider the final two phases by which these small organic molecules may have given rise to living organisms.

# 10·2 Cells and self-replicating systems evolved together to create the first life.

After the generation of numerous organic molecules such as amino acids, the second phase in the generation of life from non-life was probably the assembly of these building block molecules into self-replicating, information-containing molecules. This is where things get a bit more speculative. It's complicated enough to generate a complex organic molecule, but it's a whole lot more complicated to generate an organic molecule that can replicate itself. Researchers believe that to get to the replication phase, enzymes were required.

**Phase 2: The formation of self-replicating, information-containing molecules** Recently, researchers discovered that the nucleic acid RNA can do what proteins do—catalyze reactions necessary for replication—meaning that this single, relatively simple molecule could have been a self-replicating system and a precursor to cellular life. These findings have given rise to the so-called **RNA world hypothesis,** which proposes that the world may have been filled with RNA-based life before it became filled with the DNA-based life we see today.

These self-replicating molecules raise an important question: When exactly was the threshold between living and non-living crossed? In the RNA world, self-replicating nucleic acid molecules carried the information of how to replicate *and* served as the machinery to actually carry out the replication. Is that enough? At this point, it is reasonable to ask, again: What exactly is "life"?

Fossils of 3.4-billion-year-old cells have been found in rocks in South Africa and Australia. These cells appear to be prokaryotic cells, similar to living bacterial cells, with no nucleus, no organelles, and a circular strand of genetic information (**FIGURE 10-3**). Some even look as if they are in the process of dividing. Although, superficially, the fossilized cells look like just a couple of circles, several lines of evidence

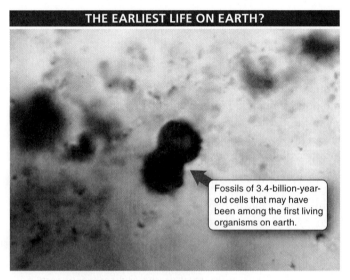

THE EARLIEST LIFE ON EARTH?

Fossils of 3.4-billion-year-old cells that may have been among the first living organisms on earth.

**FIGURE 10-3 Ancient prokaryotic cells.**

support the idea that they are indeed remnants of cells. These include: (1) the age of the rocks themselves has been reliably determined, (2) the size of the circles is similar to that of modern-day prokaryotes, and (3) the ratio of two carbon isotopes ($^{12}C/^{13}C$) is more characteristic of fossilized organisms than of typical rocks that do not contain fossils. But many questions remain. Were these cells the first living organisms on earth? And were they descendants of earlier, self-replicating molecules of RNA? Because no earlier fossils have been found, it is difficult to answer these questions.

Among humans, it's obvious whether someone is living or non-living. But this distinction becomes difficult to make when discussing simpler collections of molecules. Although no perfect, precise definition exists, most scientists agree that life is defined by two characteristics: the ability to replicate and the ability to carry out some sort of metabolism. So the self-replicating RNA molecules were right on the border, satisfying the first condition for life (replicating) but not the second (carrying out metabolism). We now explore the critical third phase in the generation of life from non-life: the development of a membrane that separated these self-replicating small molecules from their surroundings, thus making cells and facilitating metabolic activity.

## Phase 3: The development of a membrane, enabling metabolism and creating the first cells

One way that self-replicating molecules could have acquired chemicals and used them in the controlled reactions of metabolism was by packaging the molecules within membranes. As we saw in Chapter 3, cell membranes are semi-permeable barriers that separate the inside of cells from their external environment. Membranes make numerous aspects of metabolism possible. In particular, they make it possible for chemicals inside the cell to be at higher concentrations than they are outside the cell. Differences in chemical concentrations inside and outside a cell are essential to most life-supporting reactions.

So, if we could combine a self-replicating molecule and some metabolic chemicals into a unit, surrounded by a membrane, life would be possible. But how could this have happened initially? Some evidence suggests that the first cells may simply have come together spontaneously. Specifically, researchers have found that mixtures of phospholipids placed in water or salt solutions tend to spontaneously form small spherical units that resemble living cells. These units may even "sprout" new buds at their surface, appearing to divide. Because these cell-like units don't pass on any genetic material, however, they cannot be considered to be alive. But if, at some point, units like these incorporated some self-replicating molecules inside, maybe by forming around them, such **microspheres** might have been important in the third phase in the generation of life from non-life: the compartmentalization of self-replicating, information-containing molecules into cells. If this did occur, the final step in the creation of something from nothing would be complete (**FIGURE 10-4**).

The exact process by which life on earth originated is still uncertain, and research continues. What is clear, however, is

**FIGURE 10-4  How did life arise?**

that life now abounds in great diversity. In the next sections, we explore that diversity and how it is generated. We start with an investigation of what a species is and how individual species split and create additional species.

### TAKE-HOME MESSAGE 10·2

The earliest life on earth, which resembled bacteria, appeared about 3.5 billion years ago, not long after the earth was formed. Evidence supports the idea that self-replicating molecules—possibly RNAs—may have formed in earth's early environment and later acquired or developed membranes, enabling them to replicate and making metabolism possible, the two conditions that define life.

## 10·3–10·6 Species are the basic units of biodiversity.

*A world of beauty. There are more than 230,000 species of flowering plants.*

# 10·3

## What is a species?

A cat and a mouse are different things. Oprah Winfrey and Bill Gates, on the other hand, are different *versions* of the same thing. That is, although they are different individuals, both are humans. We generally distinguish between different kinds of living organisms: a rose versus a daisy, a wasp versus a fly, a snake versus a frog. Similarly, we lump some organisms into the same group and recognize them as different versions of the same thing: red roses and yellow roses, Chihuahuas and Dalmatians. How do we know when to classify two individuals as members of different groups and when to classify them as members of the same group (**FIGURE 10-5**)?

Biologists use the word **species** to label different kinds of organisms. According to the **biological species concept,** species are populations of organisms that interbreed, or could possibly interbreed, with each other under natural conditions, and that cannot interbreed with organisms from other such groups (**FIGURE 10-6**).

Notice that the biological species concept completely ignores physical appearance when defining a species and instead emphasizes **reproductive isolation,** the inability of individuals from two populations to produce fertile offspring with each other, thereby making it impossible for gene exchange between the populations to occur.

Let's clarify two important features of the biological species concept. First, it says that members of a species are actually interbreeding or *could possibly* interbreed. This emphasis means that just because two individuals are physically separated, they aren't necessarily of different species. A person living in the United States and a person living in Iceland, for example, may not be able to mate because of the distance, but if they were brought to the same location, they could. So, although the people are not in the same population, we do not consider them to be reproductively isolated. Second, our definition refers to

**FIGURE 10-5 The same but not the same.** We recognize Bill Gates and Oprah Winfrey as members of the same species, but can you tell whether the Persian buttercup and the pink rose or the Bornean orangutan and the golden lion tamarin are from the same species? (They're not.)

## WHAT MAKES A SPECIES?

**SPECIES ARE...**
- Populations of organisms that interbreed with each other
- Or could possibly breed, under natural conditions

**FIGURE 10-6 An interbreeding population.** This herd of Dall sheep lives in the remote regions of Denali National Park, Alaska.

*Even if they can be produced in captivity, organisms that do not naturally occur in the wild—such as the "zorse"—are not considered species.*

**FIGURE 10-7 Interbreeding is not enough.** This "zorse" is the offspring of a zebra and a horse, two species that will interbreed, but only in captivity.

"natural" conditions. This distinction is important because occasionally, in captivity, individuals may interbreed that would not interbreed in the wild, such as zebras and horses (**FIGURE 10-7**).

There are two types of barriers that prevent individuals of different species from reproducing: prezygotic and postzygotic barriers (**FIGURE 10-8**). (Remember, an egg that has been fertilized by a sperm cell is a *zygote*.)

## REPRODUCTIVE ISOLATION: KEEPING SPECIES SEPARATE

**PREZYGOTIC BARRIERS**
- Individuals are physically unable to mate with each other.
  - *OR*
- If individuals are able to mate, the male's reproductive cell is unable to fertilize the female's reproductive cell.

**POSTZYGOTIC BARRIERS**
- Matings produce hybrid individuals that do not survive long after fertilization.
  - *OR*
- If hybrid offspring survive, they are infertile or have reduced fertility.

**FIGURE 10-8 Barriers to reproduction.** With postzygotic barriers to reproduction, even if fertilization does occur, the resulting animal (such as the mule, on the right) is usually sterile.

**Prezygotic barriers** make it impossible for individuals to mate with each other or, if they can mate, make it impossible for the male's reproductive cell to fertilize the female's reproductive cell. These barriers include situations in which the members of the two species have different courtship rituals or have physical differences that prevent mating or fertilization.

**Postzygotic barriers** occur after fertilization and generally prevent the production of fertile offspring from individuals of two different species. These barriers are responsible for the production of **hybrid** individuals that either do not survive long after fertilization or, if they do, are infertile or have reduced fertility. Mules, for example, are the hybrid offspring of horses and donkeys, and although they can survive, they cannot produce their own offspring.

At first glance, the biological species concept seems to make it possible to determine unambiguously whether individuals belong to the same or different species. For plants and animals this is usually true. For many other organisms, however, it is impossible to apply the biological species concept. We examine those species in Section 10-5 and explore some alternative methods of identifying species. Even though it is not perfect, however, the biological species concept gives us an important tool for conceptualizing and categorizing the tremendous diversity of organisms found on earth.

---

### TAKE-HOME MESSAGE 10·3

Species are generally defined as populations of individuals that interbreed with each other, or could possibly interbreed, and that cannot interbreed with organisms from other such groups. This concept can be applied easily to most plants and animals, but is not applicable for many other types of organisms.

---

# 10·4  How do we name species?

Keeping track of a large group of anything requires an organizational system. Many libraries, for example, catalog their books by using the Dewey Decimal System, which organizes books into 10 different classes, further subdivides each class into 10 divisions, and each division into 10 sections.

With the huge number of species on earth, a classification system is particularly important. Biologists use the system developed by the Swedish biologist Carolus Linnaeus in the mid-1700s. Here's how it works (**FIGURE 10-9**). Every species is given a scientific name that consists of two parts, a **genus** (*pl.* genera) and a **specific epithet.** Linnaeus named humans *Homo sapiens,* meaning "wise man." *Homo* is the genus and *sapiens* is the specific epithet. (The genus is capitalized, and both genus and specific epithet are italicized.) The redwood tree has the name *Sequoia sempervirens.* The strength of Linnaeus's system is that it is hierarchical—that is, each element of the system falls under a single element in the level just above it.

In the Linnaean system, the species is the narrowest classification for an organism. The name for every species within a genus is unique, but many different species may be in the same genus. Similarly, many genera are grouped within a **family.** And many families are grouped within an **order.** Orders are grouped within a **class.** Classes are grouped within a **phylum.** And, as Linnaeus originally set it up, all phyla were classified under one of three **kingdoms:** the animal kingdom, the plant kingdom, or the "mineral kingdom."

Today, many of the species classifications that Linnaeus described have been revised, and some of his designations, such as the "mineral kingdom," have been left out, while two other kingdoms (fungi and protists) have been added. Also, all of the kingdoms are now classified under an even higher order of classification, the **domain.** But Linnaeus's basic hierarchical structure remains, and all life on earth is still named using this system, with all organisms belonging to one of the three domains: the bacteria, the archaea, and the eukarya.

When new species are discovered, they are given names based on the Linnaean system. The higher-level groups into which a new species should be classified are generally clear—that is, it's usually obvious whether something is in the animal or plant kingdom, or whether it is in the mammalian or amphibian class. But when scientists are assigning a specific epithet, they frequently have a little fun. In recent years, for example, an amphibian fossil was named *Eucritta melanolimnetes,* which translates roughly into "creature from the black lagoon." And in honor of Elvis Presley, a wasp species was named *Preseucoila imallshookupis.* A variety of other celebrities also have had species named after them, including the Beatles, Mick Jagger, Kate Winslet, and Steven Spielberg.

## THE ORGANIZATION OF LIFE

**DOMAIN**
Bacteria  Archaea  Eukarya

**KINGDOM**
Protista  Plantae  Fungi  Animalia

**PHYLUM**
Chordata

**CLASS**
Mammalia

**ORDER**
Perissodactyla

**FAMILY**
Equidae

**GENUS**
*Equus*

**SPECIES**
*Equus quagga*

**FIGURE 10-9 Name that zebra.** Each species on earth, like *Equus quagga,* is given a scientific name and is categorized according to hierarchical groups.

Plants and animals often are referred to by common names, which are different from their "official" names in the Linnaean system and are based on similarities in appearance. These common names, however, can cause confusion. "Fish" for example is used as part of the names of jellyfish, crayfish, and silverfish, none of which are closely related to the vertebrate group of fishes that includes salmon and tuna. Later in this chapter we'll see how an important goal of modern classification and the naming of organisms is to link an organism's classification more closely with its evolutionary history than with its physical characteristics.

### TAKE-HOME MESSAGE 10·4

Each species on earth is given a unique name, using a hierarchical system of classification. Every species falls into one of three domains.

# 10·5 — Species are not always easily defined.

Biologists, like all humans, can be biased. When investigating the natural world, for example, they often focus on plants and animals, to the exclusion of the rest of the earth's rich biodiversity. This gets them into trouble when it comes to an idea such as the biological species concept. While the biological species concept is remarkably useful when

describing most plants and animals, it falls short of representing a universal and definitive way of distinguishing many life forms (**FIGURE 10-10**).

**Difficulties in classifying asexual species** The biological species concept defines species as populations of

## THE BIOLOGICAL SPECIES CONCEPT DOESN'T ALWAYS WORK

The biological species concept is remarkably useful when describing most plants and animals, but it doesn't work for distinguishing all life forms.

**CLASSIFYING ASEXUAL SPECIES**
Asexual reproduction does not involve interbreeding, so the concept of reproductive isolation is no longer meaningful.

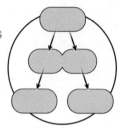

**CLASSIFYING FOSSIL SPECIES**
Differences in size and shape of fossil bones cannot reveal whether there was reproductive isolation between the individuals from which the bones came.

**DETERMINING WHEN ONE SPECIES HAS CHANGED INTO ANOTHER**
There is rarely a definitive moment marking the transition from one species to another.

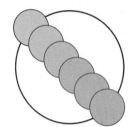

**CLASSIFYING RING SPECIES**
Two non-interbreeding populations may be connected to each other by gene flow through another population, so there is no exact point where one species stops and the other begins.

**CLASSIFYING HYBRIDIZING SPECIES**
Hybridization—the interbreeding of closely related species—sometimes occurs and produces fertile offspring, suggesting that the borders between the species are not clear-cut.

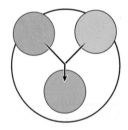

**FIGURE 10-10 A useful concept that can't always be easily applied.**

interbreeding individuals. But this is a useless distinction for the asexual species of the world. Recall from Chapter 6 that asexual reproduction, common among single-celled organisms (including all bacteria), many plants, and some animals, is a form of reproduction that doesn't involve fertilization or even two individuals. Rather, the cell (or cells) of an individual simply divides, creating new individuals. Because asexual reproduction does not involve partners or interbreeding, the concept of reproductive isolation is not meaningful and it might seem that every individual should be considered a separate species. Clearly, that's not a helpful rule to follow.

**Difficulties in classifying fossil species** When classifying fossil species, differences in the size and shape of fossil bones from different individuals can never definitively reveal whether there was reproductive isolation between those individuals. This makes it impossible to apply the biological species concept.

**Difficulties in determining when one species has changed into another** Based on fossils, it seems that modern-day humans, *Homo sapiens,* probably evolved from a related species called *H. heidelbergensis* about 250,000–400,000 years ago. This seems reasonable, until you consider that your parents—who are in the species *H. sapiens*—were born to your *H. sapiens* grandparents, who were born to your *H. sapiens* great-grandparents, and so on. If humans evolved from *H. heidelbergensiss,* at what *exact* point did *H. heidelbergensis* turn into *H. sapiens?* It may not be possible to identify the exact point at which this change occurred.

**Difficulties in classifying ring species** Living in central Asia are some small, insect-eating songbirds called greenish warblers. Unable to live at higher elevations of the Tibetan mountain range, the warblers live in a ring around it. At the southern end of the mountain range, in northwest India, the warblers interbreed with each other. Along either side of the range, the warbler population is split, and warblers on one side do not interbreed with those on the other, because the mountain range separates them. Where the two "side" populations meet up again at the northernmost end of the mountain range, in the forests of Siberia, they can no longer interbreed.

**Q** Chihuahuas and Great Danes generally can't mate. Does that mean they are different species?

What happened? Gradual changes in the warblers on each side of the mountain range accumulated so that the two populations that meet up in Siberia are sufficiently different, physically and behaviorally, that they have become reproductively incompatible. But because the two non-interbreeding populations in the north are connected by gene flow through other populations farther south, there is no exact point at which one species stops and the other begins. So where do you draw the line? The greenish warblers are just one example of more than 20 such **ring species** that have been described.

## Difficulties in classifying hybridizing species

Increasingly, **hybridization**—the interbreeding of closely related species—has been observed among plant and animal species. This phenomenon fits with the biological species concept as long as postzygotic barriers have evolved, so that the hybrids are unable to reproduce. But in some cases, such as among butterflies in the genus *Heliconius,* the hybrids have high survival rates *and* are fertile, whether interbreeding with one another or with individuals of either of the two parental species. This suggests that the borders between the species are not clear-cut.

All of these shortcomings have prompted the development of several alternative approaches to defining what a species is. These alternatives tend to focus on aspects of organisms other than reproductive isolation as defining features. The most commonly used alternative is the **morphological species concept,** which characterizes species based on physical features such as body size and shape. Although the choice of which features to use is subjective, an important aspect of the morphological species concept is that it can be used effectively to classify asexual species. And because it doesn't require knowledge of whether individuals can actually interbreed, the morphological species concept is a bit easier to use when observing organisms in the wild.

When it comes to species definitions, although the biological species concept is the most widely used and can be applied without difficulty to most plants and animals, we should not expect one size to fit all, and there will probably never be a universally applicable definition of what a species is. From asexual species to ring species to hybridizing species, the diversity of the natural world is simply too great to fit into neat, completely defined and distinct little boxes. Nonetheless, scientists can generally use a species definition that is satisfactory for a particular situation.

> ### TAKE-HOME MESSAGE 10·5
>
> The biological species concept is useful when describing most plants and animals, but it falls short of representing a universal and definitive way of distinguishing many life forms. Difficulties arise when trying to classify asexual species, fossil species, species arising over long periods of time, ring species, and hybridizing species. In these cases, alternative approaches to defining species can be used.

# 10·6 How do new species arise?

Biologists don't really have a clue about exactly how many species there are on earth. Estimates vary tremendously, from 5 million to 100 million. Biologists do know, however, the process by which all species arose.

The process of **speciation,** in which one species splits into two distinct species, occurs in two phases and requires more than just evolutionary change in a population. The first phase of speciation is *reproductive isolation,* through which two populations come to have independent evolutionary fates. The second phase is *genetic divergence,* in which two populations evolving as separate entities adapt differently to features of their separate environments, including to different predators and different types of food, thereby accumulating physical and behavioral differences over time.

The initial reproductive isolation necessary for speciation often comes about, as we've seen, when two populations are geographically separated. Although this is an effective and common way for speciation to occur, speciation can also occur without it.

**Speciation with geographic isolation: allopatric speciation** Suppose one population of squirrels is split into two separate populations because the local climate grows wetter and a river forms and splits the habitat in two. Because the squirrels cannot cross the river, the population on one side is reproductively isolated from that on the other. Over time, the two populations have different evolutionary paths as they adapt to particular features of their separate habitats, which may differ. Eventually, the two populations might genetically diverge so much that even if the river separating them disappeared and the populations came back into contact, squirrels from the two groups could no longer interbreed. In fact, two species of antelope ground squirrels formed on the north and south rims of the Grand Canyon as a result of this type of speciation, known as **allopatric speciation** (FIGURE 10-11).

Another example of allopatric speciation is seen in the various finch species of the Galápagos Islands. Individual finches from the nearest mainland, now Ecuador, originally colonized one or more of the islands. (And later, additional islands may have

## ALLOPATRIC SPECIATION

Allopatric speciation occurs when a geographic barrier causes one group of individuals in a population to be reproductively isolated from another group.

**1** INITIAL POPULATION

**2** REPRODUCTIVE ISOLATION

Suppose a river forms through the squirrels' habitat, separating the population. Because the squirrels cannot cross the river, the two groups are now reproductively isolated.

**3** GENETIC DIVERGENCE

Over time, the populations on either side diverge enough genetically that they are no longer able to interbreed.

Harris's antelope ground squirrel

White-tailed antelope ground squirrel

been colonized by birds from the mainland or from previously colonized islands.) But because the islands are far apart, the finches tended not to travel between them, and because colonization events are rare, the populations remained reproductively isolated from one another. Consequently,

14 different finch species have evolved in the Galápagos Islands, each of which specializes in eating certain of the wide range of insects, buds, and seeds found on the islands (**FIGURE 10-12**). Only one species of finch is found in mainland Ecuador.

The barrier doesn't have to be an expanse of water. A forming glacier could split a population into two or more isolated populations. Or a drop in the water level of a lake might expose strips of land that divide the lake into detached, smaller bodies of water, separating one large population of fish into two distinct populations. In each case, the result is the same: geographic isolation that enforces reproductive isolation.

Could you create a new species in the laboratory? How?

Researchers can easily create new species in the lab, using an analogous strategy. In one experiment, a single population of fruit flies (*Drosophila pseudoobscura*) was divided into two. The two populations were then maintained on different diets—one on the sugar maltose and the other on a starch-based food. After only eight generations of enforced reproductive isolation and adaptation to their differing nutritional environments, the populations had diverged sufficiently to form separate species; when the populations were mixed, the fruit flies from one population would no longer interbreed with flies from the other population.

### Speciation without geographic isolation: sympatric speciation

Speciation can also occur among populations that overlap geographically. This is called **sympatric speciation.** Among vertebrates, it is rare for populations of the same animal to become reproductively isolated when they coexist in the same area, so this method of speciation is relatively uncommon. But it *is* common among plants, and it occurs in one of two ways (**FIGURE 10-13**, page 400).

During cell division in plants (both in reproductive cells and in other cells of the plant body), an error sometimes occurs in which the chromosomes are duplicated but a cell does not divide. This creates a new cell that can then grow into an individual with twice as many sets of chromosomes as the parent from which it came. The new individual may have four sets of chromosomes, for example, while the original individual had two. This doubling of the number of sets of chromosomes is called **polyploidy.** The individual with four

**ALLOPATRIC SPECIATION: GALÁPAGOS ISLAND FINCHES**

Large cactus finch

Vampire finch

Large ground finch

Woodpecker finch

Small tree finch

Mainland finch

*Due to allopatric speciation, fourteen different species of finches have evolved in the Galápagos Islands. Only one species of finch is found on the nearest mainland.*

**FIGURE 10-12 Five of the 13 Galápagos Island finch species.** Each species of finch specializes in eating the insects, buds, and/or seeds found in its island habitat.

sets of chromosomes can no longer interbreed with individuals having only two sets, because their offspring would have three sets (two sets from the parent that had four, and one set from the parent that had two), which could not divide evenly during cell division. The individual with four sets can, however, propagate through self-fertilization or by mating with other individuals that have four sets. As a consequence, the individuals with four sets of chromosomes have achieved instant reproductive isolation from the original population and are therefore considered a new species. Although rare in animals, speciation by polyploidy has occurred several times among some species of tree frogs.

A much more common method of sympatric speciation occurs when plants of different (but closely related) species interbreed, forming a hybrid. The hybrid may not be able to interbreed with either of the parental species, but it may be able to propagate asexually—as many plants can. And meiosis often is disrupted in hybrids, causing a chromosome doubling (polyploidy). When this occurs, the hybrids with a doubled number of chromosome sets can interbreed with each other (see Figure 10-13). This method of speciation has led to the production of a large number of important crop plants, including wheat, bananas, potatoes, and coffee.

When reproductive separation occurs, for whatever reason, the new populations are isolated from each other. But because the biological species concept considers individuals from populations that could *potentially* interbreed as still belonging to the same species, speciation is not complete until sufficient differences have evolved in the two populations that they could no longer interbreed even if

Sympatric speciation results in the reproductive isolation of populations that coexist in the same area. Two scenarios, common in plants, lead to this method of speciation.

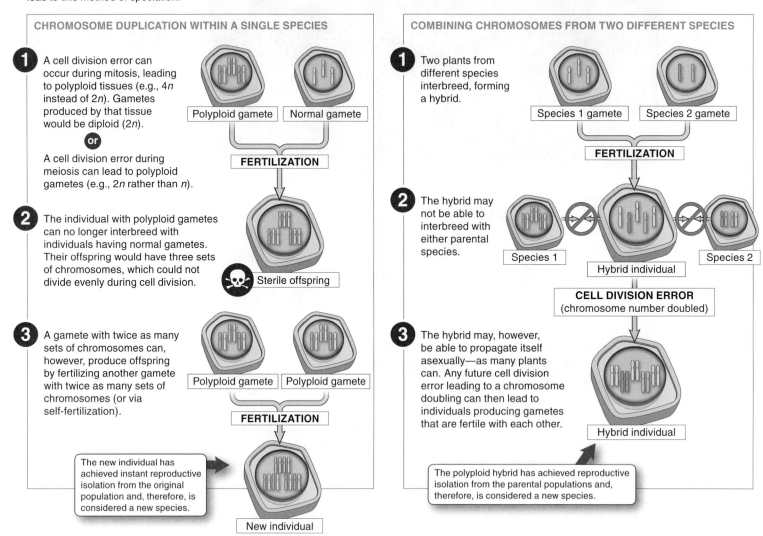

**CHROMOSOME DUPLICATION WITHIN A SINGLE SPECIES**

**1** A cell division error can occur during mitosis, leading to polyploid tissues (e.g., 4*n* instead of 2*n*). Gametes produced by that tissue would be diploid (2*n*).

**or**

A cell division error during meiosis can lead to polyploid gametes (e.g., 2*n* rather than *n*).

Polyploid gamete    Normal gamete

**FERTILIZATION**

**2** The individual with polyploid gametes can no longer interbreed with individuals having normal gametes. Their offspring would have three sets of chromosomes, which could not divide evenly during cell division.

Sterile offspring

**3** A gamete with twice as many sets of chromosomes can, however, produce offspring by fertilizing another gamete with twice as many sets of chromosomes (or via self-fertilization).

Polyploid gamete    Polyploid gamete

**FERTILIZATION**

The new individual has achieved instant reproductive isolation from the original population and, therefore, is considered a new species.

New individual

**COMBINING CHROMOSOMES FROM TWO DIFFERENT SPECIES**

**1** Two plants from different species interbreed, forming a hybrid.

Species 1 gamete    Species 2 gamete

**FERTILIZATION**

**2** The hybrid may not be able to interbreed with either parental species.

Species 1    Hybrid individual    Species 2

**CELL DIVISION ERROR**
(chromosome number doubled)

**3** The hybrid may, however, be able to propagate itself asexually—as many plants can. Any future cell division error leading to a chromosome doubling can then lead to individuals producing gametes that are fertile with each other.

Hybrid individual

The polyploid hybrid has achieved reproductive isolation from the parental populations and, therefore, is considered a new species.

**FIGURE 10-13 Speciation without geographic isolation.** Plants may genetically diverge through sympatric speciation.

they did come in contact. It is in isolation from each other, however, that two populations generally diverge into separate species—that is, the prezygotic and/or postzygotic mechanisms evolve that form the biological barriers to genetic exchange between the two new species.

Reproductive separation can occur relatively quickly—as long as it takes for a new river (or freeway) to divide one large population into two. But the genetic divergence that causes true reproductive isolation—no successful interbreeding even when individuals come into contact with each other—can

take a very long time, sometimes thousands of years. For this reason, speciation can be difficult to study and observe.

**TAKE-HOME MESSAGE 10·6**

Speciation is the process by which one species splits into two distinct species that are reproductively isolated. It can occur by polyploidy or by a combination of reproductive isolation and genetic divergence.

## 10·7–10·9
## Evolutionary trees help us conceptualize and categorize biodiversity.

*The diversity of life on earth can be thought of as branching like a tree.*

# 10·7 The history of life can be imagined as a tree.

Although Carolus Linnaeus's method for naming species was an important step in categorizing and cataloguing earth's biodiversity, his underlying assumption about where species came from was wrong. He believed, as did all biologists of his day, that all species had been created at the same time and were unchanging. For this reason, his classification of organisms into hierarchical groups was based on nothing more than his own evaluation of how physically similar various organisms appeared.

A hundred years later, however, Charles Darwin proposed and documented that species could, in fact, change and give rise to new species. With Darwin, the classification of species acquired a new goal and a more important function. In *The Origin of Species* Darwin wrote: "Our classifications will come to be, as far as they can be so made, genealogies." That is, Darwin proposed that the classifications of organisms would resemble family trees that link generations of parents and offspring over long periods of time. With these words, Darwin was the first to link classification with evolution.

Linnaean classification involved placing organisms within groups as a function of their apparent similarity with each other. The modern incarnation of Darwin's vision of classification is called **systematics** and has the broader goal of reconstructing the **phylogeny,** or evolutionary history, of

organisms. That is, through systematics, all species—even extinct species—are named and arranged in a manner that indicates the common ancestors they share and the points at which the species diverged. A complete phylogeny of all organisms is like a family tree for all species, past and present.

A phylogenetic tree not only shows the relationships among organisms but also presents a hypothesis about evolutionary history (**FIGURE 10-14**). At the beginning of life on earth there was the first living organism, one that could replicate itself. Then a **speciation event** occurred, after which the population of the first living organism split into independent evolutionary lineages. The phylogenetic tree had its first branch, and there was biodiversity. Over hundreds of millions of years, speciation events continued to occur, and today the tree has branches with millions (or possibly tens of millions) of tips that represent all the species on earth. Moving up a branch of any evolutionary tree from its trunk toward its tips, we can see when groups split, with each branching point representing a speciation event. And it doesn't stop here—speciation continues to add new branches all the time.

It is important to remember that phylogenies are hypotheses and, like any hypotheses, they are subject to revision and modification. In the next section, we explore how phylogenetic trees are constructed and look at the rich

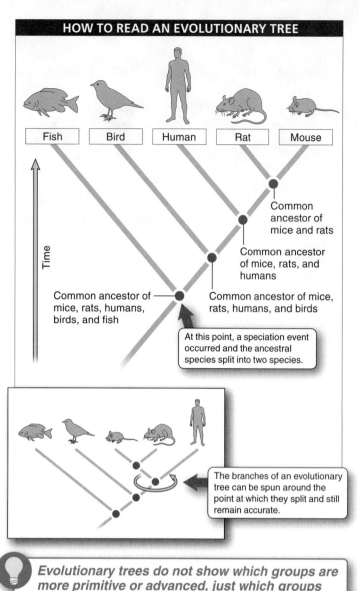

## HOW TO READ AN EVOLUTIONARY TREE

Fish | Bird | Human | Rat | Mouse

Time

Common ancestor of mice and rats

Common ancestor of mice, rats, and humans

Common ancestor of mice, rats, humans, birds, and fish

Common ancestor of mice, rats, humans, and birds

At this point, a speciation event occurred and the ancestral species split into two species.

The branches of an evolutionary tree can be spun around the point at which they split and still remain accurate.

*Evolutionary trees do not show which groups are more primitive or advanced, just which groups are more closely related to which other groups.*

**FIGURE 10-14 Mapping common ancestors.** Evolutionary trees reveal the evolutionary history of species and the sequence of speciation events that gave rise to them.

information they convey about the history of life on earth. We also investigate the modern methods of molecular systematics and see how, with new molecular and DNA evidence, biologists are revising many earlier phylogenies that were based on organisms' physical features.

### TAKE-HOME MESSAGE 10·7

**The history of life can be visualized as a tree; by tracing from the branches back toward the trunk, we can follow the pathway back from descendants to their ancestors. The tree reveals the evolutionary history of all species and the sequence of speciation events that gave rise to them.**

# 10·8

## Evolutionary trees show ancestor-descendant relationships.

Many human traditions have a sacred "Tree of Life." The bodhi tree that the Buddha sat beneath to achieve the ultimate truth is the Tree of Enlightenment; the apple tree in the Garden of Eden is the Tree of Knowledge. The history of the relationship between all organisms that constitute life on earth is another tree of life.

The trunk and branches of an evolutionary tree represent ancestor-descendant relationships that link living organisms with all life that has ever existed on earth. The evolutionary tree of life can be thought of as one giant tree, but as a

practical matter, biologists often study only particular branches. These branches can be illustrated as big trees or little trees and can be expanded to include whatever organisms the biologist is studying: the tree might, for example, include all animals or just the rodents.

In evolutionary trees, it does not matter on which side of the tree you put a particular group of organisms. Any branches can be spun around the **nodes** at which they split (see Figure 10-14). This pinwheel effect means that you cannot assume that rats are more evolutionarily advanced than mice, or vice versa. They are equally advanced in the sense that both groups derived from the same speciation event.

**Q** Are humans more advanced, evolutionarily, than cockroaches? Can bacteria be considered "lower" organisms?

Evolutionary trees tell us many things, but one thing they do not tell us is which groups are most "primitive" and which are most "advanced." This property of phylogenetic trees can serve to undermine some of our most sacred beliefs, such as the one that humans are the pinnacle of evolution. Many trees can be drawn to support this idea, including Figure 10-14. But notice that if you rotate this tree around any one of its nodes, you can get a number of different trees.

What trees do tell us is which groups are most closely related to which other groups. One of the most interesting revelations of "tree thinking" is that—despite appearances—fungi, such as mushrooms, yeasts, and molds, are more closely related to animals than to plants (**FIGURE 10-15**). We investigate this surprising fact in Chapter 11.

> "We all should know that diversity makes for a rich tapestry, and we must understand that all the threads of the tapestry are equal in value no matter what their color."
>
> — MAYA ANGELOU,
> in *Wouldn't Take Nothing for My Journey Now,* 1993

Biologists use the term **monophyletic** to describe any group in which all of the individuals are more closely related to each other than to any individuals outside that group. Monophyletic groups are determined by looking at the nodes of the trees. For example, birds and crocodiles, taken together, compose a monophyletic group because they share a more recent common ancestor (designated by node A in **FIGURE 10-16**) than either group shares with lizards or mammals. Lizards and crocodiles, taken together, do not compose a monophyletic group, because their common ancestor (at node B) is also shared by birds. But birds, crocodiles, and lizards, all taken together, *do* compose a monophyletic group, by virtue of all three sharing the common ancestor at node B. Similarly, birds, crocodiles, lizards, and mammals also compose a monophyletic group.

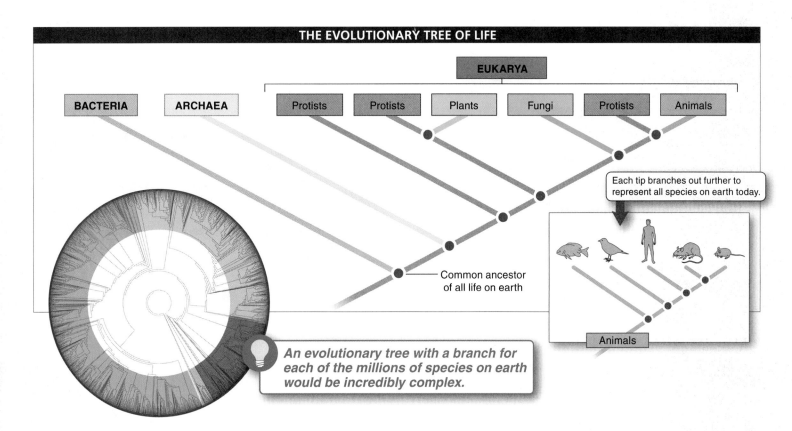

**FIGURE 10-15 A growing tree.** The evolutionary tree of life has branches with millions of tips representing all species on earth. As speciation events occur, new branches are added to the tree.

## MONOPHYLETIC GROUPS

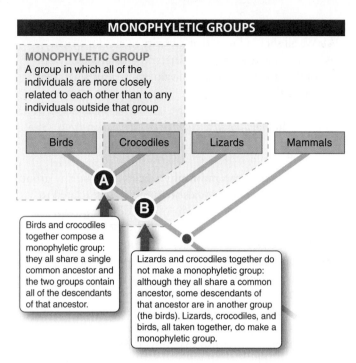

**MONOPHYLETIC GROUP**
A group in which all of the individuals are more closely related to each other than to any individuals outside that group

Birds | Crocodiles | Lizards | Mammals

**A**

**B**

Birds and crocodiles together compose a monophyletic group: they all share a single common ancestor and the two groups contain all of the descendants of that ancestor.

Lizards and crocodiles together do not make a monophyletic group: although they all share a common ancestor, some descendants of that ancestor are in another group (the birds). Lizards, crocodiles, and birds, all taken together, do make a monophyletic group.

**FIGURE 10-16 Members of a monophyletic group share a common ancestor, and the group contains all of the descendants of that ancestor.**

Notice also that in Figure 10-15, the **protists** (or "Protista"), a kingdom of mostly single-celled eukaryotes, are not actually a monophyletic group, contrary to original expectations.

## Constructing evolutionary trees by comparing similarities and differences among organisms

Reading an evolutionary tree reveals which groups are most closely related and approximately how long ago they shared a common ancestor. But how are evolutionary trees—which might hypothesize historical events, such as the cat-dog split, that happened 60 million years ago—constructed in the first place?

Until recently, these trees were assembled by looking carefully at numerous physical features of species and generating tables that compared these features across the species. **FIGURE 10-17** is a simple example of such a table, showing a clear split between the characteristics of the lion and the hyena, on the one hand, and those of the wolf and the bear, on the other. For most of the 20th century, biologists classifying organisms would often use 50 or more traits to generate a tree.

Then, beginning in the 1980s, biologists began using molecular sequences rather than physical traits to generate evolutionary trees. The rationale for this approach is that organisms inherit DNA from their ancestors and so, as species diverge, their DNA sequences also diverge, becoming increasingly different. As more time passes following the splitting of one species into two, the differences in their DNA sequences become greater. By comparing how similar the DNA sequences are between two groups, it is possible to estimate how long ago they shared a common ancestor (**FIGURE 10-18**).

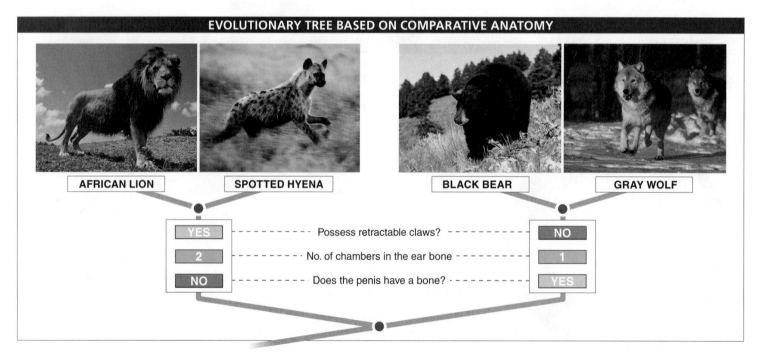

### EVOLUTIONARY TREE BASED ON COMPARATIVE ANATOMY

**AFRICAN LION** | **SPOTTED HYENA** | **BLACK BEAR** | **GRAY WOLF**

| YES | Possess retractable claws? | NO |
| 2 | No. of chambers in the ear bone | 1 |
| NO | Does the penis have a bone? | YES |

**FIGURE 10-17 Looking for clues in body structures.** Before DNA sequencing became available, physical features of species were used to determine evolutionary relatedness.

## DNA-BASED EVOLUTIONARY TREES

By comparing how similar the DNA sequences are between two groups, it is possible to estimate how long it has been since they shared a common ancestor.

Difference in DNA base pairs

Species 1 and 2 are more similar to each other than they are to either species 3 or 4. This is because they probably share a more recent common ancestor.

Species 1    Species 2    Species 3    Species 4

A
T
C
G

*The more similar their DNA sequences, the more closely two species are related.*

**FIGURE 10-18** DNA sequences reveal evolutionary relatedness.

In theory, the construction of evolutionary trees using DNA sequences is not really different from using physical traits. Each DNA base pair can be thought of as a "trait," and many, many more such "traits" can be compared.

Although most evolutionary trees are now produced by using molecular sequence comparisons, when it comes to actually naming species we still use Linnaeus's system, assigning a species name and fitting the new species within the other categories such as kingdom, class, and family. This can be difficult, because there is no objective way to assign groups above the level of species—for example, deciding whether multiple genera should be grouped in the same or different families, or whether multiple families should be grouped in the same or different orders.

### TAKE-HOME MESSAGE 10·8

**Evolutionary trees constructed by biologists are hypotheses about the ancestor-descendant relationships among species. The trees represent an attempt to describe which groups are most closely related to which other groups.**

# 10·9 ——— Similar structures don't always reveal common ancestry.

Whales and sharks have fins; bats and many insects have wings. Do bats have wings because they are most closely related to insects and inherited their wings from a common, winged ancestor, or did insects' and bats' wings evolve independently? As the evolutionary tree in **FIGURE 10-19** makes clear, insects' wings and bats' wings evolved independently—an adaptation that arose separately on more than one occasion.

The mapping of species' characteristics onto phylogenetic trees provides us with the story of evolution. Before the 1980s, biologists had tried to decode the process of evolution by comparing physical features of organisms. Then, with the advent of methods for comparing DNA sequences, tracing the history of life on earth became a more rigorous science. Let's look at a case that illuminates why the original methods were weaker (**FIGURE 10-20**).

Initially, biologists thought that African golden moles belonged in the order known as insectivores, which includes shrews, hedgehogs, and other moles. This belief seemed reasonable because the moles have many characteristics in common with these other animals: they are small, they have long, narrow snouts, their eyes are tiny, and they live in underground burrows. Biologists thought that this group of characteristics evolved just once, and that all species in the insectivore order possessed these characteristics because they inherited them from a common ancestor.

Recent evidence from DNA sequencing, however, surprised everyone who studies these animals. The DNA evidence reveals that African golden moles are more closely related to elephants than to the insectivores, including all of the other mole species! Why do they look so similar to the insectivores? Because of a phenomenon called **convergent evolution,**

## REPEATED EVOLUTION OF THE WING

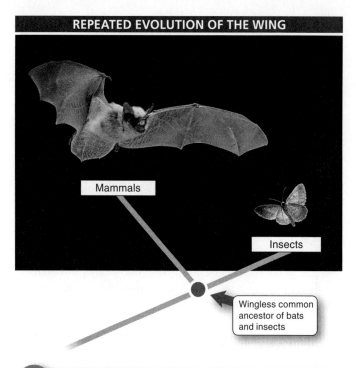

Bats and insects share a wingless common ancestor, proving that bat and insect wings evolved wings independently.

Mammals

Insects

Wingless common ancestor of bats and insects

**FIGURE 10-19** Because wings are an adaptation that has evolved independently many times in populations living in similar environments under similar selective pressures, the presence of wings does not necessarily indicate evolutionary relatedness.

Common shrew

African golden mole

African elephant

The African golden mole resembles a shrew, but DNA sequence comparisons reveal that it is more closely related to an elephant.

**FIGURE 10-20 Looks can be deceiving.** Which two of these three animals are more closely related? Can you tell just by looking?

which occurs when populations of different organisms live in similar environments and so experience similar selective forces. **Analogous traits** are characteristics (such as bat wings and insect wings) that are similar because they were produced by convergent evolution, not because they descended from a common structure in a shared ancestor. Features that are inherited from a common ancestor are called **homologous features.** All mammals have hair, for example, because they inherited this trait from a common ancestor.

Analogous features are problematic when biologists are constructing evolutionary trees, because they are the result of natural selection rather than relatedness. As a consequence, these traits should not be considered a sign of common ancestry when constructing an evolutionary tree. But how do we know whether traits are homologous or analogous? Through DNA analysis. DNA sequences do not become similar during convergent evolution, so molecular phylogenies cannot be fooled. This is an important reason why molecular-based evolutionary trees are preferable to trees based on physical features.

**Q** Humans have several fused vertebrae at the bottom of their spinal cord that look like a tiny, internal "tail." Are we evolving a tail or losing a tail?

Although analogous traits and the process of convergent evolution are not helpful in constructing evolutionary trees, that doesn't mean they are useless to biologists. Far from it. Convergence provides some of the best evidence for the power of natural selection. For example, from the 19th to the 20th centuries in industrialized parts of Europe, a large number of distinct butterfly and moth species became darker and darker. This change was not the result of their sharing a common ancestor, but instead came about because industrialization had caused an accumulation of dark soot on the bark of many trees. This soot accumulation caused light-colored moths and butterflies to stand out against the background when they landed on trees, making them an easier target for predators than dark-colored individuals. As a consequence, natural selection led to a rapid evolution of body color, with many species converging on a darker coloration at the same time.

### TAKE-HOME MESSAGE 10·9

Evolutionary trees are best constructed by comparing DNA sequences among organisms rather than comparing physical similarities, because convergent evolution can cause distantly related organisms to appear closely related, but it doesn't increase their DNA sequence similarity.

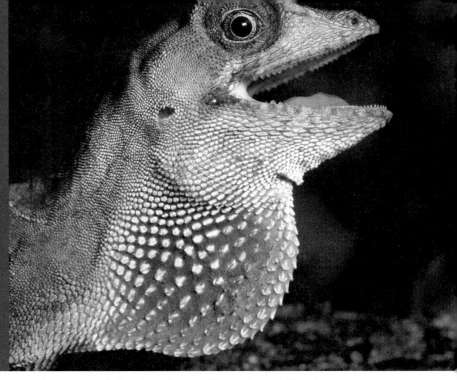

## 10·10–10·13
## Macroevolution gives rise to great diversity.

More than 400 Anolis lizard species exist. (Shown here: A blue-throated anole, Anolis chrysolepsis, *from Ecuador.*)

# 10·10  Macroevolution is evolution above the species level.

When water runs over rocks, it wears them away. The process is simple and slow, yet powerful enough to have created the Grand Canyon. To be sure, water running over rocks does not always make a Grand Canyon. Nonetheless, no additional physical processes are necessary. The process of evolution has a lot in common with a stream of water running over rocks: in the short term, it produces small changes in a population, yet the accumulation of these changes over the long term can be "canyonesque" (**FIGURE 10-21**). Let's consider some examples.

The production of 200-ton dinosaurs from rabbit-size reptile ancestors. The diversification from a single species of flowering plants into more than 230,000 species. These large-scale examples, which are the products of evolutionary change involving the origins of entirely new groups of organisms, are referred to as **macroevolution.** These examples can be contrasted with phenomena involving changes in allele frequencies in a population—referred to as **microevolution**—such as the increase in milk production in cows during the first half of the 20th century, or the gradual change in the average beak size of birds with changing patterns of rainfall (**FIGURE 10-22**).

These micro and macro events might seem like two very different processes, but they are not. Evolution, whether at the micro or macro level, is one thing only: a change in allele frequencies over time. In the short term, over one or a few generations, evolution can appear as a slight and gradual change within a species. But

*No additional processes necessary! Just as a stream of water running over rocks can create the Grand Canyon, accumulated microevolutionary changes can lead to dramatic macroevolutionary changes.*

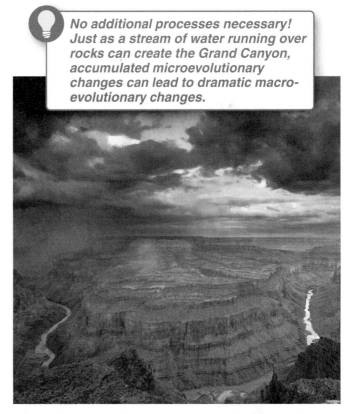

**FIGURE 10-21 From a trickle of water to the Grand Canyon.** The Grand Canyon is a product of small changes to the landscape occurring over millions of years.

## EVOLUTION: MICRO vs. MACRO

Evolution is one thing only: a change in allele frequencies within a population. But over time, these changes can lead to new species and groups of species that vary tremendously.

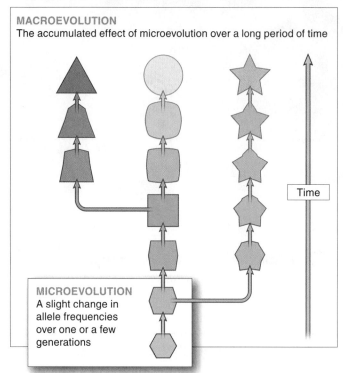

**MACROEVOLUTION**
The accumulated effect of microevolution over a long period of time

Time

**MICROEVOLUTION**
A slight change in allele frequencies over one or a few generations

**FIGURE 10-22 The scale of evolution.** Microevolutionary changes in allele frequencies in a population over time can lead to macroevolution, changes on a grand scale, including vast diversification of species.

over a longer period of time, the accumulated effects, acting continuously and combined with reproductive isolation of populations, can lead to the dramatic phenomena described as macroevolution. Just as a trickle of water, given enough time, produced the Grand Canyon, so evolution has created the endless forms of diversity on earth. In a sense, microevolution is the process and macroevolution is the result.

### TAKE-HOME MESSAGE 10·10

The process of evolution—changes in allele frequencies within a population—in conjunction with reproductive isolation is sufficient to produce speciation and the rich diversity of life on earth.

# 10·11 The pace of evolution is not constant.

If you listened in on some scientists debating their research, you might hear one side describing "evolution by jerks" and the other side making light of "evolution by creeps." They're not gearing up for a fight out in the schoolyard. They're just debating the pace of evolution.

The traditional model of how evolutionary change occurred was that populations changed slowly but surely, gradually accumulating sufficient genetic differences for speciation—hence the phrase "evolution by creeps." Spurred on by findings from the fossil record that do not always support this view, however, researchers have come to believe that evolution may often occur in a different way: brief periods of rapid evolutionary change immediately after speciation, followed by long periods with relatively little change—hence "evolution by jerks." This newer view of the pace of evolution, in which long periods of relatively little evolutionary change are punctuated by bursts of rapid change, is called **punctuated equilibrium** (**FIGURE 10-23**).

In nature, we can find examples of both gradual change and the irregular pattern of punctuated equilibrium. For many groups of organisms, such as many mollusks and mammals, there are a lot of fossils for a particular species and they reveal that it seems to have persisted with little change for very long periods of time. Then, a large number of newer mollusk or mammal fossils appear in the fossil record that are similar enough to the older fossils that they seem to be in the same evolutionary lineage, yet there are no fossils representing the gradual transition between the newer and older fossils. For other groups, such transitional fossils do exist. For example, very complete sets of fossils reveal the transitional sequence from fox-like ancient terrestrial mammals to modern whales.

Punctuated equilibrium has erroneously been portrayed as a "problem" for evolutionary biology, for two primary reasons. First, it is mistakenly believed that the rapid evolutionary changes of punctuated equilibrium are the result of some new mechanism of evolution—thus seeming to undermine

## THE TEMPO OF EVOLUTION

The pace of evolution varies for different species. Some species have evolved gradually over time, while others spend vast amounts of time with little change.

### GRADUAL CHANGE
Evolution by creeps: The pace of evolution occurs gradually in incremental steps.

Time

### PUNCTUATED EQUILIBRIUM
Evolution by jerks: Long periods of relatively little evolutionary change are punctuated by bursts of rapid change.

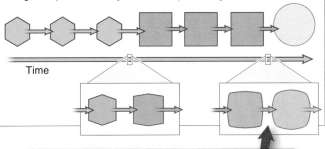

Time

Even though this period of rapid evolutionary change may cover only 1% of the species' evolutionary history, it still may cover hundreds or thousands of generations. This could take tens of thousands of years in a primate or a matter of months in bacteria.

**FIGURE 10-23 The pace of evolution can be rapid or slow.**

evolutionary theory. In actuality, although the punctuated change appears lightning-fast on a geological time scale, the per-generation rates of change necessary to produce it are relatively slow and not at all beyond the rates of change that have been measured in evolving populations.

Second, the existence of "gaps" in the fossil record seems to imply that the transitional species did not exist—again seeming to contradict evolutionary theory. To the contrary,

though, punctuated equilibrium should generate exactly the patterns found. Consider an example. In a large population, speciation is most likely to occur at the edges of the species' range. At the edges, small populations are more likely to be separated and geographically isolated from the rest of the population. The population is also likely to experience the most extreme selective pressures at the edges of its range, where the habitat differs most from the rest of its range and may represent the edge of the species' ecological tolerance. This border region is where speciation and rapid evolutionary change are most likely to occur. But because the individuals there represent only a tiny fraction of all the individuals in the population, they will make up only a tiny fraction of the fossil record.

In the end, there is no single rate at which evolution occurs across all species. Some species have spent vast periods of time with little change at all. The coelacanth species of fish, for example, seems to have undergone almost no physical changes over more than 300 million years. Other species have clearly changed gradually, over time, for their entire evolutionary history. For any organisms, the rate of evolutionary change depends on the selective forces acting on the population. Strong and directional forces may act over long periods of time and lead to rapid change, while stabilizing selection may lead to very little change.

Darwin foreshadowed the discussion about the pace of evolution in *The Origin of Species,* writing, "the periods during which species have undergone modification, though long as measured in years, have probably been short in comparison with the periods during which they retain the same form."

### TAKE-HOME MESSAGE 10·11

The pace at which evolution occurs can be rapid or very slow. In some cases, the fossil record reveals long periods with little evolutionary change punctuated by rapid periods of change. In other cases, species may change at a more gradual but consistent pace.

# 10·12 Adaptive radiations are times of extreme diversification.

Even the greatest of success stories often owe something to a bit of luck. As mammals, it might seem that we owe a little of our success to luck. Flash back to 65 million years ago. Our mammalian ancestors had their place among the organisms on

earth, but these rodent-size, insect-eating, nocturnal creatures didn't remotely possess the dominant position we hold today. It was the dinosaurs' time, and the giant reptiles dominated the earth.

All that changed in an instant when the earth was struck by an asteroid, about 6 miles (10 km) in diameter, near what is now the eastern part of Mexico. This caused the environmental conditions on earth to change quickly and drastically. We'll explore the details of this catastrophic event in the next section, but one outcome was extreme: almost all of the dinosaur species were wiped out. Our mammalian ancestors, though, survived and found themselves living on a planet where most of their competitors had suddenly disappeared. We were in the right place at the right time.

What followed was an explosive expansion of mammalian species. In a brief period of time, a small number of species diversified into a much larger number of species, able to live in a wide diversity of habitats. Called an **adaptive radiation,** such a large and rapid diversification has occurred many times throughout history (**FIGURE 10-24**).

Three different phenomena tend to trigger adaptive radiations. After one of these events, surviving species find themselves in locations where they suddenly have access to plentiful new resources.

**1. Mass extinction events.** With the near-total disappearance of the dinosaurs, a world of "opportunities" opened up for the mammals. Where previously the dinosaurs had prevented mammals from utilizing resources, mammals suddenly had few competitors. Not surprisingly, the number of mammalian species increased from a very small number—perhaps a few hundred—to more than 4,000 species in about 130 genera. This happened over about 10 million years, barely the blink of an eye by geological standards. Following other large-scale extinctions, numerous other groups that suddenly lost most of their competitors experienced similar adaptive radiations.

**2. Colonization events.** In a rare event, one or a few birds or small insects will fly off from a mainland and end up on a distant island group, such as Hawaii or the Galápagos Islands. Once there, they tend to find a large number of opportunities for adaptation and diversification. In the Galápagos, as we learned in Section 10-6, 14 finch species evolved from a single species found on the nearest mainland, 600 miles away. In Hawaii, there are several hundred species of fruit flies, all believed to have evolved from one species that colonized the islands—perhaps blown there by a storm, or carried there stuck in the feathers of a bird—and experienced an adaptive radiation.

**3. Evolutionary innovations.** In the world of computers, software developers are always looking for the "killer app"— the new application so useful that it immediately leads to huge success, opening up a large new niche in the software market or greatly expanding an already-existing niche. The first spreadsheet, email program, and web browser were all killer apps. In nature, evolution sometimes produces killer apps, too. These are innovations such as the wings and rigid outer skeleton that appeared in insects and helped them diversify into the most successful group of animals, with more than 800,000 species today (more than a hundred times the number of mammalian species). The flower is another innovation that propelled an explosion of diversity and evolutionary success of flowering plants relative to the non-flowering plants, such as ferns and pine trees. Today, about 9 out of 10 plant species are flowering plants.

## ADAPTIVE RADIATION

Adaptive radiation occurs when a small number of species diversify into a larger number of species. Three phenomena tend to trigger adaptive radiation.

**MASS EXTINCTION EVENTS**
With their competition suddenly eliminated, remaining species can rapidly diversify.

**COLONIZATION EVENTS**
Moving to a new location with new resources (and possibly fewer competitors), colonizers can rapidly diversify.

**EVOLUTIONARY INNOVATIONS**
With the evolution of an innovative feature that increases fitness, a species can rapidly diversify.

**FIGURE 10-24 A rapid diversification of species.**

### TAKE-HOME MESSAGE 10·12

Adaptive radiations—brief periods of time during which a small number of species diversify into a much larger number of species—tend to be triggered by mass extinctions of potentially competing species, colonizations of new habitats, or the appearance of evolutionary innovations.

# 10·13

## There have been several mass extinctions on earth.

If the past is a guide to the future, we know this: no species lasts forever. Speciation is always producing new species, but **extinction,** the complete loss of all individuals in a species population, takes them away. Extinction, which is always occurring, is faced by all species.

For any given time in earth's history, it is possible to estimate the rate of extinctions, and the evidence reveals that these rates are far from constant (**FIGURE 10-25**). Although the details differ in most cases, extinctions generally fall into one of two categories: "background" extinctions or mass extinctions.

**Background extinctions** are the extinctions that occur at lower rates during periods other than those of mass extinctions. Background extinctions occur mostly as the result of natural selection. Competition with other species, for example, may reduce a species' size or the range over which it can roam or grow. Or a species might be too slow to adapt to gradually changing environmental conditions and becomes extinct as its individuals die off.

**Mass extinctions** are periods during which a large number of species on earth become extinct over a relatively short period. There have been at least five mass extinctions on earth and, during each of these extinctions, 50% or more of the animal species living at the time became extinct.

There is a fundamental difference between background and mass extinctions that goes beyond differences in rates. They have different causes. Mass extinctions are due to extraordinary and sudden changes to the environment (such as an asteroid impact). As a consequence, nothing more than bad luck is responsible for the extinction of species during mass extinctions; fit and unfit individuals alike perish.

Of the five mass extinctions during the past 500 million years, the most recent is also the best understood. Sixty-five million years ago, an almost unimaginable catastrophe occurred. A massive asteroid smashed into the Caribbean near the Yucatán Peninsula of Mexico (**FIGURE 10-26**). The impact left a crater more than 100 miles wide, and probably created an enormous fireball that caused fires worldwide, followed by a cloud of dust and debris that blocked all sunlight from the earth and disturbed the global climate for months. In the aftermath of this catastrophe, about 75% of all species on earth were wiped out, including almost all dinosaurs, a tremendously successful group that had been thriving for 150 million years. (Among the rare surviving dinosaurs were those that would become the birds.)

As bad as the dinosaur-devastating asteroid event was, from a biodiversity perspective it is not the worst catastrophe in earth's history. That distinction falls to a mass extinction that took place 250 million years ago, called the Great Dying (see Figure 10-25). Although the cause is not clear—hypotheses include an asteroid impact, continental drift, a supernova (a star explosion), or extreme volcanic eruptions—more than 95% of all marine life became extinct, along with almost 75% of all terrestrial vertebrates. The causes of the three other mass

## TWO CATEGORIES OF EXTINCTIONS

Extinctions generally fall into two categories:

○ **MASS EXTINCTIONS**
A large number of species become extinct over a short period of time due to extraordinary and sudden environmental change.

○ **BACKGROUND EXTINCTIONS**
These extinctions occur at lower rates during times other than mass extinctions.

**FIGURE 10-25 Extinction never sleeps.**

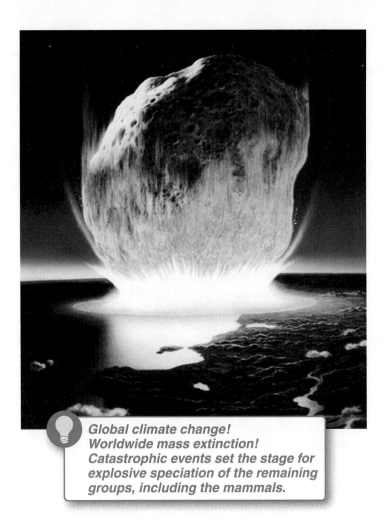

**FIGURE 10-26 Catastrophic collision.** Sixty-five million years ago, an asteroid 6 miles in diameter smashed into earth near the Yucatán Peninsula of Mexico and wiped out three-fourths of all species on earth, including almost all the dinosaurs.

*Global climate change! Worldwide mass extinction! Catastrophic events set the stage for explosive speciation of the remaining groups, including the mammals.*

extinctions also are poorly understood, but the evidence in the fossil record of the tremendous upheavals is clear.

It is interesting to note that although, from time to time, sudden and extreme events lead to mass extinctions that wipe out a large proportion of the species on earth, in every case these mass extinctions are followed by a time of explosive speciation in the groups that survive. An important question now being debated is whether we are currently in the midst of a human-caused mass extinction event. We'll explore whether this is true—and why—in Chapter 16.

> **TAKE-HOME MESSAGE 10·13**
>
> As new species are being created, others are lost through extinction, which may be a consequence of natural selection or large, sudden changes in the environment. Mass extinctions are periods during which a large number of species on earth become extinct over a short period of time. These events are usually followed by periods of unusually rapid adaptive radiation and diversification of the remaining species.

*The flower hat jellyfish (Olindias formosa), a rare species that lives in the oceans off Brazil, Argentina, and southern Japan.*

# 10·14

## All living organisms are classified into one of three groups.

Biological diversity can be humbling. At the very least, it makes it harder to believe that humans are particularly special. We are not at the center of earth's "family tree." Nor are we at its peak. We are simply one branch.

When Linnaeus first put together his system of classification, he saw a clear and obvious split: all living organisms were either plant or animal. Plants could not move and could make their own food. Animals could move but could not make their own food. So in Linnaeus's original classification, all organisms were put in either the animal kingdom or the plant kingdom (his third, "mineral" kingdom, now abandoned, included only non-living matter).

With the refinement of microscopes and subsequent discovery of the rich world of **microbes**—microscopic organisms—the two-kingdom system was inadequate. Where did the microbes belong? Some could move, but many of those could also make their own food, seeming to put them somewhere between plants and animals. And the problems didn't stop with the microbes: mushrooms and molds, among other organisms originally categorized as plants, didn't move but they didn't make their own food either—they digested the decaying plant and animal material around them.

The two-kingdom system gave way in the 1960s to a five-kingdom system. At its core, the new system was a division based on the distinction between *prokaryotic* cells (those without nuclei) and *eukaryotic* cells (those with nuclei). The prokaryotes were put in one kingdom, where the only residents were the bacteria: single-celled organisms with no nucleus, no organelles, and genetic material in the form of a circular strand of DNA. The eukaryotes—having a nucleus, compartmentalized organelles, and individual, linear pieces of DNA—were divided into four separate kingdoms: plants, animals, fungi, and protists.

The classification of organisms took a huge leap forward in the 1970s and 1980s, and the five-kingdom system had to be discarded. Until that point, organisms had been classified primarily based on their appearance. But because the ultimate goal of classification had changed to reconstructing phylogenetic trees that reflected the evolutionary history of earth's diversity, Carl Woese, an American biologist, and his colleagues began classifying organisms by their nucleotide sequences.

Woese assumed that the more similar the genetic sequences were between two species, the more closely related they were,

THE THREE DOMAINS OF LIFE

EUKARYA

BACTERIA ARCHAEA Protists Protists Plants Fungi Protists Animals

Common ancestor of all life on earth

**FIGURE 10-27 All living organisms are classified into one of three groups.**

from bacteria. The tree of life was revised to show three primary branches, called domains: the bacteria, the archaea, and the eukarya (**FIGURE 10-27**). The domain names are often capitalized: Bacteria, Archaea, and Eukarya.

Woese put the domains above the kingdom level in the Linnaean system. In Woese's new system, which is the most widely accepted classification scheme today, the bacteria and archaea domains each have one kingdom and the eukarya domain has four. Because both bacteria and archaea are microscopic, it can be hard to believe that the two domains are as different from each other as either domain is from the eukarya. However, each of the three domains is monophyletic, meaning that each contains species that share a common ancestor and includes all descendants of that ancestor. Close inspection even reveals that the archaea are more closely related to the eukarya than they are to the bacteria.

The three-domain, six-kingdom approach is not perfect and is still subject to revision. As we saw earlier, for example, within the eukarya, the single-celled protists have turned out to be much more diverse than initially thought and, problematically, are not a monophyletic group. Increasingly, it is recognized that they should be split into multiple kingdoms. Also problematic is that bacteria sometimes engage in **horizontal gene transfer,** which means that, rather than passing genes simply from "parent" to "offspring," they transfer genetic material directly into another species. This process complicates the attempt to determine phylogenies based on sequence data, because it creates situations in which two organisms might have a similar genetic sequence, not because they share a common ancestor, but as a result of a direct transfer of the sequence from one species to another.

**Q** **How can one molecule support the claim that all living organisms probably evolved from a single common ancestor?**

and he built phylogenetic trees accordingly. The only way Woese could compare the evolutionary relatedness of all the organisms present on earth today was by examining one molecule that was found in *all* living organisms and looking at the degree to which it differed from species to species. He discovered a perfect candidate for this role: a molecule called ribosomal RNA, which helps translate genes into proteins (see Chapter 5). Ribosomal RNA has the same function in all organisms on earth, almost certainly because it comes from a common ancestor. Over time, however, its genetic sequence (i.e., the DNA that codes for it) has changed a bit. Tracking these changes makes it possible to reconstruct the process of diversification and change that has taken place.

The trees that Woese's genetic sequence data generated had some big surprises. First and foremost, the sequences revealed that the biggest division in the diversity of life on earth was not between plants and animals. It wasn't even between prokaryotes and eukaryotes. The new trees revealed instead that the diversity among microbes was much, much greater than ever imagined—particularly because of the discovery of a completely new group of prokaryotes called **archaea** (*sing.* archaeon), which thrive in some of the most extreme environments on earth and differ greatly

**Q** **Are viruses alive?**

Additionally, a fourth group of incredibly diverse and important biological entities, the **viruses,** is not even included in the tree of life, because they are not considered to be living organisms. Viruses can replicate, but can have metabolic activity only by taking over the metabolic processes of another organism.

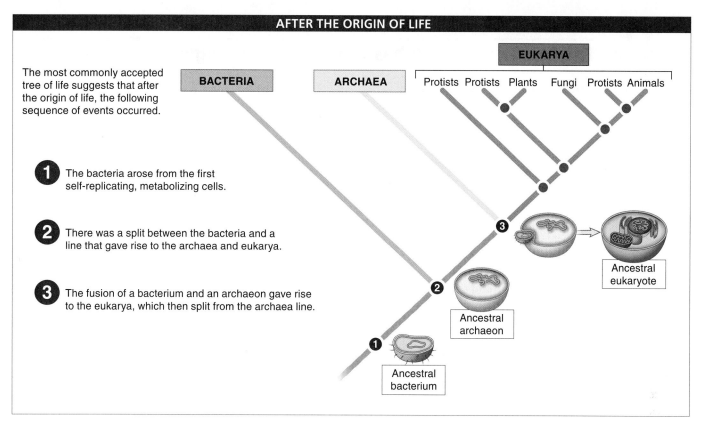

The most commonly accepted tree of life suggests that after the origin of life, the following sequence of events occurred.

**1** The bacteria arose from the first self-replicating, metabolizing cells.

**2** There was a split between the bacteria and a line that gave rise to the archaea and eukarya.

**3** The fusion of a bacterium and an archaeon gave rise to the eukarya, which then split from the archaea line.

**FIGURE 10-28  From self-replicating, metabolizing cells to complex organisms.** The biggest branches on the tree of life separate the three domains.

Their lack of metabolic activity puts viruses just outside the definition of life that we use in this book, but some scientists do view viruses as living.

The most commonly accepted tree of life suggests that, after the origin of life, the following sequence of events occurred (**FIGURE 10-28**).

1. The bacteria arose from the first self-replicating, metabolizing cells.

2. There was a split between the bacteria and a line that gave rise to the archaea and eukarya.

3. The fusion of a bacterium and an archaeon gave rise to the eukarya, which then split from the archaea line.

The next three sections introduce each of the three domains, surveying the broad diversity that has evolved within each domain and the common features that link all the members of each. Chapters 11, 12, and 13 cover the domains in greater detail.

### TAKE-HOME MESSAGE 10·14

All life on earth can be divided into three domains—bacteria, archaea, and eukarya—which reflect species' evolutionary relatedness to each other. Plants and animals are just two of the four kingdoms in the eukarya domain, encompassing only a small fraction of the domain's diversity.

# 10·15 The bacteria domain has tremendous biological diversity.

Morning breath is stinky. And learning the cause of the offensive smell may make the situation even worse. You see, when you wake up, your mouth contains huge amounts of bacterial waste products. Perhaps the only consolation is that it gives us a glimpse into just how diverse and resourceful bacteria are.

At any given time, there are several hundred species of bacteria in your mouth—mostly on your tongue—all competing for the resources you put there (**FIGURE 10-29**). Some of the bacteria are aerobic, requiring oxygen for their metabolism, and others are anaerobic. At night, because the flow of saliva slows down and the oxygen content of your mouth decreases, the anaerobic bacteria get the upper hand in terms of growth and reproduction. These bacteria metabolize food bits in your mouth, plaque on your teeth and gums, and dead cells from the lining of your mouth, breaking down proteins in these materials to use as their energy source. Because proteins are made from amino acids, some of which contain the smelly chemical sulfur, their breakdown leads to the odor in the accumulating waste products.

Once you wake up, you breathe more and produce more saliva, both of which increase the oxygen level in your mouth. This tips the battle for space and food back in favor of the aerobic bacteria. Because aerobic bacteria prefer carbohydrates as their energy source and because carbohydrates don't contain sulfur, the sulfur smell goes away as the aerobic bacteria start to outcompete the anaerobic bacteria. The aerobic bacteria are, of course, also filling your mouth with waste products, it's just that their waste products don't smell as bad.

**Q** Why is morning breath so stinky?

On a small scale, your mouth reveals some of the tremendous biological versatility of the bacteria: hundreds of species can live in a tiny area—a teaspoon of soil, for example, is home to more than a billion bacteria—they can thrive in a variety of unexpected habitats, they can utilize a variety of food sources, and they can survive and thrive with or without oxygen. Looking around the world, we find the clear dominance of bacteria. By any measure, it is their planet: the biomass of bacteria (if they were all collected, dried out, and weighed) exceeds that of all the plants and animals on earth. Bacteria live in soil, air, water, arctic ice, and volcanic vents (**FIGURE 10-30**). Many can even make their own food, utilizing light from the sun or harnessing energy from chemicals such as ammonia.

While the various species differ in many ways, the bacteria are a monophyletic group, sharing a common ancestor. For this reason, they all have a few features in common. All bacteria are single-celled organisms with no nucleus or organelles, with one or more circular molecules of DNA as their genetic material, and using several methods of exchanging genetic information. Because they are asexual—they reproduce without a partner, just by dividing—the biological species concept cannot be applied to bacteria when classifying them into narrower categories. Thus, bacteria are classified on the basis of physical appearance or, preferably, genetic sequences.

Many people think of bacteria as illness-causing organisms. However, although bacteria are responsible for many diseases, including strep throat, cholera, syphilis, pneumonia, botulism, anthrax, leprosy, and tuberculosis, disease-causing bacteria are only a small fraction of the domain, and bacteria seem to get less credit for their many positive effects on our lives. Consider that bacteria (*E. coli*) living in your gut help your body digest the food you eat and, in the process, make certain vitamins your body needs. Other bacteria

**FIGURE 10-29 Bacteria thrive on your tongue and in your mouth.**

> Bacteria have a biomass greater than that of all the plants and animals on earth. They thrive in a variety of habitats including arctic ice, volcanic vents, and the human intestine.

**FIGURE 10-30 We're outnumbered.** Shown here: bacteria thriving on the bristle of a toothbrush.

(actinomycetes) produce antibiotics such as streptomycin. Still others live symbiotically with plants as small fertilizer factories, converting nitrogen into a form that is usable by the plant. Bacteria also give taste to many foods, from sour cream to cheese, yogurt, and sourdough bread. Increasingly, bacteria are used in biotechnology—from those that can metabolize crude oil and help in the cleanup of spills to transgenic bacteria used in the production of insulin and other medical products, as described in Chapter 5.

We explore the great diversity of the bacteria in more detail in Chapter 13.

---

**TAKE-HOME MESSAGE 10·15**

All bacteria share a common ancestor and have a few features in common. They are prokaryotic, asexual, single-celled organisms with no nucleus or organelles, with one or more circular molecules of DNA as their genetic material, and using several methods of exchanging genetic information. Bacteria have a much broader diversity of metabolic and reproductive abilities than do the eukarya.

---

# 10·16 The archaea domain includes many species living in extreme environments.

In a bubbling hot spring in Yellowstone National Park, the temperature ranges from the 212° F (100° C) of boiling water down to a relatively cool 165° F (74° C) at the surface. It would seem a most inhospitable place for life. Yet researchers have found 38 different species of archaea thriving there. Perhaps even more surprising, the genetic differences among these species are more than double the genetic differences between plants and animals.

In the freezing waters of Antarctica, too, archaea abound. More than a third of the organisms in the Antarctic surface waters are archaea. In swamps, completely devoid of oxygen, and in the extremely salty water of the Dead Sea, the story is the same: where once it was assumed that no life could survive, the archaea not only exist but thrive and diversify (**FIGURE 10-31**).

If biologists ever find themselves thinking that we have a handle on the breadth of life that is possible on earth, they

**FIGURE 10-31 Archaea can thrive in the most inhospitable-seeming places.** Archaea live in a diversity of environments; they resemble bacteria but are actually more closely related to eukarya.

**Q**

*Will life be found elsewhere in the universe? How does the discovery of archaea alter that likelihood?*

would be wise to remember the archaea. Until relatively recently, our perception of life on earth was that there were bacteria and there were eukaryotes, such as the plants and animals. But in the past several decades, as researchers have explored some of the most unlikely of habitats, they have found archaea thriving.

There are entire worlds of life on earth that we never even imagined.

Analyses of genetic sequences indicate that the archaea and the bacteria diverged about 3 billion years ago and that the eukarya split off from the archaea approximately 2.5 billion years ago. The archaea are grouped in one kingdom within the domain archaea, but we have no idea how many species exist. Given that they are the dominant microbes in the deep seas, it may very well be that archaea—which we were completely ignorant of until recent decades—are the most common organisms on earth. It is still too early to tell.

We do know that, like bacteria, all archaea are single-celled prokaryotes. For that reason, under a microscope they look very similar to bacteria. Several physical features distinguish them from the bacteria, however. Specifically, the archaeal cell walls contain polysaccharides not found in either bacteria or eukaryotes. The archaea also have cell membranes, ribosomes, and some enzymes similar to those found in the eukarya.

The archaea exhibit tremendous diversity and are often divided into five groups based on their physiological features:

1. Thermophiles ("heat lovers"), which live in very hot places

2. Halophiles ("salt lovers"), which live in very salty places

3. High- and low-pH-tolerant archaea

4. High-pressure-tolerant archaea, found as deep as 4,000 meters (about 2.5 miles) below the ocean surface, where the pressure is almost 6,000 pounds per square inch (compared with an air pressure of less than 15 pounds per square inch at sea level)

5. Methanogens, which are anaerobic and produce methane

There also seem to be large numbers of archaea living in relatively moderate environments that are also commonly home to bacteria. We explore the great diversity of the domain archaea in more detail in Chapter 13.

---

**TAKE-HOME MESSAGE 10·16**

Archaea, many of which are adapted to life in extreme environments, physically resemble bacteria but are more closely related to eukarya. Because they thrive in many habitats that humans have not yet studied well, including the deepest seas and oceans, they may turn out to be much more common than currently believed.

---

# 10·17 The eukarya domain consists of four kingdoms: plants, animals, fungi, and protists.

After exploring the archaea and bacteria domains, turning to the eukarya feels like coming home. We are most familiar with this domain: all of the living organisms that we can see are eukarya. Plants, mushrooms, slime molds, insects, fish, birds, and, of course, mammals, including humans. There are three kingdoms of eukarya that can be seen with the naked eye: plants, animals, and fungi (**FIGURE 10-32**). All are made up from eukaryotic cells—they have a membrane-enclosed nucleus—and each kingdom is almost entirely multicellular.

A fourth kingdom contains the protists, which are most often invisible to the naked eye, and is a sort of grab bag

that includes a wide range of mostly single-celled eukaryotic organisms, including amoebas, paramecia, and algae. Discovery of new species of protists, as with the other microscopic organisms on earth (the bacteria and archaea), continues at a very high rate. As we noted earlier, biologists now know that the protists are not a monophyletic group and are increasingly splitting them into multiple kingdoms within the eukarya.

The eukarya split from the archaea about 2.5 billion years ago. At the time, the eukarya were single-celled and probably resembled modern protists more than they did any other modern eukarya. The split may have occurred when symbiotic

**FIGURE 10-32 All shapes and sizes.** The slime mold, passion flower, mushrooms, and an orangutan reflect the diversity of eukaryotic organisms.

bacteria became incorporated within an ancestor of the eukaryotes, resulting in what would become mitochondria. Later, approximately 1.5 billion years ago, a second important symbiosis between bacteria and eukarya resulted in what would become chloroplasts (see Chapter 3).

Because they are so much easier to see than bacteria and archaea, a disproportionate number of the named species on earth are in the domain eukarya. In fact, of the 1.5 million named species, the majority are eukarya, with about half being insects. This is more a result of the interests and biases of biologists than a reflection of the relative numbers of actual species in the world. We explore the great diversity of the eukarya in Chapters 11 (on animals) and 12 (on plants and fungi).

---

### TAKE-HOME MESSAGE 10·17

All living organisms that we can see with the naked eye (and many that are too small to be seen) are eukarya, including all plants, animals, and fungi. The eukarya are unique among the three domains in having cells with organelles.

## DO RACIAL DIFFERENCES EXIST ON A GENETIC LEVEL?

In a fraction of a second, we are aware of someone's size, gender, approximate age—and race. But what is race? Does race have biological meaning? And can humans be categorized into groups that can scientifically be called races?

**Q: WHAT IS RACE, BIOLOGICALLY SPEAKING?** The biological species concept hinges on a straightforward question: can two individuals interbreed under natural conditions? The concept of race, conversely, is not straightforward. Some biologists have proposed that races are groups of interbreeding populations that are phylogenetically distinguishable but reproductively compatible. But it is important to note that, in humans, beyond phylogenetic distinctions, religion, culture, ethnicity, and geography are common aspects of many legal, social, and political conceptions of race. So do such social constructions of "race" reflect meaningful divisions?

**Q: WHAT DO HUMAN RACIAL GROUPINGS REFLECT?** Are blacks and whites genetically different? The answer is obviously yes; black-skin alleles differ from white-skin alleles. Moreover, the prevalence of some genetic diseases varies by race and ethnicity. Sickle-cell anemia, for example, is relatively common among Africans and Southeast Asians because the genes that cause it also improve a person's resistance to malaria, a disease prevalent in Africa and Southeast Asia.

Yet Africans from the southernmost part of the continent have no greater risk of sickle-cell anemia than do the Japanese, because malaria is similarly rare in both of their homelands. Should we use this trait rather than skin color when determining a person's race? If we did, southern Africans would be grouped with the Japanese and northern Africans with southern Europeans (who share their genetic resistance to malaria and the higher risk of sickle-cell anemia).

This observation reveals an important problem with racial groupings, which DNA sequence comparisons also show: *similarity in skin color between two people doesn't tell you anything about overall genetic similarity.* There is, for example, a huge variation in blood type among Africans: some are type O, some AB, others A or B. But the same goes for Asians and Turks, Russians and Spaniards. Yet we do not group people according to blood type or any other physiological and biochemical variations.

**Q: WHAT IS THE FUTURE OF RACE?** Racial groupings such as "Hispanic" or "black" or "Asian" do not reflect consistent genetic differences. Still, race remains an important part of individual, social, cultural, and ethnic identity for many, and its use is mandated by federal agencies. For these reasons, it will remain a powerful social construct, even as scientific consensus emerges that racial groupings do not reflect meaningful genetic distinctions.

# Big Ideas in ... The Origin and Diversification of Life on Earth

**10·1–10·2 Life on earth most likely originated from non-living materials.**

- Under conditions similar to those on early earth, small organic molecules form that have some chemical properties of life.

- Self-replicating molecules—possibly RNA—may have acquired or developed membranes, enabling them to replicate and making metabolism possible, the two conditions that define life.

- However it came about, the earliest life, which resembled bacteria, appeared about 3.5 billion years ago, not long after the earth formed.

**10·3–10·6 Species are the basic units of biodiversity.**

- Species are distinct biological entities, named using a hierarchical system of classification.

- Reproductive isolation and genetic divergence together lead to speciation.

- The distinction between two species is not easily defined for all types of organisms.

**10·7–10·9 Evolutionary trees help us conceptualize and categorize biodiversity.**

- The history of life can be visualized as a tree; by tracing from the branches back toward the trunk, we can follow the pathway back from descendants to ancestors.

- A phylogenetic tree of life reveals the evolutionary history of every species and the sequence of speciation events that gave rise to them.

- Evolutionary trees are best constructed by comparing genetic similarities among organisms, because convergent evolution can cause distantly related organisms to appear much more closely related.

**10·10–10·13 Macroevolution gives rise to great diversity.**

- The process of evolution—changes in allele frequencies within a population—in conjunction with reproductive isolation is sufficient to produce speciation, diversification, and the rich diversity of life on earth.

- The pace at which evolutionary change occurs can be rapid or very slow.

- As new species are being created, others are lost through extinction, which may be a consequence of natural selection or large, sudden changes in the environment.

- Mass extinctions are usually followed by periods of unusually rapid adaptive radiation and diversification of the remaining groups or organisms.

**10·14–10·17 An overview of the diversity of life on earth: organisms are divided into three domains.**

- All life on earth can be divided into three domains, which reflect species' phylogeny, their evolutionary relatedness to each other.

- Bacteria have a much broader diversity of metabolic and reproductive abilities than eukarya.

- Archaea, many of which are adapted to life in extreme environments, physically resemble bacteria but are more closely related to eukarya.

- The eukarya are unique in having cells with organelles.

See the To-Go Guide (packaged with every new copy of this textbook) for an illustrated summary of the chapter.

# Check Your Knowledge

## GRAPHIC CONTENT

*Thinking critically about visual displays of data.*

**1.** What is the approximate rate of background extinctions over the time period shown here?

**2.** What distinguishes the mass extinctions from the background extinctions in this figure?

**3.** Why do you think the graph records the percentage of families becoming extinct rather than the percentage of species becoming extinct?

**4.** Why isn't the green line indicating the rate of background extinctions a straight line?

**5.** Rates of extinction are generally estimated based on the fossil record. Why might this not be an accurate measure of the magnitude of a mass extinction?

**6.** Conservation biologists believe it is important that we keep a close watch on the current rate of extinctions. Why might that be?

### TWO CATEGORIES OF EXTINCTIONS

**● MASS EXTINCTIONS**
A large number of species become extinct over a short period of time due to extraordinary and sudden environmental change.

**● BACKGROUND EXTINCTIONS**
These extinctions occur at lower rates during times other than mass extinctions.

The Great Dying

Extinction of almost all dinosaur species (and ¾ of all species on earth)

*Extinction rate (percentage of families that became extinct)* vs *Time (millions of years ago)*

See Answers at the back of the book.

# Key Terms in The Origin and Diversification of Life on Earth

# Multiple-Choice Questions

**1.** In a set of classic experiments performed in the early 1950s, Urey and Miller subjected an experimental system composed of $H_2$, $CH_4$ (methane), and $NH_3$ (ammonia) to electrical sparks. A few days later, they found _____ in their system.

   a) amino acids
   b) DNA
   c) microspheres
   d) cells
   e) RNA

100 HARD

13 EASY

**2.** What kind of molecule is thought to have been the first genetic material?

   a) protein
   b) DNA
   c) carbohydrate
   d) RNA
   e) microsphere

100 HARD

25
0 EASY

**3.** According to the biological species concept, species are natural populations of organisms that have the potential to interbreed and are _____ isolated from other such populations.

   a) behaviorally
   b) prezygotically
   c) postzygotically
   d) geographically
   e) reproductively

100 HARD

31
0 EASY

**4.** The classification rank that includes genera but not orders is:

   a) species.
   b) domain.
   c) class.
   d) kingdom.
   e) family.

100 HARD

50
0 EASY

**5.** Populations of *Larus* gulls around the North Pole show an unusual pattern of reproductive isolation: each population is able to interbreed with its neighboring populations, but populations separated by larger geographic distances are not able to interbreed. *Larus* gulls are an example of a(n) _____ species.

   a) ring
   b) polyploid
   c) circular
   d) Escher
   e) linked

100 HARD

62
0 EASY

**6.** In animals, it is believed that the most common mode of speciation is:

   a) autopolyploidy.
   b) chromosomal.
   c) directional.
   d) allopatric.
   e) sympatric.

100 HARD

54
0 EASY

**7.** Polyploidy:

   a) arises only when there is an error in meiosis resulting in diploid gametes instead of haploid gametes.
   b) is a common method of sympatric speciation for animals.
   c) arises when allopatric speciation causes plants to have fewer sets of chromosomes than their parent plants.
   d) is an increased number of sets of chromosomes.
   e) always results in allopatric speciation.

100 HARD

46
0 EASY

**8.** In the plant kingdom, all of the species are descended from a single common ancestor. In terms of phylogeny, what type of tree of life is this?

   a) monophyletic
   b) uniphyletic
   c) punctuated
   d) sympatric
   e) allopatric

100 HARD

27
0 EASY

**9.** Phylogenetic trees should be viewed as:

a) true genealogical relationships of species.

b) the result of vertical, but never horizontal, gene transfer.

c) intellectual exercises, not to be interpreted literally.

d) representations of allopatric speciation events.

e) hypotheses regarding evolutionary relationships of groups of organisms.

100 HARD
34
0 EASY

**10.** The difference between microevolution and macroevolution is that:

a) they take place on different time scales.

b) macroevolution occurs with physical structures, whereas microevolution occurs with physiological traits.

c) microevolution occurs with physical structures, whereas macroevolution occurs with physiological traits.

d) microevolution has been proven, whereas macroevolution is very speculative.

e) microevolution occurs in prokaryotes, whereas macroevolution takes place among eukaryotes.

100 HARD
47
0 EASY

**11.** The idea of punctuated equilibrium challenges which component of Darwin's theory of evolution?

a) steady change

b) gradualism

c) species stasis

d) Both a) and b) are correct.

e) None of the above is correct.

100 HARD
45
0 EASY

**12.** Which of the following scenarios would best facilitate adaptive radiation?

a) A population of birds native to an island archipelago is forced to relocate to the mainland by a storm.

b) A population of cheetahs goes through an event in which all genetic diversity in the population is wiped out.

c) Darker-colored moths have a selective advantage over lighter-colored moths due to industrial soot on trees.

d) A population of birds becomes stranded on an island archipelago.

e) All of the above would facilitate adaptive radiation equally.

100 HARD
78
0 EASY

**13.** The mass extinction that occurred on earth 65 million years ago was immediately followed by:

a) the rise of archaea.

b) the emergence of the first non-photosynthetic organisms.

c) the rise of the reptiles, including the dinosaurs.

d) an increase in atmospheric oxygen levels.

e) the rapid divergence and radiation of modern mammals.

100 HARD
46
0 EASY

**14.** Prokaryotes are classified into _____ domain(s).

a) 1

b) 2

c) 3

d) 4

e) 5

100 HARD
42
0 EASY

**15.** All protists are alike in that all are:

a) bacterial.

b) photosynthetic.

c) eukaryotic.

d) prokaryotic.

e) archaeal.

100 HARD
42
0 EASY

**16.** Which of the following domains share the most recent common ancestor?

a) archaea and eukarya

b) bacteria and eukarya

c) archaea and bacteria

d) None of the above; all three domains evolved from different ancestors.

e) None of the above; all three domains are equally related to each other.

100 HARD
57
0 EASY

**17.** Which of the following groups would be placed nearest the fungi in an evolutionary tree based on DNA sequences?

a) plants

b) bacteria

c) animals

d) archaea

e) protists

100 HARD
41
0 EASY

# Short-Answer Questions

1. Explain the difference between defining a species according to the biological species concept and according to the morphological species concept. Include an example of why one method might be more appropriate than the other for certain situations.

2. Evolutionary trees are important in helping to explain the history of life. Outline the difference between how evolutionary trees were constructed in the 20th century and how they are constructed today. Explain what types of information these trees can and cannot convey.

3. In mammals, the shape of the skull, as well as the shape and size of teeth, often reflects the method of feeding. A skull with a more rounded jaw and shorter, narrower teeth indicates a mammal that is well adapted to pulling bunches of leaves off trees and grinding them up. A skull with a more square jaw indicates a mammal adapted to grazing and cropping vegetation at the ground level. Explain how the adaptations of long, flat teeth and a more square jaw or muzzle could have triggered adaptive radiation in horses.

See Answers at the back of the book.
For additional study questions, go to www.prep-u.com.

# Visibility in motion

# 11

# Animal Diversification

427

# 11·1–11·3 Animals are just one branch of the Eukarya domain.

A scarlet macaw in flight (Ara macao), Tambopata National Reserve, Peru.

# 11·1

## What is an animal?

Looking at a kangaroo, it's obvious that it is an animal. The same goes for a mosquito, earthworm, or jellyfish. It gets a bit more difficult, though, when you look at a sponge. Is it an animal? Or is it a plant? Or is it something else? This is when it is helpful to have some guidelines for identifying and organizing the biodiversity we can see on earth.

For the sponge, the answer is that it is classified as an animal. Here's why. When defining what is or is not an animal, we look for certain features that are common to all animals. Surprisingly, considering the enormous differences among even the few animals mentioned above, just three characteristics are generally sufficient to define **animals** (**FIGURE 11-1**).

### WHAT IS AN ANIMAL?

**ANIMALS EAT OTHER ORGANISMS**
All animals acquire energy by consuming other organisms.

**ANIMALS MOVE**
All animals have the ability to move—at least at some stage of their life cycle.

**ANIMALS ARE MULTICELLULAR**
Animals consist of multiple cells and have body parts that are specialized for different activities.

**FIGURE 11-1 Characteristics of an animal.** Left to right: A garter snake eating a grasshopper, dolphins, and a nudibranch.

**1.** All animals eat other organisms. That is, rather than manufacturing their own food, as plants and some bacteria do through photosynthesis, animals consume plants, bacteria, other animals, fungi, or some combination of these. And because the best way to catch other organisms is to move toward them or chase after them, that brings us to the second characteristic that sets animals apart from plants.

**2.** All animals move—*at least at some stage of their life cycle.* That qualifying phrase is important because some animals move only during a period called the "larval stage" that comes early in their lives, and when they become adults they fasten themselves to a surface and stay there. If you visit a rocky seashore, you can find two of these animals— mussels and barnacles—living on rocks in the tidal zone. Organisms that are fastened in place, such as adult mussels and barnacles, are said to be **sessile,** but even animals that are sessile as adults moved when they were larvae.

**3.** All animals are multicellular, and they generally have body parts that are specialized for different activities. For example, most animals possess sensory organs (eyes, ears, nose, tongue, antennae, whiskers, and so on) that enable them to acquire information about their environment and, among other things, detect potential prey and predators.

There are a few other general features of animals—most reproduce sexually, for example—but these three are the easiest criteria to use in determining whether or not you're looking at an animal.

> ### TAKE-HOME MESSAGE 11·1
>
> **Animals are organisms that share three characteristics: all of them eat other organisms, all can move during at least one stage of their development, and all are multicellular.**

# 11·2 Four key distinctions divide the animals.

More than two-thirds of the almost two million species on earth that have been identified are animals. All of these animals share a common ancestor, an ancestral protist. (This last common ancestor to all of the animals is thought to have been a free-living unicellular organism resembling a sperm in size and shape.) With such a large number of animals to organize and identify, it is useful to divide them into a few large categories, based on their evolutionary relatedness.

Analyses of DNA and RNA sequences have helped biologists identify four major groupings into which we can classify all animals. Each of these groupings hinges on a particular adaptation and on the question of whether or not an animal has descended from an ancestor with that adaptation (**FIGURE 11-2**). At any level of classification, of course, there are additional characteristics that can be used to further subdivide and classify the animals. We'll explore many of those characteristics throughout this chapter, but first let's look at four of the key distinctions.

**1. Does the animal have specialized cells that form defined tissues?** If an animal has no tissues—groups of cells with a common structure and function—then it belongs to the phylum Porifera, better known as the sponges. Sponges are aggregations of similar cells, with little coordination of activities among the cells. All other animals have clearly defined tissues. Humans, for example, have

highly specialized cells such as skin cells, muscle cells, and sensory cells.

**2. Does the animal develop with radial symmetry or bilateral symmetry?** All animals with defined tissues develop in a shape that has some sort of symmetry: radial or bilateral. **Radial symmetry** describes animals with a body structured like a pie, such as jellyfishes, corals, and sea anemones. Slow-moving or free-floating, radially symmetrical animals have no front or back ends, and it is possible to make multiple slices, all going through the center, that divide the organism into identical pieces. In contrast, organisms with **bilateral symmetry** (such as humans, cows, and scorpions) have left and right sides that are mirror images. These animals can move adeptly, searching for food and avoiding predators.

Radial symmetry      Bilateral symmetry

Classification of animals as bilaterally or radially symmetrical is based on the pattern of symmetry seen in

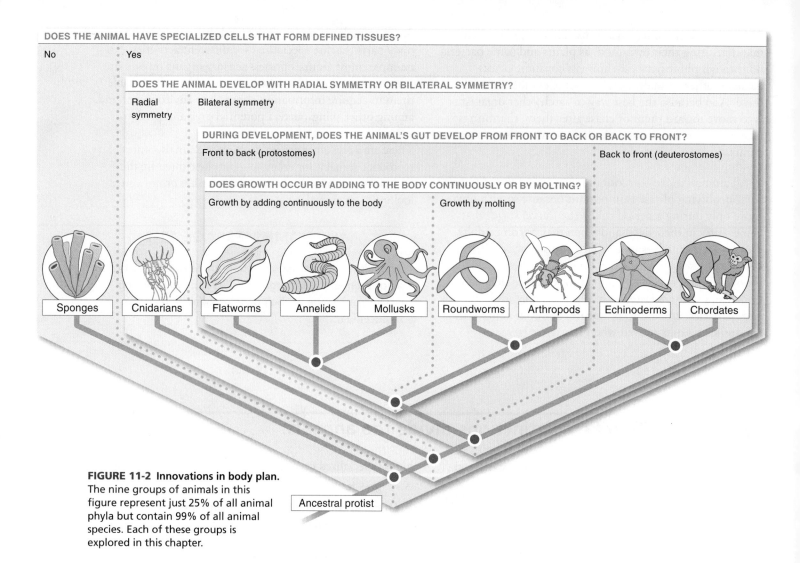

DOES THE ANIMAL HAVE SPECIALIZED CELLS THAT FORM DEFINED TISSUES?

No | Yes

DOES THE ANIMAL DEVELOP WITH RADIAL SYMMETRY OR BILATERAL SYMMETRY?

Radial symmetry | Bilateral symmetry

DURING DEVELOPMENT, DOES THE ANIMAL'S GUT DEVELOP FROM FRONT TO BACK OR BACK TO FRONT?

Front to back (protostomes) | Back to front (deuterostomes)

DOES GROWTH OCCUR BY ADDING TO THE BODY CONTINUOUSLY OR BY MOLTING?

Growth by adding continuously to the body | Growth by molting

Sponges | Cnidarians | Flatworms | Annelids | Mollusks | Roundworms | Arthropods | Echinoderms | Chordates

Ancestral protist

**FIGURE 11-2 Innovations in body plan.** The nine groups of animals in this figure represent just 25% of all animal phyla but contain 99% of all animal species. Each of these groups is explored in this chapter.

the embryo. As we'll see, in some animals, such as starfish, embryonic development is bilaterally symmetrical, but then the animal becomes radially symmetrical as an adult.

**3. During development, does the animal's gut develop from front to back or from back to front?** Early in the evolution of animals—about 630 million years ago—a major split occurred that allows us to separate the bilaterally symmetrical animals with defined tissues into two distinct lineages: the **protostomes** (which can be translated as "mouth first") and the **deuterostomes** ("mouth second"). These names refer to the way the gut develops—from front to back or from back to front. In protostomes, the gut develops from front to back, so the first opening that forms becomes the mouth of the adult animal, and the last opening becomes the anus. In deuterostomes, the gut develops from back to front. The anus is the first opening to form, and the second is the mouth. Although it's something you can see only when closely observing animals' embryonic development, the

protostome-deuterostome split is the most basic division of animals, in terms of revealing their evolutionary relatedness to one other.

**4. Does growth occur by molting or by adding continuously to the skeletal elements?** Among the bilaterally symmetrical animals that are protostomes, there is an important distinction between animals that molt and those that do not. Animals that molt (such as lobsters and insects) shed their exoskeleton (a hard outer layer) and replace it with a larger one at regular intervals during development. Other animals (such as clams, octopuses, cats, and humans) grow by adding to the size of their body in a continuous matter.

Although all animals share a common ancestor, the ancestral protist, these four distinctions help us identify and organize all of the approximately 36 phyla of living animals. And these four distinctions divide the animals into monophyletic groups. Recall from Chapter 10 that biologists use the term "monophyletic" to describe any group in which all of the individuals are more closely related to one another than to any individuals outside that group. Additional evolutionary transitions have occurred within the groups created by these four distinctions, allowing further categorization of animals into monophyletic groups. These categories help us understand the sequence of evolutionary events that have resulted in the tremendous animal diversity we see in the world today.

# 11·3

## Everything that is not extinct is evolutionarily successful.

"Never use the word higher or lower." This is a note that Charles Darwin wrote to himself in the margin as he was reading a book that proposed a theory of cosmic and biological evolution. It was an important insight, but one that can be difficult to adhere to—particularly as we turn our attention to the animals.

What is "success," evolutionarily speaking? There are two possible states for a species. It can be *extant,* meaning that it currently exists. Or it can be *extinct.* From an evolutionary perspective, every living species is successful. The fact that a species exists means it can do the things all organisms must do: find food, escape predators, and reproduce. Earthworms and tigers live very different lives, and the characteristics that work well for one species would not work at all for the other. But the earthworm is as successful at being a worm as the tiger is at being a very large cat (**FIGURE 11-3**).

Throughout his writings, Darwin emphasized adaptation among the species he described and their relationship to their environment. One species was never better or worse than another. Rather, each species was differentiated from the others, showing specializations that adapted individuals of that species to the particular niche in which they lived. Different environments posed different challenges. All extant species, as evidenced by their existence and persistence, are able to overcome those challenges.

Our tour through the animal kingdom begins with animals having a simple body plan, such as sponges, jellyfishes, and worms, continues through the more complex invertebrates, the mollusks and arthropods, and culminates in the complexity of the vertebrates—mammals (including humans), amphibians, reptiles (including birds), and fishes. Taking this path allows us to explore the major evolutionary transitions in the chronological order in which they occurred. This order does not provide any information about the relative success of any species. Nor does it imply that humans are the "crown of creation."

Although each extant species represents an evolutionary success, certain adaptations have led to the rapid and extensive diversification of particular *groups* of species. We investigate nine of the phyla with the most species (shown in Figure 11-2) and the evolutionary innovations that contributed to their success. These groups represent only a quarter of all animal phyla, but they account for about 99% of the animal species on earth today.

From an evolutionary perspective, every living species is successful.

FIGURE 11-3 Two equally successful organisms: the earthworm and the tiger.

# 11·4 Invertebrates are animals without a backbone.

*Sea nettles (Chrysaora quinquecirrha) are voracious invertebrate predators of zooplankton in Atlantic estuaries (where rivers meet the ocean), such as Chesapeake Bay.*

## 11·4 Invertebrates are the largest and most diverse group of animals.

All animals are divided into the four large categories, described in Section 11.2, which are based on patterns of evolutionary relatedness. A distinction that is perhaps more commonly used, however, is whether the animal has a backbone or not. **Invertebrates** are animals that do not have a backbone; **vertebrates** are animals that do have a backbone.

We begin our tour of the animal kingdom with the invertebrates. This group takes the prize for diversity and abundance, accounting for more than 95% of species in the animal kingdom (**FIGURE 11-4**). An example of this diversity is seen in body structure: some invertebrates (such as worms) have soft bodies, others (including snails) have soft bodies inside a shell, and still others (including insects and crustaceans such as lobsters) have a rigid external covering called an **exoskeleton.**

No invertebrate can be considered to represent the group as a whole, but the phylum Arthropoda (the **arthropods**) has the largest number of species: more than 75% of the animals on earth are arthropods. This phylum includes the most familiar invertebrates: insects, spiders, scorpions, lobsters, crabs, shrimp, and their relatives.

As "animals without backbones," the invertebrate grouping has a convenient and easy-to-apply definition, but the invertebrates are not a monophyletic group. All invertebrates are protostomes—except for one large group, whose members are all deuterostomes. Called echinoderms, and including sea stars, sea urchins, and sand dollars, this animal phylum shows the same from-back-to-front gut development as the vertebrates, and these animals are, in fact, the vertebrates' closest relatives.

The invertebrates include eight separate evolutionary lineages with an enormous diversity of body forms, body sizes, habitats, and behaviors. Sections 11-5 through 11-12 examine these lineages in detail.

### TAKE-HOME MESSAGE 11·4

Invertebrates, defined as animals without a backbone, are the largest and most diverse group of animals, comprising more than 95% of all living animal species. The invertebrates are not a monophyletic group, however, and include protostomes and deuterostomes.

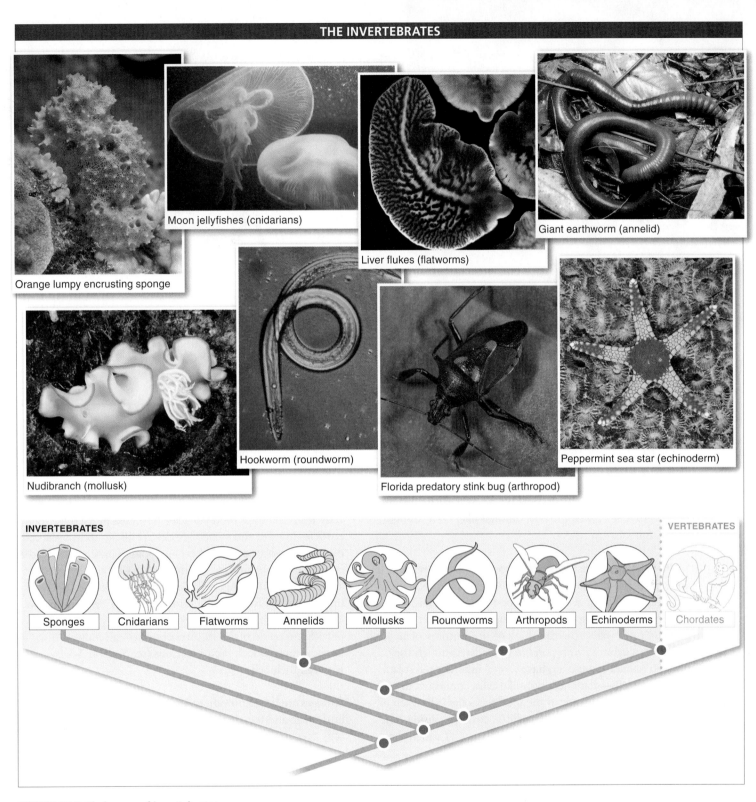

Orange lumpy encrusting sponge

Moon jellyfishes (cnidarians)

Liver flukes (flatworms)

Giant earthworm (annelid)

Nudibranch (mollusk)

Hookworm (roundworm)

Florida predatory stink bug (arthropod)

Peppermint sea star (echinoderm)

INVERTEBRATES

VERTEBRATES

Sponges | Cnidarians | Flatworms | Annelids | Mollusks | Roundworms | Arthropods | Echinoderms | Chordates

**FIGURE 11-4 Phylogeny of invertebrates.**

# 11·5–11·12 Across several evolutionary transitions, the invertebrate animals diversified.

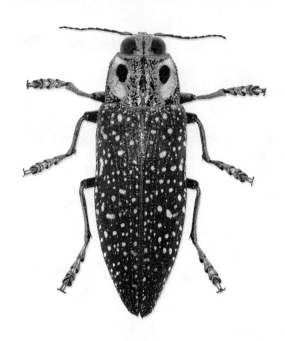

*The most diverse animals on earth? More than 25% of all described species are beetles. Shown here: a wood-boring beetle (Lampropepia rothschildi) from Madagascar.*

## 11·5

### Sponges are animals that lack tissues and organs.

In the natural world, we see many things that are hard to believe. Consider this: sponges (phylum Porifera) are animals. They are multicellular, they eat other organisms, and (for part of their life) they are mobile (**FIGURE 11-5**). Let's investigate this group that may be the least "animal-like," and simplest, of all the animals.

Sponges have no specialized tissues or organs, and most lack any symmetry. Despite their simplicity, however, sponges are remarkably efficient at obtaining food. Living as sessile suspension-feeders, sponges strain suspended matter and food particles from water. You can think of the anatomy of an adult sponge as a cylinder with pores in its walls. Although they have no tissues, sponges do have several different cell types. Epidermal cells—flattened, skin-like cells, connected to each other—cover the outside of the sponge. Collar cells, each of which has a long, whip-like flagellum, cover the inside. The middle layer is a gel, with embedded fibers that stiffen the sponge's body (**FIGURE 11-6**).

The beating of each collar cell's flagellum creates a current that carries water upward and out through the opening at the top of the sponge. That water is replaced by an inward flow of water through the pores on the sides of the sponge. The incoming water carries bacteria and algae, other microscopic organisms, and small particles of organic material into the sponge's body. The volume of

water moving through its body daily can be as much as 20,000 times the volume of the sponge.

In the "collars" of the collar cells, sticky mucus covering the cells' thin cilia traps food particles. The cilia move these food particles inward, and some particles are taken up into the cells by endocytosis (see Section 3-11). Sponges have no circulatory systems to move nutrients from one place to another. Instead, a special kind of sponge cell, called an amoebocyte cell, picks up some of the food particles from the collar cells and moves around through the sponge's body, transferring the food particles to epidermal cells. Amoebocytes can move freely because sponges lack tight junctions (see Section 3-12) between their cells.

Sponges are hermaphrodites. That is, each individual contains both male and female reproductive organs, but it produces only one kind of gamete (eggs or sperm) at a time. Sponges that are acting as males release a cloud of sperm that swim to other sponges, which are acting as females, and fertilize their eggs. The eggs develop into larvae, which are released into the water and then drift for a few days before settling on a rock or coral outcrop and developing into a sessile adult sponge.

Sponges also reproduce asexually: small buds form on the outside of the sponge and eventually break off, settle to the bottom, and grow into new sponges. Even a piece

## THE ANATOMY OF A SPONGE

Water

Epidermal cells

Collar cell

Flagellum

Pore

Amoebocyte

Gel layer

*A fist-size sponge can filter more than 1,000 gallons (5,000 liters) of water every day!*

**FIGURE 11-6 Sponge structure.**

## THE SPONGES

Purple sponge

Brown tube sponge

**COMMON CHARACTERISTICS**
- No tissues or organs
- Body consists of a hollow tube with pores in its wall
- Feed by pumping in water, along with bacteria, algae, and small particles of organic material, through their pores
- Free-swimming larvae
- Sessile as adults

**MEMBERS INCLUDE**
- About 5,000 species

**FIGURE 11-5 An overview of the sponges.**

accidentally broken off by wave action or a collision with a passing fish will grow into a new sponge. Most remarkable is the ability of sponge cells to reassemble. Picture this experiment. You put a living sponge into a food blender and puree it, then strain the liquid through a fine sieve to remove all the chunks, leaving a suspension of individual cells, which you then dump into an aquarium. Within a day you'll see small clumps of sponge forming as the individual cells move around and attach to each other when they meet, and within a week a new sponge will form. For an even more dramatic demonstration, you can repeat the experiment with two sponge species of different colors. Not only do the sponges reassemble, but the cells from each species assemble only with other cells of the same species, so you'll have two sponges, each with only the cells of its own species.

Because the dried skeletons of some sponges are soft and absorbent, humans have long made use of them. Sponges have been used as padding in helmets and as a tool for painting, for example. And in ancient Rome, a sponge on a stick was commonly used as toilet paper. Some true sponges are still marketed for use in personal hygiene, but these are far from living. They are just a matrix of soft, elastic fibers of a protein called spongin, isolated from dried sponges. Because over-fishing has dramatically reduced the populations of these sponges, they are not widely available. Rather, the "sponges" used in most kitchens and bathrooms today are nearly always produced from synthetic materials. Loofah sponges, contrary to their name, are made from plant material.

**Q** Is your kitchen sponge an animal? Is it alive?

### TAKE-HOME MESSAGE 11·5

Sponges are among the simplest of the animal lineages. A sponge consists of a hollow tube with pores in its wall; it has no tissues or organs. Sponges reproduce sexually (by producing eggs and sperm) and asexually (by budding). The fertilized eggs grow into free-swimming larvae that settle and develop into sessile, filter-feeding adult sponges.

# 11·6 Jellyfishes and other cnidarians are among the most poisonous animals in the world.

If you feel a sting when you are swimming in the ocean, you have met a jellyfish. If you are lucky, it will be one of the thousands of species that feed on tiny floating organisms called plankton, because these species have only mild stings. If it is a species that feeds on larger prey such as fish or shrimp, however, you are definitely unlucky. These jellyfishes have powerful stings that can cause extreme pain or even death.

The jellyfishes, along with sea anemones and corals, belong to the phylum Cnidaria (pronounced nigh-DARE-ee-ah). As with all animals except the sponges, the 11,000 species in this phylum have defined tissues. The cnidarians also all have a simple body plan, with radial symmetry (**FIGURE 11-7**).

There are two types of cnidarian bodies: a sessile polyp and a free-floating medusa, which is the form most people picture when they think of a jellyfish. In some species, individuals spend part of their life cycle as a polyp and part as a medusa. Other species exist only as medusas or, as in corals and sea anemones, only as polyps.

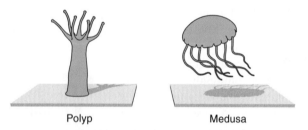

Polyp        Medusa

Cnidarians are carnivores and use their tentacles to capture and feed on a wide variety of marine organisms, from protists to fish and shellfish. Their method of capturing prey relies on a sort of weapon that is unique to the cnidarians—a stinging cell called a cnidocyst. All cnidarians have tentacles, located near their mouth, that are armed with rows of stinging cnidocysts. Each cnidocyst has a coiled thread with barbs inside and a "trigger" on the outside. When something comes in contact with the trigger, perhaps a prey item (or perhaps your leg as you swim in the ocean), the coiled thread is ejected and, like a harpoon, can penetrate the prey, often injecting a toxin (**FIGURE 11-8**). The Portuguese man-o'-war is one of the dangerously poisonous species of cnidarians. Being stung by a Portuguese man-o'-war is painful (extremely painful if the unlucky swimmer becomes entangled in the tentacles and receives hundreds of stings), but these encounters are rarely fatal.

We explore here some of the great diversity among the three major groups of cnidarians.

## THE CNIDARIANS

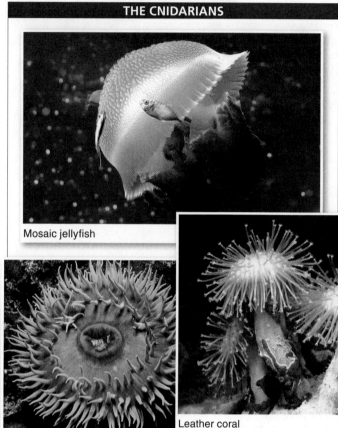

Mosaic jellyfish

Leather coral

Giant green sea anemone

**COMMON CHARACTERISTICS**
- Radially symmetrical
- Tentacles armed with rows of stinging cells, used to paralyze prey

**MEMBERS INCLUDE**
- Jellyfishes
- Sea anemones
- Corals

**FIGURE 11-7 Understanding cnidarians.**

**Corals** The corals live as small (just a few millimeters) polyps in large colonial groups. Corals secrete the chemical calcium carbonate, which gives them a hard outer structure. But if you look inside this hard external structure you'll find

## CNIDOCYST IN ACTION

Cnidarians capture prey using stinging cells called cnidocysts.

**1** Each cnidocyst has a coiled thread with barbs inside and a "trigger" on the outside.

Coiled thread

Trigger

**2** When something comes in contact with the trigger, such as a prey item, the coiled thread is ejected and can penetrate the prey, often injecting a toxin.

**FIGURE 11-8 Stinging cells are used to capture prey.**

animals that are soft-bodied, with stinging tentacles surrounding a mouth. Corals use the tentacles to catch small fish and plankton, which are directed into the mouth by the tentacles, then digested in the stomach. Corals most commonly reproduce sexually, with both external fertilization (releasing sperm and eggs into the water, where they can fuse and form larvae) and internal fertilization (in which only sperm are released and can fertilize eggs within female corals). Corals can also reproduce asexually.

Corals grow in a wide variety of shapes—from free-standing spherical or branched forms to crusts growing on rocks or other corals. A coral reef is a complex three-dimensional structure formed as polyps produce and secrete calcium carbonate. Coral reefs provide an environment that is home to a greater diversity of species than any other marine habitat. The Great Barrier Reef, located in the Coral Sea, off the coast of Queensland, Australia, extends more than 1,500 miles (about 2,600 km) from north to south. A truly awesome display of biological productivity, this reef is the largest biological structure in the world and is easily visible from the International Space Station.

**Q**

**How is global warming affecting the coral reefs of the world?**

Among the first casualties of global warming are coral reefs. Although coral polyps can catch prey using their cnidocysts, they obtain most of their nutrition from algae, called zooxanthellae, that live

symbiotically within the polyps. Each partner contributes and receives something of value in the relationship. The algae use the carbon dioxide produced by the polyps to conduct photosynthesis, and the polyps gain oxygen and nutrients produced by the algae. Strangely, when coral polyps get too hot they expel their zooxanthellae. Because the zooxanthellae are responsible for the colors of many corals, as well as much of their nutrition, when coral polyps expel their zooxanthellae, the coral turns white—a phenomenon called coral bleaching. The zooxanthellae may return to the polyps when the water cools, but repeated bleaching events often lead to the death of the coral polyps. The world's oceans are already warming, and coral bleaching events are becoming more and more frequent (**FIGURE 11-9**).

**Sea anemones** The sea anemones resemble flowers, and some of the colorful species are popular additions to saltwater aquaria. The polyp form of a sea anemone looks a bit like an upside-down medusa form of a jellyfish—the tentacles with

Healthy coral

Bleached coral

*With warming water, corals eject their photosynthetic symbionts, causing coral "bleaching." The polyps turn white and may die, endangering the coral reef and the diversity it supports.*

**FIGURE 11-9 Coral bleaching.**

their stinging cells are at the top of the anemone's body, surrounding its mouth. Sea anemones have a larval stage that swims freely, then settles on a rock and metamorphoses into the adult form. Even as adults, many sea anemones can crawl a few inches a day, and wandering sea anemones are often found in seaweed.

**Jellyfishes**  The jellyfishes (which are not fish!) range tremendously in size. The Asian giant jellyfish is more than 6 feet (about 2 m) across and weighs nearly 500 pounds (more than 200 kg). When swarms of this giant jellyfish appear, they clog the nets of fishing boats with a sticky mass of toxic stingers. At the opposite end of the size scale is the Irukandji jellyfish, about the size of a hen's egg. This species is so deadly that, in 2007, the sighting of just five Irukandji in Hervey Bay, Queensland, completely halted the filming of a major Hollywood movie (*Fool's Gold*), which had to be completed in the safety of a studio.

> ### TAKE-HOME MESSAGE 11·6
>
> Corals, sea anemones, and jellyfishes are radially symmetrical animals with defined tissues, in the phylum Cnidaria. All cnidarians are carnivores and use specialized stinging cells located in their tentacles to capture prey.

# 11·7  Flatworms, roundworms, and segmented worms come in all shapes and sizes.

It's a good thing that biologists don't run bait shops—if they did, a simple trip to buy worms for fishing would be way too complicated. The name "worm" is commonly applied to a long, skinny, slimy animal without a backbone, but you can find animals fitting that description in eight different phyla. You will never encounter most of them, and that's a good thing, because many are parasites that cause really unpleasant diseases. In this section, we consider just three phyla—**flatworms** (phylum Platyhelminthes), **roundworms** (phylum Nematoda), and **segmented worms,** or **annelids** (phylum Annelida)—that illustrate most of the diversity of the animals we call worms.

All of the worms described here have defined tissues and are protostomes—the gut develops from front to back. Unlike the cnidarians, however, the worms (along with all of the remaining animals we discuss in this chapter) develop with a body plan, characterized by bilateral symmetry, which adapts them for forward movement.

It is important to note, however, that worms do not make a monophyletic group. Their general body plan has evolved several times independently, and their evolutionary relationships are not always obvious. The segmented worms (annelids), for example, are believed to be more closely related to mollusks (including snails, clams, and octopuses) than to the roundworms. And the roundworms are believed to be more closely related to the arthropods (insects) than to either the flatworms or the segmented worms (see Figure 11-3).

**Flatworms**  Unlike the roundworms and segmented worms, flatworms have no body cavity. That is, the space between the body wall and the digestive tract is not a fluid-filled cavity. As a consequence, the flatworms have a reduced capacity for diffusion of gases and nutrients. Like the segmented worms, but unlike the roundworms, the flatworms grow by adding body mass, rather than by molting (**FIGURE 11-10**). Flatworms have well-defined head and tail regions, with clusters of light-sensitive cells called eyespots that help them orient themselves within their environment. Most flatworms are hermaphroditic, each individual producing both male and female gametes, and they engage in both sexual and asexual reproduction. Flatworms include more than 20,000 species and can be parasites (the flukes and the tapeworms) or brilliantly colored, free-living aquatic creatures.

Although many flatworms have a digestive system, one feature of the free-living flatworms (but not the annelids or roundworms) that many humans find distasteful is that their gut has only one opening, requiring them to both consume food and eliminate undigested waste through the same opening. Tapeworms, by contrast, completely lack a digestive system; these parasitic worms live in their host's gut and absorb nutrients from the host directly through their body wall.

All 5,000 species of tapeworms are parasites, and most have a two-stage life cycle that is split between two different host species. For example, the common tapeworm that infects dogs spends the other half of its life cycle in fleas, while the human pork tapeworm splits its life cycle between pigs and humans. Tapeworms have long, flat bodies made up of repeated segments, each of which is a reproductive unit. Mature tapeworms spread by breaking off segments, which are then shed in the feces of an infected individual.

## THE FLATWORMS

Tapeworm of the mammalian intestine

Liver fluke

**COMMON CHARACTERISTICS**
- Well-defined head and tail regions
- Hermaphroditic and can engage in both sexual and asexual reproduction
- Some have a single opening in the body, which serves as a mouth and an anus

**MEMBERS INCLUDE**
- Tapeworms
- Flukes

**FIGURE 11-10 Overview of the flatworms.**

**Roundworms** The roundworms, also called nematodes, are probably the most abundant animals on earth—a spoonful of garden soil contains several thousand individuals, and some species produce more than 200,000 eggs every day. Unlike segmented worms and flatworms, they grow by molting. More than 90,000 roundworm species have been named, and there may be five times as many species that have not yet been identified (**FIGURE 11-11**).

Many soil-dwelling roundworms live in the roots of plants, damaging or even killing the plant. About 15,000 species of roundworms are parasites of other animals, and roundworms are responsible for a large number of human diseases. Most roundworms are transmitted by fecal contamination of soil or food.

## THE ROUNDWORMS

Root-knot nematode

Whipworm

**COMMON CHARACTERISTICS**
- Long, narrow unsegmented body
- Bilaterally symmetrical
- Surrounded by a strong, flexible cuticle
- Must molt in order to grow larger

**MEMBERS INCLUDE**
- More than 90,000 species (scientists estimate there may be five times as many species that have not yet been identified)

**FIGURE 11-11 Roundworm diversity.**

Tiny parasitic roundworms called *filariae* are responsible for several tropical diseases, including some that have an enormous social and economic impact in certain parts of the world. Elephantiasis, for example, is a disease in which filariae transmitted by the bite of a mosquito block the lymph ducts so that fluid accumulates in the limbs or scrotum, causing grotesque swelling. These filariae occur in

India, Africa, South Asia, Pacifica, and tropical regions of the Americas.

**Segmented worms (annelids)** The segmented worms, or annelids, number about 13,000 species and are easy to recognize by the grooves running around the body that mark the divisions between segments—if you've seen an earthworm, you are familiar with those grooves. The segmented worms are protostomes with defined tissues, and they do not molt. They are organized into three different groups: marine polychaetes (pronounced POL-lee-keets), terrestrial oligochaetes, and leeches (**FIGURE 11-12**).

Polychaetes are marine worms, living on the seafloor. *Polychaete* means "many bristles," and the combination of segments and bristles makes polychaetes easy to recognize. Some species burrow through the mud and extract organic material from it. Tube worms use sand grains or limestone to make a tube in which they live, with just their waving tentacles exposed. Small particles of food are trapped in mucus on the tentacles and transported to the worm's mouth.

Earthworms, which belong to the group called *oligochaetes* ("few bristles"), are the annelid worms you are most likely to have seen. The night crawler is a typical earthworm. It gets its name from its habit of emerging from its burrow on rainy nights to crawl across the surface of the ground. More than 4,000 earthworm species have been named, ranging in size from less than half an inch (about a centimeter) to the enormous length of the giant Gippsland earthworm, 7–10 feet (2–3 m)!

Earthworms are bulk-feeders—that is, as an earthworm burrows, it consumes particles of soil and organic material. The organic material is digested as it passes through the worm's gut, and the fecal material plus the inorganic part of the soil is excreted as feces, called castings. Earthworm castings are valued by gardeners as soil supplements, and an earthworm can produce its own weight in castings every day. Earthworms' activities mix the soil components, create a more even mixture of nutrients, and expedite the breakdown of organic materials in soil, thus making the nutrients available for plants. Because they perform all of these tasks, earthworms have enormous economic value to agriculture. A government study published in 2008 estimated the economic value of earthworms in Ireland alone to be more than $1 billion per year.

Have you ever waded in a pond or swamp and emerged to find a long, dark brown worm clinging to your leg? If so, you have met the third major group of annelids, the *leeches*. Probably you pulled the leech off (not an easy thing to do, because leeches are both slippery and stretchy) and threw it as

**THE ANNELIDS**

Polychaete worm

Earthworm

Leech

**COMMON CHARACTERISTICS**
• Segmented body

**MEMBERS INCLUDE**
• Marine polychaetes (about 9,000 species)
• Earthworms (more than 4,000 species)
• Leeches (about 500 species)

**FIGURE 11-12 Annelid diversity.**

far as you could. If you had looked at it carefully, though, you'd have noticed that, like earthworms, leeches have segmented bodies.

If you have had a close encounter with a leech, you'll also have noticed that the wound it made on your leg continued to ooze blood for several hours. That's because the saliva of blood-sucking leeches contains an anticoagulant substance that prevents blood from clotting. If the leech is undisturbed while it is feeding, blood keeps flowing and the leech inflates like a small balloon.

Not all leeches are blood-suckers—in fact, more than half the leech species are predators. The horse leech, for example, which reaches a length of 8 inches (about 20 cm), feeds on smaller annelid worms, snails, and aquatic insect larvae.

### TAKE-HOME MESSAGE 11·7

Worms are found in several different phyla and are not a monophyletic group. All are bilaterally symmetrical protostomes with defined tissues. The flatworms and segmented worms (annelids) do not molt; the roundworms do. Flatworms include parasitic flukes and tapeworms, many of which infect humans. Many roundworms are parasites of plants or animals and are responsible for several widespread human diseases. Earthworms are annelids that play an important role in recycling dead plant material.

# 11·8  Most mollusks live in shells.

Just because they are related, organisms don't always share much physical resemblance. Consider these animals: A colossal predatory squid, more than 40 feet long, with eyes the size of beach balls and a razor-sharp beak. A small snail—*escargot* in French—from among the more than 700 million consumed in France each year. An oyster, with its nearly 2,000-year-old reputation as an aphrodisiac (unsupported by any scientific proof, it turns out). All of these are mollusks, among the most diverse groups of animals and occupying an impressive and impressively varied position within the human imagination.

More than 100,000 mollusk species have been named (and many more have not yet been named). The members of this large phylum (Mollusca) of animals live in the ocean, in fresh water, and on land and include many familiar creatures in addition to those mentioned above, including clams, scallops, mussels, and octopuses. They are so diverse that it is difficult to describe any single defining characteristic. The position of mollusks within the animal phylogeny, however, reflects several important characteristics that they all share: they have defined tissues, are bilaterally symmetrical, and are protostomes. Further, all mollusks grow by adding tissue, rather than by molting.

Several additional features are also common to many mollusks. These include a shell that protects the soft body, a mantle (the tissue that secretes calcium carbonate to form the shell), and a sandpaper-like tongue structure, called the radula, that is used during feeding.

There are three major groups of mollusks. The animals in these groups share most of the features common to mollusks, but with very different body plans: **gastropods, bivalve mollusks,** and **cephalopods** (FIGURE 11-13).

**Gastropods** Snails and slugs are gastropod mollusks. *Gastropod* means "belly foot," and these mollusks get their name because they have an expanded foot on the bottom of their body, which allows them to climb a vertical surface as easily as they glide across a horizontal one. Snails have a one-piece, curled shell, and slugs are snails that have reduced the shell to a tiny remnant that is not even visible because it is covered by the mantle. Found in both aquatic and terrestrial environments, snails and slugs account for three-quarters of all mollusks.

The snail's shell is its primary protection against predators, but for terrestrial slugs and sea slugs, with very little shell material, other defense mechanisms are necessary. A terrestrial slug relies on slime for defense: when a slug is attacked, it secretes slime that sticks to the predator. Worse still, anything that touches the slime coating the predator sticks to it, so a bird that attacks a slug quickly finds that its beak and face are covered by pieces of dead leaves, clods of soil, and twigs. Sea slugs have other methods of defense. Some species synthesize toxic chemicals, and others, which feed on sponges, recycle sponge toxins into their own slime. One of the most remarkable uses of another animal's defense equipment is found among sea slugs that eat sea anemones. Anemones are protected by

## THE MOLLUSKS

**COMMON CHARACTERISTICS**
- Most have a shell that protects the soft body
- Mantle (tissue that forms the shell)
- Radula (sandpaper-like tongue structure used during feeding) Found in all mollusks except bivalves

**MEMBERS INCLUDE**
- Gastropods (about 35,000 species)
- Bivalves (about 8,000 species)
- Cephalopods (about 600 species)

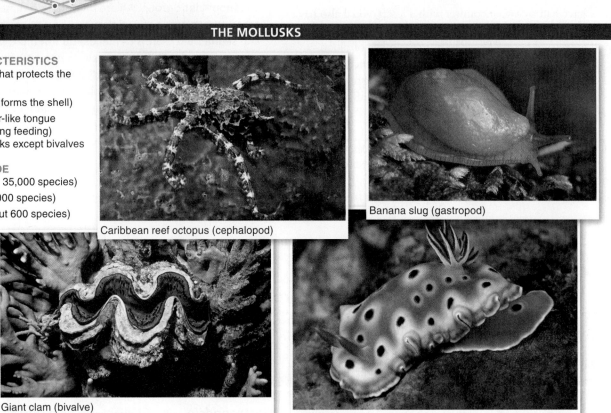

Caribbean reef octopus (cephalopod)

Banana slug (gastropod)

Giant clam (bivalve)

Nudibranch (gastropod)

**FIGURE 11-13 Mollusks: an overview.**

stinging cells, and the sea slugs that eat anemones transfer some of those cells to their own skin, enabling them to sting other creatures.

**Bivalve mollusks** The bivalve mollusks are soft-bodied animals protected by a pair of shells hinged together by a ligament. Clams, scallops, oysters, and mussels are examples of bivalves. There are about 8,000 bivalve species and most of them live in the ocean, although there are some freshwater species. Clams spend their lives buried in mud or sand, scallops live on the seafloor, and oysters and mussels fasten themselves to underwater objects, such as rocks, the pilings that support piers, and the hulls of boats. All bivalves are filter-feeders that draw a current of water in through a tube called the "incurrent siphon," across their gills, where tiny food particles are captured, and out through the "excurrent siphon."

When a grain of sand is trapped in the shell of a bivalve, the mantle may secrete layer after layer of a shell-like protein material that covers the sand grain and forms the iridescent gem called a pearl. Oysters are the best known source of pearls, but clams and other bivalve mollusks also form pearls.

**Cephalopods** The third major group of mollusks is the cephalopods (**FIGURE 11-14**). It includes 6 nautilus species and more than 600 species of squids and octopuses. The nautilus has an external shell that resembles a snail shell, except that it is not twisted. Squids have very small shells that are covered by the mantle, and octopuses have lost the shell entirely. In all these species, the tentacles appear to grow directly from the head, which explains the name: *cephalopod* translates as "head-footed."

The most obvious feature of cephalopods may be their tentacles, which they use to walk and swim and to capture prey. Squids—many of which are ferocious predators—have eight short tentacles called arms and two long, sucker-bearing tentacles that are used to capture prey. Squids are often more than 3 feet (about 1 m) long, and members of some species

**SQUIDS**
- 8 short tentacles and 2 long sucker-bearing tentacles
- Free-swimming
- About 300 species

**OCTOPUSES**
- 8 short tentacles called arms
- Bottom-dwellers, living in coral reefs and rocky coasts
- About 300 species

**NAUTILUSES**
- Chambered shell used for protection
- Free-swimming
- 6 species

**FIGURE 11-14 Cephalopod diversity.**

may exceed 40 feet (about 13 m). When a squid is in a hurry, it swims tail-end first, using jet propulsion. Water is expelled through the siphon at the tentacle end of the body, shooting the animal backward.

After a slow approach to its unsuspecting prey, a squid propels its tentacles forward with astonishing speed—accelerations of more than 800 feet (about 250 m) per second per second have been measured, the equivalent of your car accelerating from 0 to 6,000 mph in 10 seconds! The suckers on the tentacles adhere to the prey and draw it back toward the squid, where the arms take over, turning and manipulating the prey as it is bitten by the squid's sharp beak and pulled into the mouth by the tongue-like radula.

Octopuses, which can have tentacles that spread 12 feet (about 4 m), are bottom-dwellers that live in coral reefs and on rocky coasts. Because they have completely lost the shell, octopuses can squeeze through astonishingly small openings,

and their ability to escape from even carefully covered tanks is a perpetual challenge for the keepers of zoos and aquaria. Keepers are accustomed to arriving in the morning and finding an octopus wandering around the room.

In the next section we further explore the cephalopods and consider the question of whether their skills of manual dexterity make them the smartest of the invertebrates.

### TAKE-HOME MESSAGE 11·8

Mollusks are protostome invertebrates that do not molt. They are the second most diverse phylum of animals and include snails and slugs, clams and oysters, and squids and octopuses. Most mollusks have a shell for protection, a mantle of tissue that wraps around their body, and a specialized tongue called a radula.

## 11·9   Are some animals smarter than others?

Squids and octopuses are wide-ranging predators that must locate, capture, and subdue other animals so they can eat them. This is a far more active lifestyle than that of bivalves or snails, and it requires quick movements and rapid responses to stimuli. All of these features are conspicuous

elements of the behavior of an octopus as it moves through a coral reef, inserting a tentacle into every opening. In fact, as an expert multi-tasker, an octopus can capture a prey item with one of its tentacles, use a second tentacle to hold something else, and still keep the remaining six tentacles in

**FIGURE 11-15 An octopus attempts to open a jar.**

motion, searching for more prey. In addition, octopuses are very good at manipulating things with their tentacles. It doesn't take a captive octopus long to learn how to unscrew the top on a glass jar to reach a fish that's swimming inside—after all, that's just a minor variation on twisting a clam to open it (**FIGURE 11-15**).

**Q** **Are octopuses smart?** Because of their manipulative skills and easily observed exploratory behavior, octopuses are sometimes featured on TV shows and in magazine articles, where they are often described as the smartest invertebrates and their intelligence compared to that of mammals. But this is the sort of claim that must be viewed skeptically.

What is intelligence, after all? The concept of intelligence loses its meaning when animals that live in completely different worlds and respond to entirely different stimuli are compared. We can say that octopuses are very good at doing the things that octopuses need to do—searching for prey, capturing it,

and manipulating it, all within their marine environment, for example. But is following a complex sequence of actions the same thing as intelligence? Spiders, for example, construct elaborate webs that are strategically placed in the flight paths of insects; they detect the impact of an insect on the web and rush out to wrap the insect in silk before it can escape. Does that require more or less intelligence than opening a glass jar to eat a fish? And what about the red squirrels of northern Europe? Every fall they hide more than 3,000 acorns, cones, and nuts, then recover more than 80% of their hidden nuggets of food over the course of the winter. A human probably couldn't manage that.

If intelligence is defined, in part, as the ability to do some problem-solving task valued by humans, the concept of "intelligence" may actually be relevant only to humans. Applying the question of intelligence to non-human species may just be an assessment of how well they can accomplish tasks that are essential to their survival. For this reason, questions of comparative animal intelligence are not generally useful. Rather, animals' abilities should be considered as evolutionary responses to particular selective pressures imposed by their environment, and the fact that a species currently survives should be taken as an indication of some mastery of its particular niche. As stated earlier in this chapter, from an evolutionary perspective, any species alive today must be considered a success.

---

**TAKE-HOME MESSAGE 11·9**

The predatory behavior of octopuses involves exploration and manipulation, behaviors often considered as intelligent by humans. But the concept of intelligence cannot be applied objectively to other species, which have evolved in response to the selective forces at work in their own particular niches.

---

# 11·10 An external skeleton and metamorphosis produced the greatest adaptive radiation ever.

When it comes to species diversity, the insects take the cake, outnumbering all other forms of life. Flies, alone, comprise 90,000 species; bees, wasps, and ants number 150,000 species; and butterflies and moths, 120,000 species. But even those numbers are dwarfed by beetles, with 320,000 species! To put it in perspective, there are more species of beetles than there are species of plants, and nearly six times as many species of

beetles as there are species of all birds, mammals, fishes, amphibians, and reptiles combined.

All of these insects belong to the phylum of arthropods (Arthropoda), which contains protostome invertebrates that have a segmented body (i.e., with a distinct head, midsection, called the thorax, and abdomen), an external skeleton (called an exoskeleton) made of a stiff carbohydrate known as chitin, and legs with joints (think of all the joints in a lobster's claw). The four major lineages of arthropods are:

- Arachnids (including spiders, mites, ticks, and scorpions)

- Crustaceans (including lobsters, crabs, shrimp, and barnacles)

- Millipedes and centipedes

- Insects

In all, the phylum contains almost one million recognized species, or approximately 75% of all animal species and 60% of *all species* on earth (**FIGURE 11-16**). Taken together, there are more than a billion billion (1 followed by 18 zeroes) arthropods alive at any given time! And about 80% of these are insects. (Put another way, on earth, right now, there are more than 150 million insects for every single human.)

Insects and other arthropods have diversified so much that arthropod species of one kind or another thrive in nearly every habitat on earth. Insects, in particular, abound in terrestrial habitats (every species of plant, for example, is eaten by at least one insect species) and in fresh water, and crustaceans thrive across marine habitats. If diversification is a measure of evolutionary success, the arthropods are unmatched.

Two characteristics of insects that have been central to their success are flight and the way they cope with the change in body size as they grow. With flight, insects are able to avoid many predators and can efficiently search for food and for mates. Flight also enables organisms to more quickly leave one location and disperse to another that is potentially more hospitable.

Most arthropods have a modular body plan and a rugged exoskeleton, to which their muscles are attached. Like a suit of armor, the exoskeleton provides excellent protection from injury and predators. It also helps individuals conserve water and resist drying out. One challenge associated with having an exoskeleton, however, is that it prohibits

Mammals get bigger and bigger the more they eat. Why don't insects?

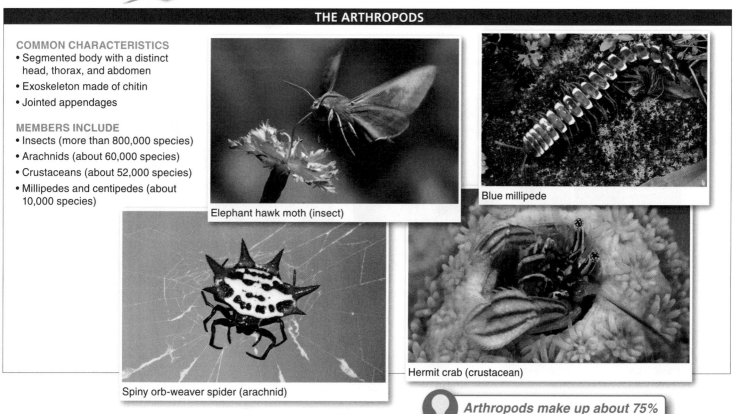

**THE ARTHROPODS**

**COMMON CHARACTERISTICS**
- Segmented body with a distinct head, thorax, and abdomen
- Exoskeleton made of chitin
- Jointed appendages

**MEMBERS INCLUDE**
- Insects (more than 800,000 species)
- Arachnids (about 60,000 species)
- Crustaceans (about 52,000 species)
- Millipedes and centipedes (about 10,000 species)

Elephant hawk moth (insect)

Blue millipede

Spiny orb-weaver spider (arachnid)

Hermit crab (crustacean)

**FIGURE 11-16 Arthropod diversity: insects, millipedes and centipedes, arachnids, and crustaceans.**

💡 *Arthropods make up about 75% of the animal species on earth.*

growth. Many arthropods, including most insects, overcome this constraint with a process of growth (shared by the roundworms, of phylum Nematoda) that is very different from the pattern seen in other animals, such as mammals. Their life is divided into three completely different stages (**FIGURE 11-17**).

**1. Larva.** An egg hatches into a **larva** (*pl.* larvae), which looks completely different from the adult form. A caterpillar is the larval stage of a butterfly or moth, and a grub is the larval stage of a beetle. A larva's job is to eat and grow large enough to enter the next life stage, the pupa. Larvae are able to grow by passing through numerous stages (usually 4–8) in which they shed their exoskeleton, increase in size, and develop a new, slightly larger exoskeleton, in the process of molting.

**2. Pupa.** Eventually, a larva completely covers itself with a casing. Covered by the case—which has different names (such as cocoon or chrysalis) in different insect groups—the insect is now called a **pupa.** In the pupa, genes are activated that code for the adult body form, and the proteins that made up the larva are broken down to amino acids, which are recycled to synthesize adult proteins and reassembled into the adult form. This rebuilding process is called **metamorphosis,** from two Greek words meaning "change

of form." The entire process requires several days, and by the end of the pupal stage, the adult insect is curled tightly inside the pupal case. The pupal case then splits and the adult squirms out. While inside the pupal case, the wings of butterflies and moths are tightly folded; once freed, they require time to expand and stiffen before the adult is ready to fly.

**3. Adult.** When the **adult** form hatches, it is at its full size. Insects do not grow after they emerge from their pupae—for this reason, some adults do not even eat. Rather than eating and growing, the job of an adult insect is reproduction.

The developmental process called **complete metamorphosis,** which includes these three stages, allows the larva and adult to act as if they were animals from different species, each optimized to perform very specialized tasks. Caterpillars eat leaves, for example, and spend their entire larval period on a single plant. In contrast, butterflies feed on nectar that they collect by flying from flower to flower. As a larva, the animal feeds and grows. As an adult, it reproduces. The genes that control the larval body form are different from the genes that determine the adult form, so natural selection can act on the larval and adult stages independently.

## METAMORPHOSIS

### COMPLETE METAMORPHOSIS

Complete metamorphosis is the division of an organism's life history into three completely different stages (occurs in 83% of insect species).

**LARVA**
An egg hatches into a larva, which eats (and eats), growing as it molts several times until large enough to enter the pupal stage.

**PUPA**
The larva encloses itself in a case, body structures are broken down, and new structures are assembled into the adult form.

**ADULT**
The adult emerges from the pupa and no longer grows. Its primary function is to reproduce.

### INCOMPLETE METAMORPHOSIS

Incomplete metamorphosis is the pattern of growth and development in which an organism does not pass through separate, dramatically different life stages (occurs in 17% of insect species).

**NYMPH**
An egg hatches into a nymph, resembling a small version of the adult (but without wings or reproductive organs), growing as it molts several times.

**ADULT**
Upon reaching adult size, the nymph stops molting and the adult forms.

 *The separation of life stages has contributed to the enormous ecological diversity of insects.*

**FIGURE 11-17 From larva to adult monarch butterfly.** Some insects have a life history divided into unique stages, each with a distinct purpose.

About 83% of insect species have complete metamorphosis, and this separation of the life stages has been an important factor in helping insects diversify into nearly 20 times as many named species as vertebrates. Among those insect species that do not undergo the dramatic changes of complete metamorphosis, typically the eggs hatch as nymphs (the juvenile form), and resemble a smaller version of the adult. These juveniles do not have wings or reproductive organs, but they live in the same habitats as adults and eat the same foods. They then undergo several molts as they grow, and stop molting when they reach adult size. This pattern of growth and development is called **incomplete metamorphosis,** and it occurs in grasshoppers and cockroaches, among many other species.

## TAKE-HOME MESSAGE 11·10

The arthropods are protostome invertebrates, and with nearly one million species (and probably at least as many more yet to be identified), they outnumber all other forms of life in species diversity. The ability to fly and the development of a body with a rugged exoskeleton have contributed to the enormous ecological diversity of insects. The life cycle of most insects includes a larval stage that is devoted to feeding and growth, a pupal stage during which metamorphosis occurs, and an adult stage in which the insect reproduces.

# 11·11 Other arthropods include arachnids, crustaceans, millipedes, and centipedes.

The arthropod phylum is so large that, even if we excluded all the insects, it would still contain more species than any other phylum. In this section we survey the other major groups of arthropods: the millipedes and centipedes, the crustaceans, and the arachnids (**FIGURE 11-18**).

**Millipedes and centipedes** The names *millipede* and *centipede* translate to "a thousand feet" and "a hundred feet" and refer to the enormous numbers of legs on these long, skinny animals. They don't really have a thousand—or even a hundred—feet, but millipedes do have more legs than

centipedes, and that is the easiest way to tell them apart. Millipedes and centipedes have long, segmented bodies that seem almost worm-like, but their jointed legs and hard exoskeletons are characteristics of arthropods. Both live among fallen leaves, especially in forests where the leaf litter layer is usually cool and moist. Millipedes feed on decaying plant material, while centipedes are predators that use fangs equipped with venom to kill insects and even snakes and small mammals. Unlike spiders and scorpions (which liquefy their prey for easy digestion), centipedes use their jaws to tear prey into pieces that are small enough to swallow.

## GROUPS OF ARTHROPODS

Japanese slave-making ant

**INSECTS**
- Three pairs of walking legs
- Legs are located on the thorax
- Life cycle consists of separate life stages

Giant desert centipede

**MILLIPEDES AND CENTIPEDES**
- Many pairs of legs (two pairs per segment in millipedes; one pair per segment in centipedes)
- Long, segmented body

A jumping spider

**ARACHNIDS**
- Usually four pairs of walking legs
- Legs are located on the thorax
- Specialized mouthparts
- Predators

Sally Lightfoot crab

**CRUSTACEANS**
- Many pairs of legs
- Usually five pairs of appendages that extend from the head
- Mostly aquatic

**FIGURE 11-18 Exploring the arthropods.**

**Arachnids** There are 60,000 species of arachnids, land-dwelling arthropods that include spiders, scorpions, mites, and ticks. Most have four pairs of walking legs and a specialized feeding apparatus. Arachnids are distinguished from other arthropods by the positioning of their legs: they have legs only on the middle section of their body (the thorax), whereas other arthropods have legs on the posterior section of their body (the abdomen) as well as on the thorax.

Arachnids are predators. Many spiders construct webs by spinning silk fibers (which consist of interconnected protein molecules). Orb-weaving spiders construct intricate spiral-shaped webs that trap (and sometimes even lure) flying insects. Spiders generally place their webs across the routes followed by flying insects, such as between plants, and when an insect gets trapped in the web, run out to wrap the prey in silk threads. Other spiders produce tangled webs, commonly called cobwebs, or funnel-shaped webs.

When a spider captures prey, it uses its fangs to inject venom that contains two types of enzymes. The first set of enzymes disrupts the prey's nervous system and paralyzes it, then other enzymes dissolve its internal organs. When the prey's innards have liquefied, the spider sucks out the contents.

The venom of most spiders is usually harmless to animals as large as humans, but two North American spiders are dangerous. The black widow spider is easily identified by a red hourglass marking on the bottom surface of the female. The brown recluse spider lacks distinctive markings, so it is harder to identify—but it is more dangerous. Both species are likely to move indoors, constructing webs in little-used spaces such as closets and cabinets and behind furniture. Brown recluse bites can create open wounds that can last for months or years and sometimes require repeated skin grafts to close.

Scorpions look dangerous to humans, but their sting is no worse than a bee's. Scorpions are lethal to insects, hunting them at night They seize prey with their claws and use their stinger only if the prey struggles. Some of the scorpion's mouth structures are covered with spines that grind the prey, while glands secrete enzymes that liquefy it and the scorpion sucks up the soup.

**Crustaceans** If arachnids are the most feared arthropods, the aquatic crustaceans—lobsters, crayfish, crabs, and shrimp—may be the most desirable arthropods, at least at meal time. In the United States, the top three fishery crops, by annual market value, are shrimp, crabs, and lobsters. There are about 52,000 species of crustaceans, and although crustaceans can look very different—from sessile barnacles to mobile, tiny krill—they have one feature in common: five pairs of appendages extending from the head. Three of these appendages are used for feeding, and two are antennae that sense the environment.

Crustaceans have many pairs of legs, and their legs are modified for many purposes. Shrimp and barnacles (which are essentially shrimp that spend their adult lives within a shell, fastened to a rock) have legs with comb-like projections that they use to capture plankton. Free-living crustaceans (shrimp, crabs, lobsters) have modified limbs in the abdominal region that are used to hold eggs or newly hatched young.

Although most crustaceans are aquatic, the inconspicuous wood lice (also known as pill bugs, sow bugs, or roly-polies) are terrestrial forms that you can find when you turn over a rock in your garden. Wood lice are herbivores and play an important role in recycling dead plant material.

---

### TAKE-HOME MESSAGE 11·11

Centipedes are predators with fangs that inject venom, and millipedes are herbivores that feed on dead plant material. Spiders and scorpions are predatory arthropods that eat insects and, occasionally, small vertebrates. Lobsters, crabs, shrimp, and barnacles are predatory marine crustaceans.

---

# 11·12 Echinoderms are vertebrates' closest invertebrate relatives and include sea stars, sea urchins, and sand dollars.

Appearances can be deceiving. Sea stars and other members of the phylum Echinodermata live in the oceans, are radially symmetrical (as adults), and are covered by spiny plates. And yet, because they are deuterostomes (their gut forms from back to front), it turns out that they are more closely related to humans than to any other invertebrate group. Echinoderms stand as an example of how classifying organisms based on evolutionary relatedness rather than physical similarity can reveal a great deal about the force of natural selection and its dramatic effects on body form in adapting populations of organisms to their environment.

The echinoderms include about 6,000 species of marine animals, most of them enclosed by a hard skeleton beneath an

## THE ECHINODERMS

Leather sea star

Sand dollar

Club sea urchin

Sea cucumber

**COMMON CHARACTERISTICS**
- Enclosed by a hard skeleton under spiny skin
- Larvae are bilaterally symmetrical and share some anatomical features with chordates
- Adults are radially symmetrical
- Undersides are covered with tube feet that aid in locomotion and grasping

**MEMBERS INCLUDE**
- Sea stars (about 1,600 species)
- Sea urchins and sand dollars (about 940 species)
- Sea cucumbers (about 1,100 species)

**FIGURE 11-19 What is an echinoderm?**

outer covering of spiny plates (**FIGURE 11-19**). Adult sea stars (starfishes and brittle stars), probably the most recognizable echinoderms, have five or more appendages evenly distributed around their body circumference. They are radially symmetrical and have no front or back ends. They move with equal ease in any direction, and their sensory organs are distributed around their circumference. Adult sea urchins and sand dollars don't have projecting arms, but they are also radially symmetrical.

Although radial symmetry is characteristic of adult echinoderms, their larvae are bilaterally symmetrical, having a left and right side of the body, and this distinguishes them from other radially symmetrical animals such as jellyfishes and the other cnidarians. Bilaterally symmetrical larvae are evidence that echinoderms evolved from bilaterally symmetrical ancestors, and their radial symmetry as adults is an evolutionary characteristic associated with their mode of locomotion and feeding specializations.

Echinoderms do not have a brain. They have a nervous system that consists of a central ring of nerves, with branches that extend into each of their appendages and help them gain information about and respond to their environment. Echinoderms creep on little tube feet that are extensions of an internal system of water-filled tubes radiating throughout the body. The undersides of a sea star's arms and of the bodies of sea urchins and sand dollars are carpeted with tube feet that extend and contract in waves, grasping and releasing the surface on which they move (the substrate) as the animal glides along.

Although each tube foot is tiny, there are thousands of them, and their combined force allows a sea star to pull the two shells of a clam or mussel apart. Once it has the shells apart, an extraordinary and unexpected thing happens. The sea star pushes its stomach out through its mouth and inserts it into the opened shells. The stomach then secretes digestive enzymes that break down the clam or mussel tissue, and the sea star absorbs the soup that results. Sea urchins feed on algae, scraping them loose from rocks with sharp tooth-like surfaces made of calcium. Sand dollars capture floating particles of algae and other organic matter by trapping them in streams of mucus, which are moved toward their mouth by beating cilia.

### TAKE-HOME MESSAGE 11·12

Because they are deuterostomes (as are vertebrates), echinoderms are the invertebrates that are the closest evolutionary relatives to the vertebrates (and other chordates). Their aquatic larvae are bilaterally symmetrical and share some anatomical features with chordates, but adult echinoderms are radially symmetrical.

*Animals in motion: these x-ray images show a tuatara (a New Zealand reptile) walking.*

# 11·13

## All vertebrates are members of the phylum Chordata.

Vertebrates are deuterostome animals, having defined tissues and bilateral symmetry. Vertebrates are all part of a phylum called Chordata (the chordates) that includes three subphyla in total; the two other subphyla are the tunicates and the lancelets (**FIGURE 11-20**). For at least a portion of their life cycle, members of all three subphyla possess four distinct chordate body structures that are found in no other animal groups (**FIGURE 11-21**).

**1.** The **notochord,** a rod of tissue extending from head to tail, is the structure that gives chordates their name (from the Latin for "cord"). The notochord stiffens the body when muscles contract during movement. The simplest chordates—those most distantly related to the vertebrates—retain the notochord throughout life. In the more complex chordates such as vertebrates, however, the notochord is present only in the early embryo and is replaced by the backbone (vertebral column) as the embryo develops.

**2.** A **dorsal hollow nerve cord** extends along the animal's back (its dorsal side) from its head to its tail. In vertebrates, this nerve cord eventually forms the central nervous system, which consists of the spinal cord and the brain. Other kinds of animals (worms, insects, and so on) also have a nerve cord, but it lies in the lower portion of the front (ventral) part of the body and is solid instead of hollow.

**3. Pharyngeal slits** are present in the embryos of all chordates, but in many chordates (including humans) the slits disappear as the animal develops. The earliest chordates were aquatic, and to breathe and feed they passed water through slits in the pharyngeal region (the area between the

back of the mouth and the top of the throat). In humans, pharyngeal slits are present only in the embryo; our gills were lost far back in evolutionary time.

**4.** A **post-anal tail** is another chordate characteristic. The posterior (back) end of the digestive system is the anus. The region of the body extending beyond the opening of the anus is referred to as "post-anal." All vertebrates have a tail in this location, but some, including humans, have a tail only for a brief period, during embryonic development.

Although all chordates share these four characteristic structures, the chordates exhibit tremendous physical and ecological diversity (**FIGURE 11-22**).

*Tunicates* (subphylum Urochordata; about 2,000 species) are invertebrate marine animals that have defined tissues, bilateral symmetry, and, as deuterostomes, from-back-to-front gut development. The adults are about the size of your thumb and look like balls of brownish green jelly. You can find them attached to docks and the mooring lines of boats. It is the free-swimming larvae of tunicates that reveal their chordate characteristics. Tunicates are filter-feeders: cilia draw a current of water through the mouth into the pharynx and out through the pharyngeal slits, and microscopic food items are trapped by the bars between the slits. Microscopic food items are trapped by the bars between the slits.

*Lancelets* (subphylum Cephalochordata; about 20 species), also bilaterally symmetrical deuterostomes, are slender, eel-like invertebrate animals, about the length of your little finger, that live in coastal waters. Unlike tunicates, adult

FIGURE 11-20 **Understanding the chordates.**

## THE CHORDATES

### COMMON CHARACTERISTICS

All chordates possess four common body structures, although in many chordates, these structures are only present during specific life stages.

- Notochord
- Dorsal hollow nerve cord
- Pharyngeal slits
- Post-anal tail

### MEMBERS INCLUDE
- Tunicates (about 2,000 species)
- Lancelets (about 20 species)
- Vertebrates (about 56,000 species)

Golden lion tamarin (vertebrate)

Compound ascidians (tunicate)

Pacific double saddle butterfly fish (vertebrate)

Thorny devil lizard (vertebrate)

## THE DISTINCT BODY STRUCTURES OF CHORDATES

The members of Chordata are defined by four distinct body structures present in all chordates but in no other animal groups.

**NOTOCHORD**
A rod of tissue extending from the head to the tail

- Simpler chordates retain the notochord throughout life
- In more complex chordates, the notochord in early embryos is replaced by a backbone

**DORSAL HOLLOW NERVE CORD**
Nerve cord that extends along the animal's back (its dorsal side)

- In vertebrates, the nerve cord eventually forms the spinal cord and brain

**PHARYNGEAL SLITS**
Slits through which water is passed in order to breathe and feed

- In many chordates (including humans), the slits disappear as the animal develops

**POST-ANAL TAIL**
Tail that extends beyond the posterior (back) end of the digestive system

- Some vertebrates (including humans) have a tail only briefly, during embryonic development

**FIGURE 11-21 Characteristic chordate body structures.**

**TUNICATES**
- Filter-feeding marine animals
- Can often be found attached to docks
- Larvae are free-swimming

**LANCELETS**
- Filter-feeding marine animals
- Live in coastal waters

**VERTEBRATES**
- The most diverse subphylum of chordates
- Range in size from the tiny hummingbird to the enormous blue whale
- Can be found in most of the habitats on earth

**FIGURE 11-22 Representing the subphyla of chordates.** Shown here (*left to right*) are a sea squirt, a lancelet, and a flap-necked chameleon.

lancelets have all the chordate characteristics. Lancelets are also filter-feeders.

*Vertebrates* (subphylum Vertebrata; about 56,000 species) are the most diverse group of chordates (**FIGURE 11-23**). Because they are deuterostomes, they are bilaterally symmetrical, with defined tissue and from-back-to-front gut development. Vertebrates differ from the other chordates in two important ways:

1. They have a backbone, formed when a column made from hollow bones (or cartilage in some organisms), called vertebrae, forms around the notochord. This backbone surrounds and protects the dorsal hollow nerve cord.

2. They have a head, at the front (anterior) end of the organism, containing a skull, a brain, and sensory organs.

The size range of vertebrates extends from tiny hummingbirds and shrews that weigh only a tenth of an ounce (about

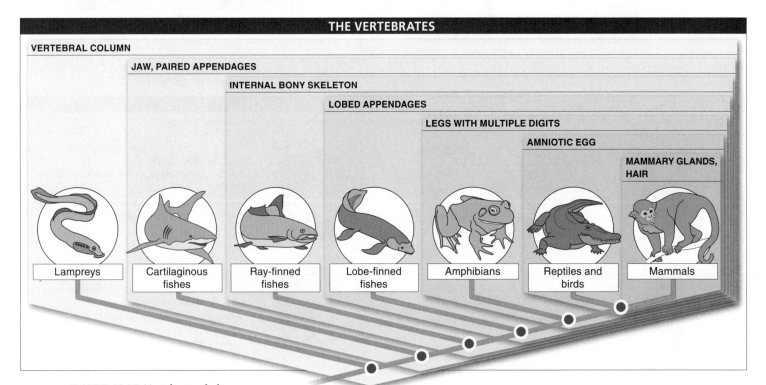

**FIGURE 11-23 Vertebrate phylogeny.**

3 grams) to blue whales, which weigh in at more than 130 tons (120,000 kg). Vertebrates move by swimming, burrowing, crawling, walking, running, and flying, and they are present in all habitats on earth. But despite all these differences, all vertebrates have the four chordate characteristics at some stage of their lives. In the remaining sections we focus on the vertebrates, which are by far the most diverse and widely dispersed group of chordates.

**TAKE-HOME MESSAGE 11·13**

All chordates have four characteristic structures: a notochord, a dorsal hollow nerve cord, pharyngeal slits, and a post-anal tail. The three subphyla of chordates are superficially very different, but are united by possessing these four structures at some stage of their life cycle.

# 11·14 The evolution of jaws and fins gave rise to the vast diversity of vertebrate species.

The earliest vertebrates were fish-like animals that lived more than 500 million years ago and did not have jaws. Two kinds of jawless vertebrates still exist: the lampreys (with 41 species) and the hagfishes (with 43 species). Lampreys are parasitic; they attach to other fishes by creating suction with a circular oral disk that is studded with sharp spines. First, the spines scrape an opening through the scales and skin. Then, a protein produced by the lamprey keeps the wound from closing, and the lamprey feeds on the blood and body fluids that seep out. Hagfishes feed on dead animals, using two spiny dental plates that they embed in their prey.

Lampreys and hagfishes are about as basic as an animal can be. And because they do not have jaws, they cannot seize prey or chew food before swallowing it. The evolution of fins and jaws is

**Q** Why are jaws needed?

what set the stage for the evolutionary explosion of diversity in the vertebrates.

Fishes swim the same way that lampreys and hagfishes swim— that is, by bending the body from side to side to create an S-shaped wave that moves from head to tail. A typical fish has seven fins serving different purposes: to drive the fish forward, to minimize rolling from side to side, and for steering and stopping. The combined effects of the seven fins allow the fish to swim rapidly in a straight line when it is in open water or to weave its way through a dense stand of plants or corals, with great maneuverability.

The evolution of fins was paralleled by the evolution of jaws, because the two structures work together. Fins get you to the organism you are going to eat, and jaws capture and kill it (**FIGURE 11-24**).

## THE EVOLUTION OF JAWS AND FINS

Oral disk

**FISHES WITHOUT JAWS OR FINS**
• Tail propels organism through water
• Feed by attaching oral disk to prey

**FISHES WITH JAWS AND FINS**
• Fins provide controlled movement through water
• Jaws lined with sharp teeth allow for seizing and chewing prey

**FIGURE 11-24 The lamprey and the shark.** The lamprey (*left*) has an oral disk that it uses for sucking in a meal. The shark has a powerful jaw lined with sharp teeth, and pairs of fins that allow it to move rapidly through the water and attack prey.

*The development of fins and jaws set the stage for the evolutionary explosion of vertebrate diversity.*

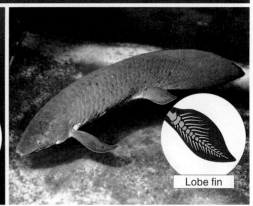

Ray fin

Lobe fin

**CARTILAGINOUS FISHES**
• Characterized by a skeleton made completely from cartilage
• About 880 species, including sharks and rays

**RAY-FINNED FISHES**
• Characterized by rigid bones and fins lined with hardened rays
• Possess a swim bladder, which aids in flotation
• About 27,000 species, including almost everything you think of as "fish"

**LOBE-FINNED FISHES**
• Characterized by two pairs of sturdy fins on the underside of their body
• 6 species of lungfish and 2 species of coelacanths

**FIGURE 11-25 Fishes with jaws: cartilaginous, ray-finned, and lobe-finned fishes.** Shown here (*left to right*) are a stingray, rainbow trout, and Australian lungfish.

The development of jaws and fins allowed great diversification in vertebrate body shapes and composition. There are two major categories of jawed fish. **Cartilaginous fishes,** with about 880 species, include the sharks and rays (**FIGURE 11-25**). Cartilaginous fishes are characterized by a skeleton made entirely of cartilage, a solid but slightly flexible connective tissue—the same tissue that gives your nose and ears their shape. **Ray-finned fishes** have a rigid skeleton made from bone, which, like cartilage, is a solid connective tissue consisting of specialized cells and an extracellular material (matrix) that the cells secrete. Bone, however, is much less flexible than cartilage because its extracellular matrix is mineralized by crystals of calcium phosphate. Ray-finned fishes have a mouth at the narrow tip, or apex, of the body, and their fins, made from webs of skin, are supported with hardened rays made of bone.

More than 97% of fishes reproduce by laying eggs, which develop and hatch outside the mother. Fertilization most commonly takes place externally, with males and females releasing their gametes into the water. However, numerous species of sharks have internal fertilization, after which the female lays the fertilized eggs. There also are some fish species, including guppies, that have internal fertilization followed by development of the eggs within the mother and live birth.

Ray-finned fishes are the largest group of jawed fishes and the most diverse group of vertebrates, with about 27,000 species. This group includes almost everything you think of as "fish," from salmon to goldfish.

An important evolutionary development in the ray-finned fishes was the swim bladder, a gas-filled organ that keeps the fish from sinking. This bladder is the evolutionary precursor of the lungs of land-dwelling (terrestrial) vertebrates. Cartilaginous fishes, which have no swim bladder, must constantly move through the water or they will sink.

A third group of jawed fishes, the **lobe-finned fishes,** are represented by just eight species. These fishes have sturdy pelvic and pectoral fins on the underside of their body (see Figure 11-25), which, unlike other fins, have a central appendage containing numerous bones and muscles that connect the fins to the body. As we see in the next section, lobe-fins were useful in initiating the move onto land.

**TAKE-HOME MESSAGE 11·14**

The development of two structures in fishes—fins and jaws—set the stage for the enormous diversity of modern vertebrates.

# 11·15 The movement onto land required lungs, a rigid backbone, four legs, and eggs that resist drying.

Movement from water onto land required four major evolutionary innovations: lungs, a backbone, four legs, and eggs that won't dry out. All four of these characteristics are necessary for a land animal, and all four evolved in predatory fishes that lived in shallow water, the immediate ancestors of terrestrial vertebrates.

**Q** **Where did legs and lungs come from?**

As we've noted, lungs developed from the swim bladder found in ray-finned fishes. All of the terrestrial vertebrates—amphibians, reptiles (including birds), and mammals—are descendants of the lobe-finned fishes that lived during the Devonian period, some 400 million years ago. At this time, the lobe-finned fishes had lungs and the rudiments of limbs, but they were still fully aquatic and living near the shore. The four sturdy fins on the underside of their body helped them move through shallow water, and lungs allowed them to breathe air when the oxygen concentration in the warm, stagnant water was low. Thus, these fishes already had some of the basic characteristics of terrestrial animals—they could walk using their fins, and they could exchange gases (getting oxygen into and carbon dioxide out of the bloodstream), using oxygen from air passing through their lungs instead of from water passing through their gills. The jointed bones in the fins of lobe-finned fishes don't look much like your arms and legs, but they are evolutionarily homologous structures.

To move onto land, a vertebrate needs more than just legs and lungs, however. It needs structural support to resist the pull of gravity. Each vertebra of a terrestrial vertebrate has projections that interlock with projections from the vertebra ahead of it and the vertebra behind it. These interlocking projections prevent the backbone from sagging under the pull of gravity, and the body weight is transmitted through the limbs to the ground.

The last innovation necessary to move onto land is an egg that resists drying out (**FIGURE 11-26**). When eggs are deposited on land, they are exposed to air and lose water by evaporation. The eggs of terrestrial animals need a waterproof covering—a membrane and a shell—to prevent them from drying out before they hatch. The appearance of eggs with a shell (about 380 million years ago) marks the final step in the evolution of entirely terrestrial vertebrates, the groups that evolved into mammals and reptiles (including birds). The eggs of terrestrial vertebrates have a water-tight membrane that keeps the embryo surrounded by a bath of fluid called amniotic fluid.

## FROM WATER TO LAND

The transition of vertebrates from life in water to life on land required overcoming three main obstacles. Four major evolutionary innovations allowed for this transition.

**PROBLEM: RESPIRATION**

Aquatic animals use gills to acquire dissolved oxygen from water. The transition onto land required the ability to breathe air.

**SOLUTION: LUNGS**
Gas exchange was transferred from gills to lungs, which evolved from the swim bladder found in ray-finned fishes.

Ray-finned fish

Early terrestrial vertebrate

**PROBLEM: GRAVITY**

The transition onto land required structural support to resist the pull of gravity.

**SOLUTIONS: LIMBS and MODIFIED VERTEBRAE**
Limbs evolved from the jointed fins found on the underside of lobe-finned fishes.

Vertebrae were modified to transmit the body weight through the limbs to the ground.

Lobe-finned fish

Early terrestrial vertebrate

**PROBLEM: EGG DESICCATION**

The transition onto land required an egg that resisted drying out when exposed to air.

**SOLUTION: AMNIOTIC EGG**
Terrestrial animals developed a waterproof eggshell, which prevents eggs from drying out before they hatch.

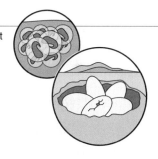

**FIGURE 11-26 How did vertebrates make the transition from life in water to life on land?**

## TAKE-HOME MESSAGE 11·15

In the transition of vertebrates from life in water to life on land, fins were modified into limbs, vertebrae were modified to transmit body weight through the limbs to the ground, and the site of gas exchange was transferred from gills and swim bladders to lungs. The only entirely new feature to appear in the early development of terrestrial vertebrates was an egg that resisted drying out.

# 11·16–11·20 All terrestrial vertebrates are tetrapods.

Although they reach 140 pounds as adults (with one recorded at more than 230 pounds!), capybara (Hydrochoerus hydrochaeris) are rodents. Shown here: parents with baby, in Brazil.

# 11·16

## Amphibians live a double life.

The ancestors of all vertebrates that live on land had lungs and four legs. Some modern vertebrates, such as whales and snakes, have evolved so that their limbs are reduced to a few shrunken and unused bones. Whales and some snakes have returned to living in water, and some salamanders have lost lungs and breathe through their skin, but these are recent changes, occurring in species whose ancestors had lungs and four legs. Taxonomically, all terrestrial vertebrates are **tetrapods** (*tetra* = four; *poda* = feet).

Terrestrial vertebrates are divided into two main groups: (1) animals (called **non-amniotes**) such as amphibians that reproduce in water and do not have desiccation-proof amniotic eggs; and (2) animals (called **amniotes**) such as reptiles and mammals that have amniotic eggs. We discuss amphibians here, then reptiles and mammals in the next two sections.

The first terrestrial vertebrates were **amphibians,** from the Greek word *amphibios,* meaning "living a double life." There

### THE AMPHIBIAN LIFE CYCLE

Amphibians, such as frogs, toads, and salamanders, are terrestrial vertebrates (tetrapods) with non-amniotic eggs. Most species live on land as adults, but develop in water.

**EGGS**
Amphibians have non-amniotic eggs, which must be laid in water to prevent desiccation.

**JUVENILES**
Amphibians spend their juvenile stage underwater and undergo metamorphosis to develop legs and lungs.

**ADULTS**
Only the adults are true land animals; however, most of the species in this group stay close to water to lay their eggs.

**FIGURE 11-27 From egg to larva to adult: the amphibian life cycle.**

are two stages in most amphibians' lives: a water-breathing juvenile form and an air-breathing adult form. The adults of many amphibian species lay their eggs in water (**FIGURE 11-27**). Most of the 6,000 species in this group are still significantly tied to life in the water, because they must always be near water to lay their eggs, which are simple structures, not unlike fish eggs. Frogs and toads make up the vast majority of amphibians (about 5,400 species). Other amphibians include salamanders (550 species) and caecilians, a group of legless, burrowing animals (170 species).

The juvenile stage of the frog, called a tadpole, lives in water, lacks legs, and eats algae. A few weeks (or up to many months in some species) after hatching, metamorphosis occurs and the tadpole develops legs, lungs, and a digestive system fit for its

adult life as a carnivore. Adult frogs have thin, moist skin through which much of their gas (oxygen and carbon dioxide) exchange with the air takes place.

The past two decades have seen a stunning decline in amphibian populations throughout the world, attributable to a combination of causes: climate change, habitat degradation, fungal diseases, and increased pollution.

> ### TAKE-HOME MESSAGE 11·16
> Amphibians are terrestrial vertebrates, but the adults of most species still lay eggs in water. The eggs hatch into aquatic juveniles.

# 11·17
## Birds are reptiles in which feathers evolved.

Soon after amniotic vertebrates appeared, two different evolutionary lineages began to diverge (**FIGURE 11-28**). One of these lineages is the mammals, the group to which humans belong. The other lineage is the one referred to in this chapter as "reptiles (including birds)." That is an awkward term, but the evolutionary tree shows why it is necessary. Surprisingly, birds (about 9,700 species) are one branch of the reptile lineage that also includes snakes and lizards (about 8,000 species), turtles (about 300 species), crocodiles and alligators (23 species), the New Zealand tuatara (2 species)—and the dinosaurs.

**Q** *Birds are reptiles?*

The characteristics that hold the bird-crocodile-dinosaur group together are mostly similarities in bones (especially the bones of the skull and legs) and DNA sequences. Reptiles are amniotes and thus have amniotic eggs. You are familiar with the hard-shelled eggs of birds, and some lizard, snake, and turtle eggs also have hard shells. Other species in these groups have eggs with a paper-like shell.

Most people find it confusing that birds are grouped with turtles, lizards and snakes, and alligators and crocodiles, and that confusion is

**THE AMNIOTIC VERTEBRATES**

REPTILES (INCLUDING BIRDS)

MAMMALS

Turtles | Snakes | Lizards | Crocodiles | Dinosaurs (EXTINCT) | Birds | Mammals

**FIGURE 11-28** Amniotic vertebrate phylogeny.

WHAT IS AN ANIMAL? | INVERTEBRATES | INVERTEBRATE DIVERSIFICATION | VERTEBRATES AND THE MOVE ONTO LAND | TERRESTRIAL VERTEBRATES

457

**REPTILES**
- Skin is covered in scales
- Body temperature is controlled by external conditions, such as air temperature (exothermic)
- Include snakes and lizards (about 8,000 species), turtles (about 300 species), crocodiles and alligators (23 species), and tuatara (2 species)

**BIRDS**
- Skin is covered in feathers, enabling flight and providing insulation
- Body temperature is maintained by heat generated from cellular respiration (endothermic)
- Include about 9,700 species

**FIGURE 11-29 Comparison of reptiles and birds.**

easy to understand (**FIGURE 11-29**). After all, birds are covered by feathers and can fly, whereas the other reptiles have bare skin and do not fly. And there is an important physiological difference as well: birds are **endotherms,** meaning they use the heat produced by cellular respiration to raise their body temperature above air temperature, whereas other reptiles are **ectotherms**—they bask in the sun to raise their body temperature and seek the shade when the air is too warm.

How can animals as different as crocodiles and hawks be closely related? During the Mesozoic era, from about 250 million to 65 million years ago, dinosaurs were the dominant terrestrial vertebrates. The fossil record shows a remarkably clear series of animal forms that bridge the transition between bare-skinned dinosaurs and feathered birds. Birds are, in fact, just one group of dinosaurs.

**Q**

**What were the first feathers used for?**

The most apparent difference between reptiles (including most dinosaurs) and birds is the presence of feathers in birds. The fossil record reveals that feathers evolved before birds, appearing in many reptile species, and that the initial evolution of feathers probably had nothing to do with flight. In fact, feathers evolved in many different kinds of dinosaur, and feathered dinosaurs continued to live for millions of years after birds appeared.

Among reptiles, many different species had feathers. A small, agile dinosaur called *Sinosauropteryx,* for example, living about 120 million years ago, had feathers that were simple spiky

*Originally, colorful feathers were used for behavioral displays. In modern birds, feathers provide insulation and aid in flight.*

**FIGURE 11-30** *Sinosauropteryx:* **an early feathered, flightless dinosaur.**

filaments on the neck, back, and tail, along with shorter filaments covering the body. These filaments were probably brightly colored and used by male *Sinosauropteryx* for courtship displays to females, as well as for aggressive displays to other males. *Sinosauropteryx* and many of the other feathered dinosaurs were latecomers, however. The pigeon-sized *Archaeopteryx*, which was among the first bird species, was already, by 147 million years ago, using its feathers to fly; by 140 million years ago, the skies were filled with birds.

If *Archaeopteryx* and a host of other early birds already had the specialized feathers needed for flight, why did later, non-flying dinosaurs still have primitive feathers? The answer is that feathers have many functions, of which flight is only one. *Sinosauropteryx*, as we saw above, probably used its feathers for courtship or territorial displays, and it may also have used them as insulation (**FIGURE 11-30**).

Another big difference between birds and reptiles is that, as noted earlier, birds use internally generated heat to maintain a high body temperature, whereas reptiles rely on the sun to heat their bodies. This evolutionary change may have been related to feathers. As feathers evolved for display and flight, they may also have provided some insulation. This insulation may then have enabled the evolution of high rates of cellular respiration and the maintenance of a high and constant body temperature.

---

**TAKE-HOME MESSAGE 11·17**

Birds are a branch of the reptile lineage but, unlike other reptiles, possess feathers and can generate body heat. The complex anatomical and physiological systems that we see in extant animals, such as feathers and endothermy in birds, are the products of hundreds of millions of years of step-by-step changes that began with simple structures. Feathers were originally colorful structures used for behavioral displays; additional functions such as insulation and flight evolved later.

---

# 11·18 Mammals are animals that have hair and produce milk.

Long before the first birds appeared, the mammals evolved from the reptiles, originally as small, nocturnal insect-eaters. These early mammals remained small and, judging from the scarce fossil record, they were never in great abundance while the dinosaurs were around. Early mammals had several features in common that are still present in all mammals today. Two important features are **hair**—dead cells filled with the protein keratin—which serves as an insulator, and **mammary glands** in female mammals, which enable them to produce nutritious, calorie-rich milk and nurse their young.

As the earliest mammals evolved, there was a gradual transition from bodies with short legs that projected out horizontally from the trunk, like the legs of an alligator, to bodies with long legs held vertically beneath the trunk, like those of a dog (**FIGURE 11-31**). These anatomical changes would have allowed the earliest mammals to run faster and farther to capture prey. But an increase in muscle activity for running had to be accompanied by an increase in the rate of cellular respiration to provide the energy to keep the leg muscles working. With this increase in metabolism, along with the evolution of hair, the early mammals could trap the heat produced and use it for temperature regulation, setting

**FIGURE 11-31 How did endothermy evolve in early mammals?**

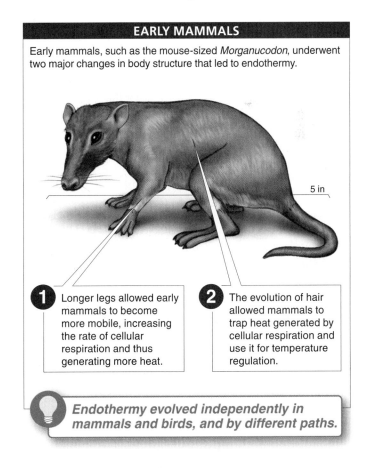

**EARLY MAMMALS**

Early mammals, such as the mouse-sized *Morganucodon*, underwent two major changes in body structure that led to endothermy.

5 in

**1** Longer legs allowed early mammals to become more mobile, increasing the rate of cellular respiration and thus generating more heat.

**2** The evolution of hair allowed mammals to trap heat generated by cellular respiration and use it for temperature regulation.

*Endothermy evolved independently in mammals and birds, and by different paths.*

Mammals are endothermic vertebrates that have hair and produce milk for their young.

**MONOTREMES**
- Females lay eggs
- Females produce milk, but do not have nipples—babies suck milk from the hairs on their mother's chest
- Only 5 species survive—the platypus and 4 species of spiny animals called echidnas

**MARSUPIALS**
- Females give birth after a short period of development
- Females of most species have a pouch where the young complete their development
- About 300 species, including kangaroos, koalas, wallabies, and possums

**PLACENTAL MAMMALS**
- Females have a placenta that provides oxygen and nutrients to embryos in the uterus
- About 4,500 species

**FIGURE 11-32 Three groups of mammals: monotremes, marsupials, and placentals.** Shown here (*left to right*) are a platypus, a kangaroo, and a giraffe.

the stage for endothermy (but through a different sequence of events than in the birds).

One feature common to most mammals is **viviparity,** or giving birth to babies ("live birth") rather than laying eggs, but it is not a defining mammalian characteristic—it isn't common to all mammals. **Monotremes** retain the ancestral condition of laying eggs. Monotremes do produce milk, but they do not have nipples—babies suck milk from the skin and hairs on their mother's chest.

Today, only five species of monotremes survive—the platypus and four species of spiny animals called echidnas. The platypus lives in streams and rivers in eastern Australia. It uses its broad, leathery, electrosensitive bill to probe under rocks and sunken logs for prey such as aquatic insect larvae and crayfish, which it finds by sensing the electrical activity of their muscles as they try to hide. The echidnas—three species in Australia and one in New Guinea—also have a leathery electrosensitive snout.

The remaining two lineages of mammals, the **marsupials** and the **placentals,** are viviparous, but the newborn young of marsupials and placentals are quite different (**FIGURE 11-32**).

Marsupials are called the "pouched mammals" because the female of most marsupial species has a pouch on her abdomen in which the young complete their development, following a short period of embryonic life in the uterus.

Placental mammals (including humans) take their name from the *placenta,* which is the structure responsible for the transfer of nutrients, respiratory gases, and metabolic waste products between the mother and the developing fetus. Dense capillary beds in the placenta send finger-like projections into the wall of the uterus, where they are surrounded by blood from the mother's circulation.

### TAKE-HOME MESSAGE 11·18

Hair and mammary glands are defining characteristics of mammals. Monotremes are egg-laying mammals. Marsupial mammals give birth after a short period of development in the uterus, and the newborn completes its development in the mother's pouch. Placental mammals have a placenta that provides oxygen and nutrients to the fetus as it undergoes a longer development in the uterus.

# 11·19

## Humans tried out different lifestyles.

The primates, the evolutionary lineage to which humans belong, originated about 55 million years ago. We can get an idea of what the ancestral primate looked like from the modern species of tree-living (arboreal) mammals commonly called prosimians—a group that includes the lemurs. Many of the anatomical characteristics of humans and the other primates can be traced to our arboreal origin. Our forward-directed eyes and binocular vision that allow us to judge distances accurately, our shoulder and elbow joints that allow our arms to rotate, and our 10 fingers and 10 toes that allow us to grasp objects—all these are traits we inherited from our arboreal ancestors (**FIGURE 11-33**).

Humans are part of the primate lineage that includes the groups commonly referred to as the New World and Old World monkeys (both of which have tails) and the apes (which lack tails). Among the apes, gibbons and orangutans live in pairs or alone, while gorillas and chimpanzees live in social groups that consist of one or more adult males and several females—the ancestral social structure for human societies. Within the apes, genetic and anatomical characteristics show that chimpanzees are our closest living relatives (**FIGURE 11-34**). Human and

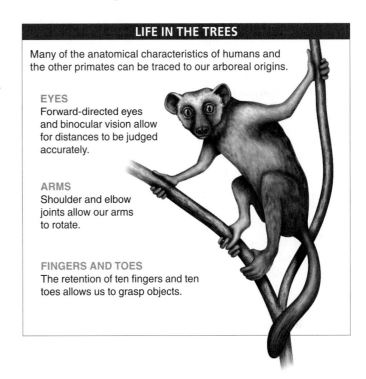

**LIFE IN THE TREES**

Many of the anatomical characteristics of humans and the other primates can be traced to our arboreal origins.

**EYES**
Forward-directed eyes and binocular vision allow for distances to be judged accurately.

**ARMS**
Shoulder and elbow joints allow our arms to rotate.

**FINGERS AND TOES**
The retention of ten fingers and ten toes allows us to grasp objects.

**FIGURE 11-33 The ancestors of modern primates lived in trees.**

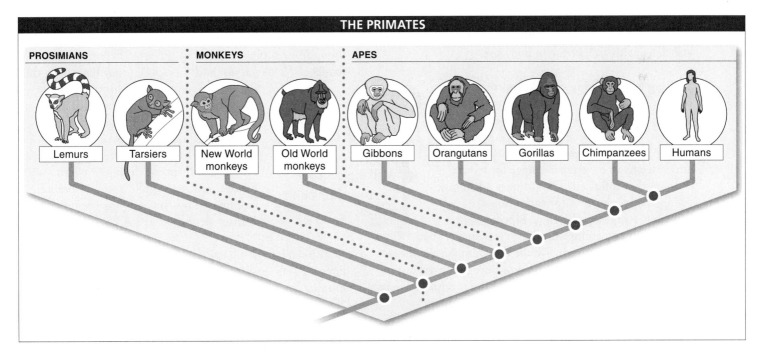

**FIGURE 11-34 Phylogeny of primates.**

chimpanzee genes are very similar: their base sequences differ by about 1%, and one-third of human and chimpanzee genes are identical.

The amount of genetic difference between humans and chimpanzees indicates that the chimpanzee and modern human lineages separated only 5 or 6 million years ago. Humans differ from chimpanzees in three major anatomical characteristics: humans are bipedal (we normally walk on two legs, whereas chimpanzees usually walk on four legs), humans are bigger than chimpanzees, and the human brain is about three times the size of the chimpanzee brain.

When we trace the appearance of these human characteristics through the fossil record, we find that humans did not become bipedal, big, and brainy all at once. Instead, the three characteristics evolved one by one. Bipedality evolved first, then brain volume increased, and finally body size increased, accompanied by a further increase in brain size.

The fossil record of human evolution looks like a bush with several episodes of branching that produced clusters of species. The first branching episode, which occurred about 4 million years ago, resulted from changes in the jaws and teeth associated with a changing diet (a shift to increased dietary flexibility, allowing consumption of a wider variety of food types). A second branching episode, about 2 million years ago, was associated with increased brain size and probably with the first use of tools. Additional branching, 200,000 to 350,000 years ago, was associated with an increase in body size.

Shifting from walking on four legs to walking on two legs required changes in several parts of the skeleton. The evolution of bipedality was not a simple process, but it was one of the first important changes in the human lineage, and it seems to have set the stage for all the changes that followed. The primary advantage of bipedal locomotion for early humans was probably energetic efficiency. Bipedal locomotion at walking speed uses less energy than quadrupedal locomotion. While the chimpanzee lineage remained in the forest, the human lineage moved out of the forest and onto the grasslands of the savannas. There are numerous hypotheses about the evolutionary advantages of bipedalism for human ancestors, including increasing the energetic efficiency of traveling and freeing the hands for carrying and for tool use. The issue is still being debated as research continues.

**Q**

What are the advantages of walking on two feet rather than four?

About 3.5 or 4 million years ago, several bipedal groups, the australopithecines, appeared (**FIGURE 11-35**). They were no larger than chimpanzees and had the same brain volume, just a bit larger than a pint jar (350–400 cc). The fossil known as Lucy was the first of many australopithecines to be discovered. An adult female, Lucy was only 3 feet (about 1 m) tall and probably weighed no more than 60 pounds (a bit less than 30 kg). The group of australopithecines to which Lucy belonged lived in grassy habitats with scattered trees; their hands and feet retained the curved finger and toe bones that are characteristic of arboreal animals.

Lucy and the other closely related species of australopithecines probably foraged on the ground for food and climbed into trees to escape predators and perhaps, at night, to sleep. With jaws and teeth much like the jaws and teeth of chimpanzees, their diet was probably also like that of chimpanzees, a mixture of leaves, soft fruits, and nuts.

Following the appearance of bipedal primates—which may have occurred in conjunction with a shift to roots rather than leaves as the secondary, or "fallback," food when fruit was in short supply—a second branching episode in human evolution originated among the australopithecines and produced the earliest species in our own genus, *Homo* (from the Latin word for "human"). These earliest humans had brain volumes about twice those of chimpanzees, but little increase in body size. The species in this radiation had markedly smaller teeth than their australopithecine ancestors, a change that might indicate they had started to use tools instead of their teeth for the initial preparation of food. Stone tools are found in the same deposits as the fossils of *Homo habilis,* and this species may have been the first to use tools. Up to this time, all of human evolution had taken place in the southern and eastern parts of Africa, but about 2 million years ago, one branch, represented by *Homo erectus,* migrated out of Africa and spread through eastern Europe and Asia, while a second branch, represented by *Homo ergaster,* remained in Africa.

Additional branching episodes in human evolution (see Figure 11-35) gave rise to several species of advanced humans, including Neandertals (*Homo neanderthalensis*), "Flores man" (*Homo floresiensis,* described in the next section), and our own species (*Homo sapiens*). This radiation coincided with an increase in body size to approximately the height and weight of modern humans and an increase in brain volume to nearly twice that of earlier ancestors. In

**GRAPHIC CONTENT!**
Thinking critically about visual displays of data.
Turn to p. 468 for a closer inspection of this figure.

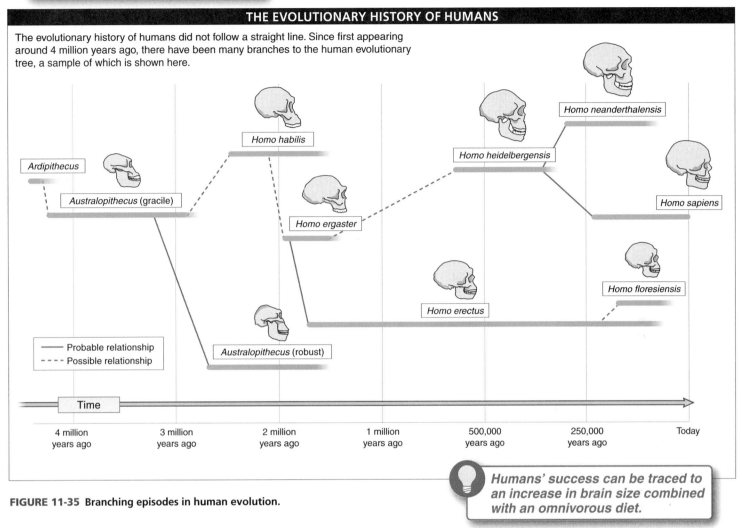

## THE EVOLUTIONARY HISTORY OF HUMANS

The evolutionary history of humans did not follow a straight line. Since first appearing around 4 million years ago, there have been many branches to the human evolutionary tree, a sample of which is shown here.

**FIGURE 11-35 Branching episodes in human evolution.**

Humans' success can be traced to an increase in brain size combined with an omnivorous diet.

addition, the body form of the species that evolved during this branching episode looked like that of modern humans, with shorter arms and longer legs.

The evolutionary lineage to which *H. sapiens* belongs became bipedal, started using tools, and developed larger brains and larger bodies. These features, along with an omnivorous diet and an increase in caloric intake associated with the development in cooking, turned out to be a winning strategy. Other lineages closely related to that of *H. sapiens* became extinct while our lineage survived.

### TAKE-HOME MESSAGE 11·19

Humans' forward-looking eyes, hands and feet with 10 fingers and 10 toes, and shoulder and elbow joints that allow the arms to rotate are characteristics retained from our arboreal ancestors. The early ancestors of humans left the trees and took up life on the ground, where they walked on two legs. Our success can be traced to an increase in brain size, possibly in conjunction with an omnivorous diet and increased caloric intake.

# 11·20 How did we get here? The past 100,000 years of human evolution.

A bit more than 100,000 years ago, a new human species evolved in Africa—the first modern *Homo sapiens*. One hundred thousand years is a ballpark figure based on a combination of molecular and fossil evidence. At that time, a small group—probably no more than 100—of these modern humans left Africa and ultimately spread across the earth.

Mitochondrial DNA shows that the initial human migration out of Africa followed three major pathways (**FIGURE 11-36**). One group turned west after leaving Africa and spread into Europe. A second group turned southeast and spread into southern Asia and through the Indo-Australian Archipelago to Australia. The third group moved northeast, populating northern Asia. This is the group that crossed the Bering Straits bridge about 15,000 years ago and spread southward through the Americas. Even as some groups migrated from Africa, many individuals remained.

There was something about the world that those *H. sapiens* groups moved into as they left Africa that we would find bizarre—*they were not alone*. Populations of at least three other species of humans (i.e., species in the genus *Homo*) were already present in some of the

areas that modern humans were moving into. That situation is totally different from the world we know today, in which we are the only human species. What would it be like to live with species that were as similar to us as dogs are to wolves, coyotes, and jackals? How would you interact with an individual who was clearly human, but not human in exactly the same way that you are?

This is the situation that the first modern *H. sapiens* encountered when they left Africa. Neandertals (*H. neanderthalensis*) had already spread across Europe and the Middle East. Neandertals were about the same size as modern humans, but they were more robust and muscular. Fossils of Neandertals often include bones that had been broken and healed. These injuries provide two types of information about Neandertals.

**Q**

Were two species of humans ever alive at the same time? If so, what happened?

**1.** Neandertals must have lived in organized groups that included a social support system, because these fossils reveal serious injuries. The injured individuals would have been incapacitated for days or even weeks until they healed, and family or clan members must have cared for them during this

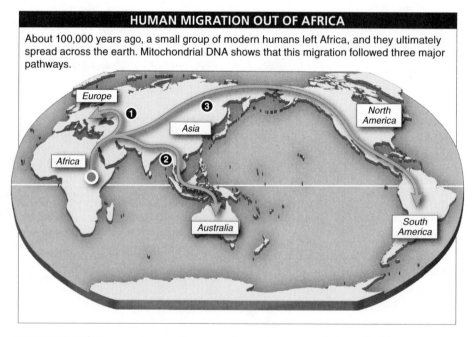

## HUMAN MIGRATION OUT OF AFRICA

About 100,000 years ago, a small group of modern humans left Africa, and they ultimately spread across the earth. Mitochondrial DNA shows that this migration followed three major pathways.

Europe **1** **3** North America

Asia

Africa **2**

Australia South America

**FIGURE 11-36** *Homo sapiens* on the move: three migratory paths out of Africa.

period. Ritual burials of Neandertals have also been found, and these provide additional support for the hypothesis that they lived in organized social groups.

**2.** The pattern of injuries found in Neandertal skeletons is distinctive—the only modern counterpart is found in professional rodeo bull and bronco riders, people who come in very close contact with large, angry animals. Neandertals probably hunted large mammals (bison, mammoths, wooly rhinoceroses) with short spears that were used for close-up jabbing instead of being thrown from a safe distance.

The *H. sapiens* who migrated through the Indo-Australian Archipelago encountered two species of humans: *H. erectus* and *H. floresiensis*. *Homo erectus* had migrated to Asia at least a million years earlier and was still present on the island of Java when modern humans arrived, about 50,000 years ago. Although *H. erectus* was the same size as modern humans, it had a smaller brain—an average brain volume of just over a quart (about 1,000 cc) for *H. erectus* compared with a quart and a half (about 1,400 cc) for *H. sapiens*. Nonetheless, *H. erectus* had technological skills. These may have been the first humans to use fire (the evidence on this is not definitive), and they almost certainly built boats that allowed them to move along coasts and from island to island.

In 2003, stone tools and fossils of a species of human only 3 feet (about 1 m) tall and with a brain volume of just over a pint (about 350 cc) were discovered on Flores Island, which lies east of Java. The paleontologists who discovered this species gave it the scientific name *Homo floresiensis* ("Flores man"), but the world press promptly called it the Hobbit because of its tiny size. When the dwarf human *H. floresiensis* lived, Flores Island was also home to a dwarf elephant, which

*H. floresiensis* probably ate, and a giant monitor lizard that probably ate *H. floresiensis*.

After modern *H. sapiens* groups spread into the areas occupied by the three other human species, these three other species disappeared. Neandertals became extinct about 30,000 years ago, *H. erectus* about 27,000 years ago, and *H. floresiensis* about 12,000 years ago. Why did these three species vanish? Broadly speaking, there are two possible explanations. They might have interbred with *H. sapiens*, in which case modern humans would carry genes from the three recently extinct species. No evidence supports that possibility, however, and it is more likely that *H. sapiens* exterminated the other species, either by monopolizing access to food and living space or by killing them in battles for food and space.

Now we are the only extant human species, a situation that is unique in the evolutionary history of humans. Ever since the first branching event, about 4 million years ago, multiple, closely related species of humans have coexisted. We are the first human species to be alone.

## TAKE-HOME MESSAGE 11·20

Modern humans (Homo sapiens) evolved in Africa between 200,000 and 100,000 years ago, and all living humans are descended from that evolutionary radiation. About 100,000 years ago, a small group of modern humans moved out of Africa, and the descendants of this group ultimately populated Europe, Asia, and the Americas. Three other species of humans were living at this time, all of which became extinct between 30,000 and 12,000 years ago, after modern humans had spread into the areas where they were living.

## WHERE ARE YOU FROM? "Recreational Genomics" and the Search for Clues to Your Ancestry in Your DNA

How much do you know about your ancestral heritage? Is it enough to know that you are Italian and Asian? More than two dozen companies now offer a promise to help you discover your genetic ancestry—usually for $100 to $1,000. Can they deliver what they promise?

**Q: ARE CLUES TO YOUR ANCESTRY HIDDEN WITHIN YOUR DNA?** The short answer is *yes*. Researchers use comparisons of DNA sequences to explore personal ancestries *within* our species, even preparing pie charts that purport to identify a person's ancestral proportions (for example, 10% West African, 70% Middle Eastern/North African, and 20% European).

**Q: HOW IS DNA USED TO BUILD A FAMILY TREE?** Most tests are based on one of three types of analysis. (1) Mitochondrial DNA tests evaluate highly variable sequences in maternally inherited mitochondrial DNA. (2) Y-chromosome tests look at variation in the paternally inherited Y chromosome. (3) Autosomal DNA tests examine highly variable sections of DNA scattered throughout the chromosomes. In each case, a sample of cells is collected (usually by swabbing the inside of the cheeks), base sequences at a number of locations (a few dozen to several hundred) are determined, and those sequences are compared with sequences from thousands of individuals in populations around the world. The matching of your sequences with those of other samples is then interpreted as potential evidence of shared ancestry.

**Q: GENETIC ANCESTRY TESTING MAY BE GOOD FUN, BUT IS IT GOOD SCIENCE?** Unfortunately, the limitations of commercially available tests of genetic ancestry are significant. These are among the most important limitations:

- High genetic variation among individuals of most populations makes it difficult to identify specific sequences that can reliably indicate membership in a population. As an analogy: if Native Americans are highly variable for a trait (e.g., height), we can't use that trait to prove someone's Native American ancestry.

- High rates of gene flow between populations reduce the reliability with which any sequence can demonstrate membership in one particular population. As an analogy: dark hair is extremely common in Asia, but having dark hair doesn't necessarily indicate Asian ancestry.

- Evaluating too few genetic loci, of which just a small number happen to be similar, can lead to the conclusion that individuals are much more genetically similar than they actually are.

- A DNA match between two individuals living today is *not* a match with an ancestor. Rather, it suggests that the two people may have inherited the DNA sequence from a common ancestor.

**Q: WHAT CAN YOU CONCLUDE?** Until we have a much more comprehensive catalog of genetic variation—including precise estimates of how much variation exists within populations and how much exists between populations—the results of genetic ancestry testing remain overly speculative and insufficiently reliable. Conclusions may be incorrect and misleading. But, because the conceptual underpinnings are solid, the future of genetic ancestry testing is promising.

# Big Ideas in ...
# Animal Diversification

## 11·1–11·3 Animals are just one branch of the Eukarya domain.

- Animals are multicellular organisms that feed on other organisms and can move during at least one stage of their life.
- Animals probably originated from an ancestral protist.
- Four key distinctions divide the extant animals: defined tissues or not, radial or bilateral symmetry, protostome or deuterostome development, and growth with or without molting.

## 11·4 Invertebrates are animals without a backbone.

- Invertebrates, defined as animals without a backbone, are the largest and most diverse group of animals, comprising 96% of all living animal species. The invertebrates are not a monophyletic group, however, and include protostomes and deuterostomes.

## 11·5–11·12 Across several evolutionary transitions, the invertebrate animals diversified.

- Sponges are among the simplest of the animal lineages. A sponge consists of a hollow tube with pores in its wall; it has no tissues or organs.
- Corals, sea anemones, and jellyfishes are cnidarians, radially symmetrical carnivorous animals with defined tissues and stinging cells called cnidocysts.
- Worms are found in several different phyla and are not a monophyletic group. All are bilaterally symmetrical protostomes with defined tissues. The segmented worms (annelids) and flatworms do not molt; the roundworms (nematodes) do.
- Mollusks, the second most diverse phylum of animals, are protostome invertebrates that do not molt.
- Arthropods are protostome invertebrates. With nearly one million species, they outnumber all other forms of life in species diversity. The ability to fly and the development of a body with a rugged exoskeleton have contributed to the enormous ecological diversity of the insects.
- Echinoderms are deuterostome invertebrates and are the closest evolutionary relatives to the vertebrates (also deuterostomes).

## 11·13–11·15 The phylum Chordata includes vertebrates, animals with a backbone.

- There are three groups of chordates, including the vertebrates, and all share four characteristic structures: a notochord, a dorsal hollow nerve cord, pharyngeal slits, and a post-anal tail.
- The development of two structures in fishes— fins and jaws—set the stage for the enormous diversity of modern vertebrates.
- In the transition of vertebrates from life in water to life on land, fins were modified into limbs, vertebrae were modified to transmit body weight through the limbs to the ground, and the site of gas exchange was transferred from gills and swim bladders to lungs. A new feature in the early evolutionary development of terrestrial vertebrates was an egg that resisted drying out.

## 11·16–11·20 All terrestrial vertebrates are tetrapods.

- Amphibians are terrestrial vertebrates (tetrapods), but most species still lay eggs in water and the eggs hatch into aquatic larvae.
- Feathers were originally colorful structures used for behavioral displays in reptiles. Their current uses in insulation and flight developed later.
- Hair and lactation are defining characteristics of mammals.
- Modern humans evolved in Africa between 200,000 and 100,000 years ago, and all living humans are descended from that evolutionary radiation.
- The evolutionary success of humans can be traced to an increase in brain size combined with an omnivorous diet.

See the To-Go Guide (packaged with every new copy of this textbook) for an illustrated summary of the chapter.

# Check Your Knowledge

## GRAPHIC CONTENT

*Thinking critically about visual displays of data.*

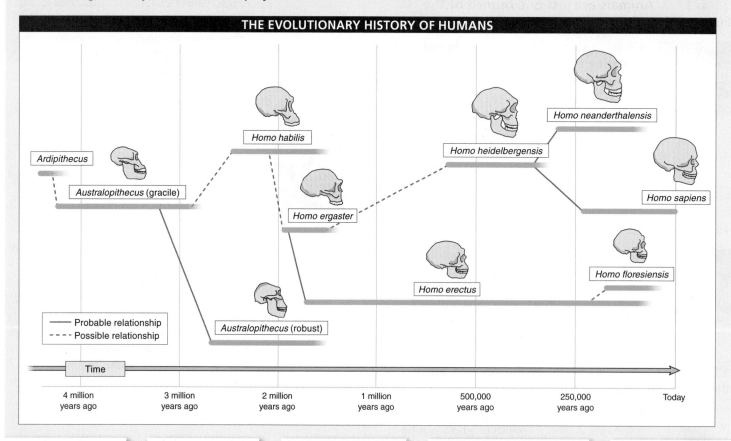

**THE EVOLUTIONARY HISTORY OF HUMANS**

Ardipithecus

Australopithecus (gracile)

Homo habilis

Homo ergaster

Australopithecus (robust)

Homo erectus

Homo heidelbergensis

Homo neanderthalensis

Homo sapiens

Homo floresiensis

—— Probable relationship
- - - Possible relationship

Time

4 million years ago | 3 million years ago | 2 million years ago | 1 million years ago | 500,000 years ago | 250,000 years ago | Today

**1.** Moving back in time from today, which species are the most likely to be the direct ancestors of modern humans? Which are not? Why?

**2.** When did the genus *Homo* first appear? How did it differ from earlier species in this evolutionary tree?

**3.** What is the difference between the dotted purple lines and the solid purple lines? Why might that information change over time?

**4.** What is the largest number of different species on the human "family tree" that were alive at the same time? When was that?

**5.** How long ago did *Homo sapiens* become the only living representative of the genus *Homo*?

See Answers at the back of the book.

## Key Terms in Animal Diversification

# Multiple-Choice Questions

1. **Which of the following is not a characteristic of all animals?**

   a) They are able to move at some point in their life.

   b) They get their energy by eating other organisms.

   c) They are multicellular.

   d) They are reproducing sexually at some point in their life cycle.

   e) All of the above are characteristics of all animals.

   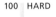
   100 HARD
   **77**
   0 EASY

2. **The lineage that first separated from the common ancestor of all animals, and retains many of those primitive features to this day, includes which of the following modern organisms?**

   a) the grasshopper

   b) the jellyfish

   c) the sponge

   d) the earthworm

   e) the sea turtle

   100 HARD
   **39**
   0 EASY

3. **Which of the following extant species is the most evolutionarily successful?**

   a) the house mouse (*Mus domesticus*)

   b) the housefly (*Musca domestica*)

   c) the Western gorilla (*Gorilla gorilla*)

   d) Irukandji, the toxic jellyfish (*Carukia barnesi*)

   e) All of the above species are equally successful because they are not extinct.

   100 HARD
   **14**
   0 EASY

4. **Sponges are sessile, meaning that they:**

   a) reproduce asexually.

   b) are parasitic and depend on their host for a constant supply of nutrients.

   c) have exoskeletons that they must shed as they grow.

   d) live within shells they find on the ocean floor.

   e) live attached to a solid structure and do not move around.

   100 HARD
   **17**
   0 EASY

5. **In cnidarians, cnidocysts are primarily used for:**

   a) creation of water flow across the body wall.

   b) formation of free-living medusas.

   c) secretion of digestive enzymes.

   d) prey capture and defense.

   e) muscular contraction during movement.

   100 HARD
   **24**
   0 EASY

6. **The mollusk's mantle is used primarily for:**

   a) feeding.

   b) gas exchange.

   c) producing the shell.

   d) excretion.

   e) reproduction.

   100 HARD
   **35**
   0 EASY

7. **Which of the following insect groups has the most species currently named by scientists?**

   a) bees, wasps, and ants

   b) beetles

   c) grasshoppers and crickets

   d) flies

   e) butterflies and moths

   100 HARD
   **36**
   0 EASY

8. **Which is the only animal phylum to have more than a million described species?**

   a) chordates

   b) arthropods

   c) flatworms

   d) nematodes

   e) mollusks

   100 HARD
   **42**
   0 EASY

**9.** The phylum Arthropoda includes all of the following kinds of animals except:

a) snails.

b) crabs.

c) crayfish.

d) butterflies.

e) scorpions.

100 HARD

53

0 EASY

**10.** Which of the following traits is unique to arthropods?

a) a ventral nerve cord

b) an exoskeleton

c) segmentation

d) a large ratio of brain size to body size

e) wheeled appendages

100 HARD

46

0 EASY

**11.** Which of the following are chordates?

a) fishes

b) humans

c) frogs

d) All of the above are chordates.

e) Only a) and c) are chordates.

100 HARD

38

0 EASY

**12.** The two most important evolutionary innovations in vertebrates, which resulted in their eventual domination among the large animals, were:

a) air-breathing lungs and a more efficient heart.

b) opposable thumbs.

c) jaws and amniotic eggs.

d) the ability to walk or fly.

e) air-breathing lungs and endothermy.

100 HARD

54

0 EASY

**13.** Which of these animals is a tetrapod that does not produce amniotic eggs?

a) salamander

b) human

c) monkey

d) elephant

e) python

100 HARD

51

0 EASY

**14.** Why is the amniotic egg considered a key evolutionary innovation?

a) It prohibits external fertilization, thereby facilitating the evolutionary innovation of internal fertilization.

b) It has an unbreakable shell.

c) It greatly increases the likelihood of survival of the eggs in a terrestrial environment.

d) It enables eggs to float in an aquatic medium.

e) It extends the time of embryonic development.

100 HARD

37

0 EASY

**15.** Which came first, the chicken or the egg?

a) The chicken, because the amniotic egg did not evolve until the first chicken appeared.

b) The egg, because the amniotic egg evolved well before the first birds.

c) The chicken, because, during speciation, the adult stage always precedes the juvenile stage.

d) The egg, because the chicken is not a real species.

e) It is impossible to determine, because eggs leave no fossils.

100 HARD

40

0 EASY

**16.** Which of the following is not a reptile?

a) sparrow

b) snake

c) frog

d) turtle

e) dinosaur

100 HARD

49

0 EASY

**17.** Marsupials and which of the following groups combine to make a monophyletic group?

a) birds

b) carnivores

c) primates

d) monotremes

e) placental mammals

100 HARD

43

0 EASY

**18.** According to the fossil record, the first humans appeared approximately _____ years ago.

a) 6 million

b) 6,000

c) 190 million

d) 100,000

e) 1 million

100 HARD

54

0 EASY

## Short-Answer Questions

1. The cheetah is an amazingly fast cat that hunts its prey on the African savannas, sprinting for short distances at speeds approaching 70 miles per hour. The sloth in the Amazon rain forests of South America is a very slow creature that hangs from a tree by all four limbs and eats vegetation. How can this difference—that the cheetah can sprint at high speeds, while the sloth moves almost in slow motion—be explained in evolutionary terms? What modifications might occur to the cheetah's ability to run if it were to occupy the same niche as the sloth, assuming the cheetah had no natural enemies there?

2. Explain why an ectotherm, such as a snake, requires only an occasional meal, while an endotherm, such as a mouse, requires more frequent meals to survive?

3. A caring pet owner brings her boa constrictor to the veterinarian because it has been more sluggish than normal and seems to be ill. The veterinarian is concerned when she sees that the owner has placed a thick sweater, specially knitted, on the snake. Why might the sweater worry the veterinarian?

4. You have a friend who confuses a terrestrial slug with a segmented worm (annelid). Describe some of the characteristics of the slug that you could point out to him that are not found in any annelids.

See Answers at the back of the book.
For additional study questions, go to www.prep-u.com.

### ABOUT THE CHAPTER OPENING PHOTO

A group of the highly sociable southern carmine bee eaters (*Merops nubicoides*) in the Luangwa Valley, Zambia.

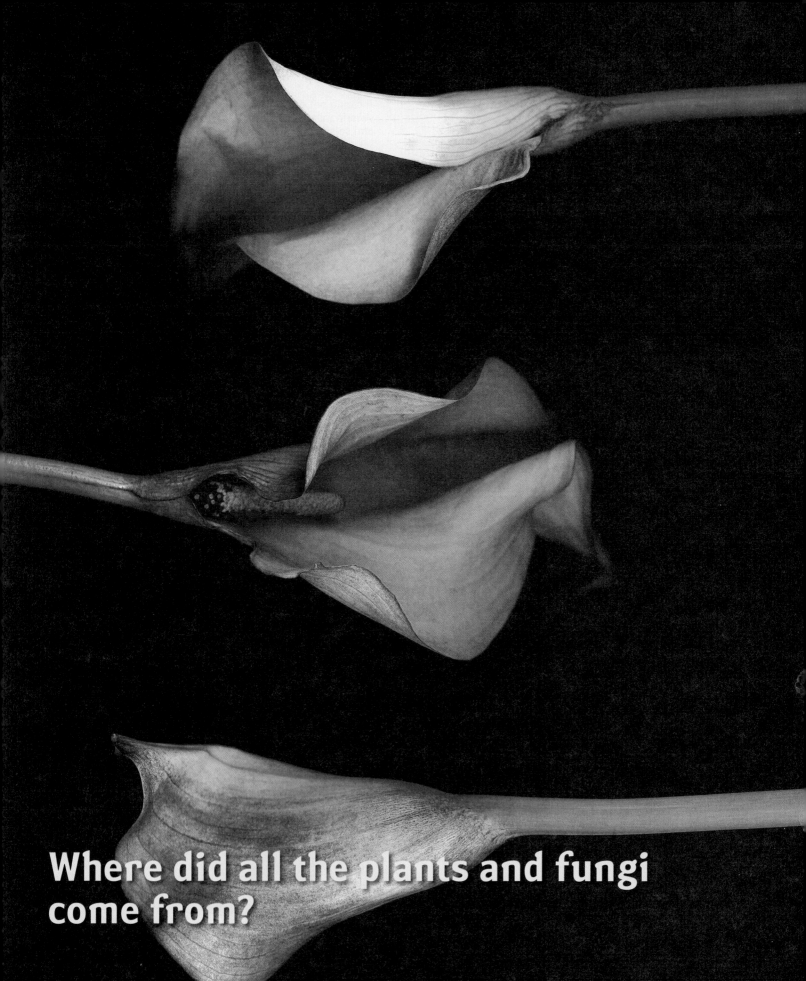

# Where did all the plants and fungi come from?

# 12

# Plant and Fungi Diversification

# 12·1 Plants are just one branch of the eukarya.

*The land plant* Psilotum nudum, *growing in Hawaii Volcanoes National Park.*

## 12·1 What makes you a plant?

An animal can move from place to place to get food or water, to avoid being eaten, or to find a mate. It's harder for a plant. It must do all the things an animal does—obtain food and water, protect itself from predators, and reproduce—*but* it can't move a centimeter in the process. Wherever a seed puts down roots, that's where the plant stays anchored for its entire life.

So, plants are organisms that are fixed in place, but that's only one aspect of being a plant. A **plant** is a multicellular eukaryote that produces its own food by photosynthesis—that is, it carries out photosynthesis, using energy from sunlight to convert carbon dioxide and water to sugar—and has an embryo that develops within the protected environment of the female parent (**FIGURE 12-1**). Plants occur almost exclusively on land, and vary in size from less than 0.04 inch (1 millimeter) to 380 feet (117 meters) tall. (Note that although there are a small number of exceptions to our definition of a plant, it applies to the vast majority of the more than 280,000 species of plants on earth today.)

There are other multicellular and photosynthetic eukaryotes on earth, including many species of algae, such as seaweed, as well as species that are the closest relatives of land plants.

### WHAT IS A PLANT?

**PLANTS CREATE THEIR OWN FOOD**
Almost all plants carry out photosynthesis, using energy from sunlight to convert carbon dioxide and water into sugar.

**PLANTS ARE SESSILE AND (MOSTLY) TERRESTRIAL**
Plants are anchored in place at their bases and occur almost exclusively on land.

**PLANTS ARE MULTICELLULAR**
Plants consist of multiple cells and have structures that are specialized for different functions.

**FIGURE 12-1 Defining characteristics of a plant.**

They differ from plants, however, in that they live only in water or on very moist land surfaces. This is in sharp contrast to most plants, which can live even in deserts.

As we have said, most plants make their own food by the process of photosynthesis. However, a plant can't live on carbohydrate alone, the product of photosynthesis. A plant needs nitrogen to build proteins, phosphorus to make ATP, and salts to create concentration gradients between the inside and outside of cells. Plants use **roots,** the part of a plant below ground, to obtain these needed substances from the soil. Above ground, plants have a **shoot** that consists of a stem and leaves. The stem is a structure that supports the main photosynthetic organ of a plant: its leaves.

When we think of plants, we often think of them as "chlorophyll-containing," but some plants have no chlorophyll—their ancestors had chlorophyll (which is why they are classified as plants), but they have lost almost all of it over evolutionary time—and they can't carry out photosynthesis. Instead, they live as parasites that steal nutrients from other plants. Dodder is an example of a parasitic plant that almost completely lacks chlorophyll and gets its sugar from the host plant it grows on (**FIGURE 12-2**). You can probably find dodder growing on plants in a nearby vacant lot or at the roadside.

Plants that carry out photosynthesis need sunlight, and that creates a challenge for some plants. Think about plants growing on the floor of a forest. It's mostly shady under the tree canopy, with just a few sunny spots where light penetrates the tree branches overhead. If a seed sprouts (germinates) in the shade, the young plant needs to

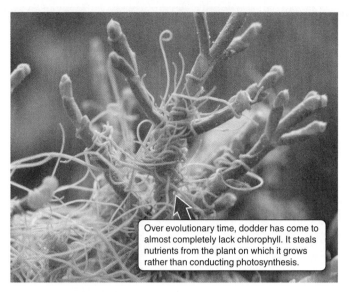

Over evolutionary time, dodder has come to almost completely lack chlorophyll. It steals nutrients from the plant on which it grows rather than conducting photosynthesis.

**FIGURE 12-2  Dodder: a parasitic, non-photosynthetic plant.**

somehow get to the nearest sunny spot. Plants can do that, of course, but not by moving. Instead, they grow toward the light (**FIGURE 12-3**). You may have seen a plant called "lucky bamboo" (it's not really a bamboo), which can grow in spirals or loops. The growers have produced these shapes by moving the light source every few days, forcing the plant to bend to follow the light. If you buy a lucky bamboo, you must continue rotating the light source or it will start to grow straight, like any other plant.

**Q** If a plant growing in the shade can't move to a new, sunnier location, how can it reach the available sunlight?

## STATIONARY LIVING CREATES CHALLENGES

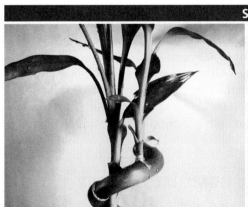

**OBTAINING FOOD**
Because plants can't move to reach sunlight, they bend in place and grow toward light.

Diploid ferns     Haploid fern

**FINDING A MATE**
Male and female plants can't meet to reproduce, so they have developed ways of getting the male gamete to the female gamete, including alternating haploid and diploid life stages, and using other organisms to transport the male gametes.

**RESISTING PREDATION**
Plants can't run from predators, so they have developed adaptations such as thorns to defend themselves.

**FIGURE 12-3  Plants must overcome the constraints of being immobile.**

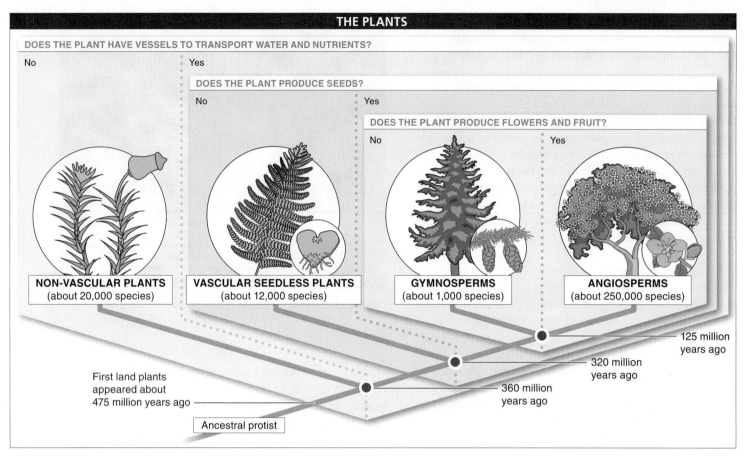

**DOES THE PLANT HAVE VESSELS TO TRANSPORT WATER AND NUTRIENTS?**

No | Yes

**DOES THE PLANT PRODUCE SEEDS?**

No | Yes

**DOES THE PLANT PRODUCE FLOWERS AND FRUIT?**

No | Yes

**NON-VASCULAR PLANTS**
(about 20,000 species)

**VASCULAR SEEDLESS PLANTS**
(about 12,000 species)

**GYMNOSPERMS**
(about 1,000 species)

**ANGIOSPERMS**
(about 250,000 species)

125 million years ago

320 million years ago

First land plants appeared about 475 million years ago

360 million years ago

Ancestral protist

**FIGURE 12-4 Phylogeny of the plants.**

Sex can also be a challenge for an organism that is anchored in place by its roots. Not surprisingly, plants have a sex life very different from that of animals. Consider humans as a typical animal. From the moment an egg is fertilized by a sperm, a human has the diploid number of chromosomes—that is, two sets of chromosomes, one set from each parent. Only the egg and sperm are haploid—an egg or a sperm has only one set of chromosomes. In plants, the haploid stage allows the male and female to reproduce, even though they will never meet because they can't move. Some plants enlist the help of animals to carry the male gamete to the female gamete.

Resisting predators is another challenge for organisms that cannot move. Running away is how most animals react to a predator, but plants have found other ways to defend themselves. Thorns, for example, are an anatomical defense that plants use to avoid predation (see Figure 12-3). Plants also use chemicals to deter predators, as many people learn when they develop an itchy rash from poison ivy.

The earliest land plants were the first multicellular organisms to live on land. They were **non-vascular,** meaning that they had no vessels to transport water and nutrients (**FIGURE 12-4**). The subsequent evolution of land plants was a series of radiations of forms with characteristics that made them increasingly independent of water. The evolutionary tree of plants shows these stages clearly: first the development of vessels to conduct water from the soil through the plant (the vascular plants), then seeds that provide nutrients to get the next generation off to a good start (the gymnosperms), and finally flowers that allow plants to entice or trick insects and birds into spreading the plant's male gametes (the angiosperms). We'll look at each of these plant adaptations in this chapter. We also explore the ecologically important kingdom of fungi. Although fungi are not plants, they are very closely associated with plants.

### TAKE-HOME MESSAGE 12·1

**Plants are multicellular organisms that spend most of their lives anchored in one place by their roots. The inability of plants to move shapes the way they obtain food, reproduce, and protect themselves from predation. Characteristics evolved that made it possible for plants to succeed on land despite their inability to move.**

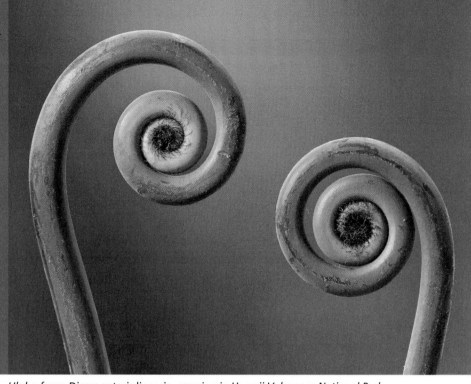

## 12·2–12·4 The first plants had neither roots nor seeds.

*Uluhe ferns,* Dicranopteris linearis, *growing in Hawaii Volcanoes National Park.*

# 12·2 Colonizing land brings new opportunities and new challenges for plants.

The aquatic ancestors of land plants were green algae, which are classified by many taxonomists as part of the plant kingdom. Like the land plants, green algae are multicellular, photosynthetic eukaryotes, but green algae live only in water or on very moist land surfaces. As water dwellers, green algae do not require specialized structures to obtain water and nutrients; water simply enters their cells by osmosis, and the nutrients they require are in solution in the water that surrounds them. Some green algae, such as sea lettuce and stoneworts, even look like land plants. But there is enough uncertainty among taxonomists about the best classification of green algae that some classify green algae within the protist kingdom, and they consider some green algae that look like slime on rocks—organisms called coleochaetes (pronounced кон-lee-oh-keets)—to be the closest relatives of plants (**FIGURE 12-5**). An individual coleochaete is about the size of a pinhead and is only one cell-layer thick. Coleochaetes can withstand exposure to air, so they survive when the water level in a lake falls and leaves them high and dry. This resistance to drying out was the first evolutionary step that plants took as they moved from water to land.

The first land plants appeared about 475 million years ago. They weren't impressive-looking, just some patches of low-growing green stems at the water's edge. They did not

Coleochaetes, a type of green alga, are the closest relative to plants. They can withstand exposure to air when water levels fall.

*Ancestors of plants began the transition from water to land with the evolution of resistance to drying.*

**FIGURE 12-5 Plants' closest relative: a green alga called a coleochaete.**

When plants emerged onto land, they faced the same two challenges that terrestrial animals faced 25 million years later.

**PROBLEM: GRAVITY**

**SOLUTION:** The earliest plants grew very close to the ground, as mosses do today, in order to resist the pull of gravity.

**PROBLEM: DESICCATION**

**SOLUTION:** Plants developed an outer waxy layer called a cuticle that covers their entire surface.

**FIGURE 12-6 Leaving the water.**

have any of the structures we associate with plants today—no roots, leaves, or flowers. But from an evolutionary perspective, those early plants were enormously important because, until terrestrial plants evolved, there was nothing on land for other land organisms to eat. Thus, the first land plants not only set the stage for the tremendous diversity of plant life that we know today but also paved the way for the evolution and diversification of land animals.

As land plants emerged, they faced the same two challenges that were to confront the first terrestrial animals, some 25 million years later: supporting themselves against the pull of gravity and reducing evaporation so they didn't dry out. The second of these problems was the more urgent. The

earliest plants did not have to grow upward—they could creep along the ground—but they *did* have to avoid drying out (**FIGURE 12-6**). The material that protects all land plants from drying is a shiny, waxy layer on the stem and leaves called the **cuticle.**

### TAKE-HOME MESSAGE 12·2

The first land plants were small, had no leaves, roots, or flowers, and could grow only at the water's edge. Nonetheless, these inconspicuous little plants set the stage for the enormous diversity of terrestrial plants and animals on earth today.

# 12·3 Mosses and other non-vascular plants lack vessels for transporting nutrients and water.

The earliest land plants were low-growing for a reason: they had no structures that could transport water and nutrients from the soil upward into the plant. The only way that these substances could move was by diffusing from one cell into an adjacent cell, and so on. Diffusion is a slow process, and plants that rely on diffusion can grow only a few centimeters tall.

Despite the limitations of diffusion, three groups of plants, all known as **bryophytes,** still use diffusion to move substances through their bodies, rather than having any sort of "circulatory system." They are the liverworts, hornworts, and mosses. More than 12,000 species of mosses grow in habitats extending from arctic and alpine regions to the tropics. Liverworts and hornworts are small (less than an

## THE NON-VASCULAR PLANTS

Moss

Liverwort

Hornwort

**COMMON CHARACTERISTICS**
• Distribute water and nutrients throughout plant by diffusion
• Release haploid spores, which grow and produce gametes
• Life cycle with multicellular haploid and diploid phases

**MEMBERS INCLUDE**
• Mosses (about 12,000 species)
• Liverworts (about 8,000 species)
• Hornworts (about 100 species)

**FIGURE 12-7 Overview of the non-vascular plants.**

inch in height), simple plants that grow in moist and shady places and resemble flattened moss. These three types of bryophyte plants are referred to as "non-vascular" because they do not have vessels to transport water and food. Water and nutrients are absorbed into the outermost layer of cells

by projections that penetrate a few micrometers into the soil. Because these projections are so short, non-vascular plants must either live in places where the soil is always moist or become dormant when the soil surface dries out (**FIGURE 12-7**).

To adapt to land, the non-vascular plants had to develop a method of reproduction that protected the plant embryo from drying out and provided it with a source of nutrients. The innovation that made this possible was a life cycle of alternating haploid and diploid generations, which is radically different from the life cycle of humans and most other animals. In animals, the haploid gametes, at fertilization, produce a new, diploid cell that becomes a multicellular organism, which, in the adult stage, starts the process all over again. In plants, however, there is an "alternation of generations."

The life cycle of a moss provides a useful example of how the alternation of generations works. When you look at a spongy mass of moss (or any other non-vascular plant), what you see is the *haploid* (or **gametophyte**) part of the life cycle, rather than the *diploid* (or **sporophyte**) part of the life cycle (**FIGURE 12-8**). For this reason, mosses are described as having a dominant gametophyte. The adult moss plants are haploid plants—that is, all the cells have only one set of chromosomes. There are male and female moss plants that have male and female reproductive structures located among the feathery leaves at the tips of the stems. Water collects here during rainstorms, allowing the sperm to "swim" (or be splashed) from the male structures to fertilize eggs in the female structures. Once the egg is fertilized in the female structure, a diploid zygote is formed, and it divides to become an embryo.

The embryo is sheltered within the female reproductive structure, which provides water and nutrients for the growing embryo. Eventually, the female structure elongates to such a degree that it breaks in two and forms a capsule extending over the top of the plant like a raised fist. Inside the capsule are haploid **spores**—single cells, containing DNA, RNA, and a few proteins. When a capsule ruptures, it releases hundreds of spores. The spores that land in moist, sheltered spots grow into new (haploid) male or female moss plants.

Some non-vascular plants are economically or ecologically important. "Peat moss" (*Sphagnum* moss), found in many parts of the world, is sold as a soil enhancer for gardening. In areas where trees are scarce, peat is dried and burned as fuel. In places with monsoon seasons, such as Malaysia and

PLANTS ARE A BRANCH OF EUKARYA

PLANTS WITHOUT ROOTS OR SEEDS

THE ADVENT OF THE SEED

FLOWERING PLANTS

PLANTS AND ANIMAL RELATIONSHIPS

FUNGI

**479**

## MOSS LIFE CYCLE

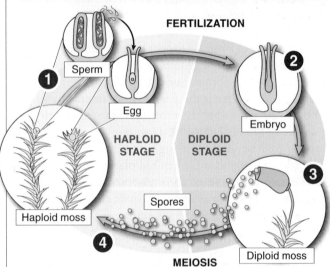

**FERTILIZATION**

Sperm

**1**

Egg

**2**

Embryo

**HAPLOID STAGE**

**DIPLOID STAGE**

Haploid moss

Spores

**4**

**3**

Diploid moss

**MEIOSIS**

**1** When there is sufficient water on the leaves—such as after rainfall—sperm swim from male reproductive structures to female reproductive structures, where they fertilize the egg.

**2** A diploid embryo forms and develops into an adult diploid moss.

**3** The diploid moss develops a capsule, which bursts and releases haploid spores.

**4** A spore lands on moist soil and grows into an adult haploid moss.

**FIGURE 12-8 Alternation of generations in mosses.** The spongy part of the moss, at ground level in the photograph, is the haploid form. Hovering above, on stalks, are the spore capsules, the diploid form.

Indonesia, peat bogs—consisting of partially decayed moss—are important in flood control, because they can absorb enormous amounts of water and release it over a period of months. Another commercial application: mounds

*Non-vascular plants such as peat moss have many important economic and ecological uses, from flood control to gardening to the production of Scotch whisky (shown here).*

**FIGURE 12-9 Uses of peat.** Peat can be dried and then burned as fuel to heat homes. In traditional Scotch whisky production, the burning of peat during the distillation process provides the smoky flavor.

of burning peat are used to dry the barley used in producing Scotch whisky. This gives Scotch its distinctive smoky taste (**FIGURE 12-9**).

Unlikely as it seems, non-vascular plants can grow in deserts, and mosses (along with lichens and cyanobacteria) are important components of the biological crust that holds desert soils in place. The crust cements the soil particles together and allows the soil to resist wind erosion, but it is extremely fragile and very slow to regenerate. When people or cattle walk over the crust, they break it into small pieces that cannot resist wind, and erosion can then strip a meter or more of soil in a few decades, leaving tree trunks supported in mid-air by their roots (**FIGURE 12-10**).

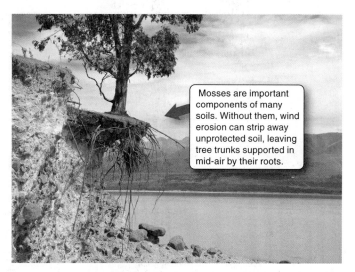

Mosses are important components of many soils. Without them, wind erosion can strip away unprotected soil, leaving tree trunks supported in mid-air by their roots.

**FIGURE 12-10 Non-vascular plants reduce erosion.** Without mosses, soils are more fragile. Shown here: a eucalyptus tree with eroded roots in New Zealand.

# 12·4 The evolution of vascular tissue made large plants possible.

What a difference some tubes make. That's all vascular tissue is—a sort of infrastructure of tubes that begins in a plant's roots and extends up its stem and out to the tips of its leaves. The evolution of vascular tissue allowed early land plants to transport water and nutrients faster and more effectively than the cell-to-cell diffusion that non-vascular plants must rely on. Vascular plants' roots penetrate far enough into the soil to reach moisture even when the soil surface is dry. Roots that reach deep into the soil also provide the support that a plant needs to grow upward without falling over. As a consequence, **vascular plants** can grow taller than non-vascular plants and are more successful in areas where the surface of the ground dries out between rains (**FIGURE 12-11**).

Ferns are the most familiar of the primitive vascular plants. During the Carboniferous period, which extended from 360 to 300 million years ago, ferns were a major component of the huge swamp forests that eventually formed the coal deposits we now mine. A related group of plants, the horsetails, like their Carboniferous ancestors, grow in wet habitats where the oxygen concentration around the roots is low because the spaces between soil particles are filled with water (rather than air). To adapt to this wet environment, horsetails developed hollow stems that allow oxygen from the air to diffuse down to the roots. The thin leaves of horsetails have a single nutrient- and water-carrying vessel extending from the base to the tip.

Simply having vessels, as horsetails do, is an important evolutionary innovation, but ferns took a further step. In addition to the central vessel in each leaflet, the leaves of ferns have vessels branching from the central vessel to the edges of the leaflet. This arrangement places a channel for the movement of water and nutrients close to each cell in the leaf.

Like non-vascular plants, horsetails and most ferns reproduce with spores. In horsetails, the central hollow stem is topped by a conical structure where the haploid spores are produced and released. Many ferns have **sporangia** (*sing.* **sporangium**) on the undersides of the leaves where spores are produced (**FIGURE 12-12**). Because the haploid gametophyte is much smaller and simpler than the diploid sporophyte, ferns are described as having a dominant sporophyte.

Horsetails and ferns are taller than non-vascular plants, so their spores can be blown by the wind when they are released, and they may settle some distance from the parent plant. This increased dispersal ability was an important adaptation—the non-vascular plants are so low-growing that wind can't play much of a role in moving their spores.

A spore that lands on moist soil grows into a tiny heart-shaped structure called a **prothallus,** which is the

## THE VASCULAR SEEDLESS PLANTS

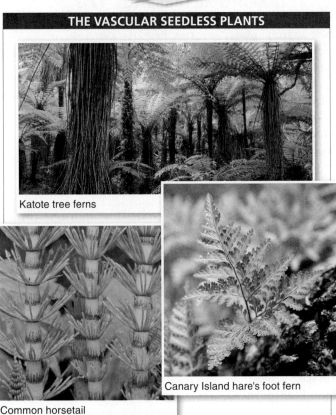

Katote tree ferns

Common horsetail

Canary Island hare's foot fern

**COMMON CHARACTERISTICS**
- Distribute water and nutrients throughout the plant with a "circulatory system" of vascular tissue
- Release haploid spores, dispersed by the wind, which grow and produce gametes
- Life cycle (unlike in animals) with multicellular haploid and diploid phases

**MEMBERS INCLUDE**
- Ferns (about 12,000 species)
- Horsetails (about 15 species)

**FIGURE 12-11 Snapshot of the vascular seedless plants: ferns and horsetails.**

free-living haploid life stage of a fern (see Figure 12-12). The prothallus produces the haploid gametes: some cells produce eggs and others produce sperm. Sperm "swim" through drops of rainwater to fertilize an egg, and the fertilized egg (a diploid zygote with two sets of chromosomes) grows into an adult fern.

## FERN LIFE CYCLE

Diploid fern                    Haploid fern

**1** Adult ferns release haploid spores, which are carried by the wind to a new location.

**2** A spore lands on moist soil and grows into a haploid fern called a prothallus.

**3** When there is sufficient water on the leaves—such as after rainfall—sperm swim from male reproductive structures to female reproductive structures, where they fertilize the egg.

**4** A diploid embryo forms and continues to grow into an adult diploid fern.

**FIGURE 12-12 Haploid and diploid life stages in ferns.** The photograph shows the sporangia, the spore-producing bodies, located on the underside of most fern leaves—on the diploid plant. The diploid and haploid plants in the fern life cycle are very different in appearance.

### TAKE-HOME MESSAGE 12·4

**Vessels are an effective "circulatory system" to carry water and nutrients up from the soil to a plant's leaves. The first vascular plants—including the earliest ferns and horsetails—were able to grow much taller than their non-vascular predecessors.**

# 12·5–12·7 The advent of the seed opened new worlds to plants.

*Forests surrounding Mistaya Lake in Banff National Park, Alberta, Canada.*

## 12·5 What is a seed?

As we've seen, the evolutionary development of vascular tissue allowed plants to grow larger. Vascular plants could colonize areas where the soil at the surface of the ground was not always wet, because their roots could penetrate the soil to reach water and nutrients. The next big innovation in plant evolution was the **seed,** an embryonic plant with its own

supply of water and nutrients encased within a protective coating (**FIGURE 12-13**).

Seeds are goodbye gifts that adult plants give to their offspring—they contain the materials needed to begin a new life. Unlike spores, which are single cells that contain only

---

### SEEDS: STRUCTURE AND GROWTH

**STRUCTURE**
Fertilization produces a diploid seed, which contains a multicellular embryo and a store of carbohydrate (endosperm) to fuel its initial growth.

Protective coating
Endosperm
Embryo

**GROWTH**
A seedling draws energy from the endosperm while it extends its leaves upward to begin photosynthesis and its roots downward into the soil to reach water and nutrients.

**FIGURE 12-13  What is a seed?** A seed is a package that contains a multicellular embryo and a store of carbohydrate, the endosperm, to fuel its initial growth. This makes many seeds desirable as food sources. Shown here are seeds for sale at a market in Jordan.

---

PLANTS ARE A BRANCH OF EUKARYA · PLANTS WITHOUT ROOTS OR SEEDS · THE ADVENT OF THE SEED · FLOWERING PLANTS · PLANTS AND ANIMAL RELATIONSHIPS · FUNGI

483

DNA, RNA, and a few proteins, seeds contain both a multicellular embryo and a store of nutrients, mostly starch. This nutritive tissue—called **endosperm** in flowering plants and trees—can fuel the seed's initial growth. A seedling draws energy from the endosperm while it extends its leaves upward to begin photosynthesis and its roots downward to reach water and nutrients in the soil. There are two modern groups of seed-producing plants: **gymnosperms** (including pines, firs, and redwoods) and **angiosperms** (all of the flowering plants and trees).

**Q**

**Do plants have sex? How are seeds formed?**

Like plants that do not produce seeds, seed plants have a life stage, called the gametophyte, that produces haploid gametes (sperm and egg). This is analogous to an individual human, which produces haploid gametes. Because the gametophyte is much smaller than the sporophyte in seed plants, these plants are considered sporophyte dominant. **Pollen grains** and **ovules** are the male and female gametophytes, respectively, of seed plants. In brief (we expand on this later in the chapter), a haploid female gamete (egg) forms inside the ovule. When a pollen grain lands near the ovule, it produces a tube that grows into the ovule. Sperm from the pollen grain move through the pollen tube into the ovule and fertilize the egg. The external layer of the ovule then forms the seed coat.

Another of the challenges that plants face is dispersing their seeds. The seed stage is the only opportunity most plants have to send their offspring away from home, and seeds and seed pods have many ways to do this. These range from the forceful send-off of exploding seed pods, to seeds that hitch a ride in or on passing animals, to those that are so small and light that they can float in water or air (such as dandelion seeds)—not to mention the use of fruits in seed dispersal. As we'll see later in the chapter, many methods for dispersing seeds have evolved.

---

**TAKE-HOME MESSAGE 12·5**

Seeds are the way that plants give their offspring a good start in life. A seed contains a multicellular embryo plus a store of carbohydrate and other nutrients. Seeds are distributed by wind, animals, or water.

---

# 12·6 With the evolution of the seed, gymnosperms became the dominant plants on earth.

Gymnosperms include four major groups: the conifers, cycads, gnetophytes, and ginkgo (just one species) (**FIGURE 12-14**). All of the 900 or more species of gymnosperms are seed-bearing plants that produce ovules on the edge of a cone-like structure. During the middle of the Mesozoic era, 160 million years ago, the forests that dinosaurs walked through consisted entirely of gymnosperms—pine trees, redwoods, cycads, and their relatives. Flowering plants (angiosperms) did not appear for another 35 million years. Indeed, when it comes to population size and habitat range, pine trees and their relatives are among the most evolutionarily successful groups of plants on earth, growing on every ice-free continent and extending from sea level to the tree line on mountains (**FIGURE 12-15**).

Pines, spruces, firs, redwoods, and their relatives are familiar to residents of the temperate regions: many of these conifers have needle-like leaves—but that is not true of all gymnosperms. The ginkgo has distinctive fan-shaped leaves, and cycads have palm-like fronds with many small leaflets (see Figure 12-15). Both are nearly identical to fossils from the Mesozoic.

**Q**

**Why does tree pollen coat the windshield of cars parked outside in the spring?**

The reproductive structures of gymnosperms—the cones—are male or female. The pine cones you are probably familiar with are the female cones, which produce the ovules and, eventually, the seeds (**FIGURE 12-16**). The male cones are smaller and release pollen that is blown by the wind, and some of it reaches the ovules, which lie beneath the protruding scales of the female cones. The quantity of pollen released by conifers is beyond imagination—pollen-heavy air in a pine forest becomes hazy, and every surface is coated with yellow. Wind dispersal is clearly an inefficient method of **pollination** (getting the pollen to the vicinity of the ovule): billions of pollen grains are wasted for each grain that lands on a female cone and produces sperm to fertilize an egg in an ovule. Nonetheless, this "brute force" method of ensuring fertilization works: for more than 200 million years, pines have been successfully pollinated by wind.

## THE GYMNOSPERMS

### COMMON CHARACTERISTICS
- Distribute water and nutrients throughout the plant with a "circulatory system" of vascular tissue
- Reproductive structures called cones produce the gametes
- Fertilization produces seeds

### MEMBERS INCLUDE
- Conifers (about 600 species)
- Cycads (about 300 species)
- Gnetophytes (about 65 species)
- Ginkgo (1 species)

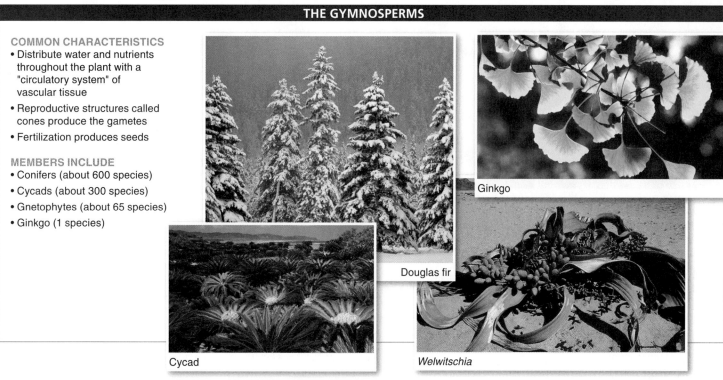

Ginkgo

Douglas fir

Cycad

Welwitschia

**FIGURE 12-14 Overview of the gymnosperms: non-flowering plants with seeds.**

## GROUPS OF GYMNOSPERMS

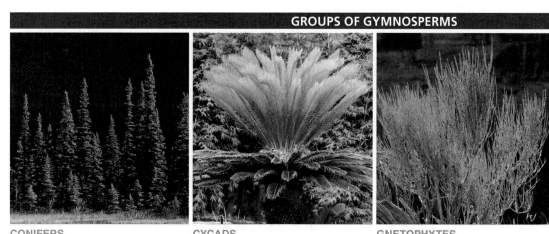

### CONIFERS
- Most commonly found in colder temperate and sometimes drier regions of the world
- Commonly have needle-shaped leaves
- Important source of timber
- Include pines, spruces, firs, cedars, hemlocks, yews, larches, cypresses

### CYCADS
- Slow-growing gymnosperms of tropical and subtropical regions
- Most resemble palm trees
- Several species are facing extinction in the wild

### GNETOPHYTES
- Composed of 3 groups: *Gnetum, Ephedra,* and *Welwitschia*
- Most gnetophyte species are *Ephedra,* a shrub-like plant sometimes used as a herbal remedy for respiratory ailments

### GINKGO
- *Ginkgo biloba* is the only remaining species
- Distinctive fan-shaped leaves
- The outer covering of the seeds emits a foul odor

**FIGURE 12-15 Major groups of gymnosperms: conifers, cycads, gnetophytes, and ginkgo.**

**MALE CONE**
The male cone releases pollen grains that require wind to reach a female cone.

**FEMALE CONE**
The female cone has ovules on the protruding scales. They produce seeds when fertilized by pollen.

 *Although billions of pollen grains are wasted for each one that fertilizes an egg, gymnosperms have successfully used wind pollination for more than 200 million years.*

**FIGURE 12-16 Cones are the reproductive structures of gymnosperms.**

When the pollen arrives at the female cone, a pollen tube forms and transports the haploid sperm to the ovule, where fertilization occurs and a diploid embryo begins to grow. The embryo develops slowly within the female cone over the course of many months, until the scales of the cone open and release the seed, ready to sprout into a new plant (**FIGURE 12-17**). With the evolution of seeds, gymnosperms developed a life history with no free-living haploid stage such as we see in non-vascular plants.

This evolutionary change in life history probably reflects the different degrees of evolutionary fitness of haploid versus diploid organisms. A haploid organism has just one copy of each chromosome, which means it has just one copy of each allele. If the haploid organism carries a defective allele for a critically important gene, the organism is doomed, because it has only that one defective version of the gene. In contrast, diploid organisms have two sets of chromosomes and thus two copies of each allele. If one copy is defective, the second copy can function as a backup, enabling the plant to produce the gene product. As a result, mutations are much less likely to be lethal for diploid organisms than for haploid organisms.

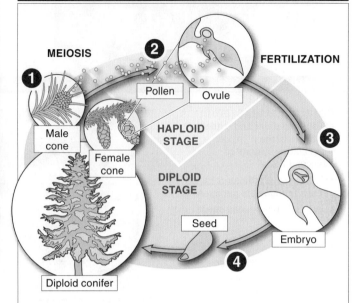

1 Male cones release pollen grains that are dispersed by the wind to ovules found beneath the scales of female cones.

2 Pollen grains release sperm that fertilize an egg within the ovule.

3 Fertilization creates a diploid embryo that matures into a seed.

4 Eventually, the seed is released from the female cone and grows into an adult tree.

 *The evolution of seeds in gymnosperms eliminated the free-living haploid life stage seen in mosses and ferns.*

**FIGURE 12-17 Assisted by the wind: the life cycle of gymnosperms.**

## TAKE-HOME MESSAGE 12·6

Gymnosperms (pine trees and their relatives) were the earliest plants to produce seeds. This mode of reproduction offered advantages over the spores of earlier plants and gave gymnosperms the boost they needed to become the dominant plants of the early and middle Mesozoic era. Gymnosperms depend on wind to carry their pollen—not an efficient method of dispersal. Conifers protect the developing seeds in the female cone.

# 12·7 Conifers include the tallest and longest-living trees.

In addition to having the widest geographic range and the greatest number of species among the gymnosperms, the cone-bearing trees—that is, the conifers—include both the tallest and the oldest living organisms on earth (**FIGURE 12-18**). The four tallest trees in the world are conifers: a coast redwood that is 380 feet (113 m) tall, a Douglas fir and a Sitka spruce, each 318 feet (97 m), and a Sierra redwood at 311 feet (95 m). But not all conifers are big: there are also miniature species of conifers, such as the shore pine, which can be just 20 centimeters tall. The oldest tree trunk belongs to a Great Basin bristlecone pine, the Methuselah tree, which has lived for more than 4,800 years. (A slightly older tree was cut down in 1964.) There's more to a tree than its trunk, though, and trunks can be replaced. That is what a Norway spruce in Sweden has been doing: the current trunk is about 600 years old, but the roots are 9,550 years old. This tree has persisted by sending up a new trunk each time the old one dies.

Trees can grow large and live to great ages because woody plants can be exceptionally strong and resistant to attack by herbivores. A cross section of a tree trunk shows the structural characteristics that allow a tree to stand erect and transport water to leaves more than 100 meters above the ground. Rigidity is provided primarily by the heartwood, which is a core of dead tissue that contains complexly cross-linked molecules.

Bark covers the outside of the tree trunk and branches. The outer layers of bark are dead tissue that can be shed without damage to the tree, protecting the living tissue from attack by plant-eating insects. Part of the success of conifers can also be traced to their ability to defend themselves by exuding a sticky pine pitch that can engulf and smother insects.

## TALLEST AND OLDEST

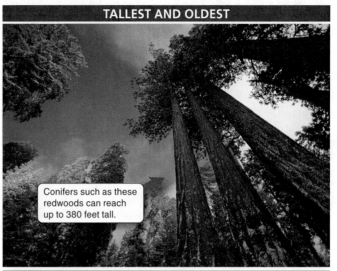

Conifers such as these redwoods can reach up to 380 feet tall.

Conifers such as these bristlecone pines can live for more than 4,800 years.

*Conifers have grown taller and reached older ages than any other plants.*

**FIGURE 12-18** Towering and ancient gymnosperms.

### TAKE-HOME MESSAGE 12·7

Conifers are the success stories among gymnosperms, with more species and a larger geographic range than all of their relatives combined. Rigidity, an exterior layer of bark, and the ability to exude sticky pitch protect conifers, allowing them to grow taller and reach older ages than any other plants.

Bleeding heart flowers, surrounded by neat, heart-shaped leaves.

# 12·8

## Angiosperms are the dominant plants today.

The appearance of flowering plants (angiosperms) about 135 million years ago set the stage for the botanical world we know today, with flowering trees, flowering bushes, and all the grasses and herbaceous (non-woody) plants we see around us. The vast majority of plants on earth are flowering plants in the angiosperm group (**FIGURE 12-19**). Many of the early flowering plants would look familiar to us, and angiosperms dominate the plant world now, with some 250,000 species, compared with approximately 1,000 species of gymnosperms.

**Flowers** come in a bewildering variety of sizes, shapes, and colors, but they all have similar structures: a supporting stem with modified leaves—the flashy petals and the sepals, which are a (usually) green wrapping that encloses the flower while it is in bud. Most angiosperms combine the male and female reproductive structures in the same flower. The male structure is the **stamen** and includes the **anther,** which produces the pollen, and its supporting stalk, the **filament.** The female reproductive structure is the **carpel.** It has an enclosed **ovary** at its base, which contains one or more ovules in which eggs develop; a stalk (the style), extending from the ovary; and a sticky tip (the stigma) (**FIGURE 12-20**).

Pollination, in angiosperms, is the transfer of pollen from the male reproductive structures to the female reproductive structures of a flower—sometimes the same flower, and sometimes a different flower (either on the same individual or a different individual). Pollination is a multi-step process

in which a pollen grain sticks to the stigma, forms a tube that grows until it reaches the ovule, and thus provides a route for sperm to travel down the tube to fertilize the egg. In the next section, we explore the tremendously rich and varied methods by which angiosperms get the male gametes to the female gametes.

Although most flowers contain both male and female structures, several thousand species of plants produce flowers that are only male or only female. Sometimes, male and female flowers are borne on different individual plants, but in quite a few angiosperm species, the same plant has some male flowers and some female flowers, often side by side. Maize (corn) is a familiar plant with male and female flowers in different places on the same plant, although you might not recognize the reproductive structures as flowers. The tassel at the top of the plant is the male flower, and the ear is formed by female flowers.

---

**TAKE-HOME MESSAGE 12·8**

Flowering plants appeared in the Cretaceous period, more than 100 million years ago, and diversified rapidly to become the dominant plants in the modern world. A flower houses a plant's reproductive structures. Most flowers have both male and female reproductive structures, but some species have flowers with only male or only female structures.

---

## THE ANGIOSPERMS

**COMMON CHARACTERISTICS**
- Distribute water and nutrients throughout the plant with a "circulatory system" of vascular tissue
- Produce flowers, which produce gametes
- Seeds are enclosed within an ovule

**MEMBERS INCLUDE**
- Flowering trees, bushes, herbs, and grasses (about 250,000 species)

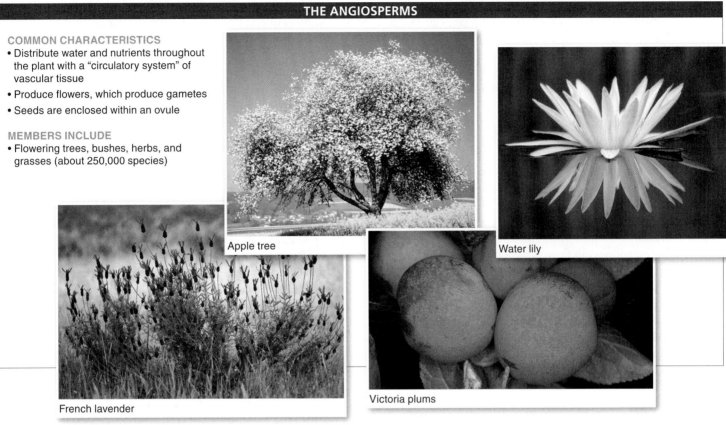

Apple tree

Water lily

Victoria plums

French lavender

**FIGURE 12-19 Snapshot of the angiosperms.**

## FLOWER STRUCTURE

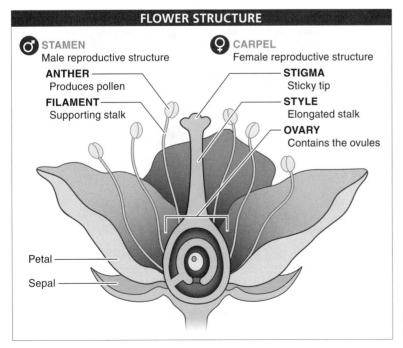

♂ **STAMEN**
Male reproductive structure

**ANTHER** — Produces pollen

**FILAMENT** — Supporting stalk

♀ **CARPEL**
Female reproductive structure

**STIGMA** — Sticky tip

**STYLE** — Elongated stalk

**OVARY** — Contains the ovules

Petal

Sepal

**FIGURE 12-20 A flower houses a plant's reproductive structures.**

PLANTS ARE A BRANCH OF EUKARYA | PLANTS WITHOUT ROOTS OR SEEDS | THE ADVENT OF THE SEED | FLOWERING PLANTS | PLANTS AND ANIMAL RELATIONSHIPS | FUNGI

489

# 12·9

## A flower is nothing without a pollinator.

Flowers may play only a tiny role in the sex lives of humans—did you give or receive any last Valentine's Day?—but they play the lead part in the sex lives of most plants. Here we examine how it is possible for individual plants, rooted in the ground in different locations, to bring their gametes together in sexual reproduction.

Picture a flower. What does it look like? Flowers differ hugely from one to another, and we all have our favorites. There are giant sunflowers and colorful roses and some rotten-smelling orchids and lilies, to name just a few. The label "flower" would seem to indicate little beyond the fact that flowers are usually brightly colored, frequently have a prominent odor, and generally sit at the end of a plant's branches. Closer inspection, however, reveals that, with very few exceptions, all flowers have the same four fundamental structures: sepals, petals, stamens, and carpels.

Fertilization occurs when the male gamete merges with the female gamete. But that's a lot easier said than done: the male gamete has to get to where the female gamete is. The first step toward this is pollination—when a pollen grain makes the journey to the stigma.

A small number of angiosperms achieve pollination by releasing tremendous amounts of pollen into the wind, as do the gymnosperms, or into water—on the slim chance that some will land on the female reproductive organs of another plant of the same species. Most angiosperms,

though, have a different way of moving pollen from the anthers of one flower to the stigma of another: they use animals to carry it (**FIGURE 12-21**).

To ensure that animals will visit a flower, picking up and delivering their pollen cargo, two strategies for achieving pollination have evolved among the flowering plants.

**Q** *Why are some flowers so flashy?*

**1. Trickery.** The plant deceives some animals into carrying its pollen from one plant to another. Some orchid species, for example, produce flowers that resemble female wasps. The mimicry is so good that male wasps mount the flower and attempt to mate with it. The male wasp twirls wildly on the flower, like a cowboy on a bucking bronco, repeatedly whacking his head against the strategically located anthers and getting pollen stuck all over his head and body. That is not enough for the plant to achieve pollination—but it's a start. If that male wasp gets fooled again by another orchid flower and mounts it in an attempt to mate, he will inadvertently deposit some of the pollen from his body onto the also strategically placed stigma of that flower. In the end, the wasp does not gain from his actions, but the orchids have used an effective system of pollination.

**2. Bribery.** The plant bribes some animals to carry its pollen from one plant to another. Rather than just using trickery, the plant offers something of value to the animal. For this

---

### STRATEGIES FOR ATTRACTING POLLINATORS

**TRICKERY**
Some plants deceive animals into carrying pollen from one plant to another. Shown here: This isn't a bee! It's actually an orchid flower. Pollen is picked up and delivered by male bees that attempt to mate with the flower.

**FIGURE 12-21 Delivering precious cargo (inadvertently).**

**BRIBERY**
Some plants offer something of value to an animal, bribing the animal to carry pollen from one plant to another. Here, a bee, covered in pollen, flies from flower to flower in search of nectar.

 *Angiosperms have developed a way to transfer pollen efficiently from the anthers of one flower to the stigma of another: get an animal to carry it!*

---

**COLORS AND PATTERNS**
**WHITE:** Nocturnal pollinators, such as moths and bats
**BRIGHT COLORS:** Visually oriented diurnal pollinators, such as
birds, butterflies, and bees

**FLOWER STRUCTURE**
**TUBE:** Pollinators with long tongues, such as moths
**INTRICATE/CLOSED:** Pollinators such as bees

**ODORS**
**SWEET:** Pollinators with a good sense of smell, such as moths,
butterflies, and bees
**STINKY:** Pollinators looking for rotten meat on which to lay eggs,
such as flies
**NO ODOR:** Pollinators with a poor sense of smell, such as birds

**NECTAR**
**ABUNDANT:** Pollinators with high energy needs, such as bees, birds,
and butterflies
**ABSENT:** Pollinators, such as flies, looking for a place to lay eggs, or
such as beetles, looking for petals, pollen, and other parts to eat

**FIGURE 12-22 Evolving together: plants and pollinators.** You can often determine the type
of animal that pollinates a flower just by examining the features of the flower.

method to work, the plant must produce (a) a sticky pollen,
(b) a flower that catches the attention of the pollinator, and,
most important, (c) something of value to the pollinator.
The payoff for the pollinator can be food, such as nutritious
nectar rich in sugars and amino acids, or perhaps a safe,
hospitable location for an insect to lay its eggs.

The variety of flower structures is tremendous. They differ in
shape, color, smell, time of day they are open, whether or not
they produce nectar, and whether or not their pollen is
edible. So, too, is the variety of pollinators: birds (mostly
hummingbirds), bees, flies, beetles, butterflies, moths, and
even some mammals (mostly bats) (**FIGURE 12-22**). In each
case, there has been strong coevolution between the plants
and their pollinators: the plants get more and more effective
at attracting the pollinators and deterring other species from
visiting the flower, while the pollinators get more and more
effective at exploiting the resources offered by the plants.
Because of this strong coevolution between the plant and

animal species, for most flowers we can determine, just by
examining the features of the flower, the type of animal that
will pollinate it.

Pollination is just one step toward fertilization. Fertilization
itself doesn't happen until the male and female sex cells meet
and fuse. We explore this process in the next section.

## TAKE-HOME MESSAGE 12·9

Angiosperms have found a way to transfer pollen from
the anthers of one flower to the stigma of another: let
an animal carry it. Flowers are conspicuous structures
that advertise their presence with colors, shapes,
patterns, and odors. Using flowers, plants are able to
trick or bribe animals into transporting male gametes to
female gametes, where fertilization can occur.

PLANTS ARE
A BRANCH OF
EUKARYA

PLANTS WITHOUT
ROOTS OR SEEDS

THE ADVENT
OF THE SEED

FLOWERING PLANTS

PLANTS AND
ANIMAL
RELATIONSHIPS

FUNGI

491

# 12·10

## Angiosperms improve seeds with double fertilization.

The process of fertilization—the fusion of two gametes to form a zygote—*begins* when a pollen grain lands on a stigma of a flower, but that is far from the end. In angiosperms, the process is called **double fertilization,** because there are two separate fusions of male nuclei (each carrying a complete set of the organism's genetic material) from the pollen grain with female nuclei in an ovule. As we'll see, double fertilization is a more efficient system than the type of fertilization in gymnosperms, because whenever an embryo is produced at fertilization (and only then), so, too, is a more substantial, ready-made food source.

When we left off in the previous section, a pollen grain—which will deliver the haploid male gamete—had just arrived on the female's stigma. The pollen grain forms a tube that extends downward through the stigma and style to the ovary, ultimately entering an ovule. As the pollen tube grows, two haploid sperm are formed within it by mitosis, and both move down the pollen tube (**FIGURE 12-23**).

While the pollen tube is growing, a cell within the ovule undergoes meiosis, forming four haploid cells, called spores. Three of these usually degenerate. The surviving haploid spore enlarges and undergoes mitosis, forming an embryo sac. A large, central cell in the embryo sac has two haploid nuclei. One of the other cells in the embryo sac is the haploid female gamete (the egg), which remains near the place where the pollen tube will be guided into the ovule by two other haploid cells within the embryo sac.

When the pollen tube reaches the ovule, one of the two sperm fuses with the egg to form a zygote. The other sperm fuses with the two nuclei in the middle of the embryo sac to form the endosperm, which has *three sets* of chromosomes (i.e., is triploid). The process is called double fertilization because two sperm enter the ovule and combine with female cells in two separate events, forming (1) a zygote (with two sets of chromosomes) and (2) an endosperm (with three sets of chromosomes).

The final steps in producing a seed occur as the diploid zygote cell undergoes multiple mitotic divisions to form an embryo, while the triploid cells multiply mitotically to produce the endosperm that will provide nutritional support for the seedling through its initial growth stages. The outer layers of the ovule form the seed coat that will protect

**FIGURE 12-23 Double fertilization fortifies the seeds of angiosperms.**

### ANGIOSPERM LIFE CYCLE

**1** Pollen grains are released from the male anthers of a flower, delivering the male haploid gamete to the stigma—the sticky tip of the female reproductive structure.

**2** The pollen grain produces a tube that extends through the stigma to the ovary. Meanwhile, a cell within the ovule undergoes meiosis, forming four haploid cells, called spores. Three of these usually degenerate.

**3** The surviving haploid spore undergoes mitosis, forming an embryo sac that contains seven haploid cells. One cell becomes the egg, another cell—with two nuclei—will form the endosperm following fertilization by a sperm cell.

**4** When the pollen tube reaches the ovule, two sperm are released. In a process called double fertilization, one of the two sperm fuses with the egg to form a zygote. The other sperm fuses with the two nuclei in the middle of the embryo sac to form the endosperm, which nourishes the embryo.

**5** The zygote and the endosperm continue to develop within the ovule, forming a seed that will eventually be released and grow into a mature plant.

the seed until it sprouts. The enclosure of the seed within the ovary is a distinction between angiosperms and gymnosperms. The seeds of a gymnosperm are unenclosed and are sometimes referred to as "naked."

**Q**

**What advantage does double fertilization give to angiosperms?**

In gymnosperms, single sperm fuses with an egg to form a zygote in an ovule that contains hundreds of other female cells that are used as a source of nutrients for the resulting embryo. By contrast, there are two important advantages that double fertilization offers angiosperms.

**1.** Double fertilization initiates formation of endosperm only when an egg is fertilized. Waiting to be sure that an egg is fertilized is a good strategy, because making endosperm is a large energy investment for a plant. Gymnosperms invest that energy up front, and nutrients in ovules that are not fertilized are wasted because those "seeds" do not contain an embryo. In contrast, angiosperms do not waste energy forming endosperm in ovules that will not contain embryos.

**2.** Angiosperms can produce smaller gametes than gymnosperms, because the large energetic reserves will be produced only *after* fertilization occurs. The small size of the male and female gametes of angiosperms ensures that seeds are produced quickly. **FIGURE 12-24** shows that as different reproductive strategies have evolved in plants, their gametes have become progressively smaller. Rapid production of seeds allows angiosperms to grow as annual plants (that is, plants that complete their life cycle from sprouting to seed production in one growing season), which is something gymnosperms cannot do.

**Q**

**Do flowers with both male and female structures fertilize themselves?**

Outbreeding, the combination of haploid cells from two different individuals, produces offspring with greater genetic diversity (that is, carrying a greater diversity of alleles) than offspring that result from inbreeding, the combination of a male and female gamete from the same individual. Angiosperms have a variety of ways to increase the chance that only sperm from another individual will fertilize the female gamete. Plants that have male and female reproductive structures in different flowers or even on different individuals, for example, increase the chance that a pollen grain landing on a stigma comes from a different plant. But many flowers with both male and female parts also have mechanisms to prevent self-fertilization. For example, the anthers may mature before the

### HAPLOID AND DIPLOID LIFE STAGES

| HAPLOID STAGE | DIPLOID STAGE |
| --- | --- |

**NON-VASCULAR PLANTS**
The majority of the life cycle is spent in the haploid stage.

**VASCULAR SEEDLESS PLANTS**
The haploid and diploid stages are both multicellular and physically independent from one another.

**GYMNOSPERMS**
The evolution of seeds almost completely eliminates the prominent haploid stage seen in mosses and ferns.

**ANGIOSPERMS**
Haploid gametes are further reduced in size, enabling more rapid seed production.

*As plants have developed different reproductive strategies, they have progressed from having a prominent haploid stage of life to simply having haploid gametes.*

**FIGURE 12-24 Overview of the haploid and diploid stages of plant life cycles.**

stigma, so the stigma is not ready to receive pollen when the anthers are active. And when the stigma becomes functional, the anthers are no longer producing pollen. Other plants use a molecular recognition system to prevent inbreeding: proteins on the surface of the stigma will not allow pollen from the same individual to form a pollen tube.

### TAKE-HOME MESSAGE 12·10

**Angiosperms undergo a process of double fertilization, which ensures that a plant does not invest energy in forming endosperm for an ovule that has not been fertilized. Angiosperms have also developed methods to reduce the occurrence of self-fertilization and increase genetic variation among offspring.**

# 12·11–12·12 Plants and animals have a love-hate relationship.

*Fruits such as these peaches are structures that contained fertilized seeds.*

## 12·11 Fleshy fruits are bribes that flowering plants pay animals to disperse their seeds.

Leaving home is an inevitable part of growing up. But as difficult as it may be, imagine you had to move away but didn't have a car or moving van—or even legs. How would you do it? This is yet another issue facing plants. Ingenious methods have evolved for transporting the male gametes to the female sex organs for fertilization, but that still results in all the fertilized eggs, and all the new offspring they develop into, living right on the female plant—or in nearly the same location, if they manage to fall off the parent plant. This is a problem because parent and offspring would end up competing for the same light, space, soil nutrients, and other resources.

The solution is the fruit, a structure that aids in dispersing seeds—the reproductive packets made up of the embryo, some food reserves, and a hard coat. The fruit usually develops from the ovary (and sometimes from nearby tissue), right around the seeds, which develop from the ovules. This should cause you to look at a field of flowers differently. Given that the ovary is part of a flower, after pollination and fertilization, that flower will turn into a fruit. Another way of looking at it is that every fruit you eat was once a flower.

Pea pods, sunflower seeds, corn kernels, and hazelnuts are fruits. These are "dry fruits," and the seeds they contain are transported by wind, water, or animals (**FIGURE 12-25**). Many angiosperms, though, make **fleshy fruits** that consist of the ovary plus some additional parts of the flower. For example, the core of an apple is the ovary and the flesh of the apple is derived from adjacent parts of the flower. Blueberries, watermelons, oranges, tomatoes, and peaches are other examples of fleshy fruits.

The fleshy part of a fruit is often larger than the ovary, and an angiosperm invests a lot of energy in producing fleshy fruits. The payoff comes when an animal eats the fruit and then defecates the seeds at a location far from the parent plant. In other words, fleshy fruit is the bait that some flowering plants use to get animals to disperse their seeds.

When you look at fruits, you can see several characteristics that help this system work.

- Fruits are colorful—red is the most common color for fruits, followed by yellow and orange. All of these colors contrast with green foliage and make fruits conspicuous.

- Fruits typically taste good—plants pour sugars into fruits, and their sweetness appeals to many animals.

- Fruit is good for animals, serving as a significant source of sugar, enabling animals to produce ATP for their needs. Additionally, fruit has other nutritional value; many birds incorporate the red and yellow colors of fruits into their own color patterns, and colorful male birds are appealing to female birds. Thus, male birds that eat fruit are likely to produce a lot of offspring who also eat fruit—a good deal for the plant.

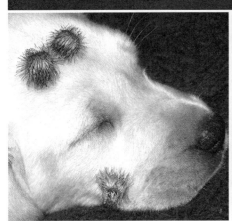

**HITCHING A RIDE**
Some seed pods have spines or projections that attach them to passing animals.

**FLYING AND FLOATING**
The structure of the seed allows it to be carried away from the parent plant by wind or water.

**PROVIDING A FOOD SOURCE**
Fleshy fruit is a form of bait that lures an animal to eat the seed and carry it far from the parent plant before eliminating it.

**FIGURE 12-25 Methods of dispersing seeds.**

**Q**

*Can seeds still sprout after being eaten by an animal?*

Of course, there's one more requirement to make fruit a successful way for a plant to disperse its seeds: the seeds must survive passage through the animal's digestive tract and be able to germinate when they come out. That's not a problem—seeds are fully viable when they emerge. It's easy enough to test this yourself. The next time you eat fresh tomatoes, check the toilet about 24 hours later. The seeds that have passed through your digestive system are still intact. If you want to test their viability, that's easy—recover a few, rinse them off, and put them on a wet paper towel in a sealed plastic bag. If you want to do a controlled experiment, you can take some seeds directly from a tomato and put them in another bag. Put both bags in a warm, dark place, and check them in a couple of days. You'll find that both sets of seeds have sprouted. (Or, you can just take our word for it.)

Seeds not only survive being eaten by animals—some seeds will not germinate *unless* they are eaten. For some plants, the pulp that surrounds the seeds in the fruit inhibits germination, and it must be digested away before the seed will sprout. The seeds of other plants have chemicals in the seed coats that must be removed by the acidity of an animal's stomach before the seeds will germinate. And there's another plus to this system: the seeds are deposited with a bit of manure that provides nutrients for the seedlings.

**TAKE-HOME MESSAGE 12·11**

Following pollination and fertilization, plants often enlist animals to disperse their fruits, which contain the fertilized seeds, depositing them at a new location where the seedlings can grow. Fruits are made from the ovary and, occasionally, some surrounding tissue.

# 12·12  Unable to escape, plants must resist predation in other ways.

The interactions of plants and animals are not simple. On the one hand, plants depend on animals to pollinate their flowers and disperse their seeds. On the other hand, many kinds of animals eat plants, and plants are vulnerable because they can't run away.

Just because plants can't run away from plant-eating animals, however, they are not defenseless. A host of defensive devices

that have evolved in plants give them some protection against being eaten. These defenses fall into two categories: anatomical structures, such as thorns, and chemical compounds, including hallucinogens (**FIGURE 12-26**).

Spines, prickles, and thorns are a common way to discourage herbivores, and some of these structures are impressively large. The acacias, a group of plants that grow as large

## PLANT DEFENSES

**ANATOMICAL STRUCTURES**
Some plants have spines, spikes, and thorns that deter predators. Some trees have thick layers of bark that are shed in order to get rid of attacking insects.

**STICKY TRAPS**
Conifers exude pitch, a sticky substance that can engulf and smother attacking insects.

Locoweed contains substances that, when eaten, can cause cattle and horses to become lethargic and stop feeding.

Stinging nettles have needle-like hairs that inject a chemical that causes a stinging sensation when contacted by an animal.

**CHEMICAL COMPOUNDS**
Some plants synthesize chemicals that induce physiological and behavioral changes in the animals eating them.

**FIGURE 12-26 Spines, sticky traps, and toxic compounds.** Plants have developed a range of defenses against predation.

bushes or small trees, are notoriously spiny. In these plants, the deterrent effects of spines are sometimes enhanced by the presence of ants that live in the swollen bases of the spines and fight off other animals—from insects to mammals, including humans—that would feed on their host acacia. Spines on palm fronds carry bacteria that infect the wounds that the spines create, providing a long-lasting reminder that it's not a good idea to try to eat a palm frond. In most plants, the surfaces of leaves and stems have microscopic hairs that protect the plant against insect herbivores by exuding a sticky fluid trap, but stinging plants such as nettles use larger hairs to inject a potent fluid into any animal that brushes against a leaf.

Many chemical defenses make plants bitter or otherwise unpalatable, but some plants play hardball by producing chemicals that affect the nervous or reproductive systems of their predators. Locoweeds, which occur all over western North America, contain alkaloids that can cause permanent damage to the nervous system. When cattle and horses eat locoweed, they tire quickly and stop eating. Other defensive chemicals are hallucinogens. Tetrahydrocannabinol, or THC, in the leaves and buds of marijuana plants disorients mammals and causes them to stop eating.

Many of the chemicals that plants produce to protect themselves have physiological effects in humans, and for thousands of years humans have used plants for medicinal purposes. The aboriginal peoples of North America and Eurasia, for example, knew that extracts of the bark from willows and aspens could relieve pain. Salicin is the compound in the bark that relieves pain, and we still use acetylsalicylic acid (aspirin) for pain relief. Opium poppies produce a more powerful pain reliever; foxglove flowers contain a compound that slows the heart rate; and a compound from the ipecacuanha plant can induce vomiting.

Plants with medicinal uses are sought after (called "bioprospecting"), and representatives of pharmaceutical companies trudge through rain forests and deserts to find them. The rewards from such a discovery can be great. For example, artemisin from Chinese sweet wormwood is effective in treating malarial infections that are resistant to conventional pharmaceutical compounds.

There's another role that chemicals can play in a plant's defense systems. Although plants can't move, they do send messages. They release volatile chemicals when they are attacked by insects, and these airborne chemicals can warn nearby plants of the impending threat or can call protective insects to their aid.

When plants are nibbled by insects, they release the chemical methyl jasmonate (MJ), which is carried with the breeze to

Some plants living in nitrogen-deficient soil have turned the tables, becoming predators on insects.

**FIGURE 12-27 Carnivorous plants feed on insects.** A true bug nymph trapped in the insectivorous northern pitcher plant, *Sarracenia purpurea,* growing in an acidic bog in Michigan.

them. Insects can also detect MJ, and insects that prey on other insects are attracted to plants emitting the chemical, because it signals the presence of prey. When corn and cotton plants are attacked by caterpillars, release of MJ summons parasitic wasps that deposit their eggs inside the caterpillars. When the eggs hatch, the wasp larvae eat the caterpillars from the inside out.

About 900 species of plants turn the tables—and eat insects. They use the protein in their insect meals to supplement the nitrogen compounds they get from the soil. Insectivorous plants are most common in boggy areas, because boggy soil often has low nitrogen concentrations (**FIGURE 12-27**). Pitcher plants are an example—their name comes from conspicuous pitcher-shaped structures that capture insects. It's easy for an insect to enter the open top of a pitcher, but hard to get out, because the inner walls are lined with hair-like structures that point downward, forcing the insect into a pool of liquid. Some pitcher plants merely absorb the nitrogen that is released when the trapped insects decay, but other species secrete digestive enzymes into the pool. These enzymes, which are basically the same as the enzymes in your stomach, digest the insects and make the nitrogen available more rapidly.

nearby plants. When these plants detect the presence of MJ, they ramp up production of their own defensive chemicals so that they are prepared if the insects move on to feast on

## TAKE-HOME MESSAGE 12·12

Plants have a wide range of defenses against herbivorous animals: from physical defenses such as thorns to chemicals that have complex effects on animals' physiology. Plants respond to insect attack by synthesizing chemicals that make the plant less palatable. Some plants living in soil that is deficient in nitrogen have switched roles, preying on insects.

## 12·13–12·15 Fungi and plants are partners but not close relatives.

*Cup fungi,* Cookeina tricholoma, *from Borneo.*

# 12·13

## Fungi are closer to animals than they are to plants.

Think about your toes for a minute. Have you ever had an itchy burning sensation between them? Almost everyone has athlete's foot at some time during his or her life. One study found that 15% of all people in the United Kingdom had this fungal infection.

Fungi make up their own monophyletic kingdom within the eukarya domain. Most fungi are multicellular, sessile decomposers (**FIGURE 12-28**). Although they were originally thought to be plants lacking chlorophyll, it turns out that they have little in common with plants. In fact, DNA sequence comparisons reveal that fungi are more closely related to animals than they are to plants (**FIGURE 12-29**). As eukaryotes, fungi have all the basic cellular components you would expect to find: nuclei, mitochondria, an endomembrane system, and a cytoskeleton. They also have cell walls, but the cell walls, instead of including cellulose, as in plants, are made of the carbohydrate chitin, a chemical also important in producing the exoskeleton of insects.

Fungi most likely arose from a unicellular, flagellated, aquatic protist more than 500 million years ago (and possibly as long as 1.3 billion years ago). Close to 100,000 species have been described, but the total number of species is estimated to be about 1.5 million. These species are

### WHAT IS A FUNGUS?

**FUNGI ARE DECOMPOSERS OR SYMBIONTS**
Fungi acquire energy by breaking down the tissues of dead organisms or by absorbing nutrients from living organisms.

**FUNGI ARE SESSILE**
Fungi are anchored to the organic material on which they feed.

**FUNGI HAVE CELL WALLS MADE OF CHITIN**
Fungi have cell walls made of chitin, a chemical that is also important in producing the exoskeleton of insects.

**FIGURE 12-28 The defining characteristics of fungi.**

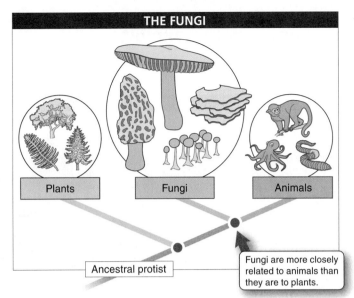

## THE FUNGI

Plants

Fungi

Animals

Ancestral protist

Fungi are more closely related to animals than they are to plants.

**FIGURE 12-29 Where the fungi fit in.**

divided into about seven phyla, but the overall phylogenetic classification is still very much in flux, with many of the relationships unresolved. Molecular evidence increasingly is guiding the process. About 98% of the described species belong to two monophyletic phyla, Ascomycota (about 64,000 species) and Basidiomycota (about 31,000 species). Besides these phyla, the two phyla with the largest number of described species are Microsporidia, with about 1,300 species, and Chytridiomycota, with about 700 species. These two groups, however, are not monophyletic.

The most commonly encountered types of fungi are (1) yeasts (the only single-celled fungi), truffles, and morels of the phylum Ascomycota; (2) mushrooms, of the phylum Basidiomycota; and (3) molds (which are not a phylogenetic grouping), such as *Penicillium*—the source of the antibiotic penicillin—of the phylum Ascomycota, and black bread mold, of the monophyletic phylum Glomeromycota (**FIGURE 12-30**).

The fungus that causes athlete's foot is multicellular and consists of thread-like structures made up of long strings of cells called **hyphae** (pronounced HIGH-fee). Because of their thinness and length, hyphae have an enormous surface area. This means they are very good at taking up nutrients from your skin, but it also makes them very vulnerable to drying out. The fungus responsible for athlete's foot grows best in moist places—like on the skin between your toes.

Unless you actually have an active athlete's foot infection right now, your most recent encounter with a fungus (or, at least, with evidence that a fungus was once present) was most likely when you last ate bread. Bread rises because yeast cells are mixed into the dough, which is then kept in a warm place while the yeast consumes sugar and produces carbon dioxide through fermentation. The carbon dioxide released by the yeast makes the dough swell into a loaf.

### TAKE-HOME MESSAGE 12·13

**Fungi are eukaryotes with the same internal cellular elements as other eukaryotes—and one distinctive feature: a cell wall formed from the carbohydrate chitin. Some fungi, the yeasts, live as individual cells; most other fungi are multicellular.**

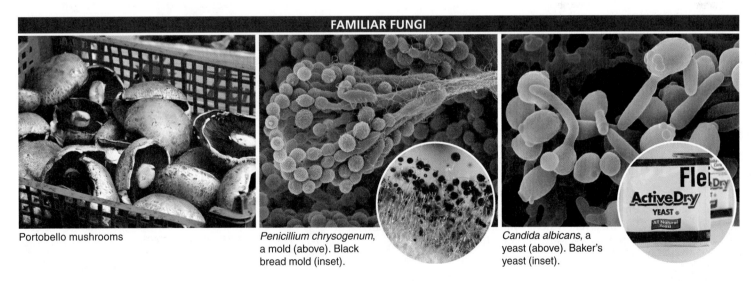

### FAMILIAR FUNGI

Portobello mushrooms

*Penicillium chrysogenum,* a mold (above). Black bread mold (inset).

*Candida albicans,* a yeast (above). Baker's yeast (inset).

**FIGURE 12-30 The fungi that humans know best.** There are more than 1.5 million fungal species, usually divided among seven phyla. Those most recognizable to humans are mushrooms of the Basidiomycota (basidiomycetes), yeasts of the Ascomycota (ascomycetes), and molds (a term that does not refer to a monophyletic group), which occur in multiple phyla.

PLANTS ARE A BRANCH OF EUKARYA · PLANTS WITHOUT ROOTS OR SEEDS · THE ADVENT OF THE SEED · FLOWERING PLANTS · PLANTS AND ANIMAL RELATIONSHIPS · FUNGI

499

# 12·14 Fungi have some structures in common, but exploit an enormous diversity of habitats.

As we've seen, most fungi are multicellular and are composed of long strings of barely visible, thread-like hyphal cells. The hyphae interconnect to form a mass of tissue called a **mycelium,** the form in which a fungus spends most of its time, usually underground or in a decaying tree or log. Unless you dug down to reach the mycelium, you would not know a fungus was there.

The structure that most people associate with fungi is the mushroom. But a mushroom is just a temporary reproductive structure (or "fruiting body"), part of a complex reproductive cycle that includes both sexual and asexual reproduction in some fungi (**FIGURE 12-31**).

1. Underground, genetically distinct haploid hyphal cells join together. But their nuclei *don't* fuse, so instead of being diploid, they are "dikaryotic," meaning that each cell in the hypha has two nuclei.

2. The dikaryotic mycelium can grow and spread for years.

3. At some point, tightly packed dikaryotic hyphae may form a mushroom.

4. The haploid nuclei in the dikaryotic hyphae fuse in some cells, putting the mushroom into a diploid state.

5. In meiosis, the diploid cells produce huge numbers of haploid spores (up to a billion in a single mushroom!), which are dispersed by wind or water or on the bodies of animals.

6. After landing in a hospitable place, the spores grow as haploid hyphae, and the cycle begins anew.

**Q** What is the largest living organism in the world?

Fungi have an unusual and effective method of getting nutrition. Unlike humans, they digest their food outside their "body." While growing underground, hyphae secrete strong enzymes that break down the organic molecules around them, and the hyphae then absorb the nutrient-rich fluid. How effective are fungi at absorbing nutrients and growing? Quite good. In fact, in eastern Oregon, a yellow honey mushroom fungus covers an area of about 4 square miles (nearly 10 square kilometers) (**FIGURE 12-32**). This fungus is estimated to be at least 2,400 years old, and it may be more than 8,000 years old, which would place it in the category of the oldest living organism. Moreover, by area, it is the largest. Huge as it is, the fungus is nonetheless hard to detect, because the

**FUNGUS LIFE CYCLE**

Dikaryotic mushroom

**DIKARYOTIC STAGE**

**DIPLOID STAGE**

Diploid nuclei

**HAPLOID STAGE**

Dikaryotic hyphae

Haploid hyphae

Haploid spores

1 Underground, distinct haploid hyphae fuse to form dikaryotic hyphae, in which single cells have two distinct nuclei.

2 The dikaryotic hyphal state can last for years, as the dikaryotic mycelium spreads throughout an area.

3 At some point, a reproductive structure—the mushroom— forms. It is produced from tightly packed dikaryotic hyphae.

4 In some cells within the mushroom, the haploid nuclei in dikaryotic hyphae fuse, becoming diploid.

5 The diploid cells within the mushroom then undergo meiosis, producing haploid spores.

6 When the spores land in a hospitable place, they start growing as haploid hyphae, and the cycle is complete.

**FIGURE 12-31 The three-stage reproductive cycle of a fungus.**

A yellow honey mushroom fungus in eastern Oregon is the largest organism by area on earth, covering four square miles, mostly underground.

**FIGURE 12-32 The largest organism in the world: a yellow honey mushroom fungus.** Shown here is just a small cluster of the fungus's temporary reproductive structures—what we call "mushrooms."

mycelium is underground. The fungus appears on the surface only as mushrooms (edible, but not particularly tasty) and as hyphae in dying and dead trees. Indeed, no one even knew that this enormous fungus existed until 2000, when Oregon foresters sought an explanation for an epidemic of dead and dying trees in the Malheur National Forest. (They discovered that the fungus was killing trees by causing their roots to rot.)

Many mushrooms are gastronomic delicacies, and commercially grown portobello and shiitake mushrooms, for example, command high prices. Truffles—the underground reproductive structures of a fungus called *Tuber*—can sell for as much as $3,500 per pound. Because truffles grow underground, there is nothing on the surface to indicate their presence, and truffle hunters use trained pigs or dogs to locate them by scent.

Fungi can grow in many different habitats because, as **decomposers,** all they need for their nutrient supply is some sort of organic material that they can break down. Fungi play an enormously important ecological role in speeding the decay of organic material in forests. They don't need light, so they can grow underground or inside dead trees and logs.

Fungi can also thrive in poorly ventilated spaces in buildings (**FIGURE 12-33**). Molds are multicellular fungi that are

responsible for many unpleasant effects. People living or working in a "sick building" can experience burning or watering eyes, a runny nose, and itchy skin—allergic reactions to the proteins in fungal spores. More severe effects, such as cancers and miscarriages, are probably produced by toxins released when the fruiting bodies of the fungi disintegrate. Curing a sick building can be so expensive that in some cases the entire building is destroyed and rebuilt.

If you break down the tissues of dead organisms, you are a decomposer, and if you break down the tissues of living organisms, you are a parasite. Just as fungi can live off dying trees, they can also grow in and on people. "Mycosis" is a general term for a disease that is caused by a fungus, such as athlete's foot, or related fungi that cause jock itch, beard itch, scalp itch, ringworm, and toenail fungus. These fungi get their nutrition by digesting some of the organic molecules of your body! And all of these fungal diseases are quite contagious, because they are spread by spores that can linger on moist surfaces and in clothing, so they tend to be a problem in places like dormitories, gymnasiums, and fitness centers.

It wouldn't be fair to mention so many undesirable effects of fungi on humans without describing some of the many beneficial fungi, particularly one that has benefited millions of humans: the fungus that produces penicillin. Alexander Fleming was a researcher studying bacteria. In 1928, just prior to taking a month-long vacation, he stacked some of his used Petri dishes containing bacterial colonies in a corner of his lab. When he returned from his vacation, he found

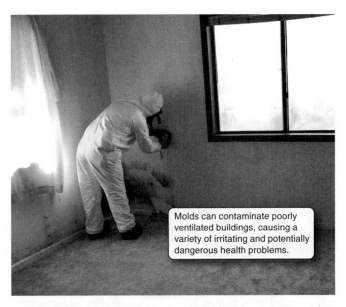

Molds can contaminate poorly ventilated buildings, causing a variety of irritating and potentially dangerous health problems.

**FIGURE 12-33 Cleaning out toxic mold.** A building overrun with mold spores may cause its occupants to become ill.

that some of the dishes had been contaminated by a fungus. Before discarding them, he noticed that near the fungal infection, the bacterial colonies were dead, while farther away from it, they continued to grow. Suspecting that the fungus might be producing something toxic to the bacteria, he cultured it and discovered that it did produce a substance that killed many different types of bacteria. He identified the fungus as *Penicillium*. Following the development by others of a method to purify the antibacterial compound produced by the fungus, penicillin became a "wonder drug," responsible for saving millions of lives by helping to treat bacterial infections.

### TAKE-HOME MESSAGE 12·14

Fungi are decomposers, and all they need to thrive is organic material to consume and a moist environment so their hyphae don't dry out. Fungi can grow almost anywhere that is moist, and they can attain enormous sizes. Fungi have complex life cycles, with both sexual and asexual phases, and the parts of a fungus that are most often visible are its temporary spore-producing bodies. Fungi can cause a variety of health problems, but also are responsible for antibacterial medicines such as penicillin.

# 12·15    Most plants have fungal symbionts.

If you examine the roots of a plant with a microscope, you will find round structures and fibers closely associated with the fine rootlets and root hairs. These are root fungi, or **mycorrhizae** (pronounced my-koh-RYE-zay). Some mycorrhizae have hyphae that press closely against the walls of root hair cells. Others send hyphae through the root cell walls and into the space between the cell wall and the plasma membrane (**FIGURE 12-34**).

Beneficial associations between roots and fungi are ancient—they have been found in 400-million-year-old fossils of early land plants—and nearly all species of modern plants have them. Lumber companies have found that they have greater success when replanting areas they have clear-cut if the conifer seedlings are first grown in soil that has been inoculated with mycorrhizal fungi. Home gardeners can buy potting soil with mycorrhizal spores or

buy spores compressed into tablets that can be placed in the soil when seedlings are transplanted. The association between mycorrhizal fungi and plants is beneficial to both partners. The mycorrhizal fungus benefits by drawing sugar from the plant, which it uses to support its own cellular respiration. The fungus extracts phosphorus and nitrogen from the soil and releases them inside the roots of the plant, allowing it to grow faster and larger. Research on the impact of root fungi on their plant hosts even demonstrated that the *offspring* of parental plants with mycorrhizae had significantly better survival than the offspring of parental plants without mycorrhizae (**FIGURE 12-35**). (In this case, unfortunately, the fungi are enhancing the growth and survival of an invasive plant called velvetleaf, which causes hundreds of millions of dollars of damage each year to corn crops in the United States.)

**MYCORRHIZAE**

Mycorrhizal fungi grow in association with the roots of plants, receiving sugar from the plant while transferring nitrogen and phosphorus from the soil to the plant.

Hyphae

Cell wall

Root hair

Plasma membrane

Some mycorrhizae have hyphae that press closely against the walls of root hair cells, while others send hyphae through the root cell walls and into the space between the cell wall and the plasma membrane.

**FIGURE 12-34 The association between mycorrhizal fungi and plants is beneficial to both.**

## SYMBIOTIC FUNGI ARE BENEFICIAL TO THEIR HOSTS

Researchers investigating the invasive velvetleaf plant evaluated how well offspring survived. They compared the survival of offspring from parents grown with and without symbiotic root fungi and reported the data below.

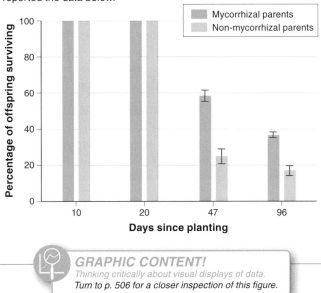

**GRAPHIC CONTENT!**
*Thinking critically about visual displays of data.*
*Turn to p. 506 for a closer inspection of this figure.*

**FIGURE 12-35 Plant partners.** Root fungi (mycorrhizae) can improve plant functioning.

Not all plants play fair, however. Some take nutrients from the fungus and give nothing in return. One of the best known of these mycorrhizal parasites is the ghost pipe, which grows in forests in the northern hemisphere (**FIGURE 12-36**). The ghost pipe is an angiosperm related to heaths and heathers. The plant's name comes from its ghostly white appearance. The roots of the ghost pipe are associated with mycorrhizal fungi that are also connected to the roots of trees. The ghost pipe obtains all its nutrients from or through the mycorrhizae: nitrogen and phosphorus are provided by the mycorrhizal fungus, and sugar travels through the fungus from tree roots to the ghost pipe.

Some plants manipulate mycorrhizae to gain a competitive advantage over other plants. Garlic mustard, for example,

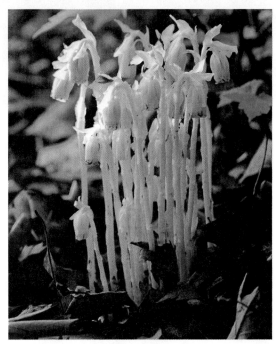

**FIGURE 12-36 Not all plants play fair.** Ghost pipe is a plant that parasitizes mycorrhizae for nutrients.

was brought to North America by European colonists to add a tangy flavor to salads. Initially it was planted in kitchen gardens, but it escaped into the wild, where it has proved itself to be an extremely successful invasive species. Garlic mustard excretes compounds from its roots that interfere with the partnership between mycorrhizal fungi and local native plants. Because the native plants *do* need mycorrhizal fungi to prosper, the destruction of these fungi by garlic mustard weakens the plants. This chemical warfare is so successful that garlic mustard has wiped out entire populations of native woodland plants and is threatening many forests in the central United States.

Fungi also form mutually beneficial relationships with chlorophyll-containing bacteria and/or algae. These two-way or three-way partnerships are called **lichens,** and they grow on surfaces such as tree trunks and rocks. The fungus is fed by its photosynthetic partner and helps it absorb water and nutrients. In addition, the fungus provides nutrients for its partners by excreting enzymes that dissolve organic material and acid that dissolves rock.

### TAKE-HOME MESSAGE 12·15

**Plants and fungi have a close and mutually beneficial association in mycorrhizae. Mycorrhizal fungi grow in intimate association with the roots of most plants, receiving sugar from the plant and transferring nitrogen and phosphorus to the plant.**

### YAMS: NATURE'S FERTILITY FOOD?

## Do you know any twins? Have you ever wondered how common twins are?

In the United States and Europe, about 12 in every 1,000 births are twin births. In Asia, the number is slightly lower, at 8 sets of twins per 1,000 births. Elsewhere, the rates are similar—except for southwest Africa, particularly in Nigeria, a place that has been called "The Land of Twins." There, the rate of twin births among the Yoruba people is more than four times that in the United States, with about 50 pairs of twins per 1,000 births. (Triplet births are unusually common, too, occurring 16 times more frequently in Nigeria than in the United States.)

**Q: WHY ARE SO MANY TWINS BORN IN NIGERIA? IS IT SOMETHING IN THE WATER?** Nope. But that's not too far off. It's the diet of the Yoruba people. A staple in their diet is the white yam, a starchy vegetable that looks a bit like a potato; many people eat yams several times each day.

**Q: WHAT'S IN THE YAMS?** Some preliminary studies suggest that an estrogen-related compound in yams is responsible for stimulating the ovaries, increasing the likelihood, in any given month, that more than one egg will be released.

**Q: HOW CAN WE BE SURE IT'S THE YAMS?** At this point, no one is certain, and the question awaits a good application of the scientific method. In one intriguing laboratory study, rats fed a diet rich in yams *doubled* their litter size. As yet, however, a definitive link between yams and increased rates of twin births in humans has not been demonstrated.

**Q: SO, WILL IT WORK FOR YOU?** It's unclear. Anecdotes abound about women having twins after purposely increasing their yam consumption. But a randomized, controlled, double-blind study has yet to be conducted. How would you set up and analyze a study like that? (It is important to note that the yams consumed in Nigeria are not the same as American yams, which are sweet potatoes and do not contain the steroid levels of yams grown in Nigeria.)

# Big Ideas in ...
# Plant and Fungi Diversification

## 12·1 Plants are just one branch of the eukarya.

- Plants are multicellular eukaryotic organisms that produce their own food by photosynthesis and have an embryo that develops within the protected environment of the female parent.
- Plants' inability to move shapes the way they obtain food, reproduce, and protect themselves from predation.
- The evolutionary history of plants reveals the progressive appearance of characteristics that allow plants to succeed on land.

## 12·2–12·4 The first plants had neither roots nor seeds.

- The first land plants were small, had no leaves, roots, or flowers, and could grow only near water.
- Non-vascular plants—mosses, liverworts, and hornworts—resemble the earliest land plants. They lack roots and vessels to move water and nutrients from the soil into the plant.
- Vessels are an effective way to carry water and nutrients from the soil to the leaves, and the first vascular plants were able to grow much taller than their non-vascular predecessors.

## 12·5–12·7 The advent of the seed opened new worlds to plants.

- A seed, which contains a multicellular embryo and a store of nutrients, is a way for plants to give their offspring a good start in life.
- Seeds are distributed by air, water, or animals.
- Gymnosperms were the earliest plants to produce seeds, which offer advantages over the spores of earlier plants, but they depend on wind to carry their pollen.
- Conifers are the most successful gymnosperms, with more species and a larger geographic range than all of their relatives combined.

## 12·8–12·10 Flowering plants are the most diverse and successful plants.

- Flowering plants (angiosperms) appeared in the Cretaceous period, over 100 million years ago, and diversified rapidly to become the dominant plants in the modern world.
- A flower advertises a plant's reproductive structures with colors, patterns, and odors which are used to trick or bribe animals into transporting male gametes to female gametes, where fertilization can occur.
- Double fertilization in angiosperms ensures that a plant does not invest energy in endosperm for an ovule that has not been fertilized.

## 12·11–12·12 Plants and animals have a love-hate relationship.

- Overcoming a challenge of immobility, plants often use the assistance of animals to disperse their fruits, which contain fertilized seeds, depositing them at a new location where the seedlings can grow.
- Fruits are made up from the ovary and, occasionally, some surrounding tissue.
- Overcoming another challenge of immobility, plants have a wide range of defenses against herbivorous animals, from physical defenses such as thorns to chemicals with complex effects.
- Some plants living in nitrogen-deficient soil have switched roles, preying on insects.

## 12·13–12·15 Fungi and plants are partners but not close relatives.

- Fungi are eukaryotes with a cell wall made of chitin. They can attain enormous sizes.
- Some fungi (yeasts) live as individual cells, but most are multicellular.
- Fungi are decomposers.
- Fungi have complex life cycles, with sexual and asexual phases, and the parts of fungi that are most often visible are their temporary spore-producing bodies.
- Mycorrhizal fungi grow in intimate association with the roots of most plants, receiving sugar from the plant and transferring nitrogen and phosphorus to the plant.

See the To-Go Guide (packaged with every new copy of this textbook) for an illustrated summary of the chapter.

# Check Your Knowledge

## GRAPHIC CONTENT
*Thinking critically about visual displays of data.*

**1.** What can you conclude from this graph?

**2.** There are error bars for the percentage of offspring surviving at 47 days and at 96 days after planting. Why are there no error bars at 10 days and 20 days after planting?

**3.** What does it mean to say that the differences in offspring survival between mycorrhizal and non-mycorrhizal parents are "statistically significant"?

**4.** What percentage of the offspring from parents with mycorrhizal fungi were surviving after 47 days? What was the percentage surviving from parents without mycorrhizal fungi?

**5.** If you could have offspring survival data for one additional point in time, when would it be (i.e., how many days after planting)? Why?

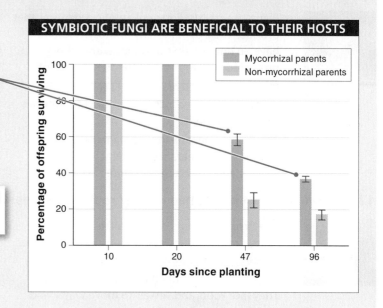

See Answers at the back of the book.

## Key Terms in Plant and Fungi Diversification

angiosperm, p. 484
anther, p. 488
bryophyte, p. 478
carpel, p. 488
cuticle, p. 478
decomposer, p. 501
double fertilization, p. 492

endosperm, p. 484
filament, p. 488
fleshy fruit, p. 494
flower, p. 488
gametophyte, p. 479
gymnosperm, p. 484
hyphae, p. 499

lichen, p. 503
mycelium, p. 500
mycorrhizae, p. 502
non-vascular plant, p. 476
ovary, p. 488
ovule, p. 484
plant, p. 474

pollen grain, p. 484
pollination, p. 484
prothallus, p. 481
root, p. 475
seed, p. 483
shoot, p. 475
sporangium, p. 481

spore, p. 479
sporophyte, p. 479
stamen, p. 488
vascular plant, p. 481

# Multiple-Choice Questions

1. In the plant kingdom, all the different species are descended from a single common ancestor. In terms of phylogeny, what type of tree is this?
   a) monophyletic
   b) unbranched
   c) biphyletic
   d) uniphyletic
   e) branched

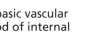

2. Which of the following is a significant new challenge that plants faced when they moved from their aquatic environment onto the land?
   a) light availability
   b) desiccation
   c) nutrient availability
   d) predation
   e) the need for osmotic regulation

3. Which is the best brief description of the vascular system of the very first terrestrial plants?
   a) The first plants developed specialized vessel cells that conducted water.
   b) The first plants did not possess a vascular system.
   c) The first plants had a very basic vascular system with a simple method of internal transport.
   d) The first plants had only long, needle-like leaves, from which water could evaporate easily.
   e) None of the above answers is correct.

4. Which of the following statements about ferns is incorrect?
   a) Their sporophyte is dominant.
   b) They require liquid water for fertilization.
   c) Their seeds are dispersed by the wind.
   d) They have vascular tissue for distributing water and nutrients throughout the plant.
   e) Their spores are contained in sporangia.

5. Mosses and ferns differ from gymnosperms and angiosperms in their reproductive strategies in which of the following ways?
   a) Mosses and ferns rely on liquid water for fertilization, whereas angiosperms and gymnosperms do not need liquid water for fertilization.
   b) Mosses and ferns have much larger seeds than do angiosperms and gymnosperms.
   c) Mosses and ferns use wind pollination, whereas angiosperms and gymnosperms use insects for pollination.
   d) Mosses and ferns are primarily diploid in their adult (reproductive) form, whereas gymnosperms and angiosperms are primarily haploid.
   e) Mosses and ferns are primarily haploid in their adult form, whereas gymnosperms and angiosperms are primarily diploid.

**6.** Which of the following is characteristic of gymnosperms?

a) They are more diverse than the angiosperms.
b) They are wind-pollinated.
c) The gametophyte generation is dominant.
d) They are water-pollinated.
e) All of the above are correct.

100 HARD

43
0 EASY

**7.** In terms of their adaptation to living on land, how are reptiles similar to the seed plants?

a) Both reptiles and seed plants became completely independent of water.
b) Reptiles eat plants.
c) Seed plants and reptiles have developed structures such as cuticles and impermeable skin to minimize desiccation.
d) Reptiles and seed plants have developed structures that house their gametes and protect them from the surrounding environment.
e) Both c) and d) are correct.

100 HARD
35
0 EASY

**8.** Angiosperms and gymnosperms differ from each other in that:

a) angiosperms have vascular systems, but gymnosperms do not.
b) angiosperms produce seeds, but gymnosperms do not.
c) angiosperms tend to rely on animal pollinators, whereas gymnosperms tend to rely on wind pollination.
d) angiosperms use pollen, whereas gymnosperms use cones.
e) angiosperms have a single fertilization process, whereas gymnosperms use double fertilization.

100 HARD
45
0 EASY

**9.** Anthers and stigmas are found on:

a) bryophytes.
b) fungi.
c) angiosperms.
d) gymnosperms.
e) all of the above.

100 HARD
40
0 EASY

**10.** Which of the following is not an example of a group of angiosperms?

a) cacti
b) cherry trees
c) pine trees
d) grasses
e) orchids

100 HARD
51
0 EASY

**11.** Over the evolutionary history of plants:

a) the sporophyte has become smaller, though more independent.
b) the gametophyte and sporophyte have grown increasingly independent of each other.
c) there has been a trend toward gametophyte dominance.
d) there has been a trend toward gametophyte dependence on the sporophyte.
e) the gametophyte has become larger, though more dependent.

100 HARD
75
0 EASY

**12.** Which of the following comparisons and contrasts between fungi and plants is incorrect?

a) Fungi cannot photosynthesize, but plants can.
b) Both fungi and plants use chitin as a structural stabilizer.
c) Fungi are heterotrophs (i.e., cannot fix carbon and so must use organic carbon for growth), but plants are not.
d) Both fungi and plants have cell walls.
e) Both fungi and plants have a sexual stage in their reproductive cycle.

100 HARD
25
0 EASY

**13.** Dispersal of fungal spores is typically done by:

a) movement of cilia.
b) insects.
c) hummingbirds.
d) wind.
e) movement of flagella.

100 HARD
25
0 EASY

**14.** You are taking a hike down a forest trail and see the familiar sight of a mushroom on the ground. This visible portion of a fungal body is the structure also referred to as a:

a) hypha.
b) fruiting body.
c) thallus.
d) spore sac.
e) mycelium.

100 HARD
26
0 EASY

**15.** In most cases, the relationship between roots and fungi in mycorrhizae can best be described as:

a) mycelium.
b) competition.
c) trickery.
d) symbiosis.
a) parasitism.

100 HARD
36
0 EASY

## Short-Answer Questions

1. What distinguishes a vascular plant from a non-vascular plant? How do these differences affect their geographic distribution?

2. How do fungi benefit plants? What benefit do the fungi derive from this relationship?

3. How does development of a fleshy fruit benefit plants?

4. If all animals became extinct, what changes would occur in plant communities? Why?

See Answers at the back of the book.
For additional study questions, go to www.prep-u.com.

Bacteria, archaea, protists, and viruses: the unseen world

# Evolution and Diversity Among the Microbes

**13·2** **There are microbes in all three domains.**

**13·1** Microbes are the simplest but most successful organisms on earth.

**13·2** Not all microbes are closely related evolutionarily.

**13·3 – 13·5** **Bacteria may be the most diverse of all organisms.**

**13·3** What are bacteria?

**13·4** Bacterial growth and reproduction is fast and efficient.

**13·5** Metabolic diversity among the bacteria is extreme.

**13·6 – 13·9** **In humans, bacteria can have harmful or beneficial health effects.**

**13·6** Many bacteria are beneficial.

**13·7** Bacteria cause many human diseases.

**13·8** Bacteria's resistance to drugs can evolve quickly.

**13·9** Sexually transmitted diseases reveal battles between microbes and humans.

**13·10 – 13·12** **Archaea exploit some of the most extreme habitats.**

**13·10** Archaea are profoundly different from bacteria.

**13·11** Archaea thrive in habitats too extreme for most other organisms.

**13·12** Much archaeal diversity has yet to be discovered.

**13·13 – 13·15** **Most protists are single-celled eukaryotes.**

**13·13** The first eukaryotes were protists.

**13·14** There are animal-like protists, fungus-like protists, and plant-like protists.

**13·15** Some protists can make you very sick.

**13·16 – 13·19** **Viruses are at the border between living and non-living.**

**13·16** Viruses are not exactly living organisms.

**13·17** Viruses are responsible for many health problems.

**13·18** Viruses infect a wide range of organisms.

**13·19** HIV illustrates the difficulty of controlling infectious viruses.

Dental plaque, a film that forms on teeth, is caused by the grow
bacterial colonies (shown here). It can generally be removed by b
if left, can lead to tooth decay.

# 13·1

## Microbes are the simplest but most successful organisms o

Humans are large organisms, and being large comes with a lot of baggage. We need a skeletal system to support our weight, and a respiratory system to take in oxygen and get rid of carbon dioxide. We need a circulatory system to move oxygen and carbon dioxide and other molecules around our bodies, and a digestive system to take in food and break it down. We even need a nervous system so that our brain knows what distant parts of our body are doing.

**Microbes**—the most abundant organisms on earth—don't have skeletal, respiratory, circulatory, digestive, or nervous systems, because they are too small to need them. In fact, "microbe" is a combination of Greek words that means "small life," and the "small" part of the name fits these organisms well. Take an amoeba, for example—it has a volume about a million billion ($10^{15}$) times smaller than a human. When you are that tiny, the force of gravity becomes trivial, so the amoeba needs no skeleton to support it. Plenty of oxygen diffuses inward across its cell membrane,

and carbon dioxide diffuses outward, so an amoeba does not need a respiratory system; and because every part of its interior is close to the body surface, it doesn't need a circulatory system to transport gases. An amoeba eats by enclosing food items in a piece of its cell membrane and digests the food with the same enzymes it uses to recycle its own proteins, lipids, and carbohydrates, so it does not need a digestive system. And no part of an amoeba is far enough from another part to require a specialized nervous system for communication.

**Q**

How can a microbe function when its body is just a single cell?

Most microbes are even smaller than an amoeba: a typical bacterium or archaeon is about one thousand million billion ($10^{18}$) times smaller than a human, and an influenza virus is about one thousand billion trillion ($10^{24}$) times smaller than you are (**FIGURE 13-1**).

They may be invisible, but microbes could never be considered unsuccessful just because of their (lack of) size. To the contrary, they are more successful than humans in almost every imaginable way.

## RELATIVE SIZES OF MICROBES

Although almost all microbes are invisible to the naked eye, they vary tremendously in size.

20× larger → 25× larger → 20× larger

| Influenza virus 100 nm | Escherichia coli 2 μm | Euglena gracilis 50 μm | Amoeba dubia 1 mm |

0 mm          0.5 mm          1.0 mm

**FIGURE 13-1** The most abundant organisms on earth are too small to see.

*If an influenza virus were the size of a pinhead, an amoeba would be the size of an oak tree.*

*Microbes are genetically diverse.* More than 500,000 kinds of microbes have been identified by their unique nucleotide sequences, and further studies will almost certainly distinguish millions of additional microbial species.

**Q** How would you decide whether a group of organisms can be considered successful?

*Microbial species live in almost every habitat on earth; among them, they can eat almost anything.* As you read these lines, more than 400 species of microbes are thriving in your intestinal tract, 500 more species thrive in your mouth, and nearly 200 species call your skin home (**FIGURE 13-2**).

## MICROBES CAN LIVE ALMOST ANYWHERE

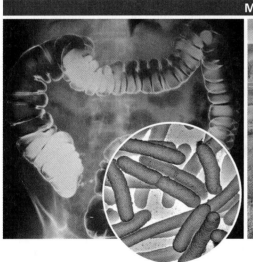

*Escherichia coli* (inset), a bacterium, lives in the human intestine.

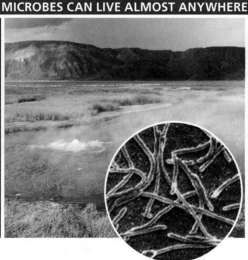

*Thermus aquaticus* (inset), a bacterium, lives in geothermal pools that can reach temperatures greater than 160° F.

Halobacteria (inset) are archaeons that live in extremely salty environments, such as the Great Salt Lake.

**FIGURE 13-2** Microbes are everywhere on earth.

*Microbes live in nearly every kind of environment, including water at temperatures of up to 750° F and as low as 5° F!*

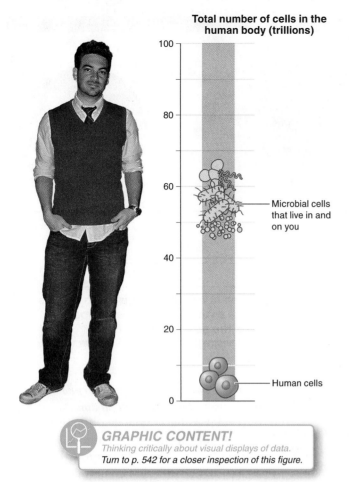

**Total number of cells in the human body (trillions)**

Microbial cells that live in and on you

Human cells

**GRAPHIC CONTENT!**
*Thinking critically about visual displays of data.*
*Turn to p. 542 for a closer inspection of this figure.*

**FIGURE 13-3 Microbe majority.** Your body has more microbial cells than human cells.

The microbes that live in you and on you eat mostly what you eat—some of the bacteria in your mouth and intestine compete with you, trying to digest your food before you can, and others use the waste products you release after you have broken down the food. Others feed on the leftovers released by the breakdown of your cells during the normal process of cell renewal.

Living conditions in the human body are relatively moderate. Other microbes inhabit some of the toughest environments on earth—in the almost boiling water of hot springs, at depths a mile below the earth's surface, and more than a mile deep in the oceans, where hydrothermal vents emit water at 400° C (750° F).

*Microbes are abundant.* Surface seawater contains more than 100,000 bacterial cells per milliliter, and diatoms (protists in the eukarya domain) are as abundant there as are the bacteria. These densities translate to about 8,000 million billion trillion ($8 \times 10^{30}$) individuals of just these two kinds of microbes in the world's oceans. Your own body is a testament to the abundance of microbes: it contains about 100 trillion cells, but only one-tenth of those cells are actually human cells—the remaining 90 trillion cells are the microbes that live in and on you (**FIGURE 13-3**). You're a minority in your own body.

> **TAKE-HOME MESSAGE 13·1**
>
> Microbes are very small, simple organisms, but they do everything that larger, multicellular organisms do. They can live anywhere, from moderate to extreme environments. There are millions of different kinds of microbes on earth, in enormous numbers.

# 13·2 Not all microbes are closely related evolutionarily.

"Microbes" is an appropriately descriptive name, but it's sloppy. We could use it to point to any one of many different types of organisms too small to see without magnification. In Chapters 3 and 10 we introduced two very different kinds of microbes: bacteria and viruses. In this chapter we explore bacteria and viruses in more detail, and also include two other kinds of microbes—the protists and the archaea. The amoeba, a kind of protist, may be familiar to you, but the archaea are probably very unfamiliar to you. That's not surprising, because archaea were only recently recognized as a distinct group of microbes. The group

doesn't even have a common name (unless you like "the group formerly known as archaebacteria"). We'll just call them archaea.

It is hard to make generalizations about microbes, because they are grouped together simply because they are small, not because they all share a recent common ancestor. In fact, microbes occur in all three domains of life—bacteria, archaea, and eukarya—and so the various types of microbes could not be more widely separated (**FIGURE 13-4**). In this chapter we focus on the tiny organisms from each of these

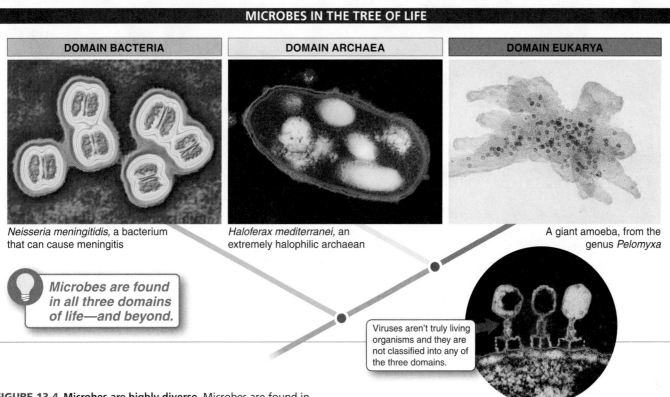

## MICROBES IN THE TREE OF LIFE

**DOMAIN BACTERIA**

*Neisseria meningitidis,* a bacterium that can cause meningitis

**DOMAIN ARCHAEA**

*Haloferax mediterranei,* an extremely halophilic archaean

**DOMAIN EUKARYA**

A giant amoeba, from the genus *Pelomyxa*

*Microbes are found in all three domains of life—and beyond.*

Viruses aren't truly living organisms and they are not classified into any of the three domains.

Bacteriophage virus

**FIGURE 13-4 Microbes are highly diverse.** Microbes are found in every domain of life.

domains. The microbes in the domains bacteria and archaea are prokaryotic, although archaea have some characteristics like those of prokaryotes and others like those of eukaryotes. Protists are the mostly microbial members of domain eukarya. And viruses, another type of microbe, are not classified into any domain at all, because they are only at the borderline of life.

### TAKE-HOME MESSAGE 13·2

Microbes are grouped together only because they are small, not because of evolutionary relatedness. They occur in all three domains of life, and also include the viruses, which are not included in any of the domains.

13·3–13·5 Bacteria
may be the most diverse
of all organisms.

Staphylococcus aureus *(green) on the surface of the small intestine.*

# 13·3
## What are bacteria?

A bacterium is the simplest organism you can imagine—and in many ways, it is the most efficient. A bacterium has a cell envelope consisting of a plasma membrane and, in most cases, a cell wall—that's what it needs to maintain conditions inside that are different from conditions outside. Inside the cell envelope it has cytoplasm—the substance that fills all kinds of cells (including your own). Proteins in the cytoplasm carry out essential functions, such as digesting molecules of food and transferring the energy gained to ATP. DNA in the cytoplasm carries the instructions for making those proteins, and messenger RNAs carry this information to ribosomes, where the proteins are synthesized. That's it—everything an organism needs, with no extra baggage.

Bacteria may be classified by their shape: some are spherical cells (known as the cocci), some are rod-shaped (the bacilli), and others are spiral-shaped (the spirilli) (**FIGURE 13-5**). Bacteria usually reproduce by binary fission, and in a few hours, a single cell can form a culture containing thousands of cells.

As a bacterial cell divides, the number of cells doubles every generation, producing a colony of cells, each of which is a clone of the original cell. Colonies of different species of bacteria look different. The familiar human intestinal

bacterium *Escherichia coli,* for example, forms beige or gray colonies that have smooth margins and a shiny, mucus-like covering. Species of *Proteus,* which are often responsible for spoiling food because they can grow at refrigerator temperatures, form colonies with a surface that looks like a contour map.

Microbiologists can often identify bacteria simply by looking at the colors and shapes of their colonies. They can get additional information by examining a single cell under a microscope, but living bacterial cells are transparent, so you can't see them with an ordinary microscope unless they have been dyed. In 1884, Hans Christian Gram, a Danish microbiologist, described a method of dyeing the cell walls of bacteria to make them visible under a microscope, and a **Gram stain** is still the first test microbiologists use when they are identifying an unknown bacterium (**FIGURE 13-6**).

*Gram-positive bacteria* are colored purple by the stain, because they have a thick layer of a glycoprotein called **peptidoglycan** on the outside of the cell wall. In *Gram-negative bacteria,* the layer of peptidoglycan is thinner and lies beneath an additional membrane, and it is not stained by the dye. Penicillin is effective in treating infections by Gram-positive bacteria because it interferes with the formation of

## BACTERIA: CLASSIFICATION AND STRUCTURE

**COCCI**
Spherical bacteria

**BACILLI**
Rod-shaped bacteria

**SPIRILLI**
Spiral-shaped bacteria

Bacterial species are commonly classified by their shape.

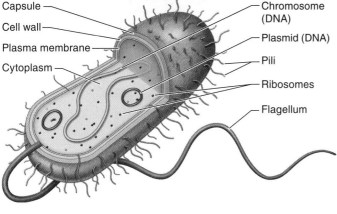

Capsule

Cell wall

Plasma membrane

Cytoplasm

Chromosome (DNA)

Plasmid (DNA)

Pili

Ribosomes

Flagellum

**FIGURE 13-5 Bacteria basics.** Bacteria are single-celled organisms that lack a nucleus.

peptidoglycan cross-links. Penicillin is less effective on Gram-negative bacteria because it does not pass through the outer membrane that covers the peptidoglycan layer.

Being able to distinguish these two groups of bacteria is a big help to microbiologists trying to identify a bacterium, but the peptidoglycan is there because it is important for the bacterial cells—not to make life easier for microbiologists. The extensive interlocking bonds of the long peptidoglycan molecules provide strength to the cell wall. Many bacteria also have a **capsule** that lies outside the cell wall. This capsule may restrict the movement of water out of the cell and allow bacteria to live in dry places, such as on the surface of your skin. In other cases, the capsule may be important in allowing the bacteria to bind to solid surfaces such as rocks or to attach to human cells.

### TAKE-HOME MESSAGE 13·3

Bacteria are efficient single-celled organisms, with an envelope surrounding the cytoplasm, which contains the DNA (they have no nuclei and no intracellular organelles). Bacterial cells undergo binary fission, and a single cell can grow into a colony of cells.

## METHODS OF IDENTIFYING BACTERIA

Some bacteria can be identified by looking at the colors and shapes of their colonies.

**GRAM-POSITIVE BACTERIA**
The glycoprotein layer is on the outside of the cell wall and can be stained with purple dye.

**GRAM-NEGATIVE BACTERIA**
The glycoprotein layer lies beneath an additional membrane and cannot be stained with the dye.

*Gram-negative bacteria—due to their cell membrane composition—are resistant to penicillin.*

**FIGURE 13-6 Bacterial IDs.** Bacteria are often identified by the appearance of an entire colony or of the cells' response to Gram staining.

# 13·4 Bacterial growth and reproduction is fast and efficient.

The time it takes for a bacterium to reproduce can be very short; most bacteria have generation times between 1 and 3 hours, and some are even shorter. *Escherichia coli,* for example, has a generation time of 20 minutes in optimal conditions, so a single *E. coli* could give rise to a population of 20 billion cells in less than 12 hours.

Bacteria carry genetic information in two structures: the chromosome and plasmids. The genes that provide instructions for all of the cell's basic life processes are usually located in a circular DNA molecule, the bacterial chromosome. Most bacteria have just one chromosome, but some have more than one. A bacterial chromosome is organized more efficiently than a eukaryotic chromosome, in two ways. First, in bacteria, the genes that code for proteins with related functions—enzymes that play a role in a pathway that breaks down food for energy, for example—are often arranged right next to one another on the chromosome. This makes it possible to efficiently control the transcription of all the genes together. Second, almost all the DNA in a bacterial chromosome codes for proteins, so bacteria do not use time and energy transcribing mRNA that will not be translated. As we saw in Chapter 5, as much as 90% of the DNA in the chromosomes of eukaryotes does not code for genes and is edited from mRNA after it has been transcribed and before it is translated into protein.

A second type of structure that carries genetic information is a circular DNA molecule called a **plasmid.** Plasmids carry genes for specific functions. For example, *metabolic plasmids* carry genes enabling bacteria to break down specific substances, such as toxic chemicals; *resistance plasmids* carry genes enabling bacteria to resist the effects of antibiotics; and *virulence plasmids* carry genes that control how sick an infectious bacterium makes its victim. Many bacteria have one or more (sometimes more than a hundred) plasmids. The strain of *E. coli* that occasionally sickens patrons of fast-food restaurants carries a virulence plasmid that magnifies the effects of a gene for a sometimes-lethal toxin. (*E. coli* strains without that virulence plasmid are normal components of the bacterial community of the human intestine.)

**Q** Would it be useful to be able to transfer genetic information from one adult human to another?

When a bacterium divides, it creates two new daughter cells, each carrying the genetic information that was present in the chromosome of the mother cell. Thus, binary fission transmits genetic information from one bacterial generation to the next (**FIGURE 13-7**). However, bacteria can also transfer genetic information laterally—in other words, not just to their "offspring," but to other individuals *within* the same generation—through any of three different processes: conjugation, transduction, and transformation (**FIGURE 13-8**).

**Conjugation** is the process by which one bacterium transfers a copy of some of its genetic information to another

**CELL DIVISION IN BACTERIA**

Bacterium cell
Chromosome (DNA)
Plasmids (DNA)

**REPLICATION**
Exact copies of the cell's chromosomal and plasmid DNA are created.

Cell elongates and begins to pinch in two.

Daughter cells are formed.

*Fission can be extremely fast—in less than 12 hours, a single* E. coli *could give rise to a population of 20 billion cells (three times the number of humans on earth)!*

**FIGURE 13-7 Binary fission.** This asexual cell division method is used by prokaryotes.

## METHODS OF GENETIC EXCHANGE IN BACTERIA WITHIN THE SAME GENERATION

**CONJUGATION**
A bacterium transfers a copy of some or all of its DNA to another bacterium, giving the second bacterium genetic information it did not have before.

Donor bacterium

Recipient bacterium

**TRANSDUCTION**
A virus containing pieces of bacterial DNA inadvertently picked up from its previous host infects a new bacterium, and passes new bacterial genes to the bacterium.

Virus

Recipient bacterium

**TRANSFORMATION**
A bacterium can take up DNA—potentially including alleles it did not carry—from its surroundings (usually from bacteria that have died).

DNA fragments

Recipient bacterium

**FIGURE 13-8** Lateral transfer of genetic information: conjugation, transduction, and transformation.

bacterium—even when the two bacteria are different species. It's very much like plugging your iPod into a computer to transfer songs from the computer to the iPod. This is not reproduction, because you start with one iPod and one computer, and that's what you have when you finish. But the iPod now contains songs that it did not have before. Plasmid transfer does exactly that for a bacterium—you still have just two bacterial cells, but the newly acquired plasmid has given the recipient bacterium genetic information that it did not have before. These genes could enable the bacterium to make an enzyme that allows it to metabolize a new chemical or to defend itself against a new antibiotic.

**Transduction** occurs when a kind of virus called a bacteriophage (one type is shown in Figure 13-4) infects a bacterial cell. The virus reproduces inside the bacterial cell, and sometimes, inadvertently, the new virus particles are filled with pieces of bacterial DNA in addition to or instead of the viral DNA. When these viruses are released and infect new bacterial cells, the bacterial DNA can be inserted into the host bacterium's chromosome, passing new bacterial genes to that bacterium.

**Transformation** is the process by which bacterial cells scavenge DNA from their environment. This DNA comes from other bacterial cells that have burst open, releasing their cellular contents. The circular chromosomes break into short lengths of DNA, which can then be taken up by living bacterial cells and inserted into their own chromosomes, potentially adding genes they did not originally have.

### TAKE-HOME MESSAGE 13·4

Bacteria grow rapidly. They have efficiently organized chromosomes—genes are organized in groups with related functions and virtually all the DNA codes for proteins. Bacteria sometimes carry genes for specialized traits on small DNA molecules called plasmids, which can be transferred from one bacterial cell to another by conjugation. DNA can also be transferred laterally between bacterial cells by transduction or transformation.

# 13·5 Metabolic diversity among the bacteria is extreme.

One important feature making bacterial diversity possible is that bacteria can metabolize almost anything. (Not all bacteria can metabolize everything. Rather, there is a huge variety of bacteria, each type with its own particular set of metabolic specializations.) Some can even use energy from light to make their own food. Microbiologists place bacteria into different "trophic" (feeding) categories that reflect their metabolic specialization.

Chemical organic feeders (**chemoorganotrophs**) are bacteria that consume organic molecules, such as carbohydrates. You probably see the products of organic

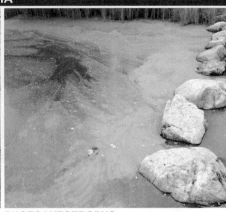

**CHEMOORGANOTROPHS**
Feed on organic molecules

**CHEMOLITHOTROPHS**
Feed on inorganic molecules (Shown here: a stream in Spain orange with oxidized iron minerals.)

**PHOTOAUTOTROPHS**
Use energy from sunlight to produce glucose via photosynthesis

**FIGURE 13-9  A bigger palate than yours.** Bacteria can metabolize almost anything.

> Bacteria are resourceful and can extract food from a huge range of sources: the sun, inorganic molecules in your drainage pipes, and even your shower curtain/door!

feeders every time you take a shower—they are responsible for the pink deposits on the shower curtain and other discolorations on the floor of the shower (**FIGURE 13-9**). Most of the bacteria that live in and on your body are also organic feeders. Some compete with you to metabolize the food you eat. Others digest things you can't digest or can't eat.

Chemical inorganic feeders (**chemolithotrophs,** meaning "rock feeders") are able to use a completely different type of food as their source of energy: inorganic molecules such as ammonia, hydrogen sulfide, hydrogen, and iron. The most common inorganic feeders are the iron bacteria responsible for the brown stains that form on plumbing fixtures in regions where tap water contains high levels of iron. Sulfur bacteria are associated with iron bacteria, and are responsible for the slimy black deposits that you will probably find if you lift the stopper out of the drain in your bathroom sink.

On a larger scale, inorganic feeders are responsible for the acidic drainage that is a by-product of mining. The desirable ore makes up only a small part of the total amount of rock that is removed from a mine. The portion that does not contain ore is discarded on the ground surface in piles called "tailings." This material is often rich in minerals such as pyrites (iron sulfides). Inorganic feeders can gain energy by oxidizing these minerals and, in the process, they release compounds that combine with rainwater to produce strong acids, such as sulfuric acid. When this acidic water drains into streams, it can kill fish and aquatic plants and insects.

"Light self-feeders" (**photoautotrophs**)—bacteria that use the energy from sunlight—contain chlorophyll and use light energy to convert carbon dioxide to glucose by photosynthesis. The floating mats of gooey green material that you see in roadside ditches are a type of photoautotroph called cyanobacteria.

The cyanobacteria living today closely resemble the first photosynthetic organisms that appeared on earth about 2.6 billion years ago. Cyanobacteria could use solar energy to build organic compounds from carbon dioxide, and in the process they broke down water molecules to release free oxygen. Before cyanobacteria, the earth's atmosphere contained no free oxygen—instead, air consisted almost entirely of nitrogen and carbon dioxide. The accumulation of oxygen released by cyanobacteria is called the **Oxygen Revolution.** Oxygen—which humans depend on—now makes up about 21% of the volume of air, and cyanobacteria still release important quantities of oxygen into the atmosphere. One of the common ways that bacteria are classified is as aerobic or anaerobic, depending on whether they require or do not require oxygen for growth—although some bacteria, called facultative anaerobes, utilize oxygen if it is present but can also switch to anaerobic respiration when oxygen is absent.

## TAKE-HOME MESSAGE 13·5

Some bacteria eat organic molecules, some eat minerals, and still others carry out photosynthesis. About 2.6 billion years ago, the photosynthesizing bacteria were responsible for the first appearance of free oxygen in the earth's atmosphere.

## 13·6–13·9 In humans, bacteria can have harmful or beneficial health effects.

*Colored scanning electron micrograph of* Escherichia coli *bacteria (blue and purple) taken from the small intestine of a child.* E. coli *are part of the normal community of organisms in the human gut.*

# 13·6

## Many bacteria are beneficial.

Do you like the taste of yogurt? You can thank bacteria for that—*Lactobacillus acidophilus* and several other species of bacteria are added to milk to create yogurt. As the bacterial cells use the milk sugar for energy, the by-product is lactic acid, which reacts with the milk proteins to produce the characteristic taste and texture of yogurt. If you buy a brand of yogurt that is labeled as containing "live cultures," you are consuming living bacterial cells as you eat the yogurt (**FIGURE 13-10**). Bacteria are also used to produce many other foods, such as cheeses, and bacteria (along with yeasts) are used in the production of beer, wine, and vinegar. Industrial microbiology is a multi-billion-dollar industry.

But you can thank bacteria for more than just tasty snacks—you owe your life to bacteria. Hundreds of species of bacteria grow in and on your body; these microbes are called your "normal flora." The normal flora take up every spot on your body that a disease-causing bacterium could adhere to, and they consume every potential source of nutrition, making it difficult for a disease-causing bacterium to gain a foothold. Thus, maintaining a robust population of these benign bacteria is your first line of defense against infection by harmful bacteria.

**Probiotic therapy** is a method of treating infections by deliberately introducing benign bacteria in numbers large

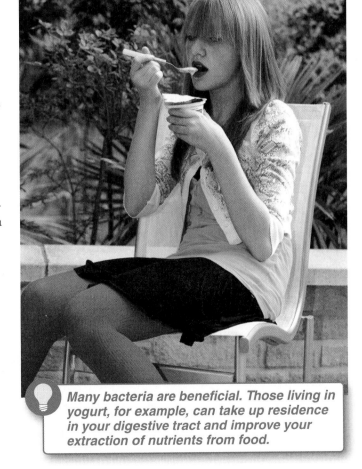

*Many bacteria are beneficial. Those living in yogurt, for example, can take up residence in your digestive tract and improve your extraction of nutrients from food.*

**FIGURE 13-10 Yogurt contains beneficial bacteria.**

enough to swamp the harmful forms. *Lactobacillus acidophilus,* which is a normal inhabitant of the human body, is used to treat gastrointestinal upsets, such as traveler's diarrhea, and urinary tract infections. In addition to growing so vigorously that it crowds out harmful bacteria, *L. acidophilus* releases lactic acid, which interferes with the growth of other bacteria and prevents them from adhering to the walls of the urinary tract and bladder.

# 13·7 Bacteria cause many human diseases.

The number of **pathogenic** (disease-causing) bacteria is very small compared with the total number of bacterial species, but some pathogens kill millions of people annually, despite advances in medicine and sanitation. Some bacteria are always pathogenic (such as those that produce cholera, plague, and tuberculosis), but others (the ones responsible for acne, strep throat, scarlet fever, and "flesh-eating" necrotizing fasciitis) are normal parts of the communities of bacteria that live in or on humans. These bacteria in our "normal flora" become pathogenic only under special circumstances.

The great cholera epidemic that devastated London in 1854 became a milestone in epidemiology (the study of the occurrence of disease outbreaks), when Dr. John Snow identified the Broad Street pump as the source of infection by mapping the pattern of deaths from cholera (**FIGURE 13-11**. When Dr. Snow persuaded the authorities to shut down the pump by removing the handle, new cases of cholera in the area dropped sharply.

Cholera epidemics are a continuing threat, especially in areas without effective sanitation. Epidemics of cholera are occurring now in parts of Iraq where the sewage systems have been destroyed and in Haiti, following the earthquake of 2010. The strains of cholera found in these areas are especially potent; as the bacteria multiply inside their hosts, they cause severe diarrhea with a massive loss of water. Victims of these severe strains of cholera are incapacitated and soon die, but before they die they release billions of cholera bacteria in diarrhea—an effective strategy for the cholera bacteria, because they contaminate the water used for drinking or bathing and rapidly infect new victims.

*Streptococcus pyogenes* is a normal part of the bacterial community of your nose and mouth. Normally it is harmless, and many people never experience any of the diseases it can cause. But when the population of *S. pyogenes*

By mapping the pattern of deaths from cholera, Dr. John Snow was able to identify the Broad Street water pump as the source of the infection.

● Deaths from cholera (size of circle proportionate to number of deaths)
◊ Street pump

**FIGURE 13-11 Contaminated!** Water from the Broad Street pump in London was the source of a cholera outbreak that killed more than 500 people.

is not held in check by competition with other members of the bacterial community, it can become a pathogen. Strep throat (more formally known as "streptococcal pharyngitis") is the most common disease caused by *S. pyogenes;* it produces a severe sore throat and a distinctive rash of red abscesses with white pus at the top of the throat (**FIGURE 13-12**). Some strains of *S. pyogenes* produce a toxin that is released into the bloodstream and produces a red rash that spreads across the skin—the disease called scarlet fever. Most threatening of all is an infection caused by strains of

Streptococcus pyogenes *is usually harmless, but some strains are responsible for strep throat, scarlet fever, and necrotizing fasciitis (flesh-eating bacteria).*

*S. pyogenes* that have a toxin that allows them to enter body tissues, where they produce necrotizing fasciitis. These strains of *S. pyogenes* are known by their well-earned name, "the flesh-eating bacteria."

### TAKE-HOME MESSAGE 13·7

Some bacteria always cause disease, and others do no harm except under certain conditions. For example, *Streptococcus pyogenes* can be harmless, but under some conditions it releases toxins that are responsible for strep throat, scarlet fever, and necrotizing fasciitis (caused by the flesh-eating strains).

## 13·8 Bacteria's resistance to drugs can evolve quickly.

Penicillin was the first antibiotic to be manufactured and used widely in the battle against illness-inducing bacteria. It came into use during the Second World War and caused a revolution in the care of the wounded. But antibiotic-resistant infections soon appeared, and the number of resistant bacteria has increased rapidly ever since (**FIGURE 13-13**). Now, 70 years after the first use of antibiotics, many bacteria are resistant to many, or even to most, antibiotics. In the United States, more people now die of *Staphylococcus aureus* infections that are resistant to many different antibiotics (MRSA infections) than die of HIV/AIDS. A few years ago, antibiotic-resistant staph infections were acquired only in hospitals, but now these infections have spread from hospitals to the community at large. Instead of contracting antibiotic-resistant staph infections in hospitals, patients (or their visitors) are now bringing the resistant bacteria into the hospital with them!

Microbes live everywhere they can, and compete for the best places to attach themselves and for the richest sources of food to eat. This competition takes many forms: rapid growth to crowd competitors out of a living space, superior ability to take in nutrients to starve competitors, and—this is the key to both the benefits of antibiotics and the problem of antibiotic resistance for humans—production of chemicals that kill other microbes, or at least stop them from growing. Antibiotics are produced by microbes to help them compete with other microbes, and most of the antibiotics we use today are derived from microbes.

Bacteria and other microbes have developed a variety of ways to resist antibiotics. For example, some bacteria pump an antibiotic out of their cells as fast as it enters, so it never reaches a lethal concentration inside the bacterial cell. Other bacteria have proteins that bind to the antibiotic molecule and block its lethal effect. Still others carry that approach a step further—they have enzymes that break down the antibiotic molecules, which are then used as fuel to help the bacteria grow faster! Antibiotic resistance within a population of bacteria can spread quickly, because many of the genes that code for resistance are on plasmids. This means that a bacterium carrying a resistance gene can transmit the gene to other bacteria by conjugation; there's no need to wait for multiple generations of natural selection

**Q** Where do antibiotics come from, and why do they so quickly lose their effectiveness?

**Q** Why is it essential to take every dose of an antibiotic prescribed by a doctor?

When an antibiotic is taken as prescribed—that is, at the times specified on the label and until all the pills have been consumed—the population of target bacteria is greatly reduced. All of the target bacteria that remain are resistant to the antibiotic, but there are not very many of them. The growth of these resistant bacteria will be held in check by competition with other types of bacteria. But,

**1** INFECTION

Patient A and Patient B both have an infection caused by a harmful strain of bacteria.

○ Harmful bacteria
○ Harmless bacteria

Patient A

Patient B

**2** ANTIBIOTICS

Both patients are prescribed an antibiotic to treat the infection, which reduces the initial number of harmful bacteria.

Patient A

Patient B

**3** OUTCOMES

**a** Patient A continues to take the antibiotic as prescribed until all of the pills have been consumed.

**b** Patient B stops taking the antibiotic before finishing all of the prescribed amount.

Patient A

Patient B

Harmful bacterial population is significantly reduced (and can be held in check by competition with other, harmless bacteria).

The harmful bacterial cells still alive are the most resistant to the antibiotic and can proliferate, leading to future health problems.

**FIGURE 13-13 Evolution of resistance.** Sixty years after the appearance of antibiotics, many bacteria are resistant to many, or even most, antibiotics.

if you stop taking the antibiotic before you have finished all of the prescription, many of the target bacterial cells will remain alive, including the ones that are most resistant to the antibiotic. These resistant cells will be the founders of a new population of bacteria in your body, so the next time you take that drug it will be ineffective. Even worse, taking antibiotics when they are not needed—to treat a viral

Livestock are given antibiotics to prevent diseases easily spread in crowded living conditions.

💡 *Agriculture in the United States uses about 25 million pounds of antibiotics each year—about eight times more than is used for all human medicine!*

**FIGURE 13-14 Antibiotics are used in agriculture.** Livestock are given antibiotics to prevent diseases easily spread in crowded living conditions.

infection, for example—selects for resistant bacteria without providing any benefit. Antibiotics have no effect on viruses.

The use of antibiotics in agriculture is another reason for the spread of antibiotic resistance. Low concentrations of antibiotics are routinely added to the feed for cattle, hogs, chickens, and turkeys. This can be beneficial in the short term, promoting growth and minimizing disease in the crowded conditions of commercial meat and milk production (**FIGURE 13-14**). But in the long run it can have disastrous consequences, as the practice can lead to selection for bacteria resistant to the antibiotics. The antibiotics can also pass through the food chain to humans. Data gathered by the Union of Concerned Scientists indicate that agriculture in the United States uses about 25 million pounds of antibiotics each year—about eight times more than is used for all human medicine!

## TAKE-HOME MESSAGE 13·8

Antibiotic resistance routinely evolves in microbes. The genes that allow bacteria to combat antibiotics are located on plasmids, and plasmid transfer allows an antibiotic-resistant bacterium to pass that resistance to other bacteria. Excessive use of antibiotics in medicine and agriculture has made several of the most important pathogenic bacteria resistant to every known antibiotic, and infections caused by these bacteria are nearly impossible to treat.

# 13·9 Sexually transmitted diseases reveal battles between microbes and humans.

To many microbes, the human genitals and reproductive tract are desirable places to find shelter, nourishment, and opportunities for reproducing and dispersing. Unfortunately, these microbes can cause problems for humans in the form of **sexually transmitted diseases (STDs).** STDs produce symptoms of varying severity, from mild to extreme discomfort, to sterility, or even death. It is estimated that more than 300 million new cases of STDs occur each year worldwide.

STDs are caused by all types of microbes—bacteria, viruses, fungi, protists—and even by some arthropods. The organisms are passed from the mucous membranes (of the genitals, anus, and mouth) of one individual to those of another during sexual contact; sometimes they are transmitted by needles used for drug injections.

Although most STDs are curable with drugs, two characteristics of STDs make them nearly impossible to completely eradicate from a population: (1) their symptoms may be mild or absent, causing many people to unwittingly pass an infection to their partners, and (2) to prevent reinfection, both partners must be treated simultaneously. Furthermore, because most microbes have such high reproductive rates, populations of a microbe can evolve quickly and become resistant to existing drugs. The treatment of STDs is one of the most pressing public health issues in the world today (**FIGURE 13-15**).

---

## TAKE-HOME MESSAGE 13·9

Sexually transmitted diseases (STDs) are caused by a variety of organisms, including bacteria, viruses, protists, fungi, and arthropods. Worldwide, more than 300 million people are newly infected each year.

---

## SEXUALLY TRANSMITTED DISEASES

| CAUSE | EXAMPLES | SYMPTOMS | TREATMENT |
|---|---|---|---|
| BACTERIUM | • Gonorrhea | Often none; sometimes painful urination, genital discharge, or irregular menstruation | Several antibiotics can successfully cure gonorrhea; however, drug-resistant strains are increasing. |
| | • Syphilis | Often no symptoms for years; eventual sores, skin rash, and if untreated, organ damage | Penicillin, an antibiotic, can cure a person in the early stages of syphilis. |
| | • Chlamydia | Often none; sometimes painful urination, genital discharge | Chlamydia can be easily treated and cured with antibiotics. |
| VIRUS | • HIV/AIDS | Initial symptoms range from none to flu-like; late stages involve severe infections and death | Currently no cure. Antiretroviral treatment can slow progression. Drug-resistant strains occur. |
| | • Genital herpes | Often none; outbreaks include sores on genitals, flu-like symptoms | Currently no cure. Antiviral medications can shorten and prevent outbreaks. |
| | • Human papilloma virus (HPV) | Often none; some types can lead to genital warts, others can cause cervical cancer | A vaccine prevents HPV, and is recommended for girls age 11–12. Warts and cancerous lesions can be removed. |
| PROTIST | • Trichomoniasis | Painful urination and/or vaginal discharge in women; often no symptoms in men | Trichomoniasis can usually be cured with prescription drugs. |
| FUNGUS | • Yeast infections | Genital itching or burning, and/or vaginal discharge in women; genital itching in men | Yeast infections can usually be cured with antifungal suppositories or creams. |
| ARTHROPOD | • Crab lice | Visible lice eggs or lice crawling or attached to pubic hair; itching in the pubic and groin area | Crab lice can be treated with over-the-counter lotions. |

**FIGURE 13-15 The most common STDs.**

## 13·10–13·12 Archaea exploit some of the most extreme habitats.

Grand Prismatic Spring hydrothermal vent in Yellowstone National Park, Wyoming. Hot springs are home to many microbes, including thermophilic archaea.

# 13·10 Archaea are profoundly different from bacteria.

When you look at the photograph of archaea in **FIGURE 13-16**, you may find it hard to believe that they are even a little bit different from bacteria, let alone *profoundly* different. Both live as single cells or colonies of cells, both are surrounded by a plasma membrane, and both have species with flagella that twirl like propellers. In fact, until the 1970s, biologists considered the archaea to be bacteria. Only then did comparisons of DNA reveal that archaea are as different from bacteria as humans are.

Those studies of archaeal nucleotide sequences stimulated other comparisons, which identified additional differences among bacteria, archaea, and eukarya. For example, the chemical compositions of the plasma membranes, cell walls, and flagella of archaea are qualitatively different from those of bacteria. And beyond the large DNA sequence differences and the differences in plasma membranes, cell walls, and flagella, a third difference reflects the position of archaea between bacteria and eukarya on the tree of life (see Figure 13-4): eukaryotes have a distinct cell nucleus that is separated from the cytoplasm by a nuclear membrane, whereas bacteria and archaea have neither a nucleus nor a nuclear membrane. The presence of a nucleus protects the chromosomes of a eukaryotic cell and allows the cell to control what molecules interact with its DNA.

*Archaea look very much like bacteria. But closer inspection—of their physiology, biochemistry, and DNA—reveals them to be profoundly different from all bacteria.*

**FIGURE 13-16 Appearances are deceiving.**

### TAKE-HOME MESSAGE 13·10

Archaea show a set of characteristics that places them between bacteria and eukaryotes on the tree of life. Archaea and bacteria may look similar, but they have large and significant differences in their DNA sequences, as well as differences in their plasma membranes, cell walls, and flagella. Furthermore, neither archaea nor bacteria resemble eukarya in one key way: only eukarya have a distinct cell nucleus and nuclear membrane.

# 13·11  Archaea thrive in habitats too extreme for most other organisms.

Archaea are famous for their ability to live in places where life would seem to be impossible, such as in water around hydrothermal vents that emerge from the seafloor, which would be boiling if it were at sea level. At 2,000 meters below the surface, the water pressure reaches 200 atmospheres, or almost 3,000 pounds per square inch, and the 400° C water coming from the vent cannot boil. Archaea thrive there. Most organisms die at temperatures between 40° and 50° C, because their protein molecules are denatured, so the ability of archaea to survive at temperatures above 100° C is truly remarkable.

Equally impressive is the ability of archaea to live in water as acidic as pH 0 or as salty as a saturated NaCl solution, and to metabolize an extraordinary range of compounds— including sulfur, iron, and hydrogen gas—to obtain energy (**FIGURE 13-17**). Some bacteria also can tolerate extreme physical and chemical conditions, and the organisms that can live in these conditions, both bacteria and archaea, are called **extremophiles** ("lovers of extreme conditions").

*Archaea in your intestine break down a chemical bond found in beans (a bond humans cannot break) … but the process generates gas, which can cause discomfort as it tries to escape.*

**FIGURE 13-18  Helping you digest tough bonds.** Intestinal archaea help you digest some chemical bonds in beans.

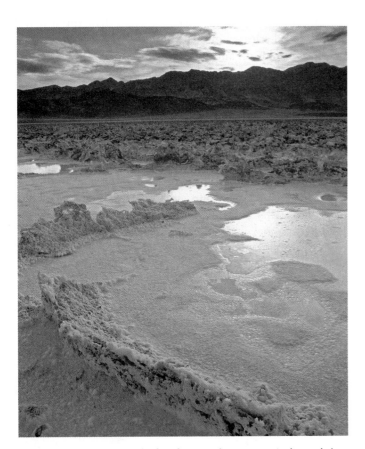

**FIGURE 13-17  Living in the harshest environments.** Archaea thrive in many extreme environments, such as this salt pond.

But not all archaea are extremophiles. Archaea live everywhere that bacteria do, and many of them live in places you would find comfortable. In fact, you are home to many archaea and, on occasion, some of them make their presence all too obvious. Beans are notorious for their tendency to produce gas in the intestine, but archaea are actually the culprits. Beans contain a couple of carbohydrates with chemical bonds that are not broken down very well by any human enzymes (**FIGURE 13-18**). Methane-producing archaea, on the other hand, have no such difficulties, producing an enzyme that targets those bonds. As a result, it is the archaea in your intestine that digest most of these carbohydrates. But in the process of breaking these bonds, they produce gases that, as they escape the digestive system, can cause considerable distress.

## TAKE-HOME MESSAGE 13·11

Archaea can tolerate extreme physical and chemical conditions that are impossible for most other living organisms, but some live in moderate conditions and even in the human intestine.

# 13·12 Much archaeal diversity has yet to be discovered.

How do you grow an organism in your lab if it requires 200 atmospheres of pressure and a water temperature of 100° C or higher, and eats a gas that is toxic to humans? It's not easy, and only about one-quarter of the identified archaeal species have been cultured in the lab. We know that the other species exist, because their transfer RNA has been identified in samples taken from the environment, and biologists are confident that thousands, perhaps millions, of species of archaea await discovery.

Extremophile archaea may have important applications in bioengineering and environmental remediation. For example, *Thermus aquaticus,* an extremophile archaeon that lives in hot springs where the water is nearly boiling, produces an enzyme important in making PCR possible (see Chapter 5), and this enzyme is now sold for laboratory use under the name *Taq* polymerase. Far from being destroyed at high temperatures, *Taq* polymerase works best at 75° C, a temperature at which most human enzymes would simply fall apart.

**FIGURE 13-19 Practical applications for archaea.** The metabolic activity of archaea has potential for use in the cleanup of oil spills.

> " He was a killer, a thing that preyed, living on the things that lived, unaided, alone, by virtue of his own strength and prowess, surviving triumphantly in a hostile environment where only the strong survive. "
> — JACK LONDON, *The Call of the Wild,* 1903

Bioengineers and biotechnologists believe that extremophile archaea and bacteria that thrive at high temperatures and pressures and metabolize toxic substances have enormous potential for industries that have to carry out activities under such conditions. Recent experiments have demonstrated the ability of some archaea to efficiently degrade hydrocarbons, making it possible for them to be used in the removal of sludge that accumulates in oil refinery tanks and, potentially, in the cleanup of contaminated environments such as oil slicks (**FIGURE 13-19**). Naturally occurring archaea may be helping to break down some of the more than 200 million gallons of oil released into the Gulf of Mexico following an oil rig explosion and subsequent massive leak in 2010. Other archaea show promise in clearing mineral deposits from pipes in the cooling systems of power plants.

## TAKE-HOME MESSAGE 13·12

Archaea are hard to study because many require extreme heat or pressure to grow, and these conditions are not easy to provide in a laboratory. But the ability of archaea to grow in such extreme conditions makes them potentially valuable for industrial and environmental purposes.

# 13·13–13·15 Most protists are single-celled eukaryotes.

*Diatoms, single-celled photosynthetic algae that form an important part of the plankton, have mineralized cell walls that provide protection and support.*

# 13·13

## The first eukaryotes were protists.

For the first 2 billion years of life on earth, organisms were really, really small—bacteria and archaea less than 10 micrometers across, one-eighth the diameter of a human hair. But in rocks about 1.9 billion years old, we find fossils of new kinds of organisms that are 10 times larger (**FIGURE 13-20**). These are a group of organisms called acritarchs (a name that can be translated as "confusing old things"). They were the first eukaryotes.

In these ancient fossilized eukaryotes, organelles are visible.

**FIGURE 13-20 Ancient protists.** These acritarch fossils represent the first eukaryotes on earth. The oldest fossils of this type were found in rock 1.9 billion years old.

The larger size is the first thing you notice about these fossil cells, but the internal changes in the cells are even more significant, because they are the basis for the success of the entire eukaryote lineage, including humans. For the first time in the history of life, cells had internal structures that carried out specific functions. These structures, the cellular organelles, perform the specialized activities that make eukaryotic cells more complex than prokaryotes.

The nucleus is an evolutionary innovation that appeared for the first time in protists. The nucleus is a region inside the cell that is enclosed by its own double-layered membrane. The nuclear membrane was probably formed by the fusion of folds of the plasma membrane that surrounded the cell. We know that many modern prokaryotes have complex infoldings of their plasma membranes—these increase the total surface area of the cell and allow better exchange of material between the cell and the external environment —and infoldings of this type may have played a role in endosymbiosis and the formation of the first organelles (see Figure 3-7). Infoldings probably fused around the DNA, for example, creating a membrane-enclosed compartment containing DNA. Thus, the cell nucleus was formed, separated by a double membrane from the cytoplasm of the cell.

Subsequently, these first nuclear membranes developed two specializations: they incorporated proteins that controlled

MICROBES ARE IN ALL DOMAINS     BACTERIAL DIVERSITY     BACTERIA AND HUMANS     ARCHAEA     **PROTISTS**     VIRUSES the movement of molecules into and out of the nucleus, and they extended outward from the nucleus to form a folded membrane called the endoplasmic reticulum. The endoplasmic reticulum is the part of a eukaryotic cell where some proteins are assembled. Further development of internal membranes produced the Golgi apparatus, where newly synthesized proteins are given some final processing steps, and sac-like structures called lysosomes, which contain enzymes that break down damaged molecules. Finally, a lineage of protists took in a guest—a bacterial cell that subsequently became specialized as the mitochondrion, the

organelle that produces most of the ATP synthesized by a eukaryotic cell.

**TAKE-HOME MESSAGE 13·13**

The nucleus is an evolutionary innovation that appeared for the first time in protists. Early protists took in a bacterial cell that subsequently became specialized as the mitochondrion, an organelle in eukaryotic cells that produces ATP.

# 13·14 There are animal-like protists, fungus-like protists, and plant-like protists.

Six major lineages of protists have been named, but even that classification does not capture the huge diversity of the group, and some of the best-known protists do not fit into any of the six lineages. Protists include forms that are very much like animals, others that seem a lot like fungi, and still others that look like plants (**FIGURE 13-21**).

**Animal-like protists** Some protists propel themselves quickly around their environment and appear to hunt for prey. These animal-like protists, which include *Paramecium*, are the ciliates, and get their name from the cilia (hair-like projections) that cover their body surfaces and propel the cells through water. *Paramecium* feeds by the process of **phagocytosis:** cilia in a funnel-shaped structure called the gullet create an inward flow of water that carries bacteria and other small particles of food with it. These particles

accumulate at the inner end of the gullet, where a portion of the plasma membrane bulges inward and eventually breaks free, forming a food vacuole that drifts into the interior of the cell. Enzymes and hydrochloric acid enter the food vacuole from the cytoplasm, and in this acidic environment the engulfed food items are broken down into molecules that diffuse from the vacuole into the cytoplasm. When all of the digestible material has been consumed, the undigested contents of the vacuole are expelled.

**Fungus-like protists** Some protists resemble fungi: living as heterotrophs (organisms that are unable to fix carbon through photosynthesis, but require carbon for growth), establishing sheet-like colonies of cells on surfaces (such as the grout in shower stalls), using spores to reproduce, and sometimes producing fruiting bodies. These

**DIVERSITY OF PROTISTS**

**ANIMAL-LIKE PROTISTS**
Some protists, such as *Trichomonas vaginalis* shown here, move around and hunt for prey like an animal.

**FUNGI-LIKE PROTISTS**
Some protists, such as the plasmodial slime mold shown here, live as heterotrophs and form sheet-like colonies of cells like a fungus.

**PLANT-LIKE PROTISTS**
Some protists, such as the kelp forest shown here, are multicellular and photosynthetic like a plant.

**FIGURE 13-21 Protists come in all shapes and sizes.**

are the slime molds. They generally spread without any individual cells moving, but rather by adding new cells at the edges of the colony. Some slime molds, however, called plasmodial slime molds, are oozing masses of gooey material that flow along a surface, engulfing bacteria, fungi, and small bits of organic material as they go. The streaming of a slime mold in its feeding phase is easy to observe with a microscope, and such a slime mold can flow around, over, or through almost anything—it can even flow through a window screen and reassemble itself on the other side! You may have seen an irregularly shaped blob of yellow material in a moist, shaded garden—this was a plasmodial slime mold.

Remarkably, a plasmodial slime mold is a very large, single cell (but has multiple nuclei). Slime molds divide by binary fission, just like other cells, but in this case a single cell may cover an area of several square centimeters! All of the nuclei in the cell undergo mitosis simultaneously.

**Plant-like protists** Other members of the protist kingdom grow in water and resemble plants. These include the protists referred to colloquially as seaweeds. The term "seaweed" generally refers to the macroscopic, multicellular marine algae, many of which are used as a source of food for humans. And although they are all protists, seaweeds encompass several groups that do not share a common multicellular ancestor. These include some red algae and some green algae (from which the land plants most likely evolved), as well as some of the brown algae—such as the giant kelp that grows in water 30 meters deep. (The seaweeds, of course, represent parts of the protist kingdom that obviously are not "microbial.")

Brown algae cover large portions of the rocks in the intertidal zone. Giant kelp, growing in temperate regions of the North and South Pacific and off the Atlantic coast of South Africa, are among the fastest growing organisms on earth, with some growing as much as 60 meters in a single year. Kelp forests are an enormously diverse habitat; more than 1,000 species of fishes, crustaceans, snails, and mammals make their homes in these marine "forests." Sea otters wrap a kelp frond around their waist when they sleep, and gray whales hide in kelp forests to escape from killer whales.

Although many protists are unicellular, there are many exceptions. As we've seen, many green and brown algae are multicellular, composed of many different cells and cell types, performing different functions. Additionally, some other protists, such as spirogyra, are colonial, living as collections of cells, each of which can carry out all of its life processes independent of the other cells.

Also among the plant-like protists are the diatoms. These organisms live in ponds, lakes, and rivers as well as in the

**FIGURE 13-22 Diatoms are aquatic photosynthetic protists.** Shown above: krill (shrimp-like crustaceans) feeding on phytoplankton (including diatoms), on the underside of ice in Antarctica.

oceans. They are so small that, for some species, 30 individuals could be lined up across the width of a human hair. A characteristic of the diatoms is that they are enclosed in a shell made of silica (**FIGURE 13-22**). Many species of diatoms float in the water, forming part of the phytoplankton—the collection of microscopic organisms that fix carbon dioxide and release oxygen. Phytoplankton can reach densities of hundreds of thousands of cells per liter, and it accounts for about one-quarter of the photosynthetic production of oxygen on earth, occupying a critical position in marine food chains. Small fishes and shrimp-like organisms called copepods feed on phytoplankton, and these small predators are eaten by larger predators, which are eaten by still larger predators. Thus the diversity of life in the marine habitat relies on diatoms and the other microbes that make up the phytoplankton.

## TAKE-HOME MESSAGE 13·14

Protists are a diverse group of mostly unicellular eukaryotic organisms. The ciliates, such as *Paramecium*, are animal-like protists. Plasmodial slime molds are fungus-like protists. Colonial protists and multicellular protists such as the giant kelp can be enormous and are plant-like in appearance.

# 13·15 Some protists can make you very sick.

A **parasite** is an organism that lives in or on another organism, called a **host,** and damages it. A parasitic protist called *Plasmodium* that is transmitted by a mosquito is responsible for a worldwide epidemic disease—malaria. Malaria occurs in tropical parts of Africa, Asia, and Latin America, and it is common in the eastern Mediterranean as well. Between 350 million and 500 million people have clinical cases of malaria, and about 1 million people die of malaria each year. Malaria is the leading cause of death for children younger than 5 years old in sub-Saharan Africa; somewhere on the African continent, about every 30 seconds, a child dies of malaria.

These are sobering statistics, and neither the incidence of malaria nor the rate of mortality has changed very much since *Plasmodium* was identified as the cause of malaria nearly a century ago. The reason for the lack of progress in combating malaria is the nature of *Plasmodium*—it has a complex development that makes it nearly invisible to the host's immune system (**FIGURE 13-23**).

The human immune system has a difficult time fighting a malarial infection because, to survive from one generation to the next, *Plasmodium* parasites go through a series of distinct developmental stages. And as the *Plasmodium* cells change from one stage to another, the parasite produces different cell surface proteins. Thus *Plasmodium* stays ahead of the human immune system by constantly changing the way it appears.

Although the immune system doesn't usually have much success fighting malarial infection, some individuals produce red blood cells that are inhospitable to the *Plasmodium* parasite, conferring a resistance to the malaria-causing protist. These individuals carry an allele, called *Hb*$^S$, which (as we learned in Chapter 7) codes for a variant of the normal adult hemoglobin that causes hemoglobin molecules to stick together in long chains that distort the red blood cell (see Figure 13-23), giving the cell a sickled shape. The sickled red blood cells are inhospitable to *Plasmodium* because they are leaky, and the substances that leak out happen to be essential to *Plasmodium*. Although carrying a copy of the *Hb*$^S$ allele confers resistance to malaria, individuals who carry two copies of the *Hb*$^S$ allele suffer from sickle-cell anemia—a severe genetic disease that is more debilitating than malaria. Individuals with the single *Hb*$^S$ allele are said to have "sickle-cell trait."

## MALARIA: A PROTIST-CAUSED ILLNESS

**1** Following the bite of a *Plasmodium*-infected mosquito, malaria-causing protists take up residence in healthy red blood cells.

*Plasmodium* parasite infecting a red blood cell

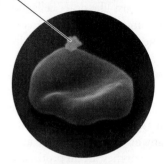

**2** Once inside a red blood cell, *Plasmodium* cells modify the cell's surface proteins, making it difficult for the immune system to fight the malarial infection.

### RESISTANCE TO MALARIA

In individuals with sickle-cell trait, the sickle shape of some red blood cells makes the cells inhospitable to the protist. This makes these individuals resistant to malaria.

**FIGURE 13-23 Malaria-infected blood cells and sickled blood cells.** The malaria parasite grows inside red blood cells and destroys them. Because sickle-cell trait changes the shape of some red blood cells, these cells resist infection by malaria parasites.

## TAKE-HOME MESSAGE 13·15

Some protists cause debilitating diseases. *Plasmodium,* the protist responsible for malaria, is one of these. *Plasmodium* has characteristics that protect it from the human immune system, but some humans do have a defense against malaria: people with sickle-cell trait make red blood cells that are inhospitable to the parasite.

## 13·16–13·19 Viruses are at the border between living and non-living.

*Colored transmission electron micrograph of T2 bacteriophage viruses (red) attacking an* E. coli *bacterium.*

# 13·16

## Viruses are not exactly living organisms.

A virus is not a cell, and that is why viruses do not fit into any of the three domains of life. A virus particle (called a virion) consists of genetic material inside a container made of protein. Some viruses also contain a few enzymes. That's all there is to a virus. The protein container is called the **capsid,** and the genetic material can be either DNA or RNA. Some viruses wrap themselves in a bit of the plasma membrane of the host cell as they are released. A virus of this type is called an enveloped virus—the flu virus is an example. Non-enveloped viruses are enclosed only by the protein container; the virus that causes the common cold is an example of a non-enveloped virus (**FIGURE 13-24**).

There is almost nothing inside the capsid of a virus except DNA or RNA. A virus particle does not carry out any metabolic processes, and it does not control the inward or

### BASIC STRUCTURE OF A VIRUS

Capsid (container made of protein)

Genetic material (DNA or RNA)

Plasma membrane (envelope)

Glycoproteins

Enveloped viruses wrap themselves in a bit of the plasma membrane of the host cell as they are released.

**ENVELOPED VIRUS**
Enclosed in the plasma membrane of a host cell

**NON-ENVELOPED VIRUS**
Enclosed only by its capsid

**FIGURE 13-24  The simple structure of viruses.** All viruses have a container (the capsid) that holds their genetic material, and sometimes the capsid is wrapped in a membrane (an envelope).

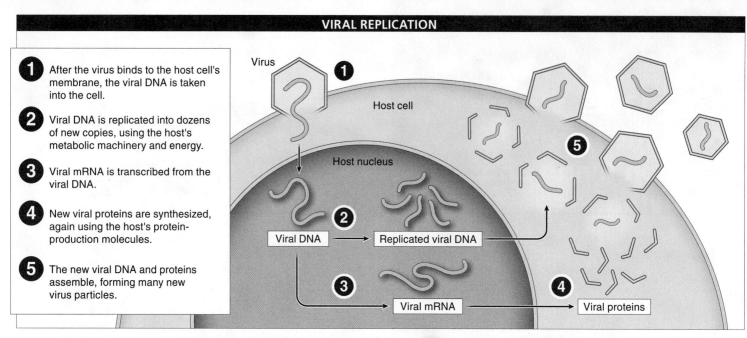

① After the virus binds to the host cell's membrane, the viral DNA is taken into the cell.

② Viral DNA is replicated into dozens of new copies, using the host's metabolic machinery and energy.

③ Viral mRNA is transcribed from the viral DNA.

④ New viral proteins are synthesized, again using the host's protein-production molecules.

⑤ The new viral DNA and proteins assemble, forming many new virus particles.

**FIGURE 13-25 Making more viruses.** A virus duplicates its own genetic material (DNA in the example shown here) by taking over the resources and metabolic machinery of a host cell.

*Because they are dependent on their hosts' metabolic machinery for replication, viruses are not considered "living."*

outward movement of molecules to make conditions inside the virus particle different from conditions outside. Viruses just wait for a chance to insert their genetic material into a living cell.

We speak of "catching" a cold, but what actually happens is that a virus catches us. Viruses identify the cells that they can infect by recognizing the structure of glycoprotein molecules on the surfaces of cells. Every cell in your body has these molecules. They are embedded in the plasma membrane and extend outward from the cell. Your immune system uses these proteins to identify the cells as part of you—as self rather than non-self—and it does not react to proteins that it recognizes as self.

Viruses have cracked the glycoprotein code, and when they find a cell with the appropriate glycoprotein on its surface, they bind to that cell's plasma membrane and insert their genetic material into the cell (**FIGURE 13-25**). Inside the host cell, the viral DNA or RNA takes over the cellular machinery and uses it to produce more viruses. Viruses carry out nearly all of their activities by hijacking materials and organelles in the host cell. Viral proteins are synthesized in exactly the same way as host cell proteins—mRNA binds to ribosomes, and tRNA matches the correct amino acid to each mRNA codon. The mRNA is produced from the virus's DNA or RNA, but all of the protein-building machinery

comes from the host cell, as does the ATP required to synthesize the new viral protein.

There are also some other types of non-living infectious agents. Prions, for example, are misfolded proteins that form plaques and interfere with normal tissue. Usually acquired through ingestion of an infected animal (or its bodily fluids), prions can cause a variety of degenerative neurological diseases in humans. One of these, Creutzfeldt-Jakob disease, is sometimes referred to as a human form of mad cow disease and is characterized by the development of holes in brain tissue, causing the tissue to appear sponge-like. It leads to dementia, memory loss, and disruption of balance and coordination. No cure or treatment is known, and few victims survive more than a year following the appearance of symptoms.

## TAKE-HOME MESSAGE 13·16

A virus is not alive, but it can carry out some of the same functions as living organisms, provided that it can get inside a cell. A virus takes over the protein-making machinery of the host cell to produce more viral genetic material (RNA or DNA) and more viral protein. The viral proteins and genetic material are assembled into new virus particles and released from the cell.

# 13·17

## Viruses are responsible for many health problems.

Many diseases are caused by viruses. Some viral diseases, such as the common cold, are not usually serious, but others have been responsible for worldwide epidemics, called pandemics. The influenza pandemic of 1918–1919 killed at least 20 million people, and possibly as many as 50 million. In the current HIV/AIDS pandemic, nearly 40 million people have been infected, with an annual mortality of about 3 million.

Herpes is a common viral infection in humans, caused by two related viruses. Oral herpes is an infection near the mouth and can cause cold sores, which last two to three weeks. More than three-quarters of all adults in the United States are infected with the virus, with more than 50 million people experiencing outbreaks each year. Genital herpes, a sexually transmitted disease, affects about one in six people between the ages of 14 and 49 in the United States. The viruses causing herpes are present in and released from sores and spread from one person to another by skin-to-skin contact, but infection can also occur when no symptoms are present. Although a variety of treatments are available for both types of herpes, including antiviral drugs that can reduce the severity and duration of symptoms, there is currently no cure.

DNA viruses have base-pair sequences that are stable over time, because the enzymes that replicate the DNA check for errors and correct them during replication. From a health perspective, the fact that DNA viruses do not change rapidly makes it much easier to fight and treat diseases caused by these viruses, using vaccines. Vaccination against a DNA virus such as smallpox, for example, provides years of protection.

RNA viruses, by contrast, change quickly, because the enzymes that carry out RNA replication do not have error-checking mechanisms. As a consequence, because the RNA-replicating enzymes make errors as they assemble new RNA molecules, RNA viruses are continuously mutating into new forms. The common influenza virus that causes outbreaks of flu every year, for example, is an RNA virus. And because flu viruses mutate so fast, the virus that causes health problems changes from one flu season to the next, necessitating a different flu vaccine every year.

**Q** Why do flu viruses change so quickly?

About every 50 years, a new variety of influenza causes a pandemic. The 20th century had three major influenza pandemics, and all of them originated in the same peculiar way. In each of these pandemics, a bird flu virus gained the ability to infect human cells by passing through pigs, and then the virus spread rapidly through the human population worldwide. The most famous of these pandemics was that of 1918–1919, mentioned above (**FIGURE 13-26**). It was called the Spanish influenza,

The influenza pandemic of 1918–1919 killed millions.

In 2002–2003, the virus that causes severe acute respiratory syndrome (SARS) infected several thousand people around the world.

**FIGURE 13-26 Viral outbreaks can threaten large populations.**

because it seemed to enter Europe through Spain, but it originated in Asia. The Asian flu and Hong Kong flu pandemics also originated in Asia, as did smaller viral outbreaks, such as Korean flu (1947), swine fever (1976), and SARS (severe acute respiratory syndrome, 2002–2003), which did not become pandemics.

# 13·18 Viruses infect a wide range of organisms.

Viruses are practically everywhere. There are viruses that infect animals, viruses that infect plants, even viruses that infect bacteria. Most viruses infect just one species, or only a few closely related species, and enter only one kind of cell in that species. But some viruses can infect a wide range of hosts—the rabies virus, for example, can infect any mammal. Pet dogs and cats are routinely immunized against rabies, but wild mammals, such as bats, skunks, and raccoons, can get rabies. That is why, if you are bitten by one of those wild animals, your treatment includes rabies shots as a precaution.

**Q** **What role does a pig play in the transmission of virus from a bird to a human?**

The glycoproteins on the surface of a virus determine which host species the virus can infect and which tissues of the host it can enter. Influenza A viruses (the ones that cause outbreaks of flu every year), for example, have two types of glycoprotein that have different functions. One glycoprotein matches that of a host cell, and allows the virus to enter the cell. The other glycoprotein allows the virus to get back out of the cell, releasing new virus particles that can infect other cells (**FIGURE 13-27**).

Influenza A is an example of a virus that can move from one species to another, and as we noted above, all of the influenza pandemics in the past century began with the transmission of a bird flu virus to humans. Here's how "species jumping" can occur.

Viruses that infect birds don't bind well to glycoproteins in human cells, making it difficult for bird flu viruses to infect humans. However, the cells of pigs have glycoproteins that allow both human and bird flu viruses to bind to them. Thus, a pig cell can be infected by a

**INFLUENZA A STRUCTURE**

SURFACE GLYCOPROTEINS

• Bind to receptors on the surface of the host cell and allow the virus to enter

• Allow the virus to break free from the host cell

*Surface glycoproteins are the "keys" that allow a virus into and out of a host cell.*

**FIGURE 13-27 A virus's surface proteins.** The influenza virus has surface glycoproteins that allow it to bind to and exit a host cell.

human flu virus and a bird flu virus at the same time. Influenza viruses are not very careful when they incorporate newly synthesized RNA into new particles, though. If RNA strands from a human flu virus and a bird flu virus happen to be in a pig cell at the same time, some virus particles released from the cell might include some RNA from the bird virus and some from the human virus. These influenza viruses are a new strain, and they may be able to infect humans (**FIGURE 13-28**).

Since 1997, one strain of bird flu has spread from Hong Kong through most of Asia and into Turkey, France, Germany, and England. This bird flu is the most recent avian influenza virus to also infect humans (not needing pigs as intermediary). The virus readily infects birds, and tens of millions of chickens, turkeys, and ducks have been killed in attempts to eradicate the infection. And because the virus's

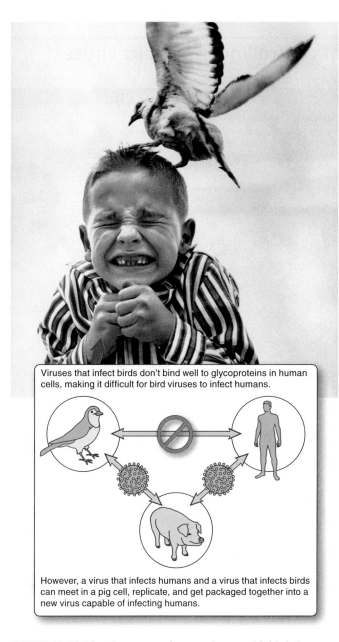

Viruses that infect birds don't bind well to glycoproteins in human cells, making it difficult for bird viruses to infect humans.

However, a virus that infects humans and a virus that infects birds can meet in a pig cell, replicate, and get packaged together into a new virus capable of infecting humans.

**FIGURE 13-28 The pig connects human viruses with bird viruses.** A virus that infects humans and a virus that infects birds can meet in a pig cell (as described in the text), replicate, and, because viral RNA replication is error-prone, get packaged together into a new virus particle capable of infecting humans.

host-entry glycoproteins do not bind well to human cells, the virus does not easily infect people and does not readily spread from human to human. So far, nearly all the people who have been infected by the avian influenza virus have contracted it through close contact with infected flocks of birds or by eating birds that had died of the viral infection. Only a half-dozen cases are known in which the virus seems to have been transmitted from one person to another. However, avian flu can be deadly when it does infect humans—more than half of the human cases reported have ended in death.

The World Health Organization and national health agencies are preparing for a worldwide pandemic of this bird flu virus that can also infect humans. Their alarm is based on a possibility involving pigs. If a pig infected with the bird flu virus is also infected with a human flu virus, and if the viruses are not careful when packing new RNA into new capsids, then a pig cell could contain RNA from both humans and birds, and some of the new viruses released from this cell could contain a mixture of RNA from the human and the bird flu viruses. A mixture of this type might produce a new form of the virus that carries both the genes that make the bird flu lethal to humans *and* the gene that codes for the host-entry glycoprotein. This virus would infect human cells and spread easily from person to person. The World Health Organization has estimated that a pandemic that started in this way could kill as many as 100 million people worldwide.

---

### TAKE-HOME MESSAGE 13·18

**Glycoproteins on the surfaces of viruses determine the types of cells a virus can invade. Most viruses infect just one species, or only a few closely related species, and enter only one kind of cell in that species.**

# 13·19

## HIV illustrates the difficulty of controlling infectious viruses.

New infectious diseases—some caused by bacteria and others by viruses—emerge quite frequently. Many of these new diseases originate in other species of animals and subsequently acquire the ability to infect humans. There are enough of these new diseases to fill the monthly issues of the journal *Emerging Infectious Diseases,* which is published by the Centers for Disease Control and Prevention. But even in that crowd, **acquired immunodeficiency syndrome (AIDS)** stands out. AIDS is caused by the **human immunodeficiency virus (HIV),** which is derived from a strain of the simian immunodeficiency virus (SIV) that jumped from chimpanzees to humans in the early 1900s. HIV has all of the characteristics that make viral diseases hard to control—plus an additional characteristic of its own.

*HIV mutates easily.* HIV is a **retrovirus,** that is, an RNA-containing virus that also contains a viral enzyme called reverse transcriptase that uses a strand of viral RNA as a template to synthesize a single strand of DNA. That DNA strand, in turn, is used as a template to make a complementary strand of DNA, and the resulting double-stranded DNA is integrated into the host cell's DNA. From here, the DNA is used by the host cell's machinery to make more viral RNA (**FIGURE 13-29**). Reverse transcriptase is so error-prone, however, that virtually every copy of HIV in an infected individual's body has a different mutation.

This enormous genetic variation in the HIV particles circulating in an infected person's body makes the infection hard to treat. Virus particles with different mutations can have different proteins on their surfaces, and these surface proteins change each time the virus replicates inside a host cell. Each new generation of HIV in the infected individual contains viruses with surface proteins that his or her immune system has never seen. Furthermore, some of the HIV mutations will confer resistance to the drugs that are being used to treat the patient, and new drugs must be used.

*HIV attacks white blood cells.* All of those problems would apply to any disease caused by a retrovirus, but HIV offers an additional challenge: it targets cells in the host's immune system, especially white blood cells, and particularly those

**FIGURE 13-29 HIV infection.** This RNA virus takes over the replicating machinery of a white blood cell to produce a new generation of HIV particles. HIV mutates rapidly during this process, because the reverse transcriptase enzyme is highly error-prone.

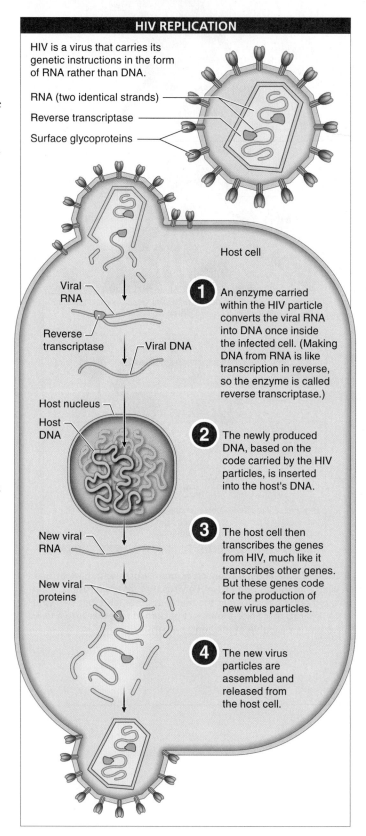

**HIV REPLICATION**

HIV is a virus that carries its genetic instructions in the form of RNA rather than DNA.

RNA (two identical strands)

Reverse transcriptase

Surface glycoproteins

Host cell

Viral RNA

Reverse transcriptase

Viral DNA

Host nucleus

Host DNA

New viral RNA

New viral proteins

**1** An enzyme carried within the HIV particle converts the viral RNA into DNA once inside the infected cell. (Making DNA from RNA is like transcription in reverse, so the enzyme is called reverse transcriptase.)

**2** The newly produced DNA, based on the code carried by the HIV particles, is inserted into the host's DNA.

**3** The host cell then transcribes the genes from HIV, much like it transcribes other genes. But these genes code for the production of new virus particles.

**4** The new virus particles are assembled and released from the host cell.

**FIGURE 13-30** HIV attacks white blood cells essential for identifying foreign invaders.

that search for and attack invading bacteria or viruses. During the incubation period, which can last for many years, HIV infects white blood cells. New white cells are produced to replace the ones killed by the virus, however, and the infected person has virtually no symptoms. Nonetheless, HIV is present in the individual's body fluids during the incubation period and can be transmitted to other individuals. HIV testing is used to detect infections during this stage.

Each time HIV infects another white blood cell, the reverse transcriptase makes errors in transcribing the RNA to DNA, and eventually one of the mutations allows the virus to bind to the glycoprotein on the surface of a specialized type of white blood cell—a bacteria- and virus-hunting white blood cell that is critically important in identifying disease-causing threats. A suitable mutation may occur in a couple of years, but more often it takes about 10 years or longer. But when it does happen, it signals a new stage in the HIV infection: the development of AIDS (**FIGURE 13-30**).

*The immune system collapses.* Normally, white blood cells all work together to identify and destroy cells that have been infected by a virus. When HIV begins to kill the cells that hunt for viruses and bacteria, the immune system begins to fail—it can no longer respond to HIV, or to any other infectious agent. Patients with AIDS develop multiple infections, bacterial and viral, as well as cancers, because they have lost the immune system cells that would normally have marked infected and cancerous cells for destruction.

---

### TAKE-HOME MESSAGE 13·19

HIV infection is especially difficult to control. Mutations change the properties of the retrovirus so that it is hard for the immune system to recognize it, and they produce variants that are resistant to the drugs used to treat the HIV infection.

## THE FIVE-SECOND RULE
### How clean is that food you just dropped?

In 2007, two students at Connecticut College decided to test the "five-second rule": the idea that it's safe to eat food you've dropped on the ground as long as you pick it up within 5 seconds. They dropped apple slices and Skittles candies on the floor of the cafeteria and a snack bar, let the foods lie there for 5, 10, 30, or 60 seconds, and then tested them for bacteria.

**Q: IS THE FIVE-SECOND RULE VALID?** The students found no bacteria on the food picked up within 30 seconds. After a minute, the apple slices had picked up some bacteria, but the Skittles had none (they found bacteria on Skittles only after 5 minutes). The students concluded that the five-second rule should get an extension, proposing that you have 30 seconds to pick up moist foods and more than a minute to pick up dry foods without risk of bacterial contamination.

**Q: SHOULD WE RELAX ABOUT EATING FOODS DROPPED ON THE FLOOR?** This question is a serious one and warrants more study. Each year in the United States there are more than 76 million cases of illness caused by contaminated food, including more than 5,000 that are fatal.

**Q: SCIENTIFIC METHOD AND DRAWING CONCLUSIONS: CAN YOU GENERALIZE FROM A SMALL NUMBER OF OBSERVATIONS TO ALL POSSIBLE SITUATIONS?** In the Connecticut College study, the students examined two food types, dropped in just two locations, and onto surfaces with unknown concentrations of bacteria. In another study, published in the *Journal of Applied Microbiology*, researchers focused on *Salmonella* bacteria. Because it has been documented that bacteria can survive on clothes, hands, sponges, cutting boards, and utensils for several days, the researchers decided to drop food on surfaces known to be covered in bacteria.

**Q: WHICH FACTOR IS MORE IMPORTANT FOR DROPPED FOOD: LOCATION OR DURATION?** Armed with slices of bologna and pieces of bread, the researchers found that when dropped on surfaces covered with *Salmonella* and left for a full minute, both types of food took up 1,500–80,000 bacteria. And although picking up the food quickly—within just 5 seconds—reduced by 90% the number of bacteria present, 150–8,000 bacteria still had time to hitch a ride on the food in that first 5 seconds.

**A GRUBBY CONCLUSION.** It's probably safe to say that if you're in the Connecticut College cafeteria, you can eat your dropped food as long as you pick it up quickly. But for all other situations, a bit more caution is advised—particularly, taking notice of the "drop zone." Unsanitary surfaces likely to have microbes will contaminate your food almost immediately. So hold on tight.

# Big Ideas in ... Evolution and Diversity Among the Microbes

## 13·1–13·2 There are microbes in all three domains.

- Grouped together because they are small, not because they are evolutionarily related, microbes are found in all three domains of life.
- Viruses, which are on the border between living and non-living, are not classified in any domain.
- Because they are so small, microbes can also be structurally simple.

## 13·3–13·5 Bacteria may be the most diverse of all organisms.

- Bacteria are single-celled organisms with no nucleus and no intracellular organelles.
- Different species of bacteria are found virtually everywhere and can eat virtually anything.
- The entire bacterial chromosome (unlike eukaryotic chromosomes) consists of functional genes, and bacteria reproduce rapidly and efficiently by binary fission.
- Genes for specialized functions, such as antibiotic resistance, are carried on small DNA molecules called plasmids that can be transferred from one bacterium to another.
- The combination of short generation times and plasmid transfer allows antibiotic resistance to spread rapidly.
- Some bacteria eat organic molecules, some eat minerals, and still others carry out photosynthesis. Some require oxygen to live, others do not.
- Photosynthesizing bacteria produced the first free oxygen in the earth's atmosphere.

## 13·6–13·9 In humans, bacteria can have harmful or beneficial health effects.

- Some bacteria always cause disease, and others do no harm except under certain conditions.
- Microbes routinely develop resistance to antibiotics.
- Excessive use of antibiotics in medicine and agriculture has made several of the most important pathogenic bacteria resistant to every known antibiotic.

- Sexually transmitted diseases (STDs) are caused by a variety of organisms, including bacteria, viruses, and protists (as well as some fungi and arthropods).

## 13·10–13·12 Archaea exploit some of the most extreme habitats.

- Although archaea resemble bacteria, they are more closely related to the eukarya.
- Archaea thrive in extreme conditions, and they can eat a variety of substances toxic to other organisms.

## 13·13–13·15 Most protists are single-celled eukaryotes.

- Most (but not all) protists are single-celled eukaryotes.
- Amoebas and *Paramecium* are animal-like protists that eat bacterial cells.
- Some protists, such as the seaweeds (a name referring to a non-monophyletic group of species), are macroscopic and multicellular. Other protist species are colonial.
- It is likely that the green algae gave rise to the land plants.
- Plasmodial slime molds are fungus-like protists, having large single cells with multiple nuclei.

## 13·16–13·19 Viruses are at the border between living and non-living.

- Viruses consist of genetic material (RNA or DNA) packed inside a protein coat.
- The virus hijacks the machinery of a cell and uses it to make more viral particles.
- RNA viruses (such as those that cause influenza) mutate rapidly.
- Human immunodeficiency virus (HIV) is an RNA retrovirus that causes AIDS, a condition in which the immune system fails, allowing infections and cancers to have much more serious impact on health and mortality.

See the To-Go Guide (packaged with every new copy of this textbook) for an illustrated summary of the chapter.

# Check Your Knowledge

## GRAPHIC CONTENT

*Thinking critically about visual displays of data.*

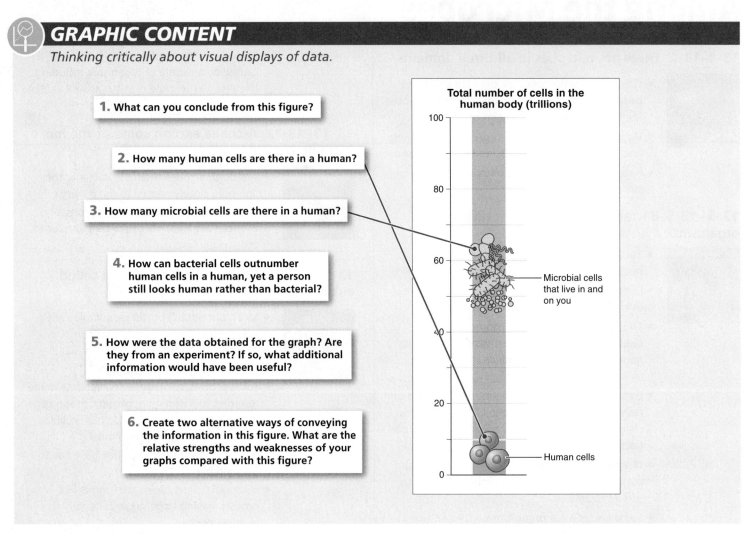

1. What can you conclude from this figure?

2. How many human cells are there in a human?

3. How many microbial cells are there in a human?

4. How can bacterial cells outnumber human cells in a human, yet a person still looks human rather than bacterial?

5. How were the data obtained for the graph? Are they from an experiment? If so, what additional information would have been useful?

6. Create two alternative ways of conveying the information in this figure. What are the relative strengths and weaknesses of your graphs compared with this figure?

**Total number of cells in the human body (trillions)**

Microbial cells that live in and on you

Human cells

**See Answers at the back of the book.**

## Key Terms in Evolution and Diversity Among the Microbes

acquired immunodeficiency syndrome (AIDS), p. 538
capsid, p. 533
capsule, p. 517
chemolithotroph, p. 520
chemoorganotroph, p. 519

conjugation, p. 518
extremophile, p. 527
Gram stain, p. 516
host, p. 532
human immunodeficiency virus (HIV), p. 538

microbe, p. 512
Oxygen Revolution, p. 520
parasite, p. 532
pathogenic, p. 522
peptidoglycan, p. 516

phagocytosis, p. 530
photoautotroph, p. 520
plasmid, p. 518
probiotic therapy, p. 521
retrovirus, p. 538

sexually transmitted disease (STD), p. 525
transduction, p. 519
transformation, p. 519

# Multiple-Choice Questions

**1.** In the human body:
  a) there are more microbial cells than human cells.
  b) archaea can survive, but bacteria are killed by the immune system.
  c) bacterial cells can survive only inside the eukaryotic cells.
  d) the diversity of bacterial cells living in the mouth is exceeded by the diversity living on the skin.
  e) the bacterial cells are larger than the eukaryotic cells.

**2.** From an evolutionary perspective, the most basic division among all organisms on earth is among:
  a) plants, animals, and bacteria.
  b) bacteria, archaea, and eukarya.
  c) plants, animals, and eukarya.
  d) bacteria, animals, and eukarya.
  e) bacteria, archaea, and protists.

**3.** Which of the following statements about prokaryotes is incorrect?
  a) Prokaryotes contain ribosomes.
  b) Some prokaryotes can conduct photosynthesis.
  c) Prokaryotes have circular pieces of DNA within their nuclei.
  d) Prokaryotes appeared on earth prior to eukaryotes.
  e) Prokaryotes contain cytoplasm.

**4.** "Establishing a phylogeny for bacteria is more difficult than establishing a phylogeny for plants or animals." This statement is:
  a) correct, because of the quick generation time of bacteria.
  b) incorrect, because lateral gene transfer makes creating a phylogeny simpler.
  c) incorrect, because the metabolic diversity of bacteria makes classification impossible.
  d) correct, because it is difficult to obtain enough morphological data from bacteria.
  e) correct, because bacteria can engage in lateral gene transfer.

**5.** A population of *Escherichia coli* can double every _____ in an ideal laboratory culture.
  a) 20 seconds
  b) 2 minutes
  c) 2 days
  d) 20 minutes
  e) 2 hours

**6.** Plasmids containing genes for antibiotic resistance can be exchanged between bacterial cells in a culture by:
  a) conjugation.
  b) artificial exchange.
  c) cloning.
  d) transduction.
  e) conduction.

**7.** Which group of organisms utilizes the largest variety of energy sources?
  a) prokaryotes
  b) fungi
  c) protists
  d) invertebrate animals
  e) vertebrate animals

**8. Which of the following statements about antibiotics is incorrect?**

a) Penicillin was the first antibiotic isolated from bacteria and was widely used to fight bacterial infections.

b) Antibiotics help microbes compete with other microbes.

c) Antibiotic-resistant microbes are selected for in humans who are taking antibiotics.

d) Antibiotics, though effective against viruses, are not effective against bacteria.

e) Antibiotics are used not just in human health care but also in agriculture.

100 HARD
51
0 EASY

**9. Which of the following domains are the most closely related, in that they share a unique common ancestor?**

a) archaea and bacteria

b) archaea and eukarya

c) bacteria and eukarya

d) None of the above; all three domains evolved from different ancestors.

e) None of the above; all three domains are equally related to one another.

100 HARD
58
0 EASY

**10. Unlike other microbes, archaea are able to thrive:**

a) in very extreme (with respect to temperature, salinity, pressure, etc.) environments.

b) in all the same places that bacteria are found.

c) in the human body.

d) in lakes and rivers (i.e., in fresh water).

e) in well-aerated soils.

100 HARD
26
0 EASY

**11. *Taq* polymerase:**

a) is produced in viruses and bacteria.

b) is denatured at temperatures above the normal human body temperature.

c) is useful in catalyzing the polymerase chain reaction.

d) can break down the sludge in oil refinery tanks.

e) can reduce the mineral deposits from rusting pipes.

100 HARD
54
0 EASY

**12. All protists are alike in that they all are:**

a) photosynthetic.

b) eukaryotic.

c) parasitic.

d) colonial.

e) multicellular.

100 HARD
34
0 EASY

**13. Which of the following is not a protist?**

a) *Paramecium*

b) slime mold

c) brown alga

d) diatom

e) All of the above are protists.

100 HARD
20
0 EASY

**14. Which of the following statements about *Plasmodium* is correct?**

a) It causes sickle-cell anemia.

b) Because it changes its surface proteins frequently, it is largely invisible to the immune system.

c) It is transmitted by *Paramecium*.

d) It confers resistance to malaria.

e) It is most common in North America.

100 HARD
65
0 EASY

**15. Many biologists do not consider viruses to be alive. Which of the following characteristics of viruses leads to this conclusion?**

a) Viruses lack a metabolic system.

b) Viruses do not respond to external stimuli.

c) Viruses are unable to reproduce on their own.

d) All of the above are correct.

e) Only a) and c) are correct.

100 HARD
56
0 EASY

**16. The genetic information in all viruses is:**

a) DNA.

b) RNA.

c) protein.

d) polymerase enzymes.

e) DNA or RNA.

100 HARD
30
0 EASY

**17. Viruses are able to infect:**

a) humans.

b) plants.

c) bacteria.

d) birds.

e) All of the above.

100 HARD
0 12 EASY

**18. Which of the following statements about HIV is incorrect?**

a) It is a virus.

b) It attacks red blood cells.

c) It mutates frequently.

d) It is derived from a simian immunodeficiency virus.

e) It contains RNA but not DNA.

100 HARD
45
0 EASY

# Short-Answer Question

**1.** Complete the chart on the characteristics of microbes.

See Answers at the back of the book.
For additional study questions, go to www.prep-u.com.

| | Prokaryotes | | Eukaryotes | Viruses |
|---|---|---|---|---|
| | Bacteria | Archaea | Protists | |
| Unicellular, colonial, or multicellular? | | | | |
| Plasma membrane? | | | | |
| Peptidoglycan in cell wall? | | | | |
| Flagella? What type? | | | | |
| Presence of nucleus? | | | | |
| Intracellular organelles? | | | | |
| Type of genetic material (DNA, RNA, or both possible)? | | | | |
| Chromosome structure (linear or circular)? | | | | |

## ABOUT THE CHAPTER OPENING PHOTO

A T4 bacteriophage virus (orange) attacking an *E. coli* bacterium. The phage consists of a large DNA-containing head and a tail composed of a central sheath with several fibers that can attach to the host cell surface. DNA from the virus is injected into the cell through the sheath. It instructs the host to build copies of the virus, which are released when the host bursts. (Colored transmission electron micrograph.)

Planet at capacity: patterns
of population growth

# 14

# Population Ecology

# 14·1–14·6 Population ecology is the study of how populations interact with their environments.

*African elephants (Loxodonta africana) at the Chobe River in Botswana.*

# 14·1

## What is ecology?

Lobster tastes delicious. Many people consider it one of the finest delicacies. It's not surprising, then, that catching and selling lobsters is big business. Generating almost $200 million a year for the State of Maine alone, it is among the most economically important businesses in New England. Now imagine for a minute that you were in charge of the lobster industry. The 6,000 lobstermen in Maine depend on your managing the industry so that not only can they catch and sell enough lobsters to survive, but that sufficient numbers of lobsters are left to ensure there will be lobsters to catch in all the years to come. How would you do it? You would be faced with some tough questions:

- How many lobsters should you allow each lobsterman to take each day? Each year?

- Should you require that certain lobsters be thrown back? If so, should they be the biggest or the smallest? Males or females?

- Is it better to increase the lobster population size or maintain it at current levels?

- But wait a minute. Forget about these obviously complex questions and start with a seemingly simpler question: how many lobsters currently live in the waters off the coast of Maine? Is it even possible to know what this number is (**FIGURE 14-1**)?

Welcome to ecology.

These difficult questions, and many others like them, are all part of **ecology,** a subdiscipline of biology defined as the study of the interactions between organisms and their

*Ecologists have a challenging task managing natural populations when it's difficult to count the number of individuals present or determine how many the environment can support.*

**FIGURE 14-1 Managing valuable resources.** Population growth models can help.

environments. But don't be fooled by the simple definition. Ecology encompasses a very large range of interactions and units of observation and is studied at different levels (**FIGURE 14-2**). These include:

### WHAT IS ECOLOGY?

Ecology is the study of interactions between organisms and their environments. It can be studied at many levels, including:

**INDIVIDUALS**
Individual organisms

**POPULATIONS**
Groups of individual organisms that interbreed with each other

**COMMUNITIES**
Populations of different species that interact with each other within a locale

**ECOSYSTEMS**
All living organisms, as well as non-living elements, that interact in a particular area

**FIGURE 14-2 From individuals to ecosystems.** Ecologists study living organisms and their relationship to the environment.

- *Individuals:* How do individual organisms respond biochemically, physiologically, and behaviorally to their environments?

- *Populations:* How do groups of interbreeding individuals change over time, in terms of growth rates, distributions, and genetic compositions?

- *Communities:* How do populations of different species in a locale interact?

- *Ecosystems:* At the highest level of organization in ecology, how do the living and non-living elements interact in a particular area, such as a forest, desert, or wetland?

> ❝ I wandered lonely as a cloud
>     That floats on high o'er vales and hills,
>     When all at once I saw a crowd,
>     A host, of golden daffodils. ❞
> —WILLIAM WORDSWORTH, *Daffodils*, 1804

We explore ecology at all of these levels throughout this and the next two chapters. We focus in this chapter on **population ecology,** a subfield of ecology that focuses on populations of organisms of a species and how they interact with the environment. We also examine the special case of human population growth and its impact on the environment, as well as the ways in which ecological knowledge can contribute to effective conservation policies. In the next chapter, we move to communities and ecosystems, where our focus will be on how species interact with each other and with their environments, and how this influences their own evolution and the environment itself.

### TAKE-HOME MESSAGE 14·1

Population ecology is the study of the interactions between populations of organisms and their environments, particularly their patterns of growth and how they are influenced by other species and by environmental factors.

# 14·2 A population perspective is necessary in ecology.

As we begin our study of population ecology, a subtle but critical shift in perspective is required. It is a shift from viewing the individual as the primary focus to viewing the population—a group of organisms of the same species living in a particular geographic region. Most ecological processes cannot be observed or studied within an individual. Rather, they emerge when considering the entire group of individuals that regularly exchange genes in a particular locale.

As an example of how a population perspective is needed, consider the case of adaptations produced by natural selection. Genetic changes don't occur within an individual. That is, a longer neck did not evolve in one individual giraffe. Instead, the genetic changes occurred within a population of giraffes over time. As a consequence of differential reproductive success among the individuals of a population with different neck lengths, there came to be more individuals with longer necks over time. Birth rates, death rates, and immigration and emigration rates, too, are features not of individuals but of populations (**FIGURE 14-3**).

In the next sections, we explore how populations grow, a question important both to the management of consumable natural resources—as in the lobster fisheries described above—and to the conservation of rare and endangered species.

*Population ecology examines features that cannot be studied on an individual organism, such as population size.*

**FIGURE 14-3 A change in perspective.** Ecology requires a focus on populations rather than on isolated individuals.

### TAKE-HOME MESSAGE 14·2

Most ecological processes cannot be observed or studied within an individual. Rather, we need to consider the entire group of individuals that regularly exchange genes in a particular locale.

# 14·3 Populations can grow quickly for a while, but not forever.

In a stable population, how many of the five million eggs laid by a female cod over the course of her life survive and grow to adulthood? How many babies that an elephant gives birth to survive to adulthood and reproduce? These may seem like difficult questions to answer, requiring complex mathematical models of population growth and special knowledge about cod or elephants. But actually, the most fundamental fact about population growth is ridiculously simple. How many of the female cod's eggs and female elephant's babies will, on average, make it to adulthood?

Just two. If a population is not growing, each individual is ultimately replacing itself with a new individual. This is true

**Q**

Who leaves more surviving offspring, a pair of elephants or a pair of rabbits?

for all species. Two of those cod eggs will make it. And two of those baby elephants will make it. (For a stable population, two survive, on average, rather than one, because both a male and a female contributed to each offspring. If each pair produced only one surviving offspring, the population would get smaller and smaller.) What happens to the rest of the eggs and offspring? Along the way from gamete to adult, they may not survive. For example, an egg may not get fertilized. Or a fertilized egg may not survive to give rise to an offspring. Predation and other factors may prevent a young offspring's surviving to adulthood. Or an adult may not find a mate. There are too many risks to name them all.

What would happen if more than two cod or two elephants survived? If there were more than one offspring per individual, the population would grow and grow and grow until the earth was covered with cod, elephants, and every other species. But that hasn't happened, and it can't. Let's investigate why.

> There is no exception to the rule that every organic being naturally increases at so high a rate that, if not destroyed, the Earth would soon be covered by the progeny of a single pair. *
> — CHARLES DARWIN, *The Origin of Species*, 1859

We start by figuring out how a population grows if each individual does more than just "replace" itself. To calculate the **growth rate** of a population—that is, the change in the number of individuals in the population in some unit of time, such as a year—we need two pieces of information: the number of individuals in the population now, $N$, and the per capita growth rate, $r$, which is simply the birth rate minus the death rate. The growth of the population in a year is the growth rate times the number of individuals present to start with, written as: $r \times N$.

If the birth rate is greater than the death rate, the population gets bigger. If the death rate is bigger than the birth rate, the population gets smaller. For example, if there are 500 individuals in a population ($N = 500$) and, over the course of a year, 125 offspring are born, the birth rate is 125/500, or 0.25 births per individual. And if 25 of the 500 individuals die during the same year, the death rate is 25/500, or 0.05 deaths per individual. In this population, then, the growth rate, $r$, is 0.25 − 0.05, or 0.20 individuals per individual per year. But how many individuals is that?

Since there were 500 individuals to start with, the population would increase by 0.20 × 500, or 100 individuals, during a year (the time period observed). After a second year, if it grew at the same rate, the population growth would be 0.20 × 600, or 120 new individuals. This would mean the population now has 720 individuals.

With the same calculations, we could predict the population size for the next 10, 50, or even 100 years (**FIGURE 14-4**). When a population's size increases at a rate proportional to its current size—in other words, the bigger the population, the faster it grows—the growth is called **exponential growth.** As the J-shaped graph reveals, the size of a population growing exponentially becomes astronomical

## EXPONENTIAL GROWTH

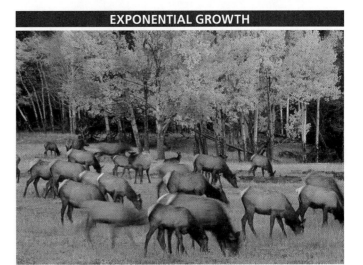

Exponential growth occurs when each individual produces more than the single offspring necessary to replace itself. According to realistic (and moderate) estimates of birth and death rates, a population of just 500 elk would grow to more than a billion individuals within 80 years and eventually would cover the earth.

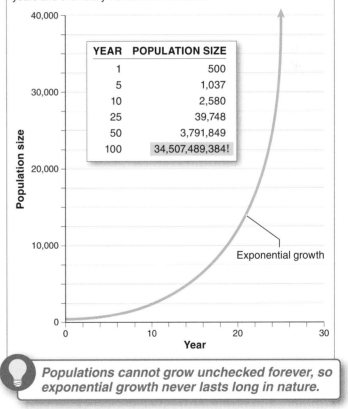

| YEAR | POPULATION SIZE |
|------|-----------------|
| 1 | 500 |
| 5 | 1,037 |
| 10 | 2,580 |
| 25 | 39,748 |
| 50 | 3,791,849 |
| 100 | 34,507,489,384! |

*Populations cannot grow unchecked forever, so exponential growth never lasts long in nature.*

**FIGURE 14-4 Unchecked growth.** Any animal or plant species that grew exponentially would eventually cover the planet.

very quickly. In 10 years, the population of 500 individuals would grow to 2,580. In 50 years, it would reach almost four million. And after 80 years, the population would surpass a billion individuals. It's clear that exponential population growth ends badly.

But the world isn't overrun with people or cod or elephants, so, clearly, exponential growth doesn't occur for long. In the next section, we explore why populations cannot keep growing unchecked forever, and we add some realism to the exponential growth description.

# 14·4 A population's growth is limited by its environment.

Life gets harder for individuals when it gets crowded. Whether you are an insect, a plant, a small mammal, or a human, difficulties arise in conjunction with increasing competition for limited resources. In particular, as population size increases, organisms experience:

- Reduced food supplies, due to competition

- Diminished access to places to live and breed, also due to competition

- Increased incidence of parasites and disease, which spread more easily when their hosts live at higher density

- Increased predation risk, as predator populations grow in response to the increased availability and visibility of prey

The limitations on a population's growth that are a consequence of **population density**—the number of individuals in a given area—are called **density-dependent factors,** and they cause more than discomfort. With increased density, a population's growth is reduced as limited resources slow it down. This ceiling on growth is the **carrying capacity, $K$,** of the environment. And as a population size approaches the carrying capacity, death rate increases, emigration rate increases (as individuals seek more hospitable places to live), and a reduction in birth rate usually occurs, too, as low food supplies result in poor nutrition, which, in turn, reduces fertility (**FIGURE 14-5**).

Here's how the carrying capacity of an environment influences a population's growth. We start with our exponential growth equation, which calculated a population's growth as $r \times N$. Then we multiply by a term, $[(K - N)/K]$, that can slow down exponential growth.

When $N$, the population size, is small relative to the carrying capacity, $K$, it means there are plenty of resources to go around. In this situation, the term $[(K - N)/K]$ is close to 1. To prove this to yourself, choose a large number for $K$, say, 100,000, and a small number for $N$, say, 1,000. In this case, the term $[(K - N)/K]$ equals $[(100,000 - 1,000)/100,000]$. This reduces to 99,000/100,000, which is the same as 99/100, or 0.99.

**DENSITY-DEPENDENT FACTORS LIMIT POPULATION SIZE**

**DENSITY-DEPENDENT FACTORS**
- Food supply
- Habitat for living and breeding
- Parasite and disease risk
- Predation risk

**FIGURE 14-5 Fighting over scarce resources.**

And so, in this situation, we are multiplying normal exponential growth by 0.99. This is almost the same as multiplying it by 1, which doesn't change it. As a consequence, the population is growing exponentially (or very, very close to exponentially).

On the other hand, if the population size, $N$, is close to the value of $K$, it means that the environment is nearly full to capacity, and the term $[(K - N)/K]$ is close to 0. Let's plug in some numbers again. We can use 100,000 again for $K$. But let's use 99,000 for $N$. In this case, $[(K - N)/K]$ equals $[(100,000 - 99,000)/100,000]$. This reduces to 1,000/100,000, which is the same as 1/100, or 0.01. In this situation, we multiply normal exponential growth by 0.01, which reduces it to almost zero. In other words, population growth slows more and more as the population size nears the environment's carrying capacity.

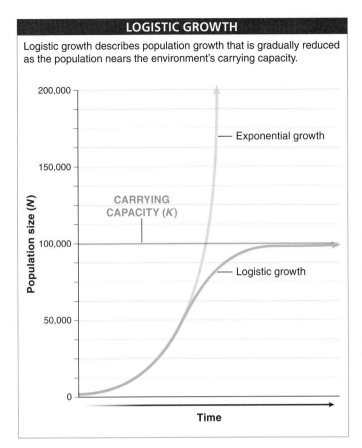

## LOGISTIC GROWTH

Logistic growth describes population growth that is gradually reduced as the population nears the environment's carrying capacity.

**FIGURE 14-6 Lack of resources limits growth.** Logistic growth is population growth that has stabilized because of limited resources.

When a population grows exponentially at first, but then slows as the population size approaches the carrying capacity, the growth is called **logistic growth** and makes an S-shaped curve (**FIGURE 14-6**). Logistic growth is a much better approximation than exponential growth of how populations grow in the real world.

**Density-independent factors** can also knock a population down, like "bad luck" limits to population growth. These forces strike populations without regard for population size, by increasing the death rate or decreasing the number and rate of offspring produced. They are mostly weather- or geology-based, including calamities such as floods, earthquakes, and fires. They also include habitat destruction by other species, such as humans. The population hit by the disaster may be at its carrying capacity, or it may be in the initial stages of exponential growth, before its growth is slowed by the carrying capacity of the environment. In either case, the density-independent force simply knocks down the size. The population then resumes logistic growth.

In an environment where these "bad luck" events repeatedly occur, a population might never have time to reach the

carrying capacity. Instead, as the series of jagged curves in **FIGURE 14-7** show, the population might perpetually be in the exponential growth portion of the logistic growth curve, with periodic massive mortality events.

The growth of populations doesn't always appear as a smooth S-shaped logistic growth curve. For some populations, particularly humans, the carrying capacity

**Q**

How many people can the earth support? Why does the number keep increasing?

## DENSITY-INDEPENDENT FACTORS LIMIT POPULATION SIZE

Density-independent forces include natural disasters such as floods, earthquakes, and fires.

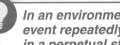 *In an environment in which a "bad luck" event repeatedly occurs, a population can be in a perpetual state of exponential growth with periodic massive mortality events.*

**FIGURE 14-7 Events can limit population growth.** Fires, floods, and earthquakes are density-independent forces that can dramatically reduce population size.

**FIGURE 14-8 Efficient crop production.** Advances in agriculture make it possible to feed the world's growing population.

Development of agricultural technologies is one example of how carrying capacity can be increased.

of an environment is not set in stone. Consider that, in 1883, an acre of farmland in the United States produced an average of 11.5 bushels of corn. By 1933, this was up to 19.5 bushels per acre. By 1992, it had increased to 95 bushels per acre. And by 2008, it had increased to 154 bushels per acre! How did this happen? The development of several agricultural technologies—including the use of vigorous hybrid varieties of corn, rich fertilizers, crop rotation, and effective pest management—has allowed farmers to produce more and more food from the same amount of land. This is just one example of how carrying capacity can be increased (**FIGURE 14-8**). Of course, over this same time, the carrying capacity has most likely been *decreased* for many other species trying to live in the same environment as humans and their crops.

---

**TAKE-HOME MESSAGE 14·4**

A population's growth can be reduced both by density-dependent factors related to crowding and by density-independent factors such as natural or human-caused environmental calamities.

---

# 14·5 Some populations cycle between large and small.

Nature is not always tidy. Exponential and logistic growth equations, for instance, help us understand the concept of population growth, but real populations don't always show such "textbook" growth patterns. Instead, they sometimes vary greatly in their rates and patterns of growth.

Locust swarms of biblical proportions certainly attest to the unpredictability of population growth (**FIGURE 14-9**). In northwest Africa, the desert locusts (migrating grasshoppers) normally live as solitary individuals in relatively small, scattered populations. In 2004, though, the population size increased rapidly, probably due to unusually good rains and mild temperatures. As the rainy season ended and green areas gradually shrank, the locusts became progressively concentrated in smaller and smaller areas. At this point, for reasons related to overcrowding but not completely understood, the locusts behaved like a mob, rather than solitary individuals. Giant swarms—some including tens of millions of insects—began flying across huge expanses of land in search of food. They completely

**FIGURE 14-9 Population explosion!** Sometimes, populations don't grow logistically, as occasionally happens for desert locusts.

Snowshoe hares (prey)
Lynx (predator)

**1** A growing hare population provides more food for the lynx, which then reproduce at higher rates.

**2** Lynx eat too many hares, thereby reducing their food source and causing their own population to crash, which enables the hare population to grow.

**FIGURE 14-10 Predator and prey.** The snowshoe hare and lynx populations rise and crash on a 10-year cycle. The population numbers are based on the number of pelts sold to the Hudson's Bay Company by trappers.

consumed giant swaths of farmland, causing more than $100 million in damage.

There is more than one way for population growth to deviate from the standard pattern. The explosive locust population growth, for example, occurs at unpredictable intervals. Another unusual pattern is the oscillations of the lynx and snowshoe hare populations of Canada. As seen in **FIGURE 14-10**, rather than undergoing smooth logistic growth, the populations of both the snowshoe hare and their predators, the lynx, cycle regularly between increases to very large numbers and crashes to much smaller numbers. Thanks to the Hudson's Bay Company, which kept detailed records for decades on the number of pelts it purchased from fur trappers, this population cycling is well documented.

Although its cause is not fully certain, to some extent, the predator and the prey cause their own cycling:

1. The hare population size grows,
2. providing more food for the lynx,
3. which then reproduce at a higher rate,
4. causing them to eat too many of the hares,
5. thereby reducing their food source,
6. causing the lynx population to crash,
7. enabling the hare population to grow, and the cycle to begin anew.

In discussing these examples of unusual population growth, we must not lose sight of the fact that, for the most part, regularly cycling populations and populations with periodic outbreaks of huge numbers are more the exception than the rule. In general, the logistic growth pattern describes populations better than any other model.

One population-growth myth that demands debunking is the story of lemmings and their supposed suicides. Do lemmings jump off cliffs when their population becomes too large? In a word: *no*. This is an urban legend. Lemming populations do occasionally experience large increases in size. Then, just as locust behavior changes when population density gets too great, lemming behavior changes, and many individuals migrate in search of less crowded habitats and more food. As they enter unfamiliar territory, some lemmings may suffer increased rates of death. But these deaths are not suicides, and they do not occur in large groups (**FIGURE 14-11**).

**Q** Do lemmings commit suicide by jumping off cliffs when their populations get too big?

**FIGURE 14-11 An urban myth debunked.** Lemmings do not jump off cliffs.

How did the myth of lemming mass suicides become so prevalent? In the filming of a 1958 Disney documentary, *White Wilderness,* many lemmings were placed on a giant snow-covered turntable so that it would appear they were migrating. In a later scene, many lemmings were filmed on a cliff overlooking a river and then chased into the water as if in a mad rush. None of these scenes were of actual lemming migrations, and in nature, lemmings never behave like this.

There are practical reasons for predicting population sizes and growth rates, as we'll see. But it can be difficult to translate simple growth models into feasible management practices.

# 14·6 "Maximum sustainable yield" is a useful but nearly impossible-to-implement concept.

Suppose you work for a logging company and are responsible for selecting which trees in an eastern hardwood forest to cut down. Or suppose you are in charge of the lobster fisheries discussed at the beginning of this chapter, responsible for deciding harvest limits. In either case, your responsibilities would be similar. In the case of the hardwood forest, how many trees would you cut down for maximum wood yield?

This is a trick question. The maximum yield on any given day would be obtained by cutting *everything* down (also known as clear-cutting). No other strategy could yield more right now. But such a harvest can be done only once, so it isn't really a management strategy. A more sensible strategy would include harvesting some individuals and leaving others for the future. With that type of strategy, which trees (or lobsters) would you harvest so as to minimally affect the

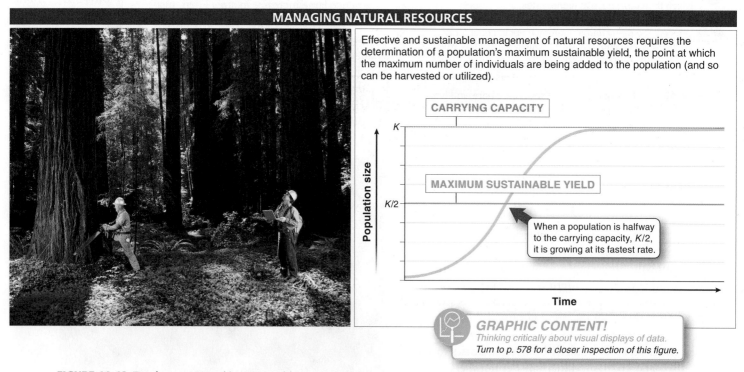

**MANAGING NATURAL RESOURCES**

Effective and sustainable management of natural resources requires the determination of a population's maximum sustainable yield, the point at which the maximum number of individuals are being added to the population (and so can be harvested or utilized).

CARRYING CAPACITY

Population size

$K$

MAXIMUM SUSTAINABLE YIELD

$K/2$

When a population is halfway to the carrying capacity, $K/2$, it is growing at its fastest rate.

Time

*GRAPHIC CONTENT!*
*Thinking critically about visual displays of data.*
Turn to p. 578 for a closer inspection of this figure.

**FIGURE 14-12 Tree harvest.** Wood is a renewable resource, but how many trees can we cut down (and which ones?) without irreversibly depleting the supply?

population's growth? Ideally, you would harvest those individuals that are no longer contributing to the population growth. But there may not be many post-reproductive individuals in the population, or it may be too hard to identify them.

What's the better solution? For long-term management, it is better to harvest some but leave others still growing and reproducing for harvest at a later time, so the population can persist indefinitely. The special case in which as many individuals as possible are removed from the population without impairing its growth is called the **maximum sustainable yield** (**FIGURE 14-12**). The value in such a harvest is that it can be carried out repeatedly, clearly yielding more in the long run than the shortsighted strategy of a complete, one-time harvest.

Your first step as manager is to determine the maximum sustainable yield for the resource, calculated as the point at which the population is growing at its fastest rate. In Figure 14-6, we can see that the population is getting larger at the fastest rate when its size is equal to half the carrying capacity. At this midpoint, scarcity of mates is not a problem, as it can be at low population levels, nor is competition a problem, as it can be near the carrying capacity—one of the reasons that the population doesn't grow at all when it reaches the environment's carrying capacity.

**Q** **Almost all natural resource managers working for the U.S. government fail to do their job exactly as mandated. Why?**

Maximum sustainable yield is a useful concept, applicable not just to tree or lobster harvesting but also to livestock, agriculture, and nearly every other natural resource. In fact, there are 31 official U.S. government agencies mandated to utilize the maximum sustainable yield concept in determining harvest levels. But here's the rub: this is nearly always an impossible task. Why is that so?

There are numerous reasons, most of which become apparent only when you leave the simplicity of the theoretician's desk and head into the messy real world. For starters, if maximum sustainable yield occurs when a population is at half its carrying capacity, do we first wait until the population stabilizes at its carrying capacity to determine what half of that carrying capacity would be? But isn't it inefficient to sit around waiting for the population to reach its carrying capacity, when we want to maintain it at half that size? Or can we just estimate carrying capacity and then calculate half of it? But, for lobsters, won't that be difficult, given that they live underwater (**FIGURE 14-13**)? Put simply, we rarely know the value of $K$.

And the problem gets worse. For one thing, with many species, not only do we not know the carrying capacity, we don't even know the number of individuals currently living. It is difficult to accurately count humans, so imagine how hard it is to count individuals of species that live underwater or fly or are microscopic.

And if we were to solve the mysteries of counting individuals and knowing the carrying capacity, we would still have to figure out whether carrying capacity is stable from year to year. If it depends on levels of resources, it may be cyclic, for instance. And even if we can figure out the carrying capacity and the population size, we would not be certain which individuals to harvest. Often, all individuals in a population are not contributing equally to population growth. The post-reproductive individuals, as mentioned above, do not contribute to this growth.

Maximum sustainable yield is a useful concept, but it is difficult to put into practice because it is hard to accurately determine population size and carrying capacity in nature.

**FIGURE 14-13 How many lobsters can be harvested sustainably?**

*The concept of maximum sustainable yield generates insights into fighting biological pests such as cockroaches.*

**FIGURE 14-14 Population explosion.** Population growth models don't just guide us to maximum production of valuable resources. They can help us efficiently reduce pest populations, too.

As a natural resource manager, you do your best, knowing that the theory behind maximum sustainable yield can almost never be put into practice perfectly.

As it turns out, harvesting natural resources for maximum efficiency generates insights into how to fight biological pests, such as cockroaches or termites (**FIGURE 14-14**). The problem is just turned on its head: which individuals in the pest population would you concentrate on killing so as to most effectively slow population growth? Those at the age of maturity, with the highest reproductive value, contribute most to the growth of the population. Similarly, because a population can still grow very rapidly with only a few males—given that any one male can fertilize a large number of females—it is effective to focus pest control on females.

Patterns of population growth and the environmental features that influence them can lead to evolutionary changes in populations, shaping features such as life span, age of reproduction, number of offspring produced, and amount of offspring care—the topics we turn to next.

A female grizzly bear and cubs in Katmai National Park, Alaska.

# 14·7–14·9 A life history is like a species summary.

## 14·7 Life histories are shaped by natural selection.

**Q**

*Do any animals mate themselves to death? If so, why?*

Some animals reproduce with a "big bang." *Antechinus* is an Australian mouse-sized marsupial, and the males are classic big-bang reproducers

(**FIGURE 14-15**). At one year of age, they enter a two- to three-week period of intense mating activity, copulating for as long as 12 hours at a time. Shortly after this, the males undergo rapid physical deterioration—they lose weight, much of their fur falls out, their resistance to parasites falls—

### VARIATION IN LIFE HISTORIES

*Antechinus*

**BIG-BANG REPRODUCTION**
- Reaches sexual maturity at one year
- Mates intensely over a three-week period
- Males die shortly after mating period
- Females usually die after weaning their first litter

Greater bulldog bat

**SLOW, GRADUAL REPRODUCTIVE INVESTMENT**
- Reaches sexual maturity at one year
- Produces about one offspring per year

House mouse

**FAST, INTENSIVE REPRODUCTIVE INVESTMENT**
- Reaches sexual maturity at one month
- Produces litters of six to ten offspring every month

**FIGURE 14-15 Reproductive strategies.** There are many strategies for reproducing, from the big bang of *Antechinus,* to the slow but sure methods of the little brown bat, to the early, fast, abundant reproduction of the house mouse.

and then they die. (Although occasionally a female will live for two or three years, most die following the weaning of their first litter.)

Other animals are a bit—but only a bit—more restrained in their reproduction. The house mouse reaches maturity in about one month and produces litters of 6–10 offspring nearly every month, sometimes generating more than a hundred offspring in its first year of life.

And some animals could not be farther from big-bang reproducers. The little brown bat is also mouse-sized, yet does not reach maturity until one year of age and typically produces only a single offspring each year. It can, however, live more than 33 years in its natural habitat.

Why all the variation? And is one strategy better than the others, evolutionarily? One of the most important recurring themes in biology is that evolution nearly always finds more than one way to solve a problem. As the marsupial mouse, house mouse, and little brown bat illustrate, there are many possible responses to the challenge of when to reproduce, how often to reproduce, and how much to reproduce in any given episode. The answers to these questions make up an organism's **life history**—the species' vital statistics, including age at first reproduction, probabilities of survival and reproduction at each age, litter size and frequency, and longevity. Life histories are a little short on warm fuzzy stories about how plants and animals live, but they tell us as much about a species as possible in a small amount of information.

Life histories vary from the big-bang reproductive strategy in *Antechinus,* in which the male's **reproductive investment**— all of the material and energetic contribution that an individual will make to its offspring—is made in a single episode, to strategies, such as that of humans, in which organisms have repeated episodes of reproduction. Plants, too, have life histories, and, like animals, they have a range of strategies. Annuals, such as corn, wheat, rice, peas, marigolds, and cauliflower, usually reproduce once and die. Perennials, such as apples, raspberries, grapes, ginger, and garlic, on the other hand, reproduce repeatedly, often living for many years.

Returning to the question posed above—which life history strategy is best?—we need to consider two questions.

**1. What is the cost of reproductive investment during any reproductive episode?** Producing offspring is risky in several ways. Gametes can be costly to produce. The act of reproducing can be risky (by increasing the likelihood of an individual's being eaten by a predator, for example). And the wear and tear of reproduction on an individual's body takes its toll. Thus, the number of offspring an organism produces can only go so high before it becomes detrimental.

An organism might produce four offspring in an episode, for example, without significantly increasing its risk of dying from the effort. But if, by chance, it produces five or six, this may take such a toll that the individual is unlikely to live to have another litter. Thus, for many organisms, a smaller litter is favored by natural selection because it enables the individual to have additional litters in the future, maximizing its *lifetime* reproductive success.

**2. What is an individual's likelihood of surviving to have future reproductive episodes?** If predation rates or other sources of mortality are very high, an individual might not be alive in one month. If so, it makes less sense to defer reproduction. Why save for a future unlikely to come?

Taken together, the answers to these questions help us understand the variety of life history strategies we see in nature.

Toward the fast extreme of reproductive strategy are rodents. They experience very high mortality rates in the wild, and natural selection has consequently favored early and heavy reproductive investment. Rodents provide

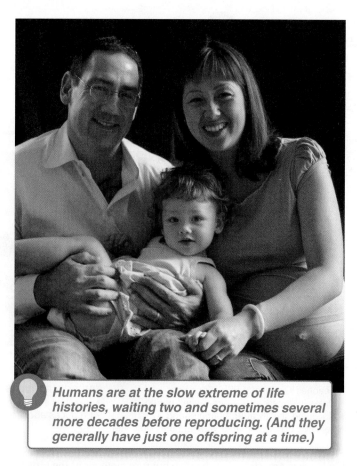

Humans are at the slow extreme of life histories, waiting two and sometimes several more decades before reproducing. (And they generally have just one offspring at a time.)

**FIGURE 14-16** With such a low probability of dying in any given year, humans can defer reproduction.

relatively little parental care but produce many litters, each with many offspring. Organisms with this type of life history tend to have relatively poor competitive ability but a high rate of population growth; each individual's likelihood of surviving is low, but so many are produced that the odds of one offspring surviving are high.

Q

**Why *do* humans put off mating so much longer than do cats or mice?**

At the slow extreme, humans have such a low probability of dying in any given year that they can defer reproduction, or at least reproduce in small amounts—once every few years—without much risk,

enabling them to have just one offspring at a time but to invest significant parental effort in maximizing its likelihood of surviving. Organisms with this type of life history generally have evolved in such a way that their competitive ability—that is, the likelihood of an individual surviving—is high, although their rate of population growth is low. That is, not many offspring are produced, but those that are have a high likelihood of surviving (**FIGURE 14-16**).

> **TAKE-HOME MESSAGE 14·7**
>
> An organism's investment pattern in growth, reproduction, and survival is described by its life history.

# 14·8 Populations can be described quantitatively in life tables and survivorship curves.

Biologists have learned some important lessons from insurance agents. Because insurance companies need to estimate how long an individual is likely to live, they invented something called the **life table** (**FIGURE 14-17**). It ought to be called a death table, though, because the table tallies the number of people in a population within a certain age group, say 10 to 20 years old, and the number of individuals within that age range who die. From these numbers, they can predict an individual's likelihood of dying or surviving within a particular age interval. For the insurance companies, making money hinges on making accurate life span predictions. For biologists, a life table is like a quick window into the lives of the individuals of a population, showing how long they're likely to live, when they'll reproduce, and how many offspring they'll produce.

From life tables, biologists create **survivorship curves,** graphs showing the proportion of individuals of a particular age that are now alive in a population. Survivorship curves

indicate an individual's likelihood of surviving through a particular age interval. And they reveal a huge amount of information about a population, such as whether most offspring die shortly after birth—think back to the five million eggs produced by some fish—or whether most survive to adulthood and are likely to live long lives, such as humans in the United States.

Three distinct types of survivorship curves are shown in **FIGURE 14-18**. The curves plot the proportion of individuals surviving at each age, across the entire range of ages seen for that species.

**Type I.** At the top of Figure 14-18, in blue, is the survivorship curve seen in most human populations, and shared by the giant tortoise. We and the tortoise have a very high probability of surviving every age interval until old age, then the risk of dying increases dramatically. Species with a type I survivorship curve, which includes most large

| LIFE TABLE | | | | |
|---|---|---|---|---|
| **AGE** (beginning of interval) | **NUMBER ALIVE** (beginning of interval) | **PROPORTION ALIVE** (beginning of interval) | **DEATHS DURING INTERVAL** | **PROBABILITY OF DYING DURING INTERVAL** |
| 0 | 210 | 1.00 | 140 | 0.67 |
| 3 | 70 | 0.33 | 28 | 0.40 |
| 6 | 42 | 0.20 | 28 | 0.67 |
| 9 | 14 | 0.07 | 10 | 0.71 |
| 12 | 4 | 0.02 | 4 | 1.00 |
| 15 | 0 | 0.00 | n/a | n/a |

In species with type II survivorship curves, individuals experience approximately the same risk of death at any age.

**FIGURE 14-17 Summarizing life and earth in a table.** Shown here: a life table for the cactus ground finch on one of the Galápagos Islands.

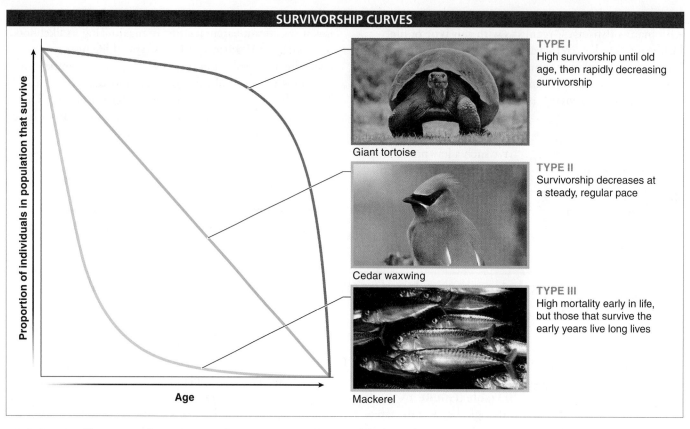

## SURVIVORSHIP CURVES

**TYPE I**
High survivorship until old age, then rapidly decreasing survivorship

Giant tortoise

**TYPE II**
Survivorship decreases at a steady, regular pace

Cedar waxwing

**TYPE III**
High mortality early in life, but those that survive the early years live long lives

Mackerel

**FIGURE 14-18 Three types of survivorship.** Survivorship curves show the proportion of individuals of a particular age that are alive in a population.

mammals, usually have a few features in common. They have few natural predators, so they are likely to live long lives. They tend to produce only a few offspring—after all, most will survive—and they invest significant time and effort in each offspring.

**Type II.** In the middle of the figure, in green, is a survivorship curve seen in many bird species, such as the cedar waxwing, and in small mammals such as squirrels. The straight line indicates that the proportion alive in each age interval drops at a steady, regular pace. In other words, with a type II survivorship curve, the likelihood of dying in any age interval is the same, whether the bird is between 1 and 2 years old, or between 10 and 11.

**Type III.** At the bottom, in purple, is a survivorship curve in which most of the deaths occur in the youngest age groups. Common in most plant and insect species, as well as in many marine species such as oysters and fishes, the type III curve describes populations in which the few individuals lucky enough to survive the first few age intervals are likely to live a much longer time. Species with this type of survivorship curve tend to produce very large numbers of offspring, because most will not survive. They also tend not to provide much parental care, if any. A classic example is the mackerel.

One female might produce a million eggs! Obviously, most of these (on average, 999,998) do not survive to adulthood, or the planet would be overrun with mackerel.

In reality, most species don't have survivorship curves that are definitively type I, II, or III. They may be anywhere in between. These three types, though, represent the extremes and help us make predictions about reproductive rates and parental investment without extensive observations of individual behavior. Survivorship curves can also change over time or location. Humans in developing countries, for example, tend to have higher mortality rates in all age intervals—particularly in the earliest intervals—relative to individuals in industrialized countries.

### TAKE-HOME MESSAGE 14·8

Life tables and survivorship curves summarize the survival and reproduction patterns of the individuals in a population. Species vary greatly in these patterns: the highest risk of mortality may occur among the oldest individuals or among juveniles, or mortality may strike evenly at all ages.

# 14·9

## There are trade-offs between growth, reproduction, and longevity.

If you were designing an organism, how would you structure its life history for maximum fitness? (Recall from Chapter 8 that fitness is an organism's reproductive output relative to that of other individuals in the population.) Ideally, you would create an organism that (1) produces many offspring (beginning just after birth and continuing every year), (2) while growing tremendously large, to reduce predation, and (3) living forever.

Unfortunately, evolution operates with some constraints. Some of these traits are just not possible, because selection that changes one feature tends to adversely affect others. When evolution increases one life history characteristic, another characteristic is likely to decrease. Life history trade-offs can be better understood in the context of the three areas to which an organism allocates its resources: growth, reproduction, and survival. When resources are limited, increased allocation to one of these areas tends to reduce allocation to one or both of the others. Some examples of the best studied trade-offs follow.

**1. Reproduction and survival.** Among big-bang reproducers such as the marsupial mouse, *Antechinus,* and salmon, investment in reproduction is exceedingly high, followed shortly by death (**FIGURE 14-19**). If individuals are physically prevented from reproducing, however, they can live many more years.

> **Q** With one simple surgical procedure, most men could add more than 10 years to their life span. Why doesn't anyone opt for it?

**2. Reproduction and growth.** Beech trees grow more slowly during years when they produce many seeds than in years when they produce no seeds. Similarly, female red deer kept from reproducing in their first year of life grow significantly larger during that year.

**3. Number and size of offspring.** The females of one lizard species, *Uta stansburiana,* can lay more eggs if the eggs are smaller, but a higher proportion of the offspring survive if the female lays larger eggs. Evolution has led to a medium-size egg that tends to maximize the number of surviving offspring.

There are many other trade-offs, such as the number of offspring produced and amount of parental care given. Additionally, among mammals, litter size increases with latitude (distance from the equator). This trend may be due to the trade-off between litter size and frequency. Closer to the equator, animals can breed for a longer portion of the year (year-round near the equator). Farther from the equator, the breeding season is shorter. To compensate for having fewer litters at higher latitudes, mammals produce more offspring per litter. Can you think of other trade-offs?

---

### TAKE-HOME MESSAGE 14·9

Because constraints limit evolution, life histories are characterized by trade-offs between investments in growth, reproduction, and survival.

---

### LIFE HISTORY TRADE-OFFS

**REPRODUCTION AND SURVIVAL**
Big-bang reproducers such as salmon make a single, exceptionally high investment in reproduction, then die shortly afterward.

**FIGURE 14-19 Life history trade-offs.**

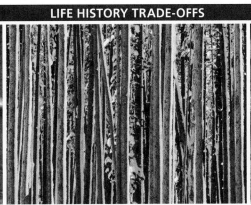

**REPRODUCTION AND GROWTH**
Beech trees grow much more slowly in the years when they produce many seeds than they do in years when they produce few seeds.

**NUMBER AND SIZE OF OFFSPRING**
Female lizards (*Uta stansburiana*) produce medium-size eggs as a compromise between many small eggs (with poor survival of offspring) and few large eggs (with better survival of offspring).

## 14·10–14·12
# Ecology influences the evolution of aging in a population.

*Bison pair in a blizzard in Minnesota.*

# 14·10
### Things fall apart: what is aging and why does it occur?

Attempting to define what was obscene, an exasperated justice of the United States Supreme Court once concluded that he might never be able to define obscenity, "but I know it when I see it." Aging is like that. Picture your grandparents or people in their seventies or eighties. Clearly they have aged. But what exactly does that mean? And does it differ from the way that people in their thirties or forties have aged? Sometimes it involves sagging skin. Sometimes it involves weakened muscles and bones. Sometimes it involves senility.

The difficulty comes when we realize that no two people age in exactly the same way, yet without fail everyone ages (**FIGURE 14-20**). So we must retreat to a somewhat vague definition. For an individual, aging is the gradual breakdown of the body's machinery. With this definition, we see how each individual can experience aging differently. But aging can be seen much more clearly by examining a population as a whole.

From a population perspective, **aging** emerges as a definitive and measurable feature: it is simply an increased risk of dying with increasing age, after reaching the age of maturity. For example, among humans, people are more likely to die between the ages of 70 and 71 than between 60 and 61. Similarly, they are more likely to die between the ages of

44 and 45 than between 34 and 35. The causes of death at these ages often differ, but the shuffle toward death becomes more and more dire, with no respite. This is the most useful definition of aging for ecologists, and it doesn't apply only to humans. Aging can be measured and assessed for any species in the same way.

Jeanne Louise Calment lived for more than 122 years, the longest life span ever documented for a human. While this is spectacularly long relative to most organisms on earth (and is significantly longer than the average human life span in the United States, 78 years), it pales in comparison with some of the longest-lived organisms. Bristlecone pine trees are the longest-lived, surviving for thousands of years. The growth rings of one bristlecone pine showed that it was more than 7,000 years old!

Among the animals, lobsters and quahogs (a type of clam) are the longest-lived invertebrates (up to 100 and 200 years, respectively). Long-lived vertebrates include striped bass, which can exceed 120 years, and tortoises, which can reach 150 years. Toward the other end of the spectrum, rodents may live just a couple of years, and fruit flies and many other insects generally live and die within the span of a few weeks. Interestingly, bird and bat species live 5–10 times longer than similarly sized rodents. These data prompt the

FIGURE 14-20 **What does "aging" mean?** Aging differs from one individual to another, but we recognize individuals of different ages without any difficulty. Shown here: group portraits of the same 16 women, taken 26 years apart. (The inscription in the upper left reads "For the future.")

questions: Why is there so much variation? And what controls how long the individuals of a species live? We explore those questions in the next section, but first we consider the more general question of why organisms age at all.

> ❝ Things fall apart; the centre cannot hold;
> Mere anarchy is loosed upon the world,
> The blood-dimmed tide is loosed, and everywhere
> The ceremony of innocence is drowned;
> The best lack all conviction, while the worst
> Are full of passionate intensity. ❞
> —WILLIAM BUTLER YEATS, *The Second Coming*, 1919

Aging, in fact, is not a mystery. The question of why we age is one of the great solved problems in biology. Oddly, though, it may also be one of the best-kept secrets in science. Let's investigate.

It is easier to *state* the evolutionary explanation for why we age than to *understand* it. But that's where we start. The explanation is based on this one fundamental fact: the force of natural selection lessens with advancing age. What does this mean?

**1.** Imagine a mutation that causes a person carrying it to die at age 10. Will that person pass the mutation on to many offspring? Of course not. The carrier of that mutant allele will die before she gets a chance to pass it on. Alternative versions of the gene that don't cause death will be the only ones that persist.

**Q** Many genetic diseases kill old people, but almost none kill children. Why not?

**2.** Now imagine a mutation that causes a person carrying it to die at the age of 150. Will that person pass it on? Yes! The carrier of this fatal allele will already have had children—passing on the mutation to them—long before she even knows she carries it. In fact, she will no doubt die long before this mutation has the opportunity to exert its disastrous effect. The same thing happens if the mutation has its negative effect at age 100 . . . or 70 . . . or even 50.

In the first example, we see that natural selection "weeds out" alleles that cause sickness and death early in life. Those alleles never get passed on (**FIGURE 14-21**). No one inherits them, and they disappear from the population without fail. Individuals with the alternative versions of the gene pass on the good versions at a higher rate. Consequently, those good alleles end up as the only versions of the gene in the population. These are the genes that we inherited from our ancestors, and this is why very few people have genetic diseases that kill them in their teens. We sometimes think of natural selection as only acting positively to increase or improve a trait—the length of a giraffe's neck, for example. But natural selection can also select *against* a trait—in this case, disease.

Natural selection isn't so effective with some other genes. The second example describes mutations that arise and only later in life make their carriers more likely to die. Examples include mutations that increase the risk of cancers or heart disease or other ailments. Unfortunately, the carriers of these mutant alleles will already have reproduced

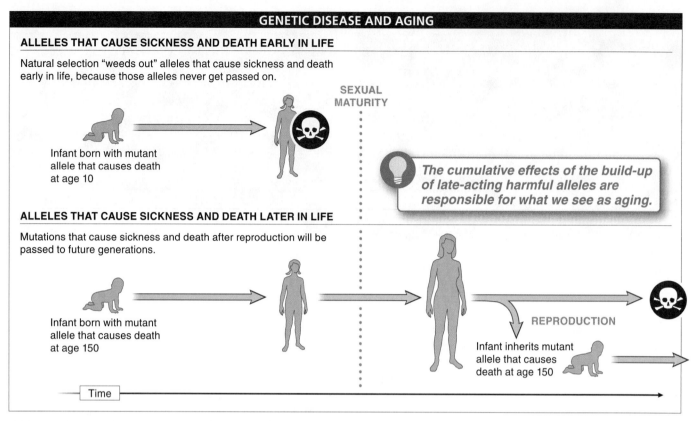

**ALLELES THAT CAUSE SICKNESS AND DEATH EARLY IN LIFE**

Natural selection "weeds out" alleles that cause sickness and death early in life, because those alleles never get passed on.

SEXUAL MATURITY

Infant born with mutant allele that causes death at age 10

The cumulative effects of the build-up of late-acting harmful alleles are responsible for what we see as aging.

**ALLELES THAT CAUSE SICKNESS AND DEATH LATER IN LIFE**

Mutations that cause sickness and death after reproduction will be passed to future generations.

Infant born with mutant allele that causes death at age 150

REPRODUCTION

Infant inherits mutant allele that causes death at age 150

Time

**FIGURE 14-21 Carrying a time bomb.** Natural selection can't weed out genetic defects that wait until we've already reproduced to harm our health. Those mutations that act early in life will be passed on to subsequent generations at a greatly reduced frequency. The cumulative effects of the buildup of late-acting harmful alleles are responsible for what we see as aging.

and passed on the alleles before they have had their negative effect. In other words, it doesn't matter whether you carry one of these mutants or an alternative allele that does not harm you, the number of offspring you produce—your **reproductive output**—is the same. Consequently, these mutants are never "cleaned" out of a population. The practical result is that we all have inherited many of these mutant alleles that have arisen over the past hundreds and thousands of generations.

These mutant alleles that have their adverse health effects later in life—after we are likely to have reproduced—are responsible for aging. The later in life they have

their effect—and therefore the more likely it is that they are passed on to offspring—the more common they will be in the population. This is why there are *many* causes of death, and why it seems as if all of our bodily systems fall apart as we get older.

**Q** A cure for cancer may be discovered, but not for aging. Why the difference?

---

**TAKE-HOME MESSAGE 14·10**

Natural selection cannot weed out harmful alleles that do not diminish an individual's reproductive output. Consequently, they accumulate in the genomes of individuals of nearly all species. This leads to the multiple physiological breakdowns that we see as aging.

# 14·11

## What determines the average longevity in different species?

Can individuals live forever? No. Just because they are not at risk from internal, genetic sources of mortality doesn't mean there aren't things in the world around them that might kill them. Predators. Lightning. Drought. Disease. The list is long. And for different species, these risks are more or less serious. Rodents, for example, are at *very* high risk of predation nearly every minute of every day, so they are likely to be dead within a few years (**FIGURE 14-22**).

Because death from some external source is likely to come so early, organisms living in high-risk worlds must reproduce early. Natural selection favors this strongly. (If they didn't reproduce early, they wouldn't leave any descendants.) Now, when a harmful mutation arises that causes its damage at an early age—say, 1 or 2 years of age—the rodent has probably already reproduced and passed the mutation on. And so that harmful allele increases in frequency in the population, causing individuals to age earlier.

Tortoises, by contrast, live in relatively low-risk worlds. Because they have armor-like shells that protect them from danger, death from external sources is low. Early, intensive investment in reproduction is not necessary and so has not been favored by natural selection. In such species, when a harmful mutation arises that has its effect at 5 or 10 years of age, the individual carrying it is likely to die before it has reproduced. Consequently, the mutation isn't passed on; natural selection weeds it out of the population.

The age at time of reproduction, then, is a key factor determining longevity. Factors that favor early reproduction also favor early aging; factors that don't favor early reproduction—or actively favor later reproduction—favor later aging. But then, what determines when an organism reproduces?

We can think of each population as evolving in a world with a specific **hazard factor** (see Figure 14-22). This factor includes

**ENVIRONMENTAL RISKS AND AGING**

Rodent

Tortoise

**HIGH HAZARD FACTOR**
- Relatively high risk of death at each age
- Individuals tend to reproduce earlier
- Earlier aging
- Shorter life spans

2 years

Longevity

**LOW HAZARD FACTOR**
- Relatively low risk of death at each age
- Individuals tend to reproduce later
- Later aging
- Longer life spans

150 years

Longevity

*In environments characterized by low mortality risk, populations of slowly aging individuals with long life spans evolve. In environments characterized by high mortality risk, the opposite occurs.*

**FIGURE 14-22 Hazard factor.** The mouse and the tortoise face different risks of dying.

the risks of death from all types of external forces. When the hazard factor is low, as it is for tortoises and humans, individuals of a species tend to reproduce later, and so natural selection can weed out all harmful mutations except those that have their effects late in life. Individuals in these species will live a long time before they succumb to aging. A high hazard factor, on the other hand, will lead to earlier reproduction and therefore early aging and shorter life span.

Which species should age more slowly in captivity, a porcupine or a guinea pig? The difference is large, revealing unambiguously that it's good to be a porcupine. Few predators want to eat you. Or rather, few predators *can* eat you. Because of their sharp quills, porcupines have evolved in a world with little risk of predation and with a lower hazard factor.

Meanwhile, guinea pigs, who have no protective quills, live with a high risk of predation. Because they've been evolving in a world characterized by a high hazard factor, they reproduce much sooner, age more quickly, and die younger. A guinea pig in captivity lives less than 10 years, whereas a captive porcupine can live 21 years. In the wild the difference is even greater: 15 years for the porcupine and only 3 or 4 years for the guinea pig.

What difference in longevity would you expect between poisonous and non-poisonous snakes, or between bats (which can fly) and mice?

### TAKE-HOME MESSAGE 14·11

The rate of aging and pattern of mortality are determined by the hazard factor of the organism's environment. In environments characterized by low mortality risk, populations of slowly aging individuals with long life spans evolve. In environments characterized by high mortality risk, populations of early-aging, short-lived individuals evolve.

# 14·12 Can we slow down the process of aging?

Life extension is possible. Real, honest-to-goodness doubling of the normal life span. Not only is it possible, researchers have demonstrated it repeatedly . . . in fruit flies.

Although the results of these life-extension studies are astonishing, the methods by which they were achieved are simple. Researchers kept several large populations of fruit flies in the laboratory, in small Plexiglas cages. The flies would feed and lay eggs on a small plate of food. Every few days, a fresh plate would be put in the cage and the old one discarded. At 2 weeks, eggs were collected and used to start the next generation of flies. Propagated in this way, the fly populations were maintained for hundreds of generations. When the longevity of flies in these populations was measured, the researchers found that the flies started to experience the physiological breakdowns associated with aging after a few weeks, and after about a month the flies died.

One researcher thought that changing the force of natural selection on the flies might produce an evolution in their aging pattern. So, instead of collecting eggs at 2 weeks to start the

**Q** Is it possible, with our current knowledge, to double longevity in humans?

new generation, he cleverly waited longer and longer (**FIGURE 14-23**). The first new generation had parents that must have lived for 3 weeks. If any fly carried a mutation that caused it to die between 2 and 3 weeks, that fly did not contribute to the next generation and that mutation was not passed on.

After several generations, eggs were collected at 4 weeks instead of 3 weeks. At this point, all the flies making up the new generation had parents that had survived for 4 weeks. Any flies that died before 4 weeks of age didn't contribute to the new generation. Natural selection was now able to reduce the prevalence of mutant genes responsible for early aging.

The experiment continued, progressively increasing the generation time to 5 weeks, then 6, 7, and ultimately 10 weeks — equivalent to decreasing the hazard factor of the fly population. At this point, only those flies with a genome free of mutations that might cause death in the first 10 weeks of life could contribute to the new generation. The net result was the creation of a

**THE EXPERIMENT**

= 500 fruit flies

Flies that didn't survive to this point do not contribute eggs (or genes) to the next generation.

Eggs

New generation

**1 INITIAL SETUP**
Start with a cage that contains a large number of fruit flies and fresh food.

**2 START NEW GENERATION**
After 2 weeks, put a fresh dish of food into the cage and collect the eggs laid on it. Transfer the eggs to a new cage. Sample some of the flies hatched from these eggs and measure their longevity.

**3 INCREASE GENERATION TIME**
Repeat the procedure, but instead of waiting 2 weeks, wait longer. Gradually increase this period until eggs are being collected only from flies that survive 10 weeks.

**THE RESULTS**

Number of flies — Longevity (days) — 0 20 40 60 80

**2-WEEK GENERATION TIME**
Avg. longevity:
**33 DAYS**

Experiment continues through 90 generations.

Number of flies — Longevity (days) — 0 20 40 60 80

**10-WEEK GENERATION TIME**
Avg. longevity:
**63 DAYS**

*Over many generations of selection for later and later age of reproduction, the average life span of the flies is doubled!*

**FIGURE 14-23 Creating longer-lived organisms through evolution in the laboratory.**

population of flies with much "cleaner" genomes. When put under ideal conditions outside the cages, these "super" flies did not experience aging until long after the original flies would have died. In fact, the average life span of the flies hatching from eggs harvested from parents of 10 weeks of age was more than 60 days—double that of the original flies!

**TAKE-HOME MESSAGE 14·12**

By increasing the strength of natural selection later in life, it is possible to increase the mean and maximum longevity of individuals in a population. This occurs in nature (as in porcupines, bats, and so on) and has also been achieved under controlled laboratory conditions.

*Earth's population is increasing by 80 million people each year. Is that sustainable?*

# 14·13–14·15 The human population is growing rapidly.

# 14·13 Age pyramids reveal much about a population.

The "baby boom." It happened half a century ago, yet young people in the United States today may end up paying a price for it. What was it? And why does it still matter? Imagine that you run a grade school. How many teachers will you need for the first grade? That all depends on how many students will be entering. If it was 50 students last year, will it be 50 students again this year? Probably. The number of students is likely to be similar from year to year. But it can change. If we set aside changes due to migration, it depends on how many babies were born six years ago. If that number was higher than in the previous year, you may need more teachers at your school. Or you may need fewer teachers if fewer babies were born.

Increasing birth rates caused the baby boom. Beginning just after World War II, and continuing through the early 1960s, families in the United States had about 30% more babies per month than they would have had if they had followed their parents' pattern. Then the birth rates began to return to their earlier

levels, where they have remained ever since. As the babies from the boom years grew to school age, schools had to increase in size to accommodate all the children. Then the schools had to downsize once the baby boomers had graduated. Now, as these individuals approach retirement age, the question of how their retirement and health care needs will be met is one of the biggest issues facing American society.

Populations often vary across space—dense in the cities and more sparse in rural areas. The baby boom illustrates that populations have an "age distribution" as well. Just as there may be more individuals in some areas and fewer in others, there may be more individuals in certain age groups and fewer in others.

Describing populations in terms of the proportion of individuals in each age group reveals interesting population features. A population can be divided into the percentages of individuals that are in specific

**Q** What is the baby boom? Why is it bad news for young people today?

age groupings, called *cohorts,* such as 0–4 years, 5–9 years, 10–14 years, and so on, in an "age pyramid." People have radically different likelihoods of dying or reproducing, for instance, depending on their age. A 10-year-old isn't going to produce any offspring, but a 30-year-old has a relatively high likelihood of reproducing. And a 10-year-old is less likely to die than an 80-year-old.

It can be useful, therefore, for a society to know the relative numbers of 10-, 30-, and 80-year-olds in its population. The age data can be used to determine whether a society would be better off investing in new schools or fertility wards or convalescent homes, among other social issues. Companies, too, rely on such demographic data to best plan their strategies for producing the goods that people will want and need.

Around the world, countries vary tremendously in the age pyramids describing their populations. If two populations are the same size but have different age distributions, they will have some very different features. Most industrialized countries are growing slowly or not at all; most of the population is middle-aged or old. In these countries, such as Norway, the age pyramid is more rectangular than pyramid-shaped (**FIGURE 14-24**). Because birth rates are

low, the bottom of the pyramid is not very wide. And because death rates are low, too, the higher age classes in the pyramid don't get significantly narrower. Instead, the cohorts remain about the same size all the way into the late sixties and seventies, at which point high mortality rates finally cause them to narrow. This gives the pyramid a more or less rectangular shape.

A more triangular shape is seen in the age pyramids of developing countries. Kenya, for example, has very high birth rates, reflected as a large base in its age pyramid (see Figure 14-24). But high mortality rates, usually due to poor health care, cause a large and continuous reduction in the proportion of individuals in older age groups. In these countries, most of the population is in the younger age groups.

Notice the "bulge" in the U.S. age pyramid (**FIGURE 14-25**). The shape of the age pyramid in the United States has economists worried that the social security system (including Social Security and Medicare) will not be able to offer older citizens sufficient benefits in the next 10–30 years. Because of the unusually large number of babies born about 50–65 years ago, an unusually large number of people are now reaching retirement age. Since the baby

**FIGURE 14-24 A visual representation of population growth.** Age pyramids go from triangular to rectangular when population growth stabilizes.

UNITED STATES OF AMERICA

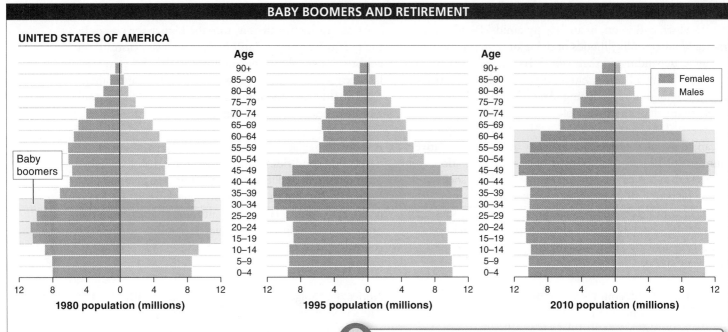

FIGURE 14-25 **The retirement boom.** As baby boomers reach retirement, the population bulge may strain the social security system in the United States.

*Because birth rates were unusually large for a brief time about 50–65 years ago, those age classes appear as a bulge moving up the age pyramid. And the younger age classes may not be large enough to support them.*

boomers were born, there haven't been any years with such a large cohort. This means that the current numbers of working individuals who contribute to the social security system are not sufficient to cover the payouts promised to the large number of retirees, and the baby boomers will be expensive to support as they reach retirement.

### TAKE-HOME MESSAGE 14·13

Age pyramids show the number of individuals in a population within any age group. They allow us to estimate birth and death rates over multi-year periods.

## 14·14 As less-developed countries become more developed, a demographic transition often occurs.

Populations can sometimes seem as strange and stubborn as individuals. Around the world, governments encourage (or sometimes almost coerce) their citizens to reduce their reproductive rates in efforts to check population growth, but they are rarely successful. Yet it seems that when governments stop trying, their countries' population growth rates slow down all on their own.

It's not that populations actually have minds of their own. Rather, many countries have undergone industrialization, with increases in health, wealth, and education. And in the process, their patterns of population growth follow common paths, marked first by periods of faster growth and later by slower growth. The sequence of changes is remarkably consistent.

Start with a country prior to industrialization. Such countries usually have high birth rates and high death rates. These result from poor and inefficient systems of food production and distribution, along with a lack of reliable medical care. Food production and health care typically improve as industrialization begins. These improvements inevitably lead to a reduction in the death rate. The birth rate, however, remains relatively high, so the country's population grows rapidly.

As industrialization continues, further changes occur. Most importantly, the standard of living increases. This results from higher levels of education and employment, and, in conjunction with improved health care, this finally causes a reduction in the birth rate. The new, lower birth rate then slows the population's growth. The progression from:

1. high birth rates and high death rates (slow population growth) *to*

2. high birth rates and low death rates (fast population growth) *to*

3. low birth rates and low death rates (slow population growth)

is called the **demographic transition** (**FIGURE 14-26**).

**Q** **Population growth is alarmingly slow in Sweden and alarmingly fast in Mexico. Why this difference?**

The demographic transition can take decades to complete, so it is not always easy to identify it as it occurs. A survey of countries around the world reveals that many are at different points along the transition. Sweden, for example, has a low fertility rate (1.7 children born per woman) and a low death rate (10.4 deaths per 1,000 people). At the other extreme, Nigeria has a very high fertility rate (5.5 children born per woman) and death rate (17.2 deaths per 1,000). Mexico, in the midst of a clear demographic transition, has a moderately high fertility rate (2.45 children born per woman) yet a very low death rate (4.7 deaths per 1,000).

In the past few decades, the demographic transition has been completed in Japan, Australia, the United States, Canada, and most of Europe, leading to a slowing of population growth. In Mexico, Brazil, Southeast Asia, and most of Africa, on the other hand, the transition is not complete and population growth is still dangerously fast.

The demographic transition illustrates how health, wealth, and education can lead to a reduction in the birth rate without direct government interventions. But because more than three-quarters of the world's population lives in developing countries and less than a quarter in the developed countries, the slowed population growth that generally accompanies the demographic transition is unlikely to be sufficient to keep the world population at a manageable level. Instead, population growth will continue to rise quickly. What are the potential consequences of such explosive growth? We explore this next.

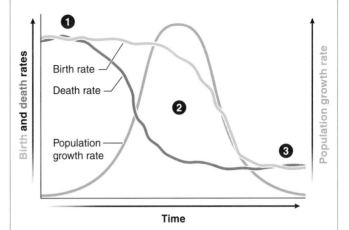

**DEMOGRAPHIC TRANSITION**

The demographic transition is a pattern of population growth that is experienced as a country industrializes. It is characterized by slow growth, then fast growth, and then slow growth again.

**1** **SLOW GROWTH**
- High birth rates and high death rates
- Populations with inefficient systems of food production and distribution, along with a lack of reliable medical care, usually have high birth rates and death rates, leading to slow population growth.

**2** **FAST GROWTH**
- High birth rates and low death rates
- As industrialization begins, both food production and health care improve, leading to a reduction in the death rate. The birth rate remains relatively high, and so the country's population grows rapidly.

**3** **SLOW GROWTH**
- Low birth rates and low death rates
- As industrialization continues, the birth rate decreases as a result of higher levels of education and employment, along with improved health care. The new, lower birth rate then slows the population's growth.

**FIGURE 14-26** With industrialization, death rates drop and, later, birth rates drop, too.

---

**TAKE-HOME MESSAGE 14·14**

The demographic transition tends to occur with the industrialization of countries. It is characterized by an initial reduction in the death rate, followed later by a reduction in the birth rate.

---

# 14·15 Human population growth: how high can it go?

Humans are a phenomenally successful species. The numbers don't lie: there are more than 7 billion people alive today (**FIGURE 14-27**). And we add 80 million people to the total each year, because birth rates greatly exceed death rates. But for all of our success, the laws of physics and chemistry still apply. We all need food for energy and space to live. We

## HUMAN POPULATION GROWTH

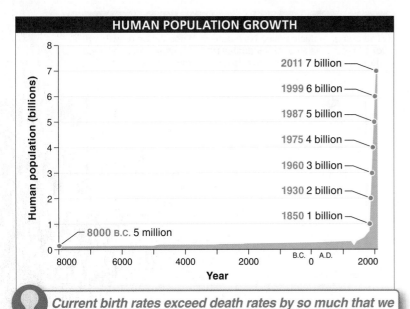

2011 7 billion
1999 6 billion
1987 5 billion
1975 4 billion
1960 3 billion
1930 2 billion
1850 1 billion
8000 B.C. 5 million

*Current birth rates exceed death rates by so much that we add 80 million people to the world population each year!*

**FIGURE 14-27** The world's human population: a slow start, but rapidly picking up steam.

need other resources, too, and we need the capability of processing and storing all of the waste our societies generate. Because of these limits to perpetual population growth, we may become victims of our own success.

This we know for certain: human population growth cannot continue forever at the current rate. As is the case for every other species, our environment has a carrying capacity beyond which the population cannot be maintained indefinitely. The question is, how high can it

go? What is the earth's carrying capacity for humans? This, unfortunately, is a difficult question to answer.

For more than 300 years, biologists have been making estimates, starting when Anton van Leeuwenhoek (the inventor of the microscope) made an estimate of just over 13 billion people. The median of all the estimates is just over 10 billion, and the United Nation suggests that the carrying capacity is somewhere between 7 and 11 billion.

There is huge variation from one biologist's estimate to the next. Why is it so hard to figure this out? After all, it's important that we know so that we can work to avoid a global catastrophe. The problem, it seems, goes back to the reasons behind the tremendous human success in the first place. We are so clever that we seem to keep increasing the carrying capacity before we ever bump into it. We do this in a variety of ways, all made possible by various technologies that we invent. In particular, we make advances on three fronts (**FIGURE 14-28**):

1. **Expand into new habitats.** With fire, tools, shelter, and efficient food distribution, we can survive almost anywhere on earth.

2. **Increase the agricultural productivity of the land.** With fertilizers, mechanized agricultural methods, and selection for higher yields, fewer people can now produce much more food than was previously thought possible.

3. **Circumvent the problems that usually accompany life at higher densities.** Public health and civil engineering advances make it possible for higher and higher densities of people to live together with minimal problems from waste and infectious diseases.

## LIMITLESS CARRYING CAPACITY?

One reason that the carrying capacity for the human population is difficult to estimate is that we can increase it in a variety of ways.

Dubai, United Arab Emirates

**EXPANDING INTO NEW HABITATS**
With fire, tools, shelter, and efficient food distribution, we can survive almost anywhere on earth.

Juehnsdorf, Germany

**INCREASING THE AGRICULTURAL PRODUCTIVITY OF THE LAND**
With fertilizers, mechanized agricultural methods, and selection for higher yields, fewer people can now produce much more food than was previously thought possible.

New York City, United States

**FINDING WAYS TO LIVE AT HIGHER DENSITIES**
Public health and civil engineering advances make it possible for higher and higher densities of people to live together with minimal problems from waste and infectious diseases.

**FIGURE 14-28** Three ways of increasing the carrying capacity for the human population.

But the question remains: how high can the population go? The most difficult problem in determining the earth's carrying capacity for humans may be assessing just how many resources each person needs. It is possible to estimate the **ecological footprint** of an individual or an entire country by evaluating how much land, food, water, and fuel, among other things, are used. This method reveals that although some countries (e.g., New Zealand, Canada, and Sweden) have more resources available than are required to support the needs of their population, resource use in the world as a whole is significantly greater than the resources available, implying that we are already at our planet's carrying capacity. The populations of many countries—including the United States, Japan, Germany, and England—currently consume an unsustainable level of resources (**FIGURE 14-29**).

> 6 6 There is in every American, I think, something of the old Daniel Boone—who, when he could see the smoke from another chimney, felt himself too crowded and moved further out into the wilderness. 9 9
> —HUBERT HUMPHREY, 38th U.S. Vice President, 1966

Another difficulty in estimating the earth's carrying capacity is that populations can and do occasionally alter their fundamental growth properties. Relatively small changes in the worldwide birth rate can have dramatic effects on the ultimate population size of the planet.

The difficulties in estimating the earth's carrying capacity illuminate an important issue. We must be more specific when we ask: how high can it go? We must add: and at what level of comfort, security, and stability, and with what impact on the other species on earth? There are trade-offs.

The ecological footprint of the 1 billion people currently living in India, for example, is much smaller than that of Japan or Norway or Australia. It is possible, we know, to live with much less impact on the environment per person. But how much do we want to sacrifice to enable a larger number to live stably on the planet? Is our goal to maximize the number? Or to reduce the number and increase the resources available to each person? Or should our goal be something else entirely? The answers to these questions will influence whether the ultimate carrying capacity is even higher than 11 billion, or perhaps lower than the current 7.1 billion people now alive.

The problem is indeed a difficult one. And it gets even more difficult because the world population currently has significant momentum. That is, even if today we immediately and permanently reduced our fertility rate to the replacement rate of just two children per couple, the world population would continue to grow for more than 40 years, putting us up to at least 8 billion. This is because there are so many young people alive that, each year, more and more of them will enter the reproductive population. As we get closer to (or exceed) the earth's carrying capacity for humans, we should be better able to recognize it, but by then it may be too late to take preemptive measures—to avoid the resource depletion that may doom us to the much more unpleasant experience of natural population controls.

## TAKE-HOME MESSAGE 14·15

The world's human population is currently growing at a very high rate, but limited resources will eventually limit this growth, most likely at a population size between 7 and 11 billion.

## LIFE HISTORY TRADE-OFFS AND A MINI-FOUNTAIN OF YOUTH

### What is the relationship between reproduction and longevity?

It's well documented that there is a trade-off between reproduction and longevity. In a wide variety of animals, decreasing reproductive effort has been shown to increase longevity.

**Q: DOES THIS HAVE ANY PRACTICAL APPLICATIONS?** Yes: spay or neuter your pet to increase its life span!

**Q: DOES THIS WORK?** Yes! Data from more than 1,000 cat autopsies showed that non-sterilized females lived a mean of 3.0 years, whereas sterilized females lived significantly longer, with a mean life span of 8.2 years. The results for males were similar. (Perhaps the most dramatic example of the reproduction-longevity trade-off is that of the marsupial mouse, *Antechinus stuartii,* in which castration led to a doubling or even tripling of the usual life span. Similarly, following castration, Pacific salmon lived up to 8 years, double their usual life span.)

**NOT THAT YOU ASKED . . .** In a study in the early 1900s of men institutionalized for mental retardation, those who were castrated lived 13 years longer (69.3 years vs. 55.7 years) than non-castrated men at the same institution, matched for age and intelligence.

**Q: WHAT IS RESPONSIBLE FOR THE TRADE-OFF BETWEEN REPRODUCTION AND LONGEVITY?** If it were a question of sterilized individuals living longer simply because they don't have to expend energy on reproduction, perhaps similar life span increases could be achieved by simply giving animals access to more energy. In practice, however, this doesn't work (and usually *decreases* life span). Rather, it seems that a significant part of the "cost" of reproduction, at least in mammals, is an increased incidence of cancer, caused by the higher levels of circulating hormones in "reproductively ready" (that is, fertile and receptive) animals.

**POSSIBILITIES TO THINK ABOUT.** The link between maintaining reproductive readiness and reduced longevity has led some researchers to contemplate the design of birth control pills that might have the additional effect of increasing longevity. Stay tuned.

# Big Ideas in ...
# Population Ecology

## 14·1–14·6 Population ecology is the study of how populations interact with their environments.

- Population ecologists study the interaction between populations of organisms and their environments, particularly patterns of growth and how they are influenced by other species and by environmental factors.

- Populations tend to grow exponentially until limited resources cause the growth to slow to logistic rates.

- A population's growth can be reduced both by density-dependent factors related to crowding and by density-independent factors, such as floods, fires, and earthquakes.

- Difficulties such as estimating population size and carrying capacity complicate the management of populations for maximum sustainable yield.

## 14·7–14·9 A life history is like a species summary.

- An organism's pattern of investment in growth, reproduction, and survival is described by its life history.

- Life tables and survivorship curves summarize the survival and reproduction patterns of the individuals of a population, varying greatly across species.

- Because constraints limit evolution, life histories are characterized by trade-offs between investments in growth, reproduction, and survival.

## 14·10–14·12 Ecology influences the evolution of aging in a population.

- Natural selection cannot weed out harmful alleles that do not diminish an individual's relative reproductive success. Consequently, these alleles accumulate in the genomes of organisms of nearly all species. This leads to the multiple physiological breakdowns that we see as aging.

- The rate of aging and pattern of mortality for a species are determined by the hazard factor of its environment.

- By increasing the strength of natural selection later in life, it is possible to increase the mean and maximum longevity of the individuals in a population.

## 14·13–14·15 The human population is growing rapidly.

- Age pyramids show the number of individuals within any age group in a population. They allow us to estimate birth and death rates over multi-year periods.

- The demographic transition tends to occur as countries industrialize. It is characterized by an initial reduction in the death rate, followed later by a reduction in the birth rate.

- The world population is currently growing at a very high rate, but limited resources will eventually limit this growth, most likely at a population size between 7 and 11 billion.

See the To-Go Guide (packaged with every new copy of this textbook) for an illustrated summary of the chapter.

# Check Your Knowledge

## GRAPHIC CONTENT

*Thinking critically about visual displays of data.*

**1.** What is the purpose of this graph?

**2.** Why does the light green line become flat after initially increasing?

**3.** When the population reaches carrying capacity it appears to persist indefinitely. How is that possible? Do organisms stop dying?

**4.** Are data presented in this graph? If so, what are they? If not, what is presented?

**5.** What data would improve this graph? Why? How might they be incorporated into the graph?

**6.** Create an alternative graph showing the relationship between population size (on the *x*-axis) and growth rate (on the *y*-axis). Label carrying capacity and the point of maximum sustainable yield on your graph.

**7.** Describe two assumptions about population growth that are implicit in this figure?

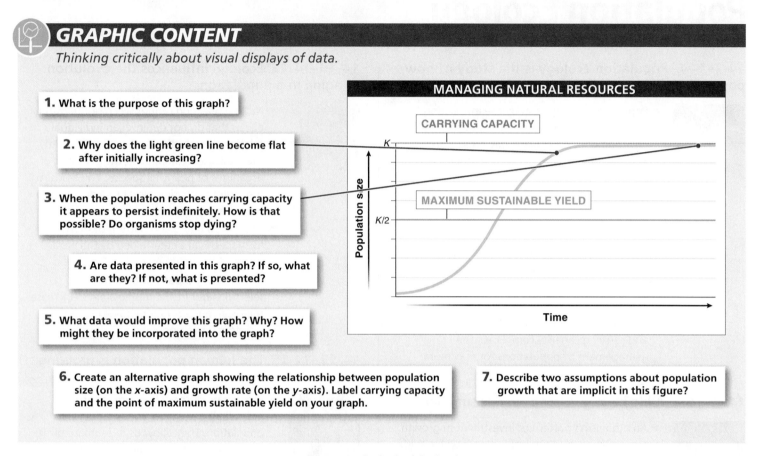

MANAGING NATURAL RESOURCES

CARRYING CAPACITY

MAXIMUM SUSTAINABLE YIELD

Population size

*K*

*K*/2

Time

See Answers at the back of the book.

## Key Terms in Population Ecology

aging, p. 564
carrying capacity, *K*, p. 552
demographic transition, p. 573

density-dependent factors, p. 552
density-independent factors, p. 553
ecological footprint, p. 575

ecology, p. 548
exponential growth, p. 551
growth rate, p. 551
hazard factor, p. 567
life history, p. 560

life table, p. 561
logistic growth, p. 553
maximum sustainable yield, p. 557
population density, p. 552

population ecology, p. 549
reproductive investment, p. 560
reproductive output, p. 566
survivorship curve, p. 561

# Multiple-Choice Questions

**1.** Ecology is best defined as the study of:

   a) the relationships between all living organisms and their environments.

   b) the relationships between parasites and their hosts.

   c) aquatic organisms.

   d) interactions between predator and prey populations.

   e) the preservation of habitats.

**2.** On average, which leaves more offspring that survive to become adults and reproduce: a pair of elephants or a pair of rabbits?

   a) The pair of elephants, because elephants live much longer and have more breeding seasons.

   b) The pair of rabbits, because they have so many more offspring per breeding season than do elephants.

   c) The pair of rabbits, because they reach sexual maturity more rapidly.

   d) The pair of elephants, because any individual elephant born is more likely to survive to become an adult and reproduce.

   e) If both populations are stable, the pair of elephants and the pair of rabbits will leave the same number of offspring that survive to become adults and reproduce.

**3.** In a population exhibiting logistic growth, the rate of population growth is greatest when $N$ is:

   a) 0.5 $K$.

   b) 0.

   c) above the carrying capacity.

   d) $K$.

   e) All of the above are correct; the rate of population growth is constant in logistic growth.

**4.** In a population, as $N$ approaches $K$, the logistic growth equation predicts that:

   a) the carrying capacity of the environment will increase.

   b) the growth rate will approach zero.

   c) the population will become monophyletic.

   d) the population size will increase exponentially.

   e) the growth rate will not change.

**5.** The number of individuals that can be supported in a given habitat is:

   a) the innate capacity for increase.

   b) one-half of the maximum sustainable yield.

   c) a density-independent effect.

   d) generally increasing over time.

   e) the carrying capacity.

**6.** Which of the following statements about maximum sustainable yield is incorrect?

   a) The maximum sustainable yield for a population is the population growth rate at $K/2$.

   b) The maximum sustainable yield for a population can be difficult to determine, because it is not always possible to accurately measure $N$.

   c) The maximum sustainable yield for a population can be difficult to determine, because it is not always possible to accurately measure $K$.

   d) The maximum sustainable yield for a population is a useful management guideline for harvesting plant products such as timber, but is not helpful for managing animal populations.

   e) The concept of maximum sustainable yield can generate useful information for fighting the growth of pest species.

**7.** Dr. David Reznick has studied life history evolution in guppies that live in streams in Trinidad. Guppies are found in two different types of habitat: sites where predation is high, and sites where predation is low. Which of the following life history characteristics would you expect to evolve in a guppy population living in a high-predation site?

a) bright colors and courtship displays

b) increased egg number

c) a female-biased sex ratio

d) increased egg size

e) delayed sexual maturation

 100 HARD  36  0 EASY

**8.** The death rate of organisms in a population exhibiting a type III survivorship curve is:

a) unrelated to age.

b) usually correlated with density-independent causes.

c) higher in the post-reproductive than in the pre-reproductive years.

d) lower after the organisms survive beyond the earliest age groups.

e) more or less constant throughout their lives.

100 HARD  48  0 EASY

**9.** Which of the following is a major trade-off in life histories?

a) size of offspring for amount of parental investment

b) size of offspring for number of reproductive events

c) growth for reproduction

d) size for life span

e) number of reproductive events for number of offspring per reproductive event

100 HARD  69  0 EASY

**10.** Natural selection:

a) does not influence aging, because aging is determined by an individual's environment.

b) cannot reduce the frequency of alleles that cause mortality among individuals who have not yet reached maturity.

c) cannot weed out from a population any alleles that do not reduce an individual's relative reproductive success, even if these alleles increase an individual's risk of dying.

d) can influence aging but not longevity.

e) leads to an increase in the frequency of any illness-inducing alleles that have their effect when an organism can reproduce.

100 HARD  39  0 EASY

**11.** Which of the following statements about the hazard factor of a population is incorrect?

a) It is a measure of organisms' risk of death from external sources.

b) It is lower for a population of porcupines than for a population of guinea pigs.

c) It is a measure of the ratio of mortality risk from external (environmental) causes relative to internal (genetic) causes.

d) It is responsible for the rate of aging among individuals in the population.

e) It is a measure of how quickly the individuals in a population age.

 100 HARD  57  0 EASY

**12.** Life extension:

a) is not possible, because natural selection cannot weed out disease-causing alleles that have an effect only after the age when reproduction is no longer possible.

b) can be achieved in species by selectively breeding individuals with the earliest age of maturity.

c) works in insects but could not work in humans.

d) has been achieved using laboratory selection for delayed reproduction.

e) None of the above are true.

 100 HARD  33  0 EASY

**13.** A population pyramid:

a) represents the number of individuals in various age groups in a population.

b) can be constructed from the data in a life table.

c) directly predicts future age distributions of the population.

d) shows the current birth and death rates of a population.

e) predicts survival and mortality rates for an individual at a given age.

100 HARD  25  0 EASY

**14.** A primary difference between the age pyramids of industrialized and developing countries is that:

a) mean longevity is significantly greater in developing countries.

b) in developing countries, much larger proportions of the population are in the youngest age groups.

c) developing countries show a characteristic "bulge" that indicates a baby boom.

d) in developing countries, females live significantly longer than males, whereas in industrialized countries the reverse is true.

e) developing countries have significantly more individuals than industrialized countries.

100 HARD  20  0 EASY

**15.** Which statement best describes expectations for the world's human population by the year 2015?

a) It will exceed 10 billion if the current rate of increase continues.

b) Negative growth in the United States and Europe will counterbalance positive growth in the developing countries.

c) It will exceed 7.5 billion if the current rate of increase continues.

d) It will drop below 6 billion if the current rate of decrease continues.

e) The problem is too complex to make any predictions.

100  HARD

33

0  EASY

# Short-Answer Questions

**1.** Explain the differences between density-dependent and density-independent factors that can limit a population's growth. Give an example of each.

**2.** Describe the characteristics of an organism with a type I survivorship curve. How would the life history change if this organism had a type III survivorship curve?

**3.** Explain why some species of organisms live very long lives and others do not. How could the life span be extended?

See Answers at the back of the book.
For additional study questions, go to www.prep-u.com.

Organisms and their environments

# 15

# Ecosystems and Communities

583

# 15·1–15·2 Ecosystems have living and non-living components.

Lesser flamingoes (Phoenicopterus minor) *feeding in Lake Nakura National Park, Kenya.*

## 15·1 What are ecosystems?

Picture a lush garden: some greenery, a bit of rotting wood, and abundant wildlife. Grazing animals abound, while predators feed on other animals and their eggs. Parasites are poised, looking for hosts, and, just below the surface, scavengers find meals among the organic detritus. It would seem to be the quintessential **ecosystem,** a community of biological organisms plus the non-living components with which the organisms interact. But now imagine that the entire scene gets up and walks away! The "camera" in your mind pulls back to reveal that the scene is playing out on the back of a beetle no more than two inches long (**FIGURE 15-1**).

The host of this mini-ecosystem is a beetle from New Guinea called the large weevil. The weevil is camouflaged from its predators by lichens—which consist of fungi and photosynthetic algae living together—while the lichens are given a safe surface on which to live. And the garden of lichens supports a wide range of other organisms, from tiny mites to a variety of other microscopic invertebrates, some free-living and others parasitic.

Not all ecosystems are the obvious assemblages of plants and animals that we usually picture—ponds, deserts, or tropical forests. A similar scene can just as easily be found in your large intestine, where several hundred bacterial species flourish. These ecosystems are contained within ecosystems that are contained within ecosystems. The

*Ecosystems are found not just in obvious places like ponds and tropical rain forests, but also in some unexpected places, like on the back of a beetle!*

**FIGURE 15-1 "Be it ever so humble . . ."** A small-scale ecosystem can exist on the large weevil of New Guinea.

scale can vary tremendously. The closer you look, the more you find.

What is important is that the two essential elements of an ecosystem are present: the biotic environment and the physical (abiotic) environment (**FIGURE 15-2**).

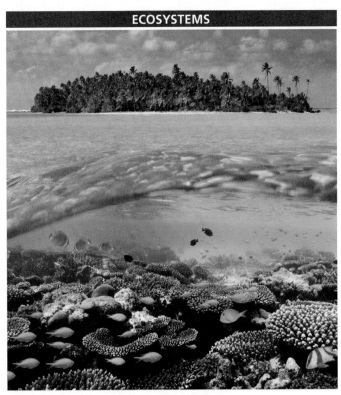

All ecosystems share two essential features:

**BIOTIC ENVIRONMENT**
- The living organisms within an area
- Often referred to as a community

**PHYSICAL (ABIOTIC) ENVIRONMENT**
- The chemical resources and physical conditions within an area
- Often referred to as the organisms' habitat

**FIGURE 15-2 What makes up an ecosystem?**

1. The **biotic** environment consists of all the living organisms within an area and is often referred to as a **community.**

2. The physical, or **abiotic,** environment, often referred to as the organisms' **habitat,** consists of:

   - the *chemical resources* of the soil, water, and air, such as carbon, nitrogen, and phosphorus, and

   - the *physical conditions,* such as the temperature, salinity (salt level), moisture, humidity, and energy sources.

Biologists view communities of organisms and their habitats as "systems" in much the same way engineers might, hence the term eco*system.* Biologists monitor the inputs and outputs of the system, tracing the flow of energy and various molecules as they are captured, transformed, and utilized by organisms and later exit the system or are recycled. They also study how the activities of one species affect the other species in the community—whether the species have a conflicting relationship, such as predator and prey, or a complementary relationship, such as flowering plants and their pollinators. On a small scale, such as the back of a beetle, making these measurements is easy. But there are also some well-studied giant ecosystems, such as the Hubbard Brook Experimental Forest in New Hampshire, which covers 7,600 acres. Regardless of size, the same principles of ecosystem study apply: observe and analyze organisms and their environments, while monitoring everything that goes into and comes out of the system.

Why should researchers bother with such a methodical—and tedious—analysis of ecosystems? This straightforward approach has led to numerous valuable discoveries, from understanding how clear-cutting of forests dramatically reduces soil quality, to understanding the link between the use of fossil fuels and the creation of destructive acid rain. As we see in this chapter and the next, humans, perhaps more than any other species in history, are significantly affecting most of the ecosystems on earth. And, in addition to improving our understanding of environmental issues, ecosystem studies can also tell us about the microbial ecosystems living inside humans and thus lead to advances in public health.

**TAKE-HOME MESSAGE 15·1**

An ecosystem is a community of biological organisms plus the non-living components in the environment with which the organisms interact. Ecosystems are found not just in obvious places such as ponds, deserts, and tropical rain forests but also in some unexpected places, such as the digestive tracts of organisms or the shell of a beetle.

# 15·2 A variety of biomes occur around the world, each determined by temperature and rainfall.

Dense vegetation surrounds you. Above you is a canopy of evergreen trees, 30–40 meters tall. Climbing vines hang from virtually all the trees. And dozens or even hundreds of species of insects are flourishing around you. Even if you've never been there, the description of a tropical rain forest is easy to recognize. But where exactly would you be if you were in this scenario? It could be South America or Africa or Southeast Asia. The species are different, but the general pattern of life forms is the same. The same holds for arctic tundra: whether you were in northern Asia or North America, the view would be similar. These are examples of the largest of the earth's ecosystems, the **biomes.**

Biomes cover huge geographic areas of land or water—the deserts that stretch almost all the way across the northern part of Africa, for example. Terrestrial (land) biomes are defined and usually described by the predominant types of plant life in the area. But looking at a map of the world's terrestrial biomes, it is clear that they are mostly determined by the weather. Specifically, when defining terrestrial biomes, we ask four questions about the weather:

## TERRESTRIAL BIOMES

Biomes are large ecosystems that cover huge geographic areas. The nine chief terrestrial biomes are determined by the temperature and amounts of precipitation in conjunction with the magnitude of seasonal variation in these factors.

FIGURE 15-3 Terrestrial ecosystem diversity.

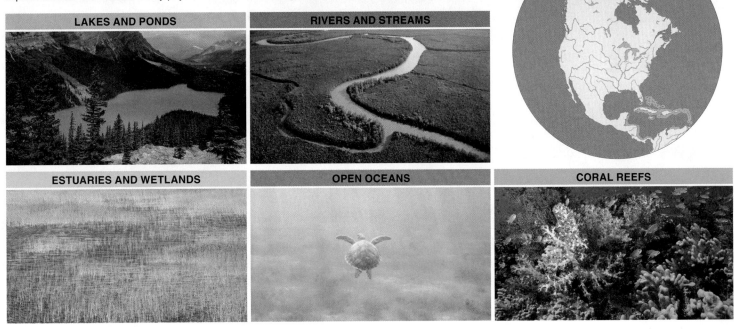

## AQUATIC BIOMES

Aquatic biomes are determined by physical features, including salinity, water movement, and depth.

**LAKES AND PONDS**

**RIVERS AND STREAMS**

**ESTUARIES AND WETLANDS**

**OPEN OCEANS**

**CORAL REEFS**

**FIGURE 15-4 Aquatic ecosystem diversity.**

1. What is the average temperature?

2. What is the average rainfall (or other precipitation)?

3. Is the temperature relatively constant or does it vary seasonally?

4. Is the rainfall relatively constant or does it vary seasonally?

By knowing the answers to these questions, we can predict, with great accuracy, the type of biome. This is because temperature and precipitation dictate the **primary productivity** levels, or the amount of organic matter produced, mainly through photosynthesis. And the numbers and types of **primary producers**—the organisms responsible for the primary productivity, such as grasses, trees, and agricultural crops—are, in turn, the chief determinants of the amount and diversity of other life in the region.

For example, where it is always moist and the temperature does not vary across the seasons, **tropical rain forests** develop. And where it is hot but with strong seasonality that brings a "wet" season and a "dry" season, **savannas** or **tropical seasonal forests** tend to develop. At the other end of the spectrum, in dry areas with a hot season and a cold season, **temperate grasslands** or **deserts** develop. **FIGURE 15-3** shows examples of the nine chief terrestrial biomes; all are determined, in large part, by the precipitation and temperature levels.

Aquatic biomes are defined a bit differently, usually based on physical features such as salinity, water movement, and depth. Chief among these environments are (1) lakes and ponds, with non-flowing fresh water; (2) rivers and streams, with flowing fresh water; (3) **estuaries** and wetlands, where salt water and fresh water mix in a shallow region characterized by exceptionally high productivity; (4) open oceans, with deep salt water; and (5) coral reefs, highly diverse and productive regions in shallow oceans (**FIGURE 15-4**).

If the terrestrial biomes of the world are determined by the temperature and rainfall amounts and seasonality, what determines those features? In other words, what makes the weather? We investigate next how the geography and landscape of the planet—the shape of the earth and its orientation to the sun, and patterns of ocean circulation—cause the specific patterns of weather that create the different climate zones and the biomes characteristic of each. Then we'll see how energy and chemicals are made available for life to flourish in these biomes.

### TAKE-HOME MESSAGE 15·2

Biomes are the major ecological communities of earth, characterized mostly by the vegetation present. Different biomes result from differences in temperature and precipitation, and the extent to which both vary from season to season.

# 15·3–15·5 Interacting physical forces create weather.

*A "weather sampler": rainbows, blue skies, and storm clouds over Utah.*

# 15·3
## Global air circulation patterns create deserts and rain forests.

> " There was a hot desert wind blowing that night. It was one of those hot dry Santa Anas that come down through the mountain passes and curl your hair and make your nerves jump and your skin itch. On a night like that every booze party ends in a fight. Meek little wives feel the edge of a carving knife and study their husbands' necks. Anything can happen. "
>
> —RAYMOND CHANDLER, "Red Wind," 1938

Temperature and rainfall, that's all. The type of terrestrial biome depends on little else besides these two aspects of the weather. But what determines the temperature and rainfall in a particular part of the world? Differences in both ultimately result from one simple fact: the earth is round. As we'll see, the earth's curvature influences temperature, and rainfall patterns are an inevitable consequence of the variation in temperature across the globe.

Wherever you are, begin walking toward the equator. As you get closer, does it get hotter or colder? Hotter, of course. Nearly everyone is aware of this universal trend (and the reverse as well: it gets colder as you move away from the equator and approach the North or South Pole). What is responsible for the increased warmth at the equator? This is most easily explained with a drawing (**FIGURE 15-5**).

### THE DISTRIBUTION OF SOLAR ENERGY

The sun shines most directly on earth's equator, leading to warmer temperatures at lower latitudes.

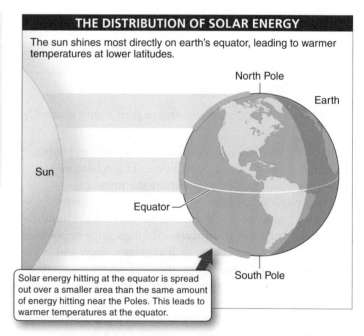

Solar energy hitting at the equator is spread out over a smaller area than the same amount of energy hitting near the Poles. This leads to warmer temperatures at the equator.

**FIGURE 15-5 The curvature of the earth.** The sun shines most directly on the equator. Closer to the Poles, the curvature of the earth causes sunlight to be spread out over a larger area, reducing its heating effect in any one spot.

The sun shines most directly on the equator. At the equator, solar energy hits the earth and spreads out over a relatively small area. Away from the equator, the earth curves toward the North and South poles. Because of this curvature, the

## FORMATION OF RAIN

**1** AIR IS HEATED AND RISES
When solar heat hits the earth, it warms the air at that point. The heated air rises.

**2** RISING AIR COOLS
As hot air rises, getting farther from the warm earth, it cools.

**3** COOLING AIR LOSES MOISTURE
Because cold air holds less moisture than warm air, clouds form and rain falls.

**FIGURE 15-6 Rainmaking.**

same amount of solar energy hitting the earth at the Poles is spread out over a much larger area and also travels a greater distance through the atmosphere, which absorbs or reflects much of the heat. With the energy dispersed over a large area, there is less warmth at any one point on the earth's surface. It's similar to the fact that at noon, the sun's rays hit the earth at a more direct angle than they do later in the day. That is why the sun provides less warmth late or early in the day. It is also why the risk of sunburn and skin cancer is greatest around noon and the nearer you are to the equator.

Global patterns of rainfall can be predicted just as easily, by taking into account that warm air holds more moisture than cold air. We'll start at the equator again. This is where the greatest warming power of the sun hits the earth, and some of that energy radiates back, warming the air. This starts a three-step process that ends with rain. (1) Hot air rises. (2) As it rises high into the atmosphere, it cools. And (3) because cool air holds less moisture, as the air rises,

clouds form, and the moisture that can no longer be held in the air falls as rain (**FIGURE 15-6**). The equator is hot, but it is also very wet.

The high, cold air in the atmosphere over the equator expands outward, to about 30° north and south. Here, the cold air, which is heavier than warm air, begins to fall downward toward the earth. As it falls, the air is warmed by heat radiated off the earth's surface. And as the air gets warmer, it can hold more and more moisture, rather than releasing it as rain. That is, moisture can be held in the air instead of being released as rain. In fact, the falling and rapidly heating air can hold so much moisture that it sometimes sucks up moisture from the land itself. For this reason, at around 30° north and south, around the world, there is very little rainfall, the ground is very dry, and deserts form (**FIGURE 15-7**). The Atacama Desert of South America,

**Q**

Nearly all of the world's deserts occur a third of the way from the equator toward the Poles, at 30° latitude. Is this just a coincidence?

## THE FORMATION OF DESERTS

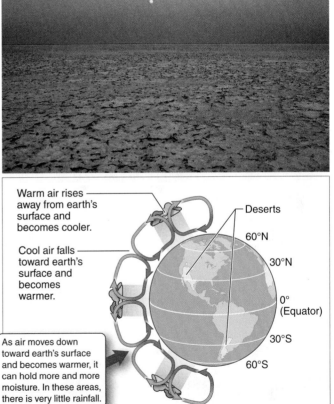

Warm air rises away from earth's surface and becomes cooler.

Cool air falls toward earth's surface and becomes warmer.

As air moves down toward earth's surface and becomes warmer, it can hold more and more moisture. In these areas, there is very little rainfall.

Deserts
60°N
30°N
0° (Equator)
30°S
60°S

**FIGURE 15-7 Desert formation.** Circulating masses of air, caused by solar energy hitting different parts of the earth at different angles, determine rainfall patterns throughout the world. The photo shows a desert in Botswana.

which lies at approximately 30° south of the equator, is an extreme example. In some parts of this large desert, *no rainfall* has ever been recorded. Other great deserts of the world are also at 30° latitude, including the Sahara, Kalahari, Mojave, and Australian deserts (although some deserts do occur at other latitudes). These areas stand in stark contrast to the equator, where it is not uncommon for an area to receive more than three meters (!) of rain in a year.

As the air falls near 30° north and south and hits the earth, it spreads equally toward the equator and toward the Poles, gradually getting warmer, until, at around 60° latitude, it begins to rise because of its accumulated heat. Again, the rising air loses its moisture as rainfall. And so at 60° latitude, two-thirds of the way toward the Poles from the equator, it's not particularly warm but there is a great deal of rain. Not

surprisingly, at these latitudes lie huge temperate forests with extensive plant growth. And finally, around the Poles, air masses again descend. And as they do, because they can hold more moisture, very little rain falls. The Poles are cold, but with little precipitation—they resemble frozen desert.

# 15·4   Local topography influences the weather.

Why is it so windy on the sidewalk around tall buildings? And why does it rain all the time on one side of some mountain ranges, while deserts form right on the other side? And is it actually warmer in the city than in the country? The answers reflect how the physical features of land, its **topography**—including features created by humans—can have dramatic but predictable effects on temperature, precipitation, and wind.

*High altitudes have lower temperatures.* With increasing elevation, the air pressure drops—this is because the weight

of the atmosphere gets less and less as altitude increases. And when pressure is lower, the temperature drops. For each 1,000 meters above sea level, the temperature drops by about 6° C. This is why the changes in weather and vegetation that you see while climbing a mountain are similar to those you would see as you moved farther and farther away from the equator—it gets colder in both cases.

*Rain shadows create deserts.* Air rises to get over the top of a mountain. But because the air cools as it rises, it can't hold much moisture. So clouds form and rain falls.

**FORMATION OF RAIN SHADOWS**

Mountain range

Rain shadow desert

**1** Wind blows from oceans toward land, rising when it hits mountains.

**2** Rising air cools and holds less moisture, leading to cloud formation and rain.

**3** Air passes over mountain top and falls, becoming warmer and increasing the moisture it can hold. This reduces rainfall and results in dry areas.

**FIGURE 15-8 The rain shadow effect.**

Asphalt, cement, and building tops absorb heat, raising the temperature.

Tall buildings force wind downward.

**FIGURE 15-9 Human engineering.** When humans alter the land, this can have unintended consequences, such as alterations in temperature and in wind speed and direction.

As the air eventually passes over the top of the mountain, it falls and warms. And because warm air holds onto moisture, there is rarely any rain on the back side of the mountain. In fact, often the air will pull moisture from the ground, intensifying the already dry conditions, creating **rain shadow** deserts (**FIGURE 15-8**). Along the west coast of the United States, the Sierra Nevada and the Cascade mountain range are responsible for the Mojave Desert in California and the Great Sandy Desert in Oregon.

 **Is it warmer or cooler in urban areas relative to nearby rural areas?**

*Asphalt, cement, and tops of buildings absorb heat, raising the temperature.* Modern landscapes also influence the weather, creating "urban heat islands." When energy from the sun hits concrete or pavement or the dark roof of a building, some of it is reflected, heating the air around it, and most of the rest is absorbed by these man-made surfaces and held until night. In the darkness, the surfaces lose heat to the sky, which is colder by comparison, through radiation. In contrast, when sunlight hits trees or other plant life, the solar energy evaporates water in the leaves. The ground surface doesn't get much hotter, and neither does the air. It's not surprising, then, that cities tend to be 1° to 6° C warmer than surrounding rural areas. And not only is it hotter in cities, but the rising warm air also alters rainfall patterns both in and around cities.

A variety of steps are being taken to create "greener" cities that absorb and release less heat from the sun, and use less energy for air conditioning. These methods include the creation of buildings with lighter-colored rooftops, the planting of trees around buildings and along roads, and the development of rooftop gardens rich with vegetation.

*Tall buildings channel wind downward.* Tall buildings are responsible for the perpetual winds you feel when walking on a city sidewalk. Here's why. Winds blow more strongly when they are higher above the earth, freed from the earth's frictional drag (from plants, dirt, rocks, and water) that slows wind as it gets near the surface. When these strong, elevated winds suddenly encounter tall buildings, they are deflected. Some of the wind goes up and over the building, but much of it is pushed downward, reaching double or even triple its initial speed by the time it reaches street level (**FIGURE 15-9**).

**Why is it so windy on streets with tall buildings?**

---

### TAKE-HOME MESSAGE 15·4

Local features of topography influence the weather. With higher altitude, the temperature drops. On the windward side of mountains, rainfall is high; on the back side, descending air reduces rainfall, causing rain shadow deserts. Urban development increases the absorption of solar energy, leading to higher temperatures, and creates wind near the bottom of tall buildings.

# 15·5 Ocean currents affect the weather.

Weather is not just affected by circulating air masses. It is also affected by the oceans. The ocean is vast and deep, warmed almost exclusively by the sun. And water is continuously moving and mixing, due to a combination of forces, including wind, the earth's rotation, the gravitational pull of the moon, temperature, and salt concentration. These forces create several large, circular patterns of flow in the world's oceans, as illustrated in FIGURE 15-10.

Much of water's effect on weather stems from its great capacity to absorb and hold heat—a 10,000 times greater capacity than air. This means that temperatures fluctuate much more in air than in water. At the beach, the air temperature can go from mild to very hot and back to mild over the course of a day, while the heat of the sun will have a tiny, almost negligible effect on the water temperature. Thus, during summers, much of the heat of the sun is absorbed and held by the ocean water in coastal towns, rather than causing hot air temperatures. Conversely, during cold winter months, heat from the water can reduce the coldness of the air.

One of the strongest ocean currents is the Gulf Stream. It travels north through the Caribbean, carrying warm water up

**Q** Why do beach communities have milder weather than inland communities (not as hot in summer, not as cold in winter)?

the east coast of the United States and then across the Atlantic Ocean toward Europe. Because the current begins in a warm part of the globe, close to the equator, it is still warm when it reaches the east coast of the United States and then Europe. The warm water also warms the climate in these areas. In fact, if it weren't for the Gulf Stream, much of Europe—given its high latitudes—would have a climate more like Canada's. Because all ocean currents in the northern hemisphere rotate in a clockwise direction, water reaching the beaches of California, unlike water reaching east coast beaches, has just come from the north, near Alaska, where it gets very cold.

Every two to seven years, a dramatic climate change driven by ocean currents, called **El Niño,** causes a sustained surface temperature change in the central Pacific Ocean (FIGURE 15-11). It is blamed for flooding, droughts, famine, and a variety of other extreme climate disruptions. We can more easily understand El Niño if we contrast it with the more common climate pattern.

*In a typical year:*

1. Steady winds blow across the Pacific Ocean from South America toward Southeast Asia and Australia.

2. These winds push the warm surface water away from the coast of South America and raise the sea level in Southeast Asia and Australia by about half a meter. The warmth of the water heats the air above it, causing the air to rise and produce great tropical rainstorms. (To experience this effect, fill your bathtub with hot water and feel the room get warmer and moister.)

3. Off the coast of South America, an upwelling of colder water from deep in the ocean replaces the water that is pushed west. The cold upwelling cools the air above it, causing extremely dry weather.

4. The upwelling of deep water brings up rich nutrients from below, enabling plankton to flourish and nourishing huge populations of fish.

*In an El Niño year:*

1. The usual South America–to–Southeast Asia winds ease just a bit. No one knows why this happens.

2. Without wind pushing it, the warm water piling up in the west flows back toward South America. This water warms the air, which rises and produces huge amounts

## OCEAN CIRCULATION PATTERNS

There are several large, circular patterns of flowing water in the oceans due to a combination of forces including wind, the earth's rotation, the gravitational pull of the moon, temperature, and salt concentration.

← Warm ocean currents
← Cold ocean currents

Beaches on the east coast of the United States have warmer water than west coast beaches at the same latitudes.

North Pacific Ocean

North Atlantic Ocean

Pacific Ocean

South Pacific Ocean

South Atlantic Ocean

Indian Ocean

**FIGURE 15-10  Ocean currents influence weather.**

El Niño occurs every two to seven years and is blamed for flooding, droughts, famine, and a variety of other extreme climate disruptions.

## USUAL CONDITIONS

**1** Steady winds blow westward across the Pacific Ocean from South America toward Southeast Asia.

**2** These winds push warm surface water away from the coast of South America, heating the air above it, which causes air to rise and produce tropical rainstorms.

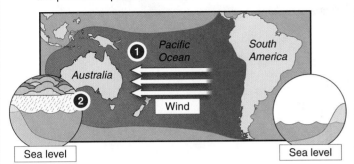

**3** Off the coast of South America, colder water upwells from the ocean depths. The cold water cools the air above, causing extremely dry weather.

**4** Water from the depths of the ocean brings up rich nutrients, enabling plankton to flourish and feed the huge populations of fish.

## EL NIÑO CONDITIONS

**1** The usual South America–to–Southeast Asia winds ease just a bit.

**2** Without the push of the wind, warm water flows back toward South America. The cooled water around Australia and the Philippines cools the air and causes dry weather that can lead to droughts.

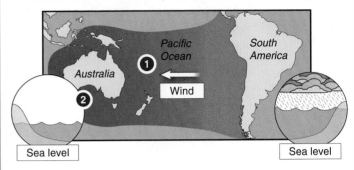

**3** Without the warm surface water being blown west, colder water from the ocean depths cannot upwell. The warm surface water warms the air, resulting in rainfall.

**4** With no upwelling of cold water and nutrients, plankton levels drop dramatically and fish stocks are wiped out.

💡 *Changes during an El Niño event start a chain reaction of unusual weather around the globe: increased rainfall in the U.S. Midwest, storms in coastal California, and warmer weather in South Africa and Japan.*

**FIGURE 15-11 Domino effect.** A reduction in the usual east-to-west ocean breeze can cause a cascade of disastrous weather.

of rainfall on the coast of South America where it is normally dry, often causing flooding.

3. Without the warm surface water being blown west, the normal upwelling of cool, deep water with high nutrient content cannot occur. The plankton levels drop to a fraction of their normal levels, and the typically rich fish stocks are wiped out.

4. Meanwhile, in the absence of warm water from the east, Indonesia, the Philippines, and Australia experience drastic climate changes. The cooler water cools the air above and causes unusually dry weather, which can lead to droughts and, in extreme years, famines.

These predictable changes during an El Niño event can start a global pattern of unusual weather. Although the

mechanisms aren't clear, in El Niño years, more rain falls in the midwestern United States and storms form off the coast of California, while South Africa, Japan, and Canada enjoy warmer weather than usual. The counterpart of the El Niño phenomenon is called **La Niña** and is characterized by the reverse effects of El Niño. During La Niña periods, ocean surface temperatures are lower than usual and the climate effects are approximately opposite to El Niño effects.

### TAKE-HOME MESSAGE 15·5

Oceans have global circulation patterns. Disruptions in these patterns occur every few years and can cause extreme climate disruptions around the world.

*The veiled chameleon (Chamaeleo calyptratus) catching a cricket.*

# 15·6

## Energy flows from producers to consumers.

All life on earth is made possible because energy flows perpetually from the sun to the earth. Looking at an ecosystem such as a desert savanna, for example—trying to understand how all of the species, from grasses and trees to birds and mammals and worms, interact with one another, and what role each plays within the system—can seem overwhelming. But if we focus on just one aspect of the ecosystem—the pathways energy takes as it flows through the system—a simple and logical order becomes clear.

The sun is where our pathway of energy flow begins. Most of the energy is absorbed or reflected by the earth's atmosphere or surface, but about 1% of it is intercepted and converted to chemical energy through photosynthesis. That intercepted energy is then transformed again and again by living organisms, making about four stops as it passes through an ecosystem. Let's examine what happens at each of the stops, known as **trophic levels** (FIGURE 15-12).

**First stop: primary producers.** When it comes to energy flow, all the species in an ecosystem can be placed in one of two groups: producers and consumers. Plants (along with some algae and bacteria) are, as we noted earlier, the *primary producers.* They convert the sun's light energy into chemical energy through photosynthesis, as discussed in Chapter 4. We use another word to describe that chemical energy: food.

**Second stop: primary consumers—the herbivores.** Cattle grazing in a field, gazelle browsing on herbs, insects devouring the leaves of a crop plant—these are the **primary consumers** in an ecosystem, the animals that eat plants. Plant material such as cellulose can be difficult to digest. Consequently, most **herbivores,** the animals that eat plants, need a little help in digesting their food. Primary consumers, from termites to cattle, often have bacteria living in their digestive system. These microorganisms benefit the organism in which they live by breaking down the cellulose, enabling the herbivore to harness the energy held in the chemical bonds of the plants' cells.

**Third stop: secondary consumers—the carnivores.** Energy originating from the sun is converted into chemical energy within a plant's tissues. The herbivore that eats the plant breaks down the chemical bonds, releasing the energy. This energy fuels the herbivore's growth, reproduction, and movement, but the energy doesn't remain in the herbivore forever. **Carnivores,** such as cats, spiders, and frogs, are animals that feed on herbivores. They are also known as **secondary consumers.** As they eat their prey, some of the energy stored in the chemical bonds of the carbohydrate, protein, and lipid molecules is again captured and harnessed for their own movement, reproduction, and growth.

**Fourth stop: tertiary consumers—the "top" carnivores.** In some ecosystems, energy makes yet another stop: the **tertiary consumers,** or "top carnivores." These are the "animals that eat the animals that eat the animals that eat the plants." They are several steps removed from the initial capture of solar energy by a plant, but the general process is the same. A top carnivore, such as a tiger, eagle, or great white shark, consumes other carnivores, breaking down their tissues and releasing energy stored in the chemical bonds of the cells. As in each of the previous steps, the top carnivores harness this energy for their own physiological needs.

This path from primary producers to tertiary consumers is called a **food chain.** We see later in this chapter why a food chain almost never extends to a fifth stop.

The food chain pathway from photosynthetic producers through the various levels of animals is a slight oversimplification. In actuality, food chains are better thought of as **food webs,** because many organisms are **omnivores** and can occupy more than one position in the chain (see Figure 15-12). When you eat a simple meal of rice and chicken, after all, you're simultaneously a carnivore and a herbivore. On average, about 30% to 35% of the human diet comes from animal products and the remaining 70% to 65% from plant products. Many other animals, from bears to cockroaches, also have diets that involve harvesting energy from multiple stops in the food chain.

In every ecosystem, as energy is transformed through the steps of a food chain, organic material accumulates in the form of animal waste and dead plant and animal matter. **Decomposers,** usually bacteria or fungi, and **detritivores,** including scavengers such as vultures, worms, and a variety of arthropods, break down the organic material, harvesting energy still stored in the chemical bonds (**FIGURE 15-13**). Decomposers are distinguished from detritivores because the decomposers are able to break down a much larger range of

---

## ENERGY FLOW WITHIN AN ECOSYSTEM

Energy from the sun is intercepted and converted into chemical energy, which passes through an ecosystem in about four stops.

Sun

**1** PRIMARY PRODUCERS
Plants convert light energy from the sun into food through photosynthesis.

**2** PRIMARY CONSUMERS
**Herbivores** are animals that eat plants.

**3** SECONDARY CONSUMERS
**Carnivores** are animals that eat herbivores.

**4** TERTIARY CONSUMERS
**Top carnivores** are animals that eat other carnivores.

The food chain is a simplified pathway. In actuality, **food webs** are often a better representation, because many organisms can occupy more than one position in the chain.

**FIGURE 15-12 Follow the fuel.**

---

### EARTH'S RECYCLERS

Mold decomposes an orange.

Compost beetles (detritivores) feed on decomposing matter.

*Decomposers and detritivores break down organic wastes, releasing chemical components that can then be reused by plants and other primary producers.*

**FIGURE 15-13 Dust to dust.** Upon dying, organisms from every level in the food chain provide sustenance for decomposers and detritivores, and important chemical components are recycled through the food chain.

organic molecules. Both groups, nonetheless, release many important chemical components from the organic material, which can eventually be recycled and used by plants and other primary producers.

Energy flows from one stop to the next in a food chain, but not in the way that a baton is passed by runners in a relay race. The difference is that at every step in the food chain, much of the usable energy is lost as heat. An animal that eats five pounds of plant material doesn't convert that into five new pounds of body weight. Not by a long shot. In the next section, we'll see how this inefficiency of energy transfers ensures that most food chains are very short.

# 15·7

## Energy pyramids reveal the inefficiency of food chains.

Look out of the nearest window. What organisms can you see? Almost without fail, you will see green plant life. Maybe some trees, possibly bushes and grasses as well. You'll have to look longer and harder to see any animals, but you'll probably see a few, most likely small animals and various insects that eat plants. On the other hand, you might stare out of the window all day and not see any animals (other than some fellow humans) that eat other animals. Why is that? And why are big, fierce animals so rare? And why are there so many more plants than animals?

The answers to these questions are closely related to our earlier observation that an animal consuming five pounds of plant material does not gain five pounds in body weight from its meal. The actual amount of growth such a meal can support is far, far less. The fraction turns out to be about 10%, and this is fairly consistent across all levels of the food chain. So the herbivore consuming five pounds of plant material is likely to gain only about half a pound in new growth, while the remaining 90% of the meal is either expended in cellular respiration or lost as feces. Similarly, a carnivore eating the herbivore converts only about 10% of the mass it consumes into its own body mass. Again, 90% is lost to metabolism and feces. Additionally, non-predatory deaths reduce the transfer of energy from one trophic level to the next. And the same inefficiency holds for a top carnivore as well. Let's explore how this 10% rule limits the length of food chains and is responsible for the rarity of big, fierce animals outside your window and across the world

**Biomass** is the total weight of living or non-living organic material in a given volume, such as a single organism, or, on a

**Q** Why are big, fierce animal species so rare in the world?

larger scale, the weight of all plant and animal matter in, say, an ecosystem. Given the 10% efficiency with which herbivores convert plant biomass into their own biomass, how much plant biomass is necessary to produce a single 1,200-pound (500 kg) cow? On average, that cow would need to eat about 12,000 pounds (5,000 kg) of grain in order to grow to weigh 1,200 pounds. But that 1,200-pound cow, when eaten by a carnivore, could only add about 120 pounds of biomass to the carnivore, and only 12 pounds to a top carnivore. That's a huge amount of plant biomass required to generate a tiny amount of our top carnivore, which explains why big, fierce animals are so rare (and why vegetarianism is more energetically efficient than meat-eating). After all, multiply that 5,000 kg of grain by several hundred (or thousands, more appropriately) to get an idea of how much plant matter would be necessary to support even a small population of top carnivores: millions of kilograms of grain can support only a few top carnivores.

How much would be required to support an even higher link on the food chain? Ten times as much, so much that there might not be enough land in the ecosystem to produce enough plant material. And even if there were, the area required would be so large that the "top, top carnivores" might be so spread out and so busy trying to eat enough that they'd be unlikely to encounter each other in order to mate. Hence, the 10% rule limits the length of food chains.

We can illustrate the path of energy through the organisms of an ecosystem with an **energy pyramid,** in which each layer of the pyramid represents the biomass of a trophic level. In **FIGURE 15-14**, we can see that for terrestrial ecosystems, the biomass (in kilograms per

## ENERGY PYRAMID: INEFFICIENT ENERGY FLOW

Only about 10% of the biomass from each trophic level is converted into biomass in the next trophic level. The rest of the available energy is lost to the environment, a consequence of several factors, including non-predatory deaths, incomplete digestion of prey/food, and respiration.

TERTIARY CONSUMERS

SECONDARY CONSUMERS

— 10% converted to biomass

PRIMARY CONSUMERS

— 10% converted to biomass

PRIMARY PRODUCERS

— 10% converted to biomass

*GRAPHIC CONTENT!*
*Thinking critically about visual displays of data.*
*Turn to p. 618 for a closer inspection of this figure.*

**FIGURE 15-14  The 10% Rule.** Inefficiencies in the transfer of energy from one trophic level to the next account for the fact that there are so many more plants than animals.

## VARIATIONS IN PRIMARY PRODUCTIVITY

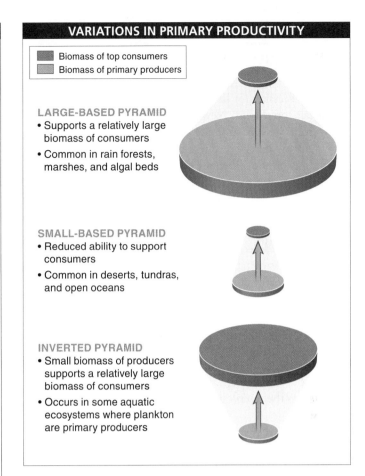

Biomass of top consumers
Biomass of primary producers

**LARGE-BASED PYRAMID**
• Supports a relatively large biomass of consumers
• Common in rain forests, marshes, and algal beds

**SMALL-BASED PYRAMID**
• Reduced ability to support consumers
• Common in deserts, tundras, and open oceans

**INVERTED PYRAMID**
• Small biomass of producers supports a relatively large biomass of consumers
• Occurs in some aquatic ecosystems where plankton are primary producers

**FIGURE 15-15  Relative biomass of producers and consumers.** Across ecosystems, there is huge variation in primary productivity.

where the primary producers are plankton. Because plankton have such short life spans and rapid reproduction rates, a relatively small biomass can support a large biomass of consumers, giving rise to an inverted pyramid (see the bottom pyramid in Figure 15-15). If you quantified the amount of energy available to consumers (rather than biomass), however, the pyramid would resemble those seen in terrestrial ecosystems.

square meter) found in photosynthetic organisms, at the base of the pyramid, is reduced significantly at each step, given the incomplete utilization by organisms higher up the food chain. **FIGURE 15-15** illustrates the huge variation in primary productivity across a variety of ecosystems. It is highest in tropical rain forests, marshes, and algal beds, and lowest in deserts, tundra, and the open ocean. In each case, the shapes of the energy and biomass pyramids are similar. With a smaller base, though, the ability of an ecosystem to support higher levels in the food chain is reduced. One dramatic exception is seen in some aquatic ecosystems

### TAKE-HOME MESSAGE 15·7

Energy pyramids reveal that the biomass of primary producers in an ecosystem tends to be far greater than the biomass of herbivores. Similarly, the biomass transferred at each successive step along the food chain tends to be only about 10% of the biomass of the organisms consumed. As a consequence of this inefficiency, food chains rarely exceed four levels.

# 15·8

## Essential chemicals cycle through ecosystems.

What is necessary for life? Energy and some essential chemicals top the list. New energy continually comes to earth from the sun, fulfilling the first need. And everything else is already here. The chemicals just cycle around and around, using the same pathway taken by energy—the food chain. Plants and other producers generally take up the molecules from the atmosphere or the soil. Then as animals consume the plants or other animals, the chemicals move up the food chain. And as the plants and animals die, detritivores and decomposers return the chemicals to the abiotic environment. From a chemical perspective, life is just a continuous recycling of molecules.

Chemicals cycle through the living and non-living components of an ecosystem. Each chemical is stored in a non-living part of the environment called a "reservoir." Organisms acquire the chemical from the reservoir, the chemical cycles through the food chain, and eventually is returned to the reservoir.

This is a useful overview, but we can get a deeper appreciation of the functioning of ecosystems and the ecological problems that can occur when they are disturbed—particularly by humans—by investigating a few of these cycles in more detail. Here we'll explore three of the most important chemical cycles: carbon, nitrogen, and phosphorus.

**Carbon** Carbon is found largely in four compartments on earth: the oceans, the atmosphere, terrestrial organisms, and fossil deposits. Plants and other photosynthetic organisms obtain most of their carbon from the atmosphere, where carbon is in the form of carbon dioxide ($CO_2$). As we saw in Chapter 4, photosynthetic organisms use carbon dioxide in photosynthesis, separating the carbon molecules from $CO_2$ and using them to build sugars and other macromolecules. Carbon then moves through the food chain as organisms eat plants and are themselves eaten (**FIGURE 15-16**). The oceans contain most of the earth's carbon. Here, many organisms use dissolved carbon to build shells (which later dissolve back into the water after the organism dies).

Most carbon returns to its reservoir as a consequence of organisms' metabolic processes. Organisms extract energy from food by breaking carbon-carbon bonds, releasing the energy stored in the bonds, and combining the released carbon atoms with oxygen. They then exhale the end product as $CO_2$.

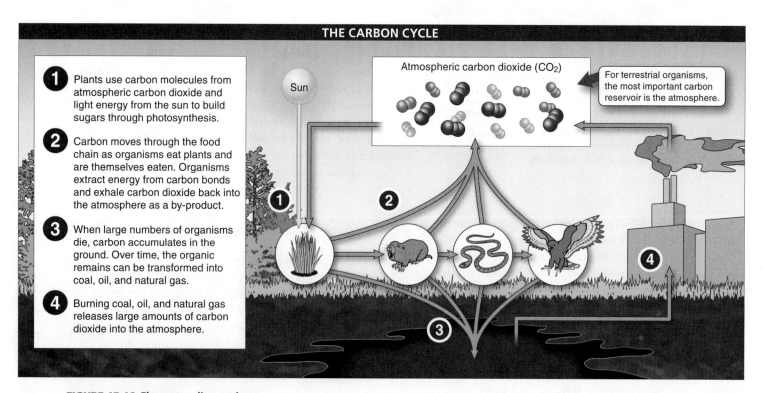

**THE CARBON CYCLE**

Atmospheric carbon dioxide ($CO_2$)

Sun

For terrestrial organisms, the most important carbon reservoir is the atmosphere.

**1** Plants use carbon molecules from atmospheric carbon dioxide and light energy from the sun to build sugars through photosynthesis.

**2** Carbon moves through the food chain as organisms eat plants and are themselves eaten. Organisms extract energy from carbon bonds and exhale carbon dioxide back into the atmosphere as a by-product.

**3** When large numbers of organisms die, carbon accumulates in the ground. Over time, the organic remains can be transformed into coal, oil, and natural gas.

**4** Burning coal, oil, and natural gas releases large amounts of carbon dioxide into the atmosphere.

**FIGURE 15-16 Element cycling: carbon.**

The burning of fossil fuels is adding significantly to the atmospheric carbon reservoir. Fossil fuels are created when large numbers of organisms die and are buried in sediment lacking oxygen. In the absence of oxygen, at high pressures, and after very long periods of time, the organic remains are transformed into coal, oil, and natural gas. Trapped underground or in rock, these sources of carbon played little role in the global carbon cycle until humans in industrialized countries began using fossil fuels to power various technologies. Burning coal, oil, and natural gas releases large amounts of carbon dioxide, thus increasing the average $CO_2$ concentration in the atmosphere—the current level of $CO_2$ in the atmosphere is the highest it has been in almost half a million years. This has potentially disastrous implications, as we'll see in Chapter 16.

Although, on average, the level of $CO_2$ in the environment is increasing, there is a yearly cycle of ups and downs in the $CO_2$ levels in the northern hemisphere. This cycling is due to fluctuations in the ability of plants to absorb $CO_2$. Many trees lose their leaves each fall and, during the winter months, relatively low rates of photosynthesis lead to low rates of $CO_2$ consumption, causing an annual peak in atmospheric $CO_2$ levels. During the summer,

**Q**

Why are global $CO_2$ levels rising?

**Q**

Global $CO_2$ levels are rising in general, but they also exhibit a sharp rise and fall within each year. Why?

trees have their leaves, sunlight is strong, and photosynthesis (consuming $CO_2$) occurs at much higher levels, causing a drop in the atmospheric $CO_2$ level.

**Nitrogen** Nitrogen is necessary to build a variety of molecules critical to life, including all amino acids, the components of every protein molecule. Just as for carbon, the chief reservoir of nitrogen is the atmosphere. But even though more than 78% of the atmosphere is nitrogen gas ($N_2$), for most organisms this nitrogen is completely unusable. The problem is that atmospheric nitrogen consists of two nitrogen atoms bonded tightly together, and these bonds need to be broken to make the nitrogen usable for living organisms. Only through the metabolic magic (chemistry, actually) of some soil-dwelling bacteria, the nitrogen-fixers, can most nitrogen enter the food chain (**FIGURE 15-17**).

These bacteria chemically convert, or "fix," nitrogen by attaching it to other atoms, including hydrogen, to produce ammonia and related compounds. These compounds are then further modified by other bacteria into a form that can be taken up by plants and used to build proteins. And once nitrogen is in plant tissues, animals acquire it in the same way they

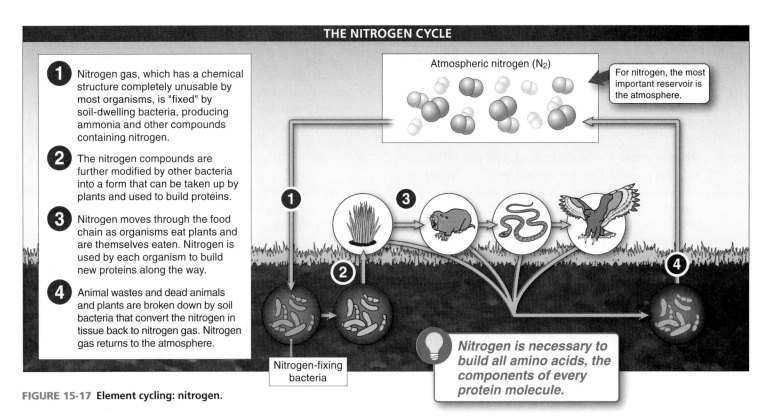

**THE NITROGEN CYCLE**

Atmospheric nitrogen ($N_2$)

For nitrogen, the most important reservoir is the atmosphere.

**1** Nitrogen gas, which has a chemical structure completely unusable by most organisms, is "fixed" by soil-dwelling bacteria, producing ammonia and other compounds containing nitrogen.

**2** The nitrogen compounds are further modified by other bacteria into a form that can be taken up by plants and used to build proteins.

**3** Nitrogen moves through the food chain as organisms eat plants and are themselves eaten. Nitrogen is used by each organism to build new proteins along the way.

**4** Animal wastes and dead animals and plants are broken down by soil bacteria that convert the nitrogen in tissue back to nitrogen gas. Nitrogen gas returns to the atmosphere.

Nitrogen-fixing bacteria

*Nitrogen is necessary to build all amino acids, the components of every protein molecule.*

**FIGURE 15-17 Element cycling: nitrogen.**

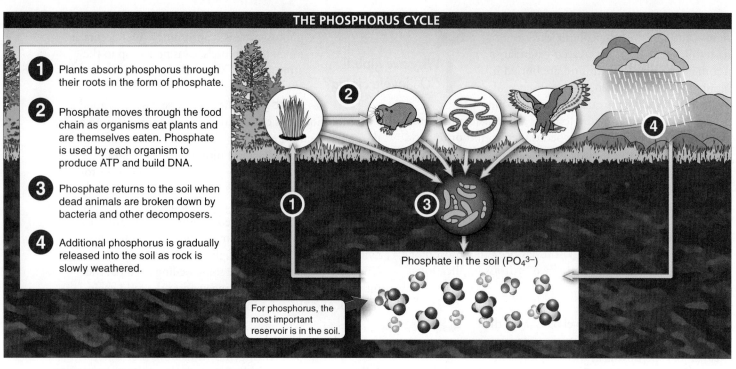

1. Plants absorb phosphorus through their roots in the form of phosphate.

2. Phosphate moves through the food chain as organisms eat plants and are themselves eaten. Phosphate is used by each organism to produce ATP and build DNA.

3. Phosphate returns to the soil when dead animals are broken down by bacteria and other decomposers.

4. Additional phosphorus is gradually released into the soil as rock is slowly weathered.

Phosphate in the soil ($PO_4^{3-}$)

For phosphorus, the most important reservoir is in the soil.

**FIGURE 15-18  Element cycling: phosphorus.**

acquire carbon: by eating the plants. Nitrogen returns to the atmosphere when animal wastes and dead animals are broken down by soil bacteria (decomposers) that convert the nitrogen compounds in tissues back to nitrogen gas.

The availability of nitrogen is like a bottleneck limiting plant growth. Because it is necessary for the production of every plant protein, and because all nitrogen must first be made usable by bacteria, plant growth is often limited by nitrogen levels in the soil. For this reason, most fertilizers contain nitrogen in a form usable by plants.

**Phosphorus**  Every molecule of ATP and DNA requires phosphorus. But no (or barely any) phosphorus is available in the atmosphere. Instead, soil serves as the chief reservoir. Like nitrogen, phosphorus is chemically converted—"fixed"—into a form usable by plants (phosphate) and then absorbed by their roots. It cycles through the food chain as herbivores consume plants and carnivores consume herbivores. As organisms die, their remains are broken down by bacteria and other organisms, returning the phosphorus to the soil. The pool of phosphorus in the soil is also influenced by the much slower process of formation of rock on the seafloor, its uplifting into mountains, and its eventual weathering, releasing its phosphorus (**FIGURE 15-18**).

Like nitrogen, phosphorus is often a limiting resource in soils, constraining plant growth. Consequently, fertilizers usually contain large amounts of phosphorus. This is beneficial in the short run, but it can have some disastrous unintended consequences. As more and more phosphorus (and nitrogen) is added to soil, some of it runs off with water and ends up in lakes, ponds, and rivers. In these habitats, also, it acts as a fertilizer, making spectacular growth possible for algae. But eventually the algae die and sink, creating a huge source of food for bacteria. The bacterial population increases and may use up too much of the dissolved oxygen, causing fish, insects, and many other organisms to suffocate and die.

 

    An atom of phosphorus 'X' had marked time in the limestone ledge since the Paleozoic seas covered the land. *Time, to an atom locked in a rock, does not pass.* The break came when a bur-oak root nosed down a crack and began prying and sucking. In the flash of a century the rock decayed, and X was pulled out and up into the world of living things. He helped build a flower, which became an acorn, which fattened a deer, which fed an Indian all in a single year.

—ALDO LEOPOLD, *A Sand County Almanac*, 1949

## EUTROPHICATION

Eutrophication—the increase in nutrients in an ecosystem, particularly nitrogen and phosphorus—often leads to the rapid growth of algae and bacteria in aquatic ecosystems. These organisms then consume much of the oxygen, leading to large die-offs of animals.

*Due to extensive runoff of fertilizers from agriculture, eutrophication now affects more than half of the lakes in Asia, Europe, and North America.*

**FIGURE 15-19** **Too much of a good thing?**

The increase in nutrients, particularly nitrogen and phosphorus, in an ecosystem, is called **eutrophication.** The consequent rapid growth of algae and bacteria, followed by large-scale die-offs of other organisms, is increasingly a problem in both small and large bodies of water, including more than half of the lakes of Asia, Europe, and North America (**FIGURE 15-19**). Lake Erie, on the U.S.-Canadian border, for example, has experienced eutrophication as a result of all the phosphorus- and nitrogen-containing wastewater that drains into the lake from the extensive surrounding farmlands. Given the lower use of fertilizers in South America and Africa, eutrophication is less common there.

### TAKE-HOME MESSAGE 15·8

Chemicals essential to life—including carbon, nitrogen, and phosphorus—cycle through ecosystems. They are usually captured from the atmosphere, soil, or water by growing organisms, passed from one trophic level to the next as organisms eat other organisms, and returned to the environment through respiration, decomposition, and erosion. These cycles can be disrupted as human activities significantly increase the amounts of chemicals added to soils for use by plants or released to the environment.

*A pair of anemone fish (Amphiprion sp.) in an anemone, Solomon Islands.*

# 15·9

## Interacting species evolve together.

There is a moth in Madagascar with a tongue that is 11 inches long! This might seem absurdly long—until the moth approaches a similarly odd-looking orchid. The orchid's flower has a very long tube, also about 11 inches long, that has a bit of nectar at the very bottom. The moth's tongue, although usually rolled up, straightens out as fluid is pumped into it and can be inserted into the long nectar tube. As its tongue reaches the bottom, the moth slurps up the nectar, gaining nourishment and energy. The moth also gets a bit of pollen stuck to it in the process, pollen that gets brushed onto the reproductive parts of the next orchid flower it visits (**FIGURE 15-20**).

**Q** **Which came first, the long-tongued moths or the long-tubed flowers?**

It's clear that having an 11-inch tongue is useful, even necessary, to extract nectar from an 11-inch nectar tube. And it's clear that putting nectar at the bottom of an 11-inch tube is a strategy to restrict access to only those pollinators that will reliably pass pollen from plant to plant of the same species. But how did such a system ever originate? Which came first, the long-tongued moths or the long-tubed flowers? Each trait only seems to make sense in a world where the other already exists. The answer is that neither came first. They both evolved—**coevolved**—together.

As a consequence of natural selection, populations of organisms commonly become better adapted to their environment. As long as there is variation for a trait, and the trait is heritable, differential reproductive success will lead to a change in the population. It is easy to imagine populations becoming more and more efficient at making use of non-living resources. But natural selection does not distinguish between biotic and abiotic resources as selective forces. Either can cause individuals with certain traits to reproduce

**FIGURE 15-20 A perfect match?** Long-tubed orchids and long-tongued moths in Madagascar each influence the evolution of the other.

at a higher rate than others, and so either can cause evolution. So small changes in the moth lead to small changes in the orchid, which select for further small changes in the moth . . . and the process continues on. In the end, species become adapted not just to their physical environment but also to the other species around them.

# 15·10 Each species' role in a community is defined as its niche.

Within a society, most humans seem to find their niche. Each person plays a particular role, defined by the nature of his or her work, activities, and interactions with others. Other species do the same thing. Within their *communities*—geographic areas defined as loose assemblages of species, sometimes interdependent, with overlapping ranges—each species has its own niche.

We can think of an organism's **niche** as its place in the community. More than just a *place* for living, however, a niche is a complete *way* of living. In other words, an organism's niche encompasses (1) the space it requires, (2) the type and amount of food it consumes, (3) the organisms for which it is a food source, (4) its influence on competitors, (5) the timing of its reproduction (its life history), (6) its temperature and moisture requirements, and almost every other aspect that describes the way the organism uses its environment and influences the other organisms in that environment (**FIGURE 15-21**).

Although a niche describes the role a species *can* play within a community, the species doesn't always get to have that exact role. Consider the rats of Boston. Until the 1990s, they lived in relative peace in the sewers beneath the city's streets. But when the city embarked on the largest underground highway construction project in U.S. history, engineers displaced and forcibly drove out thousands of rats from much of their habitat. In essence, there was suddenly an overlap between the rat niche and the human niche, and the rats were now restricted to just some portions of the sewers. As a result, the rats' **realized niche,** the narrower role that they may occupy in a community, is now just a subset of their **fundamental niche,** the full range of environmental conditions under which they can live.

It is common for species to find themselves competing with other species for parts of a niche. When this occurs, generally one of the species is restricted from its full niche. We'll examine next the various possible outcomes when such niche overlap occurs.

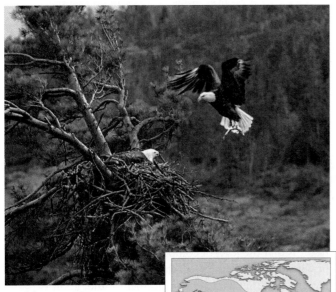

**NICHE FEATURES**
- The space an organism requires
- The type and amount of food an organism utilizes
- The timing of an organism's reproduction
- An organism's temperature and moisture requirements and other necessary living conditions
- The organisms for which it is a food source
- Its influence on competitors

☐ Bald eagle range

Constrained by competitors (or other factors), a species' realized niche may be smaller than its fundamental niche.

**FIGURE 15-21 A way of living.** "Niche" describes all of the ways in which an organism utilizes the resources of the environment.

# 15·11 Competition can be hard to see, yet it influences community structure.

Some species, particularly closely related species, have similar niches. This can lead to conflict as they try to exploit the same resources. Almost invariably, when the fundamental niches of two species overlap, competition occurs. This competition doesn't last forever, though. Inevitably, one of two outcomes occurs: competitive exclusion or resource partitioning (**FIGURE 15-22**).

**1.** In **competitive exclusion,** two species battle for resources in the same niche until the more efficient of the two wins and the other species is driven to extinction in that location ("local extinction"). In the 1930s, this was demonstrated in simple laboratory experiments using *Paramecium,* a single-celled organism. Populations of two similar *Paramecium* species were grown either separately or together in test tubes containing water and their bacterial food source. When grown separately, each species thrived. When grown together, though, one species always drove the other to extinction.

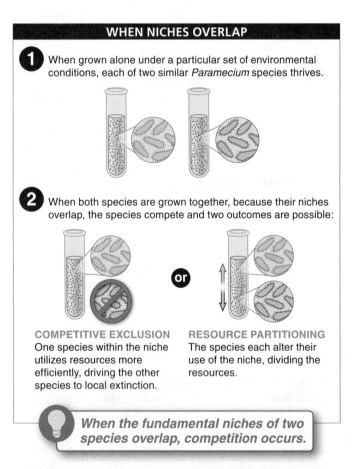

## WHEN NICHES OVERLAP

**1** When grown alone under a particular set of environmental conditions, each of two similar *Paramecium* species thrives.

**2** When both species are grown together, because their niches overlap, the species compete and two outcomes are possible:

**or**

**COMPETITIVE EXCLUSION**
One species within the niche utilizes resources more efficiently, driving the other species to local extinction.

**RESOURCE PARTITIONING**
The species each alter their use of the niche, dividing the resources.

*When the fundamental niches of two species overlap, competition occurs.*

**FIGURE 15-22 Overlapping niches: competitive exclusion and resource partitioning.**

**2. Resource partitioning** is an alternative outcome of niche overlap. Individual organisms and species can adapt to changing environmental conditions, and resource partitioning can result from an organism's behavioral change or a change in its structure. When this occurs, one or both species become restricted in some aspect of their niche, dividing the resource. In other experiments with *Paramecium,* for example, one of the two species was replaced with a third species. As in the initial experiment, each species thrived when grown alone. But when the two species were grown together in the same test tube, they ended up dividing the test tube "habitat." One species fed exclusively at the bottom of the test tube, and the other fed only at the top. Simple behavioral change made coexistence possible.

In many situations, resource partitioning is accompanied by **character displacement,** an evolutionary divergence in one or both of the species that leads to a partitioning of the niche. A clear example occurs among two species of seed-eating finches on the Galápagos Islands. On islands where both species live, their beak sizes differ significantly. One species has a deeper beak, better for large seeds, while the other has a shallower beak, better for smaller seeds, and they do not compete. On islands where either species occurs alone, beak size is intermediate between the two sizes (**FIGURE 15-23**).

Competition between species has one very odd feature: it is hard to actually see it occurring because it causes itself to disappear. That is, after only a short period of competition, one of the species becomes locally extinct or leaves the area where the niches overlap, or character displacement occurs, largely reducing the competition. In either case, the net result is that the level of competition is significantly reduced or wiped out altogether. For this reason, biologists often have to look for character displacement—the "ghost of competition past"—to identify areas where competition once occurred. Moreover, even while it is occurring, competition tends to be indirect, rather than head-to-head battles. Like a game of musical chairs, it's more a question of both species trying to use a particular resource, with one being a bit better at it.

Biologists do have the opportunity to see competition in the occasional natural experiments in which a new species is introduced to a community. This happened, for example, when the American mink, a weasel-like mammal, was introduced into Europe, where previously only the European

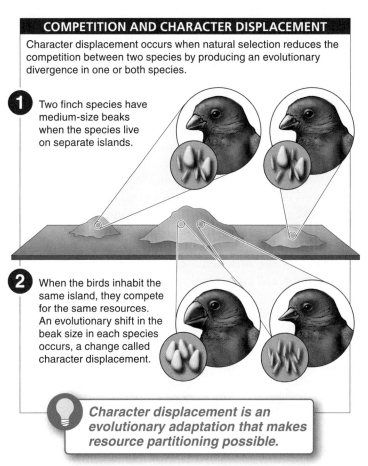

## COMPETITION AND CHARACTER DISPLACEMENT

Character displacement occurs when natural selection reduces the competition between two species by producing an evolutionary divergence in one or both species.

**1** Two finch species have medium-size beaks when the species live on separate islands.

**2** When the birds inhabit the same island, they compete for the same resources. An evolutionary shift in the beak size in each species occurs, a change called character displacement.

*Character displacement is an evolutionary adaptation that makes resource partitioning possible.*

**FIGURE 15-23 Allowing organisms to divide resources.** Character displacement is an evolutionary adaptation that makes resource partitioning possible.

mink had lived. Within 10 years, in the area where their ranges overlapped, the European mink increased in size while the American mink decreased in size, as biologists documented the evolution. In the next chapter, we'll investigate in greater depth why newly introduced species frequently win the competition with native species.

### TAKE-HOME MESSAGE 15·11

Populations with completely overlapping niches cannot coexist forever. Competition for resources occurs until one or both species evolve in ways that reduce the competition, through character displacement, or until one becomes extinct in that location.

## 15·12 Predation produces adaptation in both predators and their prey.

Some words of advice in case you ever think about quickly approaching a horned lizard: be afraid. Be very afraid. Here's why: as you get close to the lizard, it may zap you with streams of blood squirted from its eyes. In all likelihood, you will flinch. And as you flinch, the horned lizard will scurry away. The display is shocking, but the fact that evolution has produced extreme and effective anti-predator adaptations is not.

**Predation**—an interaction between two species in which one species eats the other—is one of the most important forces shaping the composition and abundance of species in a community. Predation, though, isn't restricted to the obvious interactions involving one animal chasing down and killing another. Herbivores' eating of leaves, fruits, or seeds is a form of predation, even though it doesn't

*Why do exotic species often flourish when released into novel habitats, even though natural selection has not adapted them to this new environment?*

necessarily kill the plant. And predators are not necessarily physically imposing. Each year more than a million humans die as a result of disease from mosquito bites, compared with fewer than a dozen from shark attacks.

**Q** Predators are a potent selective force: organisms eaten by predators tend to have reduced reproductive success. Consequently, in prey species, a variety of features have evolved (and continue to evolve)—including the blood-squirting-eyeball effect—that reduce their predation risk. But as prey evolve, so do predators. This coevolution is a sort of arms race with ever-changing and escalating predation-effectiveness adaptations causing more effective predator-avoidance adaptations, and vice versa. (In this light, it may seem unexpected that exotic species often flourish when released into novel habitats, even though natural selection has not adapted them to their new environment. As it turns out, just as these species are not fully adapted to their new environment, they also have few predators there. And with low predation risk, they often can flourish.) We'll examine some common adaptations of both predators and prey.

**Prey adaptations for reducing predation** There are two broad categories of defenses against predators: physical and behavioral.

**Physical defenses** include mechanical, chemical, warning coloration, and camouflage mechanisms.

**1. Mechanical defenses.** Predation plays a large role in producing adaptations such as the sharp quills of a porcupine, the prickly spines of a cactus, or the tough

armor protecting an armadillo or sow bug. These, as well as claws, fangs, stingers, and other physical structures, can reduce predation risk (**FIGURE 15-24**).

**2. Chemical defenses.** Further prey defenses can include chemical toxins that make the prey poisonous or unpalatable. Plants can't run from their predators, so chemical defenses are especially important to them. Almost all plants produce some chemicals to deter organisms that might eat them. The toxins can be severe, such as strychnine from plants in the genus *Strychnos,* which kills most vertebrates, including humans, by stimulating nonstop convulsions and other extreme and painful symptoms leading to death. At the other end of the spectrum are chemicals toxic to some insects but relatively mild to humans, such as those found in cinnamon, peppermint, and jalapeno peppers. Ironically, many plants

---

**PHYSICAL DEFENSES FOR REDUCING PREDATION**

**MECHANICAL DEFENSES**

**CHEMICAL DEFENSES**

**WARNING COLORATION**

**CAMOUFLAGE**

**FIGURE 15-24 Prey defenses: some physical means for avoiding predation.** Shown (*from top left, clockwise*) are Cape porcupine, azure poison dart frog, praying mantis, and monarch butterfly.

that evolved to produce such chemicals to deter predators are now cultivated and eaten specifically *for* these chemicals. One organism's toxic poison is another's spicy flavor.

Some animals can also synthesize toxic compounds (or perhaps more commonly can safely acquire them from the organisms they eat). The poison dart frog has poison glands all over its body, making it toxic to the touch. The fire salamander, too, is toxic, with the capacity to squirt a strong nerve toxin from poison glands on its back. Some animals, including milkweed bugs and monarch butterflies, can safely consume toxic chemicals from plants and sequester them in their tissues, becoming toxic to predators who try to eat them.

**3. Warning coloration.** Species protected from predation by toxic chemicals frequently have bright color patterns to warn potential predators. They are essentially carrying a sign that says: "Warning, I'm poisonous, so keep away." To make it as easy as possible for predators to learn, different poisonous species often have the same color patterns. It's as if devising a single common sign saying "Danger: poison" is more effective than multiple species coming up with their own specific signs.

In a clever twist on this, some species that are perfectly edible to their predators also have evolved the same bright colors, in a phenomenon known as **mimicry.** Their coloration mimics the same warning sign but without the toxins. As long as they are relatively rare compared with the toxic individuals they mimic—reducing the chance that predators might catch on to their trickery—the evolutionary ruse is quite successful.

**4. Camouflage.** An alternative to warning coloration, and one of the most effective ways to avoid being eaten, is simply to avoid being seen. An adaptation in many organisms is patterns of coloration that enable them to blend into their surroundings. Examples include insects that look like leaves or twigs and hares that are brown for most of the year but turn white when there is snow on the ground.

**Behavioral defenses** include both seemingly passive and active behaviors: hiding or escaping, and alarm calling or fighting back.

**1. Hiding or escaping.** Anti-predator adaptations need not involve toxic chemicals, physical structures, or special coloration. Many species excel at hiding or running, or both. With vigilance, it is

**Q** On islands, animals frequently have no fear of humans (in contrast to animals' more skittish behavior on the mainland). Why?

## BEHAVIORAL DEFENSES THAT REDUCE PREDATION

**HIDING OR ESCAPING**

**FIGHTING BACK**

**FIGURE 15-25 Prey defenses: behavioral means for avoiding predation.** Fish may travel in huge schools, finding safety in numbers. And fish may mob a predator, here, a shark, to make it harder for the predator to attack any one member of the school. The bird can project vomit at a predator, frightening it away.

possible to get advance warning of impending predator attacks, and then quickly and effectively avoid the predator. A variation of this strategy comes from safety in numbers: many species, including schooling fish and emperor penguins, travel in large groups to reduce their predation risk (**FIGURE 15-25**).

On many islands, animals show no fear of humans. Rather than encountering skittish lizards, for example, a human visitor to the Galápagos Islands must be careful not to step on them. The animals are not afraid because, given the small size of most islands, the number of predators is restricted. As a consequence, there has been no selection for skittishness and hypervigilance, traits that normally evolve in response to predation risk.

**2. Alarm calling and fighting back.** In many species, especially birds and mammals, individuals warn others with an alarm call. Although risky for the caller, such alarm calling can give other individuals—often close kin that are nearby—enough advance warning to escape (recall, from Chapter 9, the warning calls of Belding's ground squirrels). Some prey species also turn the tables, mobbing predators to keep them from successfully completing their task. This category might also include the blood-squirting lizard described above, or the fulmar, a seabird that defends its nest from attacks with projectile vomiting aimed at the intruder.

## Predator adaptations for enhancing predation

Just as prey use physical and behavioral features to reduce their risk of predation, predators evolve in parallel ways to increase their efficiency. As milkweed plants have evolved to produce toxic chemicals, sequestered within the structures of the plant, that kill their predators, milkweed bugs have evolved to produce toxic-avoidance methods that allow them to eat the toxic plants without suffering harm. And as prey have become better at hiding and escaping, predators have developed better sensory perception to help them detect hiding prey, and faster running ability to catch them. Predators, too, make use of mimicry. The angler fish, the tasseled frogfish, and the snapping turtle all have physical structures that mimic something—usually a food item—that is of interest to potential prey. As the prey come closer to inspect, the predator snaps them up (**FIGURE 15-26**).

**Q**

Why don't predators become so efficient at capturing prey that they drive the prey to extinction?

Although natural selection leads to predators with effective adaptations for capturing prey, the adaptations are rarely so efficient that the prey species are driven to extinction. The explanation for this is referred to as the "life-dinner hypothesis." Selection for "escape ability" in the prey is stronger than selection for "capture ability" in the predator. When, for example, a rabbit can't escape from a fox, the cost

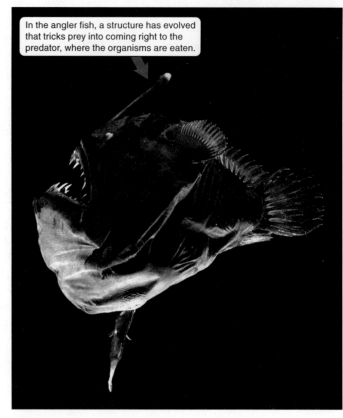

In the angler fish, a structure has evolved that tricks prey into coming right to the predator, where the organisms are eaten.

**FIGURE 15-26 A better predator.** Some adaptations enhance predation.

to the rabbit is its life—and it will never reproduce again. On the other hand, when a fox can't keep up with a rabbit, all it loses is a meal; it can still go on to capture prey and reproduce in the future. In other words, the cost of losing in the interaction is much higher for the rabbit.

---

### TAKE-HOME MESSAGE 15·12

Predators and their prey are in an evolutionary arms race; as physical and behavioral features evolve in prey species to reduce their predation risk, predators develop more effective and efficient methods of predation. The coevolutionary process can result in brightly colored organisms, alarm calling, and many types of mimicry.

# 15·13 Parasitism is a form of predation.

For most people, the word "predation" conjures images of a large animal such as a cheetah, chasing and killing another large animal such as a gazelle. But there is another world of predation, largely unseen but equally deadly: **parasitism,** defined as a symbiotic relationship in which one organism (the **parasite**) benefits while the other (the **host**) is harmed. In fact, this hidden world of parasites is thriving with activity: parasites are probably the most numerous species on earth, with three or four times as many species as non-parasites!

Two general types of parasites make life difficult for most organisms. *Ectoparasites* (*ecto* = outside) include organisms such as lice, leeches, and ticks. One species of Mexican parrot is all too aware of ectoparasites: it has 30 different species of mites living on its feathers alone. And many of the parasites even have parasites themselves! *Endoparasites* are parasites that live inside their hosts (*endo* = inside). They are equally pervasive. Endoparasites infecting vertebrates include many different phyla of both animals and protists, the single-celled eukaryotes (**FIGURE 15-27**). In all of these parasite-host interactions, as with all predator-prey interactions, the predator or parasite benefits and the prey or host is harmed.

Even though they are considered predators, parasites have some unique features and face some unusual challenges relative to other predators. The most obvious of these features is that in parasitism, the parasite generally is much smaller than its host and stays in contact with the host for extended periods of time, normally not killing the host but weakening it as the parasite uses some of the host's resources. Being located right on your food source all the time can be advantageous. But this also leads to what is perhaps the greatest challenge that parasites face: how to get from one individual host to another. A parasite can't survive long once its host dies, after all. The methods by which parasites accomplish such dispersal are unexpected and surprising. Many of their complicated life cycles involve passing through two (or more) different species (and could have come about only through coevolution with each of the host species). These life cycles are likely to give us a new appreciation for the ingenuity of these microorganisms—or rather, for the evolutionary process that has produced them. Let's look at a few representative examples.

*Case 1: Parasites can induce foolish, fearless behavior in their hosts.* Rats fear cats. During their evolution, rats have developed a protective wariness of their feline predators and

## PARASITES

Parasites are predators that benefit from a symbiotic relationship with their hosts.

**ECTOPARASITES**
Parasites that live on their host

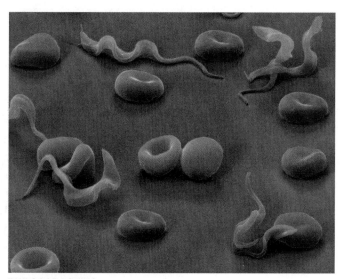

**ENDOPARASITES**
Parasites that live inside their host

 *There may be three to four times as many parasitic species as non-parasitic species on earth!*

**FIGURE 15-27 Parasites: predators dwarfed by their prey.**
A bedbug (*Cimex lectularius*), an ectoparasite of humans, is shown in extreme close-up. *Trypanosoma brucei* is an endoparasite that invades a human host and is the cause of African sleeping sickness.

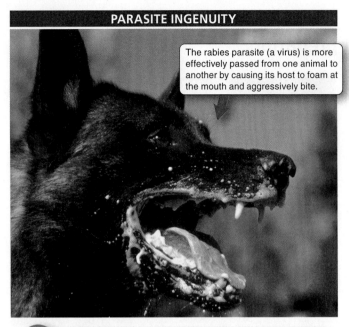

The rabies parasite (a virus) is more effectively passed from one animal to another by causing its host to foam at the mouth and aggressively bite.

 *Parasites have surprising ways of solving a challenging problem: how to get from one individual host to another.*

**FIGURE 15-28 Spreading disease.** Rabies symptoms aid the disease-causing virus in spreading to new hosts.

areas in which cats have been roaming. *Toxoplasma* is an organism that changes this. This single-celled parasite of rats must also spend part of its life cycle in cats. It does this by altering the brain of its rat host so that the rat no longer fears cats. In fact, *Toxoplasma*-infected rats not only lose their fear of cats, they become attracted to them. Is this an accident? No. This behavioral change, while quite dangerous for the rat, is exactly the change that increases the likelihood that the rat will be attacked by a cat, spreading the parasite in the process.

*Case 2: Parasites can induce inappropriate aggression in their hosts.* Rabid animals don't behave normally. They froth at the mouth and become unusually aggressive (**FIGURE 15-28**). Is this an accident? No. Rabies is caused by a virus that infects warm-blooded animals, mostly raccoons,

skunks, foxes, and coyotes. It is passed from one host to another via saliva. Inducing these "rabid" behaviors, of course, is exactly the change in behavior that will most help the virus to spread.

*Case 3: Parasites can induce bizarre and risky behavior in their hosts.* The lancet fluke is a parasitic flatworm. It has also been described as a "zombie-maker." This fluke spends most of its life in sheep and goats, but the fluke's eggs pass into snails that graze on vegetation contaminated by sheep and goat feces. Once inside the snail, the fluke eggs grow and develop, eventually forming cysts that the snail surrounds with mucus and then excretes. Continuing on their complex life cycle, the fluke cysts find their way into ants that eat the snail mucus. The flukes' journey back to a sheep or goat is now expedited by the so-called zombie-making. In an infected ant, the flukes grow into the ant's brain, altering its behavior. Whereas ants normally remain low to the ground, when infected by the lancet fluke, an ant climbs to the top of a grass blade or plant stem and clenches its mandibles on a leaf. Is this an accident? No. This behavioral change puts the ant much more at risk of being eaten by a grazing mammal—a bad outcome for the ant, but just what the parasite needs to complete its life cycle.

These parasite-induced behavior modifications are among the most dramatic, but there are many others that are more subtle yet equally effective at allowing a parasite to thrive attached to or within its host's body. This subset of predation is an active and exciting area of ecological and physiological investigation.

**TAKE-HOME MESSAGE 15·13**

Parasitism is a symbiotic relationship in which one organism benefits while the other is harmed. Parasites face some unusual challenges relative to other predators, particularly in how to get from one individual host to another, and some complex parasite life cycles have evolved.

# 15·14 Not all species interactions are negative: mutualism and commensalism.

It's easy to get the idea that all species interactions in nature are harsh and confrontational, marked by a clear winner and a clear loser. That is largely the case when it comes to competition and predation. Nonetheless, not all species

interactions are combative. Every flower you see should be a reminder that evolution produces beneficial species interactions as well. These types of interactions fall into two categories: mutualism and commensalism.

**MUTUALISM**
Interaction in which both species benefit

**COMMENSALISM**
Interaction in which one species benefits and the other neither benefits nor is harmed

**FIGURE 15-29 Not always "red in tooth and claw."** Mutualisms and commensalisms abound in nature.

## Mutualism: everybody wins

Corals that gain energy from photosynthetic algae living inside their tissues. Termites capable of subsisting on wood, but only with the assistance of cellulose-digesting microbes living in their digestive system. Flowers pollinated by animals that are nourished by nectar. Each of these relationships is an example of **mutualism,** an interaction in which both species gain and neither is harmed (**FIGURE 15-29**). Mutualism is common in almost all communities. Plants, particularly, have numerous such interactions: with nutrient-absorbing fungi, with nitrogen-fixing bacteria, and with animals that pollinate them and disperse their fruits.

Mutualistic interactions are so widespread and ecologically important that they are described in significant detail throughout the book. The relationship between plants and mycorrhizae bacteria, for example, is discussed in Section 12-15. Plants and the rich variety of mutualistic relationships they have with animal pollinators and fruit dispersers are covered throughout Sections 12-9 and 12-11. Lichens, the symbiotic association of fungi and photosynthetic algae or bacteria, are discussed in Section 12-15.

You need not look further than your own gut for an important mutualistic relationship. Living within your large intestine are huge populations of *E. coli* bacteria that synthesize a significant portion of the vitamin K—essential for several metabolic processes and blood clotting—you need each day.

## Commensalism: an interaction with a winner but no loser

Some species interactions are one-sided. The cases in which one species benefits and the other neither benefits nor is harmed are called commensal relationships, or **commensalism.** Cattle egrets have just such a relationship with grazing mammals such as buffalo and elephants (see Figure 15-29). As the large mammals graze through grasses, they stir up insects. The birds, which feed near the mammals—particularly near the forager's head—are able to catch more insects with less effort this way. The grazers are neither helped nor harmed by the presence of the birds.

---

### TAKE-HOME MESSAGE 15·14

Not all species interactions are combative. Beneficial species interactions evolve as well. In commensalism, one species benefits and the other is neither harmed nor helped. Mutualism, in which both species benefit from the interaction, is widespread and critically important to all ecosystems.

A pair of colorful ochre starfish, keystone predators within intertidal zones in the Pacific Ocean.

# 15·15

## Many communities change over time.

Human "progress" and development can completely transform an environment—turning a patch of desert into Las Vegas, for example. Urban landscapes, too, can obliterate any signs of the nature that was once there. This is why it can be surprising (and heartening) to observe what happens when humans abandon an area. Little by little, nature reclaims it. The area doesn't necessarily recover completely, and change is slow. Still, this process is almost universal and virtually unstoppable. Nature responds similarly to other disturbances, too, from a single tree falling in a forest, to a massive flood or fire, to massive volcanic eruptions.

The process of nature reclaiming an area and of communities gradually changing over time is called **succession.** It is defined as a change in the species composition over time, following a disturbance. There are two types of succession. *Primary succession* is when the process starts with no life and no soil. *Secondary succession* is when an already established habitat is disturbed, but some life and some soil remain.

**Primary succession** Primary succession can take thousands or even tens of thousands of years, but it generally occurs in a consistent sequence. It always begins with a disturbance that leaves an area barren of soil and with no life. Frequently the disturbance is catastrophic. The huge volcanic eruption on Krakatau, Indonesia, in 1883, for

example, completely destroyed several islands and wiped out all life and soil on others. Primary succession has also begun, in a less dramatic fashion, in regions where glaciers have retreated, such as Glacier Bay, Alaska. Although succession does not occur in a single, definitive order, several steps are relatively common (**FIGURE 15-30**).

- The first arrivals, or **colonizers,** in a lifeless, soil-less area are usually bacteria or fungi or other photosynthetic microorganisms, floating in on the air.

- Lichens, symbiotic associations between a fungus and a photosynthetic alga or bacterium, are also common first colonizers. They can grow on rocks and are able to generate energy through photosynthesis. As they grow, lichens change their environment, too, secreting acids that break down the rocks and release useful minerals, making the terrain more hospitable for other organisms.

- Mosses also tend to arrive in the early stages, making use of many of the nutrients freed up by lichens. They trap moisture and can provide a hospitable site for germination of the seeds of other plants.

- Following mosses, small herbs often arrive. Later, some shrubs may begin to thrive, shading out the mosses and herbs.

- Shrubs, in turn, are eventually outcompeted and shaded out by small trees.

Succession is the change in species composition in a community over time.

**COLONIZING COMMUNITY**

The first arrivals to a lifeless and soil-less area. They are good dispersers, but poor competitors, so they are gradually replaced.

**INTERMEDIATE COMMUNITIES**

A variety of species is present, typically includes prokaryotes, and species from all kingdoms of eukaryotes. Shown here: typical plant representatives in intermediate communities.

**CLIMAX COMMUNITY**

Longer-living, larger species outcompete the initial colonizers and persist as a stable and self-sustaining community.

**1** Fungi, bacteria, lichens, and seeds are often among the earliest colonizers.

**2** Mosses begin to grow and trap moisture, allowing the seeds of other plants to germinate.

**3** Small herbs arrive and shrubs grow, shading out the mosses and herbs.

**4** Small trees grow and outcompete the shrubs.

**5** Taller, faster-growing trees outcompete the smaller trees.

**PRIMARY SUCCESSION**
Primary succession begins after a disturbance leaves an area barren of soil and with no life.

**SECONDARY SUCCESSION**
Secondary succession is like primary succession with a head start. It begins when a disturbance opens up part of a community to the development and growth of species previously outcompeted by other species in the area.

 *Disturbance is a fundamental part of most ecosystems and can repeatedly set a community back to an earlier stage of succession.*

**FIGURE 15-30 The species composition of a community changes over time.**

- And the first trees generally are outcompeted by taller, faster-growing trees.

An important feature of the colonizers seen in the earliest stages of primary succession—whether plants, animals, or other species—is that while they are good **dispersers,** able to move away from their original home (hence their early arrival in a newly available locale), they are not particularly good *competitors.* That's why they are gradually replaced.

Ultimately, it is the longer-living, larger species that tend to outcompete the initial colonizers and persist within a stable and self-sustaining community, called a **climax community.** The particular species present in the climax community depend on physical factors such as temperature and rainfall.

**Secondary succession** Secondary succession is much faster than primary succession, because life and soil are already present. Rather than the thousands or even tens of thousands of years that primary succession may take, secondary succession can happen in a matter of centuries, decades, or even years. It frequently begins with organisms colonizing the decaying remains of dead organisms. This may involve fungi

establishing themselves in the decaying trunk of a fallen tree, and these being replaced over time by different species of fungi.

Or secondary succession may begin with weeds springing up in formerly cultivated land. If the weeds are allowed to grow, they are eventually outcompeted and replaced by perennial species, and then shrubs, and eventually larger trees. The process is similar to primary succession, but with a head start. Ultimately, if undisturbed, secondary succession also leads to establishment of a stable, self-sustaining climax community.

Given that primary and secondary succession both progress toward climax communities, it might seem surprising that, at any given time, so many communities are in states far from climax. This paradox reflects the fact that succession leads to climax communities in the *absence of disturbance.* But disturbance is a fundamental part of most ecosystems. Massive disturbances, such as the fires that destroyed a third of Yellowstone Park in 1988 or the volcanic eruptions of Mount St. Helens, can send a community back to square one. Minor disturbances, on

**Q** *Why doesn't succession occur on front lawns?*

the other hand, may undo just a small degree of the succession that has already occurred. Interestingly, it is at intermediate levels of disturbance that communities tend to support the greatest number of species. With very high or very low disturbance levels, the community is restricted to the top colonizers—those with the best dispersal or fastest reproductive rates—or the top competitors, respectively. At the intermediate level, a larger variety of species with intermediate combinations of features can persist.

# 15·16 Some species are more important than others within a community.

Not all species are created equal. In this chapter, we have examined the ways in which species interact with one other and with their habitats, noting the dependence of species on each other. But not all species have equal dependence on or influence over others. Within a community, the presence of some species, called **keystone species,** greatly influences which other species are present and which are not (**FIGURE 15-31**). That is, if a keystone species is removed, the species

**KEYSTONE SPECIES**

A keystone species, such as the sea star, has an unusually large influence on the presence or absence of numerous other species in the community.

COMMUNITY WITH KEYSTONE SPECIES

COMMUNITY WITH KEYSTONE SPECIES REMOVED

When sea stars were removed from an intertidal zone, species diversity decreased drastically—only mussels remained.

*Preserving just one keystone species has the effect of preserving many additional species at the same time.*

**FIGURE 15-31 Preserving biodiversity.** Keystone species can keep aggressive species in check, allowing more species to coexist.

mix in the community changes dramatically. The removal of other species, conversely, causes relatively little change.

Bison are a keystone species. Experiments were set up in which a herd of bison was allowed to graze in certain parts of a prairie in Kansas, while being excluded from other areas. The areas without the bison were soon dominated by a single species of tall grass. The grass was like a bully to other plant species. The areas where the bison grazed, on the other hand, had significantly greater species diversity. When kept in check by the grazing bison, the bully grass was unable to outcompete the other plant species, which were able to thrive.

Sea stars, too, are a keystone species (see Figure 15-31). On the rocky seashores of the Pacific Northwest, they consume mussels, which tend to crowd out other species. In a five-year study, sea stars were removed and kept out of certain areas of the intertidal zone but allowed in other areas. As with the bison study, species diversity dropped dramatically where sea stars were excluded. In fact, ultimately, just the mussels remained. In the areas with sea stars, mussels also remained but in reduced numbers. More important, 28 other species of animals and algae were also able to live there.

Suppose you are in charge of conserving biodiversity in a large region of a country. With limited funds, you cannot simply purchase and restrict access to the entire region. It

**When it comes to conservation, are some species more valuable than others? Why?**

is not possible to preserve everything. Instead, you must make difficult decisions and prioritize species for preservation. Knowing that keystone species exist can make your job considerably easier. There is real practical value in identifying keystone species and focusing conservation efforts on these. After all, in the sea star experiment, preservation of the sea stars alone would have the "side effect" of preserving 28 other species.

Keystone species make it possible to get more "bang for your buck" when your aim is to conserve biodiversity. Consequently, identifying keystone species is an important part of conservation biology. Besides bison and sea stars, some other keystone species are dam-building beavers, elephants of the African savanna, and lichens in the desert. We'll explore these questions of conservation in greater detail in the next chapter.

---

### TAKE-HOME MESSAGE 15·16

Keystone species have a relatively large influence on which other species are present in a community and which are not. Unlike other species, when a keystone species is removed from the community, the species mix changes dramatically. For this reason, preserving keystone species is an important strategy in preserving biodiversity.

### LIFE IN THE DEAD ZONE

## In boosting plant productivity on farms, we've created a "dead zone" in the Gulf of Mexico bigger than the size of Connecticut.

In the water at the mouth of the Mississippi River, each spring and summer, there is so little oxygen that virtually no life can survive.

**Q: WHICH CHEMICAL ALMOST ALWAYS LIMITS PLANT GROWTH?** Plant growth requires the production of proteins, all of which contain nitrogen. Access to nitrogen nearly always determines how much a plant grows.

**Q: IF YOU'RE A FARMER, HOW CAN YOU INCREASE THE PRODUCTIVITY OF YOUR CROPS?** Add nitrogen. This, in fact, is the chief component of all plant fertilizers.

**Q: WHERE DOES THE UNUSED NITROGEN END UP?** Spring and summer rains wash nutrients, dissolved organic materials, and other runoff (much of which comes from human sources) from the middle of the United States into nearby rivers and thus, eventually, into the Mississippi River. From there they flow into the Gulf of Mexico.

**Q: WHAT HAPPENS WHEN LARGE AMOUNTS OF NITROGEN ARE DUMPED INTO A BODY OF WATER?** Organisms such as plankton and algae living near the mouth of the Mississippi grow like crazy with all the extra nitrogen and nutrients that are washed into the Gulf of Mexico. But the rapid increase in their productivity disrupts the food chain as they decay, increasing the organic matter that sinks to the bottom and feeds the bacteria there. As the bacteria thrive, they consume ever-increasing amounts of oxygen.

**Q: WHAT HAPPENS WHEN BACTERIA USE UP TOO MUCH OXYGEN?** Excessive bacterial growth depletes the supply of dissolved oxygen in the water, starving all other life of oxygen—including fish, crabs, oysters, and shrimp, all of which die or move out of the area. This creates a "dead zone" and ruins local fisheries, with significant impact on the economies they support.

**Q: WHAT CAN YOU CONCLUDE?** Disrupting food chains, even increasing the productivity of certain trophic levels, can have unintended effects throughout an ecosystem, and even thousands of miles away.

**Q: WHAT CAN BE DONE?** Reducing and reversing the Gulf of Mexico's dead zone (and other dead zones throughout the world) can be accomplished by (1) reducing fertilizer use, (2) preventing animal wastes from getting into waterways, and (3) controlling the release of other sources of nutrients (phosphorus as well as nitrogen) from industrial facilities into rivers and streams.

# Big Ideas in ...
# Ecosystems and Communities

## 15·1–15·2 Ecosystems have living and non-living components.

- Ecosystems are the living organisms and the physical environment within specific areas.
- The largest ecosystems in the world are called biomes.
- The characteristic organisms and physical features present in each terrestrial biome depend largely on the temperature and rainfall patterns.

## 15·3–15·5 Interacting physical forces create weather.

- The greater amount of solar energy hitting earth at the equator causes global air circulation patterns that are responsible for heavy rainfall near the equator and deserts at 30° latitude, north and south.
- Global patterns of ocean circulation also influence patterns of temperature and rainfall.
- Local topography, from mountain ranges to human-made structures, can also influence rainfall, temperature, and wind patterns.

## 15·6–15·8 Energy and chemicals flow within ecosystems.

- Energy pyramids describe the energy flow through ecosystems, from primary producers to herbivores to carnivores.
- At each step of the food chain, energy capture and conversion are inefficient, with only 10% of energy transferred from one trophic level to the next.
- While all energy comes from the sun and flows in a pathway through food chains, chemicals repeatedly cycle from the physical environment through living organisms and back into the environment.

## 15·9–15·14 Species interactions influence the structure of communities.

- Interacting species in a community coevolve in a variety of ways, some antagonistic and others mutually beneficial.
- The organisms of a population fill a unique niche, defined by how they utilize the resources in their environment.
- Organisms with overlapping niches compete until character displacement or local extinction occurs.

## 15·15–15·16 Communities can change or remain stable over time.

- Most communities change over time. The patterns of succession, in which species composition changes within an ecosystem, depend on the rate and magnitude of disturbances.
- In a community, species vary in the degree to which they influence diversity. Keystone species are the most influential.

See the To-Go Guide (packaged with every new copy of this textbook) for an illustrated summary of the chapter.

# Check Your Knowledge

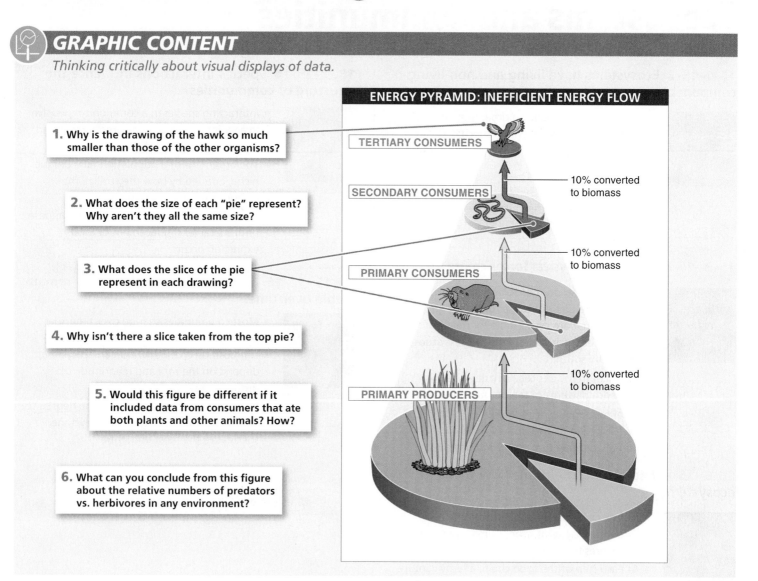

**ENERGY PYRAMID: INEFFICIENT ENERGY FLOW**

1. Why is the drawing of the hawk so much smaller than those of the other organisms?

2. What does the size of each "pie" represent? Why aren't they all the same size?

3. What does the slice of the pie represent in each drawing?

4. Why isn't there a slice taken from the top pie?

5. Would this figure be different if it included data from consumers that ate both plants and other animals? How?

6. What can you conclude from this figure about the relative numbers of predators vs. herbivores in any environment?

TERTIARY CONSUMERS

SECONDARY CONSUMERS — 10% converted to biomass

PRIMARY CONSUMERS — 10% converted to biomass

PRIMARY PRODUCERS — 10% converted to biomass

See Answers at the back of the book.

## Key Terms in Ecosystems and Communities

# Multiple-Choice Questions

1. **An ecosystem consists of _____ in a given area.**
   a) all the photosynthetic organisms
   b) all the living organisms
   c) all the abiotic factors that influence living organisms
   d) all the living organisms and non-living materials
   e) the plant life and climate

   100 HARD
   **17** 0 EASY

2. **Earth's largest terrestrial ecosystems, the biomes, are defined primarily by:**
   a) the average rainfall.
   b) the average temperature.
   c) the seasonal variability in temperature.
   d) the seasonal variability in rainfall.
   e) All of the above.

   100 HARD
   0 **11** EASY

3. **Why is it hotter at the equator than at the Poles?**
   a) The angle at which sunlight hits the earth leads to a given amount of solar energy being spread over a smaller area at the equator than at the Poles.
   b) The increased rotational speed of the earth at the Poles creates strong winds that increase radiant cooling.
   c) Greater cloud cover at the equator leads to a greater "greenhouse" effect than at the Poles.
   d) The water in the oceans has a high heat capacity, and there is significantly more water near the equator than at the Poles.
   e) All of the above are responsible for the higher average temperature at the equator than at the Poles.

   100 HARD
   **23** 0 EASY

4. **Why do beaches on the west coast of the United States have colder water than beaches on the east coast?**
   a) The ocean currents in the northern hemisphere rotate in a clockwise direction, so water off the coast of California has just come from the cold north, while water off the coast of New York has just come from the warm south.
   b) West coast beaches are not colder than east coast beaches. Ocean temperatures depend only on latitude.
   c) The Pacific Ocean is deeper than the Atlantic Ocean, so more of the sun's heat is absorbed by the colder, deep water, reducing the water temperature at the west coast beaches.
   d) Trade winds blowing north from the Tropic of Cancer in the Pacific Ocean sweep the warm air away from the west coast, taking with it much of the warmth of the water.
   e) Deep water upwellings bring cool water to the west coast beaches, while the surface currents on the east coast cause downwellings, sending warm surface water to depths where it is less likely to be lost through evaporation.

   100 HARD
   **27** 0 EASY

5. **When moist wind from an ocean blows onshore toward a mountain range:**
   a) as the air rises, it pulls moisture from the ground, causing the higher elevations to be drier.
   b) as the air goes over the top of the mountain and falls back down toward lower elevations, it holds less moisture, creating a "rain shadow" zone of unusually high precipitation.
   c) as the air rises, it holds less moisture, causing the dissipation of all clouds.
   d) as the moist air goes over the top of the mountain and falls back down toward lower elevations, it holds even more moisture, creating a "rain shadow" desert with very little precipitation.
   e) None of the above are correct.

   100 HARD
   **42** 0 EASY

6. **"Top" carnivores:**
   a) are more common than secondary consumers.
   b) rely directly on primary producers for energy.
   c) consume primarily herbivores.
   d) consume primarily carnivores.
   e) rely on symbiotic bacteria living in their digestive systems to help digest cellulose.

   100 HARD
   **26** 0 EASY

**7.** The 10% rule of energy-conversion efficiency:

a) explains why big, fierce animals are so rare.

b) explains why the biomass of herbivores must exceed that of carnivores.

c) limits the length of food chains.

d) suggests that 90% of what an organisms eats is used in cellular respiration or is lost as feces.

e) All of the above are correct.

**8.** Nitrogen enters the food chain:

a) primarily through soil-dwelling bacteria that "fix" nitrogen by attaching it to other atoms.

b) from the atmosphere when "fixed" by the photosynthetic machinery of plants.

c) when rocks dissolved by rainwater become soil, which is then utilized for plant growth.

d) through soil erosion followed by runoff into streams and ponds.

e) through methane, produced by herbivores as a by-product of the breakdown of plant material.

**9.** Which of the following statements about an organism's niche is incorrect?

a) It encompasses the space the organism requires.

b) It includes the type and amount of food the organism consumes.

c) It is not always fully exploited.

d) It may be occupied by two species, as long as they are not competitors.

e) It reflects the ways in which the organism utilizes the resources of its environment.

**10.** The "ghost of competition past" refers to the fact that:

a) competition often leads to character displacement, which remains even after direct competition is reduced.

b) competition cannot be seen in nature.

c) competition inevitably leads to the extinction of one of the competitors.

d) competition inevitably leads to the extinction of both competitors.

e) the fossil record is a record of competitive interactions.

**11.** Chemical defenses are more common among plants than animals because:

a) plants cannot move to escape predators and so must develop other deterrents.

b) the cell wall can contain the chemicals more effectively than a simple plasma membrane.

c) mechanical defenses against predators can evolve only in animals.

d) parasite loads are significantly higher in plants than in animals.

e) All of the above are correct.

**12.** Rabid animals:

a) froth at the mouth so as to increase the likelihood that the rabies-causing virus is passed on to another host.

b) have been infected by parasitic bacteria.

c) must die before their host can be passed on.

d) are infected by a viral ectoparasite.

e) often show reduced aggression, which increases their ability to get closer to other animals, thereby spreading the infection.

**13.** In a commensal relationship:

a) one species pollinates the other.

b) neither species benefits or is harmed, but the community itself benefits.

c) one species provides nutrients, usually cellulose, for the other.

d) one species benefits while the other neither benefits nor is harmed.

e) the "loser" has reduced reproductive output.

**14.** Coevolution:

a) is responsible for all the beautiful flowers in the world.

b) is responsible for nectar production by plants.

c) reveals that both biotic and abiotic resources can serve as selective forces.

d) can produce an insect with a tongue as long as its body.

e) All of the above are correct.

**15.** The chief difference between primary and secondary succession is that:

a) primary succession occurs among the plants in a habitat, whereas secondary succession occurs among the animals.

b) primary succession begins with no life or soil, whereas secondary succession begins with both.

c) secondary succession alters the biotic environment, and primary succession alters the abiotic environment.

d) primary succession occurs more quickly than secondary succession.

e) primary succession can occur in terrestrial and aquatic habitats, but secondary succession can occur only in terrestrial habitats.

**16.** **Keystone species:**

a) occur only in intertidal zones.

b) play an unusually important role in determining the species composition in a habitat.

c) can be removed from a habitat without any impact on the remaining species in the community.

d) are primary producers and therefore usually are plants.

e) are more expendable than commensal species, from a conservation perspective.

100 HARD

0 11 EASY

## Short-Answer Questions

**1.** What changes in climate occur on moving from the equator toward the Poles? Which two aspects of weather determine the type of biome found at any location?

**2.** Describe the energy flow that occurs through an ecosystem. How does this energy flow influence the number of top carnivores?

**3.** What is a niche? What is the potential fate of two species that occupy the same niche?

**See Answers at the back of the book.**
**For additional study questions, go to www.prep-u.com.**

# Human influences
## on the environment

# 16

# Conservation and Biodiversity

## 16·1–16·4
## Measuring and defining biodiversity is complex.

*Baobab trees along a dirt road in Madagascar.*

# 16·1

## Biodiversity benefits humans in many ways.

In the Pacific Northwest of the United States, a medium-size tree grows. This conifer, the Pacific yew tree (*Taxus brevifolia*), isn't the biggest tree in the forest, nor is it the most common. It's not the sort of tree that usually warrants attention in a textbook. But within the bark of the Pacific yew there is a chemical, called taxol, that has some important properties—it acts as an anti-cancer agent, interfering with the division of cells that come into contact with it (**FIGURE 16-1**). The role taxol plays for the Pacific yew is not clear; it may reduce the rate at which other organisms feed on the plant. But in humans, taxol is effective in the treatment of ovarian cancer, breast cancer, and lung cancer (generating more than $1 billion a year in pharmaceutical sales).

Consider also the following:

- The chemicals vinblastine and vincristine, isolated from the Madagascar periwinkle (*Catharanthus roseus*), are so effective in treating leukemia and Hodgkin's lymphoma that both diseases, formerly incurable, now are curable in the vast majority of people.

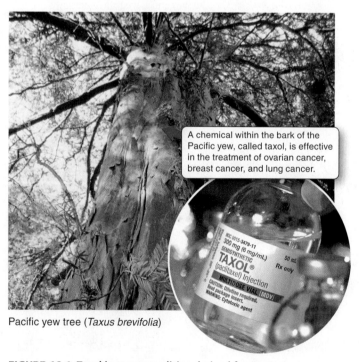

A chemical within the bark of the Pacific yew, called taxol, is effective in the treatment of ovarian cancer, breast cancer, and lung cancer.

Pacific yew tree (*Taxus brevifolia*)

**FIGURE 16-1 Taxol is cancer medicine derived from a tree.**

- The chemical ancrod, found only in the Malayan pit viper (*Agkistrodon rhodostoma*), dissolves blood clots and is effective in treating some heart attack and stroke patients.

- Epibatidine, a poisonous compound in the saliva of a small frog (*Epipedobates tricolor*) that lives in Ecuador, is 200 times more effective than morphine in relieving pain and is non-addictive.

These are just a few examples—there are many, many more—illustrating that medically important compounds often come from some unlikely organisms living in locations around the world. Usually, toxin production evolved in these organisms as a method of protecting them from other organisms. Co-opting the chemicals for human use reveals that one important value of biodiversity is as a sort of universal medicine cabinet.

But the beneficial effects of the world's biodiversity aren't limited to disease-fighting chemicals from plants and animals. This has been particularly apparent in the Gulf of Mexico during 2010 and 2011. In April of 2010, an explosion on a floating oil rig that was drilling a well about a mile under the surface of the water caused a massive leak of crude oil (**FIGURE 16-2**). Initially gushing oil at a rate of more than 2.5 million gallons per day, the well released more than 200 million gallons of oil into the Gulf over the almost three months it took to cap the well.

The oil spill caused extensive damage to the coastal and marine environments and was responsible for killing

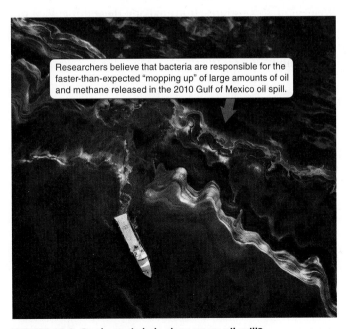

Researchers believe that bacteria are responsible for the faster-than-expected "mopping up" of large amounts of oil and methane released in the 2010 Gulf of Mexico oil spill.

**FIGURE 16-2 Can bacteria help clean up an oil spill?**

thousands of animals, including birds, dolphins, sea turtles, mollusks, and crustaceans. The longer-term consequences of the oil spill for the region—including the eight U.S. national parks threatened—are not yet clear. One bit of good news, however, came to light as scientists began asking the question: where will the oil go? Reporting in the journal *Science,* researchers using sensors at 207 locations in the Gulf were surprised to find huge regions in which the large amounts of methane released in the spill were basically gone. Oxygen levels in these regions were also very low, suggesting that bacteria in those areas—which use oxygen when consuming methane—were responsible for the rapid "mopping up" of the methane. Fragments of methane-consuming bacteria found by the researchers supported their hypothesis.

Additionally, other researchers taking samples from deepwater sites throughout the area of the spill found high concentrations of microbes (including some never before described), related to several previously observed oil-eating bacterial species. The significant drop in oxygen saturation of the water (down from 67% to 59%) suggested that the bacteria were consuming the oil more rapidly than anyone had initially expected. The researchers believe that, because the bacteria were already present in the Gulf, feeding on the slow, natural seepage of oil from the ocean bottom, they were able to respond quickly to the spill. Nonetheless, not all scientists agree with the analyses suggesting rapid bacterial consumption of the oil, and the research is continuing.

On other fronts, *Brocadia anammoxidans* is another bacterial species that may turn out to have great value to humans. This species is able to break down an important component of human waste and, more important, it can do so without oxygen. For this reason, when treating human waste with *B. anammoxidans,* it is unnecessary to aerate the sewage sludge. Normal waste treatment utilizes bacteria that need oxygen and, consequently, it requires costly, power-hungry equipment. Still other bacteria, with sulfur-metabolizing properties, are now being used to remove the stench of sewage. In these cases, as with the medicines and oil cleanups described above, the value of biodiversity to humans is utilitarian: from other species we can obtain products or make use of processes that make our lives better.

These are just a few examples of how humans use the chemicals produced naturally by plants and animals, and sometimes whole organisms themselves, in various human enterprises.

**Biodiversity** is the variety of genes, species, and ecosystems on earth. And, as we've seen from these examples, there is

**UTILITARIAN VALUE**
Biodiversity can provide materials or processes that make our lives better.

**AESTHETIC VALUE**
Biodiversity can be beautiful to look at and pleasant to experience.

**SYMBOLIC VALUE**
Biodiversity can provide imagery that conveys meaningful abstract ideas.

**NATURALISTIC VALUE**
Biodiversity can provide the satisfaction that comes from direct contact with nature.

**FIGURE 16-3 Biodiversity has value to humans in many different ways.**

tremendous value that can come from earth's biodiversity, and it often comes from unexpected places. For these reasons, the loss of biodiversity—including genes, species, and ecosystems—can be hugely detrimental to humans.

But viewing the value of biodiversity only in the context of our abilities to extract medicines or to better process the messes we make—the utilitarian view—is to take much too narrow a view. Biodiversity provides significant, perhaps less tangible, benefits by helping regulate the climate, in the form of trees that both create shade and take up carbon dioxide from the atmosphere. Biodiversity also benefits humans in numerous other ways that are difficult to quantify, but are no less valuable than medicinal plants and oil- and sludge-eating bacteria (**FIGURE 16-3**):

- *Aesthetic value.* We think of a field of flowers or a deer in the woods as beautiful to look at and pleasant to experience.

- *Symbolic value.* Humans are so connected to the natural world that images of animals and plants can convey meaningful abstract ideas: the dove as a symbol of peace; the oak as a symbol of strength; the butterfly as a symbol of transformation; the owl as a symbol of wisdom; the rose as a symbol of love.

- *Naturalistic value.* People can get great strength, solace, and satisfaction from direct contact with nature,

including the sense of awe and fascination that can come from gaining an understanding of nature's complexity and diversity.

When we recognize that people can view the same object (or organism) and value it in different ways, we are on our way to making wiser conservation decisions. This knowledge, for example, can help us anticipate potential conflict when it comes time to invest limited conservation resources in the conservation of biodiversity, and can help us seek out methods for balancing conflicting desires. We explore some strategies for making these difficult decisions later in this chapter. First, we investigate the question of where the greatest amounts of biodiversity are to be found.

### TAKE-HOME MESSAGE 16·1

Biodiversity—the variety of genes, species, and ecosystems on earth—provides significant benefits to humans in numerous utilitarian ways, such as the production of medicines and other products and processes. Biodiversity also provides important value to humans in less tangible ways, including aesthetic, symbolic, and naturalistic values.

# 16·2 Biodiversity is not easily defined.

Which habitat has greater biodiversity, one with 3 or 4 species each of birds, reptiles, mammals, plants, and insects, or one with 38 unique species of fruit flies, or one with 100 radically different species of prokaryotes? This question illustrates that it's not always clear exactly what a person means when using the term "biodiversity." In the media, discussions make it appear that there is a consensus on what biodiversity means, but there isn't.

All definitions of biodiversity refer, in a general sense, to the variety of living organisms on earth, and most reflect that biodiversity can be considered at multiple levels, from entire ecosystems to species to genes and alleles (**FIGURE 16-4**). Defining biodiversity as the variety of genes, species, and ecosystems on earth, however, is not sufficiently practical. Cataloguing all the genes in all species is not practical. And counting (or even identifying) all species is also impractical (and likely impossible). Similarly, it is difficult to determine where one ecosystem ends and another begins.

And so, returning to our original question: If you were given the task of overseeing the conservation of biodiversity in a region where these three habitats occur, which habitat would you protect first? In making your decision, you might first define biodiversity—maybe as the number of different classes or families of organisms in a habitat, or maybe as the number of distinct species in the habitat, which is, in fact, the most commonly used definition of biodiversity—to support your decision. In doing so, however, you are acknowledging that biodiversity can be defined in many ways and that the definition is not completely objective. Rather, it is a function of your values, biases, and interests. But this recognition helps us to discuss, manage, use, and conserve biodiversity in a practical and more productive manner.

**Conservation biology** is the interdisciplinary field that addresses how to understand and preserve the natural biological resources of earth, including its biodiversity at all levels (**FIGURE 16-5**). For this reason, the complexities of defining biodiversity directly affect the decisions that conservation biologists make. In this chapter, we explore how the actions of humans can have significant, even disastrous, consequences for biodiversity and natural resources, and we also explore the strategies that have been developed for effective conservation.

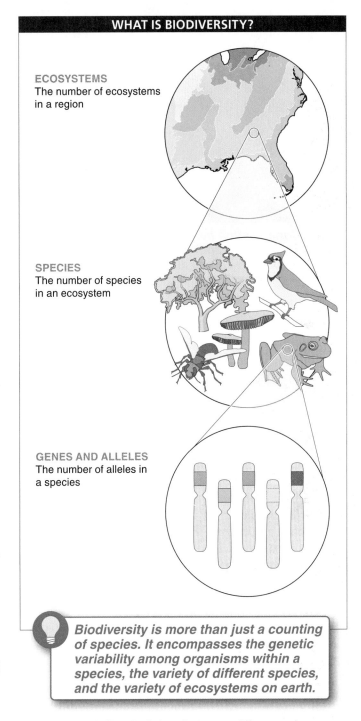

**WHAT IS BIODIVERSITY?**

**ECOSYSTEMS**
The number of ecosystems in a region

**SPECIES**
The number of species in an ecosystem

**GENES AND ALLELES**
The number of alleles in a species

*Biodiversity is more than just a counting of species. It encompasses the genetic variability among organisms within a species, the variety of different species, and the variety of ecosystems on earth.*

**FIGURE 16-4 Biodiversity is found on many different scales.**

FIGURE 16-5 Preserving the earth's natural resources.

Underpinning many of the difficulties in conservation biology is that biodiversity, as we've noted, can have value for humans in so many different ways. Throughout the chapter, we'll see how that complexity makes it difficult to balance competing interests when they are in conflict.

*Conservation biology is a multidisciplinary science that addresses how to preserve natural resources and protect biodiversity.*

## TAKE-HOME MESSAGE 16·2

**Defining biodiversity is difficult because it can be considered at multiple levels, from entire ecosystems to species to genes and alleles. In practice, it is most often defined as the number of distinct species in a habitat, which has important implications for conservation biology—the field that addresses questions of how to preserve the natural resources of earth.**

# 16·3 Where is most biodiversity?

If you were to stand at the equator in South America and identify the number of land mammals, you would count about 400 different species. If you then started walking south, away from the equator, and assessed the diversity of land mammals when you got to 20° latitude, you would find about 350 different species. Continuing your long walk south, when you got to 30° latitude you would find only 200 different species. And the trend would continue for as far south as you could go, with 100 species at 40° latitude and fewer than 50 species at 50° latitude. On a similar walk north from the equator, through North America, you would observe the same trend, with more than 300 species of land mammals at 20° latitude, reduced to only 100 species in northern Canada (**FIGURE 16-6**). Biodiversity is not evenly distributed throughout the world. And perhaps the strongest trend in biodiversity distribution is just what we've described: as you move away from the equator in either direction, diversity is reduced with increasing latitude.

The strong biodiversity gradient from the tropics to the Poles occurs not just in land mammals but in nearly any group of plants or animals observed—recent surveys have documented the trend in woody plants, herbaceous plants, and mangrove trees. Interestingly, even the diversity of organisms in the oceans follows this trend. One survey of marine copepods—tiny crustacean animals, including plankton—found the same trend moving north from the tropical region of the Pacific Ocean (80 species present) to the Arctic Ocean (fewer than 10 species). Similar patterns have been documented in fishes, mollusks, marine and terrestrial arthropods, corals, and marine protists.

**Q** Why are there more species in an acre of tropical rain forest than in an acre farther from the equator, such as in a temperate forest or prairie?

Numerous factors that influence species richness—that is, the number of species in an area—are responsible for producing the tropical-to-temperate-to-polar gradient in biodiversity. Three, in particular, play strong roles (**FIGURE 16-7**).

**1. Solar energy available.** Perhaps the simplest predictor of species diversity is the amount of solar energy available in an area. Solar energy is, after all, the fuel for all life. As we saw in Figure 15-5, the sun shines most directly

**GRAPHIC CONTENT!**
*Thinking critically about visual displays of data.*
*Turn to p. 658 for a closer inspection of this figure.*

## WHERE IS MOST BIODIVERSITY?

### LAND MAMMAL BIODIVERSITY IN THE AMERICAS

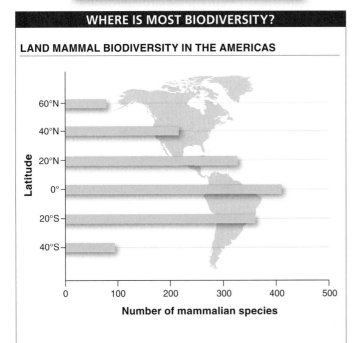

### MARINE COPEPOD BIODIVERSITY IN THE NORTH PACIFIC

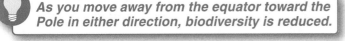

*As you move away from the equator toward the Pole in either direction, biodiversity is reduced.*

**FIGURE 16-6** Biodiversity is unevenly distributed.

on the equator, and, as we move away from the equator, the earth's curvature causes the solar energy hitting the earth to be spread out over increasingly large areas. This leads to less solar energy at any one point as latitude increases. In a variety of species of plants and animals, researchers have documented strong relationships between energy availability and species richness.

## FACTORS THAT INFLUENCE BIODIVERSITY

**SOLAR ENERGY AVAILABLE**
Greater access to solar energy, the fuel for all life, provides increased species richness.

**EVOLUTIONARY HISTORY OF AN AREA**
Communities diversify over time. The more time that passes without a climatic event, such as an ice age, the greater the diversity in an area.

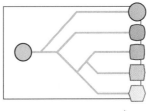

**RATE OF DISTURBANCE**
A habitat with an intermediate amount of disturbance tends to have the greatest species richness.

**FIGURE 16-7** What determines species richness in an area?

**2. Evolutionary history of an area.** Communities diversify over time. Consequently, the more time that passes, generally speaking, the greater is the diversity in an area. A high level of biodiversity in an area, however, can be knocked back down by climatic disasters such as glaciations. Thus, for example, biodiversity declines due to glaciations may have disproportionately occurred in temperate and polar regions (and on the tops of some mountains), which have been affected more frequently than tropical regions as glaciers have advanced and retreated over hundreds of millions of years.

**3. Rate of environmental disturbance.** Over time, the best competitor for a resource is expected to outcompete other species, excluding them from the community and reducing species richness. However, communities may be kept from reaching this species-depleted state. Factors such as predation, or environmental disturbances such as floods, fires, and volcanic eruptions, which may wipe out most species in an area, can prevent any one species from excluding others. For this reason, a habitat with an *intermediate* amount of environmental disturbance is expected to have the greatest species richness.

Conservation biologists are increasingly interested in **biodiversity hotspots,** those regions of the world having significant reservoirs of biodiversity that are under threat of destruction. Twenty-five biodiversity hotspots have been identified around the globe (**FIGURE 16-8**). These hotspots cover less than 1% of the world's area, but have 20% or more of the world's species. Habitats included among the biodiversity hotspots are tropical rain forests, coral reefs, and islands. While not considered a hotspot, the deepest regions of the oceans have recently proved to hold great biodiversity.

**Tropical rain forests.** These habitats are common near the equator in Asia, Australia, Africa, South America, and Central America, as well as on many Pacific islands. An important, but still poorly studied, region of tropical rain forests is the canopy, the uppermost region of a forest, where nearly all of the sunlight is intercepted. The canopy

(which can be up to 30 feet thick) is the most diverse part of the tropical rain forests; estimates suggest that as many as half of all species on earth reside in the canopy of tropical rain forests. Due to its inaccessibility (the canopy is 20–40 m above the ground), the canopy is not easily studied and the extent of biodiversity there is still not fully known.

**Coral reefs.** Built by corals, members of the animal kingdom, coral reefs occur in ocean areas with low levels of nutrients. The physical structures created by the corals provide a home and habitat for huge numbers of other species. The Great Barrier Reef of Australia (which is not listed among the 25 biodiversity hotspots, because it currently is a focus of extensive conservation efforts) is an example of just how much diversity can be supported by

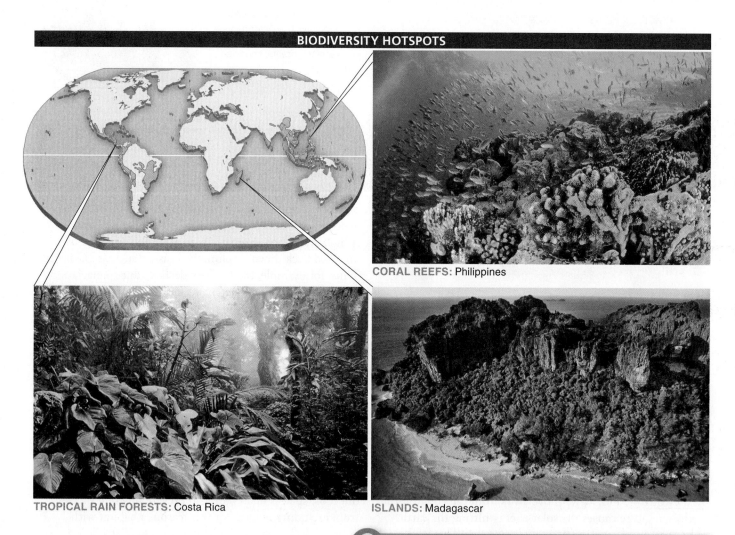

**BIODIVERSITY HOTSPOTS**

**CORAL REEFS:** Philippines

**TROPICAL RAIN FORESTS:** Costa Rica

**ISLANDS:** Madagascar

**FIGURE 16-8 Biodiversity hotspots are significant reservoirs of biological diversity.**

*There are 25 biodiversity hotspots around the world; they cover less than 1% of the world's area, but have 20% or more of the world's species.*

coral reefs. It is the largest coral reef system in the world and covers such a large area (almost the size of California) off the northeast coast of Australia that it can be seen from space. It is home to numerous species of whales, dolphins, sea turtles, sharks, and stingrays, as well as 1,500 species of fishes and 5,000 species of mollusks. Five hundred species of algae are found there, too. Above the Great Barrier Reef, more than 200 species of birds are supported by the rich diversity of life below. Other coral reefs are not as well protected by laws and conservation efforts, and the world's second largest coral reef, in New Caledonia in the southwestern Pacific, is one of the 25 biodiversity hotspots.

**Islands.** Because of their generally smaller size, which limits the population size of organisms living on them, islands can be regions where biodiversity is at risk. Madagascar, located in the Indian Ocean off the southeastern coast of Africa, is the fourth largest island in the world (just a bit smaller than Texas). It is home to 5% of the world's species, 80% of which are **endemic** (exclusively native to a place)—a higher percentage of endemic plants and animals than in any comparably sized area on earth. One of the 25 biodiversity hotspots, Madagascar is an area unusually rich in species diversity.

**Deep oceans.** The oceans that cover two-thirds of the earth's surface can be much more difficult to explore than terrestrial habitats. Deep ocean regions, in particular, are much less studied than the terrestrial parts of the world. Once thought to be completely devoid of life, the deepest regions of the oceans have recently been found to be teeming with biodiversity. One survey of the deepest parts of the Antarctic Ocean, sampling at a depth of more than 18,000 feet (6,000 m), discovered 585 new species of crustaceans alone.

---

### TAKE-HOME MESSAGE 16·3

Biodiversity is not evenly distributed over the earth. For nearly all groups of plants and animals, both marine and terrestrial, biodiversity is greatest near the equator and progressively falls toward the North and South Poles. Factors that influence the species richness in an area include the amount of solar energy available, the area's evolutionary history, and the rate of environmental disturbance. Biodiversity hotspots are regions of significant biodiversity under threat of destruction.

---

# 16·4 Island biogeography helps us understand the maintenance and loss of biodiversity.

"Can you help me exterminate all the insects on a few islands?" In 1966, when E. O. Wilson started calling exterminators in Florida, asking them to help him put tents over several islands and then pump ethyl bromide into them to kill all of the insects, he had trouble being taken seriously. Most of the exterminators opted to pass on the job, but eventually he found one (**FIGURE 16-9**).

Why would a biologist (of all people) want to wipe out all insect life on six islands? Six years earlier, Robert MacArthur and Wilson had developed a theory, and now Wilson and one of his students, Daniel Simberloff, had an idea for a controlled experiment with which they could test the theory.

MacArthur and Wilson's theory, called the "theory of island biogeography," is an attempt to explain and predict the patterns of species diversity on islands. The theory has two fundamental components (**FIGURE 16-10**).

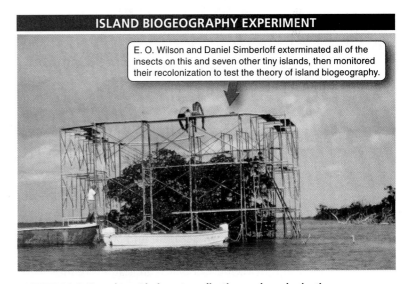

**ISLAND BIOGEOGRAPHY EXPERIMENT**

E. O. Wilson and Daniel Simberloff exterminated all of the insects on this and seven other tiny islands, then monitored their recolonization to test the theory of island biogeography.

**FIGURE 16-9 Experiment in insect eradication and recolonization: explaining patterns of species diversity on islands.**

The theory of island biogeography is designed to explain and predict patterns of species diversity on islands. It has two fundamental components:

**THE AREA EFFECT**
There is a close relationship between the number of species inhabiting an island and the island's area—the larger an island is, the more species it holds.

**Number of species on island**

**THE DISTANCE EFFECT**
There is a close relationship between the number of species inhabiting an island and how isolated an island is—the farther an island is from the mainland, the fewer species it holds.

Mainland

**Number of species on island**

**FIGURE 16-10 The lonely island.** The theory of island biogeography generates predictions about the patterns of species diversity in isolated habitats.

**1. The area effect.** MacArthur and Wilson noted that observations of islands in the West Indies had revealed a close relationship between the number of species inhabiting an island and the island's area: the larger the island, the greater the number of species present. This relationship held true for bird species as well as plants, beetles, and snails.

**2. The distance effect.** MacArthur and Wilson's observations of islands in the Pacific Ocean revealed a close relationship between how isolated an island was, measured as its distance from the nearest mainland, and the number of species that inhabited the island: the farther from the mainland the island was, the fewer species it held.

**1 RATE OF IMMIGRATION**
Islands closer to the mainland have higher rates of immigration; the mainland serves as a pool of new species. The rate of immigration of new species to the island decreases as the number of species on the island increases.

**2 RATE OF EXTINCTION**
Smaller islands have higher rates of extinction than large islands (due to smaller population sizes). The rate of extinction of species on the island increases as the number of species on the island increases.

Rate of immigration

Rate of extinction

Island close to mainland

Small island

Equilibrium number

Island far from mainland

Large island

**Number of species on island**

**3** The number of species on an island reflects a balance between immigration to the island by new species and extinction of species on the island.

**Time**

The composition of species on an island is not fixed; rather, it might change over time. An island might have five species of birds, but exactly which five species occur on the island may change over time.

**FIGURE 16-11 Same number, but possibly different species.** On islands, species composition changes over time.

From these observations, MacArthur and Wilson concluded that the number of species on an island represents a balance between the rate of immigration of new species to the island and the rate of extinction of species on the island. They proposed that the rate of immigration to an island would primarily be determined by how close the island was to the mainland, with the mainland serving as a pool of new species. And they proposed that the rate of extinction on an island would primarily be determined by the island's size: smaller islands generally support a smaller population size, making the population more vulnerable to extinction. Beyond enabling them to predict the number of species on an island, MacArthur and Wilson's theory also suggested to them that the composition of species on the island would not be fixed; rather, it might change over time. In other words, once it was "full," an island might have 10 species of birds, but exactly *which* 10 species occurred on the island could change over time (**FIGURE 16-11**).

So, in 1966, Wilson's reason for exterminating all of the insects on multiple islands was so that he could monitor the islands as they were recolonized following the exterminations. He chose the mangrove islands in the Florida Keys because they include thousands of small islands, some only yards across. He and Simberloff selected eight islands, 33–54 feet (11–18 m) in diameter. The islands varied in how far they were from the nearest mainland, with the closest just 6 feet (2 m) away and the farthest more than half a mile (1,188 m) away.

After selecting two islands as control islands, the biologists and their assistants fumigated the others, verifying that all of the insects were killed. They then surveyed the islands regularly over the next year. Here's what they found—as predicted.

1. The closer islands were recolonized more quickly than the distant islands.

2. On all of the islands, the number of insect species after fumigation returned to about the same number as before fumigation.

3. The composition of insects on each island changed over the course of the study, even on the two control islands.

The theory of island biogeography is not just about real islands—lakes and mountain tops and other geographic sites can be thought of as islands, too. The theory makes testable predictions about the biodiversity levels that occur in such areas. Perhaps most important, conservation biologists can gain insights from island biogeography into the design of nature preserves. The importance of developing nature preserves as close as possible to sources of new species, for example, is one such insight. We explore others later in this chapter, when discussing the strategies for preservation of biodiversity.

## TAKE-HOME MESSAGE 16·4

The theory of island biogeography generates predictions about the patterns of species diversity in isolated habitats (including not only islands but lakes and mountain tops, too). Both the size of the habitat and its distance from a pool of new species influence the number of species present. Ideas gained from island biogeography have important implications for the development of nature preserves.

## 16·5–16·6
## Extinction reduces biodiversity.

*The Bengal tiger has been hunted, captured, and poisoned to near extinction.*

# 16·5 There are multiple causes of extinction.

Imagine that you are on a deserted island. Alone. It would be hard and it would be lonely. The hope of rescue someday, however, would probably make things bearable. "Martha," a passenger pigeon living in the Cincinnati zoo in the early 1900s, found herself in a situation of this sort—minus the hope. Passenger pigeons in the wild had been completely wiped out by a combination of factors, including hunting of the birds for meat, loss of their habitat due to deforestation in North America, and possibly disease. And so, except for the small flock that Martha was part of, the species was gone. One by one, those birds died as well, and when the second-to-last passenger pigeon died in 1912, Martha was all alone. She lived for two more years, with no possibility for ever reproducing; she died on September 1, 1914 (**FIGURE 16-12**).

On average, species persist for about 10 million years, although some may last for hundreds of millions of years. An extinction occurs when all individuals in a species have died. And whether you look at biodiversity from a utilitarian, aesthetic, or symbolic perspective, extinction is a tragic loss.

As we discussed in Chapter 10, extinctions can be divided into two general categories: mass extinctions and background extinctions (**FIGURE 16-13**), which are distinguished by the numbers of species affected. A mass extinction occurs when a large proportion of species on earth are lost in a short period of time (typically less than two

*Once numbering in the millions, passenger pigeons were hunted to extinction. The last passenger pigeon on earth, named Martha, died at the Cincinnati Zoo in 1914.*

**FIGURE 16-12  The only way to see one today.** The passenger pigeon is a species completely lost to the earth. Shown here: a stuffed passenger pigeon from the Cincinnati Zoo. *Inset:* the aftermath of a pigeon-hunting expedition; in a single day as many as 50,000 birds could be killed.

Extinctions generally fall into two categories:

○ **MASS EXTINCTIONS**
A large number of species (even entire families) become extinct over a short period of time due to extraordinary and sudden environmental change.

○ **BACKGROUND EXTINCTIONS**
These extinctions occur at lower rates during times other than mass extinctions.

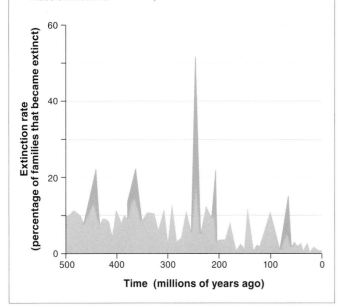

**FIGURE 16-13 Background and mass extinctions.**

million years, and in some cases much less)—such as when an asteroid struck earth 65 million years ago and about 75% of all species were wiped out. These extinctions are above and beyond the normal rate of extinctions that occur in any given period of time, referred to as the background extinction rate.

**Q**

**Does extinction only happen to weak, unfit species?**

Both background and mass extinctions result in the same outcome for the species involved, but the causes tend to differ. In mass extinctions, the particular features of a species' biology—its biochemistry, physiology, behavior—don't play a role; rather, it's more like really bad luck. Background extinctions, by contrast, tend to be a consequence of one or more aspects of a species' biology.

For any species, there is always a risk of extinction. This risk can be larger or smaller, depending on certain features. Here we look at three important aspects of a species' biology that can influence its extinction risk (**FIGURE 16-14**).

**GEOGRAPHIC RANGE:**
EXTENSIVE vs. RESTRICTED
Species restricted in their range are more vulnerable than those with extensive ranges.

**LOCAL POPULATION SIZE:**
LARGE vs. SMALL
Species with small population sizes are at increased risk of extinction.

**HABITAT TOLERANCE:**
BROAD vs. NARROW
Species with narrow habitat tolerances are at greater risk of extinction than species with broader habitat tolerances.

**FIGURE 16-14 Range, population size, and tolerance influence risk of extinction.**

**1. Geographic range: extensive versus restricted.** Species restricted in their range—including those limited to small bodies of water and those confined to islands—are more vulnerable than species with extensive ranges. The Tasmanian devil is a marsupial carnivore about the size of a dog. Although these animals once thrived in Australia, they now are confined to the island of Tasmania, smaller than the state of Maine. Unable to expand their range, Tasmanian devils are more vulnerable to extinction than if their range was not limited. The house mouse, on the other hand, is found nearly everywhere in the world and so has a much lower risk of extinction.

**2. Local population size: large versus small.** Tigers and California condors, along with *Welwitschia,* a slow-growing, long-lived plant in southwest Africa, are examples of species that—as a consequence of their small population sizes—are at increased risk of extinction. With a small population size, a species is more susceptible to extinction due to a variety of factors that can kill individuals, including fire, diseases, habitat destruction, and predation. Put simply, the more individuals that are alive, the more likely it is that some of them will survive such events. In addition, having more individuals to breed with typically leads to greater genetic variation. When a population is small, inbreeding is increased, which generally reduces the fertility and longevity of individuals.

**3. Habitat tolerance: broad versus narrow.** "Habitat tolerance" describes the breadth of habitats in which a species can survive. Some plant species, for example, can tolerate large swings in water availability or soil pH or temperature. Others are limited to very narrow habitat ranges. The now-extinct passenger pigeons could only build nests in a specific type of forest and needed large numbers of individuals, breeding communally, in order to breed successfully. As forests were cut down and as the birds were

hunted, the size of their flocks diminished. Their narrow habitat tolerance made them extremely vulnerable to extinction. In general, because species with narrow habitat tolerance cannot adapt in the face of habitat degradation and loss, they are at greater risk than species with broader habitat tolerance.

One species that is enormously successful, thanks to its extensive geographic range, large local population sizes, and broad habitat tolerance, is our own. Humans are growing and expanding their range at a staggering rate, consuming resources in an unprecedentedly voracious fashion. By any measure, humans are one of the most successful species in earth's history. This success, unfortunately, is having an increasingly negative effect on the species with which we share the planet, causing extinctions at a significantly higher rate than ever before. We next explore in more detail the conflict between humans' success as a species and the survival of other species.

---

### TAKE-HOME MESSAGE 16·5

Extinctions occur for fundamentally different reasons. Mass extinctions are the result more of bad luck than of the particulars of a species' biology, including its biochemistry, physiology, and behavior. Background extinctions tend to be a consequence of one or more features of the species' biology. Small population size, limited habitat range, and narrow habitat tolerance contribute to background extinctions.

---

# 16·6 We are in the midst of a mass extinction.

Sometimes you can be so close to something that it's difficult to really see it. That may be the situation that humans are in right now when it comes to the global loss of biodiversity. But we are becoming increasingly aware that we are in the midst of a mass extinction. In a recent survey by the American Museum of Natural History, 70% of biologists indicated that they believed this is true and that steps must be taken by governments and individuals to stop this massive loss of species.

Data on current rates of extinction in every well-studied group of plants and animals support the hypothesis that a mass extinction is under way. Among the mammals, 11% of

all species are currently *endangered,* under imminent threat of extinction, and 14% are *threatened,* characterized as vulnerable to extinction. Among birds, 4% of all species are endangered and 50% are in decline. Almost a third of all amphibian species are threatened. Among fishes, mollusks, insects, fungi, and plants, the story is the same.

Historically, background extinction rates are about one extinction per million species per year. Currently, as the numbers above suggest, extinction rates are 1,000 times (or more) greater than usual. It is difficult to know the exact magnitude of the problem because, as we saw in Chapter 10, biologists don't have much of an idea about

**DEFORESTATION**

*Deforestation in the tropics is a leading cause of biodiversity loss.*

**FIGURE 16-15** Removing habitat through deforestation.

*Bonobos, the closest living relative of humans, have been driven close to extinction, in part due to hunting.*

**FIGURE 16-16** Exploitation of species: the bonobo.

how many species there are on earth—the estimates range from 5 million to 100 million.

Unlike the last mass extinction event, in which the dinosaurs and most other species were wiped out in the wake of an asteroid's collision with earth, this current mass extinction seems to be the result of the activity of one species—humans. Ironically, it is the unprecedented success of humans that is responsible for this situation. The resource needs of so many people have inevitably interfered with the ability of other species to coexist with us.

The chief reason for the loss and impending loss of so many species is habitat loss and habitat degradation. Particularly harmful is habitat loss in earth's tropical rain forests, where biodiversity is greatest. In the past 25 years, half of the world's tropical rain forests have been destroyed, usually by burning to make way for agricultural use of the land or by logging (**FIGURE 16-15**). Urban development, too, is responsible for destruction of rain forests, as the growing human populations continue to expand. Intensive agriculture, livestock grazing, and the development of urban centers have led to the destruction and fragmentation of habitats worldwide.

The loss of biodiversity-supporting habitat is not restricted to the tropics. In the Pacific Northwest, as well, logging of forests has occurred at an unsustainable rate, leading to significant reductions in a diverse range of populations.

Beyond habitat loss, another factor that reduces biodiversity is overexploitation of species. Such overexploitation includes the killing of animals for food, pelts, tusks or other body parts, and medicinal products. Bonobos (*Pan paniscus*), for example, a type of chimpanzee and our closest living relatives, are almost never seen in the wild. As a consequence of habitat loss and their exploitation as a source of food in parts of central Africa, they have been driven to near extinction (**FIGURE 16-16**). And, as we'll see later in this chapter, the introduction of exotic species into habitats where they would not naturally occur is also having a significantly adverse effect on biodiversity. Consequences of the loss of biodiversity, beyond the values of biodiversity discussed earlier, are largely unknown, but most likely include a serious reduction in the capacity of the environment to provide clean air and water and to recover from environmental and human-induced disasters.

Later in this chapter, we discuss some strategies by which the current high rate of biodiversity loss can be reduced.

### TAKE-HOME MESSAGE 16·6

Most biologists believe that we are currently in the midst of a mass extinction, that it is the result of human activities, and that it poses a serious threat to the future survival of humans.

*Rain forest next to city sprawl, Rio de Janeiro, Brazil.*

# 16·7–16·12
# Human interference generally reduces biodiversity.

# 16·7

## Some ecosystem disturbances are reversible, others are not.

When you fly over southern New England today, you look down on an undulating carpet of forest that covers hills and valleys. Roads traverse the forest, and here and there you see towns and cities or cultivated land and pastures, but most of what you see is the tops of trees.

If you could have flown over the same route 200 years ago, things would have looked very different. In the early 1800s, most of the forest in southern New England had been cleared and the land was in cultivation. Homesteads consisting of farmhouses, barns, stables, and other outbuildings were scattered across the countryside. Stone walls extended across hills and valleys, separating cropland from pastures.

Going back farther, the same flight 400 years ago would have taken you over mostly forested land. Here and there, Indian villages would have stood among cleared fields, but most of what you saw then would have looked much like what you see today.

The change from forest 400 years ago to cleared land 200 years ago was a major ecosystem disturbance, but now the area has returned to forest. Today, the stone fences found deep in the New England woods are the only obvious trace of the agricultural history of the area. The fences still snake their way up and down hills, but now they run through solid forest instead of between fields and pastures.

By the middle 1800s, New England settlers were abandoning farms on marginal land and moving west. The fields they had cultivated began a process of ecological succession that returned them to forests (see Section 15-15).

In the first year after a field is abandoned, its bare soil provides a harsh environment for plants. Sunlight blazes down, heating the ground surface and baking moisture from the soil. It takes a tough plant to sprout and grow under those conditions, but there are plants that can do it. We call those pioneering species of plants "weeds," and we see them growing in disturbed habitats—along roadsides, in vacant lots, and even in cracks in pavement.

**Q** Once land is cultivated or developed, can it ever return to its natural state?

As weeds grow and die, year after year, their organic matter enriches the soil and helps it retain moisture. And every year, the plants grow more densely so that they shade the ground surface, which no longer gets quite so hot and dry. Additional species of plants can grow in these milder conditions. These plants are taller, so they provide still more shade and the ground becomes still cooler and more moist. At this stage, the seeds of bushes and small trees can sprout. As these woody plants grow, they make the ground cool and shady enough to allow the seeds of forest trees (oak, maple, or beech) to sprout. More saplings of forest trees sprout and grow every year, and eventually they are so closely packed that little sunshine penetrates their leafy canopies to reach ground level. At this point, the forest stops changing, because the mature trees have created conditions in which only their own seeds can sprout and grow.

An ecosystem disturbance is reversible when the disturbance alters the biotic (living organisms) and abiotic (decaying

May 1980            September 1980            July 2011

**FIGURE 16-17** **The forest returning.** Just two decades after an explosive eruption of Mount St. Helens, in Washington State, shrubs and small trees have returned to the plain at the foot of this active volcano.

organisms, soil, rocks) nature of the habitat, but does not result in the complete extinction of any species, so species can re-establish their populations. The example of New England forests is based on a disturbance that was caused by humans—clearing land for agriculture in the 1700s—but wind storms, forest fires, floods, and volcanic eruptions have been clearing forests throughout the history of life (**FIGURE 16-17**).

> “Any fool kid can step on a beetle. But all the professors in the world can't make one.”
> — ARTHUR SCHOPENHAUER, German philosopher (1788–1860)

If an ecosystem disturbance involves the complete loss of a species to extinction, however, the disturbance is irreversible. Each species is the result of a long and uninterrupted evolutionary history, involving an interplay of random and selective forces, and producing a unique genome. Once lost, a species can never be created again (**FIGURE 16-18**). For this reason, ecosystem disturbances that involve the loss of species can have more devastating consequences than those that do not lead to extinctions, no matter how great the changes to the abiotic environment might be.

**TAKE-HOME MESSAGE 16·7**

An ecosystem disturbance is reversible when the disturbance does not include the complete extinction of any species, so they can re-establish their populations. An ecosystem disturbance that involves the complete loss of a species to extinction is irreversible because a species, once lost, can never be created again.

EXTINCT            EXTINCT            ENDANGERED

Tasmanian wolves        Chinese river dolphin        Blue whale

**FIGURE 16-18** **The end of a species.**

*Once lost, a species will never exist again. The blue whale—the largest animal ever to have lived on earth—is at risk of being lost forever.*

# 16·8 Disruptions of ecosystems can be disastrous: 1. Introductions of exotic species.

In the rain and sleet of a Chicago winter, the idea of colorful birds flying about is very appealing to some people, and you might see exactly that, in the form of the monk parakeet. These small green members of the parrot family are about a foot (30 cm) from head to tail tip. And if they look out of place in Chicago, it's because they are. Monk parakeets are just one example of an **exotic species** (also called **introduced species**), which refers to species introduced by human activities, intentionally or accidentally, to areas other than the species' native range.

Native to southern South America, monk parakeets were imported to the United States as pets in the 1960s and 1970s. Some of those pet birds escaped, and free-living populations of monk parakeets now occur in 11 states across the country, including Oregon, Florida, and New York, in addition to those mentioned above, in Illinois (**FIGURE 16-19**).

The phrase "exotic species" suggests colorful birds and butterflies winging through a forest, but the ecological reality of such species can be much darker. Because although some introduced species do not cause harm to their new habitat, in many cases they can cause harm—including economic or environmental harm, or harm to human health.

Monk parakeets were imported from South America as pets. Some escaped and now make a nuisance of themselves in cities such as Chicago, where they build huge nests on utility poles, transformers, and floodlights.

**FIGURE 16-19 A parakeet in the Midwest.**

When species are introduced and cause harm, they are referred to as **invasive species.**

The appearance of these free-living monk parakeets in the United States was greeted with alarm because they are regarded as agricultural pests in South America. Records from the Inca civilization that flourished before Pizarro's invasion of Peru in 1532 attest to crop damage by monk parakeets. And when Charles Darwin visited Uruguay in 1833, he was told that monk parakeets attack fruit orchards and grain fields. Fortunately, the dire predictions that these parakeets would devastate crops in the United States have not been fulfilled as yet, but monk parakeets do make a nuisance of themselves in cities such as Chicago, where they build nests on utility poles, transformers, and floodlights. There is nothing delicate about a monk parakeet nest—it is a large mass of sticks and twigs, and a nest that has been soaked with rain is heavy enough to bring down wires, interrupting electrical service to entire neighborhoods.

While monk parakeets were initially introduced to the United States intentionally, in many cases exotic species introductions have been unintentional, as was the case with the brown tree snake, which reached Guam sometime before 1952 by stowing away in shipments of military equipment just after the Second World War. In either case, two characteristics, in particular, can lead to introduced species becoming harmful, invasive species (**FIGURE 16-20**).

1. Exotic species have no predators in their new habitat, so their populations may grow unchecked.

2. Native plants and animals have no mechanisms to compete with or defend themselves against invading exotic species.

**Q** | Why should we worry about exotic species?

Guam, which lies in the middle of the South Pacific Ocean, had no native snakes and no predators specialized to eat snakes. But Guam did have a magnificent fauna of native birds that had evolved in isolation on the island for thousands of years. Because the species of birds on Guam had never experienced predation by snakes, they had no defense mechanisms against them—and brown tree snakes have eradicated most of the native species of birds. Extraordinary efforts are being made to prevent brown tree snakes from spreading to other Pacific islands, such as Hawaii. Like Guam, these islands are home to species of birds that occur

**PROBLEM**
Exotic species can threaten native populations.

**CAUSE**
Exotic species are often introduced, both accidentally and intentionally, by humans. Once introduced, these species can multiply, unchecked by predation, overwhelming competitors and irreversibly altering ecosystems.

**STRATEGIES FOR SOLUTION**
Better regulation and restriction of intentional introductions; better vigilance against accidental introductions

**FIGURE 16-20 Understanding exotic species.**

nowhere else in the world, and the islands have no native species of snakes and no native predators of snakes.

One current promising strategy used by the U.S. Department of Agriculture is to airdrop dead mice that have tablets of acetaminophen—the active ingredient in Tylenol—packed into them. In addition to killing live animals, brown tree snakes scavenge dead animals such as mice. And it turns out that acetaminophen interferes with the oxygen-carrying capability of the snake's hemoglobin, and a child's dose of Tylenol will kill a brown tree snake. While initial results are promising, it is too early to predict the long-term success of this program.

We don't have to look beyond the borders of the continental United States, though, to find examples of exotic species that are responsible for massive ecological shifts and economic costs. Purple loosestrife is an attractive flowering plant that is native to Eurasia. It was imported to the United States in the 1800s as a garden plant. Once here, it rapidly escaped from cultivation and invaded wetlands in every state except Florida. It produces thousands of seeds a year, and also spreads by sending out underground stems. This aggressive growth overwhelms native grasses, sedges, and flowering plants, replacing diverse wetland communities with monocultures of loosestrife that provide poor-quality habitats for bog-dwelling insects, birds, reptiles, amphibians, and mammals (**FIGURE 16-21**).

The title of Most Destructive Invaders in North America may belong to the zebra mussels and quagga mussels. These thumbnail-size freshwater mussels are native to eastern Europe and western Asia. In the early 1800s, the mussels spread to western Europe, and they came to North America in the 1980s, after completion of the Saint Lawrence Seaway allowed oceangoing ships (which carried mussels in their ballast water) to travel from the Atlantic Ocean to the Great Lakes. Female mussels produce more than a million eggs per year, and the larvae settle on any solid surface. Sometimes they settle so thickly that they clog the intake pipes of water-treatment plants and factories and the cooling systems of power plants.

The ecological threat that zebra mussels represent is far more serious than the damage they cause to industrial facilities. The Great Lakes fisheries are based on game species (lake trout, salmon, walleye) and commercial species (whitefish, perch, herring), and they produce revenues of more than $1.5 billion annually. Most of the commercially valuable fish feed on small fish species, such as smelt, and the small fish feed on microscopic plants and animals (phytoplankton and zooplankton). Zebra and quagga mussels are exceptionally efficient at filtering phytoplankton and zooplankton from the

## EXOTIC SPECIES: ACCIDENTAL INTRODUCTIONS

**BROWN TREE SNAKE**
Brown tree snakes, introduced to the island of Guam, have eradicated most of the native species of birds, which had never experienced predation by snakes and had not developed defense mechanisms.

**PURPLE LOOSESTRIFE**
Purple loosestrife, introduced to the United States as a garden plant, escaped from cultivation and has invaded wetlands in every state except Florida. The aggressive plant outgrows native grasses, sedges, and flowering plants.

**ZEBRA MUSSELS**
Zebra mussels, introduced to the Great Lakes, can cause damage to industrial facilities by clogging water-intake pipes. They also deplete resources available to small fish, eliminating the food supply for larger game fish.

**FIGURE 16-21 Unwanted guests.** Exotic species sometimes get into natural habitats accidentally. The results can be devastating.

Introduced to Australia to control agricultural pests, cane toad populations exploded and have become an ecological nightmare, killing indigenous snakes, lizards, birds, and even crocodiles.

FIGURE 16-22 Cane toad infestation.

One legendary pest is the cane toad. Ironically, it was initially imported into Australia to control a native sugarcane pest, the cane beetle. Unfortunately, not only were the cane toads unsuccessful at controlling the pest populations, but the cane toad populations exploded—preying on numerous species of amphibians, reptiles, mammals, and even birds— and spread throughout much of the continent. They now number more than 200 million and have become a nuisance in much of the country, as conservation workers struggle to determine how to control their numbers (**FIGURE 16-22**). The cane toad illustrates a common problem encountered when species are intentionally introduced, say, as a method of biologically, "naturally" controlling pests: the effects of a species on an ecosystem in which it does not naturally occur are unpredictable and often terrible.

### TAKE-HOME MESSAGE 16·8

Exotic or introduced species are species intentionally or accidentally introduced into a new habitat. They are considered invasive species if they cause harm in their new habitat. Exotic species often come to be considered invasive because in a new habitat there often are no natural predators to reduce their population size and they often encounter prey that have few or no defenses. Invasive species can dominate and irreversibly alter communities and entire ecosystems.

water—so good that they are depleting the resources available to the small fish. Without enough food, the populations of small fish are crashing and that, in turn, is eliminating the food supply for the large game and commercial fish. The Great Lakes contain nearly one-fifth of the fresh water in the world, but this enormous ecosystem is being irreversibly degraded by just two exotic species.

# 16·9 Disruptions of ecosystems can be disastrous: 2. Acid rain and the burning of fossil fuels.

"What goes up must come down." That's a saying that applies to molecules just as much as it does to larger objects. The difference is that when molecules come down, they may not be in the same form as when they went up. In particular, some chemicals rise into the atmosphere as gases and return to earth as solutions of acid. Fog, sleet, snow, or rain—all forms of precipitation can be acidic.

*How can normal rain be made ten times more acidic?*

We use the term "fossil fuels" to describe oil, natural gas, and coal. These substances are composed largely of carbon and hydrogen, and they form carbon dioxide ($CO_2$) and water ($H_2O$) when they burn. Although they are referred to as hydrocarbons, fossil fuels are not *pure* hydrocarbons. In addition to carbon and hydrogen, they also contain substantial quantities of other elements. In

particular, it is the sulfur and nitrogen in fossil fuels that cause acid precipitation (**FIGURE 16-23**).

When sulfur burns, it produces the gas sulfur dioxide ($SO_2$), and nitrogen produces nitrogen dioxide ($NO_2$). These gases react with water vapor in the atmosphere to produce sulfuric acid ($H_2SO_4$) and nitric acid ($HNO_3$), and these acids fall to earth as acid precipitation (**FIGURE 16-24**).

In North America, precipitation has an average pH as low as 4.3 in some parts of the Northeast, more than 10 times as acidic as clean rain, which has a pH of about 5.6. (Clean rain is more acidic than pure water because carbon dioxide in the air dissolves in the water to form carbonic acid, making the rain slightly acidic as it breaks down.) High

1. Burning fossil fuels releases the gases sulfur dioxide and nitrogen dioxide.

2. When combined with water vapor in the atmosphere, these compounds form sulfuric acid and nitric acid.

3. The acids fall to earth as acid precipitation.

**FIGURE 16-23** Formation of acid rain.

**PROBLEM**

Acid precipitation kills plants and aquatic animals directly, and also acts indirectly via changes in soil and water chemistry.

**CAUSE**

Burning fossil fuels releases sulfur dioxide and nitrogen dioxide. The compounds form sulfuric and nitric acids when combined with water vapor.

**STRATEGIES FOR SOLUTION**

Tighter regulation and reduction of sulfur dioxide and nitrogen dioxide emissions

**FIGURE 16-24** Acid precipitation: the problem and its cause.

Pacific Ocean. Precipitation in Europe and Asia is also acidified by local and distant sources of pollution.

Both terrestrial and aquatic organisms are harmed by acid precipitation, and the effects of acid precipitation can be direct or indirect. Direct effects result from contact of living tissues with acidic water, whereas indirect effects are produced by interactions between non-living and living systems.

Dramatic evidence of the direct effects of acid precipitation on vegetation can be found in mountain forests that are often blanketed by fog. When trees are exposed to acid fog, year after year, their leaves are damaged, their rates of photosynthesis decrease, and eventually, many of the trees die (**FIGURE 16-26**).

concentrations of sulfuric and nitric acids in the atmosphere are to blame, and several factors play a role. The Northeast has more people per square mile than the rest of the United States, and that means there are more houses, factories, and automobiles burning the oil, coal, and gasoline that create acid precipitation (**FIGURE 16-25**).

But not all of the pollution that causes acid precipitation is local. In the United States, the Midwest and Southeast have a large number of electric power plants that burn coal, and wind currents carry the sulfur dioxide and nitrogen dioxide from these power plants across the Northeast. The precipitation in the western states is not as acidic as that in the Northeast (but it's still nearly 100 times more acidic than pure water). Much of this region's acid precipitation comes from sulfur dioxide and nitrogen dioxide from Asia, which are converted to acids as they blow eastward across the

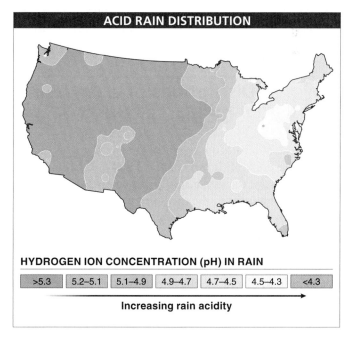

**ACID RAIN DISTRIBUTION**

**HYDROGEN ION CONCENTRATION (pH) IN RAIN**

| >5.3 | 5.2–5.1 | 5.1–4.9 | 4.9–4.7 | 4.7–4.5 | 4.5–4.3 | <4.3 |

**Increasing rain acidity**

**FIGURE 16-25** Acid precipitation across the United States. Pure water has a neutral pH of 7.

Trees exposed to acid rain and acid fog have damaged leaves and decreased rates of photosynthesis. Eventually, many of the trees die.

**FIGURE 16-26  Acid precipitation can kill trees.**

The indirect effects of acid precipitation on forests are not as conspicuous as the direct effects, but they are more far-reaching. As acid falls on the soil, it carries away calcium, potassium, magnesium, and sodium ions. These ions are essential nutrients for plants, and when they are leached out of the soil, the rate of plant growth is reduced. Some forest animals also feel the impact of this change in soil chemistry—when calcium is depleted, for example, the shells of snails are thinner than usual. As a result of the thinning of snail shells, birds that eat snails receive less calcium than normal, and these birds lay eggs with shells so fragile that they can crack while the parents are incubating them.

Lakes and streams are also affected directly and indirectly by acid precipitation, and some lakes in the northeastern United States and in northern Europe have pH values as low as 4.0–4.3, which is 1,000 times more acidic than pure water. Very few species of insects, mollusks, crustaceans, fishes, or amphibians can live in water that acidic. Still worse, acid precipitation dissolves aluminum in soil and carries the dissolved metal (aluminum ions)—which is toxic to animals—into lakes and streams, where it can kill most forms of aquatic animal life.

Acid rain is a solvable problem, because the deposition of acids from acid rain can be both prevented and cleaned up. The best approach is to prevent or reduce the emissions of sulfur dioxide and nitrogen dioxide. Preventive measures include improving energy efficiency, reducing the use of coal, switching to use of natural gas, and increasing the use of renewable energy resources, such as wind energy and solar energy. In the United States, many coal-burning plants have now installed special "scrubbers" on their smokestacks. The scrubbers extract hot gases from the power plant and function like sponges, removing sulfur dioxide before it is released into the atmosphere and converting it to a less hazardous form that can be physically removed from the tower and disposed of.

---

### TAKE-HOME MESSAGE 16·9

Burning fossil fuels releases the gases sulfur dioxide and nitrogen dioxide, and these compounds form sulfuric and nitric acids when they combine with water vapor in the atmosphere. Rain, fog, sleet, and snow that contain these acids can be more than 10 times more acidic than clean rain. Acid precipitation kills plants and aquatic animals directly, by contact with living tissues, and indirectly, through changes in soil and water chemistry.

---

# 16·10  Disruptions of ecosystems can be disastrous: 3. Global climate change and the greenhouse effect.

Not long ago, debates about global climate change focused on two questions: "Is it real?" and "Is it caused by human activities?" Now, the vast majority of scientists believe it is clear that the answer to the first question is *yes,* and that it is very likely that the answer to the second question is also *yes.* As a consequence, the scientific debate has increasingly moved to questions such as "What effect will global climate change have on the world we know?" and "How can we stop it?" Nonetheless, in part because of the significant political implications of global climate change—including, but not limited to, global warming—and because of the complex nature of the data establishing its occurrence, global climate change and the role of human activities remains a controversial topic.

Some of the evidence that the earth's atmosphere is warming comes from weather records that extend back into the 1700s. Those records show that the average temperature has increased rapidly during the past 50 years and that the 10 hottest years on record have occurred since 1990. Three centuries is a substantial time from the perspective of an individual human, but it's less than the blink of an eye in the history of the earth. We can extend the temperature records back to more than 400,000 years ago by examining ice cores from Antarctica. Although these cores do reveal short-term and long-term variations in temperature, nothing in the past 400,000 years approaches the speed of temperature change we have seen in the past 50 years.

**1** Energy from the sun passes easily through the atmosphere to warm earth's surface.

**2** Some energy is reflected back toward space and escapes the atmosphere.

**3** Some energy is absorbed by greenhouse gases and remains trapped in the atmosphere, heating the air.

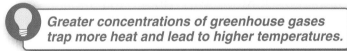

*Greater concentrations of greenhouse gases trap more heat and lead to higher temperatures.*

**FIGURE 16-27 The earth is warmed by the greenhouse effect.**

The term "greenhouse effect" describes the process by which energy from the sun warms the earth's atmosphere. The light we see lies in the visible portion of the light spectrum, and visible wavelengths of light pass easily through the atmosphere to warm the earth's surface. Some energy in the infrared part of the spectrum flows from earth back toward space, and several gases in the atmosphere absorb this energy, heating the air (**FIGURE 16-27**). These gases act like the glass panels that make up the roof of a greenhouse: they allow light energy to pass through, but they keep heat from escaping—hence the name "greenhouse gases." Carbon dioxide and methane are among the most important of the greenhouse gases. Without them, heat would escape into space and the earth's average temperature would be lower.

**Q** Why do we blame global warming on humans?

Humans have increased the amount of greenhouse gases in the atmosphere. And we've done it for primarily one reason: we rely on the burning of fossil fuels—coal, oil, gasoline, and natural gas—for much of the energy we use. The burning of fossil fuels, which have been stored deep in the earth for millions of years, releases carbon dioxide, and clearing land

and plowing soil to cultivate crops releases both carbon dioxide and methane. The same Antarctic ice cores that reveal the unprecedented rate of global warming over the past 50 years also show that the current levels of carbon dioxide and methane in the atmosphere are far higher than any that have occurred during the 400,000-year span of the cores (**FIGURE 16-28**).

The Intergovernmental Panel on Climate Change, a scientific body set up in 1988 by the World Meteorological Organization and the United Nations, has predicted an increase in the average annual temperature of 2° to 12° F (1° to 6.4° C) during the 21st century. Sea levels will rise by 7–24 inches (18–59 cm) as warmer temperatures accelerate the melting of glaciers and ice sheets near the North and South Poles. Rainfall is predicted to decrease overall, and droughts will become more frequent and more severe. The impact of droughts will be most severe in areas that are currently on the borderline of aridity, such as Australia, the American Midwest, and sub-Saharan Africa.

Both environmental and biological evidence already reveal the effects of global warming. The Arctic Ocean ice cap is melting, and the glaciers that cover Antarctica and Greenland have also been melting at unprecedented rates—and those rates are increasing (**FIGURE 16-29**). During 1996, 21.6 cubic miles (90 cubic kilometers) of the Greenland ice sheet melted, and by 2010, the annual rate of melting was more than 57 cubic miles per year—faster than in any of the 53 years over which measurements have been made. This melting will cause an increase in ocean levels, creating flooding problems for many of the world's population centers in coastal areas, including New York City, Tokyo, and Amsterdam, and in the world's small island nations.

Biological systems are also showing the effects of climate change. Trees and flowers in northern latitudes are blooming

**DISASTROUS DISRUPTIONS  INCREASED GREENHOUSE GAS EMISSIONS**

**PROBLEM**
The average temperature has increased rapidly during the past 50 years, affecting both the physical environment and the biological world.

**CAUSE**
Burning fossil fuels and clearing land to cultivate crops have significantly increased levels of greenhouse gases in the atmosphere.

**STRATEGIES FOR SOLUTION**
Reduced emissions of greenhouse gases (particularly from the burning of fossil fuels)

**FIGURE 16-28 Increasing greenhouse gases: the problem and its cause.**

## ICE CAP MELTING

1979

2007

**FIGURE 16-29 Disappearing ice pack.** The satellite image at the top shows the minimum concentration of Arctic sea ice in 1979. The lower image shows the concentration of sea ice recorded on September 14, 2007.

earlier in the spring than they used to; migratory birds are arriving at their summer ranges earlier than they did even a decade ago; and birds and butterflies are extending their geographic ranges northward.

The changes to the physical environment caused by global warming, such as the thinning and shrinking of the Arctic Ocean ice or the flooding of coastal estuaries, can have profound effects on biodiversity at the species and ecosystem levels. For example, in 2004, a research ship in the Arctic Ocean found nine walrus calves swimming alone in deep water. Normally, adult female walruses leave their babies on ice floes in shallow water while they dive to the seafloor to feed on clams and crabs, and then return to the ice to nurse their babies. But the Arctic ice cap over shallow water has melted, driving the female walruses to the remaining ice, which is in deep water. The mother walruses cannot dive deep enough to reach the sea bottom, and they seem to be abandoning their calves.

Other consequences of global warming reveal the tremendous interconnectedness of so many ecosystem elements. Many bird and butterfly species' migratory patterns have been altered as they search for lower temperatures. Many populations of small mammals living near high-altitude peaks increasingly are constrained as they attempt to move toward milder temperatures during the summer months—if they move to lower altitudes it gets hotter; moving up to higher altitudes brings them to milder temperatures, but if they are already at the peak, they've got nowhere else to go when it gets too hot. Massive croplands throughout the world may no longer be able to support the crops previously grown there.

Infectious disease experts have warned that global climate change is likely to alter the distribution of disease vectors—especially arthropods such as mosquitoes and ticks, which are particularly sensitive to temperature—changing and potentially increasing the incidence of diseases such as malaria and dengue fever.

Global warming is a difficult issue to tackle because it is so large in scope; it encompasses every biome, every ecosystem, every species, every body of water, and its solutions require individual efforts in all nations as well as international cooperation among governments. Fundamentally, however, the science is straightforward: carbon emissions must be reduced significantly. This can be done by individuals who choose to cut down on their own fossil fuel use, improve the energy efficiency of their homes and cars, and support the development of carbon-free renewable energy sources. Internationally, it can be done through efforts to reduce deforestation while increasing the replanting of forests, and by developing and implementing transportation and industrial innovations that reduce dependence on the fuels that produce greenhouse gases. Innovative technologies, such as the underground sequestration of carbon, may also contribute to the reversal of global warming.

### TAKE-HOME MESSAGE 16·10

Carbon dioxide and methane are called "greenhouse gases," because they trap heat in the atmosphere. As humans burn fossil fuels and clear forests, the concentrations of greenhouse gases have been increasing and global temperatures have been rising. Ecological changes in plant and animal communities have already been observed and are likely to become more serious unless there is a global reduction in emissions of greenhouse gases.

You are familiar with the oxygen you breathe—it comes in molecules formed by two oxygen atoms and is represented by the chemical formula $O_2$. Ozone is a different molecular arrangement of oxygen that packs three oxygen atoms into a molecule; its chemical formula is $O_3$. Ozone irritates the respiratory pathways and lungs, causing coughing and wheezing, and exposure to ozone can induce asthma attacks. If you live in a city, you are likely to hear radio and television warnings of high ozone levels during the summer. These warnings announce that ozone is expected to exceed safe levels that day. Children and adults with lung disease are advised to remain indoors during an ozone alert.

If ozone makes people sick, why would anyone worry about *depletion* of ozone in the atmosphere? Isn't that a *good* thing? The answer is, in the words of the old saying about what determines the value of real estate: "Location, location, location." At ground level, ozone is bad for you, but ozone in the upper part of the stratosphere, which is 30 miles (50 km) above the earth's surface, protects you from ultraviolet radiation, the rays that cause sunburn and skin cancer. Ozone is a Jekyll-and-Hyde molecule, and one set of environmental regulations tries to reduce the formation of ozone at ground level while a different set of regulations tries to protect ozone in the stratosphere.

There are two types of "ozone depletion": the general reduction in the amount of ozone in the stratosphere and the formation of areas with very low ozone concentration (called "ozone holes") over the North and South Poles every winter. Synthetic chemicals known as chlorofluorocarbons (CFCs) are the villains in both forms of ozone depletion. CFCs, which were developed in the 1930s, have a wide range of applications, including use as coolants in refrigeration systems. When CFC molecules leak into the atmosphere, they rise to the stratosphere, where sunlight knocks a chlorine atom off the CFC molecule. These free chlorine atoms then catalyze the breakdown of ozone to oxygen (**FIGURE 16-30**).

In the 1970s, it was noted that the amount of ozone in the stratosphere was decreasing by about 4% per decade. The cold temperatures and circular flows of air that develop over the Poles during the winter concentrate CFCs, forming ozone holes in those locations. The Antarctic ozone hole, which lasts for several months, covers the

**Should we be concerned about an ozone hole in our atmosphere?**

## DISASTROUS DISRUPTIONS — OZONE LAYER DEPLETION

**PROBLEM**
Increased levels of ultraviolet light reach the earth's surface, leading to a greater incidence of health problems in animals and decreased rates of photosynthesis in plants.

**CAUSE**
Synthetic chemicals known as chlorofluorocarbons (CFCs) leak into the atmosphere, where they cause the breakdown of ozone.

**STRATEGIES FOR SOLUTION**
Reduced production and emission of CFCs

**FIGURE 16-30 Depletion of the ozone layer: the problem and its cause.**

entire continent of Antarctica and extends northward to include the southern tips of South America and Australia (**FIGURE 16-31**). The Arctic ozone hole is smaller than the Antarctic hole and does not last as long, but it is large enough to extend southward into northern Europe, Asia, and North America.

Ozone depletion allows short-wavelength ultraviolet light (UVB light, with wavelengths of 270–315 nanometers) to reach the earth's surface. The 4%-per-decade reduction in global stratospheric ozone levels has increased the intensity of UVB radiation everywhere on earth, particularly under the ozone holes.

Not surprisingly, humans tend to think first about the effects of elevated UVB on human health. A 1% increase in UVB intensity increases the incidence of skin cancer by 2% to 3%. Exposure to UVB radiation also promotes the formation of cataracts and reduces the effectiveness of our immune system. Domestic animals suffer the same kinds of damage that humans do, and a few studies indicate that wild animals are also affected.

**Q** But we may be looking at the wrong problem. The damage to ecosystems could be a more serious consequence of ozone depletion than damage to human health. UVB radiation reduces the rate of photosynthesis by plants, and reduced photosynthesis in agricultural ecosystems means that crop yields will decline just when the worldwide cost of food is increasing.

## ANTARCTIC OZONE DEPLETION

September 1979      September 1988      September 2000      October 2010

**FIGURE 16-31 Monitoring the ozone hole.** The satellite images show changes in the ozone hole over the Antarctic during the past 30 years. The darkest blue color indicates the lowest ozone concentration.

> 💡 *Depletion of the ozone layer and the formation of an ozone "hole" over earth's polar regions have occurred as a result of human activities.*

The creation of the ozone holes is an example of how some scientifically developed industrial processes and products can have unintended and severe consequences. When CFCs were first invented, they were considered harmless, and so their use as coolants, as aerosol propellants, and in the production of materials such as Styrofoam became widespread. But the recognition that CFCs eventually reach the highest levels of the atmosphere and catalyze the destruction of the protective ozone layer prompted relatively

quick reaction: an example of the power of science to help identify environmental problems and to help find solutions (**FIGURE 16-32**). With the adoption by most countries in the 1980s of an agreement to discontinue the use of CFCs, ozone depletion was slowed, and in 2010 it was shown that the atmospheric levels of ozone had stabilized. A full recovery could occur by 2050.

It's important to note that depletion of the ozone layer is not the mechanism of global warming. Ultraviolet radiation represents less than 1% of the sun's energy. And the greenhouse gases discussed in the previous section do not reduce the atmosphere's ozone layer. Global warming and ozone depletion are largely unrelated environmental issues, linked only by their common cause: human activities. The response to ozone depletion, however, represents an encouraging example of how nations can work together to address and correct human-caused environmental problems.

## REDUCING CFCs

Products made from ozone-depleting CFCs, such as Styrofoam hamburger boxes and hair sprays with damaging propellants, have been phased out in the hope of spurring a recovery of the ozone layer.

**FIGURE 16-32 Taking simple steps to reduce ozone depletion.**

### TAKE-HOME MESSAGE 16·11

Ozone in the stratosphere prevents short-wavelength ultraviolet light (UVB) from reaching the earth's surface, but for many decades the amount of ozone was decreasing, largely due to synthetic chemicals called chlorofluorocarbons (CFCs). An increase in UVB light reaching the earth's surface can seriously damage ecosystems and has adverse effects on health, including increasing the incidence of skin cancer in humans and other animals. A decline in the production and use of CFCs is reducing atmospheric ozone depletion.

# 16·12  Disruptions of ecosystems can be disastrous:
## 5. Deforestation of tropical rain forests.

Towering trees, colorful birds and butterflies, maybe a glimpse of Tarzan swinging past on a hanging vine—that is the popular image of a tropical rain forest. And it is a reasonably accurate picture (minus the Tarzan part). But it is also a picture that is rapidly fading, as agriculture, logging, gold mines, and oil wells destroy tropical forests (**FIGURE 16-33**).

**FIGURE 16-33 Deforestation: the problem and its cause.**

Tropical rain forests grow in a region extending just a bit north and south of the equator, in South America, Africa, Asia, and Australia. As recently as a few centuries ago, that belt of rain forest covered about 6 million square miles (15.5 million square kilometers). More than half of that forest has already been destroyed, most of it in just the past 200 years (**FIGURE 16-34**).

The destruction of a tropical rain forest means an enormous loss of species, because these forests are the most diverse terrestrial habitats on earth. More than 170,000 species of plants grow in tropical rain forests, more than 60% of the total number of living plant species in the world. A survey of a Brazilian rain forest found 487 different species of trees in just 2.5 acres (1 hectare), an area less than half the size of one city block in Manhattan. To put that number in perspective, consider that only 700 species of trees are found in the 5 *billion* acres of the United States and Canada combined!

Agriculture is responsible for the greatest loss of tropical forests, and it takes a variety of forms. Sometimes, a relatively small area is affected—"slash-and-burn agriculture" refers to cutting trees, burning them, and planting crops in the newly opened area. Usually, just a few acres of forest are cleared, but the land is fertile for only two or three years after it has been cleared. Then that plot is abandoned and more forest is cleared, so the cumulative effect of slash-and-burn agriculture is substantial. At the other end of the size scale, multinational corporations clear hundreds of acres of forest to plant bananas, coffee, or oil palms, and thousands of acres of forest to create pastures for cattle.

Pollution from oil wells and mining in tropical rain forests has an impact that can extend far beyond the areas that are actually cleared. Leaking oil contaminates streams and groundwater, and acidic water drains from mines. Even worse, gold miners use mercury to extract gold, and some of

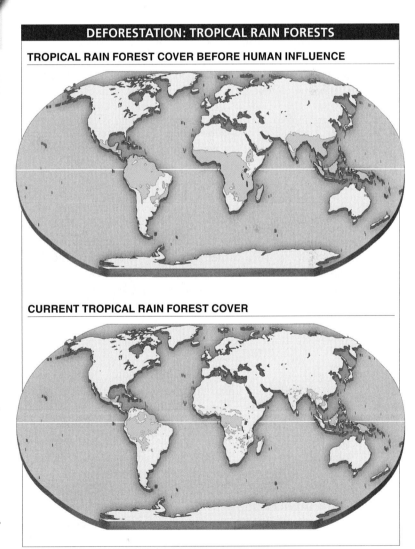

**DEFORESTATION: TROPICAL RAIN FORESTS**

**TROPICAL RAIN FOREST COVER BEFORE HUMAN INFLUENCE**

**CURRENT TROPICAL RAIN FOREST COVER**

**FIGURE 16-34 Worldwide loss of tropical rain forests.**

the mercury enters streams, where it is converted to methyl mercury—a toxic compound that accumulates in plant and animal tissue as it moves up the food chain.

Agriculture and mining both require roads to bring equipment to the sites and take crops or minerals to market. Roads have an impact on forests that greatly exceeds the relatively small area they occupy, because they make access to the forest easy. Traveling through a virgin tropical rain forest is difficult—bogs and natural tree falls often prevent travel in a straight line; slopes are often steep and the wet soil is slippery; and the ground-level vegetation can be both dense and thorny. No wheeled vehicle larger than a motorcycle can penetrate most rain forests, and traveling by motorcycle is slower than walking. These difficulties protect rain forests, but as soon as a road is bulldozed through a forest, people flock to it. They and their activities then spread into the forest on both sides of the road, creating new clearings that spread like cancers, farther and farther into the forest.

Destruction of tropical rain forests has two serious environmental impacts: reducing the earth's biodiversity and increasing the concentration of greenhouse gases in the atmosphere, which in turn affects global climate change.

**1. Reducing biodiversity.** Tropical rain forests of Africa, Asia, the Pacific, and Central and South America contain an unusually large number of species of plants and animals and probably other groups of organisms. Half of the world's biodiversity hotspots are in tropical rain forests.

**2. Increasing greenhouse gases.** Photosynthesis in tropical rain forests removes an estimated 610 billion tons (550 trillion kg) of carbon dioxide from the atmosphere each year. As we learned earlier, accumulation of carbon dioxide in the atmosphere is the major cause of global warming. Thus, the continued photosynthetic activity of tropical rain forests is important in slowing the rate of warming. The huge quantity of carbon stored in rain forests has a downside, however, because that carbon is released into the atmosphere when forests are cleared and burned. And tropical forests are being cleared at a frightening rate—more than 50,000 square miles (an area larger than the state of New York!) each year between 2000 and 2010.

Because deforestation alters characteristics of land cover, changing the proportion of solar radiation reflected and absorbed, rain forest destruction can have additional significant effects on climate—including rainfall amounts and seasonality. These changes can further accelerate climate change, with potentially significant impacts on ecosystems.

*Planting new trees can undo some deforestation damage.*

**FIGURE 16-35 Growing seedlings for reforestation.**

The problem of tropical deforestation is among the most difficult of any environmental problems to solve, and solving it will involve multiple strategies and rely on international cooperation. The most diverse areas must be identified and protected. The poverty that necessitates the destruction of tropical rain forests for human activities must be addressed. Alternative sources of food and income must be developed. Population growth must be reduced. And education about the value of preserving biodiversity is central to these issues. At present, we have not progressed much beyond identifying the problem, but that, at least, is a necessary first step (**FIGURE 16-35**).

## TAKE-HOME MESSAGE 16·12

Tropical rain forests are being destroyed at an alarming rate. This deforestation is devastating for two main reasons. First, tropical rain forests contain more species of plants and animals than all other terrestrial habitats combined, and half of the earth's biodiversity hotspots are in these forests. Second, tropical rain forests remove more carbon dioxide from the atmosphere than any other terrestrial habitat and, as a result, these forests are enormously important in limiting global warming. Deforestation further influences climate change by altering land cover characteristics.

*A worker carries a tree to plant at the edge of Maowusu Desert, China. Trees are planted to prevent the desert from expanding.*

# 16·13–16·14
# We can develop strategies for effective conservation.

## 16·13 With limited conservation resources, we must prioritize which species should be preserved.

Is it preferable to save one beautiful, well-studied bird species or 1,000 species of bacteria, none of which have even been described or named? And what about the 1,200 species of beetles in Panama that live in the evergreen tree *Luehea seemannii*, almost 200 of which live *only* in this one species of tree—how should we prioritize these species for conservation relative to, say, a single primate species? From locale to locale the particulars of these questions may change, but the underlying issue remains: in a world where species are being driven to extinction faster than we can save them, which should be singled out for preservation and which should we leave to almost certain extinction (**FIGURE 16-36**)? This is a question that biologists, policymakers, and, ultimately, citizens must address.

When setting conservation priorities, it is essential to articulate a goal. In an ideal world, the goal might be to protect "all biodiversity." Barring that, options include protecting "most biodiversity," the most diverse subset of biodiversity (in terms of genes, species, and ecosystems), or the most valuable biodiversity. There are numerous

### WHAT SHOULD WE PROTECT?

**FIGURE 16-36 Prioritizing conservation efforts.** How do we decide among the many options? Shown here: giant pandas and soil bacteria.

*Conservation efforts can focus on preserving individual species (such as orangutans) or important habitats (such as rain forests).*

**FIGURE 16-37 Which to preserve?** Limited resources and great demand make preservation choices difficult.

arguments in favor of each goal, and each has significant flaws as well. Once a goal is decided on, a related step in the decision-making process is needed—a plan that outlines priorities for achieving the goal. Frequently, such a plan involves an assessment of the degree to which various components of biodiversity are threatened. We can rank species, for example, as relatively intact, relatively stable, vulnerable, endangered, or critical. These rankings can then be used, in conjunction with measures of the biological value of the biodiversity to humans, in formulating a plan.

As we saw earlier in the chapter, biodiversity can be valued in many different ways, and its worth cannot be easily quantified or weighed. Sooner or later, many difficult and subjective decisions must be made in the goal-setting and prioritizing process. The undiscovered bacterial species may harbor the metabolic secrets to curing a devastating medical condition in humans, but the beautiful bird species may be much loved by bird-watchers or may serve as a powerful icon of strength or freedom, inspiring generations of people. The decisions are difficult, but not facing them is, in most cases, the same as making a decision.

In the United States, much conservation policy involves response to the **Endangered Species Act (ESA),** a law that defines **endangered species** as those in danger of extinction throughout all or a significant portion of their range. The law is designed to protect those species from extinction. As we noted earlier in the chapter, species that are dwindling are listed as endangered or threatened, according to an assessment of their risk of extinction. Once a species is listed, legal tools are available to help rebuild the population and

protect the habitat critical to its survival. Seemingly straightforward, the ESA has had the effect of focusing most conservation efforts on the preservation of individual species (and populations), sometimes at the expense of other elements of biodiversity and sometimes at the expense of efforts to reduce the loss of ecologically important habitats (**FIGURE 16-37**). Other difficulties are also associated with the ESA. Consider the task of determining the critical population size below which a population is endangered—is it 500 individuals or 5,000 or 50,000? With its emphasis on preservation of single species, the ESA does not effectively address ecosystem decline, which is an equally urgent and serious problem.

In spite of these difficulties, the species-focused approach to conservation has had some success in the United States. Since it became law in 1973, approximately 40 species that were listed as endangered—including the bald eagle, the peregrine falcon, the gray whale, and the grizzly bear—have been taken off the list as their numbers have recovered. In the final section, we examine other strategies that have been used in efforts to preserve biodiversity.

## TAKE-HOME MESSAGE 16·13

Effective conservation requires the setting of goals on the elements of biodiversity (genes, species, or ecosystems) to be conserved and priorities among those elements. The U.S. Endangered Species Act has focused much conservation effort on the preservation of species.

# 16·14 There are multiple effective strategies for preserving biodiversity.

In a survey asking which animal they would like to be, men and women gave very different answers. Among men, the top answer was eagle, followed by tiger, lion, and dolphin. As their first choice, women chose cat, followed by butterfly, swan, and swallow. It's just a silly survey, but the differences in male and female selections parallel the fact that people have very different preferences when it comes to preserving biodiversity. One person may view as unthinkable the loss of a particular species or habitat, while another may view other species or habitats as much more valuable.

Most approaches to conservation biology, as we've seen, have focused on the preservation of individual species. Increasingly, however, conservation biology is shifting toward the preservation of important habitats, focusing on conserving communities and ecosystems. This habitat-conserving approach also leads to the preservation of individual species, but it has the added benefit of preserving, at the same time, many different species within a habitat, including many that have not yet been identified, particularly microbes and fungi.

One example of an effort aimed at the preservation of important habitat is the World Wildlife Fund's "Global 200," an identification of all of the most biologically distinct habitats on earth. The identification of these habitats, which include terrestrial, freshwater, and marine habitats, is part of an innovative strategy, often called **landscape conservation,** that is geared toward conserving not just species but habitats and ecological processes (e.g., large-scale migrations and predator-prey interactions). In an age of scarce conservation resources and limited time, this approach prioritizes the conservation of biologically unique habitats on earth.

The habitat conservation approach is not new. The very first attempts at conservation in the United States were the creation of Yellowstone Park in 1872 and Yosemite in 1890, both large-scale efforts geared not simply toward the preservation of one or a few species but rather toward the preservation of the "wilderness." In the time since then, significant efforts have been made to establish national parks, wilderness areas, and recreation areas (**FIGURE 16-38**).

As conservation biologists have started to better understand population dynamics and biogeography, the design of natural preserves has evolved. Using insights gained from

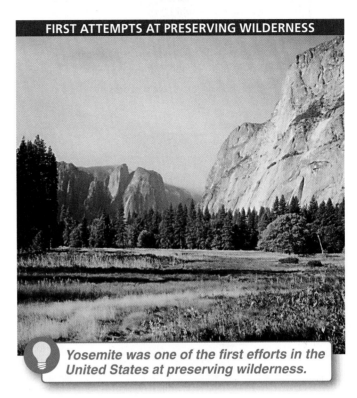

**FIRST ATTEMPTS AT PRESERVING WILDERNESS**

Yosemite was one of the first efforts in the United States at preserving wilderness.

**FIGURE 16-38 The natural beauty of Yosemite.**

the theory of island biogeography, modern preserve design focuses not simply on maximizing the variety of habitats and biodiversity preserved but on using several design features that maximize the efficiency of the preserves (**FIGURE 16-39**).

- Larger, rather than smaller, preserves (including a preference for a single, undivided preserve over several smaller preserves)

- Circular, rather than linear, preserves

- Corridors—even just relatively narrow strips of land—that connect larger natural preserves, allow gene flow, and reduce inbreeding among distinct populations in the different preserves

- Buffer zones (which permit limited amounts of human activity) around core areas (which contain the habitat to be conserved)

As conservation plans aimed at preserving habitats increase, efforts focusing on individual species still continue to be effective. Several strategies targeting individual species for

**FIGURE 16-39 Carefully planned nature preserves focus on features that maximize biodiversity.**

conservation have been particularly successful at preserving large amounts of biodiversity beyond that single species.

**1. Flagship species.** Some species, because they are particularly charismatic, distinctive, vulnerable, or otherwise appealing, can engender significant public support. Examples include the giant panda of China, the golden lion tamarin of Brazil's coastal forest, the mountain gorilla of Central Africa, the orangutan of Southeast Asia, the leatherback sea turtle, the Indian tiger, and the African elephant. Preservation of these species, given their habitat needs, can serve to preserve many other species as well.

**2. Keystone species.** As we saw in Chapter 15, keystone species have a disproportionate effect on the biodiversity of a community. Their removal can lead to massive changes in the composition of species in an ecosystem, often causing huge loss of biodiversity. Examples include kelp, California mussels, grizzly bears, beavers, and sea stars.

**3. Indicator species.** These are species whose presence within an ecosystem indicates the presence of a large range of other species. For example, the presence or absence of lichens is an indicator of air quality, because they are sensitive to sulfur dioxide, a component of industrial fumes. When lichens disappear from trees, it is usually an indicator of increased pollution, which endangers the entire ecosystem. Conversely, when lichens are present and healthy, they indicate that the ecosystem is healthy.

**4. Umbrella species.** These include wide-ranging species with such large needs for habitat and other resources that their preservation ends up protecting the numerous other species within that same habitat, without having to identify these other species as conservation targets. Umbrella species tend to be large, wide-ranging vertebrates.

For more critically endangered species and degraded habitats, all hope is not necessarily lost. Captive breeding programs and habitat restoration have had some success in bringing back biodiversity from the brink of extinction. Zoos and botanical gardens have taken the lead in many of these efforts. In the 1980s, for example, when the population size of the California condor had dropped to 22 individuals, due to poaching, lead poisoning, and habitat loss, all of the birds were caught and taken to zoos, where a captive breeding program began. With the success of these breeding programs, by 2011 the population had reached 400, including close to 200 birds in the wild and the remainder in zoos. Breeding programs are not a complete conservation solution on their own, however. It is essential that the species' habitats are not destroyed or altered if the species are to flourish again under natural conditions.

Restoration ecology, which uses the principles of ecology to restore degraded habitats to their natural state, has also been an important tool of conservation biologists. Wetlands that have been degraded by dredging and development, in particular, have benefited. Reintroduction of the native plant species and restoration of water and stream flow patterns can be instrumental in restoring these habitats.

Taken together, the strategies used by conservation biologists represent an important step toward reducing the adverse effects of one hugely successful (from a population growth perspective) species—humans—on other species. Although we are far from living with minimal disturbance of the environment in which we live, continued conservation efforts offer our best hope for a sustainable future.

**TAKE-HOME MESSAGE 16·14**

Conservation biology has focused, in the past, on preserving individual species. Increasingly, there is a shift toward the preservation of important habitats, focusing on conserving communities and ecosystems and using concepts learned from ecology and island biogeography. Several methods focusing on single species, however, remain useful, particularly when preserving the selected species requires preservation of an amount and type of habitat that simultaneously preserves many other species.

## THE PERILS OF (EXOTIC) PETS!

Some significant threats to biodiversity may come from your own neighborhood. And they're often aided by well-meaning small businesses: pet stores!

**Q:** **HOW MIGHT OTHERWISE-RESPONSIBLE CITIZENS WREAK HAVOC ON THEIR ENVIRONMENT?** The purchase of exotic, non-native animals as pets can have unintended and harmful consequences.

**Q:** **WHAT'S WRONG WITH KEEPING AN EXOTIC FISH IN YOUR AQUARIUM OR A TROPICAL BIRD IN A CAGE?** In an aquarium or in a cage, non-native species are unlikely to harm any natural environment. But the problem is that they escape, they get flushed down toilets, they're "set free" by owners who no longer want them, or, in some other way, they get into the environment. And it's not their native habitat.

**Q:** **WHAT HAPPENS WHEN NON-NATIVE SPECIES ARE INTRODUCED TO A NOVEL ENVIRONMENT?** In some cases, non-native species do no harm. But in many others, non-native or exotic species can alter habitat and crowd out, outcompete, or directly prey on beneficial native species. They can spread exotic diseases, for which native species have no resistance. And they can disrupt and harm fisheries, crops, and other valuable industries, costing millions of dollars. Consider just a few common examples.

> **FISH.** Catfish can alter river vegetation and shorelines, making them uninhabitable for native species. Lionfish from home aquariums have been released into the ocean off the coast of Florida. Expert predators, they kill a wide variety of smaller fishes, mollusks, and other invertebrates. In 2008, their population densities in some new habitats were found to have increased by 700% in just five years. China's giant snakehead fish is a voracious eater and preys on numerous species of smaller fishes and amphibians. Traced to the aquarium industry, non-native populations off the east coast of the United States have wiped out numerous populations of native species. The snakehead fish can even crawl from one pond to another, surviving as long as four days out of water! Numerous other types of popular aquarium fish, including emperor angelfish, the yellow tang, and the orbicular batfish, have been found thriving in non-native habitats in the United States.
>
> **BIRDS.** More than 150 non-native bird species have been documented in Florida, including 30 different parrot species, all of which were former pets, and many of which now cause significant crop damage, consuming fruits, berries, flowers, and seeds. Monk parakeets often build nests on electrical transformers and lampposts and have been responsible for power outages and other damage.
>
> **AMPHIBIANS AND REPTILES.** Terrapins (a type of turtle) and bullfrogs often outgrow their aquariums. If released into ponds, they can wipe out native fish, amphibian, and even small mammal populations.
>
> **PLANTS.** Numerous popular aquarium plant species, such as the fast-growing seaweed *Caulerpa,* have become invasive in the Mediterranean Sea and in Southern California. Freed from their natural predators, they thrive and have caused millions of dollars in damage.

**Q:** **WHAT CAN YOU DO?** The best advice here is to avoid purchasing exotic pets (or using non-native plant species when landscaping or planting a garden). You can also let your representatives in government know of the dangers of non-native species and the importance of restricting the import of such species into the country—an issue that is not always high on their list of priorities.

# Big Ideas in ...
# Conservation and Biodiversity

## 16·1–16·4 Measuring and defining biodiversity is complex.

- Biodiversity can be defined as the variety of genes, species, and ecosystems on earth.
- Biodiversity provides significant benefits to humans in utilitarian ways, such as the production of medicines, and provides important value in less tangible ways, including aesthetic, symbolic, and naturalistic.
- There is a strong latitudinal gradient in biodiversity, decreasing from the tropics to the Poles for nearly all groups of plants and animals.
- Biodiversity hotspots are regions of significant biodiversity under threat of destruction.
- The theory of island biogeography generates predictions about patterns of species diversity and has implications for the development of nature preserves.

## 16·5–16·6 Extinction reduces biodiversity.

- Mass extinctions tend to result from bad luck rather than from features of a species' biology, whereas background extinctions tend to be a consequence of a species' biological characteristics.
- Small population size, limited habitat range, and narrow zones of habitat tolerance contribute to background extinctions.
- Most biologists believe that we are in the midst of a mass extinction, that it is the result of human activities, and that it poses a serious threat to human survival.

## 16·7–16·12 Human interference generally reduces biodiversity.

- Because they have no natural predators in their new habitat and their prey have no defenses against them, exotic species can multiply unchecked, overwhelming competitors and dominating and irreversibly altering communities and entire ecosystems.
- Burning fossil fuels releases sulfur dioxide and nitrogen dioxide, and these gases combine with water vapor to form acid precipitation, harming plants, animals, and the abiotic environment.
- Carbon dioxide and methane are greenhouse gases, trapping heat in the atmosphere. Produced by the burning of fossil fuels and clearing of forests, greenhouse gases are increasing and global temperatures rising. Avoiding increasingly serious consequences requires a global reduction in emissions of greenhouse gases.
- Ozone in the stratosphere has been depleted through the effects of CFCs, with health consequences for humans and animals. A reduction in CFC production appears to be restoring the ozone to safer levels.
- Tropical rain forests contain more species than all other terrestrial habitats combined, and half of the earth's biodiversity hotspots are in tropical rain forests. Destruction of these forests is likely to have serious environmental consequences.

## 16·13–16·14 We can develop strategies for effective conservation.

- Effective conservation requires setting goals on which elements of biodiversity (genes, species, or ecosystems) to conserve and setting priorities among those elements.
- Although the U.S. Endangered Species Act has focused most conservation efforts on individual species, there is now a shift toward conserving communities and ecosystems. Several methods focusing on single species remain useful, however, particularly when they require preservation of an amount and type of habitat that simultaneously preserves many other species.

See the To-Go Guide (packaged with every new copy of this textbook) for an illustrated summary of the chapter.

# Check Your Knowledge

## GRAPHIC CONTENT

*Thinking critically about visual displays of data.*

1. What are the axes for these graphs? What are the variables being observed?

2. Which variable is independent? Which is dependent? Do these graphs conform to typical norms for the placement of the independent and dependent variables? Why or why not?

3. At what latitude do the largest numbers of species occur?

4. Why are data presented both for mammalian species *and* for marine copepod species (rather than just one or the other)? Why aren't data for *all* species presented?

5. At 20° north latitude, are there more mammalian or marine copepod species? What aspect of the graphs may be responsible for someone answering this question incorrectly?

6. Bar graphs often have error bars, indicating some measure of variance in the data. Why do the data bars in these graphs have no such error bars?

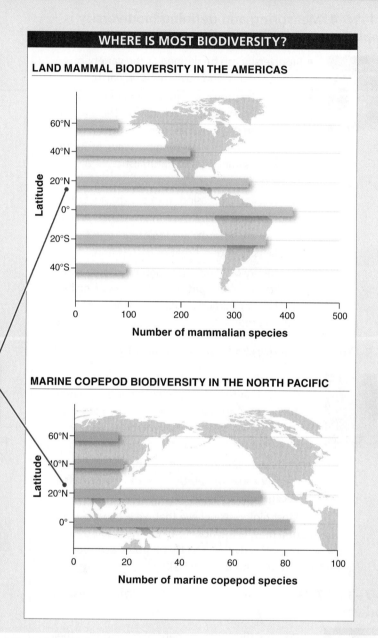

**WHERE IS MOST BIODIVERSITY?**

LAND MAMMAL BIODIVERSITY IN THE AMERICAS

MARINE COPEPOD BIODIVERSITY IN THE NORTH PACIFIC

See Answers at the back of the book.

## Key Terms in Conservation and Biodiversity

# Multiple-Choice Questions

1. **Biodiversity is considered important because of:**
   a) the direct economic benefits to humans, such as medicines and food.
   b) the symbolic value that elements of the natural world can provide.
   c) the aesthetic value it holds.
   d) its potential for helping humans understand the processes that gave rise to the diversity we see.
   e) All of the above are correct.

   100 HARD
   **40**
   0 EASY

2. **Biodiversity hotspots are defined by which two criteria?**
   a) species richness and ecosystem integrity
   b) size and distance from nearest alternative hotspot
   c) species endemism and degree of threat
   d) species richness and size
   e) ecological diversity and species diversity

   100 HARD
   **32**
   0 EASY

3. **Madagascar is unusually important to conservation because it:**
   a) has more species per unit area than any other place on earth.
   b) is the fourth largest island in the world.
   c) is home to more endangered species than any other country.
   d) has a higher percentage of endemic plants and animals than any comparably sized area on earth.
   e) is the native habitat of the rosy periwinkle, a rain forest plant that helps cure childhood lymphocytic leukemia.

   100 HARD
   **37**
   0 EASY

4. **By fumigating several islands in the Florida Keys with methyl bromide, E. O. Wilson and Daniel Simberloff demonstrated that:**
   a) the equilibrium number of insect species was a function of island size and distance from the mainland.
   b) colonization of an island by new species can no longer occur once the island is "full."
   c) for plant but not insect species, the relation of area to species number holds true for all islands.
   d) the effects of global warming will be greater on islands than on the mainland.
   e) species turnover rates are highest on small, more distant islands.

   100 HARD
   **44**
   0 EASY

5. **When an island has the equilibrium number of species:**
   a) extinction no longer occurs.
   b) immigration rates to the island are reduced to zero.
   c) both the number and composition of species on the island remain the same.
   d) both the number and composition of species on the island are changing.
   e) the composition of species on the island may change, but the number of species present remains the same.

   100 HARD

   **42**
   0 EASY

6. **Which of the following attributes of a species might render it especially susceptible to extinction?**
   a) It lives in a marine habitat.
   b) It exists as two distinct subpopulations.
   c) It is an invasive species.
   d) It is short-lived.
   e) It has a narrow dispersal range.

   100 HARD

   **46**
   0 EASY

**7.** What bird, once the most abundant in North America, was hunted to extinction by shooting and trapping during the 1800s?

a) California condor
b) blue-footed booby
c) dodo
d) passenger pigeon
e) great auk

100 HARD
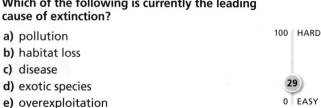
53
0 EASY

**8.** How many species do biologists estimate to be currently existing on earth?

a) fewer than 1 million
b) between 3 million and 5 million
c) more than 1 billion
d) between 5 million and 100 million
e) between 1 million and 2 million

100 HARD
45
0 EASY

**9.** Which of the following is currently the leading cause of extinction?

a) pollution
b) habitat loss
c) disease
d) exotic species
e) overexploitation

100 HARD
29
0 EASY

**10.** Exotic species can disrupt ecosystems because:

a) they frequently have no predators in their new habitat and grow unchecked.
b) they are favored by ecotourists.
c) they have no natural prey items and so must rely on humans for their survival.
d) they have better dispersal capability than endemic species.
e) All of the above are correct.

100 HARD
42
0 EASY

**11.** Even though there is a carbon cycle, carbon dioxide levels around the world seem to be rising. Which of the following best explains why this is so?

a) Animals give off carbon dioxide during their normal metabolism.
b) As the atmosphere heats up, it can contain more carbon dioxide.
c) The destruction of coral reefs leads to increased levels of carbon dioxide.
d) More carbon dioxide is being given off by ocean waters as they heat up.
e) The burning of fossil fuels releases more carbon dioxide into the atmosphere.

100 HARD
19
0 EASY

**12.** Ozone depletion is cause for concern because:

a) without atmospheric ozone, greenhouse gases can escape earth's atmosphere, leading to global warming.
b) without ozone, fossil fuels are more likely to react with water vapor in the atmosphere, increasing levels of precipitation throughout the world.
c) without the protection against solar radiation provided by ozone, rates of skin cancer are likely to increase.
d) without ozone, solar radiation can irradiate the Antarctic ice shelf, causing unsafe levels of radioactive isotopes.
e) reduced levels of ozone can increase the rate of respiratory problems in humans, particularly children.

100 HARD

0 11 EASY

**13.** Which of the following statements about habitat loss is correct?

a) Small animals are not vulnerable to habitat destruction.
b) Fragmented habitats can support only small populations of the species living in them.
c) Corridors between habitat areas are of use only to active organisms such as vertebrates.
d) Corridors can replace habitat areas for migratory species.
e) Large animals are not vulnerable to habitat destruction.

100 HARD
41
0 EASY

**14.** An umbrella species:

a) is a species whose removal will severely damage or cause the collapse of its ecosystem.
b) is a non-native species, usually introduced to an area by accident.
c) is any species that has a larger than expected effect on its ecosystem.
d) is a popular species that the public tends to rally behind and support.
e) is a species whose preservation results in the preservation of many other species as well.

100 HARD
50
0 EASY

## Short-Answer Questions

1. Where, geographically, is the greatest biodiversity? What are the primary threats to this biodiversity?

2. Why were the ecosystem disturbances created by humans in New England reversible? What changes would have made the ecosystem disturbances in New England irreversible?

3. What is the primary cause of tropical rain forest deforestation? How does this affect the biodiversity of plants and animals?

See Answers at the back of the book.
For additional study questions, go to www.prep-u.com.

Blue cliffs in the Painted Desert, Utah.

# Periodic Table

## THE PERIODIC TABLE OF ELEMENTS

**ELEMENT CATEGORIES**
- Alkali metals
- Alkaline earth metals
- Transition metals
- Lanthanides
- Actinides
- Post-transition metals
- Metalloids
- Other nonmetals
- Halogens
- Nobel gases
- Unknown chemical properties

Key:
- Atomic number
- Element symbol
- Element name
- Atomic mass

**GROUP** / **PERIOD**

| Group 1 | 2 | 3 | 4 | 5 | 6 | 7 | 8 | 9 | 10 | 11 | 12 | 13 | 14 | 15 | 16 | 17 | 18 |
|---|---|---|---|---|---|---|---|---|---|---|---|---|---|---|---|---|---|
| 1 H Hydrogen 1.00794 | | | | | | | | | | | | | | | | | 2 He Helium 4.00260 |
| 3 Li Lithium 6.941 | 4 Be Beryllium 9.012182 | | | | | | | | | | | 5 B Boron 10.811 | 6 C Carbon 12.0107 | 7 N Nitrogen 14.0067 | 8 O Oxygen 15.9994 | 9 F Fluorine 18.9984 | 10 Ne Neon 20.1797 |
| 11 Na Sodium 22.9898 | 12 Mg Magnesium 24.3050 | | | | | | | | | | | 13 Al Aluminum 26.9815 | 14 Si Silicon 28.0855 | 15 P Phosphorus 30.9738 | 16 S Sulfur 32.065 | 17 Cl Chlorine 35.453 | 18 Ar Argon 39.948 |
| 19 K Potassium 39.0983 | 20 Ca Calcium 40.078 | 21 Sc Scandium 44.9559 | 22 Ti Titanium 47.867 | 23 V Vanadium 50.9415 | 24 Cr Chromium 51.9961 | 25 Mn Manganese 54.9380 | 26 Fe Iron 55.845 | 27 Co Cobalt 58.9332 | 28 Ni Nickel 58.6934 | 29 Cu Copper 63.546 | 30 Zn Zinc 65.409 | 31 Ga Gallium 69.723 | 32 Ge Germanium 72.64 | 33 As Arsenic 74.9216 | 34 Se Selenium 78.96 | 35 Br Bromine 79.904 | 36 Kr Krypton 83.798 |
| 37 Rb Rubidium 85.4678 | 38 Sr Strontium 87.62 | 39 Y Yttrium 88.9059 | 40 Zr Zirconium 91.224 | 41 Nb Niobium 92.9064 | 42 Mo Molybdenum 95.94 | 43 Tc Technetium [98] | 44 Ru Ruthenium 101.07 | 45 Rh Rhodium 102.906 | 46 Pd Palladium 106.42 | 47 Ag Silver 107.868 | 48 Cd Cadmium 112.411 | 49 In Indium 114.818 | 50 Sn Tin 118.710 | 51 Sb Antimony 121.760 | 52 Te Tellurium 127.60 | 53 I Iodine 126.904 | 54 Xe Xenon 131.293 |
| 55 Cs Cesium 132.905 | 56 Ba Barium 137.327 | 57–71 | 72 Hf Hafnium 178.49 | 73 Ta Tantalum 180.948 | 74 W Tungsten 183.84 | 75 Re Rhenium 186.207 | 76 Os Osmium 190.23 | 77 Ir Iridium 192.217 | 78 Pt Platinum 195.078 | 79 Au Gold 196.967 | 80 Hg Mercury 200.59 | 81 Tl Thallium 204.383 | 82 Pb Lead 207.2 | 83 Bi Bismuth 208.980 | 84 Po Polonium [209] | 85 At Astatine [210] | 86 Rn Radon [222] |
| 87 Fr Francium [223] | 88 Ra Radium [226] | 89–103 | 104 Rf Rutherfordium [261] | 105 Db Dubnium [262] | 106 Sg Seaborgium [266] | 107 Bh Bohrium [264] | 108 Hs Hassium [277] | 109 Mt Meitnerium [268] | 110 Ds Darmstadium [281] | 111 Rg Roentgenium [272] | 112 Cn Copernicium [285] | 113 Uut Ununtrium [284] | 114 Uuq Ununquadium [289] | 115 Uup Ununpentium [288] | 116 Uuh Ununhexium [292] | 117 Uus Ununseptium ? | 118 Uuo Ununoctium [294] |

Lanthanides:

| 57 La Lanthanum 138.906 | 58 Ce Cerium 140.116 | 59 Pr Praseodymium 140.908 | 60 Nd Neodymium 144.24 | 61 Pm Promethium [145] | 62 Sm Samarium 150.36 | 63 Eu Europium 151.964 | 64 Gd Gadolinium 157.25 | 65 Tb Terbium 158.925 | 66 Dy Dysprosium 162.5 | 67 Ho Holmium 164.930 | 68 Er Erbium 167.259 | 69 Tm Thulium 168.934 | 70 Yb Ytterbium 173.04 | 71 Lu Lutetium 174.967 |
|---|---|---|---|---|---|---|---|---|---|---|---|---|---|---|

Actinides:

| 89 Ac Actinium [227] | 90 Th Thorium 232.038 | 91 Pa Protactinium 231.036 | 92 U Uranium 238.029 | 93 Np Neptunium [237] | 94 Pu Plutonium [244] | 95 Am Americium [243] | 96 Cm Curium [247] | 97 Bk Berkelium [247] | 98 Cf Californium [251] | 99 Es Einsteinium [252] | 100 Fm Fermium [257] | 101 Md Mendelevium [258] | 102 No Nobelium [259] | 103 Lr Lawrencium [262] |
|---|---|---|---|---|---|---|---|---|---|---|---|---|---|---|

# Answers

## Chapter 1

1. b; 2. b; 3. d; 4. e; 5. c; 6. e; 7. a; 8. a; 9. b; 10. a; 11. d; 12. c; 13. a; 14. b; 15. e; 16. a; 17. d; 18. a

### Short-Answer Questions

1. To provide proper controls, a group of participants must receive a placebo that superficially resembles the medication. The control group should be treated exactly the same as the experimental group, with one exception: the control group receives a placebo instead of the drug. This provides a basis of comparison between the experimental and control groups.

2. There are many possible answers. Four possibilities are: adding a control group; extending the test for a longer period of time; standardizing the subjects to make sure that other variables are not responsible for the weight loss; repeating the study.

### Graphic Content Questions

1. The green portion represents the proportion of all papers published in the journal *Behavioral Ecology* that had a female first author. The blue portion represents the proportion of all papers published in *Behavioral Ecology* that had a male first author. Together, the green and blue portions represent all papers published in *Behavioral Ecology* (that is, 100%).

2. The pie chart on the left presents the proportion of all papers published in *Behavioral Ecology* that had a female first author (and, in blue, the proportion that had a male first author) during the years 1997–2001, when the journal had a "single-blind" review process in which reviewers knew the authors' identities (and, presumably, their sex). The pie chart on the right presents the same information, but for the years 2002–2005, when the journal had a "double-blind" review policy in which reviewers did *not* know the authors' identities or sex.

3. When the journal changed its review policy so that reviewers did not know the sex of the authors, the proportion of accepted papers that had female first authors increased from 23.7% to 31.6%. This suggests that when reviewers knew the sex of the authors, they were less likely to accept a paper with a female first author. This would seem to indicate that the reviewers for *Behavioral Ecology* at that time were biased against female authors.

4. It could be that, beginning in 2002, there was an increase in the proportion of papers submitted to *Behavioral Ecology* that had a female first author. This explanation seems unlikely, however, given the information in the text that a similar journal that did not change its review policy had no change in the proportion of published papers with a female first author. Another possible explanation is that the proportion of women in the field of behavioral ecology increased in the early 1990s, so that, during the second period measured, 2002–2005, there were more women at a more advanced stage in their careers, so the quality of the papers they submitted increased over time.

5. The increase in proportion of published papers with female first authors was 31.6% − 23.7%, or 7.9 percentage points. It could also be described as a 7.9%/23.7% = 33% increase (if you take one-third of 23.7% and add it to 23.7%, you get 31.6%).

6. The information given in the figure tells us what percentage of the papers *published* had a female first author. We cannot tell from this information whether 50% of the papers *submitted* had female first authors. If that were true, it would be reasonable to be concerned why the proportion published wasn't also 50%. But what if only 10% of the people working as behavioral ecologists are women? If that were the case, we might expect that only 10% of the papers published in the journal would have female first authors, and we might be concerned that with such a small proportion of female behavioral ecologists, more than one-quarter of the papers published in this journal had female first authors.

7. The graphs do not prove a general bias against female scientists. Because the data are drawn from just one journal in one field of science, we cannot be certain that they reflect a similar bias across all journals in biology, or in chemistry, physics, or other scientific fields. Nonetheless, the findings do suggest such a bias may exist. It would be valuable to collect additional data on this topic across all of the sciences.

## Chapter 2

1. d; 2. e; 3. d; 4. b; 5. a; 6. c; 7. a; 8. d; 9. c; 10. b; 11. a; 12. a; 13. a; 14. a; 15. a; 16. a; 17. e; 18. e; 19. d; 20. b

### Short-Answer Questions

1. a) The sodium atom has one electron in its outermost shell; the fluorine atom has seven electrons in its outermost shell. b) An ionic bond forms between these two atoms. Neither atom has a full outer electron shell, and both are unstable. The atoms interact to achieve a full outer shell of electrons: the single electron in the outermost shell of the sodium is transferred to the outermost shell of the fluorine. Each atom becomes an ion, and the oppositely charged ions attract each other.

2. First, cohesion is a result of polar water molecules easily bonding together by hydrogen bonds; this allows one water molecule to "connect" with the next as water makes its way up from the soil to the leaves of a tall tree. Second, water has a large capacity to hold heat and resists large fluctuations in temperature; this is because added energy, in the form of heat, does not increase the movement of water molecules, as occurs in most substances. Instead, hydrogen bonds between water molecules are briefly broken and then re-form with nearby water molecules. Third, ice is less dense than water, so it floats; this is because polar water molecules are held apart by hydrogen bonds when the temperature drops (see Figure 2-15). Fourth, water is an excellent solvent because water molecules are polar. Polar molecules with charged regions are attracted to the charged regions of other polar molecules. Many substances contain polar molecules, which easily dissolve in water.

3. Water is a polar molecule and is attracted to other polar molecules, or molecules with charged regions. Oil consists of nonpolar molecules. As a result, polar water molecules will not interact with oil, and water by itself will not be able to dissolve oily dirt.

4. Lettuce and celery contain carbohydrates, which provide a source of energy or fuel for the body. Strawberries also contain carbohydrates,

but more monosaccharides, versus the complex carbohydrates of lettuce and celery. Walnuts and chicken are sources of protein. The amino acid building blocks of protein are used to build many structures and molecules required by all cells. The salad dressing contains lipids, which provide the body with energy.

**Graphic Content Questions**

1. A blue dot in a column indicates that the food summarized in that column carries the optimal level of the amino acid(s) listed for that row. A white dot indicates that the food does not contain optimal levels of that amino acid(s). For example, lentils contain optimal levels of the amino acid cysteine, but white rice does not.

2. There is a large amount of information in this figure. One very important conclusion you can draw is that some food sources of protein contain all of the essential amino acids, but others, mostly from plants, do not contain all of the essential amino acids.

3. Although neither of these foods, alone, is a complete protein, consuming them together provides all of the essential amino acids at optimal levels.

4. You don't necessarily have to avoid such a food. It is not necessary to get all of the essential amino acids from each protein source. As long as you get all of the essential amino acids from *everything* you consume at a meal, you will not have any amino acid deficiencies.

5. Apples and white rice, consumed together, would contain all of the essential amino acids. Lentils and almonds would not, since neither of these foods contains methionine and cysteine.

6. The figure doesn't specify what the "optimal levels" of an essential amino acid are. The word "optimal" seems a bit subjective, and the figure doesn't specify who determined what constitutes an "optimal" level. This could be cause for some caution in interpreting the data. In fact, though, these levels were determined by the Institute of Medicine's Nutrition Advisory Board. The Institute of Medicine is the health arm of the U.S. National Academy of Sciences, a highly respected group of the country's most distinguished scientists.

7. The amount of a food you must consume to get a sufficient amount of all the essential amino acids is not given. We can assume that a standard serving is sufficient, but it would be helpful if specific amounts were given, such as "8 ounces" of 2% fat milk, or "6 ounces" of chicken breast.

## Chapter 3

1. e; 2. b; 3. e; 4. a; 5. d; 6. d; 7. c; 8. a; 9. d; 10. d; 11. b; 12. d; 13. c; 14. e; 15. b; 16. b; 17. e; 18. e

**Short-Answer Question**

1. See Figure 3-41 on page 121.

**Graphic Content Questions**

1. The different cell types observed are the variables. For each cell type, the figure shows how many mitochondria it has.

2. It would be helpful to see data for more cell types, such as sperm cells, egg cells, or brain cells. Data on variation among cells of a particular type within an individual would also be helpful. In other words, do all liver cells have the same number of mitochondria, or does the number vary a lot? And does the number of mitochondria in a specific cell type of a person vary over time—such as when the person gets older, or in response to exercise?

3. Cells with high rates of metabolic activity seem to have more mitochondria.

4. First, perhaps "number of mitochondria per some measure of cell volume" would be better, which would be a measure of mitochondrial density. This would make it easier to compare energy-generating capacity for cells that differ greatly in size. Second, it is also possible that mitochondria can shrink or swell. If that's the case (and some researchers believe it is), maybe the energetic needs of a cell can be met by a few big mitochondria *or* by a large number of small mitochondria—in which case, the *number* of mitochondria wouldn't be the best measure. A measure of total mitochondrial *volume* might be better.

5. Some other cells with high metabolic activity might have relatively large numbers of mitochondria. Sperm cells, for example, might have larger numbers. Similarly, for one cell type, such as muscle cells, more active cells (such as leg muscles) might have more mitochondria than less used muscles (such as those controlling your baby toe).

## Chapter 4

1. a; 2. e; 3. e; 4. c; 5. a; 6. c; 7. a; 8. b; 9. d; 10. c; 11. b; 12. e; 13. b; 14. b; 15. a; 16. a; 17. e; 18. d

**Short-Answer Questions**

1. The plant can split the water molecules in the "photo" reactions of photosynthesis, producing oxygen gas, which can be used for aerobic respiration by both the plant and the mouse. The carbon dioxide produced as a by-product of aerobic respiration by the plant and the mouse can be used by the plant in the "synthesis" reactions, which ultimately produce glucose for the plant.

2. a) Water molecules are split during the "photo" reactions and oxygen gas is released. Measuring the amount of $O_2$ provides an indication of the rate of the "photo" reactions. b) Plants incorporate $CO_2$ in the "synthesis" reactions. Using the energy from the "photo" reactions, plants modify the $CO_2$ to (eventually) form glucose. The "synthesis" reactions are driven by the energy from the "photo" reactions. c) With additional $CO_2$, the rate of the "synthesis" reactions will increase, which in turn will increase the amount of energy needed from the "photo" reactions.

3. Photosynthesis involves the production of glucose, and cellular respiration involves the breakdown of glucose. Growth in plants is dependent on the rate of photosynthesis. The faster this process occurs, the more glucose is made, the more plant material can be synthesized, and the more the plant grows. If a plant is growing, the rate of photosynthesis exceeds the rate of cellular respiration.

**Graphic Content Questions**

1. The *x*-axis is the horizontal axis, on which the photosynthetic pigments are listed. The *y*-axis is the vertical axis, which shows the height of each bar and is a measure of the amount of the pigment molecule (for each pigment on the *x*-axis) present in leaves.

2. The independent variable is the time at which the researcher measured the amounts of photosynthetic pigments in leaves. The researcher controlled this, by deciding when to make the measurements. The dependent variable is the measure of the amount of the pigment molecule (for each pigment on the *x*-axis) present in leaves. Based on the information in the graph, you cannot determine how the amounts of pigment molecules were measured. Presumably, this was done by some sort of biochemical assay, but you cannot be sure. The colors of the bars give an extra bit of information, showing the color of light reflected by each of the pigments.

**3.** The source of the data is not given. In this case, this probably does not reduce the value of the information in the graph by much. It can be important to know the source of a graph's data if the graph is being used to persuade the reader to believe something controversial, or is being used to advance some sort of political or social agenda. In the figure here, it's hard to imagine why the data would not be reliable or trustworthy. However, it would be useful to know whether the researcher had the technical training and skills to carry out the data collection appropriately.

**4.** The two graphs are measures of the amounts of pigment, for three different pigments, in spring versus fall. This allows us to evaluate differences in the photosynthetic pigments present in leaves at two different times of year, when the leaves appear green and when they have the red, orange, and yellow colors characteristic of fall foliage. We can see that the amount of yellow-reflecting carotenoids is the same in spring as in fall, but the green-reflecting chlorophylls *a* and *b* are present in much smaller amounts in fall than in spring.

**5.** The colors of leaves in the fall are a consequence of reduced amounts of chlorophylls *a* and *b* in the leaves, which reduces the green reflectance, but the amounts of yellow-reflecting pigments are unchanged.

**6.** First, in the fall, leaves have many colors, including yellow. It would be helpful to see measures of the amounts of other photosynthetic pigments in leaves in the fall and spring. Which ones are responsible for the other colors? Second, as with any graph, a standard deviation or standard error bar for each bar in the graph would be useful. This would indicate how much leaves vary in the amounts of each pigment they contain. Is it a lot? Or are the amounts relatively constant from leaf to leaf? From this figure, we don't know. Third, it might also be interesting to see the same pair of plots for several different species, including some trees whose leaves change color in the fall and other trees whose leaves do not change color.

**7.** These graphs show experimental results, namely, that the relative amounts of photosynthetic pigments differ at the two time periods measured. There are no control and experimental variables, however. Rather, there are just measures of the different pigments at different times.

# Chapter 5

1. e; 2. e; 3. b; 4. c; 5. d; 6. d; 7. e; 8. b; 9. c; 10. b; 11. a; 12. e; 13. c; 14. a; 15. d

### Short-Answer Questions

**1.** Thymine makes up 30% of the bases. Cytosine and thymine have only one complementary base, so the base and its complement occur in equal numbers. Because cytosine makes up 20%, its complementary base, guanine, is also 20%. The total of cytosine and guanine is 40%, so the remaining bases total 60%. Because thymine and adenine must occur in equal numbers, half of this total (30%) must be thymine and the other half (30%) adenine.

**2.** The mRNA is UAG CUG CCU AGG. mRNA allows instructions from DNA to be delivered to ribosomes in the cytoplasm, where it directs the assembly of the protein it encodes.

**3.** The sequence of codons of the mRNA is translated into a sequence of amino acids that forms a polypeptide chain. As each codon of the mRNA is read by a ribosome, a transfer RNA with an anticodon complementary to the mRNA codon adds a specific amino acid to the growing polypeptide chain.

**4.** This is an example of a point mutation. The adenine in the fourth codon from the left has been replaced with thymine. This change results in

coding for a different amino acid, changing the type of protein produced by the gene—because the amino acid sequence has changed.

**5.** First, recombinant DNA technology is used to make certain medicines, such as insulin, growth hormone, and erythropoietin, with much less expense; this enables more people to purchase and benefit from these medicines. Second, there have been many efforts to cure some diseases caused by genetic abnormalities by using gene therapy; although this technology has not met with success, the future still holds potential. Third, genetic testing can be conducted on parents to identify whether they are at risk of producing offspring with certain genetic diseases. Prenatal genetic screening can be conducted to determine whether a baby will be born with a genetic disease. Adults can also be tested to determine whether they carry genes that put them at risk of developing a disease. With genetic screening, preventive measures and treatments can be undertaken.

### Graphic Content Questions

**1.** In the United States, 86% of the corn is genetically modified. The legend indicates that the blue portion of each pie chart represents the proportion of the crop that is genetically modified.

**2.** It is not possible to determine whether more genetically modified corn or more genetically modified cotton is produced in the United States. Although it might appear that more genetically modified cotton is produced (93% vs. 86%), that only tells you the *proportion* of each crop that is genetically modified. The figure doesn't give any information about the *total amount* produced for any of these crops. Recent data actually show that, in the United States, while corn is cultivated on 80 million acres, only about 12 million acres are used for cotton cultivation. So the absolute amount of genetically modified corn grown in the United States is much greater than the amount of genetically modified cotton.

**3.** The common theme across the three pie charts is that the vast majority of these crops, whether corn, cotton, or soybeans, are genetically modified. This may be a surprise to many people and may cause them to become interested in learning more about genetically modified crops, including whether they are nutritionally different or carry any health or environmental risks.

**4.** If absolute amounts were given, the figure would convey more information about the total size of the corn crop versus the cotton crop versus the soybean crop in the United States. While this might be interesting, it would deflect attention from the fact that the proportion of crops that are genetically modified is very high for all of these crops, independent of the total amount of each crop grown.

**5.** The information in the question suggests that the United States is relying much more heavily on genetically modified crops than is the rest of the world. This might prompt you to investigate not just why such large proportions of crops are genetically modified but why there is such a large disparity between practices in the United States and those in the rest of the world. Is it because genetically modified crops are more commonly developed in the United States? Or are there, perhaps, fewer regulations or limitations on the cultivation of genetically modified crops in the United States? Or do consumers' attitudes to genetically modified crops differ around the world?

**6.** There are several ways to present the data in the question. You could produce a series of pie charts, one for each year. Or you could produce a bar graph, with the *x*-axis indicating the year and the *y*-axis indicating the proportion of the corn crop that is genetically modified in each year, showing one bar for each year. Or you could produce a line graph with the same axes as the bar graph, but with a point for each year's proportion and a line connecting them. In each of these representations, the dramatic rate of increase would be the most obvious trend.

## Chapter 6

1. c; 2. b; 3. d; 4. d; 5. c; 6. a; 7. b; 8. e; 9. d; 10. b; 11. b; 12. c; 13. d; 14. e; 15. c; 16. a; 17. e; 18. d

### Short-Answer Questions

1. Crossing over occurs as sister chromatids of homologous chromosomes come together and swap equivalent segments containing the same genes (but possibly different alleles) during the first prophase of meiosis. This results in an uneven distribution of genetic material from maternal and paternal chromosomes. Gametes could potentially receive more genes from one (either) biological parent than the other. Also, the random division of the homologues (with respect to whether they are maternal or paternal) at anaphase I can lead to gametes having greater or less than exactly 50% maternal and 50% paternal genetic material. For every gamete having a higher percentage of its genetic material from the maternal side, there is a gamete having an equally higher percentage of paternal genetic material.

2. There cannot be any essential genetic information on the Y chromosome—because all females are able to live without a Y chromosome—but there is on the X chromosome. Receiving only the Y chromosome means that essential genetic information from the X chromosome is absent.

3. An organism is more likely to reproduce asexually in the absence of adverse environmental pressures. Should some environmental pressure develop, the organism may start reproducing sexually, producing genetically diverse offspring that are better able to survive environmental pressures. An experiment could be conducted with an organism that has the capacity to switch between sexual and asexual reproduction. This organism could be placed in an environment with ideal conditions and in another with adverse conditions. If the hypothesis is correct, the organism will reproduce asexually under ideal conditions and sexually under adverse conditions.

### Graphic Content Questions

1. The probability that a 35-year-old woman giving birth will have a baby with Down syndrome is approximately 1 in 333 (i.e., 3/1,000).

2. A woman of age 45 giving birth is unlikely to have a baby with Down syndrome. In fact, according to the data in the graph, a woman of age 45 giving birth has a 98.1% probability of having a baby that does not have Down syndrome. This is because the incidence, according to the figure, is approximately 1.9% (19 in 1,000).

3. You cannot conclude from the figure that most babies with Down syndrome are born to mothers over 35. The figure shows only, for mothers of a given age, the proportion of babies born with Down syndrome. But because relatively few of the babies born in any year are born to mothers older than 35, the total number of babies with Down syndrome in this group of mothers is small. In fact, since so many more babies are born to women under 35, it turns out that 80% of babies with Down syndrome are born to mothers under 35.

4. Because the number of births to women older than 45 is much smaller than the number of births to younger women, it becomes increasingly difficult to make accurate estimates. For this reason, researchers restrict

the data they present to situations for which they have sufficiently high confidence in the data accuracy.

5. Based on doctors' recommendations, women aged 35 and higher are likely to have amniocentesis, which will reveal whether a fetus has trisomy-21. Conversely, amniocentesis is not recommended for women under 35. As a consequence, women aged 35 and higher are more likely to learn that a baby will be born with Down syndrome and choose to terminate the pregnancy. This prenatal testing therefore reduces the incidence of babies born with Down syndrome among women 35 and over, but not among women under 35. Among women under 35, the rate of babies born with Down syndrome is much more likely to be almost the same as the incidence of fetuses having trisomy-21.

## Chapter 7

1. a; 2. b; 3. e; 4. e; 5. a; 6. c; 7. e; 8. a; 9. c; 10. e; 11. e; 12. d; 13. d; 14. b; 15. d; 16. a

### Short-Answer Questions

1. a) SMA must be a recessive genetic disorder, because the phenotype of both parents is normal—that is, they are not affected by SMA. For two unaffected parents to have children with the disease, both parents' genotypes must be heterozygous. This is an example of a disease-causing allele that doesn't always reveal itself in a population. b) Individuals who are heterozygous for a recessive genetic disorder (i.e., are carriers) will not necessarily have children with the disease. However, this example demonstrates heterozygous parents with children who inherit the disease-causing allele from both parents, resulting in SMA. The probability of having another biological child with SMA is the same as it was for each of the first two children. The probability of inheriting the recessive allele from the mother is 0.5, and the probability of inheriting the recessive allele from the father is 0.5. Therefore the overall probability is 0.25, 25%, or 1 in 4.

2. The phenotype, or color of the blooms, is affected by differences in the environment. Hydrangea flower color varies with the type of soil, because the pH of the soil affects the pigmentation of the blooms.

3. a) Genetic disorders that involve genes on the X chromosome affect men differently from women. If a female inherits an allele that produces a non-functional protein on one of her X chromosomes, she may not be affected *if* the second X chromosome has the normal allele. However, because males have only one X chromosome, they have only one chance to receive the normal allele.

b) 

|  | $X^H$ | $X^H$ |
|---|---|---|
| $X^h$ | $X^h X^H$ | $X^h X^H$ |
| Y | $X^H Y$ | $X^H Y$ |

There is a 0% chance that a son will have hemophilia. A son inherits his X chromosome from his mother. Because the mother is carrying two functional alleles, either X would provide a functional allele.

### Graphic Content Questions

1. A novelty-seeking score was measured for individuals with different genotypes. Measurement was based on the individuals' responses on a questionnaire. While this might be a fast and easy way to estimate one person's propensity for novelty-seeking relative to another person's propensity, it would have been better to observe individuals' actual behavior. If you are interested in behavior, it is always better to

see how someone behaves than to ask them how they think they would behave.

2. *CC* refers to a genotype for the promoter polymorphism −521*C/T*, which alters the transcription rate of the dopamine receptor gene. This gene influences the activity of the brain's pleasure centers. There are two different alleles, *C* and *T*, and a person can have one of each or two copies of *C* or two copies of *T*. Having one or two copies of the *T* allele (that is, *CT* or *TT*) seems to lead to reduced transcription of the dopamine receptor gene, while having the *CC* genotype is associated with greater gene activity and, hence, greater novelty-seeking behavior.

3. The data reported for the initial study provide information on how much individuals with genotypes in the two groups (first group: *CC,* second group: *CT* or *TT*) varied in their score on the novelty-seeking questionnaire (since everyone would not be expected to score exactly the same). This information is useful because it helps us see that not only was the average novelty-seeking score higher among those with the *CC* genotype, but those with the lowest novelty-seeking scores in the *CC* genotype group had higher scores than those with the highest novelty-seeking scores in the *CT/TT* group. The smaller the variation observed in each group, the more likely it is that differences between the two groups reflect true differences in the measure between the groups.

4. A meta-study combines the results of several or many studies that address a related research hypothesis. This is commonly done because the results from a group of studies can allow more accurate data analysis, whereas a single study doesn't necessarily have enough power to reveal a true effect.

5. The 11 different studies did not include exactly the same subjects. In fact, they may have used groups of subjects that differed in significant ways, such as age, sex, profession, or ancestry. They may also have measured different numbers of subjects, or they may have used slightly different versions of novelty-seeking questionnaires.

6. The results of any one study showing that individuals with the *CC* genotype score higher on a novelty-seeking questionnaire than individuals with the *CT* or *TT* genotype might be interesting, but unconvincing. Questions might remain about the population from which the subjects were chosen, the number of subjects used, the laboratory methods used in determining genotypes, the conditions under which the questionnaire was administered, and a whole range of unconscious, inadvertent, or otherwise hidden biases that the researchers may have had. But it is powerful when the results of all 11 studies that tested this hypothesis are reported and most of the studies seem to show similar results. This means that you can be more certain about the conclusions drawn in any one of the studies. It's important that *all* of the studies conducted on this hypothesis were collected and analyzed, rather than the meta-study simply selecting studies that supported or rejected the hypothesis. We can have more confidence in concluding that an individual's genotype for this gene influences novelty-seeking behavior. (In fact, though, without information about how well a person's score on this questionnaire predicts actual behavior, it might be safer to simply conclude that an individual's genotype influences his or her score on a novelty-seeking questionnaire. This conclusion gives rise to numerous new hypotheses that could be tested.)

## Chapter 8

1. e; 2. a; 3. b; 4. c; 5. d; 6. d; 7. e; 8. a; 9. e; 10. e; 11. b; 12. a; 13. c; 14. b; 15. d; 16. b; 17. b; 18. e; 19. d; 20. d; 21. c

### Short-Answer Questions

1. Evolution is a change in allele frequencies in a population. Natural selection is one of four evolutionary agents, characterized by variation, heritability, and differential reproductive success for a trait.

2. Genetic drift is a random change in allele frequencies. In small populations, random deviations from the expected genotype frequencies among offspring are less likely to be counterbalanced by deviations in the opposite direction from the expected genotype frequencies.

3. First, environments can change quickly. Natural selection may be too slow to adapt the species to a constantly changing environment. Second, variation is needed as the raw material of selection. If there is no mutation to give rise to an allele that would help make an individual more fit, the species will never be perfectly adapted. (Why, for instance, are there no mammals with wheeled appendages?) Third, evolution produces "more fit" individuals in each generation. If the optimum solution required organisms to get worse before they could evolve to "perfection," selection would never lead to such a solution.

4. In the 1940s, when penicillin was first used to treat bacterial infections, it was uniformly effective in killing *Staphylococcus aureus.* Today, more than 90% of isolated *S. aureus* strains are resistant to penicillin. Because penicillin has become such a pervasive toxin in the environment of *Staphylococcus,* natural selection has led to an increase in the frequency of alleles that make the bacteria resistant. Today, nearly a third of "staph" infections are resistant to antibiotics. The meaning of such unintentional natural selection "experiments" is clear and consistent: evolution is occurring all around us.

### Graphic Content Questions

1. The curvy line represents the average beak size of Galápagos finches, measured from 1976 to 1985.

2. The "average" beak size is calculated from measures of many birds. If more birds with large beaks die in a particular year, the average will be lower. If birds with larger beaks are healthier and reproduce at a higher rate than birds with small beaks during that year, the average will increase.

3. Each time there was a drought, the average beak size increased a bit. During the wet year, the average beak size decreased. It seems as though the environment is influencing which birds are most successful.

4. Based on this figure, it appears that when the climate gets drier, birds with larger beaks survive and reproduce better than birds with smaller beaks. Similarly, when the climate is wetter, birds with smaller beaks are more plentiful. So having information about annual rainfall would be useful in making predictions about average beak size in these finches

5. There is no "optimum" beak size. Rather, the beak size that leads to the greatest survival and reproduction varies from year to year and is dependent on the amount of rainfall.

6. Traits in a population can vary in fitness, depending on the environment. There is no such thing as a better trait or a worse trait. A trait is better or worse in a particular environment.

## Chapter 9

1. b; 2. d; 3. e; 4. b; 5. d; 6. d; 7. c; 8. a; 9. d; 10. e; 11. c; 12. a; 13. a; 14. b; 15. e; 16. b

### Short-Answer Question

1. Males provide more parental care, consistent with the two predictions about sex roles and parental investment. First, the sex that invests more

will be more discriminating. Second, members of the sex that invests less will compete among themselves for access to the higher-investing sex.

**Graphic Content Questions**

1. A female with one mate produces 60 offspring. This is the same number as for a male mating with one female. The two numbers are the same because they are measuring the same thing: one male mating with one female. Whatever the number of offspring produced, it is recorded in the left-most bar in each graph.

2. A female with two mates produces 50 offspring. It is not clear why females with two mates would produce fewer offspring than females with one mate. At least two explanations are possible. First, because these numbers are measured for different females—that is, a female's reproductive output can be measured when she has one mate or two mates, but not both—this may just be a consequence of natural variation among females. Maybe those in the two-mate group just happened to have fewer offspring, unrelated to the number of mates they had. Second, perhaps the presence of two males created stress, which somehow reduced either the female's or the males' fertility.

   For additional data, it would be useful to have error bars to determine why the number of offspring is lower for females with two mates. These would give us a sense of how much variation there is in reproductive output among females given access to two males (and among females given access to just one male). If there is much variation within those two groups, then the lower number seen here may just be an artifact of the individual females measured. In this case, if greater numbers of birds were used in the study, the average reproductive output would be the same for females given access to one mate or two mates.

3. Females with four mates would most likely still have about 60 offspring, while males with four mates would have about 150 offspring. These numbers assume the same trends shown in the graphs.

4. If a single male produces more sperm than the number of eggs a female produces, a female's having access to more sperm will not increase her number of offspring.

5. It appears that females are not limited by access to mates. Rather, they seem to be limited by their own production of eggs. Males' reproductive output, however, does seem to be constrained by the number of mates they have. If a male has more mates, he has greater reproductive output.

6. To make comparisons between the data in the two graphs, it is essential that they are measuring the same thing, using the same scales. Otherwise, it would be like comparing apples and oranges.

## Chapter 10

1. a; 2. d; 3. e; 4. e; 5. a; 6. d; 7. d; 8. a; 9. e; 10. a; 11. d; 12. d; 13. e; 14. b; 15. c; 16. a; 17. c

**Short-Answer Questions**

1. The biological species concept focuses on reproductive populations, or groups of individuals that can mate with each other and are unable to breed with other reproductively isolated groups. The morphological species concept focuses on physical characteristics of the organisms to distinguish one species from another. Many living organisms don't fit neatly into a species as defined by the biological species concept. An example is many of the living organisms that divide asexually, such as bacteria. For these organisms, because they do not require a partner to mate, reproductive compatibility is not a useful diagnostic tool for discriminating among species.

2. Biologists would rely on physical features or characteristics to create evolutionary trees and establish relationships between groups of organisms. Today, with the advent of technology, DNA sequencing (examining the base-pair sequence of DNA) is used to construct evolutionary trees. In either case, the trees define which groups of organisms are most closely related, but not which groups are more advanced than others.

3. These adaptations could be considered an "evolutionary innovation" in which a specific trait or group of traits allows an organism to fill a new niche or an unoccupied niche. This is one trigger of adaptive radiation—the huge, rapid expansion of a species, often diversifying into many more species. The specific traits of tall, flat teeth and a square jaw allowed the horse to efficiently crop grasses and vegetation from the ground. With areas of grasslands plentiful, these mammals were able to expand and flourish in this niche.

**Graphic Content Questions**

1. The blue line hovers around a rate of 10% of families becoming extinct at any time during the period recorded on the graph.

2. There are several reasons why the graph may record data on the extinction of families rather than of species. First, because the events documented in the figure occurred millions or even hundreds of millions of years ago, it is difficult to estimate the total number of species present on earth at that time. (It is nearly impossible to do this even today, when we can, at least in theory, count all of the species.) Second, it is difficult to identify whether multiple fossils belong to the same or different species. But because all of the organisms within a family share basic similarities, it is possible to identify fossils as members of the same family with much more certainty. Third, due to the limited number of fossils found, researchers are much more likely to find at least one representative from each family that has existed across time than they are to find at least one representative of each species. This enables a more accurate identification of when all species within a family became extinct (because after that point, no fossils are found of any members of the family). Still, even if no fossils of a family are found after a certain point in time, we can't be sure whether extinction occurred *or* whether the environmental situation just changed so that members of that family were less likely to be fossilized.

3. The data points for background extinction rates never get much higher than 11% or 12% of all families becoming extinct at that point. In each of the mass extinctions shown on the graph, the rate is significantly higher—usually 20% or more, except for the most recent event, which is about 15%, but still significantly higher than the typical background rate.

4. Several factors influence whether a species or a family becomes extinct. These include competition with other species, the range over which the species can grow, and other aspects of the species' environment. For this reason, conditions leading to extinction won't necessarily occur at exactly the same rate. There will be variation over time, influenced by climate, population sizes of competitors and predators, and other factors, all leading to ups and down.

5. As described in Section 8-18, fossils are left only by certain types of organisms, and under a very limited range of environmental conditions. For this reason, the magnitude of mass extinctions noted on this graph may overestimate or underestimate the rate of extinctions of the many groups of organisms that did not leave fossils. This figure is an accurate measure of overall extinction rates only if our (unstated) assumption is true: that the extinction rates of the groups of organisms not leaving

fossils were proportional to the extinction rates of the groups that did leave fossils.

6. If the current rate begins to rise above typical background rates of extinction, we might be in the midst of another mass extinction. This is important in that it may be an indicator that human activities are causing a higher than usual rate of extinctions. If this is the case, it may signal a need to reduce the human impact on extinction rates so as to reduce the likelihood of large (and maybe catastrophic, for us) changes in the species composition of earth.

# Chapter 11

1. d; 2. c; 3. e; 4. e; 5. d; 6. c; 7. b; 8. b; 9. a; 10. b; 11. d; 12. c; 13. a; 14. c; 15. b; 16. c; 17. e; 18. d

## Short-Answer Questions

1. In all animals, the ability to obtain food, whatever method that may be, has evolved through natural selection. For the sloth, it would make no sense for it to have developed the ability to spring at great speeds, because its food source (vegetation) is stationary. The cheetah, by contrast, must use high speeds to catch its prey, which also can run fast. If the cheetah were to occupy the same niche as the sloth, its ability to run would not improve its ability to obtain vegetation, so it might lose this running ability over many generations.

2. Much of the energy derived from food is utilized for heat production, so endotherms require more food to maintain a constant internal temperature. Ectotherms depend on an external heat source to maintain their body temperature, so they do not need energy derived from food to generate heat.

3. The snake, like other ectotherms, depends on an outside source of heat, so placing an insulating sweater on the snake will prevent heat from reaching its body, keeping it from warming up. For the same reason, ectotherms do not have insulating feathers or fur.

4. The slug has a radula for scraping food from surfaces, has a broad foot on which it glides, and produces a protective slime. Slugs lack the body segmentation found in segmented worms.

## Graphic Content Questions

1. We are of the species *Homo sapiens,* which has *Homo heidelbergensis* as an ancestor, which may have *Homo ergaster* as an ancestor, which may have *Homo habilis* as an ancestor, which may have *Australopithecus* (gracile) as an ancestor, which may have *Ardipithecus* as an ancestor. We seem less likely to be direct descendants of *Homo neanderthalensis* and *Australopithecus* (robust), which are dead-end branches that split off from our common ancestors and became extinct.

2. *Homo* appeared about 2.5 million years ago. It seems to have had a significantly larger skull (and so brain size) than its *Australopithecus* ancestors.

3. The solid lines represent transitions from one species to another that are better supported by fossils showing gradual changes. The dashed lines represent potential relationships that are not as well supported by transitional fossils. But the discovery of additional fossils may enable better clarification of the relationships.

4. About 1.6 million years ago, there were probably four different human species living at the same time: *Australopithecus* (robust), *Homo erectus, Homo ergaster,* and *Homo habilis.*

5. It's not long ago that our species became the only living representative of the genus *Homo.* From this figure, it appears to be approximately 10,000 years ago.

# Chapter 12

1. a; 2. b; 3. b; 4. c; 5. a; 6. b; 7. e; 8. c; 9. c; 10. c; 11. d; 12. b; 13. d; 14. b; 15. d

## Short-Answer Questions

1. Non-vascular plants lack tissue that can conduct water and dissolved nutrients, tissue that *is* found in vascular plants. Because diffusion can transport water only over very short distances, non-vascular plants are much smaller than vascular plants. Non-vascular plants must live in an environment that is wet for at least some parts of the year, and they become dormant when it is dry.

2. Mycorrhizal fungi are associated with the roots of plants, benefiting them by transporting phosphorus and nitrogen from the soil to the cells of the plant roots. In exchange, the fungus takes some sugar from the plant. This relationship enhances the overall health of the plant.

3. Plants expend considerable energy producing a fruit, which contains the seeds. Once a fruit is eaten, it is digested by the animal and the seeds are excreted unharmed, and thus relocated to an area distant from the parent plant. The plant pays the animal with a meal, and in return, the animal transports the seeds.

4. The most dramatic initial effect of the extinction of all animals would be the loss of pollinators and seed dispersers for most of the angiosperms on earth. This would lead to significant reduction in the flowering plants' rates of reproduction and population growth, if not outright extinction of animal-pollinated plants and plants with animal-dispersed seeds. Many such plants can grow vegetatively and many can self-fertilize, helping them to avoid immediate extinction, but these methods of growth and reproduction would lead to a rapid reduction in genetic diversity within populations, putting the plants at risk in changing environments. The reduction of herbivory would certainly help plants at first, reducing their mortality rates. But this benefit would be experienced by all plants, and so competition for limited resources (space, sun, soil, etc.) would probably increase.

## Graphic Content Questions

1. At the most general level, you can conclude that having mycorrhizal fungi benefits a plant. To be more precise, you can conclude that plants with symbiotic mycorrhizal fungi have offspring that are more likely to survive than the offspring of plants without the fungi. This difference in offspring survival does not become apparent until sometime between 20 and 47 days after planting.

2. It appears that for the 10-day and 20-day bars, there is no variance at all. In other words, 100% of the offspring were alive in both groups.

3. Based on the number of individuals observed in each group, the difference in survival recorded is unlikely to have occurred if there were, in fact, no true difference in survival. In other words, there is less than a 5% chance that the researchers would have found such a difference between the two groups due to random chance.

4. For the parents with mycorrhizal fungi, about 60% of their offspring were surviving after 47 days. For parents without mycorrhizal fungi, only about 30% of their offspring were surviving after 47 days.

5. Because the difference in offspring survival appeared between 20 and 47 days after planting, it would be helpful to have data for some point within that interval, probably at the mid-point of 33 or 34 days after planting. This is the likely time when some significant differences in offspring quality appeared.

## Chapter 13

1. a; 2. b; 3. c; 4. e; 5 d; 6. a; 7. a; 8. d; 9. b; 10. a; 11. c; 12. b; 13. e; 14. b; 15. e; 16. e; 17. e; 18. b

**Short-Answer Question**

| | Prokaryotes | | Eukaryotes | Viruses |
|---|---|---|---|---|
| | Bacteria | Archaea | Protists | |
| Unicellular, colonial, or multicellular? | Unicellular | Unicellular | Unicellular or colonial | Not cells |
| Plasma membrane? | Present | Present | Present | Absent |
| Peptidoglycan in cell wall? | Present | Absent | Absent | Absent |
| Flagella? What type? | Yes; spins like a propeller | Yes; spins like a propeller, but has a chemical composition different from that of bacteria | Yes; waves from side to side and has a chemical composition different from that of bacteria or archaea | No flagellum |
| Presence of nucleus? | Absent | Absent | Present | No nucleus |
| Intracellular organelles? | Absent | Absent | Present | No organelles |
| Type of genetic material (DNA, RNA, or both possible)? | DNA | DNA | DNA | DNA or RNA |
| Chromosome structure (linear or circular)? | Usually circular | Usually circular | Linear | Linear or circular |

**Graphic Content Questions**

1. Although it may be surprising, the vast majority of cells in a human body are microbial cells—probably bacterial—rather than human cells. In fact, according to this figure, 90 percent of the 100 trillion cells in a human, or 90 trillion cells, are microbial.

2. A human has about 10 trillion human cells.

3. A human has about 90 trillion microbial cells.

4. One feature of bacteria that contributes to their ability to outnumber human cells even as a human still appears human is that bacterial cells are significantly smaller than eukaryotic cells.

5. It is not clear how the cell counts were obtained. The figure would be much more helpful if it included information on how the data were collected. For example, if the cell counts for multiple humans were determined, how much variation was there among individuals, both in the total number of cells *and* in the proportion of each type?

6. There are several other ways to convey these data. Here are two examples. First, because the data include two components (number of bacterial cells and number of human cells) that make up one entity (a human), you could present them as a pie chart, with 90% of the pie in one color, representing bacteria, and 10% in another color, representing human cells. A strength of this presentation is that it conveys nicely that most of the "pie" is bacterial rather than human. A weakness is that it conveys information on how each component represents some proportion of the whole, but no information on the absolute numbers of each cell type. Second, because there is a comparison of two entities, you could present side-by-side bar graphs, with a *y*-axis labeled "Number of cells (trillions)." The bar for "human cells" would extend to 10 trillion, and the bar for "bacterial cells" would extend to 90 trillion. One strength of this approach is that it makes it easier to compare the two numbers. A weakness is that it de-emphasizes that the two components together make up the whole of how many cells there are in a human body.

## Chapter 14

1. a; 2. e; 3. a; 4. b; 5. e; 6. d; 7. b; 8. d; 9. c; 10. c; 11. e; 12. d; 13. a; 14. b; 15. c

**Short-Answer Questions**

1. Density-dependent factors are limitations on a population's growth resulting from competition among individuals for a finite amount of resources, such as food and space for living and breeding, and an increased risk of disease and predation. Density-independent factors are forces such as hurricanes and droughts that kill off many individuals in a population, regardless of the size of the population.

2. An organism with a type I survivorship curve is relatively large, has only a few offspring, invests a lot in each offspring, and lives a relatively long life. If the organism had a type III survivorship curve, it would have a high probability of dying early in life and would produce many offspring and provide very little parental care.

3. Species that lack natural enemies and live in an environment with low risk of mortality from other sources breed at a later age; late-onset genetic disorders are selected out and individuals are healthier in old age. Species that breed at an early age pass on adverse genetic traits that are expressed after their normal age of breeding; they do not live long after the normal age of breeding. If breeding is delayed, as in the fruit flies in the experiments described in the chapter, age-related disorders that interfere with breeding are selected out and the life span will be extended.

**Graphic Content Questions**

1. The graph shows how a population's size changes over time. Based on the graph's title, "Managing Natural Resources," it also is intended to reveal something about how knowledge of population growth can help resource managers be more effective. Specifically, it shows the size at which a population can most effectively be utilized for natural resources, which may involve cutting down some trees for wood or catching some lobsters for food, for example.

2. As a population gets closer to the carrying capacity of its environment, its growth slows and plateaus at that point.

3. A population can stay at the same size if the number of individuals arriving (through birth or immigration) is exactly the same as the number of individuals leaving (through death or emigration). This situation can persist indefinitely, even as organisms continue dying.

4. No data points appear on the graph. This suggests that perhaps the graph just presents a theoretical relationship. In this case, it appears to be the

relationship between population size and time, when a population is growing logistically, calculated as: $rN[(K - N)/K]$.

5. It would be helpful if some real data were presented for actual populations, showing how they grow over time. The data points could be overlaid on the graph, perhaps with points for different populations shown in different colors. This would give insight into how well the logistic growth equation predicts the actual growth of natural populations.

6. The graph described in the question will show a sort of inverted U, with the low points of growth rate at population sizes 0 and $K$, and with the maximum in the middle of the $x$-axis at $K/2$. The population grows fastest when it is at half its carrying capacity.

7. Implicit assumptions are, first, that carrying capacity does not change over time, and second, that a population's growth is restrained with perfect efficiency as it approaches its carrying capacity. In other words, growth slows exactly as required to avoid exceeding the carrying capacity—as opposed to the population growing so quickly that it overshoots its carrying capacity, after which it grows more slowly, possibly undershooting $K$, and ultimately oscillating around it.

## Chapter 15

1. d; 2. e; 3. a; 4. a; 5. e; 6. d; 7. e; 8. a; 9. d; 10. a; 11. a; 12. a; 13. d; 14. e; 15. b; 16. b

### Short-Answer Questions

1. Heated air rises and cools at the equator (0° latitude); as it cools, it cannot hold as much water and loses it as precipitation. Air moves north and south from the equator and begins to fall at about 30° latitude north and south, warming and absorbing moisture. This creates very dry conditions at these latitudes, and deserts are found there. Air continues to move toward the poles and begins to rise again at 60° latitude, cooling and losing its moisture through precipitation, making these locations very wet. Air again falls near the Poles, making these areas very dry. Temperature and amount of rainfall are the primary determinants of the kind of biome found in any terrestrial location.

2. Energy flows through each trophic level. Photosynthetic organisms, the primary producers, trap energy from the sun in the chemical bonds of food. Primary consumers, the herbivores, eat photosynthetic organisms, and these animals are consumed by secondary consumers, the carnivores. Carnivores can be consumed by tertiary consumers, the top carnivores. A natural ecosystem can support only a few top carnivores, because only about 10% of the energy is available to each subsequent trophic level.

3. An organism's niche includes the space it requires to live in, the kinds and amount of food it eats, the time of day and time of year it reproduces, the amount of moisture and temperature it requires, as well as other criteria. Two organisms that occupy the same niche compete, with two different possible outcomes: one species may drive the other to extinction (competitive exclusion), or both species may change their behavior or undergo a morphological change so that they are restricted within their own niche, dividing the resources between them (resource partitioning).

### Graphic Content Questions

1. The drawings are roughly proportional in size to the biomass of each level in the energy pyramid (i.e., primary producer, primary consumer, secondary consumer, tertiary consumer) for this particular type of ecosystem.

2. The size of the pie at each level in this diagram of energy flow reflects the biomass in an ecosystem of organisms fulfilling a particular role. The pies are different sizes because the biomass of primary producers is typically 10 times the biomass of primary consumers, and the biomass of each successive step up the pyramid is typically one-tenth of that at the previous level. This is due to inefficiencies of energy flow and conversion.

3. The pie slice, typically 10% of the size of the pie, represents the conversion efficiency of biomass at one level into biomass at the next level up the pyramid.

4. At the level of tertiary consumers, typically there is not sufficient biomass to support a viable population of quaternary consumers, given the 10% rule.

5. A secondary or tertiary consumer that also eats plants could have greater biomass than shown here because it could draw its energy not just from the animals it eats, but also from plants. The expected biomass of such a consumer would be 10% of the total biomass of the organisms it could consume. This would be much greater than for any of the consumers shown here.

6. There are likely to be many fewer predators than prey individuals (particularly herbivores), probably because the total biomass of predators is estimated to be one-tenth of the total biomass of prey organisms.

## Chapter 16

1. e; 2. c; 3. d; 4. a; 5. e; 6. e; 7. d; 8. d; 9. b; 10. a; 11. e; 12. c; 13. b; 14. e

### Short-Answer Questions

1. The greatest biodiversity, both aquatic and terrestrial, is in equatorial regions. Less biodiversity is found with increasing latitude. Habitat destruction and degradation are the greatest threats to biodiversity.

2. The biotic and abiotic nature of the habitat was altered, but species were not driven to extinction. If species had been made extinct, the ecosystem disturbances would not have been reversible.

3. Most of the deforestation of tropical rain forests is a result of clearing land to make way for agriculture; mining and pollution also contribute. The biodiversity of the planet is greatly affected by rain forest deforestation, because no other terrestrial ecosystem has the extent of biodiversity found in these forests.

### Graphic Content Questions

1. There are two graphs. Each presents data on the number of species present at different latitudes, from the equator toward the Poles. The $x$-axis shows the number of species: in the top graph, the number of mammalian species, and in the bottom graph, the number of marine copepod species. The $y$-axis shows the latitude. In the top graph, data are presented for the numbers of species at 40° and 20° south, the equator, and 20°, 40°, and 60° north. In the bottom graph, data are presented for the equator and 20°, 40°, and 60° north.

2. The independent variable is latitude. The dependent variable is species number. Typically, the $x$-axis is used for the independent variable and the $y$-axis for the dependent variable. In these graphs, the variables are placed on the opposite axes. This is because the data are overlaid on a world map, which aids in identifying exactly where in the world 20°, 40°, and 60° latitudes fall. This makes it easy to see, for example, that 20° north latitude includes Central America, while 60° north latitude cuts across southern Alaska and northern Canada.

3. For mammalian species and for marine copepods, the largest numbers of species occur at 0°, which is the equator.

4. The value of having data both for mammalian species and for marine copepod species is that it enables us to see that the trend of decreasing species diversity toward the Poles holds true for a *terrestrial* species and for a *marine* species. This suggests that the phenomenon is robust and not just a quirk of one group of organisms. It would be helpful if data were presented for *all* species, but such data do not exist. (And, as we saw in Chapter 10, for prokaryotes and many other organisms it is difficult to even define what a species is.) Only a fraction of the world's species have been described. In this figure, it is more valuable to include animal groups for which most species have been described. Also important is that the figure uses groups of organisms for which any bias in the proportion of total species described at the equator versus the proportion described at higher latitudes is unlikely.

5. There are more mammalian species than marine copepod species at 20° north latitude. A person might answer this question incorrectly because the bar representing the number of copepod species is longer than the bar representing mammalian species—but this is because the scales for the axes differ. There are more than 300 mammalian species and about 70 copepod species at 20° north.

6. The bars here represent exact counts of species rather than some repeated measure of a variable (such as height), which would have a variance associated with it.

# Glossary

*A note about notation:* The word or phrase being defined is in boldface type; it is followed by the number (in parentheses) of the chapter or chapters where it is discussed. In some cases, derivations are given. Abbreviations: Gk., Greek; Lat., Latin; *sing.,* singular; *pl.,* plural; *dim.,* diminutive (a smaller version of the object named); *pron.,* pronounced.

## A

**abiotic** (15) Relating to the physical and chemical components in an environment. These include the chemical resources of the soil, water, and air (such as carbon, nitrogen, and phosphorus) and physical conditions (such as temperature, salinity, moisture, humidity, and energy sources).

**acid** (2) Any fluid with a pH below 7.0, indicating the presence of more H⁺ ions than OH⁻ ions. [Lat., *acidus,* sour]

**acquired immune deficiency syndrome (AIDS)** (13) Infectious human disease caused by a retrovirus, HIV (human immunodeficiency virus), which compromises the immune system by attacking T cells, leaving an individual susceptible to infections, as well as cancers.

**activation energy** (2) The minimum energy needed to initiate a chemical reaction (regardless of whether the reaction releases or consumes energy).

**activator** (2) A chemical within a cell that binds to an enzyme, altering the enzyme's shape or structure in a way that causes the enzyme to catalyze a reaction.

**active site** (2) The part of an enzyme to which reactants (or substrates) bind and undergo a chemical reaction.

**active transport** (3) Molecular movement that depends on the input of energy, which is necessary when the molecules to be moved are large or are being moved against their concentration gradient.

**adaptation** (8) The process by which, as a result of natural selection, a population's organisms become better matched to their environment; also, a specific feature, such as the quills of a porcupine, that makes an organism more fit.

**adaptive radiation** (10) The rapid diversification of a small number of species into a much larger number of species, able to live in a wide variety of habitats.

**additive effects** (7) Effects from alleles of multiple genes that all contribute to the ultimate phenotype for a given characteristic.

**adenosine triphosphate (ATP)** (4) A molecule that temporarily stores energy for cellular activity in all living organisms. ATP is composed of an adenine, a sugar molecule, and a chain of three negatively charged phosphate groups.

**adult** (11) (Pertaining to insect developmental stages) In complete metamorphosis, the third and final stage of insect development.

**aging** (14) An increased risk of mortality with increasing age; generally characterized by multiple physiological breakdowns.

**alleles** (5, 7) Alternative versions of a gene. [Gk., *allos,* another]

**allopatric speciation** (10) Speciation that occurs as a result of a geographic barrier between groups of individuals that leads to reproductive isolation and then genetic divergence. [Gk., *allos,* another + *patris,* native land]

**altruistic behavior** (9) A behavior that comes at a cost to the individual performing it and benefits another. [Lat., *alter,* the other]

**amino acid** (2) One of 20 molecules built of an amino group, a carboxyl group, and a unique side chain. Proteins are constructed of combinations of amino acids linked together.

**amino group** (2) A nitrogen atom bonded to three hydrogen atoms.

**amniotes** (11) Terrestrial vertebrates—reptiles, birds, and mammals—that produce eggs (called amniotic eggs) that are protected by a water-tight membrane and a shell.

**amphibians** (11) Members of the class Amphibia; ectotherms (that is, they are cold-blooded), with a moist skin, lacking scales, through which they can fully or partially absorb oxygen. They were the first terrestrial vertebrates. The young of most species are aquatic, and the adults are true land animals. [Gk., *bios,* life + *amphi,* on both sides]

**analogous traits** (10) Characteristics (such as bat wings and insect wings) that are similar because they were produced by convergent evolution, not because they descended from a common structure in a shared ancestor. [Gk., *analogos,* proportionate]

**anaphase** (6) A phase in mitosis and meiosis in which chromosomes separate. In mitosis, it is the third phase, in which the sister chromatids are pulled apart by the spindle fibers, with a full set of chromosomes going to opposite sides of the cell. In meiosis, the homologues separate in anaphase I and the sister chromatids separate in anaphase II. [Gk., *ana,* up + *phasis,* appearance]

**anecdotal observation** (1) Observation of one or only a few instances of a phenomenon.

**angiosperms** (12) Vascular, seed-producing flowering and fruit-bearing plants, in which the seeds are enclosed in an ovule within the ovary. [Gk., *angeion,* vessel, jar + *sperma,* seed]

**animals** (11) Members of the kingdom Animalia; eukaryotic, multicellular, heterotrophic (that is, they cannot produce their own food) organisms. Many of these organisms have body parts specialized for different activities and can move during some stage of their lives. [Lat., *animal,* a living being]

**annelid** (11) Phylum of worms having segmented bodies; protostomes with defined tissues, which grow by adding segments rather than by molting. There are about 13,000 identified species of segmented worms, including earthworms and leeches.

**anther** (12) The part of the stamen, the male reproductive structure of a flower, that produces pollen. [Gk., *anthos,* blossom]

**apoptosis** (6) Programmed cell death, which takes place particularly in parts of the body where the cells are likely to accumulate significant genetic damage over time and are therefore at high risk of becoming cancerous. [Gk., *apoptosis,* falling away]

**archaea** (*sing.* **archaeon**) (10) A group of prokaryotes that are evolutionarily distinct from bacteria and that thrive in some of the most extreme environments on earth; one of the three domains of life. [Gk., *archaios,* ancient]

**arthropods** (11) Members of the invertebrate phylum Arthropoda; characterized by a segmented body, an exoskeleton, and jointed appendages. [Gk., *arthron,* joint + *pous,* foot]

**asexual reproduction** (6) A type of reproduction common in prokaryotes and plants, and also occurring in many other multicellular organisms, in which the offspring inherit their DNA from a single parent.

**atom** (2) A particle of matter than cannot be further subdivided without losing its essential properties. [Gk., *atomos,* indivisible]

**atomic mass** (2) The mass of an atom; the combined mass of the protons, neutrons, and electrons in an atom (the mass of the electrons is so small as to be almost negligible).

**atomic number** (2) The number of protons in the nucleus of an atom of a given element.

## B

**background extinctions** (10) Extinctions that occur at lower rates than at times of mass extinctions; occur mostly as the result of aspects of the biology and competitive success of the species, rather than catastrophe.

**base** (chemistry) (2) Any fluid with a pH above 7.0, that is, with more OH⁻ ions than H⁺ ions.

**base** (of DNA) (2, 5) One of the nitrogen-containing side-chain molecules attached to a sugar molecule in the sugar-phosphate backbone of DNA and RNA. The four bases in DNA are adenine (A), thymine (T), guanine (G), and cytosine (C); the four bases in RNA are adenine (A), uracil (U), guanine (G), and cytosine (C). The information in a molecule of DNA and RNA is determined by its sequence of bases.

**base pair** (5) Two nucleotides on complementary strands of DNA that form a pair, linked by hydrogen bonds. The pattern of pairing is adenine (A) with thymine (T) and cytosine (C) with guanine (G); the base-paired arrangement forms the "rungs" of the double-helix structure of DNA.

**behavior** (9) Any and all of the actions performed by an organism, often in response to its environment or to the actions of another organism.

**bilateral symmetry** (11) A body structure with left and right sides, which are mirror images. [Lat., *bi-,* two + *latus,* side]

**binary fission** (6) A type of asexual reproduction in which the parent cell divides into two genetically identical daughter cells. Bacteria and other prokaryotes reproduce by binary fission. [Lat., *binarius,* consisting of two + *fissus,* divided]

**biodiversity** (10, 16) The variety and variability among all genes, species, and ecosystems. [Gk., *bios,* life + Lat., *diversus,* turned in different directions]

**biodiversity hotspots** (16) Regions of the world with significant reservoirs of biodiversity that are under threat of destruction.

**biofuels** (4) Fuels produced from plant and animal products.

**biogeography** (8) The study and interpretation of distribution patterns of living organisms around the world. [Gk. *bios,* life + *geo-,* earth + *graphein,* to write down]

**biological literacy** (1) The ability to use scientific inquiry to think creatively about problems with a biological component, to communicate these thoughts to others, and to integrate these ideas into one's decision making.

**biological species concept** (10) A definition of species described as populations of organisms that interbreed, or could possibly interbreed, with each other under natural conditions, and that cannot interbreed with organisms outside their own group.

**biology** (1) The study of living things. [Gk., *bios,* life + *logos,* discourse]

**biomass** (15) The total mass of all the living organisms in a given area. [Gk., *bios,* life]

**biomes** (15) The major ecological communities of earth. Terrestrial biomes, such as rain forest or desert, are defined and usually described by the predominant types of plant life in the area, which are mostly determined by the weather; aquatic biomes are usually defined by physical features such as salinity, water movement, and depth. [Gk., *bios,* life]

**biotechnology** (5) The modification of organisms, cells, and their molecules for practical benefits. [Gk., *bios,* life + *technologia,* systematic treatment]

**biotic** (15) Relating to living organisms; the biotic environment, or community, consists of all the living organisms in a given area. [Gk., *bios,* life]

**bivalve mollusk** (11) Mollusks with two hinged shells; examples are clams, scallops, and oysters. [Lat., *bi-,* two]

**blind experimental design** (1) An experimental design in which the subjects do not know what treatment (if any) they are receiving.

**bond energy** (2) The strength of a bond between two atoms, defined as the energy required to break the bond.

**bottleneck effect** (8) A phenomenon by which genetic drift can occur; a sudden reduction in population size (often due to famine, disease, or rapid environmental disturbance) that can lead to changes in the allele frequencies of a population.

**bryophytes** (12) Three groups of plants (the liverworts, hornworts, and mosses) that lack vascular tissue and move water and dissolved nutrients by diffusion. [Gk., *bruon,* tree-moss, liverwort + *phytas,* plant]

**buffer** (2) A chemical that can quickly absorb excess H⁺ ions in a solution (preventing it from becoming too acidic) or quickly release H⁺ ions (to counteract increases in OH⁻ concentration).

## C

**C4 photosynthesis** (4) A method (along with C3 and CAM photosynthesis) by which plants fix carbon dioxide, using the carbon to build sugar; serves as a more effective method than C3 for binding carbon dioxide under low carbon dioxide conditions, such as when plants in warmer climates close their stomata to reduce water loss.

**Calvin cycle** (4) In photosynthesis, a series of chemical reactions in the stroma of chloroplasts, in which sugar molecules are assembled. [From the name of one of its discoverers, Melvin Calvin, 1911–1997]

**CAM (crassulacean acid metabolism)** (4) Energetically expensive photosynthesis in which the stomata are open only at night to admit $CO_2$, which is bound to a holding molecule and released to enter the Calvin cycle to make sugar during the day. In this type of photosynthesis, found in many fleshy, juicy plants of hot, dry areas, water loss is reduced because the stomata are closed during the day.

**cancer** (6) Unrestrained cell growth and division. [Lat., *cancer,* crab]

**capsid** (13) The protein container surrounding the genetic material (DNA or RNA) of a virus. [Lat., *capsa,* box, case]

**capsule** (13) A layer surrounding the cell wall of many bacteria; it may restrict the movement of water out of the cell and thus allow bacteria to live in dry places, such as the surface of the skin. The capsule contributes to the virulent characteristics of some bacteria, making them resistant to phagocytosis by the host's immune system. [Lat., *capsula,* small box or case]

**carbohydrate** (2) One of the four types of biological macromolecule, containing mostly carbon, hydrogen, and oxygen. Carbohydrates are the primary fuel for cellular activity and form much of the cell structure in all life forms. [Lat., *carbo,* charcoal + *hydro-,* pertaining to water]

**carboxyl group** (2) A functional group characterized by a carbon atom double-bonded to one oxygen atom and single-bonded to another oxygen atom. Amino acids are made up of an amino group, a carboxyl group, and a side chain.

**carnivores** (15) Predatory animals (and some plants) that consume only animals. [Lat., *carnis*, of flesh + *vorare*, to devour]

**carotenoids** (4) Pigments that absorb blue-violet and blue-green wavelengths of light and reflect yellow, orange, and red wavelengths of light. [Lat., *carota*, carrot]

**carpel** (12) The female reproductive structure of a flower, including the stigma, style, and ovary. [Gk., *karpos*, fruit]

**carrier** (7) An individual who carries one allele for a recessive trait and does not exhibit the trait; if two carriers mate, they may produce offspring who do exhibit the trait.

**carrying capacity (*K*)** (14) The ceiling on a population's growth imposed by the limitation of resources for a particular habitat over a period of time.

**cartilaginous fishes** (11) Fish species characterized by a skeleton made completely of cartilage, not bone. [Lat., *cartilago*, gristle]

**cell** (3) The smallest unit of life that can function independently; a three-dimensional structure, surrounded by a membrane and, in the case of prokaryotes and most plants, a cell wall, in which many of the essential chemical reactions of the life of an organism take place. [Lat., *cella*, room]

**cell cycle** (6) In a cell, the alternation of activities related to cell division and those related to growth and metabolism.

**cell theory** (3) A unifying and universally accepted theory in biology that holds that all living organisms are made up of one or more cells, and that all cells arise from other, preexisting cells.

**cell wall** (3) A rigid structure, outside the cell membrane, that protects and gives shape to the cell; found in many prokaryotes and plants.

**cellular respiration** (4) The process by which all living organisms extract energy stored in the chemical bonds of molecules and use it for fuel for their life processes.

**cellulose** (2) A complex carbohydrate, indigestible by humans, that serves as the structural material for a huge variety of plant structures. It is the single most prevalent organic compound on earth. [Lat., *cellula, dim.* of *cella,* room]

**central vacuole** (3) See **vacuole.**

**centriole** (6) A structure, located outside the nucleus in most animal cells, to which the spindle fibers are attached during cell division.

**centromere** (6) After replication, the region of contact between sister chromatids, which occurs near the center of the two strands. [Gk., *centron*, the stationary point of a pair of compasses, thus the center of a circle + *meris*, part]

**cephalopods** (11) Mollusks in which the head is prominent and the foot has been modified into tentacles; examples are octopuses and squids. Cephalopods have a reduced or absent shell and possess the most advanced nervous system of the invertebrates. [Gk., *kephale*, head + *pous*, foot]

**character displacement** (15) An evolutionary divergence in one or more of the species that occupy a niche that leads to a partitioning of the niche between the species. Changes in characteristics, such as behavior or body plan, of two or more very similar species that have overlapping geographic locations result in a reduction of competition between the species.

**chemical energy** (4) A type of potential energy in which energy is stored in chemical bonds between atoms or molecules.

**chemolithotrophs** (13) Bacteria that can use inorganic molecules such as ammonia, hydrogen sulfide, hydrogen, and iron as sources of energy. [Gk., *lithos*, stone, rock + *trophē*, food]

**chemoorganotrophs** (13) Bacteria that consume organic molecules, such as carbohydrates, as an energy source. [Gk., *trophē*, food]

**chitin** (2) (*pron.* KITE-in) A complex carbohydrate, indigestible by humans, that forms the rigid outer skeleton of most insects and crustaceans. [Gk., *chiton*, undershirt]

**chlorophyll** (4) A light-absorbing pigment molecule in chloroplasts. [Gk., *chloros*, pale green + *phyllon*, leaf]

**chlorophyll *a*** (4) The primary photosynthetic pigment. Chlorophyll *a* absorbs blue-violet and red light; because it cannot absorb green light and instead reflects those wavelengths, we perceive the reflected light as the color green.

**chlorophyll *b*** (4) A photosynthetic pigment similar in structure to chlorophyll *a*. Chlorophyll *b* absorbs blue and red-orange wavelengths and reflects yellow-green wavelengths.

**chloroplast** (3, 4) The organelle in plant cells in which photosynthesis occurs. [Gk., *chloros*, pale green + *plastos*, formed]

**cholesterol** (2, 3) One of the sterols, a group of lipids important in regulating growth and development; an important component of most cell membranes, helping the membrane maintain its flexibility. [Gk., *chole*, bile + *stereos*, solid + *-ol*, chemical suffix for an alcohol]

**chromatid** (6) One of the two strands of a replicated chromosome. [Gk., *chroma*, color]

**chromatin** (3) A mass of long, thin fibers consisting of DNA and proteins in the nucleus of the cell. [Gk., *chroma*, color]

**chromosomal aberration** (5) A type of mutation characterized by a change in the overall organization of genes on a chromosome, such as the deletion of a section of DNA; the moving of a gene from one part of a chromosome to elsewhere on the same chromosome or to a different chromosome; or the duplication of a gene, with the new copy inserted elsewhere on the chromosome or on a different chromosome. [Lat., *aberrare*, to wander]

**chromosome** (5) A linear or circular strand of DNA on which are found specific sequences of base pairs. The human genome consists of two copies of each of 23 unique chromosomes, one from the mother and one from the father. [Gk., *chroma*, color + *soma*, body]

**cilia** (*sing.* **cilium**) (3) Short projections from the cell surface, often occurring in large numbers on a single cell, that beat against the intercellular fluid to move the fluid past the cell. [Lat., *cilium*, eyelid]

**class** (10) In the hierarchical taxonomic system developed by Carolus Linnaeus (1707–1778), a classification of organisms consisting of related orders.

**climax community** (15) A stable and self-sustaining community that results from ecological succession.

**clone** (5) A genetically identical DNA fragment, cell, or organism produced by a single cell or organism. [Gk., *klon*, twig]

**clone library** (5) A collection of cloned DNA fragments; also known as a gene library.

**cloning** (5) The production of genetically identical cells, organisms, or DNA molecules.

**code** (5) In genetics, the base sequence of a gene.

**codominance** (7) The case in which the heterozygote displays characteristics of both alleles.

**codons** (5) Three-base sequences in mRNA that link with complementary tRNA molecules, which are attached to amino acids; a codon with yet another sequence ends the process of assembling a protein from amino acids.

**coevolution** (15) The concurrent appearance and modification over time, through natural selection, of traits in interacting species that enable each species to become adapted to the other; an example is the 11-inch-long tongue of a moth that feeds from the 11-inch-long nectar tube of an orchid.

**colonizers** (15) Species introduced into an area that has been disturbed and is undergoing the process of either primary or secondary succession. The identity of a colonizing species varies, depending on the stage and type of succession.

**commensalism** (15) A symbiotic relationship between species in which one benefits and the other neither benefits nor is harmed. [Lat., *com,* with + *mensa,* table]

**communication** (9) An action or signal on the part of one organism that alters the behavior of another organism.

**community** (in ecology) (15) The biotic environment; a geographic area defined as a loose assemblage of species with overlapping ranges.

**competitive exclusion** (15) The case in which two species battle for resources in the same niche until the more efficient of the two wins and the other is driven to extinction in that location.

**competitive inhibitor** (2) A chemical within a cell that binds to the active site of an enzyme, blocking substrate molecules from the site and thereby reducing the enzyme's ability to catalyze a reaction.

**complementarity** (6) The characteristic of double-stranded DNA that the base on one strand always has the same pairing partner, or complementary base, on the other strand.

**complementary base** (6) A base on a strand of double-stranded DNA that is a pairing partner to a base on the other strand: adenine (A) is the complementary base to thymine (T), and guanine (G) is the complementary base to cytosine (C).

**complex carbohydrate** (2) A carbohydrate that contains multiple simple carbohydrates linked together; examples are starch, which is the primary form of energy storage in plants, and glycogen, which is the primary form of short-term energy storage in animals.

**compound** (2) A substance composed of atoms of different elements in specific ratios, held together by ionic bonds. [Lat., *componere,* to put together]

**conjugation** (13) The process by which a bacterium transfers a copy of some or all of its DNA to another bacterium, of the same or another species. [Lat., *coniugatio,* connection]

**conservation biology** (16) An interdisciplinary field, drawing on ecology, economics, psychology, sociology, and political science, that studies and devises ways of preserving and protecting biodiversity and other natural resources.

**contact inhibition** (3) The limiting factor of cell growth that occurs when normal cells come into contact; in cancer cells, contact inhibition does not take place and the cells continue to divide.

**control group** (1) In an experiment, the group of subjects not exposed to the treatment being studied but otherwise treated identically to the experimental group.

**convergent evolution** (8, 10) A process of natural selection in which features of organisms not closely related come to resemble each other as a consequence of similar selective forces. Many marsupial and placental species resemble each other as a result of convergent evolution. [Lat., *con-,* together with + *vergere,* turn + *evolvere,* to roll out]

**covalent bond** (2) A strong bond formed when two atoms share electrons; the simplest example is the $H_2$ molecule, in which each of the two atoms in the molecule shares its lone electron with the other atom. [Lat., *con-,* together + *valere,* to be strong]

**critical experiment** (1) An experiment that makes it possible to determine decisively between alternative hypotheses.

**cross** (7) The breeding of organisms that differ in one or more traits.

**crossing over** (6) The exchange of some genetic material between a paternal homologous chromosome and a maternal homologous chromosome, leading to a chromosome carrying genetic material from each; also referred to as recombination.

**cuticle** (12) A waxy layer produced by epidermal cells and found on leaves and shoots of terrestrial plants, protecting them from drying out. [Lat., *cuticula, dim.* of *cutis,* skin]

**cytokinesis** (6) In the cell cycle, the stage following mitosis in which cytoplasm and organelles duplicate and are divided into approximately equal parts and the cell separates into two daughter cells. In meiosis, two diploid daughter cells are formed in cytokinesis following telophase I and four haploid daughter cells are formed in cytokinesis following telophase II. [Gk., *kytos,* container + *kinesis,* motion]

**cytoplasm** (3) The jelly-like fluid that fills the inside of the cell; in eukaryotes, the cytoplasm contains the organelles. [Gk., *kytos,* container + *plasma,* anything molded]

**cytoskeleton** (3) A network of protein structures in the cytoplasm of eukaryotes (and, to a lesser extent, prokaryotes) that serves as scaffolding, adding support and, in some cases, giving animal cells of different types their characteristic shapes. The cytoskeleton serves as tracks guiding the intercellular traffic flow and, because it is flexible and can generate force, gives cells some ability to control their movement.

# D

**daughter cells** (6) Cells produced by the division of a parent cell.

**decomposers** (12, 15) Organisms, including bacteria, fungi, and detritivores, that break down and feed on once-living organisms.

**demographic transition** (14) A pattern of population growth characterized by the progression from high birth and death rates (slow growth) to high birth rates and low death rates (fast growth) to low birth and death rates (slow growth).

**denaturation** (2) The disruption of protein folding, in which secondary and tertiary structures are lost, caused by exposure to extreme conditions in the environment such as heat or extreme pH.

**density-dependent factors** (14) Limitations on a population's growth that are a consequence of population density.

**density-independent factors** (14) Limitations on a population's growth without regard to population size, such as floods, earthquakes, fires, and lightning.

**deoxyribonucleic acid (DNA)** (2, 5) A nucleic acid, DNA carries information about the production of particular proteins in the sequences of its nucleotide bases.

**dependent variable** (1) A measurable entity that is created by the process being observed and whose value cannot be controlled; generally represented on the *y*-axis in a graph and expected to change in response to a change in the independent variable.

**desert** (15) A type of terrestrial biome; a type of dry climate, with very little rainfall, in which water loss through evaporation exceeds water gain through precipitation; typically found at 30° north and south latitude.

**desmosome** (3) Irregularly spaced connections between adjacent animal cells that, in the manner of Velcro, hold cells together by attachments but are not water-tight. They provide mechanical strength and are found in muscle tissue and in much of the tissue that lines the cavities of animal bodies. [Gk., *desmos,* bond + *soma,* body]

**detritivores** (15) Organisms that break down and feed on once-living organic matter; this group includes scavengers such as vultures, worms, and a variety of arthropods. [Lat., *detritus,* worn out + *vorare,* to devour]

**deuterostomes** (11) Bilaterally symmetrical animals with defined tissues in which the gut develops from back to front; the anus forms first, and the second opening formed is the mouth of the adult animal. [Gk., *deuteros,* second + *stoma,* mouth]

**differential reproductive success** (8) The situation in which individuals have greater reproductive success than other individuals in a population; along with variation and heritability, differential reproductive success is one of the three conditions necessary for evolution by natural selection.

**diffusion** (3) Passive transport in which a particle (the solute) is dissolved in a gas or liquid (the solvent) and moves from an area of higher solute concentration to an area of lower solute concentration. [Lat., *diffundere,* to pour in different directions]

**diploid** (6) Describes cells that have two copies of each chromosome (in many organisms, including humans, somatic cells are diploid). [Gk., *diplasiazein,* to double]

**direct fitness** (9) The total reproductive output of an individual.

**directional selection** (8) Selection that, for a given trait, increases fitness at one extreme of the phenotype and reduces fitness at the other, leading to an increase or decrease in the mean value of the trait.

**disaccharides** (2) Carbohydrates formed by the union of two simple sugars, such as sucrose (table sugar) and lactose (the sugar found in milk). [Gk., *di-,* two + *sakcharon,* sugar]

**dispersers** (15) Organisms able to move away from their original home.

**disruptive selection** (8) Selection that, for a given trait, increases fitness at both extremes of the phenotype distribution and reduces fitness at middle values.

**DNA probe** (5) A short sequence of radioactively tagged single-stranded DNA that contains part of the sequence of the gene of interest, used to locate that gene in a gene library. The probe binds to the complementary base pair on a gene in the library, which is identified by the radioactive tag on the probe.

**domain** (10) In modern classification, the highest level of the hierarchy; there are three domains, Bacteria, Archaea, and Eukarya.

**dominant** (7) Describes an allele that masks the phenotypic effect of the other, recessive, allele for a trait; the phenotype shows the effect of the dominant allele in both homozygous and heterozygous genotypes. [Lat., *dominari,* to rule]

**dorsal hollow nerve cord** (11) The central nervous system of vertebrates, consisting of the spinal cord and brain. [Lat., *dorsum,* back]

**double-blind experimental design** (1) An experimental design in which neither the subjects nor the experimenters know what treatment (if any) individual subjects are receiving.

**double bond** (2) The sharing of two electrons between two atoms; for example, the most common form of oxygen is the $O_2$ molecule, in which two electrons from each of the two atoms of oxygen are shared.

**double fertilization** (12) In angiosperms, two sperm are released by a pollen grain: one fuses with an egg to form a zygote, and the other fuses with two nuclei, forming a triploid endosperm.

**double helix** (2) The spiraling ladder-like structure of DNA composed of two strands of nucleotides; the bases protruding from each strand like "half-rungs" meet in the center and bind to each other (via hydrogen bonds), holding the ladder together. [Gk., *heligmos,* wrapping]

# E

**ecological footprint** (14) A measure of the impact of an individual or population on the environment by calculation of the amount of resources—including land, food and water, and fuel—consumed.

**ecology** (14) The study of the interaction between organisms and their environments, at the level of individuals, populations, communities, and ecosystems. [Gk., *oikos,* home + *logos,* discourse]

**ecosystem** (15) A community of biological organisms and the non-living environmental components with which they interact.

**ectotherms** (11) Organisms that rely on the heat from an external source to raise their body temperature and seek the shade when the air is too warm. [Gk., *ektos,* outside + *thermē,* heat]

**El Niño** (15) A sustained surface temperature change in the central Pacific Ocean that occurs every two to seven years; this event can start a chain reaction of unusual weather across the globe that can result in flooding, droughts, famine, and a variety of extreme climate disruptions. [Spanish, *the child;* a reference to the Christ child, because of the appearance of the phenomenon at Christmastime]

**electromagnetic spectrum** (4) The range of wavelengths that produce electromagnetic radiation, extending (in order of decreasing energy) from high-energy, short-wave, gamma rays and X rays, through ultraviolet light, visible light, and infrared light, to very long, low-energy, radio waves. [Lat., *specere,* to look at]

**electron** (2) A negatively charged particle that moves around the atomic nucleus.

**electron transport chain** (4) The path of high-energy electrons from one molecule within a membrane to another, coupled to the pumping of protons across the membrane, creating a concentration gradient that is used to make ATP; occurs in mitochondria and chloroplasts.

**element** (2) A pure substance that cannot be broken down chemically into any other substances; all atoms of an element have the same atomic number. [Lat., *elementum,* element, or first principle]

**empirical** (1) Describes knowledge that is based on experience and observations that are rational, testable, and repeatable. [Gk., *empeiria,* experience]

**endangered species** (16) As defined by the Endangered Species Act, species in danger of extinction throughout all or a significant portion of their range.

**Endangered Species Act (ESA)** (16) A U.S. law that defines "endangered species" and is designed to protect those species from extinction.

**endemic** (16) Describes species peculiar to a particular region and not naturally found elsewhere. [Gk., *en,* in + *demos,* the people of a country]

**endocytosis** (3) A cellular process in which large particles, solid or dissolved, outside the cell are surrounded by a fold of the plasma membrane, which pinches off, forming a vesicle, and the enclosed particle now moves into the cell. The three types of endocytosis are phagocytosis, pinocytosis, and receptor-mediated endocytosis. [Gk., *endon,* within + *kytos,* container]

**endomembrane system** (3) A system of organelles (the rough endoplasmic reticulum, the smooth endoplasmic reticulum, and the Golgi apparatus) that surrounds the nucleus; it produces and modifies necessary molecules, breaks down toxic chemicals and cellular by-products, and is thus responsible for many of the fundamental functions of the cell. [Gk., *endon,* within + Lat., *membrana,* a thin skin]

**endosperm** (12) Tissue of a mature seed that stores certain carbohydrates, proteins, and lipids that fuel the germination, growth, and development of the embryo and young seedling. [Gk., *endon,* within + *sperma,* seed]

**endosymbiosis theory** (3) Theory of the origin of eukaryotes that holds that, in the past, two different types of prokaryotes engaged in a close partnership and eventually one, capable of performing photosynthesis, was subsumed into the other, a larger prokaryote. The smaller prokaryote made some of its photosynthetic energy available to the host and, over time, the two became symbiotic and eventually became a single more complex organism in which the smaller prokaryote had evolved into the chloroplast of the new organism. A similar scenario can be developed for the evolution of mitochondria. [Gk., *endon,* within + *symbios,* living together]

**endotherms** (11) Organisms that use the heat produced by their cellular respiration to raise and maintain their body temperature above air temperature. [Gk., *endon,* within + *thermē,* heat]

**energy** (4) The capacity to do work, which is the moving of matter against an opposing force. [Gk., *energeia,* activity]

**energy pyramid** (15) A diagram that illustrates the path of energy through the organisms of an ecosystem; each layer of the pyramid represents the biomass of a trophic level.

**enzymatic protein** (3) See **enzyme.**

**enzyme** (2, 3) A protein that initiates and accelerates a chemical reaction in a living organism; enzymatic proteins take part in chemical reactions on the inside and outside surfaces of the plasma membrane. [Gk., *en,* in + *zyme,* leaven]

**estuary** (15) A tidal water passage, linked to the sea, in which salt water and fresh water mix; characterized by exceptionally high productivity. [Lat., *aestus,* tide]

**ethanol** (4) The end product of fermentation of yeast; the alcohol in beer, wine, and spirits. [contraction of the full chemical name, ethyl alcohol]

**eukaryote** (3) An organism composed of eukaryotic cells. [Gk., *eu,* good + *karyon,* nut, kernel]

**eukaryotic cell** (3) A cell with a membrane-surrounded nucleus containing DNA, membrane-surrounded organelles, and internal structures organized into compartments.

**eutrophication** (15) The process in which excess nutrients dissolved in a body of water lead to rapid growth of algae and bacteria, which consume much of the dissolved oxygen and, in time, can lead to large-scale die-offs. [Gk., *eu,* good + *trophē,* food]

**evolution** (8) A change in allele frequencies of a population. [Lat., *evolvere,* to roll out]

**exocytosis** (3) A cellular process in which particles within the cell, solid or dissolved, are enclosed in a vesicle and transported to the plasma membrane, where the membrane of the vesicle merges with the plasma membrane and the material in the vesicle is expelled to the extracellular fluid for use throughout the body. [Gk., *ex,* out of + *kytos,* container]

**exoskeleton** (11) A rigid external covering such as is found in some invertebrates, including insects and crustaceans. [Gk., *ex,* out of]

**exotic species** (16) Species introduced by human activities to areas other than the species' native range.

**experimental group** (1) In an experiment, the group of subjects exposed to a particular treatment; also known as the treatment group.

**exponential growth** (14) Growth of a population at a rate that is proportional to its current size.

**extinction** (10) The complete loss of all individuals in a species. [Lat., *extinguere,* to extinguish]

**extremophiles** (13) Bacteria and archaea that can live in extreme physical and chemical conditions. [Lat., *extremus,* outermost + Gk., *philios,* loving]

**F**

**facilitated diffusion** (3) Diffusion of molecules through the phospholipid bilayer of the plasma membrane that takes place through a transport protein (a "carrier molecule") embedded in the membrane. Molecules that require the assistance of a carrier molecule are those that are too big to cross the membrane directly or are

electrically charged and would be repelled by the middle layer of the membrane.

**family** (10) In the system developed by Carolus Linnaeus (1707–1778), a classification of organisms consisting of related genera.

**fatty acid** (2) A long hydrocarbon (a chain of carbon-hydrogen molecules); fatty acids form the tail region of triglyceride fat molecules.

**female** (9) In sexually reproducing organisms, a member of the sex that produces the larger gamete.

**fermentation** (4) The process by which glycolysis occurs in the absence of oxygen; the electron acceptor is pyruvate (in animals) or acetaldehyde (in yeast) rather than oxygen.

**fertilization** (6) The fusion of two reproductive cells.

**filament** (12) The supporting stalk of the anther of a stamen found in angiosperm flowers. [Lat., *filum,* thread]

**first law of thermodynamics** (4) A physical law that states that energy cannot be created or destroyed; it can only change from one form to another.

**fitness** (8) A relative measure of the reproductive output of an individual with a given phenotype compared with the reproductive output of individuals with alternative phenotypes.

**fixation** (8) The point at which the frequency of an allele in a population is 100% and, therefore, there is no more variation in the population for this gene.

**fixed action pattern** (9) An innate sequence of behaviors, triggered under certain conditions, that requires no learning, does not vary, and once begun runs to completion; an example is egg-retrieval in geese.

**flagellum** (*pl.* **flagella**) (3) Long, thin, whip-like projection from the cell body of a prokaryote that aids in cell movement through the medium in which the organism lives; in animals, the only cell with a flagellum is the sperm cell. [Lat., *flagellum,* whip]

**flatworms** (11) Worms with flat bodies that are members of the phylum Platyhelminthes; characterized by well-defined head and tail regions, with some having clusters of light-sensitive cells for eyespots. Most are hermaphroditic and are protostomes that do not molt. Examples are tapeworms and flukes.

**fleshy fruit** (12) A fruit that consists of the ovary and some additional parts of the flower; when fleshy fruits, an attractive food, are eaten by animals, the seeds may be widely dispersed.

**flower** (12) The part of an angiosperm that contains the reproductive structures; consists of a supporting stem with modified leaves (the petals and sepals) and usually contains both male and female reproductive structures.

**fluid mosaic** (3) A term that describes the structure of the plasma membrane, which is made up of several different types of molecules, many of which are not fixed in place but float, held in proper orientation by hydrophilic and hydrophobic forces.

**food chain** (15) The path of energy flow from primary producers to tertiary consumers.

**food web** (15) A more precisely described path of energy flow from primary producers to tertiary consumers than the food chain, reflecting the fact that many organisms are omnivores and occupy more than one position in the chain.

**fossil** (8) The remains of an organism, usually its hard parts such as shell, bones, or teeth, which have been naturally preserved; also, traces of such an organism, such as footprints. [Lat., *fossilis,* that which is dug up]

**fossil fuels** (4) Fuels produced from the decayed remains of ancient plants and animals; include oil, natural gas, and coal.

**founder effect** (8) A phenomenon by which genetic drift can occur; the isolation of a small subgroup of a larger population which can lead to changes in the allele frequencies of a population because all the descendants of the smaller group will reflect the allele frequencies of the subgroup, which may be different from those of the larger source population.

**fundamental niche** (15) The full range of environmental conditions under which an organism can live.

## G

**gamete** (6) Cell (often haploid) that will combine at fertilization to produce offspring; also called a reproductive cell. [Gk., *gamete,* wife]

**gametophyte** (12) The structure in land plants and some algae that produces gametes (sperm and eggs); the haploid life stage of plants and some algae, which may be either male (producing sperm) or female (producing eggs). [Gk., *gamete,* wife + *phytas,* plant]

**gap junction** (3) A junction between adjacent animal cells in the form of a pore in each of the plasma membranes surrounded by a protein that links the two cells and acts like a channel between them, allowing materials to pass between the cells.

**gastropods** (11) Mollusks that are members of the class Gastropoda; most have a single shell, a muscular foot for locomotion, and a radula used for scraping food from surfaces; examples are snails and slugs. [Gk., *gaster,* belly + *pous,* foot]

**gene** (5) The basic unit of heredity; a sequence of DNA nucleotides on a chromosome that carries the information necessary for making a functional product, usually a protein or an RNA molecule. [Gk., *genos,* race, descent]

**gene expression** (5) The process by which information in a gene's sequence is used to synthesize a gene product (commonly a protein, but also RNA).

**gene flow** (8) A change in the allele frequencies of a population due to movement of some individuals of a species from one population to another, changing the allele frequencies of the population they join; also known as migration.

**gene library** (5) A collection of cloned DNA fragments; also known as a clone library.

**gene regulation** (5) The processes by which cells "turn on" or "turn off" genes, influencing the amount of gene products formed.

**gene therapy** (5) The process of inserting, altering, or deleting one or more of the genes in an individual's cells to correct defective versions of the gene(s).

**genetic drift** (8) A random change in allele frequencies over successive generations; a cause of evolution.

**genetic engineering** (5) The manipulation of an organism's genetic material by adding, deleting, or transplanting genes from one organism to another.

**genome** (5) The full set of DNA present in an individual organism; also can refer to the full set of DNA present in a species.

**genotype** (5, 7) The genes that an organism carries for a particular trait; also, collectively, an organism's genetic composition. [Gk., *genos,* race, descent + *typos,* impression, engraving]

**genus** (*pl.* **genera**) (10) In the system developed by Carolus Linnaeus (1707–1778), a classification of organisms consisting of closely related species. [Lat., *genus,* race, family, origin]

**glycerol** (2, 3) A small molecule that forms the head region of a triglyceride fat molecule. [Gk., *glykys,* sweet + *-ol,* chemical suffix for an alcohol]

**glycogen** (2) A complex carbohydrate consisting of stored glucose molecules linked to form a large web, which breaks down to release glucose when it is needed for energy. [Gk., *glykys,* sweet + Gk., *genos,* race, descent]

**glycolysis** (4) In all organisms, the first step in cellular respiration, in which one molecule of glucose is broken into two molecules of pyruvate. For some organisms, glycolysis is the only means of extracting energy from food; for others, including most plants and animals, it is followed by the Krebs cycle and the electron transport chain. [Gk., *glykys,* sweet + *lysis,* releasing]

**Golgi apparatus** (3) (*pron.* GOHL-jee) An organelle, part of the endomembrane system, structurally like a flattened stack of unconnected membranes, each known as a Golgi body. The Golgi apparatus processes molecules synthesized in the cell and packages those molecules that are destined for use elsewhere in the body. [From the name of the discoverer, Camillo Golgi, 1843–1926]

**gonads** (6) The ovaries and testes in sexually reproducing animals. [Gk., *gonē,* offspring]

**Gram stain** (13) A test used by microbiologists in identifying an unknown bacterium; the dye that is used stains the layer of peptidoglycan outside the cell wall purple, but those bacteria in which the layer of peptidoglycan is covered by a membrane are not colored by the dye. Bacteria that take a Gram stain are known as Gram-positive bacteria, those that do not are known as Gram-negative bacteria.

**group selection** (9) The process, extremely uncommon in nature, that brings about an increase in the frequency of alleles for traits (e.g., behaviors) that are beneficial to the persistence of the species or population while simultaneously being detrimental to the fitness of the individual possessing the trait (or engaging in the behavior).

**growth rate** (*r*) (14) The birth rate minus the death rate; the change in the number of individuals in a population per unit of time.

**gymnosperms** (12) Vascular plants that do not produce their seeds in a protective structure; seeds are usually found on the surface of the scales of a cone-like structure. The gymnosperms include conifers, cycads, gnetophytes, and ginkgo. [Gk., *gymnos,* naked + *sperma,* seed]

## H

**habitat** (15) The physical environment of organisms, consisting of the chemical resources of the soil, water, and air, and physical conditions such as temperature, salinity, humidity, and energy sources. [Lat., *habitare,* to dwell or inhabit]

**hair** (11) Dead cells filled with the protein keratin that collectively serve as insulation covering the body or a part of the body; present in all mammals.

**haploid** (6) Describes cells that have a single copy of each chromosome (in many species, including humans, gametes are haploid). [Gk., *haploeides,* single]

**hazard factor** (14) An external force on a population that increases the risk of death.

**herbivores** (15) Animals that eat plants; also known as primary consumers. [Lat., *herba,* grass + *vorare,* to devour]

**heredity** (7) The greater resemblance of offspring to parents than to other individuals in the population, a consequence of the passing of characteristics from parents to offspring through their genes. [Lat., *heres,* heir]

**heritability** (8) The transmission of traits from parents to offspring via genetic information; also known as inheritance.

**hermaphrodite** (6) An organism that produces both male and female gametes. [From the names of the Greek god Hermes and goddess Aphrodite]

**heterozygous** (7) Describes the genotype of a trait for which the two alleles an individual carries differ from each other. [Gk., *heteros*, other + *zeugos*, pair]

**histones** (6) Proteins around which the long, linear strands of DNA are wrapped; serve to keep the DNA untangled and to enable an orderly, tight, and efficient packing of the DNA within the cell.

**homologous feature** (10) A feature inherited from a common ancestor. [Gk., *homologia*, agreement]

**homologous pair (homologues)** (6) The maternal and paternal copies of a chromosome. [Gk., *homologia*, agreement]

**homologous structure** (8) Body structures in different organisms that, although they may have been modified over time to serve different functions in different species, are due to inheritance from a common evolutionary ancestor.

**homozygous** (7) Describes the genotype of a trait for which the two alleles are the same. [Gk., *homos*, same + *zeugos*, pair]

**honest signal** (9) A signal, which cannot be faked, that is given when both the individual making the signal and the individual responding to it have the same interests; it carries the most accurate information about an individual or situation.

**horizontal gene transfer** (10) The transfer of genetic material directly from one individual to another, not necessarily related, individual; common among bacteria.

**host** (13, 15) An organism in or on which a parasite lives.

**human immunodeficiency virus (HIV)** (13) The virus responsible for AIDS, a deadly disease that destroys the human immune system. HIV is a retrovirus, an RNA-containing virus that is thought to have been introduced to humans from chimpanzees.

**hybridization** (10) The interbreeding of closely related species. [Lat., *hybrida*, animal produced by two different species]

**hybrids** (10) Offspring of individuals of two different species. [Lat., *hybrida*, animal produced by two different species]

**hydrogen bond** (2) A type of weak chemical bond formed between the slightly positively charged hydrogen atoms of one molecule and the slightly negatively charged atoms of another (often oxygen or nitrogen atoms). Hydrogen bonds are important in building multi-atom molecules, such as complex proteins, and are responsible for many of the unique and important features of water.

**hydrophilic** (2, 3) Attracted to water, as, for example, polar molecules that readily form hydrogen bonds with water. [Lat., *hydro-*, pertaining to water; Gk., *philios*, loving]

**hydrophobic** (2, 3) Repelled by water, as, for example, nonpolar molecules that tend to minimize contact with water. [Lat., *hydro-*, pertaining to water, Gk., *phobos*, fearing]

**hypertonic** (3) Of two solutions, that with a higher concentration of solutes. [Gk., *hyper*, above + *tonos*, tension]

**hyphae** (*sing.* **hypha**) (12) (*pron.* HIGH-fee) Long strings of cells that make up the mycelium of a multicellular fungus. [Gk., *hypha*, web]

**hypothesis** (*pl.* **hypotheses**) (1) A proposed explanation for an observed phenomenon. [Gk., *hypothesis*, a proposal]

**hypotonic** (3) Of two solutions, that with a lower concentration of solutes. [Gk., *hypo*, under + *tonos*, tension]

**I**

**inclusive fitness** (9) The sum of an individual's indirect and direct fitness.

**incomplete dominance** (7) The case in which the heterozygote has a phenotype intermediate between those of the two homozygotes; an example is pink snapdragons, with an appearance intermediate between homozygous for white flowers and homozygous for red flowers.

**independent variable** (1) A measurable entity that is available at the start of a process being observed and whose value can be changed as required; generally represented on the *x*-axis in a graph.

**indirect fitness** (9) The reproductive output that an individual brings about through apparently altruistic behaviors toward genetic relatives.

**inheritance** (8) The transmission of traits from parents to offspring via genetic information; also known as heritability.

**inhibitor** (2) A chemical within a cell that binds to an enzyme or substrate molecule and in doing so reduces the enzyme's ability to catalyze a reaction.

**innate behaviors** (9) Behaviors that do not require environmental input for their development. These behaviors are present in all individuals in a population and do not vary much from one individual to another or over an individual's life span; also known as instincts. [Lat., *innatus*, inborn]

**instincts** (9) Behaviors that do not require environmental input for their development. Instincts are present in all individuals in a population and do not vary much from one individual to another or over an individual's life span; also known as innate behaviors. [Lat., *instinctus*, impelled]

**intermediate filaments** (3) One of three types of protein fibers (the others are microtubules and microfilaments) that make up the eukaryotic cytoskeleton, providing it with structure and shape; a durable, rope-like system of numerous different overlapping proteins.

**intermembrane space** (3) In a mitochondrion, the region between the inner and outer membranes. [Lat., *inter*, between + *membrana*, a thin skin]

**interphase** (6) In the cell cycle, the phase during which the cell grows and functions; during this phase, replication of DNA occurs in preparation for cell division. [Lat., *inter*, between + Gk., *phasis*, appearance]

**intron** (5) A non-coding region of DNA.

**invagination** (3) The folding in of a membrane or layer of tissue so that an outer surface becomes an inner surface. [Lat., *in*, in + *vagina*, sheath]

**invertebrates** (11) Animals that do not have a backbone; although commonly used in organizing the animals, invertebrates are not a monophyletic group.

**ion** (2) An atom that carries an electrical charge, positive or negative, because it has either lost or gained an electron or electrons from its normal, stable configuration. [Gk., *ion*, going]

**ionic bond** (2) A bond created by the transfer of one or more electrons from one atom to another; the resulting atoms, now called ions, are charged oppositely and so attract each other to form a compound.

**isotonic** (3) Refers to solutions with equal concentrations of solutes. [Gk., *isos*, equal to + *tonos*, tension]

**isotopes** (2) Variants of atoms that differ in the number of neutrons they possess; isotopes do not vary in charge, because neutrons have no electrical charge, but the atom's mass changes with the loss or addition of a particle in the nucleus.

**K**

**karyotype** (6) A visual display of an individual's full set of chromosomes. [Gk., *karyon*, nut or kernel + *typos*, impression, engraving]

**keystone species** (15) A species that has an unusually large influence on the presence or absence of numerous other species in a community.

**kin selection** (9) "Kindness" toward close relatives, which may evolve as apparently altruistic behavior toward them, but which in fact is beneficial to the fitness of the individual performing the behavior.

**kinetic energy** (4) The energy of moving objects, such as legs pushing the pedals of a bicycle or wings beating against the air. [Gk., *kinesis,* motion]

**kingdom** (10) In the system developed by Carolus Linnaeus (1707–1778), one of the three categories—animal, plant, and mineral—into which all organisms and substances on earth were placed. In modern classification, there are six kingdoms: bacteria, archaea, protists, plants, animals, and fungi.

**Krebs cycle** (4) The second step of cellular respiration, in which energy is extracted from sugar molecules as additional molecules of ATP and NADH are formed. [From the name of the discoverer, Hans Adolf Krebs, 1900–1981]

**L**

**La Niña** (15) The counterpart of the El Niño phenomenon, characterized by the reverse effects, including a decrease in surface temperature and air pressure at the surface of the western Pacific Ocean; the resulting climate effects are approximately opposite to El Niño effects.

**landscape conservation** (16) The conservation of habitats and ecological processes, as well as species.

**language** (9) A type of communication in which arbitrary symbols represent concepts and grammar; a system of rules dictates how the symbols can be manipulated to communicate and express ideas.

**larva** (11) In complete metamorphosis, the first stage of insect development; the larva is hatched from the egg and eats to grow large enough to enter the pupa stage. The larva (for example, a caterpillar) looks completely different from the adult (a butterfly or moth). [Lat., *larva,* ghost]

**learning** (9) The alteration and modification of behavior over time in response to experience.

**lichen** (12) Symbiotic partnership between fungi and chlorophyll-containing algae or cyanobacteria, or both.

**life** (10) A physical state characterized by the ability to replicate and the presence of metabolic activity.

**life history** (14) The vital statistics of a species, including age at first reproduction, probabilities of survival and reproduction at each age, litter size and frequency, and longevity.

**life table** (14) A table presenting data on the mortality rates within defined age ranges for a population; used to determine an individual's probability of dying during any particular year.

**light energy** (4) A type of kinetic energy made up of energy packets called photons, which are organized into waves.

**linked genes** (7) Genes that are close to each other on a chromosome and so are more likely than others to be inherited together.

**lipid** (2) One of four types of macromolecules, lipids are insoluble in water and greasy to the touch; they are important in energy storage and insulation (fats), membrane formation (phospholipids), and regulating growth and development (sterols). [Gk., *lipos,* fat]

**lobe-finned fishes** (11) Fish species characterized by two pairs of sturdy lobe-shaped fins on the underside of the body.

**logistic growth** (14) A pattern of population growth in which initially exponential growth levels off as the environment's carrying capacity is approached.

**lysosome** (3) A round, membrane-enclosed, enzyme- and acid-filled vesicle in the cell that digests and recycles cellular waste products and consumed material. [Gk., *lysis,* releasing + *soma,* body]

**M**

**macroevolution** (10) Large-scale evolutionary change involving the origins of new groups of organisms; the accumulated effect of microevolution over a long period of time. [Gk., *macros,* large + Lat., *evolvere,* to roll out]

**macromolecule** (2) A large molecule, made up of smaller building blocks or subunits; four types of biological macromolecules are carbohydrates, lipids, proteins, and nucleic acids. [Gk., *macros,* large + Lat., dim. of *moles,* mass]

**male** (9) In sexually reproducing organisms, a member of the sex that produces the smaller gamete.

**mammary glands** (11) Glands in all female mammals that produce milk for the nursing of young. [Lat., *mamma,* breast]

**marsupials** (11) Mammals in which, in most species, after a short period of embryonic life in the uterus, the young complete their development in a pouch in the female. [Gk., *marsipos,* pouch]

**mass** (2) The amount of matter in a given sample of a substance.

**mass extinctions** (10) Extinctions in which a large number of species become extinct in a short period of time, usually because of extraordinary and sudden environmental change. [Lat., *extinguere,* to extinguish]

**mate guarding** (9) Behavior by an individual that reduces the opportunity for that individual's mate to interact with other potential mates.

**mating system** (9) The pattern of mating behavior in a species, ranging from polyandry to monogamy to polygyny; mating systems are influenced by the relative amounts of parental investment by males and by females.

**matrix (mitochondrial)** (3, 4) The space within the inner membrane, where the carriers NADH and FADH$_2$ begin the electron transport chain by carrying high-energy electrons to molecules embedded in the inner membrane.

**maximum sustainable yield** (14) The point at which the maximum number of individuals are removed from a population without impairing its growth rate; it occurs at half the carrying capacity.

**meiosis** (6) In sexually reproducing organisms, a process of nuclear division in the gonads that, along with cytokinesis, produces reproductive cells that have half as much genetic material as the parent cell and that all differ from each other genetically. [Gk., *meioun,* to lessen]

**Mendel's law of independent assortment** (7) Allele pairs for different genes separate independently in meiosis, so the inheritance of one trait generally does not influence the inheritance of another trait (the exception, unknown to Mendel, occurs with linked genes). [From the name of its discoverer, Gregor Mendel, 1822–1884]

**Mendel's law of segregation** (7) During the formation of gametes, the two alleles for a gene separate, so that half the gametes carry one allele, and half the gametes carry the other. [From the name of its discoverer, Gregor Mendel, 1822–1884]

**messenger RNA (mRNA)** (5) The ribonucleic acid that "reads" the sequence for a gene in DNA and then moves from the nucleus to the cytoplasm, where the next stage of protein synthesis will take place.

**metamorphosis** (11) The rebuilding of molecules from the larva stage to the adult, resulting in a change of form. *Complete metamorphosis* is the division of an organism's life history into three completely different stages; *incomplete metamorphosis* is the pattern of growth and development in which an organism does not pass through separate, dramatically different life stages. [Gk., *metamorphoun,* to transform]

**metaphase** (6) The second phase of mitosis, in which the sister chromatids line up at the center of the cell; in meiosis, the homologues line up at the center of the cell in metaphase I and the sister chromatids line up in metaphase II. [Gk., *meta,* in the midst of + *phasis,* appearance]

**microbe** (10, 13) A microscopic organism; not a monophyletic group, since it includes protists, archaea, and bacteria. [Gk., *micros,* small + *bios,* life]

**microevolution** (10) A slight change in allele frequencies in a population over one or a few generations. [Gk., *micros,* small + Lat., *evolvere,* to roll out]

**microfilaments** (3) One of three types of protein fibers (the others are intermediate filaments and microtubules) that make up the eukaryotic cytoskeleton, providing it with structure and shape, These are the thinnest elements in the cytoskeleton; long, solid, rod-like fibers that help generate forces, including those important in cell contraction and cell division.

**microsphere** (10) A membrane-enclosed, small, spherical unit containing a self-replicating molecule and carrying information, although no genetic material. Microspheres may have been an important stage in the development of life. [Gk., *micros,* small + *sphaira,* ball]

**microtubules** (3) One of three types of protein fibers (the others are intermediate filaments and microfilaments) that make up the eukaryotic cytoskeleton, providing it with structure and shape, These are the thickest elements in the cytoskeleton; they resemble rigid, hollow tubes, functioning as tracks to which molecules and organelles within the cell may attach and be moved along; also help pull chromosomes apart during cell division.

**migration** (8) A change in the allele frequencies of a population due to the movement of some individuals from one population to another; an agent of evolutionary change caused by the movement of individuals into or out of a population. [Lat., *migrare,* to move from place to place]

**mimicry** (15) The evolution of an organism to resemble another organism or object in its environment to help conceal itself from predators. [Gk., *mimesis,* imitation]

**mitochondrial matrix** (4) See **matrix.**

**mitochondrion** (*pl.* **mitochondria**) (3) The organelle in eukaryotic cells that converts the energy stored in food in the chemical bonds of carbohydrate, fat, and protein molecules into a form usable by the cell for all its functions and activities. [Gk., *mitos,* thread + *chondros,* cartilage]

**mitosis** (6) The division of a nucleus into two genetically identical nuclei that, along with cytokinesis, leads to the formation of two identical daughter cells. [Gk., *mitos,* thread + *phasis,* appearance]

**mitotic phase (M phase)** (6) The phase of the cell cycle during which first the genetic material and nucleus, and then the rest of the cellular contents, divide.

**molecule** (2) A group of atoms held together by covalent bonds. [Lat., *dim.* of *moles,* mass]

**monogamy** (9) A mating system in which most individuals mate and remain with just one other individual. [Gk., *monos,* single + *gamete,* wife]

**monophyletic** (10) A group containing a common ancestor and all of its descendants. [Gk., *monos,* single + *phylon,* race, tribe, class]

**monosaccharides** (2) The simplest carbohydrates and the building blocks of more complex carbohydrates; they cannot be broken down into other monosaccharides; examples are glucose, fructose, and galactose; also known as simple sugars. [Gk., *monos,* single + *sakcharon,* sugar]

**monotremes** (11) Present-day mammals that retain the ancestral condition of laying eggs. Monotremes are so called because they have a single duct, the cloaca, into which the reproductive system, the urinary system, and the digestive system (for defecation) open. [Gk., *monos,* single + *trema,* hole]

**morphological species concept** (10) A concept that defines species on the basis of physical features such as body size and shape. [Gk., *morphē,* shape]

**multiple allelism** (7) The case in which a single gene has more than two possible alleles.

**mutation** (5, 8) An alteration in the base-pair sequence of an individual's DNA; may arise spontaneously or following exposure to a mutagen. [Lat., *mutare,* to change]

**mutualism** (15) A symbiotic relationship in which both species benefit and neither is harmed. [Lat., *mutuus,* reciprocal]

**mycelium** (12) A mass of interconnecting hyphae that make up the structure of a multicellular fungus. [Gk., *mykes,* fungus]

**mycorrhizae** (12) (*pron.* my-ko-RYE-zay) Root fungi, that is, symbiotic associations between roots and fungi in which fungal structures are closely associated with fine rootlets and root hairs. [Gk., *mykes,* fungus + *rhiza,* root]

## N

**NADPH** (4) A molecule (nicotinamide adenine dinucleotide phosphate) that is a high-energy electron carrier involved in photosynthesis, which stores energy by accepting high-energy protons. It is formed when the electrons released from the splitting of water are passed to NADP⁺.

**natural selection** (8) A mechanism of evolution that occurs when there is heritable variation for a trait and individuals with one version of the trait have greater reproductive success than individuals with a different version of that trait.

**neutron** (2) An electrically neutral particle in the atomic nucleus. [Lat., *neutro,* in neither direction]

**niche** (15) The way an organism utilizes the resources of its environment, including the space it requires, the food it consumes, and timing of reproduction.

**node** (phylogeny) (10) The point on an evolutionary tree at which species diverge from a common ancestor. [Lat., *nodus,* knot]

**noncompetitive inhibitor** (2) A chemical within a cell that binds to part of an enzyme away from the active site, but alters the enzyme's shape so as to alter the active site, thereby reducing or blocking the enzyme's ability to bind with substrate.

**nondisjunction** (6) The unequal distribution of chromosomes during cell division; can lead to Down syndrome and other disorders caused by an individual's possession of too few or too many chromosomes.

**nonpolar** (3) Electrically uncharged.

**non-vascular plants** (12) Plants that do not have vessels to transport water and dissolved nutrients, but instead rely on diffusion; bryophytes are non-vascular plants. [Lat., *vasculum, dim.* of *vas,* vessel]

**notochord** (11) A rod of tissue from head to tail that stiffens the body when muscles contract during locomotion. Primitive chordates retain the notochord throughout life, but in advanced chordates it is present only in early embryos and is replaced by the vertebral column. [Gk., *notos,* back + Lat., *chorda,* cord]

**nuclear membrane** (3) A membrane that surrounds the nucleus of a cell, separating it from the cytoplasm, consisting of two bilayers and perforated by pores enclosed in embedded proteins that allow the passage of large molecules from nucleus to cytoplasm and from cytoplasm to nucleus; also called the nuclear envelope.

**nucleic acid** (2, 5) One of the four types of biological macromolecules; the nucleic acids DNA and RNA store genetic information in unique sequences of nucleotides.

**nucleolus** (3) An area near the center of the nucleus where subunits of the ribosomes are assembled. [Lat., *nucleolus, dim.* of *nucleus,* kernel, small nut]

**nucleotide** (2, 5) A molecule containing a phosphate group, a sugar molecule, and a nitrogen-containing molecule. Nucleotides are the individual units that together, in a unique sequence, constitute a nucleic acid.

**nucleus** (cell biology) (3) A membrane-enclosed structure in eukaryotic cells that contains the organism's genetic information as linear strands of DNA in the form of chromosomes. [Lat., *nucleus, dim.* of *nux,* nut]

**nucleus** (chemistry) (2) The central and most massive part of an atom, usually made up of two types of particles, protons and neutrons, which move about the nucleus. [Lat., *nucleus, dim.* of *nux,* nut]

**null hypothesis** (1) A hypothesis that proposes a lack of relationship between two factors. [Lat., *nullus,* none]

**nuptial gift** (9) A food item or other item presented to a potential mate as part of courtship. [Lat., *nuptialis,* of marriage]

## O

**omnivores** (15) Animals that eat both plants and other animals and thus can occupy more than one position in the food chain. [Lat., *omnis,* all + *vorare,* to devour]

**operon** (5) A group of several genes, along with the elements that control their expression as a unit, all within one section of DNA.

**order** (10) In the system developed by Carolus Linnaeus (1707–1778), a classification of organisms consisting of related families.

**organelles** (3) Specialized structures in the cytoplasm of eukaryotic cells with specific functions, such as the rough and smooth endoplasmic reticulum, Golgi apparatus, and mitochondria. [Gk., *organon,* tool]

**osmosis** (3) A type of passive transport in which water molecules move across a membrane, such as the plasma membrane of a cell; the direction of osmosis is determined by the relative concentrations of all solutes on either side of the membrane. [Gk., *osmos,* thrust]

**ovary** (12) An enclosed chamber at the base of the carpel of a flower that contains the ovules; the female gonad. [Lat., *ovum,* egg]

**ovule** (12) The structure within the ovary of flowering plants that gives rise to female egg cells. [Lat., *ovum,* egg]

**Oxygen Revolution** (13) The accumulation in the atmosphere of oxygen released by cyanobacteria and other photosynthetic organisms.

## P

**pair bond** (9) A bond between an individual male and female in which they spend a high proportion of their time together, often over many years, sharing a nest or other refuge and contributing equally to the care of offspring.

**parasite** (13, 15) An organism that lives in or on another organism, the host, and damages it. [Gk., *para,* beside + *sitos,* grain, food]

**parasitism** (15) A symbiotic relationship in which one organism (the parasite) benefits while the other (the host) is harmed.

**parent cells** (6) Cells that divide to form daughter cells, which are genetically identical to the parent cell.

**passive transport** (3) Molecular movement that occurs spontaneously, without the input of energy; the two types of passive transport are diffusion and osmosis.

**paternity uncertainty** (9) Describes the fact that among species with internal fertilization in the female, a male cannot be 100% certain that any offspring a female produces are his. [Lat., *pater,* father]

**pathogenic** (13) Disease-causing.

**pedigree** (7) In genetics, a type of family tree that maps the occurrence of a trait in a family, often over many generations.

**peptide bond** (2) A bond in which the amino group of one amino acid is bonded to the carboxyl group of another; two amino acids so joined form a dipeptide, several amino acids so joined form a polypeptide. [Gk., *peptikos,* able to digest]

**peptidoglycan** (13) A glycoprotein that forms a thick layer on the outside of the cell wall of a bacterium; in some bacteria, the layer of peptidoglycan is covered by a membrane and so is not colored by a Gram stain.

**periodic table** (2) A tabular display in which all the known chemical elements are arranged in the order of their atomic number and on the basis of other aspects of their atomic structures.

**pH** (2) A logarithmic scale that measures the concentration of hydrogen ions ($H^+$) in a solution, with decreasing values indicating increasing acidity. Water, in which the concentration of hydrogen ions ($H^+$) equals the concentration of hydroxyl ions ($OH^-$), has pH = 7, the midpoint of the scale. [Abbreviation for "power of hydrogen"]

**phagocytosis** (3, 13) One of the three types of endocytosis, in which relatively large solid particles are engulfed by the plasma membrane, a vesicle is formed, and the particle is moved into the cell.

**pharyngeal slits** (11) Slits in the pharyngeal region, between the back of the mouth and the top of the throat, for the passage of water for breathing and feeding. [Gk., *pharynx,* throat]

**phenotype** (5, 7) The manifested structure, function, and behaviors of an individual; the expression of the genotype of an organism. [Gk., *phainein,* to cause to appear + *typos* impression, engraving]

**pheromones** (9) Molecules released by an individual into the environment that trigger behavioral or physiological responses in other individuals. [Gk., *pherein,* to carry]

**phospholipid** (2, 3) A lipid that is the major component of the plasma membrane; phospholipids are structurally similar to fats, but contain a phosphorus atom and have two, not three, fatty acid chains.

**phospholipid bilayer** (3) The structure of the plasma membrane; two layers of phospholipids, arranged tail to tail (the tails are hydrophobic and so avoid contact with water), with the hydrophilic head regions facing the watery extracellular fluid and intracellular fluid.

**photoautotroph** (13) Chlorophyll-containing bacteria, or other organisms, that use the energy from sunlight to convert carbon dioxide to glucose by photosynthesis. [Gk., *phos,* light + *autos,* self + *trophē,* food]

**photon** (4) The elementary particle that carries the energy of electromagnetic radiation of all wavelengths. [Gk., *phos,* light]

**photosynthesis** (4) The process by which some organisms are able to capture energy from the sun and store it in the chemical bonds of sugars and other molecules that the plants produce. [Gk., *phos,* light + *syn,* together with + *tithenai,* to place or put]

**photosystems** (4) Two arrangements of light-absorbing pigments, including chlorophyll, within the chloroplast that capture energy from the sun and transform it first into the energy of excited electrons and ultimately into ATP and high-energy electron carriers such as NADPH. [Gk., *phos,* light + *systema,* a whole compounded of parts]

**phylogenetic tree** (5) A grouping of organisms in a hierarchical system that reflects the evolutionary history and relatedness of the organisms. [Gk., *phylon,* race, tribe, class + Lat., *genus,* race, family, origin]

**phylogeny** (10) The evolutionary history of organisms.

**phylum** (10) In the system developed by Carolus Linnaeus (1707–1778), a classification of organisms consisting of related classes. [Gk., *phylon,* race, tribe, class]

**pigment** (4) In photosynthesis, molecules that are able to absorb the energy of light of specific wavelengths, raising electrons to an excited state in the process. [Lat., *pigmentum,* paint]

**pilus** (*pl.* **pili**) (3) A thin, hair-like projection that helps a prokaryote attach to surfaces. [Lat., *pilus,* a single hair]

**pinocytosis** (3) One of the three types of endocytosis, in which dissolved particles and liquids are engulfed by the plasma membrane, a vesicle is formed, and the material is moved into the cell. The vesicles formed in pinocytosis are generally much smaller than those formed in phagocytosis. [Gk., *pinein*, to drink + *kytos*, container]

**placebo** (1) An inactive substance used in controlled experiments to test the effectiveness of another substance; the treatment group receives the substance being tested, the control group receives the placebo. [Lat., *placebo*, I shall please]

**placebo effect** (1) A frequently observed and poorly understood phenomenon in which there is a positive response to treatment with an inactive substance.

**placenta** (6) The organ formed during pregnancy (and expelled at birth) that connects the developing embryo to the wall of the uterus and allows the transfer of gases, nutrients, and waste products between mother and fetus; the placenta is so called from its shape. [Lat., *placenta*, a flat cake]

**placentals** (11) Mammals in which the developing fetus takes its nourishment from the transfer of nutrients from the mother through the placenta, which also supplies respiratory gases and removes metabolic waste products. [Lat., *placenta*, a flat cake]

**plants** (12) Members of the kingdom Plantae; multicellular eukaryotes that have cell walls made up primarily of cellulose, contain true tissues, and produce their own food by photosynthesis. Plants are sessile, and most inhabit terrestrial environments.

**plasma membrane** (3) A complex, thin, two-layered membrane that encloses the cytoplasm of the cell, holding the contents in place and regulating what enters and leaves the cell; also called the cell membrane. [Gk., *plasma*, anything molded]

**plasmid** (5, 13) A circular DNA molecule found outside the main chromosome in bacteria.

**plasmodesma** (*pl.* **plasmodesmata**) (3) In plants, microscopic tube-like channels connecting the cells and enabling communication and transport between them. [Gk., *plassein*, to mold + *desmos*, bond]

**pleiotropy** (7) A phenomenon in which an individual gene influences multiple traits. [Gk., *pleion*, more + *tropos*, turn]

**point mutation** (5) A mutation in which one base pair in DNA is replaced with another, or a base pair is either inserted or deleted.

**polar** (3) Having an electrical charge.

**polar body** (6) One of the two cells formed when a primary oocyte divides; it gets almost no cytoplasm and eventually disintegrates.

**pollen grain** (12) A structure that contains the male gametophyte of a seed plant. [Lat., *pollen*, fine dust]

**pollination** (12) The transfer of pollen from the anther of one flower to the stigma of another flower. [Lat. pollen, *fine dust*]

**polyandry** (9) A polygamous mating system in which individual females mate with multiple males. [Gk., *polys*, many + *aner*, husband]

**polygamy** (9) A mating system in which, for one sex, some individuals attract multiple mates while other individuals of that sex attract none; among the opposite sex, all or nearly all of the individuals are able to attract a mate. [Gk., *polys*, many + *gamete*, wife]

**polygenic** (7) Describes a trait that is influenced by multiple different genes. [Gk., *polys*, many + *genos*, race, descent]

**polygyny** (9) A mating system in which, among the males, some individuals attract multiple mates while other males attract none; among the females, all or nearly all of the individuals are able to attract a mate. [Gk., *polys*, many + *gune*, woman]

**polymerase chain reaction (PCR)** (5) A laboratory technique in which a fragment of DNA can be duplicated repeatedly. [Gk., *polys*, many + *meris*, part]

**polyploidy** (10) The doubling of the number of sets of chromosomes in an individual. [Gk., *polys*, many + *ploion*, vessel]

**polysaccharides** (2) Complex carbohydrates formed by the union of many simple sugars. [Gk., *polys*, many + *sakcharon*, sugar]

**population** (8) A group of organisms of the same species living in a particular geographic region.

**population density** (14) The number of individuals of a population in a given area.

**population ecology** (14) A subfield of ecology that studies the interactions between populations of organisms of a species and their environment.

**positive correlation** (1) A relationship between variables in which they increase (or decrease) together. [Lat., *com*, with + *relatio*, report]

**post-anal tail** (11) A tail that extends beyond the end of the trunk, a point that is marked by the anus; a characteristic of chordates. [Lat., *post*, after]

**postzygotic barrier** (10) A barrier to reproduction caused by the infertility of hybrid individuals or the inability of hybrid individuals to survive long after fertilization. [Lat., *post*, after]

**potential energy** (4) Stored energy; the capacity to do work that results from an object's location or position, as in the case of water held behind a dam. [Lat., *potentia*, power]

**predation** (15) An interaction between two species in which one eats the other. [Lat., *praedari*, to plunder]

**prepared learning** (9) Behaviors that are learned easily by all, or nearly all, individuals of a species.

**prezygotic barrier** (10) A barrier to reproduction caused by the physical inability of individuals to mate with each other, or the inability of the male's reproductive cell to fertilize the female's reproductive cell. [Lat., *prae*, before]

**primary active transport** (3) Active transport using energy released directly from ATP.

**primary consumers** (15) Herbivores, which consume the output of primary producers.

**primary electron acceptor** (4) In photosynthesis, a molecule that accepts excited, high-energy electrons from chlorophyll *a*, beginning the series of electron handoffs known as an electron transport chain.

**primary producers** (15) The organisms responsible for primary productivity, such as grasses, trees, and agricultural crops, which convert light energy from the sun into chemical energy (that is, food) through photosynthesis.

**primary productivity** (15) The amount of organic matter produced by living organisms, primarily through photosynthesis.

**primary structure** (2) The sequence of amino acids in a polypeptide chain.

**probiotic therapy** (13) A method of treating infections by introducing benign bacteria in numbers large enough to overwhelm harmful bacteria in the body. [Lat., *pro*, for, on behalf of + Gk., *bios*, life]

**prokaryote** (3) An organism consisting of a prokaryotic cell (all prokaryotes are one-celled organisms). [Gk., *pro*, before + *karyon*, nut, kernel]

**prokaryotic cell** (3) A cell bound by a plasma membrane enclosing the cell contents (cytoplasm, DNA, and ribosomes); there is no nucleus or other organelles.

**promoter site** (5) Part of a DNA molecule that indicates where the sequence of base pairs that makes up a gene begins.

**prophase** (6) The first phase of mitosis, in which the nuclear membrane breaks down, sister chromatids condense, and the spindle forms. In meiosis, homologous pairs of sister chromatids come together and cross over in prophase I, and the chromosomes in daughter cells condense in prophase II. [Gk., *pro*, before + *phasis*, appearance]

**protein** (2) One of the four types of biological macromolecules; constructed of unique combinations of 20 amino acids that result in unique structures and chemical behavior. Proteins are the chief building blocks of tissues in most organisms. [Gk., *proteion*, of the first quality]

**protein synthesis** (5) The construction of a protein from its constituent amino acids, by the processes of transcription and translation.

**prothallus** (12) The free-living haploid life stage of a fern; produces haploid gametes. [Gk., *pro*, before + *thallia*, twig]

**protists** (10) In modern classification, one of the four eukaryotic kingdoms; includes all the single-celled eukaryotes.

**proton** (2) A positively charged particle in the atomic nucleus; it is identical to the nucleus of the hydrogen atom, which lacks a neutron and has atomic number 1. [Gk., *protos*, first]

**protostomes** (11) Bilaterally symmetrical animals with defined tissues in which the gut develops from front to back; the first opening formed is the mouth of the adult animal. [Gk., *protos*, first + *stoma*, mouth]

**pseudoscience** (1) Hypotheses and theories not supported by trustworthy and methodical scientific studies. [Gk., *pseudes*, false + Lat., *scientia*, knowledge]

**punctuated equilibrium** (10) A theory in biology about the tempo of evolution, suggesting that in many taxa long periods in which there is relatively little evolutionary change are punctuated by brief periods of rapid evolutionary change.

**Punnett square** (7) A diagram showing the possible outcomes of a cross between two individuals; the possible crosses are shown in the manner of a multiplication table. [From the name of its designer, Reginald C. Punnett, 1875–1967]

**pupa** (11) In complete metamorphosis, the second stage of insect development, in which the larva is enclosed in a case and its body structures are broken down into molecules that are reassembled into the adult form. [Lat., *pupa*, a little girl]

**pyruvate** (4) The end product of glycolysis.

## Q

**quaternary structure** (2) Two or more polypeptide chains bonded together in a single protein; an example is hemoglobin. [Lat., *quaterni*, four each]

## R

**radial symmetry** (11) A body structure like that of a wheel, or pie, in which any cut through the center would divide the organism into identical halves. [Lat., *radius*, spoke of a wheel + Gk., *symmetria*, symmetry]

**radioactive** (2) The property of some elements or isotopes of having a nucleus that breaks down spontaneously, releasing tiny, high-speed particles that carry energy.

**radiometric dating** (8) A method of determining both the relative and the absolute ages of objects such as fossils by measuring both the radioactive isotopes they contain, which are known to decay at a constant rate, and their decay products.

**rain shadow** (15) An area in the lee of a mountain where there is no or reduced rainfall, because the air passing over the mountain falls, becoming warmer and thus increasing the amount of moisture it can hold.

**randomized** (1) Describes a manner of choosing subjects and assigning them to groups on the basis of chance, that is, randomly.

**ray-finned fishes** (11) Fish species characterized by rigid bones and a mouth at the apex of the body; they are so called because their fins are lined with hardened rays.

**realized niche** (15) The environmental conditions in which an organism is living at a given time.

**receptor-mediated endocytosis** (3) One of the three types of endocytosis, in which receptors on the surface of a cell bind to specific molecules; the plasma membrane then engulfs both molecule and receptor and draws them into the cell.

**receptor protein** (3) A protein in the plasma membrane that binds to specific chemicals in the cell's external environment to regulate processes within the cell; for example, cells in the heart have receptor proteins that bind to adrenaline.

**recessive** (7) Describes an allele whose phenotypic effect is masked by a dominant allele for a trait. [Lat., *recessus*, retreating]

**reciprocal altruism** (9) Costly behavior directed toward another individual that benefits the recipient, with the expectation that, at some later time, the recipient will behave in a similar manner, "returning the favor." [Lat., *reciprocare*, to move backward and forward + *alter*, the other]

**recognition protein** (3) A protein in the plasma membrane that provides a "fingerprint" on the outside-facing surface of the cell, making it recognizable to other cells. Recognition proteins make it possible for the immune system to distinguish the body's own cells from invaders that may produce infection, and also help cells bind to other cells or molecules.

**recombinant DNA technology** (5) Technology that depends on the combination of two or more sources of DNA into a product; an example is the production of human insulin from fast-dividing transgenic *E. coli* bacteria in which has been inserted the human DNA sequence that codes for insulin.

**recombination** (6) The exchange of some genetic material between a paternal homologous chromosome and a maternal homologous chromosome, leading to a chromosome carrying genetic material from each; also referred to as crossing over.

**replication** (6) The process in both eukaryotes and prokaryotes by which DNA duplicates itself in preparation for cell division.

**reproductive cells** (6) Haploid cells from two individuals that, as sperm and egg, will combine at fertilization to produce offspring; also called gametes.

**reproductive investment** (9, 14) Energy and material expended by an individual in the growth, feeding, and care of offspring.

**reproductive isolation** (10) The inability of individuals from two populations to produce fertile offspring together.

**reproductive output** (14) The number of offspring an individual or population produces.

**resource partitioning** (15) A division of resources that occurs when species overlap some portion of a niche in which one or more species differ in behavior or body plan in a way that divides the resources of the niche between the species.

**restriction enzymes** (5) Enzymes that recognize and bind to different specific sequences of four to eight bases in DNA and cut the DNA at that point. Restriction enzymes are important in biotechnology because they permit the cutting of short lengths of DNA, which can be inserted into other chromosomes or otherwise utilized.

**retrovirus** (13) A virus containing RNA and also reverse transcriptase, a viral enzyme, which uses the viral RNA as a template to synthesize a single strand of DNA. [Lat., *retro*, backward + *virus*, slime]

**ribonucleic acid (RNA)** (2) A nucleic acid, RNA serves as a middleman in the process of converting genetic information in DNA into protein; messenger RNA (mRNA) takes instructions for production of a given protein from DNA to another part of the cell, whereas transfer RNA (tRNA) interprets the mRNA code and directs the construction of the protein from its constituent amino acids.

**ribosomal subunits** (5) The two structural parts of a ribosome, which function together to translate mRNA to build a chain of amino acids that will make up a protein.

**ribosomes** (3) Granular bodies in the cytoplasm, released from their initial positions on the rough endoplasmic reticulum, that copy the information in segments of DNA to provide instruction for the construction of proteins.

**ring species** (10) Populations that can interbreed with neighboring populations but not with populations separated by larger geographic distances. Because the non-interbreeding populations are connected by gene flow through geographically intermediate populations, there is no clear point at which one species stops and another begins, and for this reason, ring species are problematic for the biological species concept.

**RNA world hypothesis** (10) A hypothesis that proposes that the world may have been filled with RNA-based life before it became filled with life based on DNA, the life of today.

**root** (12) The part of a vascular plant, usually below ground, that absorbs water and minerals from the soil and transports them through vascular tissue to the rest of the plant, and that anchors the plant in place. The overall structure of a plant's roots is called the root system.

**rough endoplasmic reticulum** (3) An organelle, part of the endomembrane system, structurally like a series of interconnected, flattened sacs connected to the nuclear envelope; called "rough" because its surface is studded with ribosomes. [Gk., *endon*, within + *plasma*, anything molded; Lat., *reticulum, dim. of rete*, net]

**roundworms** (11) A worm phylum with members characterized by a long, narrow, unsegmented body and growth by molting. Roundworms, also called nematodes, are protostomes with defined tissues; there are some 90,000 identified species.

**rubisco** (4) An enzyme (ribulose 1,5-bisphosphate carboxylase/oxygenase), important in photosynthesis, that fixes carbon atoms from $CO_2$ in the air, attaching them to an organic molecule in the stroma of the chloroplast. This fixation is the first step in the Calvin cycle, in which molecules of sugar are assembled. Rubisco is the most plentiful protein on earth.

## S

**saturated fat** (2) A fat in which each carbon in the hydrocarbon chain forming the tail region of the fat molecule is bound to two hydrogen atoms; saturated fats are solid at room temperature.

**savanna** (15) A type of terrestrial biome; a tropical or subtropical grassland with scattered woody plants, characterized by hot climate and distinct wet and dry seasons (rainfall is less than in the tropical seasonal forest biome).

**science** (1) A body of knowledge based on observation, description, experimentation, and explanation of natural phenomena. [Lat., *scientia*, knowledge]

**scientific literacy** (1) A general, fact-based understanding of the basics of biology and other sciences, the scientific method, and the social, political, and legal implications of scientific information.

**scientific method** (1) A process of examination and discovery of natural phenomena that involves making observations, constructing hypotheses, testing predictions, experimenting, and drawing conclusions and revising them as necessary.

**second law of thermodynamics** (4) A physical law that states that every conversion of energy is not perfectly efficient and invariably includes the transformation of some energy into heat.

**secondary active transport** (3) Active transport in which there is no direct involvement of ATP (adenosine triphosphate); the transport protein simultaneously moves one molecule against its concentration gradient while letting another flow down its concentration gradient.

**secondary consumers** (15) Animals that feed on herbivores; also known as carnivores.

**secondary structure** (2) The corkscrew-like twists or folds formed by hydrogen bonds between amino acids in a polypeptide chain.

**seed** (12) An embryonic plant with its own supply of water and nutrients, encased within a protective coating.

**segmented worms** (11) A worm phylum with members characterized by grooves around the body that mark divisions between segments. Segmented worms, also called annelids, are protostomes with defined tissues and do not molt; examples are earthworms and leeches.

**sessile** (11) Describes organisms that are fastened in place, such as adult mussels and barnacles. [Lat., *sedere*, to sit]

**sex-linked trait** (7) A trait controlled by a gene on a sex chromosome.

**sexual dimorphism** (9) The case in which the sexes of a species differ in size or appearance. [Gk., *di-*, two + *morphē*, shape]

**sexual reproduction** (6) A type of reproduction in which offspring are produced by the fusion of gametes from two distinct sexes.

**sexual selection** (8) The process by which natural selection favors traits, such as ornaments or fighting behavior, that give an advantage to individuals of one sex in attracting mating partners.

**sexually transmitted disease (STD)** (13) A disease passed from one person to another through sexual activity.

**shoot** (12) The above-ground part of a plant, consisting of stems and leaves, and sometimes flowers and fruits. The stem contains vascular tissue and supports the leaves, the main photosynthetic organ of the plant. Also called shoot system.

**short tandem repeat (STR)** (5) Region in DNA in which a short sequence (commonly four or five nucleotides) repeats over and over, a dozen or more times. Variation among individuals in the number of times the sequence repeats make it possible to use these regions as a DNA "fingerprint," unique to an individual and valuable as a forensic tool.

**sign stimulus** (9) An external signal that triggers the innate behavior called a fixed action pattern.

**simple diffusion** (3) Diffusion of molecules directly through the phospholipid bilayer of the plasma membrane that takes place without the assistance of other molecules; oxygen and carbon dioxide, because they are small and carry no charge that would cause them to be repelled by the middle layer of the membrane, can pass through the membrane in this way.

**simple sugars** (2) Monosaccharide carbohydrates, generally containing three to seven carbon atoms, which store energy in their chemical bonds and which biological cells can break down; cannot be broken down into other simple sugars; examples are glucose, fructose, and galactose.

**single-gene trait** (7) A trait that is determined by instructions on only one gene; examples are a cleft chin, a widow's peak, and unattached earlobes.

**sister chromatids** (6) The two identical strands of a replicated chromosome.

**smooth endoplasmic reticulum** (3) An organelle, part of the endomembrane system, structurally like a series of branched tubes; called "smooth" because its surface has no ribosomes. Smooth endoplasmic reticulum synthesizes lipids such as fatty acids, phospholipids, and steroids. [Gk., *endon*, within + *plasma*, anything molded; Lat., *reticulum*, dim. of *rete*, net]

**solute** (3) A substance that is dissolved in a gas or liquid; in a solution of water and sugar, sugar is the solute. [Lat., *solvere*, to loosen]

**solvent** (3) The gas or liquid in which a substance is dissolved; in a solution of water and sugar, water is the solvent. [Lat., *solvere*, to loosen]

**somatic cells** (6) The (usually diploid) cells of the body of an organism (in contrast to the usually haploid reproductive cells). [Gk., *soma*, body]

**speciation** (10) The process by which one species splits into two distinct species; the first phase of speciation is reproductive isolation, the second is genetic divergence, in which two populations evolve over time as separate entities with physical and behavioral differences.

**speciation event** (10) A point in evolutionary history at which a given population splits into independent evolutionary lineages.

**species** (10) Natural populations of organisms that can interbreed and are reproductively isolated from other such groups; in the Linnaean system, the species is the narrowest classification for an organism. [Lat., *species*, kind, sort]

**specific epithet** (10) In the system developed by Carolus Linnaeus (1707–1778), a noun or adjective added to the genus name to distinguish a species; in the name *Homo sapiens*, *Homo* is the genus name and *sapiens* is the specific epithet. [Gk., *epithetos*, added]

**spindle** (6) A part of the cytoskeleton of a cell, formed in prophase (in mitosis) or in prophase I (in meiosis), from which extend the fibers that organize and separate the sister chromatids.

**spindle fibers** (6) Fibers that extend from one pole of a cell to the other, which pull the sister chromatids apart in the anaphase stage of mitosis or the anaphase II stage of meiosis.

**sporangia** (*sing.* **sporangium**) (12) In many ferns, the structures on the underside of the leaves in which the spores are produced. [Gk., *spora*, seed + *angeion*, vessel]

**spore** (12) A reproductive structure of non-vascular and some vascular plants that have an alternation of generations; spores are typically haploid, unicellular, and develop into either a male (producing sperm) or female (producing eggs) gametophyte. The eggs and sperm produced by gametophytes unite to produce the diploid generation (sporophyte) of the plant. [Gk., *spora*, seed]

**sporophyte** (12) The multicellular diploid structure in non-vascular plants, some vascular plants, and some algae that produces asexual spores, the diploid life stage in organisms exhibiting alternation of generations.

**stabilizing selection** (8) Selection that, for a given trait, produces the greatest fitness at the intermediate point of the phenotypic range.

**stamen** (12) The male reproductive structure of a flower, consisting of a head-like anther on a stalk-like filament. There are usually several stamens in a flower. [Lat., *stamen*, thread]

**starch** (2) A complex polysaccharide carbohydrate consisting of a large number of monosaccharides linked in line; in plants, starch is the primary form of energy storage.

**statistics** (1) A set of analytical and mathematical tools designed to further the understanding of numerical data.

**stem cells** (5) Undifferentiated cells that have the ability to develop into any type of cell in the body; this property makes stem cells useful in biotechnology.

**sterol** (2) A lipid important in regulating growth and development; the sterols include cholesterol and the steroid hormones testosterone and estrogen, and all are modifications of a basic structure of four interlinked rings of carbon atoms. [Gk., *stereos*, solid + *-ol*, chemical suffix for an alcohol]

**stomata** (*sing.* **stoma**) (4) Small pores, usually on the undersides of leaves, that are the primary sites for gas exchange in plants; carbon dioxide (for photosynthesis) enters and oxygen (a by-product of photosynthesis) exits through the stomata. [Gk., *stoma*, mouth]

**stroma** (3, 4) In the leaf of a green plant, the fluid in the inner compartment of a chloroplast, which contains DNA and protein-making machinery. [Gk., *stroma*, bed]

**substrate** (2) The molecule on which an enzyme acts. The active site on the enzyme binds to the substrate, initiating a chemical reaction; for example, the active site on the enzyme lactase binds to the substrate lactose, breaking it down into the two simple sugars glucose and galactose. [Lat., *sub-*, under + *stratus*, spread]

**succession** (15) The change in the species composition of a community over time, following a disturbance.

**superstition** (1) The irrational belief that actions not related by logic to a course of events can influence an outcome.

**surface protein** (3) A protein that resides primarily on the inner or outer surface of the phospholipid bilayer that constitutes the plasma membrane of the cell.

**survivorship curves** (14) Graphs showing the proportion of individuals of particular ages now alive in a population; indicate an individual's likelihood of surviving through a given age interval.

**sympatric speciation** (10) Speciation that results not from geographic isolation but from polyploidy or hybridization and allopolyploidy; this type of speciation is relatively uncommon in animals but is common among plants. [Gk., *syn*, together with + *patris*, native land]

**systematics** (10) The modern approach to classification, with the broader goal of reconstructing the evolutionary history, or phylogeny, of organisms. [Gk., *systema*, a whole compounded of parts]

**T**

**telomere** (6) A non-coding, highly repetitive section of DNA at the tip of every eukaryotic chromosome that shortens with every cell division; if it becomes too short, additional cell division can cause the loss of functional, essential DNA and therefore almost certain cell death. [Gk., *telos*, end + *meris*, part]

**telophase** (6) The fourth and last phase of mitosis, in which the chromosomes begin to uncoil and the nuclear membrane is reassembled around them. In meiosis, the sister chromatids arrive at the cell poles and the nuclear membrane reassembles around them in telophase I; in telophase II, the sister chromatids have been pulled apart and the nuclear membrane reassembles around haploid numbers of chromosomes. [Gk., *telos*, end + *phasis*, appearance]

**tertiary consumers** (15) Animals that eat animals that eat herbivores; also known as top carnivores. [Lat., *tertius*, third]

**tertiary structure** (2) The unique and complex three-dimensional shape formed by multiple twists of the secondary structure of a protein as amino acids come together and form hydrogen bonds or covalent sulfur-sulfur bonds. [Lat., *tertius*, third]

**test-cross** (7) A mating in which a homozygous recessive individual is bred with individuals of unknown genotype, showing the dominant phenotype; this type of cross can reveal the unknown genotype by the observed characteristics, or phenotype, of the offspring.

**tetrapod** (11) An organism with four limbs; all terrestrial vertebrates are tetrapods. [Gk., *tetra-*, four + *pous*, foot]

**theory** (1) An explanatory hypothesis for a natural phenomenon that is exceptionally well supported by empirical data. [Gk., *theorein*, to consider]

**thermodynamics** (4) The study of the transformation of energy from one type to another, such as from potential energy to kinetic energy. [Gk., *thermē*, heat + *dynamis*, power]

**thylakoids** (3, 4) Interconnected membranous structures in the stroma of a chloroplast, where light energy is collected and the conversion of light energy to chemical energy in photosynthesis takes place. [Gk., *thylakis, dim. of thylakos,* bag]

**tight junction** (3) A continuous, water-tight connection between adjacent animal cells. Tight junctions are particularly important in the small intestine, where digestion occurs, to ensure that nutrients do not leak between cells into the body cavity and so become lost as a source of energy.

**tonicity** (3) For a cell in solution, a measure of the concentration of solutes outside the cell relative to that inside the cell. [Gk., *tonos,* tension]

**topography** (15) The physical features of a region, including those created by humans. [Gk., *topos,* place + *graphein,* to write down]

**total reproductive output** (9) The lifetime number of offspring produced by an individual.

**trait** (5, 8) Any characteristic or feature of an organism, such as red petal color in a flower.

**trans fat** (2) An unsaturated fat that has been partially hydrogenated (meaning that hydrogen atoms have been added to make the fat more saturated and to improve a food's taste, texture, and shelf-life). The added hydrogen atoms are in a trans orientation, which differs from the cis ("near") orientation of hydrogen atoms in the unsaturated fat. [Lat., *trans,* on the other side of]

**transcription** (5) The process by which a gene's base sequence is copied to mRNA.

**transduction** (13) A method of lateral transfer of DNA from one bacterial cell to another by means of a virus containing pieces of bacterial DNA picked up from its previous host, which infects the recipient bacterium and passes along new genes to the recipient. [Lat., *trans,* on the other side of + *ducere,* to lead]

**transfer RNA (tRNA)** (5) The type of RNA molecules in the cytoplasm that link specific triplet base sequences on mRNA to specific amino acids.

**transformation** (13) A method of lateral transfer of DNA from one bacterial cell to another in which a bacterial cell scavenges DNA released from burst bacterial cells in the environment. [Lat., *trans,* on the other side of + *formare,* to shape]

**transgenic organism** (5) An organism that contains DNA from another species.

**translation** (5) The process by which mRNA, which encodes a gene's base sequence, directs the production of a protein.

**transmembrane protein** (3) A protein that can penetrate the phospholipid bilayer of a cell's plasma membrane.

**transport protein** (3) A transmembrane protein that provides a channel or passageway through which large or strongly charged molecules can pass. Transport proteins are of a number of shapes and sizes, making possible the transport of a wide variety of molecules.

**treatment** (1) Any condition applied to the subjects of a research study that is not applied to subjects in a control group.

**triglyceride** (2) A fat having three fatty acids linked to the glycerol molecule. [Gk., *tri-* three + *glykys,* sweet]

**trophic level** (15) A step in the flow of energy through an ecosystem. [Gk., *trophē,* food]

**tropical rain forest** (15) A type of terrestrial biome, found between the Tropic of Cancer (23.5° north latitude) and Tropic of Capricorn (23.5° south latitude) and characterized by constant moisture and temperature that do not vary across the seasons; vegetation is dense.

**tropical seasonal forest** (15) A type of terrestrial biome, characterized by hot climate and distinct wet and dry seasons; trees shed their leaves in the dry season.

**true-breeding** (7) Describes a population of organisms in which, for a given trait, the offspring of crosses of individuals within the population always show the same trait; for example, the offspring of pea plants that are true-breeding for round peas always have round peas.

**turgor pressure** (3) In plants, the pressure of the contents of the cell against the cell wall, which is maintained by osmosis as water rushes into the cell when it contains high concentrations of dissolved substances. Turgor pressure allows non-woody plants to stand upright, and its loss causes wilting. [Lat., *turgere,* to swell]

**U**

**unsaturated fat** (2) A fat in which at least one carbon in the hydrocarbon chain forming the tail region of the fat molecule is bound to only one hydrogen atom; unsaturated fats are liquid at room temperature.

**V**

**vacuole (central)** (3) A membrane-enclosed, fluid-filled, multipurpose organelle prominent in most plant cells (but also present in some protists, fungi, and animals); functions vary but can include storing nutrients, retaining and degrading waste products, accumulating poisonous materials, containing pigments, and providing physical support. [Lat., *vacuus,* empty]

**variables** (1) The characteristics of an experimental system subject to change; for example, time (the duration of treatment) or specific elements of the treatment, such as the substance or procedure administered or the temperature at which it takes place. [Lat., *variare,* to vary]

**vascular plants** (12) Plants that transport water and dissolved nutrients by means of vascular tissue, a system of tubes that extends from the roots through the stem and into the leaves. [Lat., *vasculum, dim. of vas,* vessel]

**vesicle** (3) A small, membrane-enclosed sac within a cell. [Lat., *vesicula, dim. of vesica,* bladder]

**vertebrates** (11) Chordate animals that have a backbone (made of cartilage or hollow bones) and, at the front end of the organism, a head containing a skull, brain, and sensory organs.

**vestigial structure** (8) A structure, once useful to organisms, but which has lost its function over evolutionary time; an example is molars in bats that now consume an exclusively liquid diet. [Lat., *vestigium,* footprint, trace]

**viruses** (10) Diverse and important biological entities that can replicate but can conduct metabolic activity only by taking over the metabolic processes of a host organism, and therefore fall outside the definition of life. [Lat., *virus,* slime]

**viviparity** (11) The characteristic of bearing young alive, giving birth to babies (rather than laying eggs). [Lat., *vivus,* alive + *parere,* to bear]

**W**

**waggle dance** (9) Behavior of scout honeybees that indicates, by the angle of the body relative to the sun and by physical maneuvers of various duration, the direction to a distant source of food.

**wax** (2) A lipid similar in structure to fats but with only one long-chain fatty acid linked to the glycerol head of the molecule; because the fatty acid chain is highly nonpolar, waxes are strongly hydrophobic.

**X**

**X and Y chromosomes** (6) The human sex chromosomes.

# Photo Credits

Legend: *L*, left; *C*, center; *R*, right; *T*, top; *M*, middle; *B*, bottom

**CHAPTER 1    pp. 0–1:** Michael Nichols/National Geographic Stock. **p. 2:** Martin Barraud/Getty Images. **p. 3:** *Figure 11 (L to R)* AP Photo, Mike Fuentes/The New York Times/Redux Pictures, The Photo Works. **p. 4:** *Figure 1-2 (clockwise from L)* Reprinted through the courtesy of the Editors of TIME Magazine © 2008 Time Inc., Courtesy of The Economist, Newsweek/PRN/Newscom, AP Photo/Tony Gutierrez; headline used with permission of The Associated Press Copyright © 2011. All rights reserved. **p. 5:** *Figure 1-3* Walter Dawn/Photo Researchers. **p. 6:** *Figure 1-4 (TL)* Steven Puetzer/Getty Images, *(TR)* Ingolf Hatz/Corbis; *(ML)* Mark Scott/Getty Images, *(MR)* WireImage/Getty Images; *(B)* Louie Psihoyos/Corbis. **p. 7:** Frans Lanting Photography. **p. 8:** *Figure 1-6* AFP/Getty Images; *Figure 1-7* Digital Vision Photography/Veer. **p. 9:** *Figure 1-8* CDC/Photo Researchers, *(inset)* Nigel Cattlin/Alamy. **p. 10:** *Figure 1-9* LWA/Dann Tardif/Blend Images/Corbis. **p. 13:** *Figure 1-12* Peter Frank/Corbis. **p. 16:** Nigel Cattlin/Alamy. **p. 18:** *Figure 1-15* Topham/The Image Works. **p. 20:** *Figure 1-16 (L)* Courtesy of KCBD News Channel 11, *(R)* Courtesy of www.hivandhepatitis.com. **p. 22:** B. Boissonnet/age fotostock. **p. 24:** *Figure 1-20* Pacific Coast News. **p. 25:** *Figure 1-21 (L)* Image Source Black/Alamy, *(R)* Marni Rolfes. **p. 26:** *Figure 1-22* AP Photo/Paul Sakuma. **p. 27:** *Figure 1-23* Reprinted through the courtesy of the Editors of TIME Magazine © 2008 Time Inc. **p. 28:** *Figure 1-24* Waltraud Grubitzsch/epa/Corbis. **p. 29:** Patrick Landmann/Photo Researchers. **p. 30:** yellowdog/age fotostock.

**CHAPTER 2    pp. 36–37:** ©Tim Laman (www.TimLaman.com). **p. 38:** David Nicholls/Photo Researchers; *Figure 2-1 (L)* Stockfolio/Alamy, *(C)* Jeffrey Coolidge/Getty Images, *(R)* Alex Cao/Getty Images. **p. 44:** *Figure 2-10* Judith Collins/Alamy, *(inset)* Dennis Kunkel/Phototake. **p. 46:** imagewerks/Getty Images; *Figure 2-13 (L)* Kietmar Nill/Foto Natura/Minden Pictures, *(R)* R. B. Suter, Vassar College. **p. 52:** Keren Su/Getty Images. **p. 53:** *Figure 2-20* Dorling Kindersley/Getty Images. **p. 54:** *Figure 2-22 (T)* Frans Lemmens/Alamy, *(B)* Lew Robertson/Getty Images. **p. 55:** *Figure 2-25 (L)* Paul Geor/Stockxpert, *(R)* Alison Parks-Whitfield/Alamy. **p. 56:** *Figure 2-26 (T)* Lake County Museum/Corbis, *(B)* Masterfile; *Figure 2-27* Chassenet/age fotostock. **p. 57:** Jeff Foott/Discovery Channel Images/Getty Images. **p. 58:** *Figure 2-28* Bryan & Cherry Alexander Photography/Alamy. **p. 59:** *Figure 2-30* Gary Gladstone/Corbis; *Figure 2-31 (L)* Ellen Isaacs/Alamy, *(R)* Masterfile. **p. 60:** *Figure 2-32* Andrew O'Toole/Taxi/Getty Images. **p. 62:** *Figure 2-34* Hulton Archive/Getty Images. **p. 63:** Stephanie Pfriender Stylander. **p. 64:** *Figure 2-36 (T to B)* Steve Bloom Images/Alamy, Steve Gschmeissner/Photo Researchers, Dennis Kunkel/Phototake, Dennis Kunkel/Phototake/Alamy, Deco/Alamy. **p. 65:** *Figure 2-38* Michael Waring/Foodstock. **p. 67:** *Figure 2-41 (L)* Ryan McVay/Getty Images, *(R)* White Packert/Getty Images. **p. 71:** Driscoll, Youngquist, & Baldeschwieler/Caltech/Photo Researchers. **p. 74:** Gustavo Andrade/age fotostock.

**CHAPTER 3    pp. 80–81:** Marshall Sklar/Photo Researchers. **p. 82:** Alfred Pasieka/Photo Researchers; *Figure 3-1 (L)* Steve Taylor/Getty Images, *(C)* Frans Lanting/Corbis, *(R)* David Scharf/Peter Arnold Images. **p. 83:** *Figure 3-2 (T)* Clark Sumida, *(M)* Rosenfeld/age fotostock, *(B)* Clouds Hill Imaging Ltd./Corbis, *(inset)* Courtesy of Reproductive Solutions.

**p. 85:** *Figure 3-3* Dennis Kunkel/Getty Images. **p. 86:** *Figure 3-4 (L)* John Glover/age fotostock, *(C)* Juniors Bildarchiv/age fotostock, *(R)* Matt Meadows/Peter Arnold Images. **p. 87:** *Figure 3-5 (L)* Dr. Donald Fawcett/Visuals Unlimited, *(R)* Phototake/Alamy. **p. 89:** imagebroker/Alamy. **p. 90:** *Figure 3-8 (L)* Don W. Fawcett/Photo Researchers, *(R)* Caro/Alamy. **p. 93:** *Figure 13-12 (L)* Photo Researchers, *(R)* Jim Evans/Kennebec Journal & Morning Sentinel. **p. 94:** *Figure 3-13* Image Source/Veer. **p. 96:** *Figure 3-15* AP Photo/Jeffrey Boan. **p. 97:** Courtesy of Dr. G. M. Gaietta and T. J. Deerinck, CRBS/NCMIR, University of California, San Diego. **p. 101:** *Figure 3-20* Richard Keppel-Smith/Getty Images. **p. 102:** *Figure 3-21* Biophoto Associates/Photo Researchers. **p. 103:** *Figure 3-23 (L)* Chuck Brown/Photo Researchers, *(C)* Ed Reschke/Peter Arnold Images, *(R)* Carolina Biological Supply Company/Phototake. **p. 105:** Lorenz Britt/Alamy. **p. 106:** *Figure 3-25 (L)* Dennis Kunkel/Phototake, *(C)* Dennis Kunkel/Visuals Unlimited, *(R)* Courtesy of Dr. G. D. Griffin, Oak Ridge National Laboratory. **p. 107:** Dennis Kunkel/Visuals Unlimited, *Figure 3-26* Mary Ellen Mark. **p. 108:** *Figure 3-27* Phototake/Alamy. **p. 109:** *Figure 3-28* Lynne Chang/Visuals Unlimited. **p. 110:** *Figure 3-29 (L)* Dr. Yorgos Nikas/Phototake, *(R)* Juergen Berger/Photo Researchers. **p. 111:** *Figure 3-30* Dr. Donald Fawcett/Getty Images. **p. 113:** *Figure 3-32* Dr. Gopal Murti/Photo Researchers. **p. 114:** *Figure 3-33* Kevin Somerville/The Medical File/Peter Arnold Images. **p. 115:** *Figure 3-34* Dennis Kunkel/Visuals Unlimited; *Figure 3-35* Omikron/Photo Researchers. **p. 117:** *Figure 3-36* Dennis Kunkel/Visuals Unlimited. **p. 118:** *Figure 3-38* Perennou Nuridsany/Photo Researchers. **p. 119:** *Figure 3-39* Biophoto Associates/Photo Researchers. **p. 120:** *Figure 3-40* George Chapman/Visuals Unlimited. **p. 122:** Marco Garcia/Getty Images.

**CHAPTER 4    pp. 128–129:** Pete Saloutos/Corbis. **p. 130:** Inga Spence/Getty Images. **p. 131:** *Figure 4-1* Pat LaCroix/Getty Images. **p. 132:** *Figure 4-3 (TL)* David Madison/Getty Images, *(TC)* Konrad Wothe/Minden Pictures, *(TR)* Michel Gunther/Photolibrary, *(BL)* Bernhard Edmaier/Photo Researchers, *(BR)* David Stoecklein/Corbis. **p. 133:** *Figure 4-4* Michael Pohuski/Foodpix/Jupiterimages. **p. 134:** *Figure 4-5 (L)* Ed Gifford/Masterfile, *(R)* Walter Lockwood/Corbis; *Figure 4-6* David Allio/Icon SMI/Corbis. **p. 137:** Frans Lanting Photography. **p. 138:** *Figure 4-10 (L to R)* Wim van Egmond/Visuals Unlimited, Dennis Kunkel/Phototake, Dennis Kunkel/Visuals Unlimited, Norbert Wu/Minden Pictures. **p. 139:** *Figure 4-12 (T)* Value Stock Images/Fotosearch, *(B)* Dennis Kunkel/Visuals Unlimited. **p. 148:** *Figure 4-23* Wolfgang Dufner/age fotostock; *Figure 4-24 (L and R)* Dr. Jeremy Burgess/Science Photo Library. **p. 149:** *Figure 4-25 (L)* age fotostock/SuperStock, *(C)* Beth Perkins/Getty Images, *(R)* PBNJ Productions/Corbis. **p. 151:** ClassicStock/Alamy; *Figure 4-27 (L)* age fotostock/SuperStock, *(C)* Frans Lanting Photography, *(R)* Volker Steger/Photo Researchers. **p. 153:** *Figure 4-29 (L)* age fotostock/SuperStock, *(C)* Frans Lanting Photography, *(R)* A. Barry Dowsett/Photo Researchers. **p. 156:** *Figure 4-33* hybrid medical animation/Photo Researchers. **p. 159:** Martin Harvey/Getty Images; *Figure 4-36* Tobias Bernhard/Getty Images. **p. 160:** *Figure 4-38 (L)* Cephas Picture Library/Alamy, *(R)* Romilly Lockyer/Getty Images, *(inset)* SciMAT/Photo Researchers. **p. 161:** *Figure 4-39 (L)*

FoodCollection/age fotostock, *(C)* Tim Hill/Alamy, *(R)* D. Hurst/Alamy. **p. 162:** Plush Studios/Getty Images.

**CHAPTER 5    pp. 168-169:** Steven Puetzer/Getty Images. **p. 170:** Mario Tama/Getty Images; *Figure 5-1* Vasna Wilson/Virginian-Pilot. **p. 171:** *Figure 5-2 (L)* Reprinted with permission, September 2008, *Discover* magazine. Copyright © Discover Magazine. All rights reserved. *(R)* Reprinted through the courtesy of the Editors of TIME Magazine © 2008 Time Inc. *(B)* Reprinted by permission of The Wall Street Journal, © 2008 Dow Jones & Company, Inc. All rights reserved worldwide. **p. 172:** *Figure 5-3 (L)* A. Barrington Brown/Photo Researchers, *(R)* Science Source/Photo Researchers. **p. 174:** *Figure 5-5 (TL)* Courtesy of Jay Phelan, *(TC)* Sian Irvine/Getty Images, *(TR)* Eye of Science/Photo Researchers, *(BL)* Visuals Unlimited/Corbis, *(BR)* John White. **p. 177:** *Figure 5-8 (T to B)* Eye of Science/Photo Researchers, Courtesy of Jay Phelan, Sian Irvine/Getty Images, John White, Visuals Unlimited/Corbis. **p. 178:** *Figure 5-9 (T to B)* Courtesy of Jay Phelan, Eye of Science/Photo Researchers, Dr. Dennis Kunkel/Getty Images, Fancy/Veer/Corbis, Dennis Kunkel/Visuals Unlimited. **p. 180:** Denis Galante Photography, Inc. **p. 186:** *Figure 5-16* Zave Smith/Corbis. **p. 189:** © Uli Westphal 2008. **p. 190:** *Figure 5-20 (L)* Andrew Syred/Photo Researchers, *(R)* Ted Kinsman/Photo Researchers. **p. 192:** *Figure 5-22 (L)* Mika/zefa/Corbis, *(C)* Justin Pumfrey/Getty Images, *(R)* Ocean/Corbis. **p. 193:** *Figure 5-23 (L and R)* Marni Rolfes. **p. 194:** Peter Ginter/Science Faction/Corbis. **p. 198:** *Figure 5-30 (L)* John Doebley, *(R)* Ed Young/Corbis. **p. 199:** *Figure 5-31 (TL)* Martin Ruegner/Digital Vision/Getty Images, *(TR)* Scottish Crop Research Institute, Dundee, *(M and B)* Professor Peter Beyer/Humanitarian Board for Golden Rice/www.goldenrice.org. **p. 200:** *Figure 5-32 (L)* Wally Eberhart/Getty Images, *(C)* Lance Nelson/Corbis, *(R)* David Munns/Photo Researchers; *Figure 5-33* Nigel Cattlin/Visuals Unlimited. **p. 201:** *Figure 5-35* Frank Whitney/Getty Images. **p. 202:** *Figure 5-36* Dr. Garth Fletcher. **p. 203:** *Figures 5-37* and *5-38* Reuters/Corbis. **p. 205:** Karen Kasmauski/Corbis. **p. 206:** *Figure 5-39* AJPhoto/Photo Researchers, *Figure 5-40* MGM/Courtesy of Everett Collection. **p. 207:** *Figure 5-41* Robin Nelson/Photo Edit. **p. 208:** *Figure 5-42* Baylor Medical Center. **p. 210:** *Figure 5-44 (T)* AP Photo/John Chadwick, *(M)* AP Photo/Ben Margot, *(B)* AP Photo/Advanced Cell Technologies. **p. 211:** *Figure 5-45 (L)* Neville Chadwick/Science Photo Library, *(C)* Tom Grill/Corbis, *(R)* David Parker/Science Photo Library. **p. 215:** Anthony Marsland/Getty Images.

**CHAPTER 6    pp. 220–221:** Steve Schapiro/Corbis Outline. **p. 222:** Dennis Kunkel Microscopy, Inc./Visuals Unlimited. **p. 223:** *Figure 6-1* Dr. Peter Landsorp/Visuals Unlimited; *Figure 6-2* Barcroft/Fame Pictures. **p. 225:** *Figure 6-4* Dennis Kunkel Microscopy. **p. 229:** Indeed/Getty Images; *Figure 6-7 (L to R)* Dan Guravich/Photo Researchers, Biology Media/Photo Researchers, SPL/Photo Researchers, Alex Blackburn Clayton/Photolibrary. **p. 230:** *Figure 6-8* Micro Discovery/Corbis. **p. 231:** *Figure 6-9* David Scharf/Getty Images. **p. 232:** *Figure 6-11* Courtesy of T. Wittman. **p. 235:** *Figure 6-13* GJLP/CNRI/Science Photo Library. **p. 236:** *Figure 6-15* Kevin Laubacher/Taxi/Getty Images. **p. 237:** Dr. Yorgos Nikas/Photo Researchers. **p. 238:** *Figure 6-17* AP Photo. **p. 243:** *Figure 6-21* Dr. David Phillips/Visuals Unlimited. **p. 245:** *Figure 6-23* Margot Granitsas/Photo Researchers. **p. 246:** *Figure 6-25* James H. Robinson/Science Photo Library/Photo Researchers; *Figure 6-26* Dr. Kari Lounatmaa/Photo Researchers. **p. 247:** Cormac Hanley/Getty Images; *Figure 6-27* Biophoto Associates/Photo Researchers. **p. 249:** *Figure 6-29 (L)* Frans Lanting Photography, *(C)* imagebroker/Alamy, *(R)* Michael Patrick O'Neill/Alamy. **p. 250:** Visuals Unlimited/Corbis. **p. 251:** *Figure 6-31* CNRI/Photo

Researchers; *Figure 6-32* Saturn Stills/Science Photo Library/Photo Researchers. **p. 252:** *Figure 6-33 (L)* Mika/zefa/Corbis, *(R)* L. Willatt/East Anglian Regional Genetics Service/Photo Researchers. **p. 256:** Michael Elins.

**CHAPTER 7    pp. 262–263:** Exactostock/SuperStock. **p. 264:** Mitsuaki Iwago/Minden Pictures; *Figure 7-1* Felix Wirth/Corbis. **p. 266:** *Figure 7-4* Newspix/Rex USA. **p. 267:** *Figure 7-5 (row 1)* Les3photo8/Dreamstime.com, Custom Medical Stock Photo/Alamy, *(row 2)* Geoff du Feu/Alamy, Gabriela Medina/Getty Images, *(row 3)* Kevin Winter/Getty Images, Kevin Winter/Getty Images, *(row 4)* Dorling Kindersley/Getty Images, Beyond Fotomedia GmbH/Alamy. **p. 269:** *Figure 7-7* James King-Holmes/Photo Researchers. **p. 270:** *Figure 7-8 (L)* blickwinkel/Alamy, *(R)* Sally Reed/age fotostock. **p. 272:** *Figure 7-10* Reinhard Kunkel/Photolibrary. **p. 274:** Inspirestock/Corbis. **p. 276:** *Figure 7-14* Richard Ellis/Alamy. **p. 278:** *Figure 7-16* Courtesy of Michael Comte. **p. 279:** Photodisc. **p. 280:** *Figure 7-18* Rene Maltete/Gamma Rapho. **p. 283:** *Figure 7-22* Big Cheese Photo LLC/Alamy. **p. 285:** *Figure 7-23* Wellcome Images. **p. 287:** *Figure 7-25* Tommy Moorman. **p. 288:** *Figure 7-26 (T)* Juniors Bildarchiv/age fotostock, *(B)* Juniors Bildarchiv/Alamy. **p. 289:** Joel Meyerowitz/Gallerystock. **p. 291:** *Figure 7-28* Armando Gallo/Retna Ltd./Corbis. **p. 292:** Patrick J. Miller.

**CHAPTER 8    pp. 298–299:** Chris Johns/National Geographic Stock. **p. 300:** Frans Lanting Photography; *Figure 8-1* Graphic Science/Alamy. **p. 303:** Frans Lanting Photography. **p. 304:** *Figure 8-3 (L to R)* Leonard de Selva/Corbis, Sheila Terry/Photo Researchers, The Granger Collection, New York, The Granger Collection, New York, Mary Evans Picture Library/Alamy. **p. 305:** *Figure 8-5* Getty Images. **p. 308:** *Figure 8-7 (L)* The Granger Collection, New York, *(R)* Louie Fasciolo/Bernard J. Shapero Rare Books. **p. 309:** Manfred Danegger/Corbis; *Figure 8-9* Renee Lynn/Corbis. **p. 312:** *Figure 8-11 (L)* Juniors Bildarchiv/Alamy, *(R)* blickwinkel/Alamy. **p. 313:** *Figure 8-12* J. D. Talasek/Phototake. **p. 314:** *Figure 8-13* Lester V. Bergman/Corbis. **p. 315:** *Figure 8-14* Ross Warner/Alamy. **p. 317:** *Figure 8-16 (TL)* Annie Katz/Getty Images, *(TC)* Catherine Ledner/Getty Images, *(TR)* Deanne Fitzmaurice/Corbis, *(BL)* Jason Homa/Getty Images, *(BC)* Paul Costello/Getty Images, *(BR)* Jacob Langvad/Stone/Getty Images; *Figure 8-17* Max Nash/AFP/Getty Images. **p. 318:** *Figure 8-18* Benn Mitchell/Riser/Getty Images. **p. 322:** Terry Andrewartha/NPL/Minden Pictures. **p. 323:** *Figure 8-22 (L and R)* Lynn Johnson/National Geographic Stock. **p. 324:** *Figure 8-23* J. Sneesby/B. Wilkins/Getty Images. **p. 326:** *Figure 8-26* Patricia McDonough. **p. 327:** *Figure 8-27* Andrew Sacks/Getty Images. **p. 328:** *Figure 8-28* Walter B. McKenzie/Getty Images. **p. 328:** *Figure 8-29* Tom & Pat Leeson. **p. 329:** *Figure 8-30* RubberBall/Alamy. **p. 330:** *Figure 8-31* Steve Allen/Getty Images. **p. 331:** Constantinos Petrinos/NPL/Minden Pictures. **p. 332:** *Figure 8-32 (L)* Jason Edwards/National Geographic/Getty Images, *(R)* Kjell Sandved/Visuals Unlimited. **p. 334:** *Figure 8-36 (T)* Joe McDonald/Visuals Unlimited, *(BL)* G. Armistead/Vireo, *(BC)* Aaron French, *(BR)* Jack Jeffrey. **p. 335:** *Figure 8-37 (TL)* Joe McDonald/Visuals Unlimited, *(TC)* Martin Pot, *(TR)* Dave Watts/Alamy, *(BL)* Kim Taylor/Minden Pictures, *(BC)* Morales/age fotostock, *(BR)* John W. Warden/age fotostock. **p. 336:** *Figure 8-38 (TL)* Jose I. Castro, *(TR)* Laboratory of Evolutionary Morphology, *(BL)* Courtesy of Bradley Smith, University of Michigan, *(BR)* S. K. Ackerley, Department of Integrative Biology, University of Guelph, Guelph, ON, Canada; *Figure 8-39 (L to R)* PBNJ Productions/PBNJ/age fotostock, Juniors Bildarchiv/age fotostock, Malcolm Schuyl/Alamy, Juniors Bildarchiv/age fotostock. **p. 337:** *Figure 8-40* Adrian Warren/Ardea; *Figure 8-41 (T)* Michael Durham/Minden Pictures/Getty Images, *(B)* Kevin Schafer/Getty Images.

**p. 338:** *Figure 8-42 (L to R)* Getty Images, Arco Images/Alamy, PhotoSpin/Alamy, Danita Delimont/Alamy, Ulrike Klenke and Zeev Pancer, Center of Marine Biotechnology, Baltimore, MD.

**CHAPTER 9    pp. 346–347:** Harald Lange/zefa/Corbis. **p. 348:** DLILLC/Corbis; *Figure 9-1* AP Photo/Independence Daily Reporter/Rob Morgan. **p. 349:** *Figure 9-2* Radius Images/Alamy; *Figure 9-3* Isabel M. Smallegange. **p. 350:** *Figure 9-4 (L)* Juniors Bildarchiv/Alamy, *(TR)* Andrew Darrington/Alamy, *(BR)* ARCO/H. Reinhard/age fotostock. **p. 352:** *Figure 9-6 (T)* Sharon Dominick/Getty Images, *(BL)* Michael Fogden/Bruce Coleman Inc., *(BR)* Lew Robertson/Corbis. **p. 355:** tbkmedia.de/Alamy; *Figure 9-8* Dr. Theodore Evans. **p. 357:** *Figure 9-9* Pete Cairns/NPL/Minden Pictures. **p. 360:** *Figure 9-12* Courtesy of Gerald Carter. **p. 361:** *Figure 9-13* Ben Cranke/Getty Images. **p. 363:** *Figure 9-14* PA Wire URN:5620400 (Press Association via AP Images). **p. 365:** Tsuneo Hayashida/Nature Production/Minden Pictures; *Figure 9-16* David M. Phillips/Photo Researchers. **p. 367:** *Figure 9-18 (L)* Suzi Eszterhas/Minden Pictures, *(C)* Gregory Basco/Visuals Unlimited, *(R)* Frans Lanting/Corbis. **p. 368:** *Figure 9-19 (L)* Universal Pictures/Photofest, *(R)* Summit Entertainment/Photofest. **p. 369:** *Figure 9-20* Michael Fogden/Photolibrary. **p. 370:** *Figure 9-21 (TL)* Anthony Mercieca/Photo Researchers, *(TR)* Dr. John Alcock/Visuals Unlimited, *(BL)* Joe McDonald/Corbis, *(BR)* Miroslav Beneda/Alamy. **p. 372:** *Figure 9-22* Thomas Marent/Visuals Unlimited; *Figure 9-23* Heidi & Hans-Juergen Koch/Minden Pictures. **p. 373:** *Figure 9-24* Yva Momatiuk & John Eastcott/Minden Pictures. **p. 374:** *Figure 9-25* M. Botzek/age fotostock. **p. 376:** *Figure 9-27 (L)* Yva Momatiuk & John Eastcott/Minden Pictures, *(R)* Frans Lanting Photography. **p. 377:** Juniors Bildarchiv/Alamy. **p. 378:** *Figure 9-28 (L)* Thomas Marent/Minden Pictures, *(C)* David Ponton/Design Pics/Corbis, *(R)* Jupiterimages/Creatas/Alamy; *Figure 9-29 (L)* Scott Camazine/Alamy, *(C)* Ocean/Corbis, *(R)* Michael Newman/PhotoEdit. **p. 379:** *Figure 9-30* Werner Bollmann/age fotostock. **p. 380:** Titus Lacoste/Getty Images.

**CHAPTER 10    pp. 386–387:** Kellar Autumn, Lewis & Clark College. **p. 388:** Frans Lanting Photography. **p. 389:** *Figure 10-1* Peter Sawyer, NMNH, Smithsonian Institution; *Figure 10-2* Roger Ressmeyer/Corbis. **p. 390:** *Figure 10-3* Courtesy of Andrew Knoll. **p. 392:** Photodisc; *Figure 10-5 (L to R)* Peter Turnley/Corbis, Steve Sands/New York Newswire/Corbis, Plantography/Alamy, The Collection by Jan Smith Photography/Alamy, Frans Lanting Photography, Frans Lanting Photography. **p. 393:** *Figure 10-6* Galen Rowell/Corbis; *Figure 10-7* European Pressphoto Agency B.V.; *Figure 10-8 (L)* Juniors Bildarchiv/Alamy, *(R)* Masterfile. **p. 398:** *Figure 10-11 (L)* John Cancalosi/age fotostock, *(R)* Wildlife/Peter Arnold Images. **p. 399:** *Figure 10-12 (TL)* Minden Pictures/SuperStock, *(TC)* Tui De Roy/Minden Pictures, *(TR)* Hiroya Minakuchi/Minden Pictures, *(BL)* David Hosking/Alamy, *(BC)* Tui De Roy/Minden Pictures, *(BR)* Michael Woodruff. **p. 401:** Fred Habegger/Grant Heilman Photography. **p. 404:** *Figure 10-17 (L to R)* DLILLC/Corbis, Nigel Dennis/age fotostock, James Gritz/Getty Images, Altrendo/Getty Images. **p. 406:** *Figure 10-19* Michael Durham/Minden Pictures; *Figure 10-20 (L)* David Hosking/Alamy, *(C)* Galen Rathbun, California Academy of Sciences, www.afrotheria.net, *(R)* IFA-Bilderteam/Jupiterimages. **p. 407:** Michael & Patricia Fogden/Corbis; *Figure 10-21* David Edwards/Getty Images. **p. 412:** *Figure 10-26* Reuters/Corbis. **p. 413:** Frans Lanting Photography. **p. 416:** Radius Images/Alamy. **p. 417:** *Figure 10-30* Steve Gschmeissner/Photo Researchers; *Figure 10-31* John Wang/Getty Images, *(inset)* Derek Lovley/Kazem Kashefi/Photo Researchers. **p. 419:** *Figure 10-32 (L to R)* Michael & Patricia Fogden/Corbis, from *Botanica* by Howard Schatz ©Schatz Ornstein

2005, Adam Jones/Photo Researchers, Frans Lanting/Corbis. **p. 420:** Lawrence Migdale/Photo Researchers.

**CHAPTER 11    pp. 426–427:** Frans Lanting Photography. **p. 428:** Frans Lanting Photography; *Figure 11-1 (L)* Bill Beatty/Visuals Unlimited, *(C)* Stephen Frink/zefa/Corbis, *(R)* Reinhard Dirscherl/Visuals Unlimited. **p. 431:** *Figure 11-3 (L)* Jef Meul/Minden Pictures, *(R)* Frans Lanting Photography. **p. 432:** Courtesy of Jay Phelan. **p. 433:** *Figure 11-4 (top row, L to R)* Barry Brown/Visuals Unlimited, Dirk Wiersma/Photo Researchers, Sinclair Stammers/Photo Researchers, Wayne G. Lawler/Photo Researchers, *(bottom row, L to R)* Matthew Oldfield/Scubazoo/Photo Researchers, Arthur Siegelman/Visuals Unlimited, Gerry Bishop/Visuals Unlimited, Brandon Cole/Visuals Unlimited. **p. 434:** Poul Beckmann. **p. 435:** *Figure 11-5 (L)* Juan Carlos Calvin/Photolibrary, *(R)* Robert Yin/Corbis; *Figure 11-6* Doug Perrine/NPL/Minden Pictures. **p. 436:** *Figure 11-7 (T)* Melissa Fiene Photography, *(BL)* Brandon Cole/Visuals Unlimited, *(BR)* Michael Pitts/NPL/Minden Pictures. **p. 437:** *Figure 11-9 (T)* ©Tim Laman (www.TimLaman.com), *(B)* Howard Hall/Photolibrary. **p. 439:** *Figure 11-10 (L)* Alamy, *(R)* Sinclair Stammers/Photo Researchers; *Figure 11-11 (T)* CNRI/Photo Researchers, *(B)* LSHTM/Photo Researchers. **p. 440:** *Figure 11-12 (T)* Ross & Diane Armstrong/Minden Pictures, *(M)* Konrad Wothe/Minden Pictures, *(B)* ARCO/H. Reinhard/age fotostock. **p. 442:** *Figure 11-13 (TL)* David B. Fleetham/SeaPics.com, *(TR)* Gerry Ellis/Minden Pictures, *(BL)* Jeff Rotman/Getty Images, *(BR)* Norbert Wu/Getty Images. **p. 443:** *Figure 11-14 (L)* David B. Fleetham/Photolibrary, *(C)* Reuters/Corbis, *(R)* Norbert Wu/Getty Images. **p. 444:** *Figure 11-15* Tomasz Machnik. **p. 445:** *Figure 11-16 (TL)* blickwinkel/Alamy, *(TR)* George Grall/Getty Images, *(BL)* Scott Leslie, *(BR)* Chris Newbert/Minden Pictures. **p. 446:** *Figure 11-17 (L to R)* Frans Lanting/Corbis, Alamy, Alamy, Nigel Cattlin/Visuals Unlimited, Satoshi Kuribayashi/Nature Production/Minden Pictures. **p. 447:** *Figure 11-18 (L to R)* Satoshi Kuribayashi/Nature Production/Minden Pictures, Gerold & Cynthia Merker/Visuals Unlimited, Mark Moffett/Minden Pictures, Gerald & Buff Corsi/Visuals Unlimited. **p. 449:** *Figure 11-19 (T)* Brandon Cole/Visuals Unlimited, *(ML)* Jeffrey L. Rotman/Corbis, *(MR)* Stephen Frink/Corbis, *(B)* Frans Lanting Photography. **p. 450:** Frans Lanting/Corbis. **p. 451:** *Figure 11-20 (TL)* Frans Lanting Photography, *(TR)* Ross Armstrong/Photolibrary, *(BL)* Art Wolfe/Getty Images, *(BR)* Images&Stories/Alamy. **p. 452:** *Figure 11-22 (L)* Georgette Douwma/Science Photo Library/Photo Researchers, *(C)* Natural Visions/Alamy, *(R)* Frans Lanting/Corbis. **p. 453:** *Figure 11-24 (L)* Reinhard H/Photolibrary, *(inset)* Darlyne A. Murawski/Getty Images, *(R)* Masa Ushioda/SeaPics.com. **p. 454:** *Figure 11-25 (L)* Paul Souders/Corbis, *(C)* Tim Fitzharris/Minden Pictures, *(R)* Rudie Kuiter/OceanwideImages.com. **p. 456:** Luciano Candisani/Minden Pictures; *Figure 11-27 (L)* Stephen Dalton/Minden Pictures, *(C)* Michael Redmer/Visuals Unlimited, *(R)* McDonald Wildlife Photography, Inc. **p. 458:** *Figure 11-29 (L)* John Cancalosi/Photolibrary, *(R)* © Tim Laman (www.TimLaman.com); *Figure 11-30* © Chuang Zhao and Lida Xing. **p. 460:** *Figure 11-32 (L)* Dave Watts/Alamy, *(C)* Martin Rugner/Photolibrary, *(R)* Joe McDonald/Visuals Unlimited. **p. 466:** Ariel Skelley/Blend Images/Corbis.

**CHAPTER 12    pp. 472–473:** Katinka Matson. **p. 474:** Frans Lanting/Corbis; *Figure 12-1 (L)* David Pearson/Alamy, *(C)* Wendy Foden, *(R)* David Evans/National Geographic Stock. **p. 475:** *Figure 12-2* David Cavagnaro/Visuals Unlimited; *Figure 12-3 (L)* Zanetta Hardy: sxc.hu, *(C)* Chad Ehlers/Alamy, *(R)* Herbert Hopfensperger/age fotostock. **p. 477:** Frans Lanting Photography; *Figure 12-5* Charles F. Delwiche. **p. 478:** *Figure 12-6 (L)* David Muench/Corbis, *(R)* Gallo Images/Getty Images. **p. 479:** *Figure 12-7*

(T) Alistair Dove/Alamy, (BL) Jacques Rosès/Photolibrary, (BR) Daniel Vega/age fotostock. **p. 480:** *Figure 12-8* Barry Turner/Alamy; *Figure 12-9* (T) The Irish Image Collection/Design Pics/Corbis, (B) D Johnston & Company (Laphroaig) Ltd. **p. 481:** *Figure 12-10* David Wall/Alamy. **p. 482:** *Figure 12-11* (T) Frans Lanting Photography, (BL) Gerry Ellis/Minden Pictures, (BR) Pascal Goetgheluck/Science Photo Library/Photo Researchers; *Figure 12-12* (L) Snowflake Studio/Getty Images, (R) Visuals Unlimited/Corbis. **p. 483:** Jeff Vanuga/Corbis; *Figure 12-13* Courtesy of Kevin Phelan. **p. 485:** *Figure 12-14* (TL) Design Pics Inc./Alamy, (TR) Bohemian Nomad Picturemakers/Corbis, (BL) Takahiro Miyamoto/Getty Images, (BR) Jouan & Rius/naturepl.com; *Figure 12-15* (L to R) Thomas Kitchin & Victoria Hurst/AllCanadaPhotos.com, Alan & Linda Detrick/Photo Researchers, Scott S. Warren/National Geographic Stock, blickwinkel/Alamy. **p. 486:** *Figure 12-16* (L) Wil Meinderts/Minden Pictures, (R) Pam Moore, East Stroudsburg, PA. **p. 487:** *Figure 12-18* (T) Jim Brandenburg/Minden Pictures, (B) Philippe Bourseiller/Getty Images. **p. 488:** Andy Small/Corbis; *Figure 12-19* (TL) blickwinkel/Alamy, (TR) Frans Lanting Photography, (BL) Bas Van Den Boogaard/Foto Natura/Minden Pictures, (BR) Roger Standen/Science Photo Library/Photo Researchers. **p. 490:** *Figure 12-21* (L) James L. Castner/Visuals Unlimited, (R) Topic Photo Agency/age fotostock. **p. 491:** *Figure 12-22* (top row, L to R) Craig K. Lorenz/Photo Researchers, Andrew Darrington/Alamy, Amanda Jastremski/National Geographic Stock, Mark Moffett/Minden Pictures, (BL) Ch'ien Lee/Minden Pictures, (BR) Sharkawi Che Din/Alamy. **p. 492:** *Figure 12-23* Paul Souders/Corbis. **p. 494:** Dezorzi/Corbis. **p. 495:** *Figure 12-25* (L) Scott Camazine/Photo Researchers, (C) James L. Stanfield/National Geographic Stock, (R) Michael Thirnbeck/Getty Images. **p. 496:** *Figure 12-26* (TL) Beverly Joubert/National Geographic Stock, (TR) Wildlife GmbH/Alamy, (M) Howard Grey/Getty Images, (BL) Gertrud & Helmut Denzau/naturepl.com, (BR) Dr. Keith Wheeler/Getty Images. **p. 497:** *Figure 12-27* Ed Reschke/Peter Arnold Images. **p. 498:** Frans Lanting Photography; *Figure 12-28* (L) Jan Vermeer/Foto Natura/Minden Pictures, (C) Frans Lanting Photography, (R) Glen Threlfo/Minden Pictures. **p. 499:** *Figure 12-30* (L) imagebroker.net/SuperStock, (C) Visuals Unlimited/Corbis, (inset) Gregory G. Dimijian/Photo Researchers, (R) Visuals Unlimited/Corbis, (inset) W. H. Freeman and Company. **p. 500:** *Figure 12-31* Ingo Arndt/Minden Pictures. **p. 501:** *Figure 12-32* Joerrn M/Photolibrary; *Figure 12-33* D. Hurst/Alamy. **p. 502:** *Figure 12-34* Visuals Unlimited/Corbis. **p. 503:** *Figure 12-35* Paul Redfearn, Ozarks Regional Herbarium, Missouri State University; *Figure 12-36* Jim Brandenburg/Minden Pictures. **p. 505:** Pius Otomi Ekpei/AFP/Getty Images.

**CHAPTER 13    pp. 510–511:** Biozentrum, University of Basel/Photo Researchers. **p. 512:** Derren Ready, Wellcome Images. **p. 513:** *Figure 13-1* (L to R) Dr. Gopal Murti/Visuals Unlimited, Dr. Terry Beveridge/Visuals Unlimited, Dennis Kunkel Microscopy, Astrid & Hanns-Frieder Michler/Photo Researchers. **p. 513:** *Figure 13-2* (L) CNRI/Photo Researchers, (inset) Phototake/Alamy, (C) Federico Veronesi/Getty Images, (inset) Agriculture et Agroalimentaire Canada, (R) Brad Mogen/Visuals Unlimited, (inset) Dennis Kunkel/Visuals Unlimited. **p. 514:** *Figure 13-3* Courtesy of Jay Phelan. **p. 515:** *Figure 13-4* (L) Alfred Pasieka/Photo Researchers, (C) Eye of Science/Photo Researchers, (R) Biodisc/Visuals Unlimited/Alamy, (B) Eye of Science/Photo Researchers. **p. 516:** Visuals Unlimited/Corbis. **p. 517:** *Figure 13-5* Dennis Kunkel/Alamy. **p. 517:** *Figure 13-6* (L) mediacolor/Alamy, (R) Raymond Otero/Visuals Unlimited. **p. 520:** *Figure 13-9* (L) Dodie Ulery, (C) Thomas Dressler/age fotostock, (R) Reino Hanninen/Alamy. **p. 521:** Stephanie Schuller/Photo Researchers; *Figure 13-10* Mediablitz Images. **p. 522:** *Figure 13-11* Justinc. **p. 523:** *Figure 13-12* ImageState Royalty Free/Alamy. **p. 524:** *Figure 13-14* AGStockUSA/Alamy. **p. 526:** Frans Lanting Photography; *Figure 13-16* B. Boonyaratanakornkit and D. S. Clark, G. Vrdoljak/EM Lab, University of California Berkeley/Visuals Unlimited. **p. 527:** *Figure 13-17* Russ Bishop/age fotostock; *Figure 13-18* Tim Mantoani/Masterfile. **p. 528:** *Figure 13-19* Salvador de Sas/epa/Corbis. **p. 529:** Steve Gschmeissner/Photo Researchers; *Figure 13-20* The Microfossil Image Recovery and Circulation for Learning and Education (MIRACLE) Project, The Micropalaeontology Unit of the Earth Sciences Dept., UCLA. **p. 530:** *Figure 13-21* (L) Visuals Unlimited/Corbis, (C) C-FH/Grant Heilman Photography, (R) Norbert Wu/Minden Pictures. **p. 531:** *Figure 13-22* Paul Nicklen/National Geographic Society/Corbis, (inset) Wim van Egmond/Visuals Unlimited. **p. 532:** *Figure 13-23* (T) Peter Arnold/Alamy, (M) Eye of Science/Photo Researchers, (B) Phototake/Alamy. **p. 533:** D. Simon/Photo Researchers; *Figure 13-24* (L) SPL/Photo Researchers, (R) A. Barry Dowsett/Photo Researchers. **p. 535:** *Figure 13-26* (T) Science Source/Photo Researchers, (B) Reuters/Corbis. **p. 537:** *Figure 13-28* Bettmann/Corbis. **p. 539:** *Figure 13-30* Scott Camazine/CDC/Science Photo Library/Photo Researchers. **p. 540:** Tommy Moorman.

**CHAPTER 14    pp. 546–547:** Frans Lanting Photography. **p. 548:** Frans Lanting Photography; *Figure 14-1* Barrett & MacKay/age fotostock. **p. 550:** *Figure 14-3* Frans Lanting Photography. **p. 551:** *Figure 14-4* Juan Carlos Muñoz/age fotostock. **p. 552:** *Figure 14-5* Hal Beral/Visuals Unlimited. **p. 553:** *Figure 14-7* John McColgan, Bureau of Land Management, Alaska Fire Service. **p. 554:** *Figure 14-8* Bill Stormont/Corbis; *Figure 14-9* Reuters/Pierre Holtz/Landov. **p. 555:** *Figure 14-10* Lynn M. Stone/naturepl.com; *Figure 14-11* Tom McHugh/Science Photo Library/Photo Researchers. **p. 556:** *Figure 14-12* Dirk Anschutz/Getty Images. **p. 557:** *Figure 14-13* Martin Bwenetti/AFP/Getty Images, (inset) Andrew J. Martinez/Photo Researchers. **p. 558:** *Figure 14-14* National Museum of Natural History, Smithsonian Institution. **p. 559:** Alaska Stock Images/National Geographic Stock; *Figure 14-15* (L) Dave Watts/naturepl.com, (C) Frans Lanting Photography, (R) Stephen Dalton/NHPA. **p. 560:** *Figure 14-16* Greg Katsoulis. **p. 562:** *Figure 14-18* (T) Frans Lanting Photography, (C) George Grall/National Geographic Stock, (B) Ken Lucas/Visuals Unlimited. **p. 563:** *Figure 14-19* (L) Gunter Marx Photography/Corbis, (C) Frans Lanting Photography, (R) Gary Meszaros/Photo Researchers, (inset) iStockphoto.com/jeridu. **p. 564:** Jim Brandenburg/Minden Pictures. **p. 565:** *Figure 14-20* Them #6, photography, 78 × 200 cm, 1998–2000 (left-hand original, 1973), ©Hai Bo, Courtesy of Pace Beijing. **p. 567:** *Figure 14-22* (L) Renaud Visage/Getty Images, (R) W. Perry Conway/Corbis. **p. 570:** Gao Shanyue/Xinhua Press/Corbis. **p. 574:** *Figure 14-28* (L) Yadid Levy/Anzenberger Agency/Jupiterimages, (C) Sean Gallup/Getty Images, (R) Michael S. Yamashita/Getty Images. **p. 575:** *Figure 14-29* Edward Burtynsky. **p. 576:** Elke Hesser/Getty Images.

**CHAPTER 15    pp. 582–583:** Frans Lanting Photography. **p. 584:** Frans Lanting Photography; *Figure 15-1* © Tim Laman (www.TimLaman.com). **p. 585:** *Figure 15-2* Steve Allen/Getty Images. **p. 586:** *Figure 15-3* (TL) Frans Lanting Photography, (TC) Gerry Ellis/Minden Pictures, (TR) Fritz Poelking/Photolibrary, (ML) John Warburton-Lee Photography/Alamy, (MC) Michael Melford/Getty Images, (MR) Frans Lanting Photography, (BL) Dan Lamont/Corbis, (BC) Peter Arnold/Alamy, (BR) Frans Lanting/Corbis. **p. 587:** *Figure 15-4* (TL) James Randklev/Corbis, (TR) Frans Lanting/Corbis, (BL) Momatiuk–Eastcott/Corbis, (BC) Troy Plota/Getty Images, (BR) Mark Conlin/V&W/Image Quest Marine. **p. 588:** Steven Besserman. **p. 589:** *Figure 15-6* Mike Boyatt/AGStockUSA; *Figure 15-7* Frans Lanting Photography. **p. 590:** *Figure 15-8* Jacques Descloitres,

MODIS Land Rapid Response Team, NASA/GSFC. **p. 591:** *Figure 15-9 (L)* Courtesy of NASA/Goddard Space Flight Center Scientific Visualization Studio, *(R)* Sandy Felsenthal/Corbis. **p. 594:** Scott Linstead/FN/Minden Pictures. **p. 595:** *Figure 15-13 (L)* AJPhoto/Photo Researchers, *(R)* Eye of Science/Photo Researchers. **p. 601:** *Figure 15-19* AP Photo/Gautam Singh. **p. 602:** Norbert Wu/Minden Pictures; *Figure 15-20* Mitsuhiko Imamori/Minden Pictures. **p. 603:** *Figure 15-21* Jim Brandenburg/Minden Pictures. **p. 606:** *Figure 15-24 (TL)* Anthony Bannister, Gallo Images/Corbis, *(TR)* Thomas Marent/Minden Pictures, *(BL)* Beata Slonecka/National Geographic My Shot/National Geographic Stock, *(BR)* Michael & Patricia Fogden/Minden Pictures. **p. 607:** *Figure 15-25 (T)* Fred Bavendam/Minden Pictures, *(B)* John Downer/naturepl.com. **p. 608:** *Figure 15-26* E. Widder/HBOI/Visuals Unlimited. **p. 609:** *Figure 15-27 (T)* Stephen Dalton/Photo Researchers, *(B)* Eye of Science/Photo Researchers. **p. 610:** *Figure 15-28* Biosphoto/Heuclin Daniel/Peter Arnold Images. **p. 611:** *Figure 15-29 (T)* Mitsuhiko Imamori/Minden Pictures, *(B)* Arco Images GmbH/Alamy. **p. 612:** Michael Melford/National Geographic Stock. **p. 614:** *Figure 15-31* Siede Preis/Getty Images. **p. 617:** NASA Image by Robert Simmon, based on Landsat data provided by the UMD Global Land Cover Facility.

**CHAPTER 16    pp. 622–623:** Emmanuel Coupé, Greece. **p. 624:** DeAgostini/Getty Images; *Figure 16-1* Inga Spence/Getty Images, *(inset)* Bristol-Myers Squibb. **p. 625:** *Figure 16-2* Daniel Beltrá. **p. 626:** *Figure 16-3 (L to R)* Jupiterimages/Getty Images, Jeff Vanuga/Corbis, Jack Phelan, François Gohier/Photo Researchers. **p. 628:** *Figure 16-5* Dennis

MacDonald/Photolibrary. **p. 630:** *Figure 16-8 (T)* blickwinkel/Alamy, *(BL)* Michael & Patricia Fogden/Corbis, *(BR)* Frans Lanting Photography. **p. 631:** *Figure 16-9* Courtesy of Edward O. Wilson. **p. 632:** *Figure 16-10* Danny Lehman/Corbis. **p. 634:** Gallo Images/Corbis; *Figure 16-12* Tom Uhlman/Alamy, *(inset)* Libraries of Saginaw, MI, Local History and Genealogical Collection. **p. 637:** *Figure 16-15* Gerry Ellis/Minden Pictures; *Figure 16-16* Frans Lanting Photography. **p. 638:** Frans Lanting Photography. **p. 639:** *Figure 16-17 (L and C)* USGS Photo by Harry Glicken, *(R)* Barry O'Neil; *Figure 16-18 (L)* E. J. Keller/National Zoo/Smithsonian Institution, *(C)* Mark Carwardine/NPL/Minden Pictures, *(R)* Denis Scott/Corbis. **p. 640:** *Figure 16-19* Jany Sauvanet/Photo Researchers, *(inset)* Gale S. Hanratty/Alamy. **p. 641:** *Figure 16-21 (L)* Michael & Patricia Fogden/Minden Pictures, *(C)* Michael P. Gadomski/Photo Researchers, *(R)* Peter Yates/Photo Researchers. **p. 642:** *Figure 16-22* Jack Picone/Alamy. **p. 643:** *Figure 16-23* Anil Risal Singh–UNEP/Peter Arnold Images. **p. 644:** *Figure 16-26* Will & Deni McIntyre/Corbis. **p. 646:** *Figure 16-29 (T and B)* NASA/Goddard Space Flight Center Scientific Visualization Studio–Thanks to Rob Gerston (GSFC) for providing the data. **p. 648:** *Figure 16-31 (from L)* NASA Goddard Space Flight Center Scientific Visualization Studio, *(far R)* NASA; *Figure 16-32* AP Photo/Elise Amendola, *(inset)* New Line/Photofest. **p. 650:** *Figure 16-35* Frans Lanting/Corbis. **p. 651:** Jason Lee/X01757/Reuters/Corbis; *Figure 16-36 (T)* Mitsuaki Iwago/Minden Pictures, *(B)* Dennis Kunkel Microscopy. **p. 652:** *Figure 16 37 (L)* Frans Lanting Photography, *(R)* Florian Schultz/www.visionsofthewild.com. **p. 653:** *Figure 16-38* Tommy Moorman. **p. 656:** Corbis/SuperStock.

# Index

Bottleneck effects, **314**–315, *315*
Boyer, Herbert, 205
Breeding. *See* Artificial selection
*Brocadia anammoxidans,* 625
Bryophytes, **478**–480, *479, 480, 481*
Buffer zones, 653, *654*
Buffers, **51**, *51*
Buffon, Georges, 303, *304*
Buildings, and weather, 591, *591*
Buttercup, Persian, *392*
Butterflies
   convergent evolution of, 406
   glasswing (*Cithaerias aurorina*), *300*
   monarch, *446, 606,* 607

## C

C4 photosynthesis, **148**–150, *149, 150*
Cacti, golden barrel, *149*
Caecilians, *456,* 457
Calment, Jeanne Louise, 564
Calvin cycle, **146**–147, *147*
CAM photosynthesis, *149,* **150**
Camels, 298–299
Camouflage, *606,* 607
Cancer, **234**–236. *See also* Diseases and genetic
      disorders; Tumors
   and apoptosis, 230
   and gap junctions, 106
   and metastasis, 234, *235*
   and mutations, 189–190
   skin, 647
   and telomeres, 223
   treatments for, 235–236, *236*
*Candida albicans,* 499
Capsids, **533**
Capsules, **517**, *517*
Captive breeding, 654
Capybara (*Hydrochoerus hydrochaeris*), *456*
Carbo-loading, 52, 54, *54*
Carbohydrates, **52**–56, *53*
   and the Calvin cycle, 146–147, *147*
   complex, 54–55
   as fuel, 52–53
   indigestible, 56
   monosaccharides, 53–54
   synthesis of, 116, 146–147, *147*
Carbon, electronic structure of, 42, *42,* 44
Carbon cycle, *598,* 598–599
Carbon dioxide
   and the carbon cycle, *598,* 598–599
   and cellular respiration, 151–152, *152,* 154,
      *154*
   and deforestation, 650
   and the greenhouse effect, 644–646, *645*
   and photosynthesis, 137–138, *138*
Carboxyl groups, **64**, *64*
Carnivores, **594**, *595*
Carotenoids, **141**, *141*
Carpels, **488**, *489*
Carriers, **278**
Carrying capacity, *K,* **552**–553, *553*
   increasing, *574,* 574–575
Cartilaginous fishes, **454**, *454*
Catalysts. *See* Enzymes
Caterpillars, *446, 446*
Cats
   parasites of, 609–610
   pigmentation of, 287, *288*
*Caulerpa,* 656
Causation, 25
Cell components, 107–121, *121*
   cell walls, 109–110
   chloroplasts, 109–110
   cytoskeleton, 109–110

endomembrane system, 109–110
   evolution of, 529–530
   Golgi apparatus, 109–110
   lysosomes, 109–110
   mitochondria, 109–110
   nucleus, 107–108
   vacuoles, 109–110
Cell cycle, *226,* **226**–227
Cell division, 222–228. *See also* Meiosis; Mitosis
   and chromosomes, 224
   in eukaryotes, *226,* 226–227
   in prokaryotes, 225, *225*
   and replication, 227–228, *228*
   and telomeres, 222–223, *223*
Cell membranes, 89–96
   faulty, 93–94
   fingerprints of, 95–96
   function of, 91–92
   origin of, 391, *391*
   structure of, 89–91
Cell theory, 84
Cell walls, **85**, *85*
   in plants, *118,* 118–119
Cells, 80–81, **82**–88
   connections between, 105–106, *106*
   discovery of, 82–84
   eukaryotic, 84, 86–88
   parent *vs.* daughter, 225, *225*
   prokaryotic, 84–85
   somatic *vs.* reproductive, 226, 237
Cellular respiration, **131**, 151–158, *158*
   anaerobic, 159–160, *160*
   and ATP synthesis, 156–158
   and the electron transport chain, 156–158
   and glycolysis, 152–153
   and the Krebs cycle, 154–156
   overview of, 151–152, *152*
Cellulose, 54, **56**, *56*
Centipedes, *447, 447*
Central vacuoles, *119,* **119**–120
Centrioles, *232,* **233**
Centromeres, *226,* **227**
Cephalopods, **441**, *442,* 442–443, *443*
   intelligence of, 443–444, *444*
Chameleon (*Chamaeleo calyptratus*), *594*
Chance. *See* Probability
Chaparral, *586*
Character displacement, **604**, *605*
Charts. *See* Visual displays of data
Cheetahs, *159,* 314–315, *315*
Chemical cycles, *598,* 598–601, *599, 600*
Chemical defenses, *606,* 606–607
Chemical energy, **133**, *133*
Chemical-induced mutations, 192
Chemolithotrophs, **520**, *520*
Chemoorganotrophs, **519**–520, *520*
Chemotherapy, 235–236
Chimpanzees, *1, 348, 461,* 461–462
   bonobos, *637, 637*
Chitin, **56**, *56*
   in fungi, *498, 498*
Chlorofluorocarbons, 647–648
Chlorophyll, **139**, *139*
   and photosynthesis, *142,* 142–143
Chlorophyll *a,* **140**–141, *141*
Chlorophyll *b,* **140**–141, *141*
Chloroplasts, 88, *120,* 120–121, **139**
   and endosymbiosis, 86–88, 121
   photosynthesis in, 139, *139*
Cholesterol, *61,* **61, 92**
   and receptor-mediated endocytosis, 102–104,
      *103*
Chordata [phylum], 450–453, *451, 452. See also*
     Lancelets; Tunicates; Vertebrates

Chorionic villus samples (CVS), 207, 252
Chromatids, **233**
Chromatin, **108**, *108*
Chromosomal aberrations, **190**, *191*
Chromosomes, **175**, *175*
   abnormal numbers of, 250–255
   circular *vs.* linear, 224, *224*
   and karyotyping, 250–253
   and polyploidy, 398–400
Chytridiomycota [phylum], 499
Cilia, **110**, *110*
Ciliates, 530, *530*
Claims, testing of, 3
Clams, *442, 442*
Class, *394, 395*
Classification, 413–415. *See also* Phylogenies
   nomenclature for, 394–395, *395*
"Clever Hans," 18, *18*
Climax communities, *613, 613*
Clone libraries, **197**, *197*
Clones, **196**
Cloning, **196**–197, *197,* 209–210
Cnidaria [phylum], *436,* 436–438, *437. See also*
     Corals; Jellyfishes; Sea anemones
Cnidocysts, *436, 437*
Cocci, 516, *517*
Cockroaches, *447,* 558, *558*
Code, genetic, **175**, 183, *183*
Codominance, **280**, *280*
Codons, **183**, *183*
Coelacanth, 409
Coevolution, 491, *491,* **602**, **602**–603, *603*
   and predation adaptations, 605–608, *606,*
     *607*
Cohen, Stanley, 205
Cohesion, **36**–37, 47, *47*
Coleochaetes, *477, 477*
Colonizers, **612**–613
Commensalism, 610–**611**, *611*
Communication, **377**–379
   evolution of, 377–379
   and honest signals, 379
   by plants, 496–497
Communities, 584–**585**, *585,* 602–615
   adaptations for predation in, 605–608, *606,*
     *607*
   climax, 613, *613*
   coevolution in, *602,* 602–603
   commensalism in, 610–611, *611*
   competition in, *604,* 604–605, *605*
   mutualism in, 610–611, *611*
   niches in, 603, *603*
   parasitism in, *609,* 609–610, *610*
   and population ecology, 549, *549*
   succession in, 612–614, *613*
Comparative anatomy, 335–337, *336*
Competition
   in communities, *604,* 604–605, *605*
   sexual, 369–371
Competitive exclusion, **604**, *604*
Competitive inhibitors, **69**
Complementarity, **227**
Complementary bases, **227**
Complete metamorphosis, *446,* **446**–447
Complex carbohydrates, **54**–55, *55*
Compounds, **44**
   and bonding, 43–45
Concentration, and enzymes, 69, *69*
Concentration gradients
   and active transport, 101, *101*
   and diffusion, 97–98
   and potential energy, 132
Conclusions, and the scientific method,
     13–15

Condors, California, 636, 654
Cones, 484, *486*
Coniferous forests, *586*
Conjugation, **518**–519, *519*
Conservation biology, **627**–628, *628,* 651–655
   landscape, *653,* 653–655, *654*
   strategies for, 651–652
Control groups, **16**
Convergent evolution, **336**–337, *337,* **405**–406,
     *406*
Coral reefs, *437, 437,* 587, *587*
   as biodiversity hotspots, *630,* 630–631
   bleaching of, 437, *437*
Corals, *436,* 436–437, *437*
Corn, *149*
   and artificial selection, 198, *198*
   flowers of, 488
   transgenic, 199–201, *200, 201*
Correlation, positive, **25**
Corridors, 653, *654*
Cotton, transgenic, 199, *200*
Courtship rituals, 369–371, *370*
Covalent bonds, *43,* **43**–44, *45*
Coyotes
   communication in, *378*
   parasites of, 609–610
Crabs, *349, 445, 447,* 448
Cranes, red-crowned (*Grus japonensis*), courtship
     rituals of, *365*
Crassulacean acid metabolism (CAM)
     photosynthesis, *149,* **150**
Crayfish, 448
Crick, Francis, 172, *172,* 227
Crick, Odile, 172
Crickets, *594*
   reproductive investment of, 368, *369*
Critical experiments, **12**–13
Crocodiles, 457–459
Crosses, **268**
   test-, 276, *276*
Crossing over, *240,* **241**
   as source of variation, 244, *244*
Crustaceans, *445, 447,* 448
Cup fungi (*Cookeina tricholoma*), *498*
Cuticle, 478, **478**
Cuvier, Georges, 303, *304*
Cyanide poisoning, 158
Cycads, 484, *485*
Cytokinesis, **227**
   in meiosis, *241,* 242
   in mitosis, *226,* 227, *233*
Cytoplasm, **84**, *85,* 109–110
Cytosine (C), *71, 72,* 173, *173*
Cytoskeleton, *109,* **109**–110

## D

Darwin, Charles, 303, *304,* 305
   and adaptation, 323, 431
   and agricultural pests, 640
   and altruism, 355–356
   aboard HMS *Beagle,* 305–307
   and behavior, 353
   and biogeography, 334
   and evolution, 308, 318, 339
   and fitness, 322
   and natural selection, 318, 326
   and the origin of life, 388
   and the pace of evolution, 409
   and perfectibility, 325
   and species classification, 401
   and variation, 317
Darwin, Erasmus, 303
Dating, radiometric, **331**–332, *332*
Daughter cells, **225**, *225*

prepared learning in, 351–352, *352*
reciprocal altruism in, 361
sex determination in, 247–248, *248*
sexual selectiveness in, 367–369, *368*
success of, 636
vestigial structures in, 336
vision of, 140
Huxley, Thomas, 319
Hybridization, **397**
Hybridizing species, *396*, 397
Hybrids, **394**
Hydrogen bonds, **44–45**, *45*
in DNA, 173, *173*
Hydrogen ions, 50, *50*
Hydrogenation, *60*, 60–61
Hydrophilic molecules, **57, 89**
Hydrophobic molecules, **57, 89**
Hydroxide ions, 50, *50*
Hypertonic solutions, **99**, *99*
Hyphae, **499**
Hypotheses, **9**–10
null, 10
and theories, 15
Hypotonic solutions, **99**, *99*

**I**

Ice caps, and global climate change, 645–646, *646*
Immigration, and island biogeography, *632*, 632–633
Immune system
and lysosomes, 113
and recognition proteins, 92
Immunization, and autism, 26–28
Inbreeding, 493
Inclusive fitness, **358**
Incomplete dominance, **279**–280, *280*
Incomplete metamorphosis, *446*, 446–**447**
Independent variables, **23**, *23*
Indicator species, 654
Indirect fitness, **358**
Inefficiency of energy conversion, 133–135
in food chains, 596–597, *597*
Inheritance, 317–**318**, *318, 319*
Inhibitors, *69*, 69–70
Innate behaviors, *350*, 350–**351**
Insects, 444–447, *445, 446*
Instincts, **351**
Insulin
and biotechnology, 205–206, *206*
and exocytosis, 104
Intelligence, 443–444
Intermediate filaments, **109**, *109*
Intermembrane space, *111*, **112**
Interphase, **226**
in meiosis, 239–240, *240*
in mitosis, *226, 227*, 232, *232*
Introduced species, **640**. *See also* Exotic species
Introns, **177**
Invagination, **88**, *88*
Invasive species, 502, 503, **640**–642, *641, 642*
Invertebrates, *432*, **432**–433
Annelida (segmented worms), *438*, **440**, 440–441
Arthropoda (insects, millipedes, centipedes, arachnids, crustaceans), 444–448, *445, 446, 447*
Cnidaria (corals, sea anemones, jellyfishes), *413, 436*, 436–438, *437*
Echinodermata (sea stars, sea urchins, sand dollars), 448–449, *449*
Mollusca (mollusks), 441–443, *442, 443*
Nematoda (roundworms), *438, 439*, 439–440
Platyhelminthes (flatworms), *438, 439*

Porifera (sponges), 429, 434–435, *435*
Ionic bonds, **44**, *44, 45*
Ions, *42*, *42*
Island biogeography, *631*, 631–633, *632*
and landscape conservation, 653–654, *654*
Islands, as biodiversity hotspots, *630*, 631
Isotonic solutions, **99**, *99*
Isotopes, *40*, *40*
Ivy, *149*

**J**

Jackson brothers, *220–221*, 238
Jeffreys, Alec, 211
Jellyfishes, *413*, 436, *436*, 438
Junctions between cells, 105–106, *106*
"Junk" DNA, 177–178

**K**

Kangaroos, *460*
Karyotypes, **250**, **250**–253, *251*
Kelp, *530*, 531
Kennedy brothers, 238, *238*
Keystone species, **614**, **614**–615
and landscape conservation, 654
Kin selection, **356**–359, *357, 358*
Kinetic energy, *132*, **132**–133
King protea (*Protea cynaroides*), *137*
Kingdom, **394**, *395*
Krebs cycle, **154**–156, *155*
Krill, 448

**L**

La Niña, **593**
Lactic acid, 159–160, *160*, 521
*Lactobacillus acidophilus*, 521–522
Lactose, 54
Lakes and ponds, 587, *587*
Lamarck, Jean-Baptiste, 303, *304*
Lampreys, 453, *453*
Lancelets, 450, *452*
Lancet flukes, 610
Landscape conservation, **653**, **653**–655, *654*
Language, *378*, **378**–379
Larvae, **446**, *446*
Lavender, *489*
Learned behaviors, 351–353, *352*
Learning, **351**
prepared, 351–352, *352*
via scientific method, 7–8
Leeches, *440*, 440–441
Leeuwenhoek, Anton van, 574
Legs, *455*, *455*
Lemmings, and population ecology, 555, 555–556
Lemurs, 461, *461*
Lichens, **503**
as colonizers, 612
mutualism in, 611
Life, **388**
classifications of, *413*, 413–415, *414*
origin of, 388–391
Life cycles
of angiosperms, 490–493, *492, 493*
of ferns, 481–482, *482, 493*
of fungi, 500, *500*
of gymnosperms, 484, 486, *486, 493*
of insects, *446*, 446–447
of mosses, 479, *480, 493*
Life histories, 559–563, **560**
and life tables, 561, *561*
and natural selection, *559*, 559–561, *560*
and survivorship curves, 561–562, *562*
and trade-offs, 563, *563*

Life tables, **561**, *561*
Light energy, *140*, **140**–141
Lignin, 118
Lilies, calla, *472–473*
Line graphs, *22, 23*, *23*
Linked genes, *291*, *291*–292
Linnaeus, Carolus, 394, 401, 413
Lions, kin selection in, 359
Lipids, **57**–62, *58*
and energy storage, 57–58
and fats, 58–61
and hormones, 61–62
synthesis of, 114–115
Lipoproteins, 66
Litter sizes, 563
Liverworts, 478–479, *479*
Lizards, *407*, 457–459
defense mechanisms of, 605
life histories of, 563, *563*
thorny devil, *451*
Lobe-finned fishes, **454**, *454*
Lobsters, 448
longevity of, 564
and population ecology, *548*, 548, 557, *557*
Locoweeds, 496
Locusts, and population ecology, *554*, 554–555
Logistic growth, *553*, *553*
Longevity, 567–568
Loosestrife, purple, 641, *641*
*Luehea seemannii*, 651
Lungs, 454–455, *455*
Lyell, Charles, 304, *304*
Lynxes, and population ecology, 555, *555*
Lysosomes, **112**–113, *113*

**M**

MacArthur, Robert, 631–633
Macaws, scarlet (*Ara macao*), *428*
Mackerel, survivorship curves of, 562, *562*
Macroevolution, **407**–412. *See also* Evolution
and adaptive radiations, 409–410, *410*
and mass extinctions, *411*, 411–412
pace of, 408–409, *409*
scale of, 407–408, *408*
Macromolecules, **52**
Males, **365**
Malthus, Thomas, 307
Mammals, *459*, 459–560, *460*
adaptive radiation of, 409–410
Mammary glands, **459**
Marijuana, 496
Marsupials, **460**, *460*
and biogeography, 334–335, *335*
Mass, **39**
Mass extinctions, 410, *411*, **411**–412, 634–635, *635*
anthropogenic, 411, 636–637
Mastodons, extinction of, 303, *304*
Mate guarding, 371–372, *372*
Mating systems, **373**, **373**–375, *374*
and sexual dimorphism, 375–376, *376*
Matrix, *111*, **112**
Maximum sustainable yield, 556, 556–558, *557*, **557**, *558*
Mechanical defenses, 606, *606*
Medicines. *See* Drugs
Meiosis, 237–246
details of, 239–242, *240–241*
overview of, 237–238, *238*
segregation of alleles in, 270–271, *271*
Membranes. *See* Cell membranes; Nuclear membranes
Mendel, Gregor, 268–270, *269*, 289, 350

Mendel's law of independent assortment, 289–**290**, *290*
Mendel's law of segregation, **271**, *271*
Messenger RNA (mRNA), **179**, *179*
in transcription, 180–182, *181*
in translation, 182–185, *184*
Metabolic pathways, 68
anaerobic, 159–160, *160*
Metabolism, 68
of bacteria, 519–520, *520*
Metamorphosis, *446*, **446**–447
Metaphase, **233**
in meiosis, 240–241, *241, 242*
in mitosis, *232*, 233
Methane, covalent bonds in, *43, 44*
Mice
fitness in, 322–323, *323*
house, 559
life histories of, *559*, 560
longevity of, 567, *567*
Microarrays, 185–186
Microbes, **413**, **512**–515. *See also* Archaea; Bacteria; Protists
classification of, 514–515, *515*
diversity among, 414, 512–514, *513*
Microevolution, **407**, *408*
Microfilaments, *109*, *109*–110
Microspheres, **391**
Microsporidia [phylum], 499
Microtubules, **109**, *109*
Migration, **315**–316, *316*
Miller, Stanley, *389*, 389–390
Millipedes, *445*, 447
Mimicry, **607**, 608
Mink, as exotic species, 604–605
"Missing links," 333, *333*
Mites, 448, *609*
Mitochondria, **88**, **110**–112, *111*
and ATP synthesis, *156*, 156–158
and endosymbiosis, 86–88, 112
Mitochondrial matrix, *156*, **156**–157
Mitosis, **226**, **227**, 229–236
and cell replacement, 229–230
details of, *232–233*, 232–234
overview of, 231, *231*
Mitotic phase, **226**, *226*, 227
Molds. *See* Fungi; Slime molds
Molecular biology, 337–338, *338*
Molecules, **43**–45
hydrophilic *vs.* hydrophobic, 57, 89
macro-, 52
models of, 43
polar, 45
polar *vs.* nonpolar, 89
self-replicating, 390–391, *391*
Moles, classification of, 405, *406*
Mollusca [phylum], 441–443, *442, 443*. *See also* Mollusks
Mollusks, 441–443, *442, 443*
Molting, 430
Monkeys, 461, *461*
communication in, 378
prepared learning in, 351–352, *352*
reciprocal altruism in, 360, *361*
Monogamy, **373**–375, *374*
Monophyletic groups, **403**–404, *404*, 430–431
Monosaccharides, *52*, **52**–53
Monotremes, **460**, *460*
Morels, 499
*Morganucodon*, 459
Morphological species concept, **397**
Mosquitoes, 324
Mosses, 478–480, *479, 480, 481*
as colonizers, 612